The Evolution of Mating Systems
in Insects and Arachnids

In the broad array of human diversity, George Eickwort stood out as one of those who was both delightful and productive to be with. Whether as a student, a teacher, a co-researcher, or whatever else, he worked hard while always managing to be both stimulating and entertaining to those around him. I always felt I was fortunate to have him as a student and friend. He is much missed.

C. D. Michener

We dedicate this volume to the memory and inspiration of George Eickwort.
J. C. Choe & B. J. Crespi

The Evolution of

Mating Systems in Insects and Arachnids

Edited by

JAE C. CHOE Museum of Zoology, University of Michigan, USA
Seoul National University, Korea

and BERNARD J. CRESPI Simon Fraser University, Canada

CAMBRIDGE
UNIVERSITY PRESS

PUBLISHED BY THE PRESS SYNDICATE OF THE UNIVERSITY OF CAMBRIDGE
The Pitt Building, Trumpington Street, Cambridge CB2 1RP, United Kingdom

CAMBRIDGE UNIVERSITY PRESS
The Edinburgh Building, Cambridge CB2 2RU, United Kingdom
40 West 20th Street, New York, NY 10011–4211, USA
10 Stamford Road, Oakleigh, Melbourne 3166, Australia

First published in 1997

Printed in the United Kingdom at the University Press, Cambridge

Typeset in MT Ehrhardt 9/12pt

A catalogue record for this book is available from the British Library

Library of Congress Cataloguing in Publication data

The evolution of mating systems in insects and arachnids / edited by
    Jae C. Choe and Bernard J. Crespi.
        p.   cm.
    Includes index.
    ISBN 0 521 58029 3 (hardback). – ISBN 0 521 58976 2 (pbk.)
    1. Insects – Behavior – Evolution.   2. Arachnida – Behavior –
    Evolution.   3. Sexual behavior in animals.   I. Choe, Jae C.
    II. Crespi, Bernard J.
QL496.E95   1997
595.7056 –dc20   96–25108 CIP

ISBN 0 521 58029 3 hardback
ISBN 0 521 58976 2 paperback

Contents

Contributors

RICHARD D. ALEXANDER
Museum of Zoology, The University of Michigan, Ann Arbor, MI
    48109-1079, USA.

GÖRAN ARNQVIST
Department of Animal Ecology, University of Umea, S-901 87
    Umea, Sweden.

WILLIAM D. BROWN
Batiment de Biologie, Institut de Zoologie et d'Ecologie Animale,
    University of Lausanne, 1015 Lausanne, Switzerland.

JAE C. CHOE
Museum of Zoology, The University of Michigan, Ann Arbor, MI
    48109-1079, USA, and Department of Biology, Seoul National
    University, Kwanak-Gu, Shillim-Dong San 56-1, Seoul 151-742,
    Korea.

STEVEN G. COMPTON
Department of Pure and Applied Biology, University of Leeds,
    Leeds LS2 9JT, UK.

JAMES M. COOK
Department of Biology, NERC Centre for Population Biology,
    Imperial College at Silwood Park, Ascot, Berkshire SL5 7PY,
    UK.

JOHN R. COOLEY
Museum of Zoology, The University of Michigan, Ann Arbor, MI
    48109-1079, USA.

BERNARD J. CRESPI
Department of Biological Sciences, Simon Fraser University,
    Burnaby, British Columbia V5A 1S6, Canada.

JANIS L. DICKINSON
Hastings Reservation, University of California, 38601 E. Carmel
    Valley Road, CA 93924, USA.

GARY N. DODSON
Department of Biology, Ball State University, Muncie, IN 47306,
    USA.

WILLIAM G. EBERHARD
Biologia, Universidad de Costa Rica, Ciudad Universitaria, Costa
    Rica.

OLA M. FINCKE
Department of Zoology, University of Oklahoma, Norman, OK
    73069, USA.

MICHAEL D. GREENFIELD
Department of Entomology, University of Kansas, Lawrence, KS
    66045, USA.

H. C. J. GODFRAY
Department of Biology, NERC Centre for Population Biology,
    Imperial College at Silwood Park, Ascot, Berkshire SL5 7PY,
    UK.

DARRYL T. GWYNNE
Department of Zoology, Erindale Campus, University of Toronto,
    Mississauga, Ontario LFL 1C6, Canada.

CHARLES S. HENRY
Department of Ecology and Evolutionary Biology, Box U-4375,
    Neagleville Road, University of Connecticut, Storrs, CT 06268,
    USA.

EDWARD ALLEN HERRE
Smithsonian Tropical research Institute, Unit 9408, APO AA
    34002-0948, USA.

ROBERT R. JACKSON
Department of Zoology, University of Canterbury, Private Bag
    4800, Christchurch, New Zealand.

FINN KJELLBERG
Centre National de la Recherche Scientifique, CEFE, Louis
    Emberger, Route de Mende, BP 5051, 34033 Montpellier Cedex,
    France.

WALTER D. KOENIG
Hastings Reservation, University of California, 38601 E. Carmel
    Valley Road, CA 93924, USA.

JAMES E. LLOYD
Department of Entomology and Nematology, Bldg. 970, Hull
    Road, PO Box 110620, University of Florida, Gainesville, FL
    32611-0620, USA.

DAVID C. MARSHALL
Museum of Zoology, The Universityof Michigan, Ann Arbor, MI
    48109-1079, USA.

ERIK PETERSSON
USDA ARS, Insect Attractants, Behavior and Basic Biology
    Research Laboratory, 1700 SW 23rd Drive, PO Box 14565,
    Gainesville, FL 32604, USA.

P. LARRY PHELAN
Department of Entomology, Ohio Agricultural Research and
    Development Center, Wooster, OH 44961, USA.

SIMON D. POLLARD
Department of Zoology, University of Canterbury, Private Bag
    4800, Christchurch, New Zealand.

RONALD L. RUTOWSKI
Department of Zoology, Arizona State University, Tempe, AZ
  85287-1501, USA.

TODD E. SHELLY
Organization for Tropical Studies, Box 90633, Durham, NC
  27708-0633, USA.

LEIGH W. SIMMONS
Department of Environmental and Evolutionary Biology,
  University of Liverpool, PO Box 147, Liverpool L69 3BX,
  UK.

JOHN M. SIVINSKI
USDA ARS, Insect Attractants, Behavior and Basic Biology
  Research Laboratory, 1700 SW 23rd Drive, PO Box 14565,
  Gainesville, FL 32604, USA.

JONATHAN K. WAAGE
Department of Ecology and Evolutionary Biology,
  Box G-W, Brown University, Providence, RI 02912,
  USA.

STUART A. WEST
Department of Biology, NERC Centre for Population Biology,
  Imperial College at Silwood Park, Ascot, Berkshire SL5 7PY,
  UK.

TIMOTHY S. WHITTIER
Hawaiian Evolutionary Biology Program, University of Hawaii,
  Honolulu, HI 96822, USA.

GERALD S. WILKINSON
Department of Zoology, University of Maryland, College Park,
  MD 20742-4415, USA.

DAVID W. ZEH
Department of Biology, University of Houston, Houston, TX
  77204-5513, USA.

JEANNE A. ZEH
Department of Biology, University of Houston, Houston, TX
  77204-5513, USA.

MARLENE ZUK
Department of Biology, University of California, Riverside, CA
  92521, USA.

Acknowledgements

Chapters in this book were peer-reviewed. We are grateful to P. Adler, J. Alcock, R. Alexander, G. Arnqvist, W. Bell, W. Brown, W. Cade, J. Cooley, J. Dickinson, G. Dodson, W. Eberhard, S. Frank, M. Greenfield, G. Gries, D. Grimaldi, D. Gwynne, C. Henry, A. Herre, R. Lederhouse, D. Marshall, T. New, L. Phelan, M. Ridley, R. Rutowski, J. Seger, T. Shelly, A. Sih, L. Simmons, J. van den Assem, C. Wiklund, S. Wilcox, G. Wilkinson, M. Zuk, and those who wish to remain anonymous for sharing their expertise. For inspiration and guidance, we thank R. D. Alexander, W. G. Eberhard, Jane Goodall, W. D. Hamilton, B. Hölldobler, C. D. Michener, M. J. West-Eberhard, and E. O. Wilson. We are also grateful to R. D. Alexander and the University of Michigan Museum of Zoology, and Bobbi Low and the Michigan Society of Fellows, for financial and emotional support throughout the project. For enthusiasm and dedication, we thank our editor, Tracey Sanderson. And for love, patience, and support we thank our families, friends, students, and cats; they taught us the true meanings of cooperation.

Introduction

JAE C. CHOE AND BERNARD J. CRESPI

If there is conflict of interest between parents and children, who share 50 per cent of each others' genes, how much more severe must be the conflict between mates, who are not related to each other?

Richard Dawkins (1976, p. 151)

Sexual behavior and social behavior are profoundly alike in that both involve one set of individuals more or less willingly providing a limiting resource to another set (Queller 1994). Thus, in sexual interactions females provide resource-rich ova and other parental investment to males, and in social interactions workers provide labor to queens. In both situations, the parties are virtually always in conflict over the allocation of the resources, but their interests also partly coincide: eggs must be fertilized and offspring produced, and a new generation of reproductives must be successfully protected and reared. The complex mixtures of conflict and cooperation that thus typify sex and sociality make them among the most endlessly fascinating and difficult topics in ecology and evolution.

This book, and its companion (Choe and Crespi 1997), explore the intricacies of sexual and social competition. We have drawn together, for each of these topics, a set of authors whose expertise is both taxon-deep and broadly based in the theory that guides interpretation of natural history. Our goal has been to bring theory and observation together, to find parallels and convergences between disparate taxa, and to sketch out the patterns of engagement that will allow us to understand how conflicts and confluences of interest evolve together. Insects and arachnids display the most impressive diversity of mating and social behavior among animals, and we have chosen them as our subjects and teachers.

The chapters in this volume investigate sexual competition, and the variety of ways in which males and females pursue, persuade, manipulate and control, and help one another. The evolutionary study of this topic has a peculiar history. Despite repeated awakenings due to Fisher (1915, 1930) and Bateman's (1948) elegant experimental illustration of sexual selection at work, the theory of sexual selection proposed by Darwin (1871) remained largely dormant

for nearly a century. Whether it was put on hold because of the formidable theoretical challenge by Wallace and his followers, or it required the proper social atmosphere provided by the women's rights movement (Cronin 1991) or liberal politics to resurface, it was the inspiration of individual and gene selectionist thinking in the 1960s (Hamilton 1964; Lack 1966; Williams 1966) that resurrected it. Once reborn, it grew into a bewildering diversity of theoretical and empirical research, and today it is the most active topic in behavioral ecology, comprising over 20% of all research in the field (Gross 1994).

Ernst Mayr (1982) asserted that '[o]ur understanding of the world is achieved more effectively by conceptual improvements than by the discovery of new facts, even though the two are not mutually exclusive'. Since Hamilton, Lack and Williams renewed our thought program, a new generation of behavioral ecologists raised as individual selectionists began viewing reproductive strategies from the perspectives of individuals striving to maximize their inclusive fitness. Both proximate and ultimate mechanisms responsible for generating diversity in mating patterns of a population were investigated using cost–benefit analysis of individuals' reproductive decisions.

One of the first conceptualizations of mating behavior based on individual selectionist thinking was the polygyny threshold model (Orians 1969). The model predicts that females should select mates according to the quality of the male's territory and the potential for paternal care. Incorporating the various mate-acquiring tactics of males, Emlen and Oring (1977) proposed one of the most widely accepted classification of mating systems. They recognized that male mating success is influenced by ecological factors such as the distribution of resources, females included, in time and space. Their scheme has generated a host of testable hypotheses in relation to sexual selection theory, and many subsequent field and laboratory studies have investigated the relation between ecological variables and particular mating systems (see, for example, Blum and Blum 1979; Thornhill and Alcock 1983; Rubenstein and Wrangham 1986).

Inspired by Williams' (1966) question, 'Why are males masculine, females feminine and occasionally vice-versa?', Trivers (1972) formulated the concept of relative parental investment by considering not only the gametic size difference but also the difference in the total reproductive energy expenditure between the sexes. In most species, females invest more heavily and thus become a limiting resource for which males compete. How females and, in some cases, males choose their mates has been an active area of research (see, for example, Bateson 1983), and causes and consequences of sexual dimorphism have also been issues central to the study of mating systems and sexual selection (see, for example, Alexander et al. 1979; Shine 1989).

Viewing the male–female phenomenon in terms of sexual conflict (see, for example, Parker et al. 1972; Alexander and Borgia 1979; Parker 1979) has been instrumental in understanding why the sexes differ in morphology and behavior, and in determining which sex is in control of mating events. Such conceptualization has prompted behavioral ecologists to go back to the field with a fresh outlook and observe mating episodes, keeping in mind the possibilities of sperm competition (Parker 1970; Smith 1984; Birkhead and Møller 1992) and postcopulatory female choice (Thornhill 1983; Eberhard 1985, 1996). Molecular genetic analyses using DNA fingerprinting techniques are now routinely employed to assign paternity (Hadrys and Siva-Jothy 1994; Scott and Williams 1994), which allows assessment of the connections between mating behavior and its genetic consequences.

This volume is intended to update the synthesis of insect mating systems by Thornhill and Alcock (1983) and complement a recent synthesis of the study of sexual selection by Andersson (1994). The volume comprises chapters addressing the evolution of mating systems in as wide a range of insects and arachnids as we could assemble. Although each chapter addresses some aspect of sexual selection theory, the volume as a whole adopts a taxon-centered approach, which was predicted by Wilson (1989) to be an increasingly important approach in biology. We have challenged each of the authors to: (1) provide a comprehensive review of mating systems in a group of insects or arachnids of their expertise, and discuss the evolution

of mating systems and intersexual conflict; (2) discuss intrinsic and extrinsic factors responsible for the observed reproductive strategies in their taxon from a comparative perspective; and (3) identify new research areas and useful taxa, and furnish the most promising questions for future studies. The chapters encompass the full range of diversity of theoretical issues related to mating systems, including male–male competition, mate choice, sex ratios, sexual dimorphism, and conflict between the sexes. Each of these facets of mating systems reflects in some way their core, the transfer of limiting resources between the sexes, and the remarkably complex forms of cooperation and conflict of interest that ensue.

LITERATURE CITED

Alexander, R. D. and G. Borgia. 1979. On the origin and basis of the male–female phenomenon. In *Sexual Selection and Reproductive Competition in Insects*. M. S. Blum and N. A. Blum, eds., pp. 417–440. New York: Academic Press.

Alexander, R. D., J. L. Hoogland, R. D. Howard, K. M. Noonan and P. W. Sherman. 1979. Sexual dimorphisms and breeding systems in pinnipeds, ungulates, primates and humans. In *Evolutionary Biology and Human Social Behavior*. N. A. Chagnon and W. Irons, eds., pp. 402–435. North Scituate: Duxbury Press.

Andersson, M. 1994. *Sexual Selection*. Princeton: Princeton University Press.

Bateman, A. J. 1948. Intra-sexual selection in *Drosophila*. *Heredity* 2: 349–368.

Bateson, P., ed. 1983. *Mate Choice*. Cambridge University Press.

Birkhead, T. R. and A. P. Møller. 1992. *Sperm Competition in Birds*. London: Academic Press.

Blum, M. S. and N. A. Blum, eds. 1979. *Sexual Selection and Reproductive Competition in Insects*. New York: Academic Press.

Choe, J. C. and B. J. Crespi, eds. 1997. *The Evolution of Social Behavior in Insects and Arachnids*. Cambridge University Press.

Cronin, H. 1991. *The Ant and the Peacock*. Cambridge University Press.

Darwin, C. 1871. *The Descent of Man and Selection in Relation to Sex*. London: John Murray.

Dawkins, R. 1976. *The Selfish Gene*. Oxford University Press.

Eberhard, W. G. 1985. *Sexual Selection and Animal Genitalia*. Cambridge, Massachusetts: Harvard University Press.

–. 1996. *Females in Control: Sexual Selection by Cryptic Female Choice*. Princeton: Princeton University Press.

Emlen, S. T. and L. W. Oring. 1977. Ecology, sexual selection and the evolution of mating systems. *Science* (Wash., D.C.) **197**: 215–222.

Fisher, R. A. 1915. The evolution of sexual preference. *Eugenics Rev.* 7: 184–192.

Fisher, R. A. 1930. *The Genetical Theory of Natural Selection*. Oxford:

Clarendon Press.

Gross, M. R. 1994. The evolution of behavioural ecology. *Trends Ecol. Evol.* **9**: 358–360.

Hadrys, H. and M. T. Siva-Jothy. 1994. Unravelling the components that underlie insect reproductive traits using a simpler molecular approach. In *Molecular Ecology and Evolution: Approaches and Applications.* B. Schierwater, B. Streit, G. P. Wagner and R. DeSalle, eds., pp. 75–90. Basel: Birkhauser Verlag.

Hamilton, W. D. 1964. The genetical evolution of social behaviour. *J. Theor. Biol.* **7**: 1–52.

Lack, D. 1966. *Population Studies of Birds.* Oxford: Clarendon Press.

Mayr, E. 1982. *The Growth of Biological Thought.* Cambridge, Massachusetts: Harvard University Press.

Orians, G. H. 1969. On the evolution of mating systems in birds and mammals. *Am. Nat.* **103**: 589–603.

Parker, G. A. 1970. Sperm competition and its evolutionary consequences in the insects. *Biol. Rev.* **45**: 525–567.

Parker, G. A. 1979. Sexual selection and sexual conflict. In *Sexual Selection and Reproductive Competition in Insects.* M. S. Blum and N. A. Blum, eds., pp. 123–166. New York: Academic Press.

Parker, G. A., R. R. Baker and V. G. F. Smith. 1972. The origin and evolution of gamete dimorphism and the male–female phenomenon. *J. Theor. Biol.* **36**: 529–553.

Queller, D. C. 1994. Male–female conflict and parent-offspring conflict. *Am. Nat.* **144**: S84–S99.

Rubenstein, D. I. and R. W. Wrangham. 1986. *Ecological Correlates of Social Evolution.* Princeton: Princeton University Press.

Scott, M. P. and S. M. Williams. 1994. Measuring reproductive success in insects. In *Molecular Ecology and Evolution: Approaches and Applications.* B. Schierwater, B. Streit, G. P. Wagner and R. DeSalle, eds., pp. 61–74. Basel: Birkhauser Verlag.

Shine, R. 1989. Ecological causes for the evolution of sexual dimorphism: a review of the evidence. *Q. Rev. Biol.* **64**: 419–461.

Smith, R. L., ed. 1984. *Sperm Competition and the Evolution of Animal Mating Systems.* New York: Academic Press.

Thornhill, R. 1983. Cryptic female choice and its implications in the scorpionfly *Harpobittacus nigriceps. Am. Nat.* **122**: 765–788.

Thornhill, R. and J. Alcock. 1983. *The Evolution of Insect Mating Systems.* Cambridge, Massachusetts: Harvard University Press.

Trivers, R. L. 1972. Parental investment and sexual selection. In *Sexual Selection and the Descent of Man.* B. Campbell, ed., pp. 136–179. Chicago: Aldine.

Williams, G. C. 1966. *Adaptation and Natural Selection.* Princeton: Princeton University Press.

Wilson, E. O. 1989. The coming pluralization of biology and the stewardship of systematics. *BioScience* **39**: 242–245.

1 · Evolutionary perspectives on insect mating

RICHARD D. ALEXANDER, DAVID C. MARSHALL AND JOHN R. COOLEY

We need a new theory of mating systems . . . [one] that incorporates the
conflicting interests of males and females, and the factors determining which
sex is in control, in order to predict patterns of male–female pairing.

Gross, 1994

Given that females, to one extent or another, subvert male interests by the
internal manipulation of ejaculate, it is not inconceivable that males will have
evolved little openers, snippers, levers and syringes that put sperm in the
places females have evolved ('intended') for sperm with priority usage –
collectively a veritable Swiss Army Knife of gadgetry!

Lloyd, 1979

ABSTRACT

The male–female interaction is an asymmetrical, usually
obligate mutualism in which there are conflicts of interest
whenever multiple potential partners that vary in quality
are available for either sex. Understanding male–female
confluences and conflicts of interest is required to explain
the sexual sequence and how it evolves. Mating interactions
involve multiple steps or stages, distinguishable because of
differences arising out of changes in selection that occur
during the sequence. Sexual selection and competition
take several different forms, which must be understood
before accurate interpretations can be made of mating
events in any particular case.

Sexual selection guided primarily by male–female conflicts of interest can result in resolvable evolutionary chases
that lead to evolutionarily stable strategies but perhaps
more frequently lead to chases that tend to be unending
(Parker 1979). In the latter, successful changes in one sex
lead to countering evolutionary changes in the other sex,
so that conflict is repeatedly exacerbated by evolutionary
changes in the participants that enhance their ability to
overcome countering traits evolving in the other sex. In
contrast, sexual selection controlled primarily by female
choice should yield directional changes which, if long-
term, have some likelihood of becoming Fisherian runaway selection because females favor males with extreme
traits. When conflict is a trivial component of such interactions, runaway selection will tend to accelerate, then reach
an equilibrium, because the external environment imposes
penalties on the individuals with extreme traits, and these
penalties counteract the benefits of the traits in mating
success (Fisher 1958). It is evidence of a conflict-of-interest
chase when males evolve to make extreme traits less expensive (thereby less valuable to females in choosing males) and
also evolve modifiable handicaps that can be adjusted
according to condition or available resources (Hill 1994).

With runaway evolution of traits wholly involved in
mating success, such as some courtship or copulatory
structures and behaviors, costs from the external environment may be minimal or absent from the start because
costs are mainly from caloric or genetic consequences of
step-by-step evolutionary changes per se (i.e. not from the
external environment). The resulting long-term directional
change can take the form of a runaway process if females
gain from favoring males best able to achieve fertilization
(Eberhard 1985); but such changes may also represent
chases based on continuing conflict.

Males and females each evolve to control fertilization at
some expense to one another. The uncoupling of insemination and fertilization that characterizes most or all copulatory acts (Eberhard 1985) is evidence of male–female
conflict. Males are expected to value zygotic success less
than do females because of their concern for the success
of their own sperm; hence they are less concerned than
the female with timing of fertilization and placement of
zygotes. Uncoupling is thus in the female's interests
because she is concerned with timing of fertilization in

4

relation to appropriate placement of zygotes, and because she may gain from additional matings. It is difficult to imagine that male and female interests in this regard ever have been identical.

For various reasons insects appear less likely than birds and mammals to generate directional changes in female choice that can result in runaway processes. Female insects often choose passively or indirectly (for example by mating with the fastest male in a lek) and they may mate with males that meet minimal criteria as a result of thresholds generated independently of the variation in quality of males available to any particular female. Females can also choose 'best-of-n' males directly, by elevating internally generated thresholds with successive matings. To the extent that adult females can compare males directly, they can also select extreme males in groups.

It is evidently important to distinguish between evolutionary changes in courtship, copulation, genitalia, and titillatory devices that are owing to continuing conflict, and those owing to females evolving structures and behaviors that match or accept – and therefore favor – traits of males showing superiority in mating effort (Eberhard 1985). In the latter case, both sexes change so as to yield better matches between the shapes of structures or the nature of actions used in the sexual sequence, rather than one sex changing to increase the match and the other sex evolving to decrease it or not evolving at all in that context.

Courtship during copulation is a consequence of the uncoupling of insemination and fertilization; thus it could not have evolved, and could not be maintained, except as a consequence of a male–female conflict of interest. If copulatory courtship involves females evolving to favor certain males, rather than males overcoming female resistance (Eberhard 1985), then the apparent universality of uncoupling of insemination and fertilization means there must be no instances in which male and female interests are both served by recoupling insemination and fertilization.

INTRODUCTION

The male–female reproductive interaction is unique among social interactions, and is characterized by seven features: (1) it is an asymmetrical, intraspecific mutualism that is typically obligate; (2) it involves multiple potential mates for each sex; (3) it most often takes place between non-relatives or distant relatives (therefore it tends not to be directly nepotistic); (4) the two sexes play predictably different roles; (5) the interests of the interactants overlap, but they are not identical; (6) neither interactant overwhelmingly controls all aspects of the mating process; and (7) prospective mates are able to exploit the interaction at one another's expense. These features generate a complex mix of cooperation and conflict. The male–female interaction resembles the investments of partners in a temporary social reciprocal interaction, or in a nepotistic interaction in which the reproductive benefits of investment in another individual are realized indirectly, via the descendants of the investor. Unlike in either of these interactions, however, the male and female share reproductive interests in their jointly produced offspring. Because there are partial and obligate overlaps and conflicts of interest, the male–female interaction also resembles the parent–offspring interaction (Dawkins 1976; Parker 1979; Queller 1994), except that parents tend to be considerably more dominant over individual offspring than members of one sex are over members of the other (Trivers 1972; Alexander 1979). If female and male interests were wholly coincident, the study of mating systems would be of little theoretical interest.

With respect to natural selection the hostile forces of nature are predators, parasites, pathogens, food shortages, climate and weather (Darwin 1859). With respect to sexual selection (Darwin 1871), the hostile forces are (1) members of the other sex that reduce (or deny) the reproductive success of an individual by (a) rejecting it as a mate or (b) coercing it against its interests (for the latter, see Smuts and Smuts 1993; Clutton-Brock and Parker 1995); and (2) members of the same sex that compete for mates. Conflict and competition occur because some potential mates are likely to be better than others, and individuals gain from mating with the highest-quality partner available (Andersson 1994). Overlaps of reproductive interest between partners can be eliminated as a result of the appearance of a better potential mate for either.

The importance of intersexual conflicts and competition in shaping the mating sequence is compounded by males and females having evolved different life strategies. Evidently, an original disruptive selection on smaller and larger gametes was caused by differences in success before and after zygote formation, respectively (Parker et al. 1972), and was followed by a long history of multiple mates being available for both sexes. As a consequence, the two sexes produce different-sized gametes, invest differently at different stages of offspring production, and have correspondingly divergent life strategies (Bateman 1948; Trivers 1972;

Alexander and Borgia 1979). Females can be defined as the sex that invests more prezygotically in the individual gametes (ova), evidently owing to a long history of tending to invest more postzygotically in the individual offspring (although females do not invariably do so now). Males, in contrast, have evolved to invest more extensively in collections of gametes (i.e. ejaculates rather than individual gametes), hence they typically exert more effort in securing multiple mates and thereby potentially fertilizing more ova. This divergence of life patterns has caused males (generally) to seek control of fertilization in the interest of fertilizing more ova. In contrast, females (generally) seek control of fertilization in the interests of (1) restricting advantageously the paternity of offspring, and (2) controlling the timing and context of production of zygotes so as to increase the likelihood of survival of offspring in ways that typically go beyond, or oppose, the interests of the male. Exceptions are species in which males (secondarily) invest more in offspring (zygotes) than do females.

The male–female difference is evidence that dramatic effects can arise from even slight conflicts of interest if such conflicts are long continued (Parker *et al.* 1972; Haig 1993). The ultimate consequences of the male–female divergence, discussed by Bateman (1948), Trivers (1972), and Alexander and Borgia (1979) are that competition and conflict between males and females are inevitable (Borgia 1979; Clutton-Brock and Parker 1995), and that in most species females are limiting for males rather than vice versa (see, for example, Emlen and Oring 1977; Andersson 1994).

In this chapter we explore sexual conflicts and confluences of interest, and develop a theoretical framework for understanding stages of the mating sequence. Using this framework, we develop general predictions about the role of sexual conflict in trait evolution and speciation, and attempt to test alternative theories.

COMPONENTS OF THE MATING SEQUENCE

For sexual animals with internal fertilization the mating (reproductive) sequence can be divided into eight components, not all of which are represented in all animal groups. As with all biological phenomena, it is useful to consider the events of this sequence as (1) the product of past evolution and (2) what selection has to work on presently. Distinctness of events in such a sequence suggests the extent to which selection has operated on them differently.

Our discussion is biassed towards insects, especially towards the pterygote clade (winged and secondarily wingless insects), and, with respect to examples, towards certain groups of Orthoptera and Homoptera with long-range acoustical signals (see also Gwynne and Morris 1983; Otte 1977; Bailey and Ridsdill-Smith 1991; Alexander in prep.). Because we are most concerned with male–female interactions, we do not discuss direct male–male competition, oviposition, or parental care by the female alone.

(1) Rapprochement or pair formation

This phase involves (a) members of one or both sexes actively seeking one another in the appropriate habitat, (b) members of both sexes gathering at some locale as a result of extrinsic stimuli such as the odor of oviposition plants or visual patterns in the environment, or (c) members of one sex signaling ('calling') to members of the other sex. During the different events of rapprochement, one sex may perceive the other through a single sense, either or both sexes may be guided by localized extrinsic stimuli, or neither sex may perceive either the other or any particular aspect of the environment, even though one or both may be searching for the other.

(2) Courtship

The courtship phase may be defined structurally as that part of the mating sequence, typically but not exclusively prior to copulation, when *both* male and female are in range of the other through at least one sense and are responding to one another, at least one of them (the courter) sexually and positively. In functional definitions we generally assume that one individual is exerting effort toward reducing the other individual's resistance to copulation either (1) by signaling the presence of a courter (usually the male), causing a change in motivational state or 'willingness', as by accelerating a general physiological change to mating readiness, independent of the particular male; or (2) by advertising positive qualities that may bias the other individual's choice toward the particular courting individual. Watson and Lighton (1994) include in the functions of courtship: (1) closing the distance between male and female (attractant courtship: including our rapprochement); and (2) influencing the female's choice of courting male (persuasive courtship: largely, our second function of courtship). In general, and for reasons well reviewed, males tend to court and females to choose (Bateman 1948; Trivers 1972; Alexander and Borgia 1979).

Our functional definition would include in courtship not only the usual close-range interactions, but also exchanges of acoustical or visual (and possibly olfactory) signals that occur during later stages of rapprochement in insects such as katydids (while the pair is still outside tactual range) (Spooner 1968) and fireflies (Lloyd 1983). In these insects females go only part way towards the signaling males during rapprochement and then begin answering the males' signals while waiting for the males to approach them. Courtship would also include behaviors, primarily by the male, that facilitate insemination and fertilization through altering the female's behavior after engagement of the genitalia. Courtship evolves to end when actions of the courting individual can no longer affect the likelihood of his paternity, either because paternity is decided or because the female controls remaining events leading to fertilization and he cannot influence her. As Eberhard (1985) points out, to the extent that female choice can occur after onset of copulation, or that coupling does not automatically ensure insemination or fertilization, there might be 'copulatory courtship' or even postcopulatory courtship.

(3) Copulation

During this phase the genitalia are engaged, and intromission may be involved. One sex or the other may be resisting, but the pair is mechanically or physically coupled. In many sexual invertebrates and vertebrates (e.g. all primitively wingless insects) coupling does not occur because spermatophores (sperm clumps or drops, usually encased) are transferred indirectly. All winged and secondarily wingless insects, however, actually copulate (Alexander 1964), although odonate males must first load sperm into an intromittent organ evolved on the second and third abdominal sternites, separate from the opening of the genital tract (Corbet 1962). Spermatophores occur generally, not only among both vertebrates and invertebrates that transfer sperm indirectly, but also in copulating forms, including nearly all anciently derived pterygote groups (Kristensen 1981). In many forms, (e.g. the Orthoptera Ensifera), only devices that thread the spermatophore tube enter the female's body; the bulb of the spermatophore remains external.

(4) Insemination

In this phase sperm enter the body (usually the spermatheca) of the female, either during or following copulation. In arthropods that do not copulate directly (gonopore-to-gonopore), insemination can be under the female's control and occur even in the absence of the male, for example in Collembola in which spermatophores are placed in 'fields' and picked up by females independently of the presence of males (Alexander 1964, in prep.; Kristensen 1981).

(5) Postcopulatory and intercopulatory events

Alcock (1994) reviews what he terms 'postinsemination associations' and various hypotheses offered for them, focussing on the mate-guarding hypothesis (Parker 1970). Included are such behaviors as holding and guarding of the female by the male, formation of mating plugs, manufacture and installation of spermatophylaxes (Sakaluk 1984) and prolonged copulation. These behaviors may take place either following insemination or while insemination is proceeding, the latter for example after copulation in insects such as *Gryllus* that copulate via spermatophores with long tubes through which the sperm pass during a period following the actual coupling (Alexander and Otte 1967a; Loher and Rence 1978). Postcopulatory holding includes the evolution of extended periods of coupling maintained by males that are able to prevent disengagement of the genitalia (review by Thornhill and Alcock 1983, p. 345). In some insects, postcopulatory or intercopulatory behavioral interactions, which are appropriately seen as a part of the mating sequence, continue until fertilization has occurred (Sivinsky 1984). During the same periods, females may engage in various activities that influence paternity (reviewed by Eberhard 1985).

(6) Fertilization

In this phase sperm penetrate eggs and zygotes are produced. In pterygotes, fertilization tends to occur at the time of oviposition, the sperm stored in an internal spermatheca. Special mechanisms bring the sperm and the egg together as the egg passes the opening of the spermathecal tube.

(7) Cooperative parental care

In this phase, both sexes invest in rearing their jointly produced offspring. Examples are some monogamous carrion beetles (Milne and Milne 1976), cockroaches, and dung beetles (Thornhill and Alcock 1983; Matthews and Matthews 1978). Presumably confidence of paternity is high in all such cases, with postcopulatory guarding or holding virtually certain unless the pair is isolated from possibilities of philandering. In some cases, as in *Anurogryllus* (and probably other genera of the typically deep-burrowing, parental Brachytrupinae, such as *Gymnogryllus*: R. D. Alexander, unpublished), males provide extensive burrows stocked with food which are sometimes commandeered by the female through maneuvers associated with the copulatory act (Alexander and Otte 1967a). These burrows are subsequently used to rear the brood (West and Alexander 1963; Walker 1983).

(8) Bonding

Bonding ceremonies are a component of long-term cooperative parenting. Such ceremonies, well-known in birds and mammals, may be erroneously interpreted as courtship because they include courtship-like behavior (cf. Daly and Wilson 1983). They may, however, occur outside oestrus or circumovulatory periods, and they may involve little or no evidence of sexual interest by either partner. On the other hand, they can also involve actual copulation, or behaviors otherwise restricted to events surrounding fertilization, which, in bonding, have nothing directly to do with zygote formation (as, for example, in humans). Evidently, no-one has reported bonding ceremonies in insects; this may not be surprising, since individual recognition seems not to have been demonstrated in insects, unlike in many birds and mammals. If bonding has evolved in insects, likely candidates are cockroaches, beetles, and termites that remain monogamous for long terms; one might examine the details of tandem behavior of male and female termites after rapprochement (cf. Thornhill and Alcock 1983, p. 241).

HOW SELECTION WORKS ON THE MATING SEQUENCE

Conflicts and confluences of interest between male and female

Despite the fact that for both sexes any sexual interaction is better than none (= no reproduction, excepting nepotism to collateral relatives), divergences in the ways the two sexes deal with (1) competitors of the same sex and (2) zygotes mean that, at every point in the sexual sequence, there are potential conflicts as well as confluences of interest between male and female. These conflicts and confluences, which are always mediated by ecological factors, determine the manner of evolution of the mating sequence (a) by determining the nature and intensity of sexual competition within the sexes and (b) by causing evolutionary chases in trait changes (see below).

Conflicts between the sexes should be absent only in cases of lifetime monogamy with little or no opportunity for reproductively profitable philandering, and consistently equal investment in the offspring by the sexes. Once copulation has begun, conflicts will be much reduced in species in which females typically mate but once (e.g. screwworms: Leopold 1976) or mate multiply with but a single male (e.g. probably in some termites), even if males are polygynous and do not participate in care of the offspring. Single copulations by females, however, do not eliminate conflicts of interest if copulations that are terminated 'prematurely' cause females to remate (Thornhill and Alcock 1983). Moreover, because remating by females can reduce the parentage of the first male (e.g. in Tettigoniidae: Gwynne 1988), repeated mating is a form of female choice, hence it might be expected to be employed in some cases in which females normally mate but once (Walker 1980; Thornhill and Alcock 1983; Eberhard 1985; Westneat *et al.* 1990). Males with opportunities to be polygynous may still gain from controlling events of mating in ways that conflict with the interest of even a wholly monogamous female (e.g. minimizing time or investment on any one female).

Fertilization (phase 6) represents the culmination and *raison d'être* of the sexual part of the reproductive sequence (i.e. phases 1–6). To some degree the evolution of the entire sequence can be interpreted as competition between the sexes for control of fertilization. Control of the final aspects of fertilization by the female increases her control of which male sires her offspring, therefore control over all aspects of sexual selection. Similarly, any degree of male control of fertilization thwarts female choice and serves the male's interests against those of the female. As Parker (1979, p. 149) notes, '...the asymmetry of aims [of the sexes] may ultimately be a much more important determinant of the evolutionary outcome than selection intensity, though the result must depend on the interactions between the two'.

Conflicts of interest and evolutionary chases

In whatever respects male and female interests differ there will be evolutionary chases (or races) between the two, which may be unending. Parker (1979, pp. 124–5) describes the asymmetry of this conflict as follows.

. . . consider a case in which a characteristic yielding a mating advantage to males causes some disadvantage (cost) to the females with which they mate. The female will always benefit from a mating with a male possessing the characteristic, provided that the cost is infinitesimal, if this means that some of her sons will inherit the advantage (see Fisher 1930; O'Donald 1962; Maynard-Smith 1956). Similarly for the male, if the costs are felt by his own progeny via the damage to his mate, then his mating advantage must be correspondingly greater than for zero costs. Hence as the cost increases there will be two thresholds of cost, one for the male and one for the female, beyond which the male characteristic (or a mating with a male possessing the characteristic) becomes disadvantageous. If these thresholds differ, then sexual conflict exists when conditions lie between the two, for example, when the characteristic is favorable to males but not to females.

Parker (1979) divided complex evolutionary chases between males and females into resolvable chases, which lead to an eventual evolutionarily stable strategy, and unresolvable chases which do not.

. . . if there is nothing one sex can do to avoid disadvantages inflicted by the other, then evolution simply favors making the best of things. Alternatively, there appear to be some instances where each sex can 'retaliate' against the other and where benefits become conditional on the strategy of one's opponent. This can lead to complex and sometimes apparently unresolvable 'evolutionary chases'.

He notes that unresolvable chases can occur when the game involves costs that are independent of the opponent. Thus:

. . . a male might have an anatomical feature . . . which causes enhanced success against female rejection, and vice versa. The cost of the morphological specialization is thus constant and independent of conflicts, though *gains* from conflicts do depend on the level of cost 'chosen'. This sort of game could be fundamentally a rather important one, especially for prey-predator systems as well as sexual conflicts.

Thus, if females evolve to resist males (or to be selective in ways that exclude certain males or cost all males in time and effort), males will evolve to overcome resistance of females (see, for example, Arnqvist and Rowe 1995; Jorma-lainen and Merilaita 1995). Each change in one sex that helps its members with the change secure their own interests to a greater degree is likely to be countered by changes in the other sex. Directions of change will not be entirely predictable because changes in strategy by each sex will be responses to alterations of behavior or morphology in the other sex that may occur by chance or for reasons independent of the interactions between the sexes (hence Parker's word 'chase'). Any change in one sex may incidentally thwart members of the other sex in serving their interests, and initiate responses that will also be unpredictable, except that they will be favored if they counter the changes in the first sex.

Complexity of structure and function, seemingly an inevitable development when male and female interests conflict, will tend to accelerate evolutionary change in mating traits, and increase unpredictability, by providing ever more different avenues of effective change. Thus, when sexual conflict drives the evolution of traits, great complexity and diversity are expected both between and within species, in the latter case depending on degrees of isolation among conspecific populations.

Evolutionary chases between the sexes will result not only in complexity but in 'matches' between the complex genitalia of the female and the male, in ways that may sometimes deceive observers into thinking that some kind of cooperative or convergent evolution has taken place, when in reality each sex may have been evolving to thwart the other, one changing first and the other following so as to counteract the change. The clue to this interpretation will often be evidence of forcing or manipulation by one sex, usually the male.

Because females are generally more limiting for males than vice versa, forcing or coercive mating behavior (cf. Smuts and Smuts 1993; Clutton-Brock and Parker 1995) tends to evolve more in males than in females. Even though females are sometimes able to counter male coercive tactics, the general evolutionary trend has been toward greater male control of events prior to fertilization (for a reversal in insects, see Simmons and Bailey 1990). This is because males tend to gain more from evolving risky coercive mating strategies than females gain from evolving to counter them. Parker (1979) notes that males are likely to persist more in an attempted mating than females can afford to resist: 'An important condition in the "war of attrition" game is that the costs to each opponent are set by the opponent willing to persist least' (see also Parker 1983). We do not argue that the female is designed to thwart every action of the male, but that she is designed to prevent the male from controlling fertilization completely.

Evidence supporting the concept of persistent conflicts of interest between males and females can be found in the nearly universal 'uncoupling' (Eberhard 1985) of insemination and fertilization in organisms that copulate, as compared with many kinds of non-copulating organisms (e.g. some fish and amphibians) in which sperm are deposited directly on eggs, causing immediate fertilization. When insemination and fertilization are 'uncoupled', insemination of a female by a male does not guarantee that his sperm fertilize the eggs at that time, or even at all. Females gain from maintaining the distinctness of these processes because (1) they retain the ability to determine the appropriate (for the zygotes) time and context for future reproductive events such as oviposition, and (2) they may retain the ability to use the sperm of different males. Both of these female advantages thwart male interests. Delay of fertilization means increased potential for a male's gametes to be supplanted by those of other males. It is thus unlikely that males and females will ever share exactly the same interests regarding fertilization, especially when males are polygamous or nonparental. Males, then, should evolve to place their gametes directly on eggs, except in the absence of male–male competition (e.g. extreme isolation of monogamous pairs, in which timing and placement of zygotes is equally critical for both sexes). Given the above characterization of male–female interests, the apparent absence of such 'direct fertilization' in copulating species is likely explainable by females evolving counteradaptations that reduce the effectiveness of male traits for controlling fertilization. If females always gained from favoring males that tended to serve their own interests by placing the sperm closer to the eggs, then the uncoupling of insemination and fertilization could not persist. Copulatory courtship (Eberhard 1985) is a consequence of the uncoupling of insemination and fertilization; it could not have evolved, and could not be maintained, except as a consequence of a male–female conflict of interest. Further, if copulatory courtship involves females evolving to favor certain males, rather than males overcoming female resistance, there must not be any instance in which both male and female interests are served by recoupling insemination and fertilization.

EVOLUTION OF COMPONENTS OF THE MATING SEQUENCE

Evolution of rapprochement

The evolution of the initial coming together of the sexes, and its developmental background in insects, has been shaped by confusing signals or noise from other sources, predators and parasites attracted by signals, sexual selection, and changes in patterns of parental investment. With respect to signal evolution we consider only the first three of these four factors.

Long-range signals and resources
Sexual selection can occur during rapprochement as soon as traits of either sex become involved in the search. Long-range signals, for example, may vary in range, precision, and amounts and kinds of interruption. Even ability to be in a certain habitat at a certain time is a trait on which sexual selection can act.

There are three possible situations with respect to whether or not males signaling at long range have resources to offer responding females: (1) the male has no resources except genes; (2) the male has mobile gifts such as nutritious gland secretions, large spermatophylaxes, or other kinds of food offerings; and (3) the male has stationary resources (such as burrows or crevices) that can be used by the female as: (a) shelter, (b) oviposition sites, (c) sources of stored food (including the male's own body), or (d) locations suitable for rearing young, usually including all of the above. All of these variations occur in the acoustical Orthoptera and Homoptera, and appear to have influenced the forms of acoustical signaling and the female's approach to the male.

When males with long-range signals (as opposed to males that search for females) have no resources but genes, males tend to develop dense aggregations called leks (more specifically non-resource-based leks: Alexander 1975) to which females are attracted. This situation has evidently been promoted by females (by their refusal to mate except when multiple males are present) because it provides them with the possibility of securing a high-quality mate with the least expense (Wrangham 1980; Bradbury 1981; Reynolds and Gross 1990). Once a female has approached such an aggregation, she may signal no further, requiring the male to locate her and then court her, sometimes extensively, or she may signal nearby calling males (e.g. in *Magicicada*: Alexander *et al.* ms.). Presumably, there is a better chance that a more vigorous male will locate a female sooner.

When the male's resource is movable, as with glands and spermatophylaxes, females may approach males producing long-range signals but stop and engage in an exchange of signals, eventually requiring the male to approach the female rather than vice versa. This system fosters competition among males in the vicinity, several or many of whom may attempt to disrupt the communication between male and female, replace the signaling male, and establish themselves as the successful copulating individual (Spooner 1968; Otte 1970). This system may sometimes protect the female from parasites or predators attracted to the male's calling song (see, for example, Walker 1964; Cade 1975; Bell 1979; Burk 1982; Heller and von Helversen 1993). If the spermatophylax becomes valuable enough, males become limiting and coy, and females compete for them (Gwynne 1981; Gwynne and Simmons 1990; Will and Sakaluk (1994), however, failed to obtain evidence that the large spermatophylax of the cricket *Gryllodes sigillatus* has reproductive value for the female).

When the male's resource is a burrow or other stationary resource the female approaches him directly and may copulate quickly. In the Ensifera, in which the female mounts the male in copulation, she may walk directly into the mating position. In certain Brachytrupinae (Gryllidae) the female sometimes copulates only after entering the burrow and establishing herself in a position from which she can retain ownership of the burrow; after mating, the female usurps the male's burrow and its stored food and uses them to rear offspring (West and Alexander 1963; Alexander and Otte 1967a; Walker 1983).

The nature of a male's resources may thus affect the evolution of the mating sequence; presumably, long-range signaling is always associated with some kind of stationary resource, whether it be a safe haven (for the female), an oviposition site, a group of males from which to select a mate, or food, including glandular secretions, spermatophores, wings, or the male's blood (Alexander 1964; Alexander and Otte 1967b). Luring sequences can continue only so long as the appropriate resource remains part of the male's portfolio.

Long-range signals and risks

When long-range signals are involved in rapprochement, there are two general situations: (a) signaling is more dangerous or expensive than moving to the signaler; or (b) moving to the signaler is more dangerous or expensive than signaling. In each case it appears that the female has assumed the less risky task, so that the male is required to perform the more expensive task (Alexander and Borgia 1979; Sakaluk 1990; Bailey 1991). This outcome is possible because females invest more in individual offspring; hence they are able to be choosier (Trivers 1972). When investment patterns are reversed, so, correspondingly, are the patterns of communication and coyness (Williams 1966; Trivers 1972; Alexander 1974; Morris 1979). When long-range signals are acoustical or visual they tend to be made by males rather than females. Such signals are calorically expensive and also easily followed by predators and parasites, which only need to hear them and not to decode them. Males are more likely to accept these costs. Apparently, when long-range signals are made by females rather than males, they are likely to be chemical signals because chemical signals are more difficult for predators to notice; thus, males are taking the larger risk. With chemical signals predators are more likely to have to evolve to respond to the actual signals, in contrast to an acoustical frequency channel that can carry a wide range of different sound patterns.

Williams (1992) doubts that most chemicals produced by females have evolved as chemical signals, referring to 'the female pheromone fallacy'. He expresses skepticism about females signaling to males at long range, and argues that female-produced pheromones are not designed as signals; male pheromones, he argues further, function in courtship but not in long-range signaling. He suggests that females do not typically have organs designed to disseminate pheromones, that female pheromones are not typically species-specific, and that only minute amounts of chemical are released compared with male pheromones and alarm pheromones (Greenfield (1981) reviews these questions). Numerous female insects, however, have prominent scent-producing glands, some of them protrusible, and many females signal using special positions of the abdomen (Jacobson 1972; Roelofs 1975; Hölldobler and Wilson 1990; see also Phelan, this volume). Moreover, many scents produced by females are indeed species-specific; that these may be '...precise blends of a number of components' (Roelofs 1975) rather than separate individual chemicals does not alter this point. Neither Williams (1992) nor Hammerstein and Parker (1987) cites the argument of Alexander and Borgia (1979) that females signal with pheromones because pheromones are less available to predators, so that travel to the signaler is more dangerous in such species than is the signaling. Production of small amounts of pheromone is consistent with minimizing risk of predation, and possibly with mate choice by females (Greenfield 1981). Nevertheless, predators do sometimes locate pheromonal signals of prey. Thus, Thornhill and Alcock (1983, p. 127) relate a case in which a predacious clerid beetle responds to the sex pheromone of tunneling female *Dendroctonus* bark beetles (Vité and Williamson 1970). It appears that male bark beetles alighting on a tree infested with burrowing females are at greater risk than the signaling females in their burrows, which can account for females continuing to produce pheromones despite the predator having evolved to detect them.

Long-range signals and confluences of interest

The effectiveness of a calling signal is determined by the number of sexual partners that respond positively to it. This means, first, that such signals will be most effective if produced when the greatest numbers of potential sexual partners are active and willing to respond, so there will be restriction of calling to daily periods when, for example, predators are least dangerous. Second, it means that signals that carry farther will be more effective. So there will be directional selection on the intensity of male rapprochement signals: in the case of acoustical signals, loudness, or carrying potential for the female. Insofar as pitch (frequency or kilohertz) is concerned, the auditory organs that evolve in conjunction with calls lacking in melody but temporally patterned (all acoustical insects) tend to hear some frequencies as 'louder' than others. As might be expected, the auditory organs of a species tend to be most sensitive to the particular frequencies present in the calls of conspecifics (Ewing 1984; Huber and Thorson 1985). The calls and the auditory organs are tuned together so that conspecific calls can be perceived at the lowest possible level, or such that the range of the call is maximized (Walker 1957; Gerhardt 1994; Forrest 1994). In this feature conflict of interest between males and females is only expected if variation in frequency or frequency spectrum in the male's call involves a cost that prevents the male from achieving an extreme that allows the female to use the frequency trait in mate choice (for possible examples, see Bailey 1985; Bailey et al. 1990).

There are two aspects of temporal patterning in a signal: the particular species-specific pattern, and the overall continuity or 'uninterruptedness' of signaling ('uninterruptedness' does not refer to call differences such as those between chirping and trilling, but to signals not broken as by disturbances). We hypothesize that an uninterrupted flow of signaling attracts more sexual partners than a frequently interrupted or broken pattern, for two possible reasons. First, if females move only toward the loudest call, then any break in the call will presumably cause a female to turn temporarily toward the next loudest call (see, for example, Minckley 1995). The longer the break lasts, the more likely it will be that the female will have moved to a point where the second male's call is louder than that of the first male, so that the female will continue to respond to it even when the first male resumes calling.

Second, amounts and kinds of interruption in a male's calling may give a female information enabling her to discriminate against a frequently broken pattern. Unbroken calling patterns could indicate that (1) the male's vicinity is safe, lacking predators and parasites that could interrupt the male and endanger the female; (2) the male has access to a safe retreat or burrow and takes greater risks; (3) the male is unusually healthy or capable (not only expending more calories but also less susceptible to predation); (4) there is an abundant food supply; or (5) the male is old and can afford greater risks. Singing is extremely costly, because of high caloric costs (Prestwich and Walker 1981), because parasites (Cade 1975) and predators (Burk 1982) are attracted, and because a male cannot easily forage while calling. These kinds of information may be so important that females have evolved positive responses to uninterrupted or rhythmically interrupted calling, and negative ones to irregularly broken calling, over and above the purely mechanical effects described earlier (see, for example, Hedrick 1986).

The effective range of an acoustical call depends not merely on its intensity, and its frequency in relation to the response curve of the hearing organs of conspecifics, but also on its structure: on the rates and patterning of pulses and chirps within it (Bailey 1985). Fainter signals are harder to recognize because, with greater distance from the signaler, the signal's distinctiveness becomes less apparent (Simmons 1988; Forrest 1994). Thus, a signal's range may be increased by selection causing a signal's pattern to be relatively invariant. This effect can occur whether or not there are other similar signals in the acoustical environment. In human communication, a call for help is less likely to be identified as such as distance from it increases, whether or not there are other calls (similar or not) in the environment at the time. With greater distance from the caller, there is also likely to be an increasing number and variety of potentially confusing or obscuring signals. These signals may be those of related or similar species, or they may come from any acoustical source at all. When a call has a complex within-chirp pattern that is demonstrably not necessary to attract the female from long distance, as Walker (1957) showed with the snowy tree cricket (*Oecanthus fultoni*), it is reasonable to wonder if the within-chirp pattern becomes important to the female at close range. In the bladder cicada, *Cystosoma saundersii*, song frequency works at long distance, and pattern up close (Doolan and MacNally 1981).

When songs of species are confusingly similar – as is often true when two species are newly sympatric and synchronic – females presumably gain when their auditory tympana and central nervous systems are tuned to match the frequencies and pulse patterns, respectively, of the males most different from confusing species (Otte, 1992). Males, in turn, will change to match the most common frequency tuning of female auditory tympana and the most common pulse rate or pattern receptivity (see Fig. 1-1). Both sexes will then stabilize on the same ranges in these features. On the other hand, males will continue to evolve to produce more intense (longer-range) songs and steadier songs, including any profitable deceptive elements if possible. The parts of the song in which there is the least conflict of interest between male and female are those expected to be under stabilizing selection. Alexander (1962a) identifies the particular elements in different kinds of cricket songs that are expected to be under stabilizing and directional selection, respectively, by contrasting the kinds of changes that occur within species repertoires, and between species in the same signaling context.

We believe that the effects just described will cause the temporal patterns of calling signals in insects and anuran amphibians to evolve in roughly the same way that the auditory organs and sounds are frequency-tuned to one another. Loudness in calling increases signal range as a result of directional selection on the male, while both frequency (kHz) and consistency in temporal patterning maximize signal range as a result of stabilizing selection on both sexes.

Evolution of courtship

When males change from calling (rapprochement) signals to courtship, they tend to shift to less intense or different signals (Alexander 1962a, 1975; Thornhill and Alcock 1983), either because such changes render males less available to competitors and predators, or because the female is being asked to do something different from rapprochement, or both. The combination of these shifts and tendencies is surely responsible for the ability and tendency of biologists to make confident distinctions between rapprochement and courtship. Nevertheless, there is a long history of confusion on these topics, partly because it has commonly been assumed that female selection of males terminates when copulation is initiated (but see Thornhill and Alcock 1983; Eberhard 1985; LaMunyon and Eisner 1993), and partly because courtship in humans (almost unavoidably used as

a frame of reference) has often been regarded as beginning with the initial actions of pair formation. Moreover, confusion has existed about which aspects of human sexually significant interactions parallel sexual interactions of organisms such as insects, and which of them parallel bonding ceremonies in organisms such as birds and mammals in which both sexes participate in parental care.

Courtship behaviors usually serve one of two functions (mentioned above): to signal the arrival of a ready mate (and cue subsequent reproductive behaviors on the part of the courted individual), or to advertise qualities that may bias the choice of the courted individual toward the courter. The former function need not involve conflicting interests between male and female, while the latter function invariably does.

Evolution of genitalia and copulation

The above assumptions about sexual conflict lead us to predict that, whenever such conflicts of interest are driving genitalic evolution, genitalic diversity and complexity will tend to be exaggerated when (1) females mate multiply, (2) sperm precedence is not complete, (3) fertilization is separated from insemination, (4) males cannot entirely sequester females following insemination, and (5) males do not participate in parental care. Moreover, particular species may embark upon sequences of evolutionary change that lead to increasing conflict between individual males and females (see, for example, melanopline grasshoppers below) and accelerate changes relating to copulation in entire species groups. Changing the above circumstances may reduce the extent of complexity and diversity in genitalia, copulatory and postcopulatory holding devices, and behaviors associated with postcopulatory guarding. Complexity and diversity in such attributes are expected to be minimal when male and female participate about equally in parental care and remain monogamous either for the rearing of an entire brood or for life.

The above predictions probably fit most insect species, herd-living mammals, polygynous birds that lack male parental care, and other forms with the appropriate conditions of life. In contrast, genitalia are virtually absent in some birds, and in termites in which the reproductive pair is isolated for their adult lives; in some of the latter sperm are not motile (Sivinski 1984). These situations remain to be clarified, but may stem either from reduced conflict (as in termites) or from special kinds of conflicts (as in birds: Isabel Constable, unpublished manuscript).

FEMALE CHOICE AND THE MATING SEQUENCE

If males have heritable variations in quality with fitness consequences for females, and if females have heritable differences in tendencies to mate with the particular males that tend to maximize the females' reproduction (Maynard Smith 1987), then selection on females will lead them to make pre-mating 'choices' among males in several ways (Otte 1974; Janetos 1980; Maynard Smith 1987). With respect to choice we use the terms 'direct' and 'indirect' to describe the manner in which the 'best' male is identified to the female: choice is indirect if processes such as male–male competition allow the female to mate with the best male without need for relative comparisons.

1. A female may choose directly among males, judging traits such as size, color, fighting or courting activities, the size or quality of a copulatory gift proffered, or a signal quality such as continuity of song that suggests a safe retreat.

2. A female may favor a particular male indirectly, rather than directly, in one of four ways:
    a. By being attracted to the loudest or most easily located signals, without comparing males (i.e. with regard to nearness).
    b. By mating with a male who removed other males from contention by dominating in a contest.
    c. By favoring a male on the basis of some trait or resource: for example, the female may feed on a glandular secretion or other male gift as she copulates, breaking off the mating when the gift is gone and thereby incidentally disfavoring a male with an inadequate gift.
    d. By favoring a male on the basis of his performance with respect to some environmental cue: for example, females may mate with whatever male happens to be in the right place at the right time, and he may be there because of particular sensory, thermoregulatory, or neuromuscular capabilities.

Female choice and runaway selection

We suggest that insects tend not to show extremes of directional sexual selection comparable to the widely studied sexual ornaments of (especially) birds, and that this restriction arises from a difference in the kinds of female choice employed by birds and insects (we do not intend here to refer to visual ornaments *per se*, but to the concept of ornaments having evolved in fashions parallel to the sexually selected ornaments of birds; excluded are traits evolved directionally as a consequence of male–male competition, such as beetle horns). Often, it may not be necessary to invoke direct comparison (direct choice of 'best-of-*n*'), especially involving learning and memory, to explain mate choice in insects. For example, if females mated with particular kinds of males are more likely to remate than females mated with some other kind of male (e.g. heterospecific versus conspecific matings), the variation may be owing to the female's ability to identify incomplete or 'imperfect' matings, rather than a 'best-of-*n*' choice. Other mechanisms, such as male–male interference, could also give the impression of 'best-of-*n*' choice: Wilkinson and Reillo (1994) found that males of stalk-eyed flies that had wider eyes also had more females in their aggregations. However, it does not necessarily follow that females actively choose, as, for example, by moving from group to group and stopping more often or for longer periods where they perceive more widely spaced eyes on the male. One must first eliminate the possibility that the contests between males, described by the authors, result in males with wider eyes being located where resources perceived by females are more plentiful or where more females are present for incidental reasons. Goulson *et al.* (1993) found that female death watch beetles are more likely to accept larger (heavier) males; it is not necessary, however, to argue that they are comparing males and choosing after the comparison: they may be using a threshold (see below).

Instead of using 'best-of-*n*' comparisons, insects appear more often to use tests involving sets of minimal criteria ('threshold choice') that can be applied even to lone individuals, minimal here meaning only that the criteria are the minimum needed to cause female acceptance. This is what Janetos (1980) called 'fixed threshold' or 'fixed threshold with last chance option' (the latter referring to cases in which a threshold of acceptance drops as time passes without opportunity to mate; threshold can also rise after mating without eliminating further mating). For example, female periodical cicadas fly into huge leks of singing males, attracted by the males' songs, and become motionless there, apparently mating with the first male to locate them and engage in at least a minimal set of courtship behaviors (Marshall and Cooley ms.; Alexander *et al.*, ms.). Even in insects that mate in huge aggregations the 'minimal criterion' may involve chiefly being in the right place at the right time (Otte's (1974) 'female accepts the most available male', p. 413).

Animals are restricted to the use of thresholds when mate criteria are developed in the individual without involvement of social (extrinsic) stimuli (what some authors intend by the term 'hard-wired'); such mate criteria work only if the internal developmental program yields a uniform response across the breeding population (Alexander 1969, 1990). Unlike birds and mammals, individual insects frequently do not encounter members of the other sex before beginning to signal or respond sexually. Most insect juveniles, for example, do not have learning experiences with their parents, as do virtually all mammals and birds, that can influence how they select mates as adults. Thus, in temperate climates, at least, all adults of one generation are usually dead before juveniles of the next generation hatch. Even when adults remain after the next generation hatches, unless there are parent–offspring interactions, possibilities of juveniles learning from adults are practically non-existent. In a high proportion of insect species a significant proportion of individuals may never encounter another individual, even of their own generation, before mating; insects in which this is the case must be prepared to perform without social experience. Adult life is often extremely short, and mating may occur soon after eclosion, which may be evidence that for various reasons, including some of the above, learning has been unlikely to yield net benefits. Insect juveniles often live so differently, and are so different from adults, that learning experiences useful in adult sexual behavior as a result of juvenile socialization seem unlikely. Finally, there seems to be no evidence from actual studies that insects show learned modifications of sexual behavior as a result of social interactions with other individuals; individuals reared in isolation seem to perform in sexual contexts in ways indistinguishable from those reared with social experience (see, for example, Alexander 1969). Although insects with migratory and non-migratory phases change some aspects of reproductive behavior as a result of crowding or isolation during rearing (see, for example, Kennedy 1961), it is not obvious that even adult insects learn socially in ways that influence sexual behavior. For example, Goulson *et al.* (1993) recorded no sequential changes (i.e. changes consistent with learning) in the behavior of females exposed successively to different males.

All of the consequences of an absence of social learning about mate choice are by no means clear. Differences in the modes of development of variable traits of males, and of female responses to such variable traits, can affect the possible kinds of runaway sexual selection (see, for example,

Moore and Moore 1988). Females of many birds and mammals appear able to learn about male traits during successive encounters of potential mates and other conspecifics so as to compare the range of variation in any group of males and choose the extreme regardless of precisely where it may fall (Lande's (1981) 'relative' and 'open-ended' preferences, the latter yielding a so-called super-optimal stimulus: see, for example, Williams 1992). This mode was that described in Trivers' (1972) version of runaway sexual selection. When females generate their preferences without learning from external or social stimuli, they necessarily enter into mating with either a threshold or a preference for some *particular* state of the male trait (Lande's (1981) 'absolute' preference). In such a case females can still select the 'best-of-*n*' mate in the following ways: (1) indirectly, for example by favoring the fastest male in a mating flight (as in honey bees: Page 1980; Thornhill and Alcock 1983); (2) by sensing multiple males simultaneously (rather than successively) and responding only to the extreme individual (Otte 1974); (3) by copulating multiply using minimal criterion choice and *raising* mating thresholds in a way that modifies the minimal criteria upward with each successive mating; or (4) by favoring the extreme males incidentally. The last case assumes that the female threshold may be set beyond the range of available males, and that females in such situations can *lower* their criteria and accept males less extreme than the female's (otherwise) 'absolute' minimum preference. This is the 'fixed threshold with last chance option' of Janetos (1980). In this last case, the female's original criterion may or may not exceed the range of male variation in the trait; if it does, the male trait will evolve to match the females' preference and then stop, unless somehow the females are again moved to even higher extremes of preference: for example, according to Lande (1981) by genetic drift. We do not regard this as nearly so likely as the kind of runaway selection described by Trivers (1972), in which females can evolve an ability to choose the extreme of male traits in long-lasting directional selection. The kind of choice excluded from this list is successive comparisons of possible mates causing learned changes in the female's willingness to mate with certain kinds of males, or in her mate criteria, with the result that she mates with only one male or a specific subset of males.

Following is a list of differences between female choice using minimal criteria or 'threshold' choice (TC) and the 'best-of-*n*' mate (BN); all assume females as the limiting sex. These criteria seem to suggest that insects are more likely to possess TC systems. They predict generally that, because internal development of mate criteria without social experience is more likely in insects, (1) extreme polygyny is less likely, (2) single matings by females are more likely, and (3) degrees and diversities of ornamentation should be less extreme.

1. In TC systems, females are expected more often to mate only once because there will be fewer easy possibilities of identifying better males following mating (but see threshold changes, above).
2. A female using BN is not expected to mate only with the first male she encounters.
3. In TC systems females are not expected to reject a male at first and then return to choose him after examining other males; this should happen frequently in at least some BN systems.
4. Directional selection on male traits will be less intense in TC systems because usually a higher proportion of males will possess the minimal criteria.
5. Assuming that females tend to agree on 'best' male features, the degree of polygyny will tend to be higher in BN systems because more females will exert effort to mate with the extreme male(s); extreme polygyny tends to increase sexual dimorphism and extreme traits in the more variable sex (for a likely exception, see Robertson 1986).
6. The degree of polygyny in BN systems will increase as the number of males available for comparison increases; this will not always happen in TC systems.
7. If it is typical for each male to be encountered only once, and one at a time, the system is likely to be TC, because females will be unable to make comparisons.
8. If development of choice is entirely internal (there are no social possibilities), the system is likely to be TC or indirect BN choice; minimal criteria can be generated internally, but ability to choose extreme males, which may differ from situation to situation, is unlikely to be generated internally.
9. When individuals mate across multiple years (or bouts), there are more likely to be learning opportunities that promote appearance of BN systems.
10. With successive years (or bouts) of mating experience, females should become measurably better at picking the extreme male only in BN systems; and long-lived iteroparous males can continue to develop extremeness in behavioral traits, including learning about fighting and competition.

11. Memory and individual recognition are likely in BN systems, unlikely in TC systems.
12. If all of the acceptable males in a TC system are removed, females are expected to reject the remaining males (unless the female employs a 'last-chance option'); if the extreme male is removed in a BN system, females should mate readily with the next (second) ranked (extreme) male.

Changes in females' ability to choose during the mating sequence

A female choosing a male, in whatever fashion, is in a curious bind. Early in the sexual sequence she has less information about a male on which to base a decision to accept or reject. As the sequence advances it becomes more costly and sometimes less possible for her to reject him. As she acquires information that would make exercise of choice significant, she may lose the ability to make the choice, even though as the interaction proceeds into closer proximity and greater physical intimacy she is also likely to be presented with a more rapid flow of information about the male. Rejection is obviously easiest (and involves the least investment) during rapprochement, particularly in species in which the male produces long-range signals so that the female is aware of him before he is aware of her. At the other extreme, rejection is probably most difficult (and costly) after insemination (Eberhard (1985) and Birkhead and Møller (1993) review possible ways to reject a mate late in the mating sequence). Thus a female gains from securing as much information as she can about a male from the nature of his long-range signals, while retaining, if possible, control in late stages, such as the ability to keep sperm from fertilizing her eggs even if they have already been deposited in her body. When possible, males may be expected to evolve to exploit this situation, as by manipulating information in long-range signals (e.g. uninterruptedness) and evolving to reduce the female's ability to reject sperm following insemination.

With respect to the actions of the female, one could divide copulations into three kinds: (1) those in which the female never rejects the male upon initiation of copulation; (2) those in which the female rejects the male only in particular situations, such as when only a single male is present and she gains from stimulating lek formation; and (3) those in which the female tries, at least briefly, to reject every male that initiates copulation. In the end, the female must accept the sperm of some male, so whenever males tend to interact with females one at a time, and not as simultaneous competitors, it also behooves females either to (a) create situations in which males compete simultaneously (e.g. by advertising ovulation, forcing ostentatious courtship chases, or only mating when or where multiple males are present) or (b) store a male's sperm in such a fashion as to be able to use either it or the sperm of a subsequent male should the subsequent male somehow be judged a better father for her offspring. LaMunyon and Eisner (1993) believe they have found such a case, stating that 'In our judgment postcopulatory sperm sorting in *Utetheisa* is exercised by the female herself'. At the least, females that store sperm can reduce one male's success in favor of that of another by mating again.

Under these circumstances males are expected to evolve to (1) defeat or avoid simultaneous competitors and (2) strive to bring about fertilization of the female's eggs by (a) placing the sperm as close to the eggs as possible or on them (hence, in part, the evolution of insemination and copulation: Alexander 1964, and see below), (b) positioning the sperm so as to reduce or preclude supersedure by other males (Smith 1984), (c) removing the sperm of previous males (as in Odonata: Waage 1979), (d) preventing the female from copulating after insemination by guarding or holding her (Alcock 1994), or (e) copulating multiple times or putting more sperm in each ejaculate. It seems unlikely that either sex can secure complete control over fertilization indefinitely, although females in species in which indirect sperm transfer has reached the point where the female alone is responsible for insemination may represent such a case.

The contrast between luring acts and coercive copulatory acts

With respect to differences in conflicts of interest, we will consider two divergent examples within the Orthoptera, illustrating two major kinds of insect mating acts: luring or persuasive acts, and coercing or forcing and manipulative acts. These two examples illustrate opposite extremes on the scale of sexual conflict. Evidently, luring copulatory acts involving transfers of spermatophores are primitive in the Pterygota, having evolved from the persuasive indirect acts of thysanuran-like ancestors. Seizing and forcing acts in Pterygota, involving either spermatophores or free sperm, are apparently all derived, often independently (Alexander 1964, in prep.), implying a general trend toward use of force by male insects.

A luring act

All field crickets studied in the widespread and speciose genus *Gryllus* copulate with the female mounting the male and the male reaching up from underneath to connect the genitalia and insert the tube of the spermatophore into the spermathecal tube of the female (Alexander 1960; Alexander and Otte 1967a). This luring act can be terminated at any moment by the female simply walking off the male's back. *Gryllus* males are also unusual among Gryllidae in lacking dorsal chemical or other gifts that could keep the female in the copulatory position for long periods. Not surprisingly, *Gryllus* mating is a brief act, usually lasting 20–30 seconds, and the male genitalia of all *Gryllus* are not only simple but essentially identical in the 69 known species found over almost all the world (Alexander 1964, 1990). On a spectrum between the two theoretical extremes of an entirely luring or enticing act and one involving force, manipulation, or coercion at every stage, these field crickets would be near the first end. Evidence of conflicts of interest between males and females of *Gryllus* is relatively slight, and exists mostly with respect to behaviors immediately prior to mating and during postmating behavioral guarding of the female by the male. Nevertheless, the long spermathecal tube of some grylline females, and the correspondingly long spermatophore tube of the male, in one species at least twice the length of the female's body (D. Otte, personal communication), suggest an unending evolutionary chase (cf. Nillsson 1988).

Numerous field crickets in other genera have both long-range calling signals and grasping genitalia (Alexander 1962b; Alexander and Otte 1967a; D. Otte, unpublished). In such cases, however, the female mounts the male and the copulatory act continues to be a luring act until the genitalia are engaged; the long cerci and antennae are thus retained (they appear inevitably to be reduced significantly when male-mounting acts evolve). The grasping and holding aspects are apparently all (still) assignable to the postinseminatory guarding situation, even though they appear to have facilitated copulation that (later) begins end-to-end in some cases (Alexander 1964), and were probably the forerunner of male-above copulation of the caeliferan (grasshopper) type in which the male reaches below the female's abdomen to attach the genitalia.

A coercive act

In contrast to field cricket males, all grasshoppers, apparently, have evolved an act (evidently anciently derived from a cricket-like luring act: Alexander 1964, in prep.) in which the male seizes the female from above and then passes the abdominal terminalia under the end of her abdomen to engage the genitalia. In species in which the male seeks the female visually and pounces on her (e.g., species of Melanoplini: Acrididae (Cantrall and Cohn 1972; Cohn and Cantrall 1974; Otte 1970)), the female presumably has little or no information about the male at the moment he pounces upon her. A female thus pounced upon by a male is also more likely be non-receptive because she is not mature or has already mated. In striking contrast to females in the field cricket genus *Gryllus*, then, the female melanopline's interests are apt to be strongly divergent from those of the male, at least at first contact. 'In all species [of Cyrtacanthacridinae and Catantopinae observed: 13] males appear to approach females stealthily and to jump onto the females without warning...Struggling between male and female occurs frequently, sometimes resulting in the separation of the pair, other times in copulation.' (Otte 1970, p.113).

The male grasshopper not only mounts the female but employs genitalia that seize and hold the female's genitalia during a protracted copulatory act that in its length is almost surely contrary to the female's interests. Conflict of interest can be estimated by the extent of female efforts to pull free, which, for example, we have observed repeatedly in cicadas (Alexander *et al.* ms.); we expect such efforts to be universal in species in which the male genitalia are employed to prolong copulation significantly. As expected, grasshopper genitalia tend to be complex and species-specific, and also vary unpredictably – and sometimes dramatically – among small populations that have been allopatric for differing lengths of time (see, for example, Cantrall and Cohn 1972; Cohn and Cantrall 1974). As also expected, genitalia of grasshoppers that pounce on females without prior acoustical or visual signals have been highly useful to taxonomists but *Gryllus* genitalia have not (Alexander 1962b, 1964, 1990); the same is true for the genitalia of other Ensifera in which the male grasps and holds the female. Indeed, the general usefulness of insect (pterygote) male genitalia in distinguishing sister species, and in all aspects of taxonomy, is probably attributable in large part to conflicts of interest between males and females, and resultant unending evolutionary chases. If long continued, such conflicts need not be large to cause quite large differences (Parker 1979; Haig 1993). Obviously, from the nature of this argument, there is no reason to expect universal genitalic differences between sister species or uniformity within species. Indeed, genitalia should vary within

species, although their complexity may still yield species-specific differences.

How luring and forcing acts have evolved

Aside from earlier speculations about stationary and movable resources, the particular historical reason why the kind of difference between cyrtacanthacridine grasshoppers and field crickets exists between specific groups is moot; in this case, the ancestral caeliferan (grasshopper) line may have lost long-range signaling, or else possessed only short- or intermediate-range signals that operated when another sensory cue was also available (e.g. an acoustical signal between individuals within view of one another, or vibrations produced after the male has mounted the female). The long-range acoustical rapprochement signals of some grasshoppers evidently evolved later from short-range visual signals involving leg and wing movements that became acoustical (Otte 1970). The interpretation that the ancestor of grasshoppers probably had a short- or intermediate-range male signal derives from: (1) the indication that ancestors of grasshoppers copulated female-above and were in the luring or attracting mode (these are universal correlates); and (2) the independent evolution of both stridulatory and auditory structures in Caelifera and Ensifera (Alexander 1964, in prep.). After an early stage when females approached and mounted males, male caeliferans evidently came to gain by seeking out females and then by leaping on them and holding them. Cricket males, however, continued a luring or attracting mode in achieving the copulatory position, and most retained long-range signals.

In some respects a luring mode is incompatible with modifying the onset of copulation to include a male seizing or grasping mode. From an initial luring or enticing act, one can envision the addition of more benefits or gifts (e.g. secretions or spermatophylaxes) or the incorporation of force, but, even considering what is known of alternative mating strategies, it is not easy to visualize increases in both gifts and force used simultaneously or for the same function. This incompatibility may indicate a significant divergence point in evolutionary history, sometimes, at least, caused by the addition of manipulatory or holding acts by the male near the end of copulation, which then tend to involve events earlier and earlier in the mating act until the entire act is changed. This evolutionary sequence has been repeated independently many times in the Pterygota. It is represented within the Gryllinae in modern crickets (Alexander 1964; Alexander and Otte 1967a). Thus, some crickets have evolved holding devices on the genitalia that function after copulation has begun, and that replace postcopulatory guarding. Such a change could have been a repeated transitional route to copulations initiated by the male seizing and holding the female. Loss of long-range signals in the early caeliferan line could have occurred during a stage in which the two sexes aggregated using extrinsic stimuli, as is evidently happening in some Gryllidae (Alexander and Otte 1967b). Subsequently, aggregations may have become the main attractant for females in some species, and in others the males may have begun seeking females outside aggregations and leaping upon them; both methods of rapprochement are represented among Caelifera today (Otte 1970).

Not all polygynous males lacking parental care and transformed from signalers into seekers and seizers have genitalia of comparable complexity and diversity. Sometimes it may be difficult to compare quite different kinds of genitalia with respect to complexity, and that fact may account for some of the variation in interpretations, as may differences in phylogenetic precursors when the above selective forces began to operate. Sometimes the relevant variations may not be in genitalia *per se*, but in some other structure or activity associated with copulation, such as claspers, the spermatophore and its attachment devices, or mate-guarding actions. Spermatophore attachment devices should be predictable from knowledge of the structure of the devices (molds) that produce them, but such molds may not be hardened and pigmented, and as a result they may be difficult to understand morphologically. To some extent, the same problem exists with respect to the internal genitalia of females, as compared to the intromittent and clasping devices of males.

Some variations in complexity and diversity among groups may occur because of differences in the intensity of competition among males: as in the extent to which there is simultaneous competition over copulation. In some cases males may compete by fighting at the copulatory site, or they may compete using structures (genitalic and otherwise) that differentially enable them to initiate copulation. Finally, there may be variation in how much of the control of fertilization is effected by males sequestering or holding females after copulation. Females may vary in how quickly after copulation they oviposit (or give birth), and they may also vary in the number of times they copulate, or in how well they can control fertilization following insemination (which is related to environmental factors that influence how important it is for them to

delay oviposition or to exert effort in hiding or placing the eggs). All of these factors can influence the extent to which selection favors elaboration of the genitalia in male–female races to control fertilization, and all must be taken into account as the relevant comparisons are developed.

Presumably, rapid directional changes are more likely in genitalic evolution when there are no long-range signals. The genitalia of grasshoppers in which males pounce on females without prior signals, for example, may be expected to change rapidly even in the absence of interspecies interference, causing divergence even during allopatry. In effect, such genitalia are *always* under the kind of selection that works on long-range calling signals only following establishment of new sympatry between species with confusingly similar songs. If species are largely defined by genitalic differences in such groups, then many more species, many more allopatric species, and many more species with small ranges are likely to be encountered. Such a difference seems to exist between the Ensifera with long-range acoustical signals and the Caelifera that lack them. Caelifera with long-range acoustical signals, however, tend to compare in these regards with Ensifera rather than with the other Caelifera (Cantrall and Cohn 1972; Cohn and Cantrall 1974; Otte 1970).

The role of the spermatheca in genitalic evolution

Particular directions in the evolution of genitalia and sperm storage devices influence future directions of change. Thus, as female insects evolved sperm storage structures and the ability to restrict egg fertilization to the moment when the micropyle is held against the spermathecal tube opening, males evidently could gain only via those sperm that happened to end up in the evolving spermatheca. This situation greatly influenced the subsequent evolution of the male genitalia, and female responses to them.

The evolution of spermathecae resulted, generally, in sperm precedence tending to favor the last male to inseminate the female – the last male to place sperm in the spermatheca – rather than the first inseminator, as is more likely in mammals that do not store sperm. This situation may sometimes place the female in the position of being able to ensure that the last male with which she mates fertilizes most or all of her eggs. At least potentially, it gives her an opportunity to choose among males, even after multiple copulations. Evidence of sperm mixing in the spermatheca, however, tends to argue against total female control of paternity since mixing nullifies the outcome of sperm competition. Competition among males is expected to result in such mixing, whereas females may be expected to evolve means of giving precedence to the last sperm to enter the spermatheca. This is so because a fully inseminated female is only likely to mate with additional males when they possess attributes superior to those that have already inseminated her (Sherman *et al.* 1988).

Evolution of spermathecae and delayed fertilization places males in a position of being able to control fertilization only by somehow causing the female not to mate again, or to oviposit soon after copulation, and it must have led to a variety of postinsemination techniques (Alcock 1994) as well as the kinds of male structures such as are used by damselfly males to clean out the sperm of previous males from the female's spermatheca (Waage 1979). In some cases it may have led to a nullification of genitalic evolutionary races, placing emphasis instead on males' ability to control females after insemination, as in *Gryllus* (Alexander 1960; Alexander and Otte 1967a).

Genitalic complexity and copulatory courtship

We have argued above that genitalic evolution may often be explained as the result of sexual conflict over control of fertilization. Eberhard (1985) argues that genitalia have evolved under the influence of female choice, with the principal advantage from genitalic changes being that males with the changes sired more offspring because they caused females to do things that increased the likelihood that those particular males' sperm would fertilize eggs. This postinsemination, internal sexual selection scenario is one way that conflicts of interest between male and female could be characterized, and one way they could be resolved. But Eberhard essentially restricts his hypothesis to female choice that can lead to Fisherian runaway sexual selection, in the sense that females choose males solely on the basis that their sons will also be capable of titillating females more effectively and thereby fertilizing more eggs. The conflict-of-interest race emphasized in this essay does not preclude any kind of 'female choice', but it ties male–female conflicts of interest about the use and structure of genitalia (and other holding devices and guarding behavior) to the question of who controls fertilization, and it seems unlikely to lead to runaway sexual selection. In this hypothesis, information that the female is securing about a male with whom she is mating, or who has just inseminated her, is not restricted to the degree or ways in which

his genitalia stimulate her internal (or external) parts, but to any and all attributes that might indicate the kind of offspring he will produce. Moreover, copulatory courtship can also evolve as a facilitator of insemination as precopulatory courtship presumably does. As with the act of capturing the female, the act of coupling the genitalia does not necessarily insure insemination, and the possibility must be considered that variations in success of insemination result from force or coercion rather than titillation.

The problem of 'matches' between male and female structures

Eberhard (1985) notes that females sometimes have special structures that appear to have evolved to fit the males' clasping devices. He believes these cannot be explained as outcomes of sexual conflict, but that the match evolved through the female changing to fit the male or both sexes changing so as to fit each other. Several possible explanations, however, are consistent with the conflict of interest hypothesis, and require only that the male evolves to fit the female and for some reason she does not or cannot prevent this outcome.

First, there is no incompatibility between some male and female devices evolving to fit one another, and others evolving so as to look as though they fit but actually resulting from conflicts. Not all events in the mating sequence involve conflicts of interest, but long-term tracking of one structure by another, even when conflicts of interest are involved, can give the impression of a mutually beneficial fit. An excellent example may be found in the bewildering complexity of the mammalian placental–uterine interface, long interpreted as serving cooperative homeostatic functions; this complexity may instead have resulted from protracted conflict between mother and fetus over resource allocation (Haig 1993).

Second, devices that male claspers fit may have existed in more or less their present form since before the male began to use them, so that we have only to ask why the female did not alter them to prevent use by the male; some other conflicting function may be one answer. An example of a structure difficult to change may be the constriction behind the pronotum in many carabid beetles, used for example by the male of *Pasimachus punctulatus* to hold the female during courtship and copulation by grasping her there with his long mandibles (Alexander 1959).

Third, the male's and female's interests may be convergent by the time (in the sexual sequence) that the claspers are used. Thus, females may sometimes be injured or killed if a male whose genitalia are coupled with hers is torn off her body by another male; if the male has grasped her with, for instance, his mandibles as well, this result may be prevented.

This list of possibilities is surely not exhaustive, but it demonstrates that even seemingly perfect correspondences between male and female parts need not jeopardize the conflict of interest hypothesis. We think it also places the burden of proof on investigators who hypothesize mutually beneficial fits of male and female genitalia.

Genitalic complexity and courtship complexity
Eberhard (1985) shows what he calls a 'weak correlation' between complex courtship and simple genitalia (and vice versa). He developed these data to test the question whether interspecific interactions or intraspecific sexual selection is responsible for species differences in courtship and genitalia. In his version of the latter hypothesis (pp. 82, 146–7), females get so much information prior to copulation that they do not gain as much by seeking additional information during or after copulation (Thus, Eberhard says, p. 82: 'Females of species that discriminate among males on the basis of premating signals have already discriminated strongly using nongenitalic cues before copulation begins; they may sometimes use additional genitalia criteria and sometimes not' and, p. 147: 'In effect, they [the genitalia] could be sheltered from sexual selection by prior strong screening according to other criteria').

These statements would suggest that, for example, because *Gryllus* females do not terminate copulations once they are initiated (i.e. because they have already 'discriminated strongly'), males do not gain from evolving genitalia that prevent termination of copulation. But males in many genera of crickets with complex, species-distinctive genitalia signal during courtship as extensively and complexly as do *Gryllus* males (Alexander 1962a,b, 1964; Alexander and Otte 1967a; Otte and Alexander 1983); they, however, remain coupled after attachment of the spermatophore, and the complex genitalia are used to hold them during postcopulation (and in some other genera have, presumably from this beginning, evolved to be attached at the onset of copulation). But *Gryllus* males may have simple genitalia because they have a luring copulatory act, not because they have complex songs or courtship. Cicadas have long- and short-range acoustical signals, but the males mount the females and have complex genitalia that apparently enable them to hold the female for hours (Alexander and Moore 1962; Alexander 1968).

Although a female responding to a long-range male signal and then to a bout of complex courtship doubtless obtains more information about the male prior to copulation than does a female abruptly seized from above by a male of which she was previously unaware, it seems unlikely that any female secures sufficient information to forego additional selection of males at any time prior to fertilization. One might as well say that when females locate males by responding to long-range calling signals, extensive or complex courtship is unlikely; however, numerous examples exist of species with both calling signals and complex courtship.

Conflicts should, by definition, become reduced when a female has received sufficient information to choose a given male, but one cannot thereby assume that females in species with relatively complex courtship receive sufficient information about potential mates. The generalization may be that conflicts over copulation should be reduced in species in which females have the most control over continuation of the mating sequence from the start of rapprochement until the point of genitalic contact (that is, when males must lure females until the beginning of copulation). Eberhard's female choice hypothesis requires that females not suffer significant direct costs as a result of male adaptations that increase control of fertilization. However, we propose that these costs can instead result in selection favoring females who regain control from males. This leads to a prediction opposite to the one described by Eberhard regarding genitalic complexity and courtship complexity: species with luring acts (and generally more complex courtship), such as *Gryllus* species, perhaps should be predicted to have been *more* likely to undergo the kind of selection on male genitalia that Eberhard (1985) describes, because the luring nature of the act reduces conflicts that might otherwise constrain the evolution of female choice on genitalia. In other words, females who control the initiation of copulation are less likely to benefit from evolving to thwart males who are better able to ensure their own paternity.

If a female has this control, one can assume that she will less often attempt to terminate a given copulation before insemination, and manipulating or grasping genitalia in males will be less likely to evolve in response to such a threat. However, this does not imply that males of any such species might not begin to evolve grasping devices to prolong copulation, or thwart female interests involving any later event in which male and female interests diverge. Once a mating sequence begins to take the form of a lure, it is likely to continue to evolve in that direction unless changing to a coercive act does not interfere with the success of the lure. Major change in the genitalia can be effected through postcopulatory holding, and change in the species' way of life. If, for example, the sexes begin to assemble for reasons other than male signals, males might then gain by moving to nearby females and taking the initiative in copulation. This could lead to evolution of seizing and grasping actions and devices, as evidently happened in the Caelifera (Otte 1970; Alexander in prep.). Once grasping devices have been evolved to function late in copulation or in postcopulatory mate guarding, they can be employed earlier and earlier in the mating sequence.

Attended and unattended spermatophores

The most extreme cases of female control of insemination appear to be those in which females independently take up spermatophores after males have produced them and left the area (references in Alexander 1964; Kristensen 1981). Because there is so little opportunity for manipulation of females by males, and because females pick up the spermatophores (presumably after inspection), little conflict of interest is expected at the point of insemination, and spermatophore complexity designed to thwart female interests is not expected. Similarly, greater spermatophore complexity is expected in species in which males are present and have the opportunity to coerce females into accepting spermatophores. Indeed, spermatophores do seem to be less complex in those species in which males and females are not in contact with one another during insemination, and more complex in those species in which the sexes are in contact (Eberhard 1985, fig. 3.5, p. 48).

Eberhard attempts to explain the pattern of spermatophore complexity as support for a female choice hypothesis, stating that complex spermatophore devices for copulatory courtship '... would be superfluous in species without male–female contact since only receptive females ever bring their genitalia into contact with spermatophores' (Eberhard 1985). Eberhard's overall theory, however, is based on the idea that male genitalic devices are initially selected to stimulate females in ways that will promote postinseminatory events in the sexual sequence, such as effective transfer of sperm to storage sites, non-receptivity to future mates, maturation of eggs, and oviposition. His hypothesis does not give a reason for the decision to take up a spermatophore being different from the decision to copulate with a male (although he notes that a female 'does not touch her genitalia to the [unattended] spermatophore

unless she is receptive', implying that males attending spermatophores must overcome some female reluctance); thus, the hypothesis does not predict a difference in complexity between the two classes of spermatophores discussed above, and might predict that spermatophores picked up by the female when she is alone should possess complicated stimulatory devices. Although Eberhard eventually uses the term 'persuasive' for structures such as 'any characteristic of a spermatophore, such as a guiding spine or an injecting mechanism, that makes it more likely that the female's genitalia will actually take up the spermatophore', earlier he says that 'Males that do not contact females cannot *oblige them to be inseminated...*' and 'males will sometimes *maneuver less receptive females* into positions in which spermatophores are in the near vicinity of the female's genitalia...' and 'Both attended and unattended spermatophores can be sensed by *marginally receptive females*, but only attended spermatophores are likely to stimulate and/or fit into such a female's genital area' (emphasis added). Spermatophores left unattended by males for later pickup by females do not seem different from male genitalia with respect to variations capable of initiating female choice and runaway selection. All of the statements quoted from Eberhard (1985, p. 153) appear to fit a conflict-of-interest hypothesis and to argue in a general way against Eberhard's explanation.

The copulatory courtship hypothesis and the conflict-of-interest/control of fertilization hypothesis both predict erratic, unpredictable directions and rates of evolution. The courtship model involves males racing to keep up with ever-changing female preferences. The conflict-of-interest hypothesis involves males evolving to coerce females, and females evolving to evade coercion. Eberhard's model almost exclusively requires the operation of 'best-of-*n*' choice, because threshold decision processes do not readily yield the open-ended preference that most easily leads to runaway selection. Eberhard is explicit (p. 71) that his model includes runaway selection: 'A key aspect of this model of genitalic evolution (and of runaway evolution by female choice in general) is the arbitrary nature of the cues used by females to discriminate among males'.

In insects 'best-of-*n*' choice mechanisms can only operate in female choice of genitalia if females copulate with a set of males and differentially favor the sperm on the basis of genitalic differences. The hypothesis of conflict over control of fertilization does not require large numbers of species in which females mate but once, but if females in large numbers of species do mate only once, the generality of Eberhard's (1985) hypothesis of copulatory courtship is reduced.

The question of the evolution of genitalia must ultimately be settled by understanding the extent to which male and female interests differ over use of male genitalia. Eberhard's hypothesis requires that females suffer only minimal costs when males evolve more effective ways of ensuring paternity; otherwise these costs will constrain the evolution of female preference for those males, and females will be selected instead to counter males and enhance their own control. For all the above reasons, and because we regard runaway evolution as less likely in insects than in vertebrates, we doubt the generality of Eberhard's hypothesis that insect genitalia evolve as a result of female choice.

Forced copulation and female control of fertilization

Controlling the final aspects of fertilization does not appear an unlikely evolutionary possibility for female insects; indeed, in some Hymenoptera females can control whether or not eggs being laid are fertilized, evidently by contracting or relaxing a sphincter that is responsible for whether or not sperm are released from the spermathecal tube (Gerber and Klostermeyer 1970; Werren 1980). It is possible that females could use such a device to control whether or not the sperm of any particular copulating male enters or leaves the spermatheca. A female may be inseminated by a male that (1) has traits indicating high quality prior to mating (see earlier), (2) is extremely adept at securing matings, or (3) is extremely good at forced mating. If there is no evidence of females of a given species having evolved to prevent certain males' sperm from entering the spermathecal tube, then it is likely that (1) the males have some exceedingly unusual kind of intromittent organ that is effective at thwarting female efforts to exert control or (2) it is in the interest of a female to accept the sperm of any male that has actually been able to couple with her. This rather startling hypothesis implies that, for whatever reasons, female choice is largely finished prior to intromission or insemination. It also implies that the evolution of male genitalia is a matter of male–male competition (e.g. getting the sperm closer and closer to eggs, removing sperm, and plugging the female's reproductive tract) and that females have evolved to accept the sperm of the male most able to get his sperm into their genital tracts

and nearest to their eggs, once the competition has become internal to the female.

If these arguments are correct, then one wishes to predict the kinds of animals in which females would be most likely to resist male attempts to control fertilization. One may at first imagine that such control should occur in species in which (1) males give no parental care, (2) males are plentiful and vary greatly in worth as mates, and (3) males are very effective at forced copulation. But this combination of conditions is paradoxical. When there is no paternal care, then one of the best signs of a male's quality as a mate is his ability to get sperm into females, and, if forced copulation is highly successful, then males who inherit the ability to be good at forcing copulation will tend to be reproductively successful. Females might be expected to resist forced copulation, even as strenuously as possible (without sacrificing actual reproduction in the effort), because presumably a copulation that must be forced is in some way contrary to their interests (see, for example, Cooley 1995). When force becomes a principal indicator to the female of male quality, females are not expected to exert as much effort to prevent a forced copulator's sperm from fertilizing their eggs once they have been placed in the reproductive tract. In other words, in the particular species in which a female's ability to control fertilization by rejecting certain males' sperm would appear at first to have its greatest potential benefit, the benefit is largely canceled by the fact that the quality of males is determined primarily by their ability to get sperm into females when they are resisting (the situation that in general would be expected to cause intense selection for postinsemination female choice). However, if only because non-parental males serving their own interests almost inevitably reduce the reproductive success of their individual mates, females may be expected to resist yielding complete control of fertilization to males.

What about species in which there is much paternal care, and paternal care is dependent on exclusive mating rights (or, in insects, birds, and reptiles, last mating rights) with the female, and high confidence of paternity? If males cannot identify or respond to variations in females' ability to control the fate of sperm once they have entered the female's body, mate-guarding (and, in some animals, infanticide when guarding fails) will be the only successful strategy for the male in controlling fertilization. If a female can obtain parental care by a superior father not good at causing or forcing mating, she may be in a position to gain by using his paternal care and the sperm of a superior male or the sperm of a male good at forcing matings. Eberhard (1985) reviews the evidence that a large amount of differential paternity occurs as a result of events during copulation, and sometimes following insemination.

Origin of copulation

The above arguments about sexual conflict help us to reconstruct the evolution of internal fertilization. Males' efforts to control fertilization in ancestral external fertilizers (to outcompete other males in the vicinity) presumably caused them to gain from placing the sperm or spermatophore closer and closer to the female's genital opening, nearer and nearer to the time of oviposition, finally inserting the sperm into the female's body before oviposition (Eberhard 1985). Thus began the process whereby females withheld the eggs at the time of copulation, males evolved longer intromittent organs, and both sexes added to the complexity of the devices they used in copulation and insemination; both sexes thereby were exerting effort to control fertilization of the eggs. Evolution toward internal fertilization surely was also abetted, at least in some cases, by a female tendency to protect eggs, especially as mating began to occur in terrestrial environments. Did side pockets in the female genital tract (as in salamanders: Salthé 1967; Boisseau and Joly 1975), and eventually spermathecae, evolve because sperm survived longer in such sites, and because females possessing such structures could time fertilization and oviposition to the best reproductive advantage? Or did such structures evolve because females were initially sequestering sperm so as to control which males fertilized their eggs?

Uncoupling could have been furthered (and maintained) by females evolving to favor males that placed sperm where the female gained from having them (e.g. in spermathecae). In such case, females would continue to disfavor males that tended to place their sperm in more direct contact with the eggs. This disagreement would tend to cause an unending evolutionary race.

THE MATING SEQUENCE AND SPECIATION

Calling (rapprochement) signals have been discussed extensively in connection with speciation because they tend to be species-specific among sympatric synchronic species, for reasons that are not entirely clear. Here we discuss evolutionary changes in signals, their consequences for speciation, and why character displacement is infrequently documented.

Despite growing acceptance of arguments that species signaling differences may often be due more to sexual selection within species than to interactions between members of different species, and therefore reproductive isolation (see, for example, West-Eberhard 1983, 1984), it is tempting to assume that interspecies interactions are the reason for widespread signal uniformity within species, and similarly for the consistency of differences between species. For insects, however, another explanation must be dismissed first. When social learning is absent, the only way that members of a breeding population can identify one another reliably (and, incidentally, populations can be potentially panmictic) is through an evolutionary tracking of signal structure and signal response capabilities. There may be brief periods of directional selection when evolutionarily divergent populations with confusingly similar signals initially become sympatric and synchronic. However, unusual signals and unusual responses will be disfavored even when there are not confusingly similar signals from other species. Internally regulated development of signals and responses, and the resultant stabilizing selection, can lead to convergences between male and female; sometimes, the same or linked genes may influence both signal structure and signal response (Alexander 1962a; Hoy 1974; Pires and Hoy 1992). Nevertheless, only arguments including species interactions, such as character displacement, can explain *both* species distinctiveness among signals of sympatric and synchronic species *and* lack of species distinctiveness among signals of similar allopatric or allochronic species.

Male visual and acoustical calling signals, more than other signals, emphasize the puzzle of apparent shifts between rapid directional selection around the time of speciation coupled with intense stabilizing selection at other times. Only male calls tend to be structurally complex, or patterned temporally, spatially, or both. Long-range visual and acoustical answering signals of females tend to be simple and nondescript in structure, with their precision often a matter of timing in relation to the male's signals. When females are the limiting sex they evidently gain little by advertising more than their sexual receptiveness. The apparent uniformity of long-range chemical signals seems less puzzling because fairly simple changes may yield great differences between species and there are probably few chemical attributes that can change on more or less continuous scales as can rates in the patterning of acoustical and visual signals. Probably, quick changes upon sympatry following speciation depend on multiple chemicals being present in a signal (Roelofs 1975), especially when their proportions and importance already vary within the species (as in different signals used in different parts of the mating sequence). Much of the apparent difference between chemical signals and acoustical and visual signals in this respect may also be illusory, or result from greater difficulty in humans comparing chemical signals. Whatever the case, principles that apply to visual and acoustical signals probably also apply to chemical signals, so we can use one kind of signal to make some arguments applying to all three kinds. In the rest of this discussion we will emphasize acoustical signals, as they occur in crickets, katydids, grasshoppers, and cicadas.

Various authors have made claims, not necessarily compatible with one another, about the importance of sexual selection (1) in initiating speciation, or (2) in cementing species differences immediately following speciation, and (3) on character displacement. Thus, West-Eberhard (1983, 1984) argues that sexual (or social) selection is a principal initiator of population divergences that later lead to speciation. Paterson (1993) argues that sexual signals are under stabilizing selection, diverge during the allopatry that leads to or allows speciation, and prior to sympatry between the newly divergent populations, and that this divergence occurs without interaction between the two evolving species. Otte (1994) in general agrees with Paterson, expecting that the acoustical environment in general causes song differences to begin in allopatry, and that this initial divergence facilitates subsequent divergence under sympatry and therefore accelerates speciation (D. Otte, personal communication). Otte attributes somewhat more significance to the general acoustical environment during allopatry than we do. Our position stems from a consideration of allopatric or allochronic species that lack any song differences, or apparently significant song differences (e.g. *Gryllus veletis*, *G. pennsylvanicus* and *G. campestris* (Alexander 1957, 1969); *G. firmus* and *G. bermudensis* (Alexander ms.); *G. rubens* and *G.* undescr. [integer of authors] (Walker 1974); *Oecanthus quadripunctatus*, *O.* [dwarf quadripunctatus] (Walker and Rentz 1967); *O. pini* and *O. laricis*; *O. nigricornis* and *O.* [undescribed on willow] (Walker 1963); *Magicicada septendecim* and *M. tredecim*; *M. cassini* and *M. tredecassini*; *M. septendecim* and *M. tredecim* (Alexander and Moore 1962)).

Species-specific patterns in calling songs may be the best example of traits that evolve towards commonality of interests between males and females. Traits that evolve under similar selection should not vary within populations and should not diverge quickly in allopatry. In contrast, traits that enable the male to act in his own interests and contrary to those of the female, or vice versa, involve continuing conflicts of interest between male and female, and thus result in unending evolutionary chases. Examples are changes in a male's genitalia or other features that enable him to capture the female for mating, hold her more securely during copulation, or hold her longer after copulation against her interests. Such traits should evolve rapidly within populations and differ extensively and in many and unpredictable ways among newly allopatric populations. They can be responsible for speciation because once they diverge far enough to prevent copulation between species they cannot be bypassed. Thus, there are likely several times as many grasshopper (caeliferan) species in North America as crickets and katydid (ensiferan) species. Most of those grasshopper species are concentrated in the Melanoplini (D. Otte, personal communication), where the greatest evidence exists of conflict of interest between males and females over use of the male genitalia (Otte 1970). Allopatry of forms with genitalic differences is also likely greater in the Melanoplini than in other Saltatoria. Thus, when genitalic evolution appears to result in large extent from male–female conflicts, correlating with use of force by males, genitalia are more complex and diverse, and speciation appears to have been more rapid. If Fig. 1-1 portrayed genitalic evolution in melanopline grasshoppers, rates of change between stages 1 and 6 would be more alike, individual populations would show greater variation, and the genitalic changes themselves might often have been causal in speciation.

Eberhard (1985, 1993a,b) has argued that the genitalia of male animals evolve in ways that benefit females as well as males, and as a result they are extensively involved in female choice even following insemination and evolve rapidly on that account. We have argued (above) that animal genitalia typically evolve rapidly because they involve conflicts of interest between male and female with respect to the details of copulation. We also suspect that character displacement is a frequent or virtually universal aspect of the evolutionary divergence of similar species newly returned to sympatry (as well as a continuing aspect of competition among species within communities, often when they are not related or highly similar overall). We also suspect that this divergence in sympatry, rather than in allopatry, accounts for most of the divergence between

Stage 1. A single species with the female-attracting parts of its song under stabilizing selection

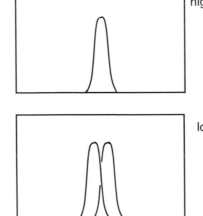

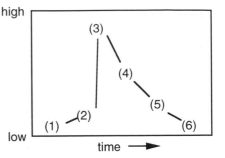

Presumed rates of evolution at each of the six stages of song divergence. Stages 1 and 6 will evolve slowly because, on average, environments will have been effectively stable for longer. Stage 2 will evolve more rapidly because populations in allopatry are likely to have different environments. Stage 3 will evolve fast, and faster than 4 and 5, because selection favoring divergence, and minority extremes, will be strongest then.

Stage 2. Two allopatric populations with songs diverged slightly through drift, pleiotropy, or slight differences in the acoustical environment

Stage 3. Sympatry occurs after postmating difficulties have appeared; song extremes at points (a) and (b), respectively, are favored in the two populations

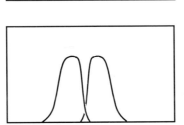

If stages 2 and 3 are heavily represented among allopatric populations, and very little of stage 6, the hypothesis that most divergence occurs during allopatry is denied. Moreover, if songs are so sensitive to acoustical overlap that selection is effective in allopatry, in the absence of extensive overlap with the songs of other species, then it should be that much stronger when species with closely similar songs become sympatric and synchronic.

Stage 4. Strong selection for divergence in song; strong selection favoring divergence as a result of interference and wasted effort because of overlap in songs between incompatible populations.

Stage 5. Two sympatric species with songs still overlapping; selection for divergence (directional selection) still strong

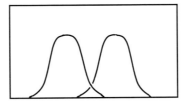

In this hypothesis, stages 1, 2, and 6 occupy the vast majority of time; 3, 4, and 5 occur so swiftly as to be observed only rarely. If stages 3, 4, and 5 are passed through very rapidly, this model may be able to account for all observations to date.

Stage 6. Songs of sympatric species have become non-overlapping; selection for divergence is essentially absent; stabilizing selection is strong within each species.

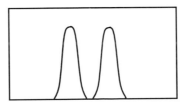

Fig. 1-1. A character-displacement model to explain song relationships among similar species that are either allopatric or sympatric. The crucial question is when sympatry appears after allopatric divergence.

similar species in sexual characters, therefore that it tends to be an effect of speciation rather than a cause, and occurs late in speciation rather than early.

If sexual selection in allopatry were a frequent initiator of speciation, as opposed to being primarily incidental to speciation or occurring near the end of the process, we should find such divergence on a large scale among recently allopatric populations. If divergence under sexual selection is sufficient to cause speciation, we should find numerous sympatric cognate species differing only in premating sexually selected traits: that is, lacking postmating differences that could maintain the interbreeding failure. If reproductive character displacement is prevalent, sympatric related species should always differ in sexually selected traits but allopatric forms similarly diverged in other respects should not (similarly diverged is a crucial phrase here, for in the alternative hypothesis divergence would be unaffected by sympatry). We should also find that recently formed species which overlap geographically only partially should be more different in the regions of overlap than in the regions of allopatry, although this requirement has to be restricted. Our argument depends on such character displacement happening so quickly that it is rarely observed in the form of great divergence in regions of overlap of newly formed species than in their regions of allopatry. Fig. 1-1 shows the process.

With respect to their roles in these kinds of events, there are several different kinds of traits under sexual selection. Traits that attract the other sex and are apparently under directional selection (such as bright coloration or tail length) are probably the sort West-Eberhard would regard as central to her argument. They are also the ones most likely to become involved in runaway selection. These are also traits (1) for which confusion among different species is likely minimal, because the signals are not broadcast great distances, (2) which tend to be attention-getters, with additional traits immediately in view, and (3) for which there are not large numbers of species signaling in the same modality (as is possible with acoustical signals). Such traits should diverge rapidly in allopatry, but because they can be bypassed during the mating sequence they are not likely to cause speciation in the absence of significant postmating differences that cause hybrids to be at a disadvantage. If these traits are involved in causing speciation one should frequently find newly formed species differing in these traits but able to produce hybrids in induced matings which survive and reproduce as well as the parental forms in the field.

Traits that attract the other sex and are evidently under stabilizing selection, including aspects of the pattern of acoustical signals or the molecular structure of chemical signals, are the kind that Paterson might include in his arguments. They are signals for which the channels are likely to be noisy because (1) they are broadcast great distances, (2) they tend to operate alone without additional confirming traits that are brought immediately into sensory range (for example, when an animal sees a bright spot, its attention is at the same time drawn to the rest of the individual's attributes), and (3) they can easily be patterned almost endlessly in the same modality. Such traits should diverge slowly in allopatry but may diverge quickly in sympatry because they will be temporarily placed under intense directional selection. If these traits diverge primarily in sympatry, one should expect to find evidence of character displacement between partially overlapping species unless there is good reason to invoke quick changes that also spread quickly to the species limits outside sympatry. Specifically, among singing insects one finds that species that breed in the same places at the same times never have the same songs; instead one finds dramatically different signals no matter how similar the species are in other regards. Species that do not breed in the same places, or in other cases at the same times, sometimes do have the same songs, or songs that overlap almost completely in their attributes, sometimes when the species involved are not even similar. The paradox is that, despite dramatic differences between the songs of sympatric, synchronic sister species, and the frequent absence of such differences between allopatric or allochronic sister species, species that partly overlap in breeding places and times do not show greater song differences in the regions of such overlap that are obvious and on the same axes of variation as the species differences (Alexander 1979, pp. 119–120). This fact has impressed workers in this arena such as Alexander (1968), Walker (1974), Otte and Alexander (1983), and Thomas and Alexander (1962) who have documented them in tapes of, overall, perhaps a thousand species of Gryllidae, Tettigoniidae, and Cicadidae with overlapping ranges. These cases are from published and unpublished reviews of, usually, the vast majority of acoustical Orthoptera and Cicadidae traced geographically throughout numerous regions including virtually all of North America and Australia.

Howard (1993) thoroughly and carefully reviewed the topic of character displacement. He concluded, certainly without disagreement from any of the above authors, that character displacement has not been studied well enough to show that it is rare or absent. However, the discovery, from meticulous studies, of even large numbers of subtle or cryptic cases (Benedix and Howard 1991) will not resolve the question of why obvious cases of the sort expected from the particular and dramatic differences among the songs of sympatric and synchronic species remain undiscovered. The reason may be that suggested in Fig. 1-1: that they are extremely short-lived.

ACKNOWLEDGMENTS

We thank Deborah Ciszek, Richard Connor, Bernie Crespi, William G. Eberhard, Richard D. Howard, James E. Lloyd, Dan Otte, and the participants of a 1996 UMMZ graduate seminar in behavioral and evolutionary ecology for criticizing the manuscript.

LITERATURE CITED

Alcock, J. 1994. Postinsemination associations between males and females in insects: the mate-guarding hypothesis. *Annu. Rev. Entomol.* **39**: 1–21.

Alexander, R. D. 1957. The taxonomy of the field crickets of the eastern United States (Orthoptera: Gryllidae: *Acheta*). *Ann. Entomol Soc. Am.* **50**: 584–602.

–. 1959. The courtship and copulation of *Pasimachus punctulatus* Haldemann (Coleoptera: Carabidae). *Ann. Entomol. Soc. Am.* **52**: 485.

–. 1960. Aggressiveness, territoriality, and sexual behavior in field crickets (Orthoptera: Gryllidae). *Behaviour* **27**: 130–223.

–. 1962a. Evolutionary change in cricket acoustical communication. *Evolution* **16**: 443–467.

–. 1962b. The role of behavioral study in cricket classification. *Syst. Zool.* **11**: 53–72.

–. 1964. The evolution of mating behaviour in arthropods. In *Insect Reproduction*. K. C. Highnam, ed., pp. 78–94. (Symp. No. 2, R. Entomol. Soc. London.) Dorking, Surrey: Bartholomew Press.

–. 1968. Life cycle origins, speciation, and related phenomena in crickets. *Q. Rev. Biol.* **43**: 1–41.

–. 1969. Arthropods. In *Animal Communication*. T. A. Sebeok, ed., pp. 167–216. Bloomington, Indiana: Indiana University Press.

–. 1974. The evolution of social behavior. *Annu. Rev. Ecol. Syst.* **5**: 325–383.

–. 1975. Natural selection and specialized chorusing behavior in acoustical insects. In *Insects, Science, and Society*. D. Pimentel, ed., pp. 35–77. New York: Academic Press.

–. 1979. *Darwinism and Human Affairs*. Seattle: University of Washington Press.

–. 1990. *Gryllus nigrohirsutus*, n. sp. from Korea, with a review of the genus *Gryllus*. *Great Lakes Entomol.* **24**: 79–84.

–. In prep. Insect mating, phylogeny and wings.

Alexander, R. D. and G. Borgia. 1979. On the origin and basis of the male-female phenomenon. In *Sexual Selection and Reproductive Competition in Insects*. M. F. Blum and N. A. Blum, eds., pp. 417–440. New York: Academic Press.

Alexander, R. D. and T. E. Moore. 1962. The evolutionary relationships of 17-year and 13-year cicadas, and three new species (Homoptera, Cicadidae, *Magicicada*). *Univ. Mich. Mus. Zool. Misc. Publ.* **121**: 1–59.

Alexander, R. D. and D. Otte. 1967a. The evolution of genitalia and mating behavior in crickets (Gryllidae) and other Orthoptera. *Univ. Mich. Mus. Zool. Misc. Pub.* **133**: 1–62.

–. 1967b. Cannibalism during copulation in the Brown Bush Cricket, *Hapithus agitator* (Gryllidae). *Fla. Entomol.* **50**: 79–87.

Alexander, R. D., R. A. Sanders, R. D. Howard, D. C. Marshall, J. C. Wilkins and D. R. Karra. ms. Acoustical behavior in *Magicicada* leks. Unpublished manuscript.

Andersson, M. 1994. *Sexual Selection*. Princeton: Princeton University Press.

Arnqvist, G. and L. Rowe. 1995. Sexual conflict and arms races between the sexes: a morphological adaptation for control of mating in a female insect. *Proc. R. Soc. Lond.* B **261**: 123–127.

Bailey, W. J. 1985. Acoustic cues for female choice in bushcrickets (Tettigoniidae). In *Acoustic and Vibrational Communication in Insects*. K. Kalmring and N. Elsner, eds., pp. 101–110. Berlin: Verlag Paul Parey.

–, ed. 1991. *Acoustic Behaviour of Insects: An Evolutionary Perspective*. London: Chapman and Hall.

Bailey, W. J. and J. Ridsill-Smith, eds. 1991. *Reproductive Behaviour of Insects: Individuals and populations*. London: Chapman and Hall.

Bailey, W. J., R. J. Cunningham and L. Lebel. 1990. Song power, spectral distribution and female phonotaxis in the bushcricket *Requena verticalis* (Tettigoniidae: Orthoptera): active female choice or passive attraction. *Anim. Behav.* **40**: 33–42.

Bateman, A. J. 1948. Intra-sexual selection in *Drosophila*. *Heredity* **2**: 349–368.

Bell, P. D. 1979. Acoustic attraction of herons by crickets. *J. N. Y. Entomol. Soc.* **87**: 126–127.

Benedix, J. N. Jr. and D. J. Howard. 1991. Calling song displacement in a zone of overlap and hybridization. *Evolution* **45**: 1751–1759.

Birkhead, T. and A. Møller. 1993. Female control of paternity. *Trends Ecol. Evol.* **8**: 100–104.

Boisseau, C. and J. Joly. 1975. Transport and survival of spermatozoa in female Amphibia. In *The Biology of Spermatozoa*. E. S. E. Hafez and C. G. Thibault, eds., pp. 94–104. Basel: Kanger.

Borgia, G. 1979. Sexual selection and the evolution of mating systems. In *Sexual Selection and Reproductive Competition in Insects*. M. S. Blum and N. A. Blum, eds., pp. 19–80. New York: Academic Press.

Bradbury, J. W. 1981. The evolution of leks. In *Natural Selection and Social Behavior*. R. D. Alexander and D. W. Tinkle, eds., pp. 138–169. New York: Chiron Press.

Burk, T. 1982. Evolutionary significance of predation on sexually signaling males. *Fla. Entomol.* **65**: 90–104.

Cade, W. 1975. Acoustically orienting parasitoids: fly phonotaxis to cricket song. *Science (Wash., D.C.)* **190**: 1312–1313.

Cantrall, I. J. and T. J. Cohn. 1972. Melanoploid genitalia and mechanical isolation. *Proc. Intl. Study Conf. Current and Future Problems of Acrididology*, London, 1970, C. F. Hemming and T. H. C. Taylor, eds., pp. 35–44. London: Centre for Overseas Pest Research.

Clutton-Brock, T. H., and G. A. Parker. 1995. Sexual coercion in animal societies. *Anim. Behav.* **49**: 1345–1365.

Cohn, T. J. and I. J. Cantrall. 1974. Variation and speciation in the grasshoppers of the Conalcaeini (Orthoptera: Acrididae: Melanoplinae): the lowland forms of western Mexico, the genus *Barytettix. San Diego Soc. Nat. Hist. Mem.* **6**: 1–131.

Cooley, J. R. 1995. Functional hypotheses concerning the mating behavior of *Lytta nuttali* Say (Coleoptera: Meloidae) in the Front Range of central Colorado. *J. Coleopt. Soc.* **49**: 132.

Corbet, P. S. 1962. *A Biology of Dragonflies.* Chicago: Quadrangle Books.

Daly, M. and M. Wilson. 1983. *Sex, Evolution, and Behavior*, 2nd edn. Boston, Mass.: PWS Publishers.

Darwin, C. R. 1859. *The Origin of Species.* London: John Murray.

–. 1871. *The Descent of Man, and Selection in Relation to Sex.* London: John Murray.

Dawkins, R. 1976. *The Selfish Gene.* New York: Oxford University Press.

Doolan, J. M. and R. C. MacNally. 1981. Spatial dynamics and breeding ecology in the cicada *Cystosoma saundersii*: the interaction between distributions of resources and intraspecific behavior. *J. Anim. Ecol.* **50**: 925–940.

Eberhard, W. G. 1985. *Sexual Selection and Animal Genitalia.* Cambridge, Mass.: Harvard University Press.

–. 1993a. Copulatory courtship and genital mechanics of three species of *Macrodactylus* (Coleoptera: Scarabeidae: Melolonthinae). *Ethol. Ecol. Evol.* **5**: 19–63.

–. 1993b. Evaluating models of sexual selection: genitalia as a test case. *Am. Nat.* **142**: 564–571.

Emlen, S. T. and L. W. Oring. 1977. Ecology, sexual selection, and the evolution of mating systems. *Science (Wash., D.C.)* **197**: 215–223.

Ewing, A. W. 1984. Acoustic signals in insect sexual behavior. In *Insect Communication.* T. Lewis, ed., pp. 223–240. New York: Academic Press.

Fisher, R. A. 1930. *The Genetical Theory of Natural Selection.* Oxford: Clarendon Press.

–. 1958. *The Genetical Theory of Natural Selection*, 2nd edn. New York: Dover.

Forrest, T. G. 1994. From sender to receiver: Propagation and environmental effects on acoustic signals. *Am. Zool.* **34**: 644–654.

Gerber, H. S. and E. C. Klostermeyer. 1970. Sex control by bees: a voluntary act of egg fertilization during oviposition. *Science (Wash., D.C.)* **167**: 82–84.

Gerhardt, H. C. 1994. Selective responsiveness to long-range acoustic signals in insects and anurans. *Am. Zool.* **34**: 706–714.

Goulson, D., M. C. Birch, and T. D. Wyatt. 1993. Paternal investment in relation to size in the deathwatch beetle, *Xestobium rufovillosum* (Coleoptera: Anobiidae), and evidence for female selection for large mates. *J. Insect Behav.* **6**: 539–547.

Greenfield, M. D. 1981. Moth sex pheromones: an evolutionary perspective. *Fla. Entomol.* **64**: 4–17.

Gross, M. R. 1994. The evolution of behavioural ecology. *Trends Ecol. Evol.* **9**: 358–360.

Gwynne, D. T. 1981. Sexual difference theory: Mormon crickets show role reversal in mate choice. *Science (Wash., D.C.)* **213**: 779–780.

–. 1988. Courtship feeding in katydids benefits the mating male's offspring. *Behav. Ecol. Sociobiol.* **23**: 373–377.

Gwynne, D. T. and G. K. Morris, eds. 1983. *Orthopteran Mating Systems: Sexual Competition in a Diverse Group of Insects.* Boulder, Colorado: Westview Press.

Gwynne, D. T. and L. W. Simmons. 1990. Experimental reversal of courtship roles in an insect. *Nature (Lond.)* **346**: 171–174.

Haig, D. 1993. Genetic conflict in human pregnancy. *Q. Rev. Biol.* **68**: 495–532.

Hammerstein, P. and G. A. Parker. 1987. Sexual selection: games between the sexes. In *Sexual Selection: Testing the Alternatives.* J. W. Bradbury and M. B. Andersson, eds., pp. 119–142. Chichester: John Wiley.

Hedrick, A. V. 1986. Female preferences for calling bout duration in a field cricket. *Behav. Ecol. Sociobiol.* **19**: 73–77.

Heller, K. and D. von Helversen. 1993. Calling behavior in bush-crickets of the genus *Poecilimon* with differing communication systems (Orthoptera: Tettigonioidea, Phaneropteridae). *J. Insect Behav.* **6**: 361–388.

Hill, G. E. 1994. Trait elaboration via adaptive mate choice: sexual conflict in the evolution of signals of male quality. *Ethol. Ecol. Evol.* **6**: 351–370.

Huber, F. and J. Thorson. 1985. Cricket auditory communication. *Scient. Am.* **253**: 60–68.

Hölldobler, B. and E. O. Wilson. 1990. *The Ants.* Cambridge, Mass.: Harvard University Press.

Howard, D. J. 1993. Reinforcement: origin, dynamics, and fate of an evolutionary hypothesis. In *Hybrid Zones and the Evolutionary Process.* R. G. Harrison, ed., pp. 46–69. New York: Oxford University Press.

Hoy, R. R. 1974. Genetic control of acoustic behavior in crickets. *Am. Zool.* **14**: 1067–1080.

Jacobson, M. 1972. *Insect Sex Pheromones.* New York: Academic Press.

Janetos, A. C. 1980. Strategies of female mate choice: a theoretical analysis. *Behav. Ecol. Sociobiol.* **7**: 107–112.

Jormalainen, V. and S. Merilaita. 1995. Female resistance and duration of mate-guarding in three aquatic peracarids (Crustacea). *Behav. Ecol. Sociobiol.* **36**: 43–48.

Kennedy, J. S., ed. 1961. *Insect Polymorphism*. (Symposia of the Royal Entomological Society of London, no. 1). London: Royal Entomological Society.

Kristensen, N. P. 1981. Phylogeny of insect orders. *Annu. Rev. Entomol.* **26**: 135–157.

LaMunyon, C. W. and T. Eisner. 1993. Postcopulatory sexual selection in an arctiid moth (*Utethesia ornatrix*). *Proc. Natl. Acad. Sci. U.S.A.* **90**: 4689–4692.

Lande, R. 1981. Models of speciation by sexual selection on polygenic traits. *Proc. Natl. Acad. Sci. U.S.A.* **78**: 3721–3725.

Leopold, R. A. 1976. The role of male accessory glands in insect reproduction. *Annu. Rev. Entomol.* **21**: 199–221.

Lloyd, J. E. 1979. Mating behavior and natural selection. *Fla. Entomol.* **62**: 17–34.

–. 1983. Bioluminescence and communication in insects. *Annu. Rev. Entomol.* **28**: 131–160.

Loher, W. and B. Rence. 1978. The mating behavior of *Teleogryllus commodus* (Walker) and its central and peripheral control. *Z. Tierpsychol.* **49**: 225–259.

Marshall, D. C. and J. R. Cooley. ms. Mate choice in periodical cicadas (Homoptera: Cicadidae: *Magicicada*).

Matthews, R. W. and J. R. Matthews. 1978. *Insect Behavior*. New York: John Wiley.

Maynard Smith, J. 1956. Fertility, mating behaviour and sexual selection in *Drosophila subobscura*. *J. Genet.* **54**: 261–279.

–. 1987. Sexual selection – a classification of models. In *Sexual Selection: Testing the Alternatives*. J. W. Bradbury and M. B. Andersson, eds., pp. 9–20. New York: John Wiley.

Milne, L. J. and M. Milne. 1976. The social behavior of burying beetles. *Scient. Am.* **235**: 84–89.

Minckley, R. L., M. D. Greenfield and M. K. Tourtelot. 1995. Chorus structure in tarbush grasshoppers: Inhibition, selective phonoresponse, and signal competition. *Anim. Behav.* **50**: 579–594.

Moore, A. J. and P. J. Moore. 1988. Female strategy during mate choice threshold assessment. *Evolution* **42**: 387–391.

Morris, G. K. 1979. Mating systems, paternal investment and aggressive behavior of acoustic Orthoptera. *Fla. Entomol.* **62**: 9–17.

Nilsson, L. A. 1988. The evolution of flowers with deep corolla tubes. *Nature (Lond.)* **334**: 147–149.

O'Donald, P. 1962. *Genetic Models of Sexual Selection*. Cambridge University Press.

Otte, D. 1970. A comparative study of communicative behavior in grasshoppers. *Univ. Mich. Mus. Zool. Misc. Pub.* **141**: 1–168.

–. 1974. Effects and functions in the evolution of signaling systems. *Annu. Rev. Ecol. Syst.* **5**: 385–417.

–. 1977. Communication in Orthoptera. In *How Animals Communicate*. T. A. Sebeok, ed., pp. 334–361. Bloomington, Ind.: Indiana University Press.

–. 1992. Evolution of cricket songs. *J. Orthopt. Res.* **1**: 25–49.

–. 1994. *The Crickets of Hawaii*. Philadelphia, PA: Acad. Nat. Sci. of Philadelphia.

Otte, D. and R. D. Alexander. 1983. The Australian crickets (Orthoptera: Gryllidae). *Acad. Nat. Sci. Philadelphia Monogr.* **22**: 1–477.

Page, R. E. Jr. 1980. The evolution of multiple mating behavior by honey bee queens (*Apis mellifera* L.) *Genetics* **96**: 263–273.

Parker, G. A. 1970. Sperm competition and its evolutionary consequences in the insects. *Biol. Rev.* **45**: 525–568.

–. 1979. Sexual selection and sexual conflict. In *Sexual Selection and Reproductive Competition in Insects*. M. S. Blum and N. A. Blum, eds., pp. 123–166. New York: Academic Press.

–. 1983. Arms races in evolution – an ESS to the opponent-independent costs game. *J. Theor. Biol.* **101**: 619–648.

Parker, G. A., R. R. Baker and V. G. F. Smith. 1972. The origin and evolution of gamete dimorphism and the male-female phenomenon. *J. Theor. Biol.* **36**: 529–553.

Paterson, H. E. H. 1993. *Evolution and the Recognition Concept of Species*. Baltimore, Maryland: Johns Hopkins University Press.

Pires, A. and R. R. Hoy. 1992. Temperature coupling in cricket acoustic communication. *J. Comp. Physiol.* A**171**: 79–82.

Prestwich, K. N. and T. J. Walker. 1981. Energetics of singing in crickets: effect of temperature in three trilling species (Orthoptera: Gryllidae). *J. Comp. Physiol.* B**143**: 199–212.

Queller, D. C. 1994. Male-female conflict and parent-offspring conflict. *Am. Nat.* **144**: S84–S99.

Reynolds, J. D. and M. R. Gross. 1990. Costs and benefits of female mate choice: Is there a lek paradox? *Am. Nat.* **136**: 230–243.

Robertson, J. G. M. 1986. Male territoriality, fighting and assessment of fighting ability in the Australian frog *Uperoleia rugosa*. *Anim. Behav.* **34**: 763–772.

Roelofs, W. L. 1975. Insect communication – Chemical. In *Insects, Science, and Society*. D. Pimentel, ed., pp. 79–99. New York: Academic Press.

Sakaluk, S. K. 1984. Male crickets feed females to ensure complete sperm transfer. *Science (Wash., D.C.)* **223**: 609–610.

–. 1990. Sexual selection and predation: balancing reproductive and survival needs. In *Insect Defenses*. D. L. Evans and J. O. Schmidt, eds., pp. 63–90. Albany, NY: SUNY Press.

Salthé, S. N. 1967. Courtship patterns and phylogeny of the Urodeles. *Copeia* **1967**: 100–117.

Sherman, P. W., T. D. Seeley, and H. K. Reeve. 1988. Parasites, pathogens, and polyandry in social Hymenoptera. *Am. Nat.* **131**: 602–610.

Simmons, L. W. 1988. The calling song of the field cricket, *Gryllus bimaculatus* (DeG.): Constraints on transmission and role in intermale competition and female choice. *Anim. Behav.* **36**: 380–394.

Simmons, L. W. and W. J. Bailey. 1990. Resource influenced sex roles of zaprochiline tettigoniids (Orthoptera: Tettigoniidae). *Evolution* **44**: 1853–1868.

Sivinski, J. 1984. Sperm in competition. In *Sperm Competition and the Evolution of Animal Mating Systems*. R. L. Smith, ed., pp. 85–115. New York: Academic Press.

Smith, R. L., ed. 1984. *Sperm Competition and the Evolution of Animal Mating Systems*. New York: Academic Press.

Smuts, B. B. and R. W. Smuts. 1993. Male aggression and sexual coercion of females in nonhuman primates and other mammals: Evidence and theoretical implications. *Adv. Stud. Behav.* **22**: 1- 63.

Spooner, J. D. 1968. Pair-forming acoustic systems of phaneropterine katydids (Orthoptera: Tettigoniidae). *Anim. Behav.* **16**: 197–212.

Thomas, E. S. and R. D. Alexander. 1962. Systematic and behavioral studies on the meadow grasshoppers of the *Orchelimum concinnum* group (Orthoptera: Tettigoniidae). *Univ. Mich. Mus. Zool. Occas. Paper* **626**: 1–31.

Thornhill, R. and J. Alcock. 1983. *The Evolution of Insect Mating Systems*. Cambridge, Mass.: Harvard University Press.

Trivers, R. L. 1972. Parental investment and sexual selection. In *Sexual Selection and the Descent of Man 1871–1971*. B. Campbell, ed., pp. 136–179. Chicago: Aldine.

Vité, J. P. and D. L. Williamson. 1970. *Thanasimus dubius*: prey perception. *J. Insect Physiol.* **16**: 233–239.

Waage, J. K. 1979. Dual function of the damselfly penis: Sperm removal and transfer. *Science (Wash., D.C.)* **203**: 916–918.

Walker, T. J. 1957. Specificity in the response of female tree crickets (Orthoptera, Gryllidae, Oecanthinae) to calling songs of the males. *Ann. Entomol. Soc. Am.* **50**: 626–636.

–. 1963. The taxonomy and calling songs of United States tree crickets (Orthoptera: Gryllidae: Oecanthinae). II. The *nigricornis* group of the genus *Oecanthus*. *Ann. Entomol. Soc. Am.* **56**: 772–789.

–. 1964. Experimental demonstration of a cat locating orthopteran prey by the prey's calling song. *Fla. Entomol.* **47**: 163–165.

–. 1974. Character displacement and acoustic insects. *Am. Zool.* **14**: 1137–1150.

–. 1983. Mating modes and female choice in short-tailed crickets (*Anurogryllus arboreus*). In *Orthopteran Mating Systems: Sexual Competition in a Diverse Group of Insects*. D. T. Gwynne and G. K. Morris, eds., pp. 240–267. Boulder, Colorado: Westview Press.

Walker, T. J. and D. C. Rentz. 1967. Host and calling song of dwarf *Oecanthus quadripunctatus* Beutenmuller (Orthoptera: Gryllidae). *Pan-Pac. Entomol.* **43**: 326–327.

Walker, W. F. 1980. Sperm utilization strategies in nonsocial insects. *Am. Nat.* **115**: 780–799.

Watson, P. J. and J. B. Lighton. 1994. Sexual selection and the energetics of copulatory courtship in the Sierra dome spider, *Linyphia litigosa*. *Anim. Behav.* **48**: 615–626.

Werren, J. H. 1980. Sex ratio adaptations to local mate competition in a parasitic wasp. *Science (Wash., D.C.)* **208**: 1157–59.

West, M. J. and R. D. Alexander. 1963. Sub-social behavior in a burrowing cricket *Anurogryllus muticus* (DeG). (Orthoptera: Gryllidae). *Ohio J. Sci.* **63**: 19–24.

West-Eberhard, M. J. 1983. Sexual selection, social competition, and speciation. *Q. Rev. Biol.* **58**: 155–183.

–. 1984. Sexual selection, competitive communication, and species-specific signals in insects. In *Insect Communication*. T. Lewis, ed., pp. 284–324. New York: Academic Press.

Westneat, D. W., P. W. Sherman and M. L. Morton. 1990. The ecology and evolution of extra-pair copulations in birds. *Curr. Ornithol.* **7**: 331–369.

Will, M. W. and S. K. Sakaluk. 1994. Courtship feeding in decorated crickets: is the spermatophylax a sham? *Anim. Behav.* **48**: 1309–1315.

Williams, G. C. 1966. *Adaptation and Natural Selection*. Princeton University Press.

–. 1992. *Natural Selection: Domains, Levels, and Challenges*. New York: Oxford University Press.

Wilkinson, G. S. and P. R. Reillo. 1994. Female choice response to artificial selection on an exaggerated male trait in a stalk-eyed fly. *Proc. R. Soc. Lond.* B **255**: 1–6.

Wrangham, R. W. 1980. Female choice of least costly males: A possible factor in the evolution of leks. *Z. Tierpsychol.* **54**: 357–367.

2 · Sexual selection by cryptic female choice in insects and arachnids

WILLIAM G. EBERHARD

ABSTRACT

Females can have important, but often underestimated effects on the likelihood that any given copulation will result in fertilization of their eggs. Examples of at least 15 different processes that occur during or after intromission, and in which females can selectively favor one male's chances of paternity over those of another, are given for insects and arachnids.

Two general phenomena suggest that such 'cryptic' female choice is not a rare biological curiosity, but rather a widespread and common phenomenon. Male courtship behavior in insects and spiders often (probably usually) begins or continues after intromission has occurred; the most likely explanation of this otherwise paradoxical behavior is that it serves to influence cryptic female choice. In addition, male seminal products in insects and ticks commonly influence several aspects of female reproductive physiology. Several kinds of evidence indicate that these male products have evolved under sexual selection, usually as triggering mechanisms rather than as nutrients. They may constitute 'chemical genitalia' that influence cryptic female choice.

Several widely held ideas about sexual selection and courtship should be modified to take into account the fact that females are less passive in male–female interactions than has previously been supposed.

INTRODUCTION

The classical view of sexual selection by female choice presumes that females selectively screen males prior to copulation. The precopulatory nature of female choice was emphasized by Darwin:

... [what] is required [for sexual selection by female choice to operate] ... is that choice should be exerted before the parents unite (1871, p. 694).

The common use of the term courtship has a similar precopulatory emphasis: 'Overtures to mating ... broadly conceived, [it] encompass[es] all behavior patterns of pre-copulation, pair formation, and pair bonding' (Immelmann

and Beer 1989, p. 62). I will argue that these conceptions are incomplete, and need to be changed. Female choice between males probably often occurs after the female has already allowed the male to insert or attach his genitalia.

The morphology of males and females in almost all animals dictates that copulation does not by itself automatically result in fertilization of eggs, because the male almost never places his sperm directly on the female's eggs (Eberhard 1985). With a few possible exceptions (e.g. some Strepsiptera: Lauterbach 1954), this is certainly true for insects and arachnids. In most species, several female-controlled processes, including sperm transport, sperm storage, oogenesis, ovulation, oviposition, and restraint from remating must all occur after a copulation has begun if that copulation is to result in fertilization of all of the female's eggs. There are also several additional female processes of possibly less general application in other species (see list below). It is thus possible that females sometimes exercise choice among potential fathers of their offspring by biasing the performance of these processes, rather than (or in addition to) biasing their acceptance of copulation with different males. Table 2-1 gives a conservative list of 15 different mechanisms; several other possiblities are mentioned in the text. Studies of insects and spiders demonstrate that females use many of these mechanisms selectively. Such female biases could result in sexual selection on males: any male better able to obtain a favorable response from a female would have a competitive advantage (in terms of offspring sired) over other males that had sexual access to the same female.

Thornhill (1983) coined the term 'cryptic female choice' for biases of this sort. He had discovered a tendency in females of the scorpionfly *Harpobittacus nigriceps* to lay more eggs immediately following some copulations (those with larger males) than after others. The word 'cryptic' refers to the fact that the bias would not be noticed by classic, Darwinian measures of 'overt' female choice that only take into account a male's success in achieving copulations.

Although cryptic female choice is clearly a theoretical possibility, it has not been clear whether it is common in nature. Some ambitious claims have been made for its general importance (Eberhard 1985), but it has generally been ignored in recent discussions of sexual selection (Maynard Smith 1987, 1991; Kirkpatrick and Ryan 1991; Andersson 1994). This chapter establishes that females often have the wherewithal with which to exercise cryptic female choice, describing the multitude of female mechanisms by which a male's chances of paternity can be increased or decreased even after copulation has begun. Concrete examples are given in which many of these mechanisms are used selectively (or at least in contexts that suggest selectivity). Two lines of evidence that cryptic female choice has been a widespread, important factor in the evolution of insects and arachnids are then discussed. Finally, relationships between cryptic female choice and male–female conflicts of interest are discussed. Most of the ideas discussed here are present in more detail elsewhere (Eberhard 1996), including evidence from many other groups (including ctenophores, nematodes, mammals, birds, fish, barnacles and snakes). The reader is referred to this more complete discussion for further arguments, examples, and details. My aim here is not to attempt a final synthesis of the importance of cryptic female choice. The field is far too young for such an undertaking to be worthwhile. Instead, I will attempt to demonstrate how a variety of disparate phenomena are compatible with and suggestive of cryptic female choice, in an attempt to alert readers to the possible importance of previously neglected or underestimated phenomena.

MECHANISMS OF CRYPTIC FEMALE CHOICE

It has been increasingly recognized that complex and generally very poorly understood processes such as sperm movement and survival within the female can have important effects on male reproductive success. Probably at least partly as a result of the still prevalent 'myth of the passive female' (Batten 1992), emphasis has generally been placed on those aspects of these processes that are perceived to be under male control. Indeed, the very name usually applied to this general topic, 'sperm competition', emphasizes possible male rather than female roles in determining the outcomes of male competition.

But in fact most of these crucial processes, including discarding or destroying sperm, transporting sperm to storage, maintenance of sperm in storage, and movement of both sperm and eggs to fertilization sites, are largely if not completely under the ultimate control of the female rather than the male. It is, after all, *her* body in which these events occur. Differential or 'discriminating' responsiveness on her part, or morphological or physiological features that favor the reproduction of some conspecific males over that of others, can result in sexual selection on males.

Even in cases in which males appear to be in direct control over a particular process (e.g. sperm removal: Waage 1979, 1984, 1986; Ono *et al.* 1989; Yokoi 1990; Gage 1992), males are able to exercise control of paternity only if female characteristics permit. For instance, a long, thin tortuous spermathecal duct, or a rough, irregular interior surface of the spermatheca with nooks and crannies where sperm can lodge, or the use of bursal sperm rather than spermathecal sperm for fertilizations, or ready remating after some copulations but not after others, or greater reluctance to oviposit after some copulations than after others, could all change the reproductive payoffs for male attempts to remove the sperm of previous males from storage sites in the female.

The behavior, morphology, and physiology of the female specify the 'rules of the game' for competing males, because they determine that some variants of male behavior and morphology will be more reproductively successful than others. The potential roles of females in determining how sexual selection acts on males, and the role of cryptic female choice in this process, deserve attention. Instead of the usual custom of taking female characteristics that affect male success as 'givens', the possibility that they have evolved in order to produce different effects on different males should also be considered.

Take for instance, mating in the dryomyzid fly *Dryomyza anilis* (Otronen 1984, 1989, 1990). Males fight with each other on and near oviposition sites (carrion), where they mount females. Once a pair forms, the flies move a short distance away, where copulation occurs. After an approximately 1 min intromission, the male withdraws his genitalia, and performs several 'tapping sequences' in which he taps rhythmically on the female's external genitalia with his genitalic claspers and presses her abdomen with his hind legs. Each tapping sequence ends with a longer clasper contact (unless the female is resisting; see below). After several tapping sequences (range 8–31), the female returns to the oviposition site, generally still guarded by the mounted male. Experimental manipulations showed that larger numbers of tapping sequences result in higher paternity frequency for the male in the eggs laid when the female moves back onto the carrion to oviposit.

Before beginning to oviposit, the female discharges a droplet about the size of a single ejaculate, which contains sperm. After the female has laid several eggs, the male appears to induce her to move away again by vibrating his wings, and by attempting to copulate (M. Otronen, personal communication). They remate, and the male again performs several postcopulatory tapping sequences. Up to six cycles of copulation–oviposition occur. Eggs laid earlier in a given bout of oviposition are more likely to be sired by the guarding male than are those later in that bout.

Females sometimes (but not always) resisted intromission and tapping, by lowering the abdomen to make genitalic contact more difficult or impossible, or by shaking their bodies vigorously. Such resistance was sometimes effective, as males often abandoned females after repeated resistance. Larger males were able to perform more tapping sequences than smaller males, and larger females, which resisted more often, spent less time in copulation bouts. The strength of female resistance also varied.

Female resistance to copulation and tapping may be favored by natural selection favoring more rapid oviposition (see discussion of male–female conflicts at the end of this chapter) (see also Alexander et al., this volume; Arnqvist, this volume). However, in addition, there is a correlation between the number of tapping bouts and the male's paternity, perhaps due to the female discarding more or less of the current male's sperm in the preoviposition droplet (Otronen and Siva-Jothy 1991). Thus the female resistance to tapping also results in cryptic sexual selection that favors larger males. Cueing sperm-use patterns on the number of postcopulatory genitalic tapping sequences seems arbitrary with respect to natural selection on females, and seems instead likely to have evolved as a mechanism to discriminate among males. In addition, a further bias (also favoring larger size) is imposed by direct male–male battles for control of oviposition sites (Otronen 1984). Still further female-imposed biases may result from variations in the intensity of female resistance. In addition, larger males achieved more paternity when the number of tapping sequences was held constant (Otronen 1995).

Female traits that result in cryptic choice could originally evolve, as suggested in this example, either under natural selection or sexual selection. A trait that was established by natural selection could subsequently undergo further evolution due to any one of several possible benefits from sexual selection. For instance, a female tendency to resist extended copulation bouts could have been originally favored by a slight reduction in predation, or an increase in oviposition rates (see Alexander et al., this volume), and then further accentuated by the benefit of increasing the genetic representation in her offspring of larger or more vigorous males, or those with a greater tendency to perform more tapping sequences. Selection on female traits resulting from any one of these benefits could result in sexual selection by cryptic female choice on males. Both natural selection and sexual selection could be involved in the maintenance of particular male and female traits.

There are many ways in which a female could cryptically bias the reproductive success of a male with which she has made genitalic contact. Table 2-1 lists 15 female-controlled processes that occur during or after copulation, and which could possibly influence a male's reproductive success. Each process is illustrated by a sample species of insect or arachnid (a more complete listing, including other insects and spiders as well as other animals, totalled over 100 species (Eberhard 1996)). There are also additional mechanisms which could bias male paternity success, but for which I have not found concrete examples in insects or arachnids: selective abortion of developing offspring (known in some mammals and perhaps widespread in plants: Stephenson and Bertin 1983; Queller 1987, 1994; Eberhard 1996); less efficient use of sperm for fertilizing eggs after some copulations (sperm of male *Drosophila melanogaster* that transfer esterase 6 in the semen may be used less efficiently: Gilbert 1981); failure to ovulate (suspected in some mammals, and probably cued by male seminal products which have been under sexual selection by cryptic female choice in some insects and ticks: see next section); failure to develop eggs (probably also cued by seminal products in some insects and ticks); and biased fusion with the nuclei of different sperm that enter the same egg (suspected in a ctenophore: Carré and Sardet 1984).

Some of these processes are actually composites of several subprocesses, each potentially subject to independent female control. For instance, in *Aedes aegypti* oviposition (no. 6 in Table 2-1) involves searching for the host, feeding, digesting the food, mobilizing reserves to the ovaries, oogenesis, searching for an appropriate oviposition site, and laying the eggs. In the moth *Helicoverpa zea* failure to remate (no. 5) can involve inhibition of release of attractant pheromone, inhibition of production of the pheromone, resorption of the pheromone already produced, and rejection of approaching males by kicking, and each process has separate controls (Kingan et al. 1993a,b). Different male traits could presumably affect different female processes independently.

HOW COMMON IS CRYPTIC FEMALE CHOICE?

It is clear that females do not lack opportunities and means to exercise cryptic choice. The question remains, however, of whether or not the cases of cryptic female choice in Table 2-1 are isolated biological 'freaks' that represent an uncommon phenomenon of limited importance. At the opposite extreme, they may be the tip of an iceberg, representing a large, pervasive, but little-studied pattern in animal evolution. In the sections that follow I present evidence that the second possibility may be closer to the truth.

Copulation as courtship

Perhaps the most convincing evidence favoring the widespread nature of cryptic female choice is the frequent occurrence of male copulatory courtship (courtship behavior performed during or after copulation). Male courtship is generally characterized as stereotyped behavior performed during interactions with females, repeated both within and between interactions, and designed to stimulate the female. It is presumed to function by inducing the female to respond in a way that increases the male's chances of reproduction with her. Usually male courtship is thought to function by inducing the female to allow the male to copulate with her. But male courtship (behavior with the characteristics just mentioned) often occurs during or following copulation, when it obviously does not function to allow intromission to occur; instead it is likely to function to induce further responses by the female after copulation has begun, that also improve the male's chances of reproduction.

Several other male characteristics could also influence such female responses: courtship prior to copulation (see, for example, Crews 1984); genitalic stimulation during copulation (see, for example, Eberhard 1985, 1991); or chemical induction of female responses via seminal products or other nuptial gifts (see next section). Therefore male copulatory courtship constitutes a conservative indication that males have been under selection to influence cryptic female choice. In a survey of the copulation behavior of insects and spiders, in which relatively conservative criteria were used to distinguish courtship from other types of behavior (Eberhard 1994), copulatory courtship occurred in 81% of 131 species, 79% of 102 genera and 76% of 49 families. A similar survey of previously published accounts of insect behavior, in which there were several reasons to suspect that copulatory courtship was under-reported, also gave surprisingly high frequencies (36% of 302 species, 34% of 231 genera, 43% of 102 families) (Eberhard 1991). The behavior patterns used by males in copulatory courtship are diverse, and include biting, licking, regurgitating on, rubbing, tapping, kicking, shaking, squeezing, rocking, waving to, vibrating, lifting, and singing to the female, often in rhythmic and stylized fashion (Fig. 2-1). The implication is that males have often been under selection to court females to induce them to respond in ways which further the male's reproduction – even after copulation has begun.

These estimated frequencies did not include behavior involving male genitalia such as repeated intromissions, rhythmic pushing, and thrusting. Experimental studies of mammals (especially rodents) have shown that these aspects of male copulatory behavior affect several female reproductive processes in ways favoring the male's chances of paternity (see, for example, Dewsbury 1989; summary in Eberhard 1996), so the estimated frequencies are probably underestimates.

Several additional data confirm that the male behavior classified as copulatory courtship indeed functions as courtship and is not just reproductively insignificant 'twitching' by highly excited males. Of a total of 279 different copulatory courtship behavior patterns observed in the two surveys, 40% were also performed by the same males prior to copulation, constituting apparent classical precopulatory courtship. Male copulatory courtship also differed intragenerically, as is expected to often be the case if it is under sexual selection, in each of the 20 genera in which more than a single species was observed (Eberhard 1994, 1996). It also differed intraspecifically in three of the four species that have been observed at geographically separated sites (Wcislo et al. 1992; Eberhard 1994). Copulatory courtship is not limited to arthropods, as there are scattered observations of copulatory courtship in a wide range of other groups, ranging from nematodes to mammals (Eberhard 1996).

It is worth mentioning in passing that the details of copulatory courtship often seem to have little relationship to male size or vigor (see, for example, Fig. 2-1). The presumed advantage to females of using these characters thus seems more likely to result from obtaining good attractiveness genes from the male (via a Fisherian process) than good viability genes (see Eberhard 1996).

Table 2-1. *Mechanisms for cryptic female choice in insects and arachnids, with an illustrative species for each*

(1) Sometimes discard sperm of current male: *Dryomyza anilis* (Diptera).

Comments: The number of series of postcopulatory genital taps by the male was positively correlated with the proportion of the offspring he sired from the eggs that the female laid immediately after copulation. Apparently the correlation resulted from a reduction in the amount of the current male's sperm that was included in the droplet of sperm that females discarded immediately before beginning to oviposit (Otronen 1990; Otronen and Siva-Jothy 1991).

(2) Sometimes discard sperm of previous male: *Paraphlebia quinta* (Odonata).

Comments: Males are dimorphic; females emitted droplets of sperm more often during copulations with those males with hyaline wings than with males with dark wing tips. Judging by the fact that the phase of copulation in which sperm emission occurs was followed by an 'inactive' phase similar to that during which other male odonates transfer sperm, the emitted sperm was probably from previous males (E. Gonzalez, manuscript in preparation). Further work is necessary to determine the roles of males and females in sperm emission, and to establish more securely that emitted sperm was from previous males.

(3) Sometimes prevent complete intromission: *Anastrepha suspensa* (Diptera).

Comments: Male song produced by wing fanning while mounted on the female often continued until the male had clamped the tip of the female ovipositor (the aculeus) and had threaded his long aedeagus at least part way into the female's long vagina. Those males that produced less intense songs, and those whose wings were clipped and thus produced much weaker songs, were more often rejected (were dislodged, usually owing to vigorous female movements, after at least one second of mounting) (Webb *et al.* 1984). Sometimes a male was dislodged after he had clamped the female and inserted his genitalia part way through her vagina (J. Sivinski, personal communication).

(4) Sometimes fail to transport sperm to storage : *Neriene litigiosa* (Araneae).

Comments: Greater male 'copulatory vigor' (higher rate of intromissions during preinsemination copulations, lower percentage of intromission attempts that failed, greater length of preinsemination copulation behavior) correlated positively with a higher percentage of paternity when the female had mated with another male previously, as often occurs in the field (Watson 1991). Female control may have been exercised by altering the opening of valve-like structures in the ducts leading to her spermathecae.

(5) Sometimes remate with another male: *Centris pallida* (Hymenoptera).

Comments: After copulation and insemination, the male courted the female briefly by stroking her vigorously with his legs, abdomen, and antennae, and producing a rasping, rattling song. When males were prevented from performing this courtship, females remated much more readily (Alcock and Buchmann 1985).

(6) Sometimes reduce rate of oviposition: *Harpobittacus nigricornis* (Mecoptera).

Comments: Females laid more eggs in the three hours immediately following copulation with a large male than following copulation with a small male. Because females of this species remate frequently, the difference in oviposition rates probably has important consequences for male reproductive success (Thornhill 1983).

(7) Sometimes forcefully terminate copulation prematurely: *Zorotypus barberi* (Zoraptera).

Comments: Sperm transfer (which can be observed through the male's transparent abdominal cuticle) occurred at a relatively predictable stage of copulation. Females frequently terminated copulation prematurely (27.9% of 247 copulations) by pulling away abruptly, often with the male in vigorous pursuit with his intromittent organ still exposed. In 74.2% of these cases termination occurred within the 5 s just before sperm transfer was expected to begin or the 5 s after it had begun (Choe 1995). The possibility that females were selective (were more likely to interrupt the copulations of some males rather than those of others) was not investigated.

(8) Sometimes impede plugging of reproductive tract: *Agelena limbata* (Araneae).

Comments: After insemination, the male deposited a mass of material covering the entrances of the female insemination ducts. If the mass was large enough to fill the cavity on the female's external genitalia (epigynum) where the insemination ducts open, it prevented subsequent males from successfully inseminating her. When smaller quantities were applied (especially frequent in copulations with smaller males), subsequent males were able to pry off the plug, inseminate the female, and sire substantial proportions of her offspring (Masumoto 1991, 1993). The cavity on the female's epigynum thus had a selective effect on paternity, favoring larger-sized males.

(9) Sometimes impede plug removal by the male: *Phidippus johnsoni* (Araneae).

Comments: The male often deposited a plug in one or both insemination duct openings after inseminating the female (55% of 44 females had plugs after a single mating) (Jackson 1980). These plugs adhered only loosely, however, and a subsequent male was often able to remove the plug and introduce sperm of his own, resulting in appreciable paternity in the subsequent clutch of eggs (Jackson 1980). Longer copulations more often resulted in removal of plugs. The duration of copulation (and thus the probability of plug removal) was apparently determined by the female, as the female always initiated termination. The possibility that females selectively favored some males over others was not checked.

Table 2-1. (cont.)

(10) Sometimes remove spermatophore before sperm transfer is complete: *Gryllus bimaculatus* (Orthoptera).

Comments: Sperm were gradually taken into the female's spermatheca during about the first 60 min after the male's spermatophore was attached to the her (the male remained nearby). The female sometimes turned and removed the spermatophore (and ate it) before it had emptied its sperm into her reproductive tract. Early spermatophore removal occurred more often when the male was smaller (as soon as only 2 min after it was attached (Simmons 1986). Females remated frequently; a male's chances of paternity were apparently a function of the proportion of his sperm in her spermatheca, so female reduction of sperm transfer probably reduced a male's chances of siring offspring.

(11) Biased use of stored sperm: *Chorthippus parallelus* (Orthoptera).

Comments: When females of two different geographic subspecies were each mated to one male of their own subspecies and one of the other, males of the same subspecies sired more of the offspring (Bella *et al.* 1992) (the order of matings was varied; paternity was determined by examining the chromosomes of embryos 9-day old). Both hybrid mortality and differences in copulation duration were tentatively eliminated as possible explanations of the biases.

(12) Move a previous male's sperm to a site where the current male can manipulate them *Metaplastes ornatus* (Orthoptera).

Comments: The male induced emptying of previously deposited sperm from the female's spermatheca by rubbing his genitalia in the female's oviduct, apparently thereby mimicking the stimuli produced by an egg moving down the oviduct to be fertilized (von Helversen and von Helversen 1991). After rubbing the oviduct several times, the male pulled the distal portion of the female's reproductive tract from her body, turning it inside out. The female then ingested the exposed sperm from a previous male before pulling her oviduct back inside her body. Experimental rubbing in the oviduct to mimic male movements resulted in sperm transfer to the oviduct (von Helversen and von Helversen 1991).

(13) Sometimes increase the difficulty of sperm transfer by changes in morphology: *Nephila clavipes* (Araneae).

Comments: Copulation with a newly molted virgin female lasted about 48 hr, but sperm were transferred within the first 3 h. If copulation was artificially interrupted (as could happen in nature if a second, larger male arrived, or if the female moved away), second matings resulted in about twice as large a percentage of the offspring being sired by the second male (Christenson 1990) (possible female biasses in deserting males were not checked for). The female's genital ducts harden gradually over the space of several days, and the ducts harden more rapidly when the female has been mated. Insertions of male genitalia and contractions of male genital membranes were less rapid and less regular during copulations with females with hardened ducts. It is thought that differences in the degree of hardening of the female's ducts are causally related to the differences in paternity (Christenson 1990).

(14) Sometimes resist male manipulations that result in discharge of his spermatophore: *Bothriurus flavidus* (Scorpiones).

Comments: After seizing the female with his pedipalps and searching for an appropriate substrate, the male deposits a spermatophore and pulls her forward so that a portion of the spermatophore enters her gonopore. He then pushes her backward so that the spermatophore pivots, discharging its contents of sperm into her reproductive tract. Females not infrequently resisted this final backward push, after having cooperated in all preceding steps (Peretti 1996). The male sometimes repeatedly attempted to push the female, which sometimes led to the female stinging him, finally causing him to release her. The possiblity that females resisted some males more than others was not checked.

(15) Sometimes invest less in each offspring: *Requena verticalis* (Orthoptera).

Comments: The size but not the number of the eggs laid after a mating was increased if the amount of spermatophylax material ingested by the female was greater (Gwynne 1988a,b). Larger eggs produced offspring that were more likely to survive the following winter. The female response is thought to be due to nutrient donations by the male. Male reproduction is probably limited by competition with other males for access to females, and male attract females with calling songs (Gwynne 1986).

Effects of male seminal products on female reproductive processes

There is a substantial literature on the physiological effects of male seminal products on females in insects and, to a lesser extent, in ticks. Many of the data come from implantation experiments, and from injections of extracts of male accessory glands or testes into the female. The general pattern, with more than 50 species now studied, is that products of the male reproductive organs often induce female responses such as oviposition, resistance to further mating, and, less often, sperm transport (reviews in insects: Leopold 1976; Chen 1984; Raabe 1986; Gillott 1988; Kaulenas 1992; in ticks: Diehl and Aeschlimann 1982; Connat *et al.* 1986; Oliver 1986; summary: Eberhard 1996). These same female responses can serve as mechanisms

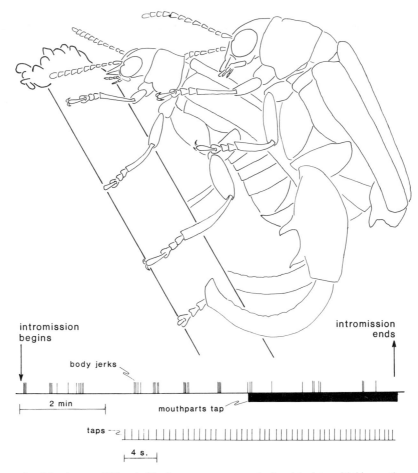

intromission
begins

intromission
ends

body jerks

2 min

mouthparts tap

taps

4 s.

Fig. 2-1. A copulating male of the chrysomelid beetle *Megalopus armatus* taps on the female's elytra with his mouthparts, one aspect of copulatory courtship. The graphs below show one complete copulation (upper graph), and a small portion of the latter portion of another. During the first several minutes of copulation the male occasionally jerks the female's entire body dorsally and forward (and probably stridulates) ('body jerks' on the graph). Later the male leans ventrally and taps the female's dorsum with complex, stereotyped movements of his maxillary and labial palps (bottom graph). The male's giant hind legs, which function as weapons in male–male combat, play no role in courtship, either before or during copulation. The apparent importance of details of the movements of the male mouthparts in this species illustrates the frequent uncoupling between the stimuli the male provides during copulatory courtship and his size and vigor. This uncoupling suggests that the benefit that females gain from discriminating between males is through good attractiveness genes (e.g. through a Fisherian process) rather than through good viability genes (after Eberhard and Marin 1996).

for cryptic female choice, so differences in the quantities and active properties of male seminal products could be subject to cryptic female choice on males.

Several aspects of the male seminal products with these effects on females, and details of how they act on females, indicate that they have evolved under sexual selection by cryptic female choice rather than or in addition to natural selection. Males appear to compete via their abilities to induce more favorable responses from females. Other data

indicate that these male products generally function as hormonal or triggering substances, rather than as nutrient donations to the female (see also Gwynne, this volume). Most of this section is dedicated to demonstrating these points.

It has been generally supposed in the physiological literature that natural selection on females rather than sexual selection on males is responsible for females using stimuli from chemicals received from the male during copulation to trigger reproductive processes. It is clearly advantageous, for instance, for a female to refrain from ovipositing before mating, and then to begin ovipositing afterward. The expected female responses to copulation favored by natural selection (e.g. sperm transport, oogenesis, oviposition, resistance to further mating attempts) could also be subject, however, to sexual selection. In any species with polyandrous females in which these female responses were not always complete (for example, the female did not always lay her entire set of mature eggs soon after copulation), a male's reproduction could be furthered if he were able to increase or accentuate the responses in his mates. And if male abilities to trigger responses differed, a female could gain by raising her response thresholds so that only males delivering superior stimuli triggered her maximum responses. The female could benefit either through increased stimulatory abilities in her sons ('good attractiveness genes') or, if male stimulatory abilities were linked to viability traits, through improved viability in her offspring ('good viability genes'). Thus the original female responses, resulting from natural selection, could subsequently lead to either runaway sexual selection as envisaged by Fisher (1958) or other forms of female choice.

What evidence can be used to decide whether sexual selection on male seminal products has occurred, or whether female responses to these products have resulted only from natural selection on females? Several points on which the two hyotheses give contrasting predictions are listed below, and brief descriptions of particular illustrative cases are given. A more thorough discussion is given in Eberhard (1996).

'Invasive' action of male products

Under natural selection, the evolution of a female response to a chemical cue in male semen would presumably often begin with chemical receptors in the female's reproductive tract that responded to the presence of the male product by sending a message to some more central coordinating site. In contrast, sexual selection on male abilities to induce female responses would often favor more direct action on the female. If, for instance, the male products could pass through the walls of the female's reproductive tract, enter her body cavity, and act directly on target organs such as her ovaries or her nervous system, the male could increase the chances that a favorable response would occur. In effect, by invading the female's body the male could bypass possible female recalcitrance at intermediate steps. Females could subsequently evolve traits that reduced the male's control of their own bodies. The final balance of male and female influences is not easily predicted for any given case (see section on male–female conflict; see also Alexander et al., this volume). Nevertheless, an invasive action of male substances would be a sign of sexual selection on males.

Although the target organs of many male seminal products have yet to be determined (Chen 1984), the most frequent site of action of male seminal substances in insects is thought to be the female's central nervous system (Gillott 1988). Two especially well studied cases illustrate this type of action. Direct recordings of nervous activity in the isolated last abdominal ganglion of the silk moth *Bombyx mori* showed that a male factor (presumably the one that stimulates the female to oviposit) acts directly on this ganglion (Yamaoka and Hirao 1977). The last abdominal ganglion innervates both the female's genitalia and her abdominal muscles. The male factor apparently alters the ionic permeability of the neural sheath of the ganglion; the subsequent change in the composition of the fluid bathing the interior of the ganglion results in the expression of the oviposition reflex.

The invasive nature of the semen of the male house fly, *Musca domestica*, is especially dramatic. Apparently one male product acts to open a hole in the female's reproductive tract, allowing other seminal components direct access to her body cavity (Leopold et al. 1971a,b). Presumably these include the male accessory gland substance (or substances) which inhibit(s) female remating and stimulate(s) oviposition (Riemann et al. 1967; Riemann and Thorson 1969). Radioactive labeling experiments showed that male products reached the female head and thorax only ten minutes after mating began (Leopold et al. 1971b). A combination of ligation, labelling, and decapitation experiments suggested that the male products inhibiting female sexual receptivity act directly on the female's brain (Leopold et al. 1971a,b).

Male substances may also act outside the female's reproductive tract in many other species. Most experimental demonstrations of the effects of male accessory gland substances on female oogenesis, oviposition and receptivity have involved either implantation of portions of male reproductive tract, or injection of extracts of his glands or their secretions into the female's body cavity, rather than into her reproductive tract (40 different species are listed in Eberhard (1996)). Riemann et al. (1967) and Loher (1984) argue that, owing to diffusion problems, female receptors on the inner surface of the her reproductive tract are unlikely to respond to substances injected into her body cavity (see also Smith et al. (1990), on the relative impermeability of the spermatheca of the blowfly *Lucilia cuprina*). If the female tract is indeed often relatively impermeable to these substances, then the implantation and extract injection studies imply that female receptor sites exist outside of the interior of her reproductive tract in these species. Such widespread invasive male action would constitute especially strong evidence favoring the sexual selection over the natural selection hypothesis to explain the action of seminal products on female reproductive physiology and behavior. The females seem in danger of losing control of their own bodies!

Multiplicity of cues for a single female process

If females evolved under natural selection to use male products associated with insemination as cues, then one might expect the female to often use one particular cue or substance to trigger any particular response (e.g. ovulation, resistance to remating). Aside from the advantage of redundancy, there is no reason to expect a multiplicity of cues within a species. Selection for efficient coordination might be expected to produce evolutionarily conservative control systems. By minimizing her dependence on different stimuli from the male, a female would be able to exercise more precise control over her own crucial reproductive processes, and thus adjust them in her own best interests.

If, on the other hand, male sexual secretions evolve under sexual selection by cryptic female choice, one possible mechanism by which a male could escalate or increase the effect of his signal would be by adding further stimulatory substances to his semen. Thus the sexual selection hypothesis predicts a frequent (though not necessarily universal) evolutionary proliferation of male substances that trigger a given female process, instead of single substances. Well-studied insects such as *Drosophila* offer the best material for testing these predictions, since the chances are greater that complexity in male stimuli, if it exists, will have been discovered.

In *D. melanogaster* there are multiple male stimuli that switch off female remating, supporting the sexual selection hypothesis. One seminal factor is a product of the male accessory (paragonial) gland that has a short-term effect (Burnet *et al.* 1973; Scott 1986; Chen *et al.* 1988; Kalb *et al.* 1993). Another stimulus that switches off females comes from the sperm, or some other product produced by the testes acting in the female sperm storage organs. This second factor has both an ephemeral, short-term effect on female receptivity to males (Scott and Richmond 1985; Kalb *et al.* 1993), and a long-term effect (Manning 1967).

Kalb *et al.* (1993) suggested that perhaps the short-term accessory gland cue was needed because sperm storage was delayed, but this idea does not fit well with the times involved: maximum sperm storage occurs about five hours after copulation (Gilbert 1981), but the short-term effects of the accessory gland product are still strong 24 hours after copulation (Kalb *et al.* 1993).

In addition, when females were mated with males lacking both sperm and an accessory gland product that reduces female receptivity without reducing her attractiveness to males (esterase 6), some females still did not remate (Scott 1986). Thus there is yet another male stimulus (or stimuli) associated with mating which induces females to reject further mating attempts.

Further cues may also be involved. Copulation must last 9–13 min to switch on the female rejection behavior involving ovipositor extension, but the female is induced to produce an antiaphrodesiac pheromone and decrease the amount of sex attractant produced after only 3–4 min (Tompkins and Hall 1981). In sum, the male's participation in the control of different subprocesses involved in switching off female receptivity in *D. melanogaster* is quite complex. Similar multiplicity occurs in seminal effects on induction of oviposition in this species (Chen 1984).

Graduated female responses to quantitative variation in male cues

If females use chemical cues from males to sense that mating has occurred, one would expect natural selection to often produce all-or-none switches rather than graduated responses to the range of dosages transferred by normal males. This is because a female has either mated or she has not, and she is not expected to respond partly or 'half-heartedly' to a normal mating. Only if lower amounts of the male cue were correlated with insufficient numbers of sperm or some other male product crucial for female reproduction might partial female responses be favored. Sperm numbers are often far in excess of those needed by females (Parker 1970; Drummond 1984; Ridley 1988), so this correlation may often be absent.

Sexual selection, on the other hand, could favor increases in the amounts transferred to the female, by favoring increases in the intensity of male chemical signals, and subsequent raising of female response thresholds. This would be similar, in a vocal display, to selection favoring louder singing because of female preference for louder songs. Thus quantitative, dosage-dependent female responses whose maxima fall within or above the range of normal amounts transferred during copulation are predicted (though not for all cases) by the sexual selection hypothesis. Female response maxima above the normal range of amounts are not expected under natural selection.

The data needed to test these predictions are available for only a few species. A dosage-dependent female response is particularly clear in the cricket *Acheta domesticus* because the active male factor has been identified. The male transfers a prostaglandin synthetase (or synthetase complex) associated with his sperm, which acts on arachidonic acid (produced by the female and stored in her spermatheca) to produce prostaglandin, a hormone which stimulates oviposition. When gravid females were injected with different quantities of prostaglandin, larger numbers of eggs were laid on the following day by females receiving larger doses (Destephano and Brady 1977). Increases in female responses extended to doses larger than those associated with normal matings (Murtaugh and Denlinger 1982, 1984). Because the oviposition response is rapid rather than long-term, and these crickets remate readily (Murtaugh and Denlinger 1985), it is not likely to be a naturally selected adaptation to substandard numbers of sperm or male nutrients. When doses of the male factor were varied in association with the other stimuli associated with normal copulations, females also responded in a dosage-dependent manner. Leaving a spermatophore attached for only 2–3 min instead of 10 min sharply reduced the number of eggs laid, both in the short term and in the long term (Murtaugh and Denlinger 1985). A study of female behavior preceding and during oviposition (Destephano *et al.* 1982) showed that both the duration of oviposition and the completion of different components of a complete oviposition were apparently proportional to the dosage of prostaglandin.

Interspecific diversity of male products

Under natural selection there is no apparent reason to expect chemical messengers in the male semen to evolve and diversify particularly rapidly. In fact, they might be expected to be relatively conservative, because messenger substances are isolated from most effects of habitat change, and act after any species-isolating mechanisms that reduce hybridization (owing to natural selection against hybrids) have had their effects.

The predicted trend under sexual selection is the opposite, because rapid divergent evolution is common in sexually selected characters (West-Eberhard 1983). The inexorable competition between males for access to female gametes, the evolutionarily ephemeral nature of advantages in this competition, and the resulting advantages to females of altering their criteria to favor particular males, would continue to favor new seminal products or new variations of old products that induced favorable female responses.

The strongest tests of these contrasting predictions are comparisons of closely related species, such as those in the same genus. By far the best studied genus in this respect is *Drosophila*; the available data give quite dramatic demonstrations of interspecific diversity. In some species the male products have been characterized chemically. One male substance that increases female fecundity in *D. melanogaster* is a 36 amino acid peptide (Chen *et al.* 1988); one that has the same effect in *D. funebris* is apparently a glycine-carbohydrate derivative (Baumann 1974). The accessory gland of a third species, *D. nigromelanica*, completely lacks a sex peptide, but has a very large amount of glutamic acid, which may function as a neurotransmitter (Chen and Oechslin 1976).

Another type of data involves comparing the major protein components in the male accessory glands of related species. The major protein components of accessory gland secretions in eight *Drosophila* species differed in molecular mass and electrical charge (Fig. 2-2), as well as isoelectric points. There were even clear differences between three sibling species, *D. melanogaster*, *D. simulans* and *D. mauritiana* (Chen and Oechslin 1976; Chen 1984; Chen *et al.* 1985; Stumm-Zollinger and Chen 1988). These differences are especially dramatic because females of these species are virtually indistinguishable on the basis of their external morphology, and male identification depends on differences in the genitalia.

Bownes and Partridge (1987) found a similar diversity in electrophoretic analyses of proteins in the accessory glands of *D. melanogaster* and *D. pseudoobscura* (not a single one of the eight major polypeptides of *D. melanogaster* was present in *D. pseudoobscura*). Although the functions of these proteins are not known, it seems likely that many are transferred to the female during copulation or are associated with such substances (e.g. precursors).

When the presence and absence of β-esterases in the male's ejaculatory bulb, and the sensitivity of these substances to inhibition by alcohol (a trait of uncertain biological significance in this context) were determined in 93 *Drosophila* species, and compared with a probable phylogeny, extensive species-group heterogeneity was apparent (Johnson and Bealle 1968). The electrophoretic mobilities

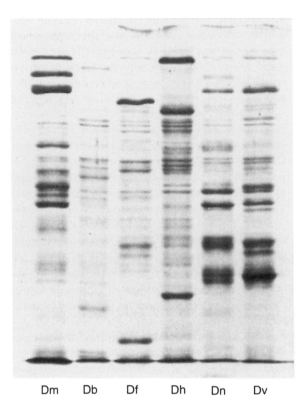

Dm   Db   Df   Dh   Dn   Dv

Fig. 2-2. Electrophoretic separation of proteins in the male accessory glands of six different species of *Drosophila*. The diversity of products in even closely related species such as these is in accord with the idea that sexual selection has acted on the products of male accessory glands. It is probable that the products of male reproductive tracts often act as 'chemical genitalia' to influence cryptic female choice when transferred to females (from Chen 1984).

of β-esterases also varied widely among species (Johnson and Bealle 1968). Intraspecific differences also occurred, as 'numerous' electrophoretic types of β-esterases occurred in different *D. aldrichi* males (Johnson and Bealle 1968).

An additional type of evidence related to divergence comes from tests of cross-reactivity. When the females of one species receive male products from a different species, their responses or lack of responses, in terms of oviposition or sexual receptivity, can give a conservative indication of differences between male products and/or female responsiveness to male products. The sex peptide PS-1 from the accessory glands of male *D. funebris* reduced the receptivity of conspecific females by 57–58%, but had no effect on female *D. melanogaster* (Baumann 1974).

PS-1 and PS-2 peptides, which reduce receptivity and stimulate oviposition in *D. funebris*, are apparently not even present in the male accessory glands of several other species of *Drosophila* (Chen in Bodnaryk 1978). Ovulation in female *D. melanogaster* was induced more frequently by injections of accessory gland secretions of conspecific males (39%) than by similar injections from five other *Drosophila* species (0.8–9.4%) (Chen *et al.* 1985). The very closely related species *D. melanogaster*, *D. simulans* and *D. mauritiana* showed more cross-reactivity in the ability to stimulate virgin females to ovulate than with more distantly related *Drosophila* (Chen *et al.* 1985; Stumm-Zollinger and Chen 1988). Even so, there were also differences between the effects of cross-injections compared with conspecific injections on both ovulation and female receptivity (Stumm-Zollinger and Chen 1988).

The effects on females were asymmetrical in one pair of species. Female *D. pulchrella* were less likely to ovulate when injected with an extract of the accessory glands of *D. suzukii* males than when injected with a similar extract from conspecific males, but *D. suzukii* females responded equally to injections of extracts from both species of male (Fuyama 1983). These species are closely related, and can produce fertile hybrids (Fuyama 1983).

Mimicry of female messenger molecules

Under the natural selection hypothesis, the origin of female use of male products as reproductive cues would occur when the female began to sense and respond to the presence of a substance that was present in the ejaculate and that functioned in some other context (e.g. nutrition for the sperm, lubricant, buffer to counteract adverse conditions within the female ducts). It is unlikely that such male substances would tend to resemble molecules such as hormones or neurotransmitters used by the female as messengers within her own body. Only if such female messenger molecules were present in the male, and were present in his ejaculate, either because they 'leaked' into the semen in large enough quantities to be sensed by the female, or because they also had some positive effect on sperm nutrition, survival or movement, would such substances be possible candidates as cues for the female. Such effects have not, to my knowledge, ever been observed.

In contrast, if males are under selection to influence and manipulate female responses, such female messenger molecules could be especially powerful tools in the semen. By introducing such a substance into the female, a male could take advantage of her pre-existing responses to such molecules. He might also avoid having his message muted or possibly ignored by the female at higher levels such as her central nervous system. Use of such substances could thus be an especially effective 'sensory trap' (West-Eberhard 1983; Ryan 1994; Christy 1995), especially since hormone molecules tend to be evolutionarily conservative (presumably because they often have many pleiotropic effects). This is another 'aggressive' characteristic like the invasive traits discussed above.

Testing these hypotheses can only be approximate. Chemical mimicry would not necessarily evolve under sexual selection in all groups, since its appearance could depend on imponderables such as the occurrence of appropriate mutations in males. The predictions are only approximate: mimicry completely absent or very rare under natural selection; and mimicry more common but possibly sporadic under sexual selection. In addition, the chemical nature of male accessory gland products is known for only a few species of insect (Gillott 1988). Even when a male product has been isolated, lack of demonstration that it resembles a female molecule may simply mean that no attempt has been made to compare it with female molecules.

Nevertheless, some striking resemblances between male seminal signals and female messenger molecules are known. Perhaps the clearest cases are the crickets *A. domesticus* and *Teleogryllus commodus*, in which males transfer an enzyme catalyzing the production of prostaglandin. The seminal fluid of the silkworm *Bombyx mori* also contains prostaglandin (Setty and Ramaiah 1979, 1980). Females of both *T. commodus* (Ai *et al.* 1986) and *B. mori* (Setty and Ramaiah 1980) also produce prostaglandins of their own. The widespread occurence of prostaglandins and their varied physiological effects (Brady 1983; Loher 1984) make it very probable that the males of these species are 'mimicking' female hormone molecules.

A suggestion that males also use juvenile hormone or its analogs comes from studies of moths. Adult female moths synthesize juvenile hormone, which can have effects on pheromone release activities (Cusson and McNeil 1989). The males of several saturniid species produce and store large quantities of juvenile hormone (JH) in their abdomens (references in Webster and Cardé 1984), and in *Hyalophora cecropia* the male accumulates large quantities of JH in his accessory glands, and transfers it to the female during copulation (Shirk *et al.* 1980). After mating, female calling behavior is reduced. Webster and Cardé (1984) showed that JH induced both termination of calling behavior, and the gradual elimination of attractant pheromone from the female's glands in the moth *Platynota stultana* (in a different family). They speculated that the function of JH transfer in the seminal fluid of saturniid and other moths may be to induce the switch from virgin to mated behavior in the female.

Summary of evidence for natural selection on females vs. sexual selection on males

For each of the five different topics, the available evidence fits more easily with the predictions of the sexual selection hypothesis than with those of the natural selection hypothesis. The data on the invasive action of male products are particularly clear: large numbers of species are included and the differences between the predictions are relatively sharp. The existence of such 'aggressive' signals by males suggests that partial rather than complete female responses to male seminal signals have been common in the past (and perhaps in the present). There are no data directly favoring an additional alternative, that of direct male–male competition (e.g. the seminal products of one male deactivating those of another). It must be noted, however, that such effects have probably seldom been searched for.

It is important to keep in mind that the discussion here is focussed on the *origins* of different male traits. It is possible that natural selection often also contributes to the subsequent maintenance of traits that originally appeared through sexual selection. Females can come to depend on male cues. For instance, females of the bug *Rhodnius prolixus* are apparently unable to move sperm into the spermatheca unless products from the male accessory gland are present (Davey 1958). Thus disentangling the effects of natural and sexual selection on maintenance will sometimes be difficult (see Fincke *et al.*, this volume). The combination of effects could explain the persistence of female use of male cues if there were periods during which there was no variability among males of her species for such cues (and thus no sexually selected payoff), or when females of a species mated with only a single male (and there was thus no sexual selection by cryptic female choice on males).

It is also important to note that natural selection operating in a different context, that of male–female conflicts (Alexander *et al.*, this volume) could produce predictions similar to those of the sexual selection hypothesis. The trends described above thus do not rule out this type of natural selection.

An alternative mechanism: nutritional contributions to the female

Some male seminal products may have effects on female reproduction because of their nutrient value to the female. For instance, male substances may increase egg production by providing the female with materials with which she can synthesize additional eggs. Alternatively (as has generally been supposed in the literature on the physiological effects of substances from the male reproductive tract on females), the male products may have a non-nutritive, triggering or hormone-like effect on the female. Both types of effect could result in sexual selection by cryptic female choice acting on males. The distinction between the two mechanisms may have important consequences, however, with respect to both the intensity of sexual selection acting on males to provide seminal products and the reproductive benefits to descriminating females (see Gwynne, this volume).

A male's ability to provide nutrient effects would presumably often be associated with his abilities to accumulate resources. If the donation was especially difficult for the male to obtain, his reproduction could be limited less by competition with other males for access to females, and more by his ability to amass resources. This would result in a reduction in the intensity of sexual selection on males. In extreme cases, sexual selection could even cease to act on the male, and females would begin to compete for access to males, as in the katydid *Kawanaphila nartee* (Gwynne and Simmons 1990).

With respect to possible direct female benefits, male nutrient transfer could favor a selective female through increases in her own reproduction. For instance, a female with a higher threshold for induction of refractory behavior would obtain more nutrients from males. Non-nutrient effects would not be expected to confer direct benefits of this sort, except in cases of female dependence on male cues mentioned above. They might in fact result in net losses, since the female reproductive behavior that would be in a male's best interests might differ from that in the best interests of the female. For instance, in a species in which females remate repeatedly and there is not complete first-male sperm precedence, a male's best interests might be served by inducing massive oviposition immediately after he copulated, with perhaps reduced attention to the quality of the oviposition site; the female might be better served by more gradual oviposition.

If male contributions are nutritive, a selective female might also gain indirectly through better viability genes in her offspring. Increased male ability to provide nutrients would presumably often be associated with superior viability characters such as better foraging behavior, feeding behavior, digestion, etc. In contrast, increased male ability to produce substances with hormone-like effects would not necessarily be associated with better male abilities to accrue resources, unless for some reason the hormone-like messenger molecules were especially costly. Improvement in hormone-like activity would usually amount to better signaling ability *per se*, and a female discriminating in favor of such a male would obtain good attractiveness genes for her male offspring.

The two mechanisms thus differ both in the intensity of expected sexual selection on male abilities to transfer seminal products to the female, and in the type of expected payoff to more discriminating females. Distinguishing between the two mechanisms can be tricky. For instance, several authors have deduced a possible nutrient function after finding that when the male is labeled radioactively, his labeled seminal products move out of the female's reproductive ducts and into her ovaries, where they are incorporated into developing eggs (Friedel and Gillott (1977) on the grasshopper *Melanoplus sanguinipes*; Huignard (1983) on the beetle *Acanthoscelides obtectus*; Sivinski and Smittle (1987) on the fly *Anastrepha suspensa*; Bownes and Partridge (1987) on *Drosophila melanogaster*; Markow and Ankney (1988) on 13 other *Drosophila* species; and Bowen *et al.* (1984) on the katydid *Requena verticalis*). This technique may be overly sensitive, however. Subsequent analyses have shown that in at least the first four of these species the amounts of male material are so small that significant nutrient effects are unlikely (Sivinski and Smittle 1987; Cheeseman and Gillott 1989; Kaulenas 1992). For instance, about 5 µg of protein are contributed to the eggs by a male *M. sanguinipes* grasshopper, compared with about 1.4 g of protein from the female. Only if the male substance were both crucial, even in minute quantities, and difficult or impossible for the female to produce in such amounts (neither has been demonstrated) would the nutrition hypothesis be likely to apply.

It has also been argued that the nutrition hypothesis is demonstrated when the magnitude of the male's effect on female reproduction is greater when the female has been fed poorly than when she has been fed well. Even such a demonstration of a nutritional effect is not enough, however, to show that a male seminal product has been under selection as a nutritional contribution to the female. The nutrition may be an incidental side effect of a selectively more important triggering effect. For instance, larger male nutrient donations to the female via larger spermatophores resulted in increases in egg production in the butterfly *Danaus plexippus*, but also increased the length of the period during which the female was refractory to further mating (Oberhauser 1989). Variations in the female response via decreased receptivity probably have a greater impact on the male's reproduction than those via the increased egg production resulting from nutrient contributions (Oberhauser 1992). Thus selection for nutrient contributions cannot be established until the entire spectrum of possible effects on the female that could favor a male's reproduction (e.g. induction of sperm transport, reduction in propensity to remate, induction of cooperation in copulation in the form of reduced resistance, longer copulations, relaxation of internal barriers that result in more sperm or more effective plugs being deposited: in fact a majority of the mechanisms listed in Table 2-1). Even for a species like *R. verticalis*, in which several lines of evidence point to a nutrient function for the spermatophore (Gwynne 1984, 1986, 1988a,b), judgement must still be reserved.

Several types of data give relatively strong evidence of hormone-like effects on females. I will limit myself here to one or two illustrative examples for each (see Eberhard 1996 for more examples, and a discussion of limitations of both the arguments and the data).

1. The male product is also synthesized by the female, where it functions as a hormone or other messenger or triggering substance. Or the male product is a direct precursor of such a substance. An example, discussed above, comes from the crickets *Acheta domesticus* and *T. commodus* (Destephano and Brady 1977; Loher 1979; Loher *et al.* 1981), in which males transfer prostaglandin synthetase to the female.

2. The male product acts directly on a female effector organ, instead of being incorporated into eggs. For instance, a male accessory gland substance stimulates an abdominal ganglion in the female of both the silkmoth *B. mori* (Yamaoka and Hirao 1977) and the mosquito *Aedes aegypti* (Gwadz 1972) (inducing, respectively, oviposition and resistance to genitalic coupling by further males).

3. Non-nutritional mimics of stimuli produced by male products induce the same female responses. For instance, mechanical stimulation of the female reproductive tract by objects mimicking spermatophores

increase oogenesis in roaches (*Nauphoeta*, *Leucophaea*, and *Diploptera*: Engelmann 1970) and in ticks (*Ornithodoros* and *Argas*: Connat *et al.* 1986; Leahy and Galun 1972).

4. Alteration of the portion of the female nervous system on which the male product is thought to act abolishes the female response, even though the same amount of nutrition is transferred to her by the male. For example, inhibition of pheromone release in the moth *Epiphyas postvittana* results from transection of the ventral nerve cord (Foster 1993), and inhibition of female receptivity in the housefly *Musca domestica* follows destruction or removal of the brain (Leopold *et al.* 1971a).

5. The effect of the substance contributed by the male lasts for much longer than the substance itself does inside the female. In *Drosophila melanogaster*, for instance, the male protein *msP 355a* is thought to increase oviposition, an effect that can last up to weeks. However *msP 355a* is detectable in the female, with a sensitive technique employing monoclonal antibodies, for only about 4–6 hours after copulation (Monsma *et al.* 1990).

6. The effect on female oviposition and oogenesis is too rapid to be due to nutrition. For example, the average number of eggs laid in the 24 h after treatment went from 2 after a control injection to 114 eggs after an injection of male substance in the cricket *Acheta domesticus* (Destephano and Brady 1977; Destephano *et al.* 1982).

7. Male supplies of the substance that elicit the female response are apparently not limiting, since males are able to transfer enough substance to elicit normal, complete female responses in copulations with several different females in rapid succession (such limitation has been adduced to favor a nutrient function; see, for example, Gwynne 1993). In the hessian fly, *Mayetiola destructor*, for instance, a male can generally mate with 15–20 females in quick succession before his ability to induce oviposition begins to decline, and he can still reduce female receptivity after mating with 45 different females during a single morning (Bergh *et al.* 1992), many more copulations than a male is likely to perform in nature (Bergh *et al.* 1990).

One further type of data suggest non-nutritional effects. In some groups the male products of one species have no effect on the reproduction of females of closely related species. Such specificity occurs, for example, in inhibition of female remating by female muscoid flies in the genera *Stomoxys*, *Phormia*, *Musca*, and *Scatophaga* (Morrison *et al.* 1982). Unless the constituents of eggs (or at least those contributed by the male) are sharply different in such species, differences of this sort are not expected under the nutrient hypothesis.

This discussion is incomplete. It omits additional species for which there is similar evidence in favor of non-nutritional male effects on female reproduction, and others for which there is no evidence one way or the other on these points. The nutritional hypothesis can be saved from some types of data by *ad hoc* arguments. For instance, rapid female responses to male nutrient donations (no. 6) could occur if these trigger the female to invest in eggs other nutrients that she has been withholding (for example, for maintenance functions). Even here, however, the male nutrients would also be acting as signals, and sexual selection could act to improve their signaling abilities (see Eberhard 1996 for a more complete discussion).

There are, to be sure, a few groups in which the available data indicate that males are selected with respect to the nutritional qualities of their seminal products (for example, *Pieris* butterflies: Rutowski *et al.* 1987; *Requena* and *Kawanaphila* katydids: Gwynne 1984, 1986, 1988a,b; Simmons 1990; Gwynne and Simmons 1990). In other species selection may favor male ability to transfer poisonous or defensive substances (see, for example, Doussourd *et al.* 1988; LaMunyon and Eisner 1993 on the moth *Utetheisa ornatrix*). Nevertheless, the general trend seems clear. Male seminal products often, and probably usually, act on females in ways that seem more hormone-like than nutritional.

Other evidence

Several additional types of data also support the idea that sexual selection by cryptic female choice is widespread (discussed in detail in Eberhard 1996).

In many species the duct leading to the female's spermatheca is long, thin, and has a complex, tortuous shape (Fig. 2-3). Spermathecal ducts reach apparently ridiculous extremes, as for instance in some cassidine beetles where they are more than 20 times the length of the female's entire body (D. Windsor, personal communication). The frequently tortuous morphology of spermathecal ducts suggests that it has often been advantageous to females to increase the difficulty of either entry into the spermatheca

(e.g. limit the entrance of male genitalia, his sperm, or other components of his semen), or exit from the spermatheca (avoid having too many sperm move out to fertilize eggs in the oviduct). Comparative data indicate that difficulty of entry has been more important than difficulty of egress in the evolution of tortuous spermathecal ducts (Eberhard 1996) (Fig. 2-3). Generalizing from these data, the selective advantage to females that explains the widespread trend toward tortuous spermathecal ducts is probably related to limiting entry of sperm into the spermatheca. Limiting male access makes sense as a mechanism for discriminating among males, and such discriminations can result in cryptic female choice (see numbers 4 and 13 in Table 2-1). Alternative hypotheses exist, such as limiting the access of venereally transmitted infections that are limited to or for some other reason must pass through the spermatheca, but I know of no evidence supporting them. Even if difficulty of access arose for other reasons, it could nevertheless result in selection via cryptic female choice favoring improvements in male abilities to gain access.

Another body of evidence supporting the widespread existence of cryptic female choice is related to the evolution

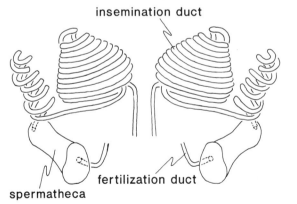

Fig. 2-3. A 'ridiculous' extreme in tortuous female spermatheca ducts in the linyphiid spider *Labulla thoracica*. Male access to a female's spermathecae is through the long, coiled and twisted 'insemination' ducts; sperm exit from the spermatheca to the oviduct, where fertilization occurs, through the short, simple 'fertilization' ducts. Thus the advantage to the female of having tortuous spermathecal ducts, which are common in many other groups, is apparently related to limiting male access. Such limitation makes sense as a mechanism for discriminating among males, and can lead to cryptic female choice (see numbers 4 and 13 in Table 2-1) (after Milledge 1993).

of male genitalia. When compared with many other body parts, male genitalia in most insects and arachnids (and many other groups) tend to undergo relatively rapid divergent evolution, often having different forms even in closely related species. I have argued previously that the reason for this pattern, which may be one of the most widespread trends in animal evolution, is that male genitalia often function as internal courtship devices that influence cryptic female choice (Eberhard 1985). If this argument is correct (see criticisms by Alexander *et al.*, this volume, and responses in the section on male–female conflict, below), then the widespread nature of rapid divergence in male genitalia constitutes further evidence that cryptic female choice is also very widespread. Further empirical data have accumulated since the original proposal, and in general they have given it further support.

One particularly dramatic confirmation is a sperm precedence study using males with different genitalic morphologies, and experimental modification of male genitalia in the tortoise beetle *Chelymorpha alternans* (Rodriguez 1994; Rodriguez *et al.* 1996). The length of a male genitalic sclerite (the flagellum), which is extremely long in these beetles (approximately three times the length of the male's body), has a strong effect on a male's chances of paternity if the female mates repeatedly (as occurs in nature) (see discussion of this work by Dickinson, this volume).

Other studies have tested the prediction that since the intensity of sexual selection by cryptic female choice is likely to be higher in species in which females mate with more males, male genitalia should diverge more rapidly in such species. This prediction was borne out in a comparison of 21 species of bees in four families: the males had more complex internal sacs in species in which published accounts indicate that females mate with several males (Roig-Alsina 1993). A similar correlation was found using 130 species of primates (Dixson 1987).

A final type of evidence favoring cryptic female choice is not related to genitalia, but to biases in paternity. It seems likely (though not altogether necessary: see Eberhard 1996) that if cryptic female choice is responsible for biasing paternity in favor of certain males, the bias will often favor the same males that are also favored by overt, precopulatory female choice. This concordance is especially likely if the selective advantage favoring female discrimination is better viability genes. Concordance between overt female choice and paternity biases that are probably due to cryptic female choice mechanisms occur in primates and birds (Tutin 1979; Lifjeld *et al.* 1993). Data from

the beetle *Tribolium castaneum* also point in the same direction, although less conclusively. Males whose pheromones were more attractive to females also achieved higher proportions of paternity when mated to females that were mated previously with another male (Lewis and Austad 1994). The number of copulations with each male was not determined, however, so the paternity bias could also have been due to overt female choice (Lewis and Austad 1994).

RELATIONSHIP BETWEEN CRYPTIC FEMALE CHOICE AND MALE–FEMALE CONFLICT

Female choice of any sort (cryptic or not) inevitably results in a conflict between the reproductive interests of males and females (Eberhard 1996; Alexander *et al.*, this volume). As illustrated in the discussion of *Dryomyza anilis* above, there are ways that a female might benefit from failing to cooperate completely and not allowing a male to fertilize all her eggs. If interactions with males reduce a female's production of offspring directly (for example, by interfering with oviposition), natural selection could favor the female's ability to facultatively resist males indiscriminately, no matter what their phenotypes; this would result in male–female conflict in the context of natural selection, as discussed by Alexander *et al.* (this volume) and Arnqvist (this volume). Female resistance to males would be expected to often be only partly effective, because selection on males will favor their abilities to overcome female resistance, both when there is competition with other males for access to the eggs of the same female (sexual selection on the male) and when no other male has access to the female (natural selection on the male).

A second type of possible benefit to the female can result from sexual selection, when her resistance is less than perfectly effective, and when some types of males are better than others at overcoming it. By biasing paternity in favor of such males, the female could produce genetically superior offspring. Female 'resistance' in this context might better by characterized as 'screening of males'. Such screening could result in sexual selection on males, favoring those with the good viability genes or the good attractiveness genes needed to pass female attempts to screen them.

The differences between naturally selected male–female conflict and sexual selection can be restated in terms of female cooperation. Direct benefits to females via natural selection on the ability to prevail in male–female conflict situations would tend to lead to facultative female mechanisms for categorical lack of cooperation. Lack of completely effective resistance mechanisms would be due to the abilities of males to overcome them, rather than to female cooperation. Indirect benefits in the context of sexual selection, on the other hand, could often lead to *selective cooperation* with some males and not others.

Females of a species could be subject to both types of selection simultaneously, and the balance between the two could change. The male–female conflict and sexual selection hypotheses generate many similar predictions (see section on seminal products above, and Alexander *et al.*, this volume) and can be difficult to distinguish. Is there any way to determine which type of selection has more often been responsible for the evolution of male–female interactions in insects and spiders?

At first glance, resolution of this question seems difficult: precise quantification of the benefits to females in nature is often not feasible, and the degree of female cooperation would seem difficult to determine. For many of the traits discussed above, such as male accessory gland products in the semen, there are simply no reliable data available, rendering attempts to discuss possible conflict (see, for example, Chapman *et al.* 1995) inconclusive.

Nevertheless, there are several sources of relevant data. They indicate that sexual selection has been more important than naturally selected male–female conflict. The details of male–female interactions in some species allow one to infer that sexual selection is more likely. For instance, the apparent effect of postcopulatory courtship on sperm-dumping in *Dryomyza anilis* (Otronen and Siva-Jothy 1991) is probably due to sexually selected benefits to the female, since it is hard to imagine a naturally selected advantage to the female from selective sperm dumping. Similar arguments could be made for several of the groups in Table 2-1 (e.g. *Paraphlebia*, *Centris*, *Harpobittacus*, *Neriene*, *Gryllus*). In addition, the often seemingly arbitrary copulatory courtship movements (e.g. stroking the female's dorsum), and their relatively gentle effects on the female (Eberhard 1994) do not have the physically forceful nature one would expect of direct male–female conflict.

The details of female morphology can also reveal the degree of female cooperation. They suggest that females often tend to cooperate selectively, although both selective cooperation and indiscriminate resistance probably occur. Males of many animals have specialized, often species-specific grasping organs (both genitalic and non-genitalic) that can hold and potentially manipulate the female. In

some species the area of the female's body that is grasped is not modified in any way to resist such contact. Some examples in which females show no modifications at all include the front legs of *Microhexura* and *Euagrus* spiders (Coyle 1981, 1986); the caudal filaments of calanoid copepods (Blades 1977; Blades and Youngbluth 1979); the antennae of sminthurid collembolans (Massoud and Betsch 1972); the thorax of *Macrodactylus* beetles (Eberhard 1993a); the ovipositor of *Ceratitis* flies (Eberhard and Pereira 1994); and the abdominal sternites in sepsid flies (Eberhard and Pereira 1996) (for other examples, see Eberhard 1985).

In other species, the area where the female is grasped by the male is modified, and the form of the modification is revealing. The female generally has grooves, cavities, and slits into which the male structures fit, thus mechanically *aiding* any male that fits (Fig. 2-4). Examples include slots in the genital cavity of *Craspedosoma* millipedes (Tadler 1993); pits in the abdomen of *Dictyna* spiders (Huber 1995); grooves in the abdomen of *Leioproctus* and *Callonychium* bees (Toro and de la Hoz 1976; Toro 1985); pits in the eyes and head of *Epigomphus* dragonflies (Corbet 1962); and indentations in the sides of the thorax of *Cicindella* tiger beetles (Freitag 1974) (see summary in Eberhard 1985). In these cases both the male grasping structure and the female grooves, etc. are often species-specific in form.

In still other species, male and female roles in grasping are reversed: the female grasps the male. Contrary to predictions of the male–female conflict hypothesis, these male structures are also often elaborate and species-specific. Examples mentioned in Eberhard (1985) include erigonine spiders, *Argyrodes* spiders, *Chordeuma* millipedes, *Meleoma* lacewings, *Ectobius* roaches, *Collops* beetles and schizomid arachnids.

Conspicuous by their absence are the 'aggressive' female resistance structures that would be expected if natural selection due to male–female conflict were usually more important than sexual selection. The only likely case I know involves the spines near the genital opening of the female water strider *Gerris incognitus* that have been shown experimentally to impede male copulation attempts (Arnqvist and Rowe 1995). Natural selection on female water striders to resist mating attempts may be unusually strong, owing to intense predation (Arnqvist, this volume); and even here it is not certain whether the female structure may not favor some conspecific males over others. The possibility that females of *G. incognitus* may also gain indirect benefits via the production of sons better able to overcome this barrier is yet to be tested.

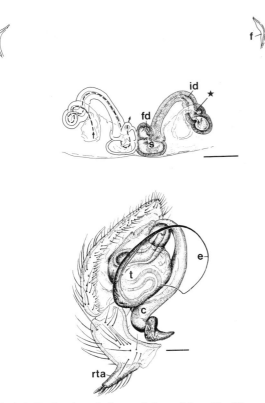

Fig. 2-4. The female genitalic morphology of the spider *Dictyna uncinata* suggests selective cooperation with males, rather than the indiscriminate resistance to male manipulations that would be expected if male–female conflict were responsible for rapid divergent genitalic evolution, as argued by Alexander *et al.* (this volume). The female (above) has a pair of pockets or foveae (f) anterior to the openings of her insemination ducts (id). The male genitalia or palp (below) has a spur-like prominence (rta). When the tip of the rta is inserted into the fovea during preliminary scraping movements on the female's abdomen, the male's palp is locked into position on the female's abdomen so that the embolus (e) can be inserted into her insemination duct. In one species group of *Dictyna* the male rta is especially long, and the female foveae are correspondingly distant from the entrances to the insemination ducts. In another species group the rta is absent, and the female lacks foveae. Far from opposing the hooking action of the male's rta, the female's foveae *facilitate* engagement; but her cooperation is selective, as it is available only for those males with appropriate morphology (from Huber 1995).

In sum, female morphology in areas grasped by males indicates that selection on females has more often favored selective cooperation than indiscriminate resistance. If the resolution of conflict in grasping situations is similar to that in others, such as male seminal products that

influence the female's reproductive behavior and physiology (I know of no *a priori* reason to doubt this, but no conclusive data are available), natural selection due to male–female conflict has generally been less important than sexual selection.

In addition to these empirical arguments, one theoretical consideration militates against the importance of forceful male–female conflict in many possible cryptic female choice contexts. A male that is attempting to achieve genitalic coupling with a female can gauge her resistance, and tell when he succeeds. He can thus harass her or otherwise increase his efforts until he has succeeded. But a male attempting to induce female responses such as sperm transport, ovulation, or rejection of further males, is in a poorer position to judge the female's compliance. He will thus be unable to utilize the efficient 'escalate until succeed' tactic (see, for example, Clutton-Brock and Parker 1995).

Alexander *et al.* (this volume) propose that male–female conflict has been more important than sexual selection by cryptic female choice in the evolution of insect male genitalia, partly because they doubt whether runaway sexual selection is likely to occur in insects. A careful consideration of their arguments is useful for several reasons. For biologists more accustomed to thinking in terms of male–female conflicts than sexual selection by female choice, a direct contrast between our views may help clarify the concept of cryptic female choice. In addition, although cryptic choice in general is not more tightly linked to runaway selection than to female choice without a runaway effect (see Eberhard 1996), there is reason to think that the two often go together. I will focus on points of more general interest, and avoid disagreements that stem from simple misinterpretations of what I have written previously.

Alexander *et al.* assert that runaway selection is unlikely in insects. They support this conclusion with speculations about how female insects choose males, and theoretical considerations of how changes are produced during the runaway process. Concerning female choice mechanisms, Alexander *et al.* argue that insect females seldom use comparative or 'best-of-*n*' criteria, and that instead 'insects *appear* more often to use tests [in mate choice] involving sets of minimal criteria' (p.14, emphasis added), citing only a single unpublished manuscript on one genus of cicadas. Later they give a list of contrasts between threshold and comparative female choice mechanisms; without offering supporting data, they state that the list 'seem[s] to suggest that insects are more likely to possess TC [threshold choice] systems' (p.15). In the same vein, they argue that in most or all insects females generate their preferences without learning from external or social stimuli.

Their discussion also fails to take into account the fact that copulation itself can be a social learning experience. A female that mates more than once can use comparisons of successive copulations to modify both the criteria by which she judges males, and her tendency to use the sperm of any given male to fertilize her eggs via cryptic female choice mechanisms (Table 2-1). Mechanisms of comparison could involve classic neural memory, and different degrees of triggering of female reproductive physiological responses. At one point Alexander *et al.* appear to reverse themselves, and to recognize the possibility that females could compare successive copulations; they state, however (with no supporting arguments or data) that in such cases 'extremes in ornamentation...are not expected' (p. 20). In sum, the arguments offered by Alexander *et al.* concerning comparative female choice mechanisms in insects do not provide convincing reason for rejecting the possiblity of runaway processes in insects.

As an alternative to comparative female choice mechanisms, Alexander *et al.* mention Lande's (1981) absolute female preference model. They argue that this kind of choice is unlikely to result in runaway evolution, because the extreme males can [only] be favored incidentally, assuming females 'can *lower* their criteria and accept males less extreme than the female's (otherwise) "absolute" minimum preference' (p. 15), and that 'the male trait will evolve to match the female's preference and then stop, unless somehow the females are again moved to even higher extremes of preference' (p. 15).

This discussion makes three important mistakes that lead to underestimates of the likelihood of runaway sexual selection. First, as is true for many models, male traits and female preferences are treated unidimensionally. Ignored is the fact that both male displays and female criteria in a given species vary in different social, ecological, morphological, and physiological contexts (see Jackson 1981, 1992; Endler 1994; Endler and Houde 1995; Andersson 1994 for changes in both male behavior and female criteria in different ecological contexts; see also Petrie and Jennions (1997) for a summary). This variability increases the chances that variant male stimuli will occur, and that they will be able to elicit favorable female responses.

Also neglected is the fact that animal nervous systems are highly interconnected (see, for example, Stein and Hart 1983; Prosser 1991; Harris-Warwick *et al.* 1992; Dusenbury 1992), so that neurons in a female decision center are likely to receive input (both directly and indirectly) from a variety of male-generated stimuli, and females can use multiple criteria for making a single decision. This interconnectedness will also increase the chances that variant male stimuli can elicit favorable female responses. It is likely that males sometimes use female responses that evolved in other contexts to induce females to respond to them, as has been argued in the the widely discussed 'sensory trap' and 'sensory bias' models of sexual selection (see, for example, West-Eberhard 1983; Eberhard 1993a,b; Ryan 1994; Andersson 1994; Christy 1995). As I have noted above, such sensory traps are especially likely to be available to males to trigger cryptic female choice mechanisms.

Finally, Alexander *et al.* (this volume) may consider the runaway process unlikely because they get the sequence of changes expected to occur in runaway processes backwards. The most likely sequence does not start with the male waiting for accidental changes to alter female preferences, but rather with some males winning out in intense competition by devising a new way to stimulate females so as to increase favorable responses (probably most often by playing to pre-existing female tendencies to respond that were established by natural or by sexual selection). This change in males can followed by changes (often reductions) in female tendencies to respond to male stimuli. The female changes can be favored when they increase the chances that the female's sons will be sired by the favored males. As the new male trait spreads in the population, selection in favor of still further male abilities to elicit favorable female responses will increase in intensity. Any change in the set of female responsiveness traits will constitute a new substrate, to which the males can play in a new round of evolution.

Turning to topics related more specifically to genitalia, Alexander *et al.* argue that female structures such as pits, pores, and grooves that I have cited (above; see also Eberhard 1985) as evidence against the male–female conflict hypothesis because they are often species-specific in form, that lack any other obvious functional significance, and that appear to *aid* males (see, for example, Fig. 2-4), are nevertheless consistent with the conflict-of-interest hypothesis. They do this, however, by failing to deal with the data in question. At first they mischaracterize the male and female structures as merely 'fit[ting] one another' (p. 20). This leads them to discuss an inappropriate test case, the beetle *Pasimachus punctulatus*, in which the female structure is neither species-specific nor without other possible functional significance. In no place do they deal with the female traits I have cited that contradict their predictions.

Another set of pertinent data omitted by Alexander *et al.* are the experimental demonstrations that apparently coercive male structures (clasping organs) can function to induce favorable responses from females after the female has been seized (see Loibl 1958 and Krieger and Krieger-Loibl 1958 on male abdominal clasping organs in dragonflies; Belk 1984 on male antennal appendages in fairy shrimps), or that they are used in ways suggesting a stimulatory function (see Lorkovic 1952 on the genitalic claspers of a butterfly).

The discussion of the proposed correlation between greater diversity of genitalic morphology and more coercive copulations also omits much conflicting evidence. The correlation is documented by comparing crickets of the genus *Gryllus* with some groups of grasshoppers, and is extended in the next sentence to cover the male genitalia of pterygote insects in general ('probably attributable in large part to conflicts of interest...' (p.18)), without a single supporting citation. Such a correlation does not occur in some major insect groups. For instance, in many moths females produce long-range attractant pheromones, and a male cannot even find a female, much less coerce her to initiate copulation, unless she first cooperates; yet male moths usually have extremely diverse and complex genitalia (see, for example, Tuxen 1970). There are many other examples in which males do not forcefully coerce females to initiate mating, and male genitalic structures are nevertheless usually species-specific: numerous flies in which females must fly to swarms and enter them before the male can clasp her (e.g. many mosquitoes, midges, empidids) (McAlpine *et al.* 1987); zorapterans in which females control both initiation and termination of copulation (Choe 1995 and this volume); lampyrid beetles in which the female can only be found after she has actively signaled to passing males (Lloyd 1979 on behavior; Green 1957 on genitalia); salticid spiders that easily reject attempts both to initiate and to prolong copulation by simply decamping (see, for example, Jackson 1980, 1992 on behavior; Kaston 1948 on genitalia); *Argyrodes* and *Leucauge* spiders in which the larger female seizes the male prior to mating with her chelicerae (rather than vice versa) and can easily break off copulation by releasing him (Lopez and Emerit 1979 and W. G. Eberhard and B. A. Huber (in prep.) on behavior; Exline and Levi 1962 and Levi 1980 on genitalia). A survey of other groups (e.g. many other insects, belonging to at least six different orders, that mate at leks or landmark

encounter sites: Thornhill and Alcock 1983) would almost certainly yield many additional examples.

Only if these initially 'luring' copulations could be shown to later involve coercion, with the male overcoming barriers imposed by indiscriminate female resistance (rather than by selective female cooperation) could the conflict-of-interest hypothesis be saved from this objection based on data on female behavior. The single documentation of this sort by Alexander *et al.* (this volume) is an unpublished manuscript on a cicada, cited to support a prediction made for grasshoppers (p.18). Although pertinent data have yet to be assembled for most of the groups just listed, there evidence in several (zorapterans and the spiders) that females are cooperating selectively; females of these groups apparently control termination as well as initiation of copulation.

In general, I fail to be convinced by a discussion of genitalic evolution that overlooks so many pertinent data, especially when many of the data have been previously assembled and related to the points under discussion (Eberhard 1985: pp. 60–64, 173, 182–183).

In summary, the available evidence that can discriminate between the sexual selection by cryptic female choice and the male–female conflict hypotheses is more generally in accord with the former. It should be re-emphasized, however, that a completely convincing resolution of this controversy must await future studies. Doubts about the validity of the argument of Alexander *et al.* that runaway selection is uncommon in insects do not mean that the male–female conflict hypothesis they propose is necessarily incorrect, or that both cannot occur. The emphasis on female morphology above is partly due to the fact that other types of data are generally unavailable. For most species and for most contexts, the kinds of data needed to discriminate between sexual selection by cryptic female choice and naturally selected male–female conflicts are not yet available.

CONCLUSIONS

This chapter presents evidence that cryptic female choice occurs in insects and arachnids. Cryptic choice is relatively certain in at least a few well-studied species. More controversially, the chapter also argues that cryptic female choice may be extremely widespread. If cryptic female choice is widespread in insects and arachnids, there are several important consequences. The classic Darwinian criterion for female choice on the basis of the male's ability to induce the female to allow him to copulate should be modified so as to include the male's abilities to induce critical postintromission events in the female. Commonly used concepts of male courtship (see, for example, Immelmann and Beer 1989) should be revised to include male behavior that influences all female processes leading up to fertilization of the egg and the production of offspring, rather than just those leading to copulation. Sexual selection itself should probably be redefined to emphasize that males are not competing for access to females, but for access to their eggs.

There are also practical consequence for those concerned with the evolution of male display behavior and morphology. If females are exercising cryptic choice, then attempts to estimate a male's reproductive success by measuring his copulation success may give misleading results. Determination of paternity, rates of oviposition and remating, etc., will be needed for more confident estimates of male reproductive success. This of course is unfortunate, because such studies are generally more costly and more difficult technically than studies of copulation success.

However, the payoffs for attention to cryptic female choice may be considerable. It offers possible explanations for otherwise puzzling behavior (such as copulatory courtship), physiology (such as the invasive action of male accessory gland products on female reproductive processes), and morphology (such as the elongate genitalic flagellum of male tortoise beetles). Cryptic female choice links evolutionary theory with a large body of studies on reproductive physiology which has heretofore been largely devoid of a theoretical basis for interpretation (Eberhard and Cordero 1995). A series of questions that were previously of only incidental interest in physiological studies emerge as being of special evolutionary interest. This link promises to foster the growth of a field of 'evolutionary reproductive physiology' related to male–female interactions, similar to other compound fields such as ecological physiology and evolutionary medicine.

The existence of cryptic female choice does not negate or diminish the importance of either classical, overt precopulatory female choice, or direct male–male conflicts. Rather, the different processes may work in concert. Males of many species may be 'tested' sequentially for their abilities to physically overcome other males, to induce females to allow them to initiate copulation, and then to induce females to respond behaviorally and physiologically in ways that increase their likelihood of paternity.

In a more general context, recognition that females can and do often play active and important roles in determining paternity even after the male has achieved intromission is another stage in the gradual abandonment of 'the myth of the passive female' (Batten 1992), which has long dominated both theoretical and practical studies of male–female interactions.

ACKNOWLEDGEMENTS

I thank R. D. Alexander and M. J. West-Eberhard for extensive comments on a preliminary draft, and the Smithsonian Tropical Research Institute and the Vicerrectoría de Investigacion of the Universidad de Costa Rica for financial support.

LITERATURE CITED

Ai, N., S. Komatsu, I. Kubo and W. Loher. 1986. Manipulation of prostaglandin-mediated oviposition after mating in *Teleogryllus commodus*. *Int. J. Invert. Reprod. Devel.* **10**: 33–42.

Alcock, J. and S. L. Buchmann. 1985. The significance of post-insemination display by male *Centris pallida* by its male (Hymenoptera: Anthophoridae). *Z. Tierpsychol.* **68**: 231–243.

Andersson, M. 1994. *Sexual Selection*. Princeton: Princeton University Press.

Arnqvist, G. and L. Rowe. 1995. Sexual conflict and arms races between the sexes: a morphological adaptation for control of mating in a female insect. *Proc. R. Soc. Lond.* **B261**: 123–127.

Batten, M. 1992. *Sexual Strategies*. New York: G. P. Putnam's Sons.

Baumann, H. 1974. Isolation, partial characterization and biosynthesis of the paragonial substances, PS-1 and PS-2 of *Drosophila funebris*. *J. Insect Physiol.* **20**: 2181–2194.

Belk, D. 1984. Antennal appendages and reproductive success in the Anostraca. *J. Crust. Biol.* **4**: 66–71.

Bella, J. L., R. K. Butlin, C. Ferris and G. M. Hewitt. 1992. Asymmetrical homogamy and unequal sex ratio from reciprocal mating-order crosses between *Chorthippus parallelus* subspecies. *Heredity* **68**: 345–352.

Bergh, J. C., M. O. Harris and S. Rose. 1990. Temporal patterns of emergence and reproductive behavior of the Hessian fly, *Mayetiola destructor* (Diptera: Cecidomyiidae). *Ann. Entomol. Soc. Am.* **83**: 998–1004.

Bergh, J. C., M. O. Harris and S. Rose. 1992. Factors inducing mated behavior in female Hessian flies (Diptera: Cecidomyiidae). *Ann. Entomol. Soc. Am.* **85**(2): 224–233.

Blades, P. 1977. Mating behavior of *Centropages typicus* (Copepoda: Calanoida). *Mar. Biol.* **40**: 57–64.

Blades P. and M. J. Youngbluth. 1979. Mating behavior of *Labidocera aestiva* (Copepoda: Calanoida). *Mar. Biol.* **51**: 339–355.

Bodnaryk, R. P. 1978. Structure and function of insect peptides. *Adv. Insect Physiol.* **13**: 69–132.

Bowen, B. J., C. G. Codd and D. W. Gwynne. 1984. The katydid spermatophore (Orthoptera: Tettigoniidae): Male nutritional investment and its fate in the mated female. *Austr. J. Zool.* **32**: 23–31.

Bownes, M. and L. Partridge. 1987. Transfer of molecules from ejaculate to females in *Drosophila melanogaster* and *Drosophila pseudoobscura*. *J. Insect Physiol.* **33**: 941–947.

Brady, U. E. 1983. Review: Prostaglandins in insects. *Insect Biochem.* **13**(5): 443–451.

Burnet, B., K. Connolly, M. Kearney and R. Cook. 1973. Effects of male paragonial gland secretion on sexual receptivity and courtship behaviour of female *Drosophila melanogaster*. *J. Insect Physiol.* **19**: 2421–2431.

Carré, C. and C. Sardet. 1984. Fertilization and early development in *Beroe ovata*. *Devel. Biol.* **105**: 188–195.

Chapman, T., F. L. Liddle, J. M. Kalb, M. F. Wolfner and L. Partridge. 1995. Cost of mating in *Drosophila melanogaster* females is mediated by male accessory gland products. *Nature (Lond.)* **373**: 241–244.

Cheeseman, M. T. and C. Gillott. 1989. Long hyaline gland discharge and multiple spermatophore formation by the male grasshopper, *Melanoplus sanguinipes*. *Physiol. Entomol.* **14**: 257–264.

Chen, P. S. 1984. The functional morphology and biochemistry of insect male accessory glands and their secretions. *Annu. Rev. Entomol.* **29**: 233–255.

Chen, P. S. and A. Oechslin. 1976. Accumulation of glutamic acid in the paragonial gland of *Drosophila nigromelanica*. *J. Insect Physiol.* **22**: 1237–1243.

Chen, P. S., E. Stumm–Zollinger, T. Aigaki, J. Balmer, M. Bienz and P. Bohlen. 1988. A male accessory gland peptide that regulates reproductive behavior of female *D. melanogaster*. *Cell* **54**: 291–298.

Chen, P. S., E. Stumm–Zollinger and M. Caldelari. 1985. Protein metabolism of *Drosophila* male accessory glands-II. *Insect Biochem.* **15**: 385–390.

Choe, J. 1995. Courtship feeding and repeated mating in *Zorotypus barberi* (Insecta: Zoraptera). *Anim. Behav.* **49**: 1511–1520.

Christenson, T. E. 1990. Natural selection and reproduction: a study of the golden orb-weaving spider, *Nephila clavipes*. In *Contemporary Issues in Comparative Psychology*. D. A. Dewsbury, ed., pp. 149–174. Sunderland, MA: Sinauer.

Christy, J. A. 1995. Mimicry, mate choice, and the sensory trap hypothesis. *Am. Nat.* **146**: 171–181.

Clutton-Brock, T. H. and G. A. Parker. 1995. Sexual coercion in animal societies. *Anim. Behav.* **49**: 1345–1365.

Connat, J.-L., J. Ducommun, P. A. Diehl and A. Aeschlimann. 1986. Some aspects of the control of the gonotrophic cycle in the tick *Ornithodoros moubata* (Ixodoidea, Argasidae). In *Morphology, Physiology, and Behavioral Biology of Ticks*. J. R. Sauer and J. A. Hair, eds., pp. 194–216. New York: John Wiley & Sons.

Corbet, P. 1962. *A Biology of Dragonflies*. London: Withersby.

Coyle, F. A. 1981. The mygalomorph spider genus *Microhexura* (Araneae, Dipluridae). *Bull. Am. Mus. Nat. Hist.* **170**: 64–75.

–. 1986. Courtship, mating, and the function of male-specific leg structures in the mygalomorph spider genus *Euagrus* (Araneae, Dipluridae). *Proceedings of the Ninth International Congress of Arachnology, Panama 1983*. W. G. Eberhard, Y. D. Lubin and B. C. Robinson, eds., pp. 33–38. Washington, D. C.: Smithsonian Institution Press.

Crews, D. 1984. Gamete production, sex hormone secretion, and mating behavior uncoupled. *Hormones Behav.* **18**: 22–28.

Cusson, M. and J. N. McNeil. 1989. Involvement of juvenile hormone in the regulation of pheromone release activities in a moth. *Science (Wash., D. C.)* **24**: 210–212.

Darwin, C. [1871] 1960. *The Descent of Man and Selection in Relation to Sex*. New York: Modern Library.

Davey, K. G. 1958. The migration of spermatozoa in the female of *Rhodnius prolixus* Stal. *J. Exp. Biol.* **35**: 694–701.

Destephano, D. B. and U. E. Brady. 1977. Prostaglandin and prostaglandin synthetase in the cricket, *Acheta domesticus*. *J. Insect Physiol.* **23**: 905–911.

Destephano, D. B., U. E. Brady and C. A. Farr. 1982. Factors influencing oviposition behavior in the cricket, *Acheta domesticus*. *Ann. Entomol. Soc. Am.* **75**: 111–114.

Dewsbury, D. 1989. Copulatory behavior as courtship communication. *Ethology* **79**: 218–234.

Diehl, P. A. and A. Aeschlimann. 1982. Tick reproduction: oogenesis and oviposition. In *Physiology of Ticks*. F. D. Obenchain and R. Galun, eds., pp. 277–350. New York: Pergamon Press.

Dixson, A. 1987. Observations on the evolution of genitalia and copulatory behaviour in primates. *J. Zool. (Lond.)* **213**: 423–443.

Drummond, B. A. 1984. Multiple mating and sperm competition in the Lepidoptera. In *Sperm Competition and the Evolution of Animal Mating Systems*. R. L. Smith, ed., pp. 291–371. New York: Academic Press.

Dusenbury, D. B. 1992. Sensory ecology. New York: W. H. Freeman.

Dussourd, D. E., K. Ubik, C. Harvis, J. Resch, J. Meinwald and T. Eisner. 1988. Biparental defensive endowment of eggs with acquired plant alkaloid in the moth *Utetheisa ornatrix*. *Proc. Natl. Acad. Sci. U.S.A.* **85**: 5992–5996.

Eberhard, W. G. 1985. *Sexual Selection and Animal Genitalia*. Cambridge, Mass.: Harvard University Press.

–. 1991. Copulatory courtship in insects. *Biol Rev.* **66**: 1–31.

–. 1993a. Copulatory courtship and genital mechanics of three species of *Macrodactylus* (Coleoptera: Scarabeidae: Melolonthinae). *Ethol. Ecol. Evol.* **5**: 19–63.

–. 1993b. Evaluating models of sexual selection: genitalia as a test case. *Am. Nat.* **142**: 564–571.

–. 1994. Copulatory courtship in 131 species of insects and spiders, and consequences for cryptic female choice. *Evolution* **48**: 711–733.

–. 1996. *Female Control: Sexual Selection by Cryptic Female Choice*. Princeton: Princeton University Press.

Eberhard, W. G. and C. Cordero. 1995. Sexual selection by cryptic female choice on male seminal products – a new bridge between sexual selection and reproductive physiology. *Trends Ecol. Evol.* 10: 493–496.

Eberhard, W. G. and M. C. Marin. 1996. The sexual behavior of the beetle *Megalopus armatus* (Coleoptera: Chrysomelidae: Megalopinae). *J. Kansas Entomol. Soc.*, in press.

Eberhard, W. G. and F. Pereira. 1994. Functions of the male genitalic surstyli in the Mediterranean fruit fly, *Ceratitis capitata* (Diptera: Tephritidae). *J. Kansas Entomol. Soc.* 66: 427–433.

–. 1996. Functional morphology of male genitalic surstyli in the dungflies *Achisepsis diversiformis* and *A. ecalcarata* (Diptera: Sepsidae). *J. Kansas Entomol. Soc.*, in press.

Endler, J. A. 1994. Multiple-trait coevolution and environmental gradients in guppies. *Trends Ecol. Evol.* 10: 22–29.

Endler, J. A. and A. E. Houde. 1995. Geographic variation in female preferences for male traits in *Poecilia reticulata*. *Evolution* 49: 456–468.

Engelmann, F. 1970. *The Physiology of Insect Reproduction*. New York: Pergamon Press.

Exline, H. and H. W. Levi. 1962. American spiders of the genus *Argyrodes* (Araneae: Theridiidae). *Bull. Mus. Comp. Zool.* 127: 75–204.

Fisher, R. 1958. *The Genetical Theory of Natural Selection*. New York: Dover.

Foster, S. P. 1993. Neural inactivation of sex pheromone production in mated lightbrown apple moths, *Epiphyas postvittana* (Walker). *J. Insect Physiol.* 39: 267–273.

Freitag, R. 1974. Selection for non-genitalic mating structure in female tiger beetles of the genus *Cicindela* (Coleoptera: Cicindelidae). *Can. Entomol.* 106: 561–568.

Friedel, T. C. and C. Gillott. 1977. Contribution of male-produced proteins to vitellogenesis in *Melanoplus sanguinipes*. *J. Insect Physiol.* 23: 145–151.

Fuyama, Y. 1983. Species-specificity of paragonial substances as an isolating mechanism in *Drosophila*. *Experientia* 39: 190–192.

Gage, M. J. G. 1992. Removal of rival sperm during copulation in a beetle, *Tenebrio molitor*. *Anim. Behav.* 44: 587–589.

Gilbert, D. G. 1981. Ejaculate esterase 6 and initial sperm use by female *Drosophila melanogaster*. *J. Insect Physiol.* 27: 641–650.

Gillott, C. 1988. Arthropoda – Insecta. In *Reproductive Biology of Invertebrates*. K. G. Adiyodi and R. G. Adiyodi, eds., pp. 319–471. New York: John Wiley & Sons.

Green, J. W. 1957. Revision of the nearctic species of *Pyractomena* (Coleoptera: Lampyridae). *Wasmann J. Biol.* 15: 237–284.

Gwadz, R. W. 1972. Neuro-hormonal regulation of sexual receptivity in female *Aedes aegypti*. *J. Insect Physiol.* 18: 259–266.

Gwynne, D. T. 1984. Courtship feeding increases female reproductive success in bushcrickets. *Nature (Lond.)* 307: 361–363.

–. 1986. Courtship feeding in katydids (Orthoptera: Tettigoniidae): investment in offspring or in obtaining fertilizations? *Am. Nat.* 128: 342–352.

–. 1988a. Courtship feeding in katydids benefits the mating male's offspring. *Behav. Ecol. Sociobiol.* 23: 373–377.

–. 1988b. Courtship feeding and the fitness of female katydids (Orthoptera: Tettigoniidae). *Evolution* 42(3): 545–555.

–. 1993. Food quality controls sexual selection in Mormon crickets by altering male investment. *Ecology* 74: 1406–1413.

Gwynne, D. T. and L. W. Simmons. 1990. Experimental reversal of courtship roles in an insect. *Nature (Lond.)* 346: 172–174.

Harris-Warwick, R. M., R. Nagy and M. B. Musbaum. 1992. Neuromodulation of stomatogastric networks by identified neurons and transmitters. In *Dynamic Biological Networks*. R. M. Harris-Warwick, E. Marder, A. I. Selverson and M. Moulins, eds., pp. 87–137. Cambridge, Massachusetts: MIT Press.

Huber, B. A. 1995. The retrolateral tibial apophysis in spiders – shaped by sexual selection? *Zool. J. Linn. Soc.* 113: 151–163.

Huignard, J. 1983. Transfer and fate of male secretions deposited in the spermatophore of females of *Acanthoscelides obtectus* Say (Coleoptera Bruchidae). *J. Insect Physiol.* 29: 55–63.

Immelmann, K. and C. Beer. 1989. *A Dictionary of Ethology*. Cambridge, Mass.: Harvard University Press.

Jackson, R. R. 1980. The mating strategy of *Phidippus johnsoni* (Araneae, Salticidae): II. Sperm competition and the function of copulation. *J. Arachnol.* 8: 217–240.

–. 1981. Relationship between reproductive security and intersexual selection in a jumping spider *Phidippus johnsoni* (Araneae: Salticidae). *Evolution* 35(3): 601–604.

–. 1992. Conditional strategies and interpopulation variation in the behaviour of jumping spiders. *N.Z. J. Zool.* 19: 99–111.

Johnson, F. M. and S. Bealle. 1968. Isozyme variability in species of the genus *Drosophila*. V. Ejaculatory bulb esterases in *Drosophila* phylogeny. *Biochem. Gen.* 2: 1–18.

Kalb, J. M., A. J. DiBenedetto and M. F. Wolfner. 1993. Probing the function of *Drosophila melanogaster* accessory glands by directed cell ablation. *Proc. Natl. Acad. Sci. U.S.A.* 90: 8093–8097.

Kaston, B. J. 1948. Spiders of Connecticut. *Bull. St. Geol. Nat. Hist. Surv. Conn.* 70: 1–874.

Kaulenas, M. S. 1992. *Insect Accessory Reproductive Structures*. New York: Springer-Verlag.

Kingan, T. G., A. K. Raina and P. Thomas-Laemont. 1993a. Control of reproductive behavior of female moths by factors in male seminal fluids. In *ACS Conference Proceedings Series Pest Management: Biologically Based Technologies*. R. D. Lumsden and J. L. Vaughn, eds., pp. 117–120. New York: American Chemical Society.

Kingan, T. G., Thomas-Laemont, P. A. and Raina, A. K. 1993b. Male accessory gland factors elicit change from 'virgin' to 'mated' behaviour in the female corn earworm moth *Helicoverpa zea*. *J. Exp. Biol.* 183: 61–76.

Kirkpatrick, M. and M. Ryan. 1991. The evolution of mating preferences and the paradox of the lek. *Nature (Lond.)* 350: 33–38.

Krieger, F. and E. Krieger-Loibl. 1958. Beitrage zum Verhalten von *Ischnura elegans* und *Ischnura pumilio* (Odonata). *Z. Tierpsychol.* 15: 82–93.

LaMunyon, C. W. and T. Eisner. 1993. Postcopulatory sexual selection in an arctiid moth (*Utetheisa ornatrix*). *Proc. Natl. Acad. Sci. U.S.A.* **90**: 4689–4692.

Lande, R. 1981. Models of speciation by sexual selection on polygenic traits. *Proc. Natl. Acad. Sci. U.S.A.* **78**: 3721–3725.

Lauterbach, G. 1954. Begattung und Larvengeburt bei den Strepsipteren zugleich ein Beitrag zur Anatomie der *Stylops*-Weibchen. *Z. Parasitenkunde* **16**: 255–297.

Leahy, Sr. M. G. and R. Galun. 1972. Effect of mating on oogenesis and oviposition in the tick *Argas persicus* (Oken). *Parasitology* **65**: 167–178.

Leopold, R. A. 1976. The role of male accessory glands in insect reproduction. *Annu. Rev. Entomol.* **21**: 199–221.

Leopold, R. A., A. C. Terranova and E. M. Swilley. 1971a. Mating refusal in *Musca domestica*: Effects of repeated mating and decerebration upon frequency and duration of copulation. *J. Exp. Zool.* **176**: 353–360.

Leopold, R. A., A. C. Terranova, B. J. Thorson and M. E. Degrugillier. 1971b. The biosynthesis of the male housefly accessory secretion and its fate in the mated female. *J. Insect Physiol.* **17**: 987–1003.

Levi, H. W. 1980. The orb-weaver genus *Mecynogea*, the subfamily Metinae and the genera *Pachygnatha*, *Glenognatha* and *Azilia* of the subfamily Tetragnathinae North of Mexico (Araneae: Araneidae). *Bull. Mus. Comp. Zool.* **149**: 1–74.

Lewis, S. M. and S. Austad. 1994. Sexual selection in flour beetles: the relationship between sperm precedence and male olfactory attractiveness. *Behav. Ecol.* **5**: 219–224.

Lifjeld, J. T., P. O. Dunn, R. J. Robertson and P. T. Boag. 1993. Extra-pair paternity in monogamous tree swallows. *Anim. Behav.* **45**: 213–229.

Lloyd, J. E. 1979. Sexual selection in luminescent beetles. In *Sexual Selection and Reproductive Competition in Insects*. M. S. Blum and N. A. Blum, eds., pp. 293–342. New York: Academic Press.

Loher, W. 1979. The influence of prostaglandin E2 on oviposition in *Teleogryllus commodus*. *Entomol. Exp. Appl.* **25**: 107–119.

–. 1984. Behavioral and physiological changes in cricket females after mating. *Adv. Invert. Reprod.* **3**: 189–210.

Loher, W., I. Ganjian, I. Kubo, D. Stanley-Samuelson and S. S. Tobe. 1981. Prostaglandins: their role in egg-laying of the cricket *Teleogryllus commodus*. *Proc. Natl. Acad. Sci. U.S.A.* **78**: 7835–7838.

Loibl, E. 1958. Zur Ethologie und Biologie der deutschen Lestiden (Odonata). *Z. Tierpsychol.* **15**: 54–81.

Lopez, A. and M. Emerit. 1979. Donnes complementaires sur la glande clypeale des *Argyrodes* (Araneae, Theridiidae): utilisation du microscope electronique a balayage. *Rev. Arachnol.* **2**: 143–153.

Lorkovic, Z. 1952. L'accouplement artificiel chez les Lépidoptères et son application dans les recherches sur la fonction de l'appareil génital des insectes. *Physiol. Comp. Oecol.* **3**: 313–319.

Manning, A. 1967. The control of sexual receptivity in female *Drosophila*. *Anim. Behav.* **15**: 239–250.

Markow, T. A. and P. F. Ankney. 1988. Insemination reaction in *Drosophila*: found in species whose males contribute material to oocytes before fertilization. *Evolution* **42**: 1097–1101.

Massoud, Z. and J.-M. Betsch. 1972. Étude sur les insectes collemboles, II: les caractères sexuels secondaires des antennes des Symphypléones. *Rev. Ecol. Biol. Sol.* **9**: 55–97.

Masumoto, T. 1991. Males' visits to females' webs and female mating receptivity in the spider *Agelena limbata* (Araneae: Agelenidae) *J. Ethol.* **9**: 1–7.

–. 1993. The effect of the copulatory plug in the funnel-web spider, *Agelena limbata* (Araneae: Agelenidae). *J. Arachnol.* **21**: 55–59.

Maynard Smith, J. 1987. Sexual selection – a classification of models. In *Sexual Selection: Testing the Alternatives*. J. W. Bradbury and M. B. Andersson, eds., pp. 9–20. New York: John Wiley & Sons.

–. 1991. Theories of sexual selection. *Trends Ecol. Evol.* **6**: 146–151.

McAlpine, J. F., B. V. Peterson, G. E. Shewell, J J Teskey, J. R. Vockeroth and D. M. Wood. 1987. *Manual of Nearctic Diptera.*, vols. 1 and 2. Ottawa: Research Branch Agriculture Canada.

Milledge, F. 1993. Further remarks on the taxonomy and relationships of the Linyphiidae, based on the epigynal duct conformation and other characters (Araneae). *Bull. Br. Arachnol. Soc.* **9**: 145–156.

Monsma, S. C., H. A. Harada and M. F. Wolfner. 1990. Synthesis of two *Drosophila* male accessory gland proteins and their fate after transfer to the female during mating. *Devel. Biol.* **142**: 465–475.

Morrison, P. E., K. Venkatesh and B. Thompson. 1982. The role of male accessory-gland substance on female reproduction with some observations of spermatogenesis in the stable fly. *J. Insect Physiol.* **28**: 607–614.

Murtaugh, M. P. and D. L. Denlinger. 1982. Prostaglandins E and F_2 in the house cricket and other insects. *Insect Biochem.* **12**: 599–603.

–. 1984. Regulation of long-term oviposition in the house cricket, *Acheta domesticus*: roles of prostaglandin and factors associated with sperm. *Arch. Insect Biochem. Physiol.* **6**: 59–72.

–. 1985. Physiological regulation of long-term oviposition in the house cricket, *Acheta domesticus*. *J. Insect Physiol.* **31**: 611–617.

Oberhauser, K. S. 1989. Effects of spermatophores on male and female monarch butterfly reproductive success. *Behav. Ecol. Sociobiol.* **25**: 237–246.

–. 1992. Rate of ejaculate breakdown and intermating intervals in monarch butterflies. *Behav. Ecol. Sociobiol.* **31**: 367–373.

Oliver, J. H. 1986. Induction of oogenesis and oviposition in ticks. In *Morphology, Physiology and Behavioral Biology of Ticks*. J. R. Sauer and J. A. Hair, eds., pp. 233–247. New York: John Wiley & Sons.

Ono, T., M. T. Siva-Jothy, and A. Kato. 1989. Removal and subsequent ingestion of rivals' semen during copulation in a tree cricket. *Physiol. Entomol.* **14**: 195–202.

Otronen, M. 1984. The effect of differences in body size on the male territorial system of the fly *Dryomyza anilis*. *Anim. Behav.* **32**: 882–890.

–. 1989. Female mating behaviour and multiple matings in the fly *Dryomyza anilis*. *Behaviour* **111**: 77–97.

–. 1990. Mating behavior and sperm competition in the fly, *Dryomyza anilis*. *Behav. Ecol. Sociobiol.* **26**: 349–356.

–. 1995. Fertilization success in the fly *Dryomyza anilis* (Dryomyzidae): effects of male size and mating situation. *Behav. Ecol. Sociobiol.* **35**: 33–38.

Otronen, M. and M. T. Siva-Jothy. 1991. The effect of postcopulatory male behaviour on ejaculate distribution within the female sperm storage organs of the fly, *Dryomyza anilis* (Diptera: Dryomyzidae). *Behav. Ecol. Sociobiol.* **29**: 33–37.

Parker, G. A. 1970. Sperm competition and its evolutionary consequences. *Biol. Rev.* **45**: 525–567.

Peretti, A. V. 1996. Análisis del comportamiento de transferencia espermática de *Bothriurus flavidus* Kraepelin (Scorpiones, Bothriuridae). *Rev. Soc. Entomol. Arg.* **55**, in press.

Petrie, M. and M. D. Jennions. 1997. Variation in female mate choice: causes and consequences. *Anim. Behav.*, in press.

Prosser, C. L. 1991. *Neural and Integrative Animal Physiology*. New York: John Wiley & Sons.

Queller, D. C. 1987. Sexual selection in flowering plants. In *Sexual Selection: Testing the Alternatives*. J. W. Bradbury and M. B. Andersson, eds., pp. 165–179. New York: John Wiley & Sons.

–. 1994. Male-female conflict and parent-offspring conflict. *Am. Nat.* (*Suppl.*) **144**: 584–599.

Raabe, M. 1986. Insect reproduction: regulation of successive steps. *Adv. Insect Physiol.* **19**: 29–154.

Ridley, M. 1988. Mating frequency and fecundity in insects. *Biol. Rev.* **63**: 509–549.

Riemann, J. G. and B. J. Thorson. 1969. Effect of male accessory material on oviposition and mating by female house flies. *Ann. Entomol. Soc. Am.* **62**: 828–834.

Riemann, J. G., D. J. Moen and B. J. Thorson. 1967. Female monogamy and its control in houseflies. *J. Insect Physiol.* **13**: 407–418.

Rodriguez, V. 1994. Fuentes de variación en la precedencia de espermatozoides de *Chelymorpha alternans* Boheman 1854 (Coleoptera: Chrysomelidae: Cassidinae). Masters thesis, Universidad de Costa Rica.

Rodriguez, V., W. G. Eberhard and D. Windsor. 1996. Longer genitalia confer a sexually selected advantage in a beetle. *Proc. R. Soc. Lond.* B, submitted.

Roig-Alsina, A. 1993. The evolution of the apoid endophallus, its phylogenetic implications, and functional significance of the genital capsule (Hymenoptera, Apoidea). *Boll. Zool.* **60**: 169–183.

Rutowski, R., G. W. Gilchrist and B. Terkanian. 1987. Female butterflies mated with recently mated males show reduced reproductive output. *Behav. Ecol. Sociobiol.* **20**: 319–322.

Ryan, J. J. 1994. Mechanisms underlying sexual selection. In *Behavioral Mechanisms in Evolutionary Biology*. L. Real, ed., pp. 190–215. Chicago: University of Chicago Press.

Schaible, U., C. Gack and H. F. Paulus. 1986. Zur Morphologie, Histologie und biologischen Bedeutung der Kopfstrukturen mannlicher Zwergspinnen (Linyphiidae: Erigoninae). *Zool. Jb. Syst.* **113**: 389–408.

Scott, D. 1986. Inhibition of female *Drosophila melanogaster* remating by a seminal fluid protein (esterase 6). *Evolution* **40**: 1084–1091.

Scott, D. and R. C. Richmond. 1985. An effect of male fertility on the attractiveness and oviposition rates of mated *Drosophila melanogaster* females. *Anim. Behav.* **33**: 817–824.

Setty, B. N. Y. and T. R. Ramaiah. 1979. Isolation and identification of prostaglandins from the reproductive organs of male silkmoth, *Bombyx mori* L. *Insect Biochem.* **9**: 613–617.

–. 1980. Effect of prostaglandins and inhibitors of prostaglandin biosynthesis on oviposition in the silkmoth *Bombyx mori*. *Ind. J. Exp. Biol.* **18**: 539–541.

Shirk, P. D., G. Bhaskaran and H. Roller. 1980. The transfer of juvenile hormone from male to female during mating in the Cecropia silkmoth. *Experientia* **36**: 682–683.

Simmons, L. W. 1986. Female choice in the field cricket *Gryllus bimaculatus* (De Geer). *Anim. Behav.* **34**: 1463–1470.

–. 1990. Nuptial feeding in tettigoniids: male costs and the rates of fecundity increase. *Behav. Ecol. Sociobiol.* **27**: 43–47.

Sivinski, J. and B. Smittle. 1987. Male transfer of materials to mates in the Caribbean fruit fly, *Anastrepha suspensa* (Diptera: Tephritidae). *Fla. Entomol.* **70**: 233–238.

Smith, P. H., C. Gillott, L. Barton Browne and A. C. M. van Gerwen. 1990. The mating-induced refractoriness of *Lucilia cuprina* females: manipulating the male contribution. *Physiol. Entomol.* **15**: 469–481.

Stein, D. S. and R. Hart. 1983. Brain damage and recovery: problems and perspectives. *Behav. Neurobiol.* **37**: 185–222.

Stephenson, A. G. and R. I. Bertin. 1983. Male competition, female choice, and sexual selection in plants. In *Pollination Biology*. L. Real, ed., pp. 109–149. New York: Academic Press.

Stumm-Zollinger, E. and P. S. Chen. 1988. Gene expression in male accessory glands of interspecific hybrids of *Drosophila*. *J. Insect Physiol.* **34**: 59–74.

Tadler, A. 1993. Genitalia fitting, mating behaviour and possible hybridization in millipedes of the genus *Craspedosoma* (Diplopoda, Chordeumatida, Craspedosomatidae). *Acta Zool.* **74**: 215–225.

Thornhill, R. 1983. Cryptic female choice and its implications in the scorpionfly *Harpobittacus nigriceps*. *Am. Nat.* **122**: 765–788.

Thornhill, R. and J. Alcock. 1983. *The Evolution of Insect Mating Systems*. Cambridge, Mass.: Harvard University Press.

Tompkins, L. and J. C. Hall. 1981. The different effects on courtship of volatile compounds from mated and virgin *Drosophila* females. *J. Insect Physiol.* **27**: 17–21.

Toro, H. 1985. Ajuste genital en la cópula de *Callonychium chilense* (Hymenoptera, Andrenidae). *Rev. Chil. Entomol.* **12**: 153–158.

Toro, H. and E. de la Hoz. 1976. Factores mecánicos en la aislación reproductiva de *Apoidea* (Hymenoptera). *Rev. Soc. Entomol. Arg.* **35**: 193–202.

Tutin, C. E. G. 1979. Mating patterns and reproductive strategies in a community of wild chimpanzees (*Pan troglodytes schweinfurthii*). *Behav. Ecol. Sociobiol.* **6**: 29–38.

Tuxen, S. 1970. *Taxonomist's Glossary of Genitalia of Insects*. Darien, Conn.: S-H Service Agency.

von Helversen, D. and O. von Helversen. 1991. Pre-mating sperm removal in the bushcricket *Metaplastes ornatus* Ramme 1931 (Orthoptera, Tettigonoidea, Phaneropteridae). *Behav. Ecol. Sociobiol.* **28**: 391–396.

Waage, J. 1979. Dual function of the damselfly penis: sperm removal and transfer. *Science (Wash., D. C.)* **203**: 916–918.

–. 1984. Sperm competition and the evolution of odonate mating systems. In *Sperm Competition and the Evolution of Animal Mating Systems*. R. L. Smith, ed., pp. 251–290. New York: Academic Press.

–. 1986. Evidence for widespread sperm displacement ability among Zygoptera (Odonata) and the means for predicting its presence. *Biol. J. Linn. Soc.* **28**: 285–300.

Watson, P. J. 1991. Multiple paternity as genetic bet-hedging in female sierra dome spiders, *Linyphia litigiosa* (Linyphiidae). *Anim. Behav.* **41**: 343–360.

Wcislo, W. T., R. L. Minckley and H. C. Spangler. 1992. Pre-copulatory courtship behavior in a solitary bee, *Nomia triangulifera* Vachal (Hymenoptera: Halictidae). *Apidologie* **23**: 431–442.

Webb, J. C., J. Sivinski and C. Litzkow. 1984. Acoustical behavior and sexual success in the Caribbean fruit fly, *Anastrepha suspensa* (Loew) (Diptera: Tephritidae). *Environ. Entomol.* **13**: 650–656.

Webster, R. P. and R. T. Cardé. 1984. The effects of mating, exogenous juvenile hormone and a juvenile hormone analogue on pheromone titre, calling and oviposition in the omnivorous leafroller moth (*Platynota stultana*). *J. Insect Physiol.* **30**(2): 113–118.

West-Eberhard, M. J. 1983. Sexual selection, social competition, and speciation. *Q. Rev. Biol.* **58**: 155–183.

Whitehouse, M. E. A. and R. R. Jackson. 1994. Intraspecific interactions of *Argyrodes antipodiana*, a kleptoparasitic spider from New Zealand. *N. Z. J. Zool.* **21**: 253–268.

Yamaoka, K. and T. Hirao. 1977. Stimulation of virginal oviposition by male factor and its effect on spontaneous nervous activity in *Bombyx mori. J. Insect Physiol.* **23**: 57–63.

Yokoi, N. 1990. The sperm removal behavior of the yellow spotted longicorn beetle *Psacothea hilaris* (Coleoptera: Cerambycidae). *Appl. Entomol. Zool.* **25**: 383–388.

3 · Natural and sexual selection components of odonate mating patterns

OLA M. FINCKE, JONATHAN K. WAAGE AND WALTER D. KOENIG

ABSTRACT

Traditionally, students of odonate reproductive behavior have focussed on how males compete for access to mates and fertilizations. This tendency has yielded considerable information on male reproductive strategies and on the proximate and ultimate mechanisms involved in male–male competition, but has left numerous gaps in our knowledge of other aspects of odonate mating systems.

We review relevant aspects of odonate biology and examine the extent to which current data on mating patterns support predictions arising from sexual selection theory. Although long-term studies offer some such support, they also indicate that natural selection for longevity and stochastic factors such as weather play critical roles in influencing reproductive success. Relatively little of the variance in male reproductive success in odonates has been traced to variance in male phenotype.

We emphasize the role of females as determinants of odonate mating patterns and discuss sexual conflicts of interest over mating, fertilization, and oviposition decisions. Finally, we explore ways in which natural selection underlies female mating decisions and how larval and adult ecology interact to influence adult reproductive behavior.

INTRODUCTION

The study of mating systems is the study of the behavioral, physiological, and ecological factors that underlie predictable patterns of male and female interactions during reproduction. Much of the literature on mating systems emphasizes male–male competition and its effects on male morphology and behavior. However, fertilization success of males cannot be explained solely by pre- and post-copulatory interactions of males or their gametes (see Alexander *et al.*, this volume). There are at least three reasons for this in odonates.

First, the unique copulation process in odonates requires cooperation on the part of a female. Males may be able to take a female in tandem (grasp the female's head or prothorax with specialized anal appendages), but she must bend her abdomen to engage the male's penis. Second, sperm competition does not take place in a neutral arena; it occurs within females. Males invade this arena by removing sperm of prior mates as part of copulation (Waage 1979a, 1986a; Siva-Jothy 1987a). If females are able to differentially use the sperm from individual males, then male copulatory behavior or mating order alone would not be sufficient to determine paternity. Third, females can negate whatever males accomplish through sperm removal and replacement simply by leaving a male's territory and mating with another male before ovipositing any eggs. These three factors demonstrate that the role of females in odonate mating systems goes well beyond simply being a limiting factor in male reproductive competition.

RELEVANT ODONATE BIOLOGY

Odonates are well suited for interspecific comparisons, field studies, and enclosure studies. They show considerable variation in reproductive behavior within and among species. Adults are often relatively easy to catch, making it easy to quantify their behavior and perform long-term studies on marked individuals. Because their lives are divided into three distinct stages, selection pressures on different life-history episodes can be studied.

The larval stage of odonates lasts from several months to several years, during which time they are aquatic predators. They emerge from this stage as winged, visually oriented predators and enter a period (lasting a few days to several weeks) of feeding and sexual maturation called the teneral stage, usually, but not always, away from oviposition areas. Adult reproductive lifespans range from days to months. Females mature batches of eggs and usually spend the time between reproductive episodes away from the water.

Odonates are one of the taxa for which the mechanisms and associated morphology for sperm competition have been most clearly identified (Siva-Jothy 1984, 1987a;

Waage 1984a, 1986a; Miller 1990, 1991a). Copulation involves a two-stage transfer of sperm, first within the male from the penis to the secondary storage vesicle and then to the female using copulatory apparatuses that are not homologous to other insect reproductive organs. The odonate penis consequently is used for two distinct functions: to transfer sperm to a female's storage organs and to remove or displace sperm from these organs. The success of these behaviors results in considerable sperm precedence by males that are the last to mate with females prior to oviposition (Fincke 1984a; McVey and Smittle 1984; Wolf *et al.* 1989; Michiels 1992; Hadrys *et al.* 1993).

Because males may mature faster than females and are usually able to mate daily whereas females often require at least several days to mature eggs, operational sex ratios for most species are male-biassed. Male agonistic behavior takes the form of chases and ritualized fights and attempts to displace males in tandem with females. Competition among males for fertilizations is common and often fierce, resulting in a diverse array of mate-guarding behaviors (Waage 1979b, 1984a; Sherman 1983; Tsubaki *et al.* 1994) and alternative mate-finding tactics (Waltz 1982; Fincke 1985, 1992a; Forsyth and Montgomerie 1987; Waltz and Wolf 1988, 1993). Shifts from one tactic to another are typically conditional on age, energy stores, and habitat structure or density, but may also be phenotypically fixed as has been shown in *Mnais* damselflies (Watanabe and Taguchi 1990; Watanabe 1991; Nomakuchi 1992).

Odonates include two major groups, the damselflies (suborder Zygoptera) and dragonflies (suborder Anisoptera). These groups show many intriguing patterns of convergent and divergent evolution in morphology and behavior. For example, zygopterans copulate for a minimum of 1–2 min and oviposit into plant tissues, whereas anisopteran copulations may be as short as a few seconds and most females deposit eggs directly onto the water surface. Males of both taxa use the penis to remove or displace sperm. However, the morphologically more complex anisopteran penis is homologous to the temporary sperm-holding vesicle in Zygoptera, whereas the zygopteran penis with its sperm-removal morphology is homologous to a supporting member of the anisopteran copulatory complex (Waage 1984a). Interestingly, the families with the most complex courtship and territorial behavior (Calopterygidae, Pseudostigmatidae and Libellulidae) are quite remote phylogenetically and very different morphologically, suggesting that mating behavior is plastic and may be subject to relatively few phylogenetic constraints. Information on phylogenetic patterns of male sperm competition and mate guarding can be found in Waage (1984a, 1986a).

THE CLASSIFICATION OF ODONATE MATING SYSTEMS

Odonate reproductive behavior can be organized according to a variety of variables. Although varying widely in emphasis, most such classifications are organized around the continuum of resource monopolization proposed by Emlen and Oring (1977). The basic premise is that the distribution and abundance of either oviposition sites, females, or both determines the degree and type of competition for matings that is profitable for males. As oviposition areas used by females become too large or widely dispersed for an individual male to monopolize, or as females become more synchronous in receptivity, searching for mates becomes more profitable for males than localized defense of encounter sites. Where oviposition sites are smaller relative to a male's patrol flight ability and more spatially clumped, and as females become more asynchronous in receptivity, defense of oviposition sites becomes more profitable. Between these two extremes, males should control female encounter sites if females are found at such sites more predictably than at oviposition areas. If females oviposit immediately after mating, males that compete by mate-searching should remain in tandem following mating until their mates oviposit in order to avoid losing their sperm investment. In contrast, an oviposition site that attracts many females should be more valuable to a territorial male than any given mate. Therefore, territorial males generally use non-contact guarding of females to ensure paternity, a behavior that offers greater opportunity to chase intruders from oviposition sites they control.

Several predictions emerge concerning the expected strength of sexual selection in different odonate mating systems. For example, because fewer males can monopolize sites than can monopolize females themselves, competition among males for mates and sexual dimorphism in traits functioning in male–male competition should increase from mate-searching systems to those in which males defend oviposition sites. Sexual selection on males should increase with increasing asynchrony in female receptivity since this offers males more opportunity to monopolize multiple mates. Finally, because the operational sex ratio of odonates tends to increase with population density (Fincke 1994a), male–male competition should also increase with density.

None the less, the expected overall strength of sexual selection in different mating systems is unclear. Female mate choice, to the extent that it occurs, may increase in tandem with male–male competition, leading to strong sexual selection in species with male resource defense. Alternatively, female choice of males may be greatest where males do not control oviposition sites, encounters between the sexes are fairly frequent, and consequently male–male competition is relatively weak (Conrad and Pritchard 1992). If this is true, the strength of female choice and male–male competition oppose each other and there should be moderate sexual selection across all levels of male resource monopolization.

Although much is known concerning the factors influencing the mating patterns of adult odonates, the role of at least one prominent feature of their life history – larval ecology – has thus far been neglected (Buskirk and Sherman 1985). For example, three species of tropical pseudostigmatid damselflies overlap in their use of tree-hole oviposition sites in Panama, but males of only the largest species, *Megaloprepus coerulatus*, defend this resource (Fincke 1984b). Based on the Emlen and Oring (1977) paradigm, all three species should be territorial since tree holes are limiting and easily defendable. The solution to this paradox appears to be interspecific larval competition that prohibits territoriality from being profitable for the two smaller *Mecistogaster* species (Fincke 1992b). Although male *Mecistogaster* stand to gain as many matings as do *Megaloprepus coerulatus* by defending large holes in tree-fall gaps, survivorship of *Mecistogaster* larvae is rarely greater than one per hole. Thus, although males of both genera are capable of defending the oviposition site, the pay-off of tree-hole defense differs between the two taxa, not in the number of matings obtained, but in the number of offspring produced.

SEXUAL SELECTION

Evidence from odonate mating patterns

Many aspects of odonate biology suggest strong sexual selection. They exhibit a wide variety of mating patterns from female monogamy (Rowe 1978; Fincke 1987) to mate-searching and territorial polygyny. Many species are sexually dimorphic. Males in some species fight for access to mates through defense of territories. Finally, males often compete for fertilizations by sperm competition and pre- and postcopulatory guarding (Waage 1984a). These features render odonates well suited to testing sexual selection theory, even though none provides unambiguous evidence of ongoing sexual selection.

Color and size dimorphism illustrate this problem. Most odonate males and females differ in coloration and patterns of the thorax and abdomen, which in some species functions in sexual recognition (Corbet 1962; Moodie 1995). However, if males and females generally occupy different foraging habitats, natural selection may explain much of the observed sexual dimorphism in body coloration (Hafernik and Garrison 1986). Males of a few species have distinct wing patterns or color patches on the body that are displayed during courtship. Although wing dimorphisms may have a sexual selection function in male–male competition or female choice, most of the evidence to date suggests that they aid in species or sexual recognition rather than in sexual selection (Buchholtz 1951; Waage 1975, 1979c; Fincke 1984b; De Marchi 1990).

Similarly, most odonates exhibit sexual size dimorphism, with females typically being slightly larger than males (Anholt *et al.* 1991). Since larger females can usually carry more eggs than smaller females, this suggests that natural selection on females to increase clutch size has been greater than sexual selection favoring large males. Male-biassed size dimorphism is rare in odonates, and in only a few species has size been shown to increase mating efficiency (see below).

Nor is variation in mating success sufficient evidence for sexual selection. If success at defending territories or accumulating matings over time is mostly due to foraging success or predator avoidance, and if much of that success is due to the timing of weather events during maturation or adulthood, then reproductive success will primarily be due to a combination of natural selection and chance (Fincke 1986a; Koenig and Albano 1986; Michiels and Dhondt 1991a).

Finally, although the widespread occurrence of sperm removal and postcopulatory guarding in odonates provides unambiguous evidence for sexual selection in the past, these behaviors are not sufficient to demonstrate current sexual selection. For that, one must be able to detect and correlate variation in sperm displacement or guarding duration with fertilization success among individual males.

Evidence from long-term studies

Does the opportunity for sexual selection vary with the ability of males to monopolize resources or females, as

Table 3-1. *Studies of lifetime reproductive success in odonates*

Species	References
Zygoptera (damselflies)	
Enallagma hageni	Fincke 1982, 1986a, 1988
E. boreale	Anholt 1991; Fincke 1994a
Coenagrion puella	Banks and Thompson 1985, 1987;
	Thompson 1987, 1989, 1990;
	Harvey and Walsh 1993
Ischnura gemina	Hafernik and Garrison 1986
I. graellsii	Cordero 1992a
Argia chelata	Hamilton and Montgomerie 1989
Anisoptera (dragonflies)	
Erythemis simplicicollis	McVey 1988
Libellula luctuosa	Moore 1989, 1990
Plathemis lydia	Koenig and Albano 1987
Nannophya pygmaea	Tsubaki and Ono 1987
Sympetrum rubicundulum	Van Buskirk 1987
Sympetrum danae	Michiels and Dhondt 1991a[a]

[a] Enclosure study.

predicted by Emlen and Oring (1977)? Because daily variation in male mating success is typically much higher than variation over the lifetime of individuals (Fincke 1982, 1988; Banks and Thompson 1985; Koenig and Albano 1987; McVey 1988), studies of lifetime reproductive success are best used to answer this question. Odonates have provided fertile ground for measuring lifetime reproductive success with at least 12 species, evenly divided between damselflies and dragonflies, studied to date (Table 3-1).

An index of the relative opportunity for sexual selection relative to natural selection can be obtained by partitioning total variance in lifetime reproductive (LRS) or mating (LMS) success (Arnold and Wade 1984a,b). For example, lifetime mating success (LMS) can be partitioned into variance in lifespan (days alive), mating efficiency (mates per day), and female reproduction (eggs fertilized per mating). Of these three episodes, only mating efficiency is likely to primarily represent sexual selection, and thus the proportion of the total variance in LRS accounted for by variation in mating efficiency provides an estimate of the relative importance of sexual selection.

Unfortunately, quantifying and partitioning variance in reproductive success entails numerous pitfalls (Sutherland 1985, 1987; Koenig and Albano 1986; Wade 1987; Grafen 1988). First, because most odonates have mechanisms for sperm competition, mating success may not reflect fertilization

success. Second, sampling must be unbiassed. This is particularly important for intersexual comparisons, because differential dispersal or other ecological dissimilarities can result in sex-related differences in reproductive success that are easily misconstrued as being due to sexual selection. Even within members of a sex, demographic parameters may vary significantly within a season, and thus the apparent importance of sexual selection may depend on what part of the breeding season is sampled (Fincke 1988).

Third, natural and sexual selection are not easy to separate, especially if male behavior affects both mating success and offspring survivorship (Fincke 1992a). Dividing up behaviors influencing fitness into ever finer episodes can help distinguish components representing natural and sexual selection and thus alleviate this problem, but at the cost of increasing the total proportion of variance attributable to covariance components that are themselves difficult to interpret. For example, Koenig and Albano (1987) found the most important component of the opportunity for selection in male *Plathemis lydia* was lifespan, explaining 27% of the total variance in LRS. However, they partitioned the remaining variance into multiple episodes, many of which entailed aspects of male mating efficiency such as visits per day and matings per hour. Consequently, covariance components represented 52% of the total variance, making the results difficult to interpret and compare with other studies.

Despite these difficulties, some trends predicted by sexual selection are qualitatively supported by long-term studies on odonates. For example, the majority of lifetime studies of odonates show greater variance in lifetime mating (LMS) or reproductive success (LRS) for males than females, consistent with odonates having polygamous mating systems (Waage 1984a; Conrad and Pritchard 1992). However, there are several exceptions.

In the territorial dragonfly *Erythemis simplicicollis*, the proportion of total variance in male LRS due to sexual selection was high, but the total variance in male LRS was slightly less than that of female LRS (McVey 1988). Similarly, variance in mating success among males and females did not differ between male and female *Ischnura gemina* (Hafernik and Garrison 1986), although the total selection on males was slightly higher than that on females. This species is particularly interesting because its mating pattern is closer to serial monogamy than to polygyny. Males are not territorial and spend a long time *in copula* and tandem mate-guarding, obtaining on average only one mating every four days. This suggests that male copulatory and postcopulatory behavior may constrain sexual selection on males.

Studies partitioning variance in LMS or LRS into episodes of selection on lifespan and mating efficiency support the prediction that the importance of sexual relative to natural selection increases across species as males are better able to monopolize reproductive resources. For example, male mating efficiency accounted for 61% of the variance in male LRS in the territorial *Erythemis simplicicollis* (McVey 1988) compared with only 39% of male LRS in the non-territorial *Enallagma hageni* (Fincke 1988). Similarly, although variance in lifespan accounted for only about a quarter of the total variance in male LRS in territorial *Plathemis lydia* (Koenig and Albano 1987), it accounted for 78% of the variance in LMS in *Coenagrion puella* (Banks and Thompson 1985), a species in which males compete by mate-searching and remain with females during oviposition. Although these comparisons are promising, more lifetime studies are needed on understudied odonates including gomphids, aeshnids, and species breeding in streams and in the tropics.

In summary, the numerous long-term studies of odonates offer at least weak support for the hypothesis that sexual selection on males increases as the ability of males to monopolize fertilizations increases. These studies are considerably more consistent on two other scores. First, they unambiguously demonstrate that survivorship is an important predictor of reproductive success for males and females regardless of the male mating pattern (Fincke 1982, 1986a, 1988; Banks and Thompson 1985; Koenig and Albano 1987; Tsubaki and Ono 1987; McVey 1988; Hamilton and Montgomerie 1989; Michiels and Dhondt 1991a; Cordero 1992a; Harvey and Walsh 1993). Second, they indicate that weather is a major constraint on sexual selection in odonates, with much of the variance in male and female reproductive success being explained by environmental conditions such as the number of sunny days occurring over an adult's lifetime (Thompson 1990; Tsubaki and Ono 1987). Indeed, Michiels and Dhondt (1991a) accurately estimated *Sympetrum danae* lifetime reproductive success by using only the lifetime number of sunny days along with the mean number of matings or ovipositions observed per sunny day.

Role of size and mass in male reproductive success

Even in the absence of any heritable phenotypic differences among males, the male-biassed operational sex ratios that are typical of most odonate populations and the limited number of favored oviposition sites in territorial species would alone result in considerable variation in male mating success. To the extent that the opportunity for sexual selection is realized, we expect reproductively successful males to differ from unsuccessful ones. Although heritability of traits has only been studied in a few odonates (Fincke 1988; Cordero 1992b), we assume that phenotypic differences among males generally reflect genetic differences.

Body size and mass are the most commonly studied phenotypic trait in adult odonates. Although correlated, there is an important distinction between these characters. 'Size' refers to the length or area of a body part, which in insects does not change after emergence as an adult. In contrast, mass may vary over the lifespan of an adult.

There are several ways in which correlating either of these traits with reproductive success can be misleading. First, body size of emerging adults typically decreases over a reproductive season (Banks and Thompson 1985; Tsubaki and Ono 1987; Fincke 1988; Michiels and Dhondt 1989), indicating the need to control for date of emergence when determining the effect of male size on reproductive success. Similarly, because the mass of an individual can change, it is important to weigh individuals at comparable stages of their lives; for example, all females should be measured either when gravid or after oviposition. Second, if dispersal is size- or age-dependent (Anholt 1990; Michiels and Dhondt 1991b), a false relationship between size and survivorship might be found. Third, we know little about how selection for size acts on females and even less about how it acts on larvae, whose growth determines adult size. The result of these difficulties is that it is often hard to interpret selection on size or mass in odonates unless large samples sizes are available and extraordinary care is taken in acquiring the data. Even then, the difficulties in dividing episodes of selection means that interpretations are often ambiguous.

Given these problems, it is a challenge to test even the most basic predictions of sexual selection theory using comparative data. An attempt to do so using data from eleven studies is summarized in Table 3-2.

Prediction 1: size should be a better predictor of male mating efficiency in species with male-biased compared with female-biased size dimorphism
Three of four (75%) populations in which size dimorphism was male-biased yielded evidence for sexual selection

Table 3-2. *Relationship between size dimorphism, resource defense polygamy, and selection on male mating efficiency with respect to size in odonates*

Species	Larger sex	Exhibits resource defense?	Selection on size relative to mating efficiency	Reference
Coenagrion puella	Females	No	Smaller more successful	Banks and Thompson 1985
Enallagma hageni	Females	No	Stabilizing	Fincke 1988
E. boreale	Females	No	Smaller more successful	Anholt 1991
Pyrrhosoma nymphula	Females	No	Not detected	Gribbin and Thompson 1991
Megaloprepus coerulatus	Males	Yes	Mated males larger	Fincke 1992a
Libellula luctuosa	Males	Yes	Mated males larger	Moore 1990
Orthetrum chrysostigma	Males	Yes	Mated males larger	Miller 1983
Pachydiplax longipennis	Males	Yes	Not detected	Dunham 1993[a]
Plathemis lydia	Females	Yes	Not detected	Koenig and Albano 1987
Sympetrum rubicundulum	Females	Yes	Not detected	Van Buskirk 1987
S. danae	Females	No	Mated males larger	Michiels and Dhondt 1991[a]

[a] Enclosure study.

acting via male size compared to one of seven (14%) populations in which size dimorphism was female-biased (Fisher exact test, $p = 0.085$). Excluding the two enclosure studies listed in Table 3-2, the count becomes three of three (100%) male-biassed versus zero of six (0%) female-biased populations yielding evidence for sexual selection acting via male size ($p = 0.011$). This difference remains significant even if the two studies of *Enallagma*, both of which yielded similar results, are combined ($p = 0.018$).

Prediction 2: selection favoring large males should be more likely to occur in species exhibiting resource defense polygamy

Evidence for this prediction includes the extensive comparative analysis by Anholt *et al.* (1991) demontrating that there is less female-biased size dimorphism in species exhibiting resource defense. Using the studies summarized in Table 3-2, three of six populations with resource defense (50%) show evidence for sexual selection acting via male size compared with one of five (20%) that do not. Considering only field studies, the count is three of five versus zero of four. This difference is not significant (Fisher exact test, $p = 0.12$), although the trend is in the expected direction.

These results are at best preliminary: besides the problems in measuring selection already discussed, patterns of resource defense within a population can differ according to ecological conditions. However, the results at least

suggest that selection studies in odonates generally support basic predictions of sexual selection theory, despite the numerous difficulties inherent in such comparisons.

One potential reason why the relationship between large size and male success in contest competition is not clearer is that large size may counter the advantages territorial males gain from competitive flight maneuvers and endurance during territorial fights. The result would in some cases be stabilizing selection on size, as found by Moore (1990) among mated males of the territorial *Libellua luctuosa*. This possibility is also suggested by studies demonstrating the importance of factors related to energy reserves and flight capability as opposed to size or mass *per se*. Marden (1989), for example, found that male *Plathemis lydia* with high flight-muscle mass to body mass ratios experienced greater short-term mating success; Marden and Waage (1990) found that fat reserves of male *Calopteryx maculata* correlated with their ability to win territorial contests and thus presumably influenced their reproductive success.

These studies are significant given the importance of energetically expensive flight for patrolling, fighting, courtship, and other activities related for territorial defense (Fried and May 1983; Vogt and Heinrich 1983; Singer 1987; Rüppell 1989a; Marden and Waage 1990; May 1991). Energy reserves may even play a role in determining the alternative mating tactics chosen by males, as suggested by the observation that shifts from territorial to non-territorial behavior occur in old age or when energy stores are low (Waltz 1982; Forsyth and Montgomerie 1987; Waltz and Wolf 1988, 1993).

We suggest four areas of research that should provide a more balanced view of size and mass in relation to odonate mating patterns. (1) Does selection on size or mass differ for males and females in the same population? (2) Is selection on size or mass the same relative to both survivorship and mating efficiency? (3) what are the exact mechanisms that cause size to be related to survival or mating success? (4) What factors during larval development and the teneral stage affect adult size and mass? Selection relative to size or mass may relate to multiple stages of the life cycle rather than just to adult survival and reproduction.

COURTSHIP AND SEXUAL CONFLICT

Male–female interactions and reproductive conflicts

Much of the literature on odonate mating patterns has focussed on males, with females often ignored or considered as passive partners. As a result, the process of odonate reproduction is often considered to end with mating. However, the inability of males to force females to mate, combined with the fact that many females may be able to fertilize all of their eggs with stored sperm from a single mating (Grieve 1937; Fincke 1987), opens the door for females to exert considerable control over reproductive patterns and for sexual conflict over when, where, and with whom to mate. For example, female *Ischnura verticalis*, in which a single mating supplies enough sperm to fertilize her eggs for life, rarely remate despite frequent encounters with receptive males (Fincke 1987). Although the male's penis morphology suggests that sperm removal by males could occur (Waage 1984a), female monogamy constrains the potential for sperm competition in this system.

Given that female odonates cannot be forced to mate and mating is often costly in terms of time, why do most female odonates mate multiply? This question can be viewed from two complementary perspectives.

First, females may remate in order to gain direct benefits through increased quality or quantity of sperm, nuptial gifts, increased access to high-quality oviposition sites, or some form of male protection (Walker 1980; Waage 1984a; Fincke 1986b). There is little evidence that females remate in order to insure against inviable sperm, to obtain superior male genotypes, or to obtain resources in the form of nuptial gifts. However, despite sperm storage, females may not always obtain sufficient sperm to last throughout their lifetime in a single mating, at least when copulations are extremely short or are interrupted, as occurs frequently in anisopterans. Thus, females may sometimes remate in order to replenish diminishing sperm stores (Fincke 1987; Miller and Miller 1989). More important is almost certainly access to high-quality oviposition sites (Fincke 1992a; Siva-Jothy *et al.* 1996; Tsubaki *et al.* 1996). Male protection of females against drowning also occurs in at least one species, *Enallagma hageni*, where females oviposit under water and guarding males help females that get stuck resurfacing, thereby decreasing the mortality risk of oviposition (Fincke 1986b). If females benefit from remating by acquiring access to better oviposition sites or better sires, males are likely to gain by advertising the quality of their territories or their own physical or genetic quality so as to 'entice' females to mate with them.

Alternatively, a female may gain nothing from remating other than minimizing the cost of harassment by a persistent male. To the extent that males ignore female rejection signals, males can be considered to be trying to 'coerce' unwilling females into mating. Mating in order to minimize harassment may be widespread in odonates (Waage 1979b, 1984a; Koenig 1991; but see Fincke 1997), and is considered in detail in the following section.

Avoidance of male harrassment

Male harassment of females while either ovipositing or searching for an oviposition site may considerably reduce oviposition efficiency. For example, *Erythemis simplicicollis* copulate for 19 s and undisturbed ovipositions average 39 s. However, in areas of high male density, oviposition duration averaged 112 s for females copulating once and ranged up to 780 s for a female who mated seven times and finally left the water, apparently without ever completing oviposition (Waage 1986b). Increased time at oviposition sites increases the risk of predation to females (Convey 1992; Rehfeldt 1992). In extreme cases, male harassment can even injure or kill females in tandem with males (Rüppell and Hadrys 1988).

It is therefore not surprising that females avoid harassment from mate-searching males in a variety of ways. Many oviposit at times and in places where males are rare or absent (Corbet 1962; Koenig 1991). Some female Zygoptera submerge to oviposit, a behavior that serves several functions (Corbet 1962) including avoiding males, who cannot interfere as long as a female is under water (Fincke 1986b). Female *Platycypha caligata* synchronize their above-water oviposition, permitting them to use a male's territory without first mating with the owner (Martens and Rehfeldt 1989). Similarly, females of several species parasitize the guarding behavior of territorial males attending other females with which the males are mated (Waage 1979b; Koenig 1991).

Yet another mechanism by which females may avoid male harassment is to mimic males in coloration or behavior or both. In many coenagrionids, the thorax and ventral abdominal color of some females, called heteromorphs, is distinct from that of males whereas other females, called andromorphs, are male-like in body coloration (Johnson 1975). Andromorphs may avoid unwanted attention from males if they are less likely to be recognized as potential mates (Robertson 1985). The costs to andromorph females may be higher predation rates (Robertson 1985) or the risk of not mating at all when male density is low (Hinnekint 1987). Interestingly, andromorphs appear to be more frequent in some high-density populations than in low-density ones (Forbes *et al.* 1995).

Although plausible, few data support the hypothesis that andromorphs benefit by lessened harassment from males or that fitness correlates differ between morphs (Cordero 1992a; but see Fincke 1994b). Lifetime reproductive success studies of three coenagrionids (Thompson 1989; Fincke 1994a) indicated no statistically significant differences between female morphs in mating frequency, survivorship, or the number of eggs laid. When andromorphs are in the majority, they are actually more likely to be recognized as mates than are heteromorphs (Forbes 1991a). Recent evidence indicates that mate recognition by males is modified by their past experience (Moodie 1995).

Female rejection signals and male reactions

A potentially more efficient means of avoiding male harassment is for females to signal their willingness to mate and to repel males with refusal displays. If such cues or displays are effective, they can minimize time and energy losses to both sexes and, at least in some *Ishnura* spp., the possibility of being eaten by an unreceptive female (Robertson 1985; Cordero 1992c; Fincke 1994a). Willingness to mate in some species may involve actively seeking males (Kaiser 1985; Moodie 1995) or giving a characteristic wing display (Waage 1984b; Fincke 1987). In *Aeschna cyanea* and *Ischnura verticalis*, female developmental color changes correlate

with receptivity and males are differentially found in areas where there are many females that have recently attained sexual maturity (Kaiser 1985; Fincke 1987). However, males have not as yet been shown to respond differentially to this color cue.

Females of many species indicate their unwillingness to mate by directly resisting (Bick and Bick 1963) and dislodging males following capture (Rüppell 1989b), holding onto perches (Fincke 1986b; Oppenheimer and Waage 1987), 'shaking' (Forbes *et al.* 1995), or curling their abdomen and spreading their wings (Fincke 1987; Utzeri 1988; Gorb 1992). Females typically give such displays following oviposition, and males that take such females in tandem release them relatively quickly (Kaiser 1985; Fincke 1986b).

To what extent are females that have sufficient sperm able to repel males? In general, it benefits a male to ignore a female's rejection signal whenever she has eggs to lay as long as the cost of influencing her mating decision in his favor is less than the benefits of success. The likelihood that this will be the case is in turn influenced by female oviposition behavior, particularly submerged oviposition, which decreases male control over sites and thus the degree to which he can gain matings by harassing ovipositing non-mates.

Consider *Calopteryx maculata* and *C. dimidiata*, two broadly sympatric congeners that share male-defended oviposition sites (Waage 1980, 1984b, 1988). *Calopteryx maculata* females oviposit at the water surface and depend on male guarding for prolonged oviposition and often on mating with the resident male for access to oviposition sites. In contrast, *C. dimidiata* females oviposit only while submerged, increasing the possibility that a female can oviposit without mating with the territory owner. Correlated with this difference, *C. dimidiata* males stop courting when females give refusal displays whereas *C. maculata* males ignore refusal displays and persist in courting, successfully copulating with the female 30% of the time (Waage 1984b).

Conflict of interest between the sexes does not end once a female is taken in tandem. Females can refuse to mate whereas males can persist in holding females. This can result in tandem pairs remaining intact for extended periods of time without successfully mating. For example, male *Megaloprepus coerulatus* have been observed to hold unwilling females in tandem, ultimately failing to mate, for 1.5 h (Fincke 1984b) (Fig. 3-2). Experimentally, this has been demonstrated in *Calopteryx maculata* by hand-pairing males with females made 'unwilling' by experimentally

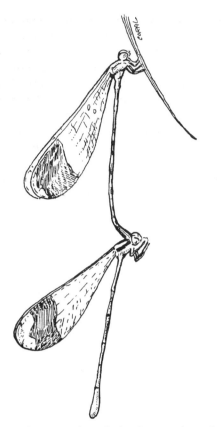

Fig. 3-1. *Megaloprepus coerulatus*: the female may resist mating once in tandem by remaining rigid in response to a male 'jerking' on her pronotum.

stiffening their abdomens. Pairs remained in tandem for up to 5 min, over three times the length of normal copulation (Oppenheimer 1991).

From the male's point of view, the costs of such a war of attrition include the risk of wasting sperm prematurely transferred to their secondary genitalia, losing additional matings, or even losing their territory (Waage 1979b). The potential benefit is that females unable to dislodge males and seeking to maximize oviposition efficiency may ultimately save time and energy by copulating. However, the extent to which males successfully obtain fertilizations by outlasting females otherwise unwilling to mate is unknown.

Courtship displays: a role in female choice?

Courtship displays are found in a variety of odonates, including damselflies in the Calopterygidae, Chlorocyphidae,

Coenagrionidae, Eupaeidae, Hemiphlebidae and Pseudo-stigmatidae, as well as several libellulid dragonflies (Jacobs 1955; Corbet 1962, 1980; Miller 1991a). Courtship displays have traditionally been thought both to aid in species recognition and reproductive isolation (natural selection function) and to provide females with information about the quality of potential mates (sexual selection function).

Although species and sex recognition in odonates is generally believed to be based on visual recognition of color, shape, and behavior, the importance of such cues probably depends at least in part on mechanical isolating mechanisms. For example, some species of damselflies that lack courtship and are similar in color and morphology avoid interspecific pairings by mechanical incompatibility, whereas species with courtship and associated differences in wing and body coloration often lack species-specific differences in morphology used in tandem formation and copulation (Waage 1975; Tennessen 1982; Robertson and Paterson 1982). Cases of the latter include sympatric *Calopteryx* species that have similar anal appendages and genitalia and that form interspecific tandems when hand-paired (Oppenheimer 1991). Species isolation would seem to be a likely function of courtship in these species (Waage 1975). However, courtship in species such as *Calopteryx* that are highly territorial and mate frequently could, in addition, provide females with the time and information needed for them to assess the quality and condition of potential mates.

Unfortunately, although female mate choice based on courtship cues seems possible, demonstrating its existence in odonates has proved surprisingly difficult. Female reproductive decisions in at least two detailed studies of *Calopteryx maculata* were consistent with choice based on territory quality or minimizing time spent at the water and failed to to unequivocally demonstrate female mate choice based on male courtship or appearance (Oppenheimer 1991; Fitzstephens 1994). However, recent experimental work by Hooper (1994) and M. Siva-Jothy (unpublished data) on *Calopteryx splendens* indicates that males that obtain copulations have significantly darker and more homogeneous wing spots and perform more displays than males that fail to mate, after holding territory quality and male wing length constant.

These findings offer the first positive evidence in odonates for active female mate choice based at least in part on courtship displays and suggests that, although subtle, similar examples may be more common than currently thought. Good candidates for additional studies are territorial species with courtship where male display may correlate with male quality, females have the time and opportunity to assess male quality, and the physiological challenges of territoriality place a premium on females choosing genetically high-quality mates. A promising way to proceed would be to carefully monitor the reaction of receptive virgin and non-virgin females to artificially modified secondary sexual characteristics of males displayed during courtship.

Regardless of the role that courtship plays in female mate choice, it clearly provides males with information helping them to determine their mating and reproductive options. For example, rejection displays from courted *Calopteryx maculata* females indicates that copulation is possible but that the female is unlikely to subsequently oviposit, diminishing the male's opportunity for fertilizations (Oppenheimer 1991). A similar interaction occurs in some coenagrionids, where males elicit information about a female's willingness to mate by jerking on her pronotum. Males transfer sperm in preparation for copulation only if the female indicates receptivity by raising her abdomen to tap on his genitalia (Robertson and Tennessen 1984; Fincke 1984b). Males adjust their mating behavior to potential fertilization returns in other ways as well. For example, 'satellite' males persuing alternative reproductive tactics, at risk that females will remate with a territorial male before oviposition, mate for longer than territory-holders (Siva-Jothy 1987b; Siva-Jothy and Tsubaki 1989a,b; Fincke 1992a).

In summary, refusal and courtship displays in odonates often involve a complex set of interactions and conflicts. Viewing pair formation and courtship as a two-way process in which both males and females acquire information subsequently influencing their behavior may aid us to better understand their roles in odonate reproduction.

THE ROLE OF NATURAL SELECTION IN ODONATE MATING PATTERNS

A general conceptual view of odonate reproductive decisions

Fig. 3-2 presents a conceptual view of the ways in which the factors influencing male and female reproductive decisions may interact (see also Michiels and Dhondt 1991a). Most work on odonate mating patterns relates primarily to the hexagonal boxes representing sexual competition. Relatively unstudied are the factors related to growth and survivorship.

Natural selection underlies the type of mating pattern and potential for sexual selection in at least two major ways. First, factors affecting survivorship, oviposition efficiency, and the quality of oviposition sites may constrain female choice of males. Second, the operational sex ratio, temporal synchrony of females, and spatial distribution of ovipositing females – factors ultimately determining the degree of male–male competition – are affected both by weather and by natural selection on larvae, tenerals, and non-reproductive adults. In this section we address female decision-making, arguing that determining the time and location of reproduction may generally be more important to female fitness than choice of mate based on a male's phenotype.

Female mating decisions: what do they choose?

Females control mating by a series of decisions about where, when, and how often to mate and oviposit (Koenig 1991). Choosing which male to mate with, or active female choice, is the only one of these reproductive decisions directly constituting sexual selection.

Ways in which female odonates could exercise active mate choice include (1) refusing to copulate with a particular male, (2) varying the number of fertilized eggs for a given male, (3) choosing to remate before ovipositing, and (4) differentially utilizing sperm of multiple mates. Despite all these possibilities, evidence among odonates for female choice among male phenotypes in a way that might result in sexual selection is lacking (Conrad and Pritchard 1992). Instead, female mating decisions are generally based on natural selection considerations such as maximizing survivorship, oviposition efficiency, and the quality of resources controlled by males (Ubukata 1984; Waage 1987; Wolf and Waltz 1988; Michiels and Dhondt 1990; Martens 1991; Wildermuth and Spinner 1991; Fincke 1992a; Wildermuth 1992, 1993).

In *Plathemis lydia*, females show strong preferences for time and location of oviposition and regularly refuse matings, but do not appear to choose among males on any consistent basis. In fact, 20% of females visiting ponds do not mate at all before oviposition (Koenig 1991), suggesting that many females with sperm stored from prior matings prefer not to mate if they can avoid it. This conclusion is supported by *Cordulia aenea*, in which females prefer to oviposit in areas where interference from males is least (Ubukata 1984), and species like *Calopteryx maculata* (Waage 1979b) and *Platycyhpa caligata* (Rehfeldt 1989) in which females are able to gain access to oviposition sites and

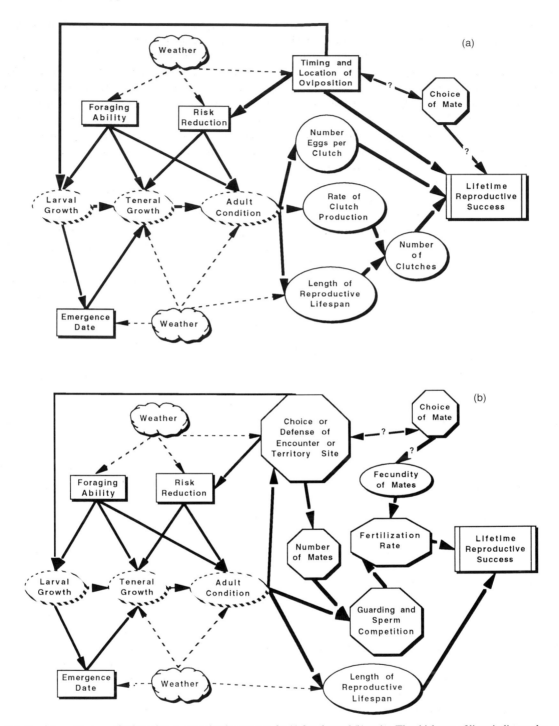

Fig. 3-2. Overview of factors affecting odonate reproductive success for (a) females and (b) males. The thickness of lines indicates the presumed importance of a particular causal factor. Octagons represent behaviors that are the primary focus of mating system classifications based on sexual selection. Other shapes represent additional life history, behavioral, and environmental factors.

male guarding without mating with the resident male. Such parasitism of mate-guarding behavior is facilitated by the tendency for females of some species to be attracted to ovipositing conspecifics (Waage 1979b, 1987; Martens 1989; Moss 1992), a behavior that may allow females to efficiently find and exploit high-quality oviposition sites without mating with the resident male.

In *Enallagma ebrium*, males heavily parasitized by larval water mites are less likely to mate, but this is apparently due to their inability to find mates rather than rejection by females (Forbes 1991b). Heavily parasitized females in this species were also not rejected as mates by males even though they were less fecund than unparasitized females (Forbes And Baker 1991). Given the current interest of the role that parasitism plays in sexual selection, there is clearly plenty of room for additional work along these lines in odonates.

Although female odonates rarely choose directly among males, intense male–male competition for control of resources attractive to females in many species means that females may exercise indirect choice of mates by selecting high-quality, limiting oviposition sites. Female *Megaloprepus coerulatus*, for example, mate only at tree-hole oviposition sites defended by large males, but they apparently do not discriminate among males on the basis of size, as indicated by their willingness to mate with small satellite males that take them in tandem at territories defended by larger males (Fincke 1992a).

As in other taxa, differentiating between female choice of oviposition site and female choice of males is difficult. Moore (1990), for example, tried to differentiate the role of these mechanisms in determining sexual dimorphism in *Libellula luctuosa*. Females visited and rejected several males before mating and varied the duration of oviposition following matings. Such behavior is necessary, but not sufficient, for sexual selection by female choice to be occurring. As in *Plathemis lydia*, females could be both rejecting and accepting males arbitrarily with respect to their phenotype (Koenig 1991). An alternative explanation for Moore's (1990) results is that females are choosing among oviposition sites rather than territorial males *per se*. Over a two-year period, for example, three of the 11 territories accounted for 62% of all observed matings (Moore 1989). At high male density, females visited fewer territories and oviposited for shorter durations, as expected if they were minimizing the risks of male harassment during oviposition. Nevertheless, among males defending a given territory, those with larger wing bands had greater mating success, suggesting that females might additionally be discriminating among male phenotypes.

Direct female choice of mates should evolve when the benefits of choosing outweigh the costs involved in discriminating among possible mates or mating opportunities. Its apparent rarity in odonates suggests that in general, whom a female mates with may be less important than minimizing the risks associated with oviposition and gaining access to resources controlled by males.

Why choose oviposition sites?

Territorial males, and females of all species, should select oviposition sites favoring larval survivorship and growth (Wolf and Waltz 1988; Michiels and Dhondt 1990). A major assumption of odonate biology is that females use these criteria for choosing oviposition sites. Unfortunately, this is difficult to test with stream or pond species, because larvae disperse away from the oviposition site and development time often exceeds a year. However, several studies that have addressed this assumption suggest that it is valid. *Calopteryx splendens xanthostoma* females prefer sites in areas of high stream flow, where eggs develop faster and have lower mortality than in slower-flowing water (Gibbons and Pain 1992; Siva-Jothy *et al.* 1995). Female *Megaloprepus coerulatus* place more eggs in large tree holes, which support a greater number of offspring to emergence, than in small holes, which typically produce only a single adult. Males discover oviposition sites independently of the presence of females, but defend tree holes at sites that predictably attract females only if the holes are also high-quality larval habitats (Fincke 1992a).

Much of the data on oviposition-site choice is also consistent with the hypothesis that, within a given habitat, sites are selected for their proximate value as safe or efficient oviposition sites in addition to their ultimate value as a good place for egg survival and larval growth. Separating these fitness consequences can be difficult. For example, when given a choice among different-sized sites within or between territories, female *Calopteryx maculata* tend to mate and oviposit at larger vegetation clumps independent of the identity of the resident male (Waage 1987; Alcock 1987; Meek and Herman 1991). Females may prefer large sites because they are better places for the development of eggs or larvae. Alternatively, females may have an easier time locating large sites and remain longer at them once they find them, thus maximizing oviposition efficiency (J. K. Waage, personal observations). Testing these alternatives will require considerably more data on the fate of eggs deposited at different sites and on oviposition safety and efficiency as factors determining the attraction and retention of females at particular sites.

Integrating larval ecology with adult behavior

The interplay between selective pressures during the larval and adult life-history stages represents the largest gap in our understanding of odonate mating patterns, making it impossible to test the basic assumption that successful adult reproductive behavior translates into more or higher-quality offspring. This is unfortunate, because selection at the larval stage is likely to have significant consequences for patterns of adult behavior (Johnson 1991; Fincke 1994c).

For example, interspecific larval competition in tropical guilds of tree-hole-breeding damselflies affects synchrony of emergence and favors differing female and male reproductive strategies among the species (Fincke 1992a,b). In *Megaloprepus coerulatus*, selection against the longer developmental time required to produce a large male appears to balance sexual selection favoring large adult males (O. M. Fincke, unpublished manuscript). Similarly, selection for large adults may have resulted in the abnormally high propensity for cannibalism among larvae in this tree-hole-breeding species (Fincke 1994c, 1996). High rates of larval cannibalism may be one factor selecting for an adult male that continues to attract mates to tree holes already saturated with his offspring. Because of cannibalism the number of matings obtained by territorial males – the usual indicator of fitness – is not a good predictor of surviving offspring (Fincke 1992a).

This limitation can now be overcome with molecular fingerprinting techniques that make it possible to measure adult fitness in terms of offspring surviving to emergence and to look for parent–offspring correlations in successful traits or behaviors (see, for example, Hadrys *et al.* 1993). Good candidates for such work would be odonates whose adult and juvenile stages can be studied in large flight cages over small ponds and damselflies whose larval habitats are small and discrete, such as those breeding in tree holes.

SUGGESTIONS FOR FUTURE RESEARCH

Odonates have long fascinated biologists by their obvious and diverse mating behaviors as well as by their mastery of the aquatic and aerial worlds. Numerous questions and challenges remain for future odonatologists.

First, most studies to date have been done on territorial species and species in which individuals tend to remain fairly localized. Less amenable to study are the more wide-ranging and strong-flying anisopterans such as the Aeshnidae, Cordulegasteridae, Gomphidae, Libellulidae and Petaluridae, which are difficult to capture without harming and tend to leave the area once marked. These families, and tropical ones, represent important gaps in our perspectives on odonate mating patterns and are in need of more study.

Second, long-term studies of marked females that focus on the fitness consequences of mating multiply and of resisting mating attempts by males are critical to understanding how selection on females underlies the evolution of male mating strategies. Third, direct tests of female mate choice by manipulation of male phenotypes are needed to unambiguously assess the relative importance of choice of site versus choice of sire. Fourth, studies measuring fitness in terms of the quality and quantity of surviving larvae would link the complex life history of odonates with their reproductive behavior. Finally, the behavioral diversity of this group combined with modern molecular methods make odonates ripe for phylogenetic analyses of mating and female oviposition patterns.

ACKNOWLEDGEMENTS

We thank M. Siva-Jothy for access to unpublished data, G. Johannes, B. Crespi, and J. Choe for comments on the manuscript, and especially Bernie and Jae for their patience. Our special thanks go to G. and J. Bick, P. S. Corbet, B. Kiauta, P. Miller, and F. A. Pitelka, whose earlier work and kind support has greatly helped our own research. We gratefully acknowledge previous support from the NSF as well as from the Max-Planck-Institut für Verhaltensphysiologie at Seewiesen (JKW), the American Philosophical Society, the Smithsonian Tropical Research Institute, and the University of Oklahoma (OMF).

LITERATURE CITED

Alcock, J. 1987. The effects of experimental manipulation of resources on the behavior of two calopterygid damselflies that exhibit resource-defense polygyny. *Can. J. Zool.* **65**: 2475–2482.

Anholt, B. R. 1990. Size-biased dispersal prior to breeding in a damselfly. *Oecologia (Berl.)* **83**: 385–387.

–. 1991. Measuring selection on a population of damselflies with a manipulated phenotype. *Evolution* **45**: 1091–1106.

Anholt, B. R., J. H. Marden, and D. M. Jenkins. 1991. Patterns of mass gain and sexual dimorphism in adult dragonflies (Insecta: Odonata). *Can. J. Zool.* **69**: 1156–1163.

Arnold, S. J. and M. J. Wade. 1984a. On the measurement of natural and sexual selection: theory. *Evolution* **38**: 709–719.

–. 1984b. On the measurement of natural and sexual selection: applications. *Evolution* **38**: 720–734.

Banks, M. J. and D. J. Thompson. 1985. Lifetime mating success in the damselfly *Coenagrion puella*. *Anim. Behav.* **33**: 1175–1183.

–. 1987. Lifetime reproductive success of females of the damselfly *Coenagrion puella*. *J. Anim. Ecol.* **56**: 815–832.

Bick, G. H. and J. C. Bick. 1963. Behavior and population structure of the damselfly *Enallagma civile* (Hagen) (Odonata: Coenagrionidae). *Southwest. Nat.* **8**: 57–84.

Buchholtz, C. 1951. Untersuchungen an der Libellengattung *Calopteryx* Leach unter besonderer Berücksichtigung ethologischer Fragen. *Z. Tierpsychol.* **8**: 273–386.

Buskirk, R. E., and K. J. Sherman. 1985. The influence of larval ecology on oviposition and mating strategies in dragonflies. *Fla. Entomol.* **68**: 39–51.

Conrad, K. F. and G. Pritchard. 1992. An ecological classification of odonate mating systems: the relative influence of natural, inter- and intra-sexual selection on males. *Biol. J. Linn. Soc.* **45**(3): 255–269.

Convey, P. 1992. Predation risks associated with mating and oviposition for female *Crocothemis erythraea* (Brulle) (Anisoptera: Libellulidae). *Odonatologica* **21**: 343–350.

Corbet, P. S. 1962. *A Biology of Dragonflies*. London: Witherby Ltd.

–. 1980. Biology of odonata. *Annu. Rev. Entomol.* **25**: 189–217.

Cordero, A. 1989. Reproductive behaviour of *Ischnura graellsii* (Rambur)(Zygoptera:Coenagrionidae). *Odonatologica* **18**: 237–244.

–. 1992a. Density-dependent mating success and colour polymorphism in females of the damselfly *Ischnura graellsii* (Odonata: Coenagrionidae). *J. Anim. Ecol.* **61**: 769–780.

–. 1992b. Morphological variability, female polymorphism and heritability of body length in *Ischnura graellsii* (Rambur) (Zygoptera: Coenagrionidae). *Odonatologica* **21**: 409–419.

–. 1992c. Sexual cannibalism in the damselfly species *Ischnura graellsii* (Odonata: Coenagrionidae). *Entomol. Gen.* **17**: 17–20.

De Marchi, G. 1990. Precopulatory reproductive isolation and wing colour dimorphism in Calopteryx *splendens* females in southern Italy (Zygoptera: Calopterygidae). *Odonatologica* **19**: 243–250.

Dunham, M. L. 1993. Fighting and territorial behavior in the dragonfly *Pachydiplax longipennis*. Ph.D. dissertation, Brown University.

Emlen, S. T., and L. W. Oring. 1977. Ecology, sexual selection, and the evolution of mating systems. *Science (Wash., D.C.)* **197**: 215–223.

Fincke, O. M. 1982. Lifetime mating success in a natural population of the damselfly, *Enallagma hageni* (Walsh) (Odonata: Coenagrionidae). *Behav. Ecol., Sociobiol.* **10**: 293–302.

–. 1984a. Sperm competition in the damselfly *Enallagma hageni* (Odonata: Coenagrionidae), and the benefits of multiple matings for males and females. *Behav. Ecol. Sociobiol.* **10**: 293–302.

–. 1984b. Giant damselflies in a tropical forest: reproductive biology of Megaloprepus coerulatus with notes on *Mecistogaster* (Zygoptera: Pseudostigmatidae). *Adv. Odonatol.* **2**: 13–27.

–. 1985. Alternative mate-finding tactics in a non-territorial damselfly (Odonata: Coenagrionidae). *Anim. Behav.* **33**: 1124–1137.

–. 1986a. Lifetime reproductive success and the opportunity for selection in a nonterritorial damselfly (Odonata: Coenagrionidae). *Evolution* **40**: 791–803.

–. 1986b. Underwater oviposition in a damselfly (Odonata: Coenagrionidae) favors male vigilance, and multiple mating by females. *Behav. Ecol. Sociobiol.* **18**: 405–412.

–. 1987. Female monogamy in the damselfly *Ischnura verticalis* Say (Zygoptera: Coenagrionidae). *Odonatologica* **16**: 791–803.

–. 1988. Sources of variation in lifetime reproductive success in a nonterritorial damselfly (Odonata: Coenagrionidae). **In**: *Reproductive Success*. T. H. Clutton-Brock, ed., pp. 24–43. Chicago: University of Chicago Press.

–. 1992a. Consequences of larval ecology for territoriality and reproductive success of a neotropical damselfly. *Ecology* **73**: 449–462.

–. 1992b. Interspecific competition for tree holes: Consequences for mating systems and coexistence in Neotropical damselflies. *Am. Nat.* **139**: 80–101.

–. 1994a. Female colour polymorphism in damselflies: failure to reject the null hypothesis. *Anim. Behav.* **47**: 1249–1266.

–. 1994b. On the difficulty of detecting density-dependent selection on polymorphic females of the damselfly *Ischnura graellsii*: failure to reject the null. *Evol. Ecol.* **8**: 328–329.

–. 1994c. Population regulation of a tropical damselfly in the larval stage by food limitation, cannibalism, intraguild predation and habitat drying. *Oecologia (Berl.)* **100**: 118–127.

–. 1996. Larval behaviour of a giant damselfly: territoriality or size dependent dominance? *Anim. Behav.*, **51**: 77–87.

–. 1997. Conflict resolution in the Odonata: implications for understanding female mating patterns and female choice. *Biol. J. Linn. Soc.*, in press.

Fitzstephens, D. M. 1994. Color as a reliable signal of fighting ability in male damselflies, *Calopteryx maculata*. Ph.D. dissertation, Michigan State University.

Forbes, M. R. L. 1991a. Female morphs of the damselfly *Enallagma boreale* Selys (Odonata: Coenagrionidae): a benefit for androchromatypes. *Can. J. Zool.* **69**: 1969–1970.

–. 1991b. Ectoparasites and mating success of male *Enallagma ebrium* damselflies (Odonata: Coenagrionidae). *Oikos* **60**: 336–342.

Forbes, M. R. L. and R. L. Baker. 1991. Condition and fecundity of the damselfly, *Enallagma ebrium* (Hagen): the importance of ectoparasites. *Oecologia (Berl.)* **86**: 335–341.

Forbes, M. R. L., J. M. L. Richardson and R. L. Baker. 1995. Frequency of female morphs is related to an index of male density in the damselfly *Nehalennia irene* (Hagen). *Ecoscience* **2**: 28–33.

Forsyth, A. and R. D. Montgomerie. 1987. Alternative reproductive tactics in the territorial damselfly *Calopteryx maculata*: sneaking by older males. *Behav. Ecol. Sociobiol.* **21**: 73–81.

Fried, C. S. and M. L. May. 1983. Energy expenditure and food intake of territorial male *Pachydiplax longipennis* (Odonata: Libellulidae). *Ecol. Entomol.* **8**: 283–292.

Gibbons, D. W. and D. Pain. 1992. The influence of river flow rate on the breeding behaviour of *Calopteryx* damselflies. *J. Anim. Ecol.* **61**: 283–289.

Gorb, S. 1992. An experimental study of the refusal display in the damselfly *Platycnemis pennipes* (Pall.) (Zygoptera: Platycnemididae). *Odonatologica* **21**: 299–307.

Grafen, A. 1988. On the uses of data on lifetime reproductive success. In: *Reproductive Success*. T. H. Clutton-Brock, ed., pp. 472–486. Chicago: University of Chicago Press.

Gribbin, S. D. and D. J. Thompson. 1991. The effects of size and residency on territorial disputes and short-term mating success in the damselfly *Pyrrhosoma nymphula* (Sulzer) (Zygoptera: Coenagrionidae). *Anim. Behav.* **41**: 689–695.

Grieve, E. 1937. Studies on the biology of the damselfly *Ischnura verticalis* Say, with notes on certain parasites. *Entomol. Am.* **17**: 121–153.

Hadrys, H., B. Schierwater, S. L. Dellaporta, R. DeSalle and L. W. Buss. 1993. Determination of paternity in dragonflies by random amplified polymorphic DNA fingerprinting. *Molec. Ecol.* **2**: 79–88.

Hafernik, J. E. and R. W. Garrison. 1986. Mating success and survival rate in a population of damselflies: results at variance with theory? *Am. Nat.* **128**: 353–365.

Hamilton, L. D. and R. D. Montgomerie. 1989. Population demography and sex ratio in a neotropical damselfly (Odonata: Coenagrionidae) in Costa Rica. *J. Trop. Ecol.* **5**: 159–171.

Harvey, I. F. and K. J. Walsh. 1993. Fluctuating asymmetry and lifetime mating success are correlated in males of the damselfly *Coenagrion puella* (Odonata: Coenagrionidae). *Ecol. Entomol.* **18**: 198–202.

Hinnekint, B. O. N. 1987. Population dynamics of *Ischnura elegans* (Vander Linden) (Insecta: Odonata) with special reference to morphological colour changes, female polymorphism, multiannual cycles and their influence on behaviour. *Hydrobiologia* **146**: 3–31.

Hooper, R. 1994. Sexual selection in a damselfly: female perspectives. Ph.D. dissertation, University of Sheffield.

Jacobs, M. E. 1955. Studies on territorialism and sexual selection in dragonflies. *Ecology* **36**: 566–586.

Johnson, C. J. 1975. Polymorphism and natural selection in ischnuran damselflies. *Evol. Theory* **1**: 81–90.

Johnson, D. M. 1991. Behavioral Ecology of Larval Dragonflies and Damselflies. *Trends Ecol. Evol.* **6**: 8–13.

Kaiser, H. 1985. Availability of receptive females at the mating place and mating chances of males in the dragonfly *Aeschna cyanea*. *Behav. Ecol. Sociobiol.* **18**: 1–7.

Koenig, W. D. 1991. Levels of female choice in the white-tailed skimmer *Plathemis lydia* (Odonata: Libellulidae). *Behaviour* **119**: 193–224.

Koenig, W. D. and S. S. Albano. 1986. On the measurement of sexual selection. *Am. Nat.* **127**: 403–409.

–. 1987. Lifetime reproductive success, selection, and the opportunity for selection in the white-tailed skimmer *Plathemis lydia* (Odonata Anisoptera). *Evolution* **41**: 22–36.

Marden, J. H. 1989. Bodybuilding dragonflies: costs and benefits of maximizing flight muscle. *Physiol. Zool.* **62**: 505–521.

Marden, J. H. and J. K. Waage. 1990. Escalated damselfly territorial contests are energetic wars of attrition. *Anim. Behav.* **39**: 954–959.

Martens, A. 1989. Aggregation of tandems in *Coenagrion pulchellum* Van der Linden 1825 during oviposition (Odonata; Coenagrionidae). *Zool. Anz.* **223**: 124–128.

–. 1991. Plasticity of mate-guarding and oviposition behaviour in *Zygonyx natalensis* (Martin) (Anisoptera: Libellulidae). *Odonatologica* **20**: 293–302.

Martens, A., and G. Rehfeldt. 1989. Female aggregation in *Platycypha caligata* (Odonata: Chlorocyphidae): a tactic to evade male interference during oviposition. *Anim. Behav.* **38**: 369–374.

May, M. L. 1991. Dragonfly flight – power requirements at high speed and acceleration. *J. Exp. Biol.* **158**: 325–342.

McVey, M. E. 1988. The opportunity for sexual selection in a territorial dragonfly, *Erythemis simplicicollis*. In: *Reproductive Success*. T. H. Clutton-Brock, ed., pp. 44–58. Chicago: University of Chicago Press.

McVey, M. E. and B. J. Smittle. 1984. Sperm precedence in the dragonfly *Erythemis simplicicollis*. *J. Insect. Physiol.* **30**: 619–628.

Meek, S. B. and T. B. Herman. 1991. The influence of oviposition resources on the dispersion and behaviour of Calopterygid damselflies. *Can. J. Zool.* **69**: 835–839.

Michiels, N. K. 1992. Consequences and adaptive significance of variation in copulation duration in the dragonfly *Sympetrum danae*. *Behav. Ecol. Sociobiol.* **29**: 429–435.

Michiels, N. K. and A. A. Dhondt. 1989. Effects of emergence characteristics on longevity and maturation in the dragonfly *Sympetrum danae* (Anisoptera: Libellulidae). *Hydrobiologia* **17**: 149–158.

–. 1990. Costs and benefits associated with oviposition site selection in the dragonfly *Sympetrum danae* (Odonata: Libellulidae). *Anim. Behav.* **40**: 668–678.

–. 1991a. Sources of variation in male mating success and female oviposition rate in a nonterritorial dragonfly. *Behav. Ecol. Sociobiol.* **29**: 17–25.

–. 1991b. Characteristics of dispersal in sexually mature dragonflies. *Ecol. Entomol.* **16**: 449–460.

Miller, P. L. 1983. The duration of copulation correlates with other aspects of mating behaviour in *Orthetrum chrysostigma* (Burmeister) (Anisoptera: Libellulidae). *Odonatologica* **12**: 227–238.

–. 1990. Mechanisms of sperm removal and sperm transfer in *Orthetrum coerulescens* (Fabricius) (Odonata: Libellulidae). *Physiol. Entomol.* **15**: 199–209.

–. 1991a. The structure and function of the genitalia in the Libellulidae (Odonata). *Zool. J. Linn. Soc.* **102**: 43–74.

–. 1991b. Pre-tandem courtship in *Palpopleura sexmaculata* (Fabricius) (Anisoptera: Libellulidae). *Not. Odonatol.* **3**: 99–101.

Miller, P. L. and A. K. Miller. 1989. Post-copulatory 'resting' in *Orthetrum coerulescens* (Fabricius) and some other Libellulidae: time for 'sperm handling'? (Anisoptera). *Odonatologica* **18**: 33–41.

Moodie, M. N. 1995. Evolution of female color morphs in damselflies. M.S. thesis, University of Oklahoma, Norman.

Moore, A. J. 1989. The behavioral ecology of *Libellula luctuosa* (Burmeister) (Odonata: Libellulidae): III. Male Density, OSR, and male and female mating behavior. *Ethology* **80**: 120–136.

–. 1990. The evolution of sexual dimorphism by sexual selection: the separate effects of intrasexual selection and intersexual selection. *Evolution* **44**: 315–331.

Moss, S. P. 1992. Oviposition site selection in *Enallagma civile* (Hagen) and the consequences of aggregating behaviour (Zygoptera: Coenagrionidae). *Odonatologica* **21**: 153–164.

Nomakuchi, S. 1992. Male reproductive polymorphism and form-specific habitat utilization of the damselfly *Mnais pruinosa* (Zygoptera: Calopterygidae). *Ecol. Res.* **7**: 87–96.

Oppenheimer, S. D. 1991. Functions of courtship in *Calopteryx maculata* (Odonata: Calopterygidae): an experimental approach. Ph.D. dissertation, Brown University.

Oppenheimer, S. D. and J. K. Waage. 1987. Hand-pairing: a new technique for obtaining copulations within and between *Calopteryx* species (Zygoptera: Calopterygidae). *Odonatologica* **16**: 291–296.

Rehfeldt, G. E. 1989. Female arrival at the oviposition site of *Platycypha caligata* (Selys): temporal patterns and relation to male activity (Zygoptera: Chlorocyphidae). *Adv. Odonatol.* **4**: 89–94.

–. 1992. Aggregation during oviposition and predation risk in *Sympetrum vulgatum* L. (Odonata: Libellulidae). *Behav. Ecol. Sociobiol.* **30**: 317–322.

Robertson, H. M. 1985. Female dimorphism and mating behaviour in a damselfly, *Ischnura ramburi* : females mimicking males. *Anim. Behav.* **33**: 805–809.

Robertson, H. M. and H. E. H. Paterson. 1982. Mate recognition and mechanical isolation in *Enallagma* damselflies (Odonata: Coenagrionidae). *Evolution* **36**: 234–250.

Robertson, H. M. and K. J. Tennessen. 1984. Precopulatory genital contact in some Zygoptera. *Odonatologica* **13**: 591–595.

Rowe, R. J. 1978. *Ischnura aurora* (Brauer), a dragonfly with unusual mating behaviour (Zygoptera: Coenagrionidae). *Odonatologica* **7**: 375–383.

Rüppell, G. 1989a. Kinematic analysis of symmetrical flight manoeuvers of Odonata. *J. Exp. Biol.* **144**: 13–42.

–. 1989b. Fore legs of dragonflies used to repel males. *Odonatologica* **18**: 391–396.

Rüppell, G. and H. Hadrys. 1988. *Anax junius* (Aeshnidae) – Sexual male competition and oviposition behaviour. *Publ. Wiss. Film. Sekt. Biol.*, Ser. 20, No. E 2998, pp. 1–12.

Sherman, K. J. 1983. The adaptive significance of postcopulatory mate guarding in a dragonfly, *Pachydiplax longipennis*. *Anim. Behav.* **31**: 1107–1115.

Singer, F. D. 1987. The behavioral and physiological ecology of dragonflies. Ph.D. dissertation, University of Minnesota.

Siva-Jothy, M. T. 1984. Sperm competition in the family Libellulidae (Anisoptera) with special reference to *Crocothemis erythrae* (Brulle) and *Orthetrum cancellatum* (L.). *Adv. Odonatol.* **2**: 195–207.

–. 1987a. The structure and function of the female sperm-storage organs in libellulid dragonflies. *J. Insect. Physiol.* **33**: 559–567.

–. 1987b. Variation in copulation duration and the resultant degree of sperm removal in *Orthetrum cancellatum* (L.) (Libellulidae: Odonata). *Behav. Ecol. Sociobiol.* **20**: 147–151.

Siva-Jothy, M. T. and Y. Tsubaki. 1989a. Variation in copula duration in *Mnais pruinosa pruinosa* Selys (Odonata: Calopterygidae) 1. Alternative mate-securing tactics and sperm precedence. *Behav. Ecol. Sociobiol.* **24**: 39–45.

–. 1989b. Variation in copula duration in *Mnais pruinosa pruinosa* Selys (Odonata: Calopterygidae) 2. Causal factors. *Behav. Ecol. Sociobiol.* **25**: 261–268.

Siva-Jothy, M. T., D. W. Gibbons and D. Pain. 1995. Female oviposition-site preference and egg hatching success in the damselfly *Calopteryx splendens xanthostoma. Behav. Ecol. Sociobiol.* **37**: 39–44.

Sutherland, W. J. 1985. Chance can produce a sex difference in variance in mating success and explain Bateman's data. *Anim. Behav.* **33**: 1349–1352.

–. 1987. Random and deterministic components of variance in mating success. In: *Sexual Selection: Testing the Alternatives.* J. W. Bradbury and M. B. Andersson, eds., pp. 209–220. Chichester: John Wiley & Sons.

Tennessen, K. J. 1982. Review of reproductive isolating barriers in Odonata. *Adv. Odonatol.* **1**: 251–265.

Thompson, D. J. 1987. Lifetime reproductive success of females of the damselfly *Coenagrion puella. J. Anim. Ecol.* **56**: 815–832.

–. 1989. Lifetime reproductive success in andromorph females of the damselfly *Coenagrion puella* (L.) (Zygoptera: Coenagrionidae). *Odonatologica* **18**: 209–213.

–. 1990. The effects of survival and weather on lifetime egg production in a model damselfly. *Ecol. Entomol.* **15**: 455–462.

Tsubaki, Y. and T. Ono. 1987. Effects of age and body size on the male territorial system of the dragonfly, *Nannophya pygmaea* Rambur (Odonata: Libellulidae). *Anim. Behav.* **35**: 518–525.

Tsubaki, Y., M. T. Siva-Jothy and T. Ono. 1994. Recopulation and post-copulatory mate guarding increase immediate female reproductive output in the dragonfly *Nannophya pygmaea* Rambur. *Behav. Ecol. Sociobiol.*, **35**: 219–225.

Ubukata, H. 1984. Oviposition site selection and avoidance of additional mating by females of the dragonfly *Cordulia aenea amurensis* Selys (Odonata: Corduliidae). *Res. Popl. Ecol.* **26**: 285–301.

Utzeri, C. 1988. Female 'refusal display' versus male 'threat display' in Zygoptera: is it a case of intraspecific imitation? *Odonatologica* **17**: 45–54.

Van Buskirk, J. 1987. Influence of size and date of emergence on male survival and mating success in a dragonfly *Sympetrum rubicundulum*. *Am. Midl. Nat.* **118**: 169–176.

Vogt, F. D. and Heinrich, B. 1983. Thoracic temperature variations in the onset of flight in dragonflies (Odonata: Anisoptera). *Physiol. Zool.* **56**: 236–241.

Waage, J. K. 1975. Reproductive isolation and the potential for character displacement in the damselflies, *Calopteryx maculata* and *C. aequabilis* (Odonata: Calopterygidae). *Syst. Zool.* **24**: 24–36.

–. 1979a. Dual function of the damselfly penis: sperm removal and transfer. *Science (Wash., D.C.)* **203**: 916–918.

–. 1979b. Adaptive significance of postcopulatory guarding of mates and nonmates by male *Calopteryx maculata* (Odonata). *Behav. Ecol. Sociobiol.* **6**: 147–154.

–. 1979c. Reproductive character displacement in *Calopteryx* (Odonata: Calopterygidae). *Evolution* **33**: 104–116.

–. 1980. Adult sex ratios and female reproductive potential in *Calopteryx* (Zygoptera: Calopterygidae). *Odonatologica* **9**: 217–230.

–. 1984a. Sperm competition and the evolution of odonate mating systems. **In**: *Sperm Competition and the Evolution of Animal Mating Systems*. R. L. Smith, ed., pp. 251–290. New York: Academic Press.

–. 1984b. Female and male interactions during courtship in *Calopteryx maculata* and *C. dimidiata* (Odonata: Calopterygidae): influence of oviposition behaviour. *Anim. Behav.* **32**: 400–404.

–. 1986a. Evidence for widespread sperm displacement ability among Zygoptera (Odonata) and the means for predicting its existence. *Biol. J. Linn. Soc.* **28**: 285–300.

–. 1986b. Sperm displacement by two libellulid dragonflies with disparate copulation durations (Anisoptera). *Odonatologica* **15**: 429–444.

–. 1987. Choice and utilization of oviposition sites by female *Calopteryx maculata* (Odonata: Calopterygidae): I. Influence of site size and the presence of other females. *Behav. Ecol. Sociobiol.* **20**: 439–446.

–. 1988. Reproductive behaviour of the damselfly *Calopteryx dimidiata* Burm. (Odonata: Calopterygidae). *Odonatologica* **17**: 365–378.

Wade, M. J. 1987. Measuring sexual selection. **In**: *Sexual Selection: Testing the Alternatives*. J. W. Bradbury and M. B. Andersson, eds., pp. 197–208. Chichester: John Wiley & Sons.

Walker, W. F. 1980. Sperm utilization strategies in nonsocial insects. *Am. Nat.* **115**: 780–799.

Waltz, E. C. 1982. Alternative mating tactics and the law of diminishing returns: the satellite threshold model. *Behav. Ecol. Sociobiol.* **10**: 75–83.

Waltz, E. C. and L. L. Wolf. 1988. Alternative mating tactics in male white-faced dragonflies (*Leucorrhinia intacta*): plasticity of tactical options and consequences for reproductive success. *Evol. Ecol.* **2**: 205–231.

–. 1993. Alternative mating tactics in male white-faced dragonflies: Experimental evidence for a behavioural assessment ESS. *Anim. Behav.* **46**: 323–334.

Watanabe, M. 1991. Thermoregulation and habitat preference in two wing color forms of *Mnais* damselflies (Odonata: Calopterygidae). *Zool. Sci.* **8**: 983–989.

Watanabe, M., and M. Taguchi. 1990. Mating tactics and male wing dimorphism in the damselfly, *Mnais pruinosa costalis* Selys (Odonata: Calopterygidae). *J. Ethol.* **8**: 129–138.

Wildermuth, H. 1992. Visual and tactile stimuli in choice of oviposition substrates by the dragonfly *Perithemis mooma* Kirby (Anisoptera: Libellulidae). *Odonatologica* **21**: 309–321.

–. 1993. Habitat selection and oviposition site recognition by the dragonfly *Aeshna juncea* (L.): An experimental approach in natural habitats (Anisoptera: Aeshnidae). *Odonatologica* **22**: 27–44.

Wildermuth, H. and W. Spinner. 1991. Visual cues in oviposition site selection by *Somatochlora arctica* (Zetterstedt) (Anisoptera: Corduliidae). *Odonatologica* **20**: 357–368.

Wolf, L. L. and E. C. Waltz. 1988. Oviposition site selection and spatial predictability of female white-faced dragonflies (*Leucorrhinia intacta*) (Odonata: Libellulidae). *Ethology* **78**: 306–320.

Wolf, L. L., E. C. Waltz, K. Wakeley, and D. Klockowski. 1989. Copulation duration and sperm competition in white-faced dragonflies (*Leucorrhinia intacta*; Odonata: Libellulidae). *Behav. Ecol. Sociobiol.* **24**: 63–68.

4 · Sexual selection in resource defense polygyny: lessons from territorial grasshoppers

MICHAEL D. GREENFIELD

ABSTRACT

Sexual selection in resource defense polygyny systems is characterized by male–male competition for valuable resource patches, female settlement among patches based on both resource quality and male quality, and positive correlations between resource quality and male quality. These correlations confound the processes of sexual selection and pose special challenges for their study. *Ligurotettix*, an unusual genus of gomphocerine grasshoppers in which males defend individual host shrubs as mating territories, have proved to be useful species for investigating these issues.

Female *Ligurotettix* generally settle on certain host shrubs whose foliage represents high-quality food, which may promote egg development. Males are usually found on the same set of host shrubs that harbor females. The settlement of adult males, which occurs prior to the seasonal appearance of adult females, is based on the expected value of a shrub as both a female encounter site and a food resource whose consumption could increase male competitive abilities. Males compete for exclusive residence at valuable shrubs by means of early adult maturation, searching mechanisms for finding the shrubs, and aggression. Loud acoustic signaling by the males attracts females and influences their initial settlement among these shrubs.

Despite high levels of aggression, *Ligurotettix* males congregate on the most valuable host shrubs. Females may prefer congregated males *per se*, possibly as a means of reducing the costs of selecting males and locating resources. Some findings suggest that males display mutual attraction and exploit this possible female preference. In addition, male congregations probably form at sites where resource quality and female number are sufficiently high that it would not always be adaptive for a male to depart when confronted by aggression.

INTRODUCTION

Darwin (1871) pointed out that characters specifically associated with mating evolved by processes distinct from natural selection. He referred to these processes as intrasexual selection, commonly manifested as male–male competition, and intersexual selection, generally female choice. Our currently prevailing scheme (see Kirkpatrick and Ryan 1991; Andersson 1994) further divides female choice into 'direct selection' of males based on material (non-genetic) benefits and 'indirect selection' based on genetic factors.

The various processes of sexual selection co-occur and are confounded in many species (Harvey and Bradbury 1991). Consequently, those mating systems in which specific processes of sexual selection can be viewed in isolation are attractive to investigators for practical reasons. Systems that include several co-occurring processes are more typical, however, and they also offer the opportunity and challenge to study the relative importance of each process and their interdependence. Resource defense polygyny systems are particularly valuable in this regard, as males are expected to compete for encounter sites, and females, in addition to moving toward these sites, might select certain males based on material benefits, genetic factors, or both. Moreover, the material benefits (e.g. parental care, nutritional spermatophores) that a male offers can reflect his success in competition for valuable food resources or encounter sites, and outcomes of these contests may be determined by chance events, such as a male's access to food during development, as well as by genetic differences.

In this chapter I present findings on sexual selection in an unusual genus of territorial grasshoppers, *Ligurotettix* (Orthoptera: Acrididae: Gomphocerinae), found in the deserts of southwestern North America. I use the resource defense mating systems in *Ligurotettix* to examine the hypotheses (1) that males compete to settle on valuable patches and increase their mating success via such competition;

(2) that a male's competitiveness and attractiveness to females are reinforced by (receive positive feedback from) resources on a valuable patch; and (3) that female choice modifies the mating success that a male would accrue solely from his settlement on a given patch. Because the conditions and resources expected to maximize male and female reproductive success may not coincide, I discuss the possibility of sexual conflict (see Alexander *et al.*, this volume) in resource defense systems.

NATURAL HISTORY OF *LIGUROTETTIX* GRASSHOPPERS

Host plant associations

The genus *Ligurotettix* includes two recognized species, both of which exhibit oligophagy on desert shrubs and male defense of individual host shrubs as mating territories. *Ligurotettix coquilletti* McNeill is found in the Sonoran and Mohave Deserts and is generally associated with creosote bush (*Larrea tridentata*; Zygophyllaceae). *Ligurotettix planum* (Bruner) occurs in the Chihuahuan Desert and is nearly restricted to tarbush (*Flourensia cernua*; Asteraceae). Most of the information on *L. coquilletti* reported here stems from populations studied at Deep Canyon, California, from 1982 to 1994. Material on *L. planum* comes mostly from populations studied at Rodeo, New Mexico, from 1984 to 1988 and 1993 and at K-Bar, Big Bend National Park, Texas, from 1991 to 1993. At these locations *L. coquilletti* and *L. planum* fed only on *Larrea* and *Flourensia*, respectively. More complete descriptions of the geographic ranges and host plant associations of *Ligurotettix* are given in Rehn (1923), Otte (1970, 1981), Otte and Joern (1975, 1977), Chapman *et al.* (1988), and Greenfield and Shelly (1990).

Life history and demography

Ligurotettix are univoltine grasshoppers that overwinter as diapausing eggs in the desert soil (Greenfield and Shelly 1990). Nymphs hatch in early spring and migrate to their host plants. At Deep Canyon, California, male *L. coquilletti* require 60–90 days to progress through five nymphal instars (Wang 1990). Adult *L. coquilletti* are present from June to October, and longevities of individual adults may exceed two months (Greenfield and Shelly 1985). Both species are protandrous; in *L. coquilletti* the peak of male eclosion precedes that of females by 3–6 weeks (Wang *et al.* 1990). Male *L. coquilletti* also live longer than females, and

the vast majority of adults found at the end of the season are males (Greenfield and Shelly 1985; Wang *et al.* 1990).

Immature and adult *Ligurotettix* of both sexes are usually found on their host plants (Shelly and Greenfield 1989; Greenfield and Shelly 1990). Oviposition is the only activity that regularly occurs away from *Larrea* or *Flourensia* shrubs. *Ligurotettix coquilletti* females deposit up to six eggs per pod and can oviposit as often as every four days (Wang and Greenfield 1994), presumably the time necessary for maturing the ultimate oocytes in each ovariole.

Population densities of *L. coquilletti* in *Larrea* stands at Deep Canyon, California, and of *L. planum* in *Flourensia* stands at Rodeo, New Mexico (Shelly and Greenfield 1989), and K-Bar, Texas (Greenfield and Minckley 1993), averaged 200–400 adults ha^{-1} in midsummer. Adult sex ratios (male:female) were approximately 1.5:1 during periods of peak population density.

Courtship and mating behaviors

Pair formation in *Ligurotettix* begins with movement of a female towards a shrub harboring a stridulating male. Once on the same shrub, however, the female never approaches the male. Rather, the male quickly jumps towards the female, probably attracted by visual stimuli identified with her movement. While approaching the female, the male often continues to stridulate and may also display stereotyped hindleg motions (tibial extension, femur-tipping, etc.; see Otte 1970) that do not produce audible sound. The female may perform various silent hindleg motions at this time, but the relationship between these signals and acceptance or rejection of the male (or female?) is not clear. When a male contacts a female, he usually mounts her, but actual mating follows only one of every three or four mountings (Greenfield and Shelly 1985; M. D. Greenfield, unpublished data). If mating occurs, it lasts approximately 10 and 15 min in *L. coquilletti* and *L. planum*, respectively (Greenfield and Shelly 1985; Shelly and Greenfield 1989) and results in transfer of a small spermatophore with no external spermatophylax. There is no evidence for forced copulation, and if a female does not curve her abdomen downward, the mounting is quickly terminated. Males do not guard females before or after mating.

Male and female *Ligurotettix* do not exhibit courtship behaviors until approximately five days after the adult molt (Greenfield and Shelly 1985; Minckley and Greenfield 1995). This delay may be under endocrine control (cf. Blondheim and Broza 1970; but see Pener 1986).

Average male mating frequencies, inferred from systematic observations of natural populations (Greenfield and Shelly 1985; Shelly and Greenfield 1989), are approximately once every 5–10 days. Males resume sexual advertisement immediately after copulation and have been observed to remate (with a different female) on the same day. Females may also remate, but their schedule and precise incidence of multiple mating are unknown. Courtship and mating are restricted to midday hours on hot, sunny days.

Acoustic signaling

Long-range sexual advertisement by *Ligurotettix* males is accomplished via femurotegminal stridulations. These calls influence the settlement of females among host plants – male territories – by occasionally eliciting phono-taxis from receptive individuals moving between shrubs or returning from the ground to shrubs following morning oviposition activity (Shelly and Greenfield 1991; Greenfield 1992; Minckley and Greenfield 1995). In *L. coquilletti* the calls can be perceived by females, and by other males, up to 15 m from their source (Bailey *et al.* 1993).

Calling in *L. coquilletti* is a series of 20 ms clicks, which are a sufficiently conspicuous feature of the Sonoran Desert that the species has earned the epithet 'desert clicker' (Ball *et al.* 1942). Clicks average 60 dB in amplitude at 1 m (dB re 20 μPa); click rates vary from 10 to 60 min^{-1} both within and between individuals. The insects' repertoire includes single clicks made by movement of one hindleg, serial (2–4) clicks made by movements of one or alternate hindlegs in rapid succession, and double clicks produced by simultaneous movement of both hindlegs. Serial and double clicks tend to be made during aggression (see below) and are 3–6 dB louder than single clicks.

Advertisement calls in *L. planum* are 350 ms 'rasps' delivered at 5–10 min^{-1} (Greenfield and Minckley 1993). During aggression (see below) *L. planum* usually produce a louder, more complex call in which 2–60 'shucks' precede, and sometimes follow, a terminal rasp (Greenfield and Minckley 1993).

Aggressive and territorial behaviors

As many as 80% of males in *L. coquilletti* populations and 50% of *L. planum* males actively defend the shrubs on which they feed, perch, signal and mate (Shelly and Greenfield 1989). Defense is both passive, via the advertisement calls, which also serve as territorial signals, and active, via chasing and fighting (Shelly and Greenfield 1985; Wang and Greenfield 1991). If another conspecific male intrudes on a

defended shrub, the resident usually detects him visually and approaches. The intruder may depart at this juncture, but if he does not, the two insects typically exchange calls and visual hindleg displays. In *L. planum* this phase is a highly stereotyped series of shuck + rasp calls produced in an antiphonal 'acoustic duel' (Fig. 4-1) (see Greenfield and Minckley 1993; cf. Young 1971). Antiphony may increase the reliability with which calls can be mutually assessed, because hearing is impaired (masked?) during calling (Greenfield and Minckley 1993; cf. Hedwig 1990 and Eiríksson 1992, who notes that acridid males call more interruptedly in high population density). If the contest is not resolved via these exchanges such that one individual – the one who sustains a lower signal rate – departs or becomes silent, fighting ensues. Fighting may include kicking, grappling, and biting, but injuries occur only rarely in these conflicts.

Defense of host plants is accompanied by a high degree of site fidelity. In *L. coquilletti* populations at Deep Canyon, California, males remained associated with given *Larrea* shrubs for an average of 21 days. In populations of *L. planum* at Rodeo, New Mexico, males only remained on given *Flourensia* shrubs for an average of five days, but this lower value reflects a larger proportion of non-territorial and highly transient males in this species (Shelly and Greenfield 1989). Female site fidelity in both species is considerably lower. Both *L. coquilletti* and *L. planum* males defend only some of the available shrubs within a host-plant stand. In high populations this trait generates congregations that persist despite the aggressive behavior between males. The sets of *Larrea* and *Flourensia* shrubs that are defended tend to remain consistent from year to year (Greenfield *et al.* 1987, 1989a,b; Shelly *et al.* 1987).

INTRASEXUAL SELECTION IN *LIGUROTETTIX*

In resource defense polygyny systems, intrasexual selection may largely consist of competition between males vying to remain on and exclude each other from resource patches where females can be (sequentially) encountered at high rates. Such appears to be the case in *Ligurotettix*, as males located on certain host shrubs fight more regularly, encounter more females over the course of a season, and mate more frequently than males situated elsewhere (Greenfield and Shelly 1985; Shelly and Greenfield 1985). Competition to defend particular shrubs and skewed female distributions occur because of marked differences in food quality between shrubs, a limited number of high-quality shrubs, and requirements of females for food at high-quality shrubs. These factors have been demonstrated by feeding experiments and manipulations of populations in the field.

Assessment of resource patches

In *L. coquilletti*, females normally increase in mass by as much as 150–200% after eclosion. The rate at which mass increases and the mass eventually attained, however, will depend on the *Larrea* shrubs that the insects feed on (Greenfield *et al.* 1987, 1989b). Field and laboratory feeding trials showed that females gained little or no mass when confined to the foliage of certain *Larrea* shrubs. These (low-quality) shrubs seldom harbor females or territorial males and are termed 'avoided'. Low growth rates result

(a)

during feeding on foliage from avoided shrubs because of the combined effects of low rates of consumption and low rates of conversion of digested food into mass gain (Greenfield *et al.* 1989b). Both deterrence of feeding and reduction of food conversion are caused by toxic compounds present in higher concentrations in avoided than in 'preferred' *Larrea* shrubs (Greenfield *et al.* 1987; Chapman *et al.* 1988; see also Rhoades 1977). The structures and modes of action of these secondary chemicals, however, remain incompletely known. It is possible that much of the improved growth on preferred (high-quality) shrubs is translated into faster egg maturation, larger eggs, and/or a higher proportion of complete (six eggs) egg pods.

Because of the importance of feeding on the appropriate foliage while an adult, female *L. coquilletti* may be strongly selected to find and settle on high-quality shrubs. This prediction could not be tested via observations of natural populations, because *Larrea* shrubs differed in food quality as well as in the number and quality of signaling males that they harbored. These two potential attractions to females were disentangled as follows. First, an experiment was performed in which one and only one randomly selected signaling male was placed on each of 12 avoided and 12 preferred shrubs in a 0.25 ha plot (Shelly *et al.* 1987). All other males were removed from the plot, and the settlements of females initially present and those released into the plot at central locations were monitored for seven weeks. Females arrived in equal numbers at avoided and preferred shrubs, but they were much more likely to remain for extended periods (≥ 3 days) on preferred ones.

Males placed on preferred shrubs may have signaled more vigorously than those placed on avoided shrubs (see following section). Thus, the skewed pattern of female

(b)

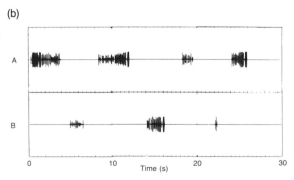

Fig. 4-1. (a) Two male *Ligurotettix planum* posturing during a contest over ownership of a *Flourensia* shrub (K-Bar, Big Bend Nat. Park, Texas; July 1992). (b) Oscillograms of shuck + rasp calls produced antiphonally by two male *L. planum* (A and B) engaged in an acoustic duel (K-Bar, Big Bend Nat. Park, Texas; August 1991).

settlement seen in the experiment above was not necessarily due to resource differences. Consequently, the experiment was repeated save that signaling males were replaced by loudspeakers broadcasting identical digitized recordings of male signals (Fig. 4-2a) (see Shelly and Greenfield 1991). Again, females arrived equally at avoided and preferred shrubs but tended to remain on the latter. These latter findings indicated more definitively that the long-term settlement of females was influenced by food quality, despite their failure to distinguish among shrubs prior to arrival (see Waage 1987 for experiments on selection of oviposition sites by female odonates).

A corollary of the above prediction is that males should assess resource patches by criteria similar to those evaluated by females, since by doing so, males would maximize their encounter rates with females (see Alcock 1990 for experiments on such tactics by male odonates). Several points suggest that this expectation occurs. Feeding trials showed that *L. coquilletti* males fed more readily on foliage from the same *Larrea* shrubs as females did (Greenfield *et al.* 1987, 1989b), and territorial males settled on preferred shrubs at the beginning of the season before the appearance of adult females (Wang *et al.* 1990). In the experiment in which randomly selected males were placed singly on avoided and preferred shrubs (Shelly *et al.* 1987), males remained much longer on preferred ones whether or not female adults, or nymphs, were present.

Assessment of resource patches by *L. planum* is probably based on principles similar to those in *L. coquilletti*. Both female and male *L. planum* tested in feeding trials consumed more foliage taken from preferred than avoided *Flourensia* shrubs (Greenfield and Shelly 1990). However, neither further nutritional studies nor manipulative experiments separating the influences of resources and males on female settlement have been performed in this species.

Although the basic factors influencing territorial defense of resource patches in *Ligurotettix* are now clarified, some critical questions remain. Females gain more mass and do so faster because of their assessment of host shrubs, but it is uncertain what this portends for actual fitness benefits, i.e. the number and quality of eggs laid. It is also unclear whether the mating success that a male can achieve on a given shrub is linearly related to his female encounter rate, which can be computed as the summation of 'female-days' (= $\sum_i d_i$, where d_i = number of days that female i remains on the shrub) (see Wang *et al.* 1990). If females tend to mate shortly after reaching a shrub, the higher $\sum_i d_i$ on preferred shrubs would not necessarily yield higher male mating success, because female arrival rates at avoided and preferred shrubs are equivalent.

The above treatment implicitly assumes that males benefit from assessment of resource patches solely because encounter rates with receptive females will be higher on some patches. However, males also benefit directly by consuming resources in high-quality patches. This benefit may lead to increased competitive abilities, which enhance the fitness benefits accrued through higher female encounter rates, and it could reinforce the selection pressure on males to make accurate assessments. In *Ligurotettix*, males situated on high-quality shrubs will avoid dietary toxins and consequently may be physiologically able to perform the activities involved in locating females quickly and fighting intruding males effectively. This would constitute positive feedback and may explain why *L. coquilletti* males show a relatively stronger aversion to consuming foliage from avoided *Larrea* shrubs during feeding trials than females do (Greenfield *et al.* 1989b).

Competition for resource patches

Three primary factors affect the ability of *Ligurotettix* males to compete for exclusive residence on high-quality resource patches: date of adult maturation, efficiency of finding and remaining on the best host shrubs, and aggression. In cases where two or more factors are independent, overall competitive success may approximate a multiplicative combination of success in each independent phase (see Wang and Greenfield 1994).

Timing

In *L. coquilletti*, early adult maturation affords a male access to the best female encounter sites (Wang *et al.* 1990). When males who eclose early move onto *Larrea* shrubs, they can generally remain on high-quality shrubs without contest, because few other males are yet mature. If an intruding male disputes ownership of the shrub later in the season, the previous resident – usually the individual who matured earlier – generally wins (Wang and Greenfield 1991). This rule presents a dilemma, though, because the number of high-quality shrubs is limited and the period during which receptive females are available is short. Under such circumstances, in which the loser is not expected to receive a future reward, both contestants would be expected to fight aggressively rather than obey a convention such as

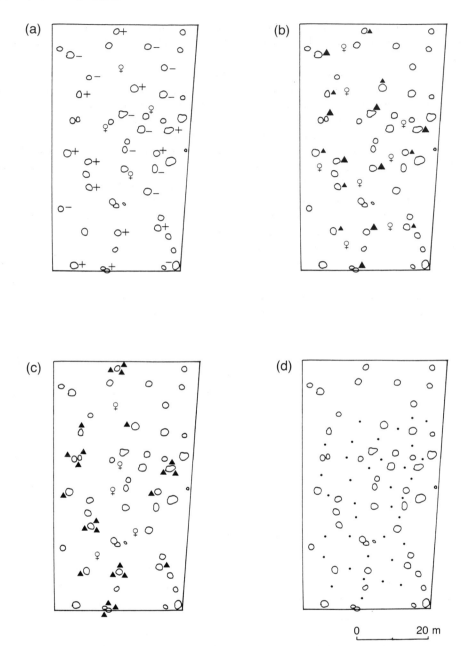

Fig. 4-2. Designs of sample replicates in experiments testing the phonotaxis of *Ligurotettix coquilletti* in a 0.25 ha plot at Deep Canyon, California (1986–1988). The elliptical outlines represent *Larrea* shrubs; most remaining area is bare ground. (a) Deployment of loudspeakers in test of female attraction and retention on shrubs of differing quality. Single loudspeakers broadcasting identical male calls at each of 12 preferred shrubs (+) and at each of 12 avoided shrubs (−); positions (♀) at which females were released. (b) Deployment of loudspeakers in test of female attraction and retention by male calls of different amplitude. Single loudspeakers broadcasting 64 dB (SPL at 1 m) calls at each of eight preferred shrubs (▲) and 56 dB calls at each of another eight preferred shrubs (▴); positions (♀) at which females were released. (c) Deployment of loudspeakers in test of female attraction and retention by single versus grouped male calls. All loudspeakers (▲) broadcast identical calls and were situated at preferred shrubs only; positions (♀) at which females were released. (d) Positions (·) at which males were released in test of mutual attraction, via phonotaxis, during settlement on territories.

previous residence (see Grafen 1987). Perhaps this implies that early eclosers are distinguished by actual fighting abilities that are currently unrecognized.

Adult eclosion date in *L. coquilletti* can be influenced by features of the site at which the egg developed and of the *Larrea* shrub on which nymphal development occurred. Soil humidity and temperature may affect the date of egg hatch, and nymphs that hatch early may reach the adult molt sooner. Nymphs that feed on foliage from preferred *Larrea* shrubs will also mature more rapidly (Greenfield *et al.* 1989b; Wang 1990; see Fincke 1992 for influences of larval ecology on male territoriality in odonates).

Advantages that males accrue by early adult molts, however, may be canceled by mortality before the presence of receptive females and by losing to late-eclosing males in the sperm competition arena (see Greenfield and Pener 1992). The first problem does not materialize in *L. coquilletti*, however, because lengthy adult male lifespans ensure that most early-eclosing individuals survive until the female population peaks (Wang *et al.* 1990). Lack of information on sperm precedence and female remating schedules prevents evaluation of the second anticipated difficulty.

Search

Ligurotettix coquilletti find high-quality *Larrea* shrubs during two distinct episodes. Most nymphs are found on preferred shrubs; it is inferred that they locate these sites via trial-and-error searching shortly after hatching (Wang and Greenfield 1994). Although adults eclose on the high-quality shrubs on which they developed as nymphs, males undertake a second episode of trial-and-error searching after they molt. This reshuffling may be a means of locating the best shrubs, an objective more likely to be achieved by (winged) adults than by nymphs. In both trial-and-error episodes, success may be determined by locomotory ability, the 'program' with which azimuths of successive intershrub moves are chosen, and memory (see below). Gustatory facility for assessing shrub quality accurately may complement a male's overall ability to find high-quality shrubs.

In addition to finding and settling on the best shrubs, males in both *L. coquilletti* and *L. planum* commit visual cues associated with these sites to long-term memory and may use this information to return quickly towards their shrubs if they depart for any reason (Greenfield *et al.* 1989a). Possibly, this homing ability has been selected for because it affords an insect the opportunity to relocate a valuable resource patch – the same one – without the burden of another trial-and-error search. The apparent

inability of *Ligurotettix* to judge shrub quality from a distance would make homing behavior particularly adaptive.

Aggression

Among *L. planum* males, success in aggression over shrub ownership is determined by signaling rate in acoustic duels, previous residence, and size, in that order of priority (Greenfield and Minckley 1993). Both signaling prowess and size may be influenced by prior nutrition (cf. Marden and Waage (1990) for findings in male odonates). In *L. coquilletti*, size does not appear to be a significant factor in aggression (Shelly and Greenfield 1985; Wang and Greenfield 1991) and signaling characteristics have not been fully investigated.

Ligurotettix planum males who are defeated in aggressive encounters tend to depart and move among *Flourensia* shrubs as 'transients', whereas defeated *L. coquilletti* males usually remain on the same *Larrea* shrub as 'satellites' (Shelly and Greenfield 1989). This difference may reflect smaller (more defensible) shrub size, shorter distances between neighboring shrubs, and greater shrub : insect population ratios in the *Flourensia* – *L. planum* system. Satellite *L. coquilletti* males move and stridulate little or not at all, and they mate infrequently despite high female encounter rates (Greenfield and Shelly 1985). Low mating success occurs because the approach of a satellite towards a female usually attracts the dominant male present on the shrub, who then either mates or chases the satellite. Some subordinate *L. coquilletti* males remain as signalers on low-quality *Larrea* shrubs, sites where they will encounter few aggressive males, but few females either. Signaling on low-quality *Larrea* shrubs may alternate with satellite behavior over the course of a subordinate male's lifespan (Shelly and Greenfield 1985).

Satellite and transient behaviors in *Ligurotettix* are mating strategies alternative to territoriality. Data on male mating success suggest that these alternatives are probably 'conditional' as opposed to 'individual mixed strategies' or 'genetic polymorphisms' (see Rubenstein 1980). *Ligurotettix* males suffer reduced mating rates when behaving as satellites or transients (Greenfield and Shelly 1985; Shelly and Greenfield 1989), indicating that these behaviors and territoriality are not being maintained at equilibrium frequencies within individual repertoires. Moreover, *L. coquilletti* males who consistently behave as satellites are not compensated by extended longevity, implying that their lifetime mating success is likewise reduced. The apparent disparity between fitnesses conferred by satellite and territorial behaviors indicates that these behaviors probably do not represent different genetic morphs existing at equilibrium frequencies within the population. It is therefore assumed that males adopt satellite or transient strategies because of some environmentally induced condition, such as poor nutrition or parasite infection, that would render futile any attempt at dominant, territorial behavior. However, efforts to ascertain the condition(s) responsible have not been rewarded (Greenfield and Shelly 1990); the only obvious distinction of alternative strategists is that they eclose late in the season (see Choe 1994 and this volume on a similar distinction of subordinates in Zoraptera). Whereas feeding on low-quality host shrubs might generate subordinate characteristics, it should be noted that in *L. coquilletti* both satellites and dominant territory owners are generally found on high-quality shrubs during nymphal as well as adult stages (Wang 1990).

Each of the factors determining male competitive ability in *Ligurotettix* may be influenced by genetic and environmental variance. Parental location of oviposition sites conducive to early hatch, finding high-quality host shrubs, and rapid nymphal development may be heritable, but expression of each could also be influenced by stochastic, environmental events. The relative contributions of these factors and the extent to which each is heritable or environmentally influenced await resolution.

FEMALE CHOICE IN *LIGUROTETTIX*

In a resource defense polygyny system, females may move toward, remain in, and mate on a resource patch because of attributes of the male(s) present as well as the quality of the resource. Alternatively, females may visit several resource patches sequentially, but only mate with males at one or a subset of the patches. Females may also control utilization of stored sperm such that their eggs are fertilized by only a specific portion of the sperm transferred in several matings (see Eberhard, this volume). With the stipulation that selection of males or their gametes is non-random, any of these possibilities would constitute female choice and could supplement the differential male reproductive success generated by variation in male competitive abilities for finding and defending valuable encounter sites.

Indirect selection

At a rudimentary level, female choice may operate in *Ligurotettix* simply by selection of those males who call (see Kriegbaum and von Helversen 1992 for similar

findings in the gomphocerine grasshopper *Chorthippus biguttulus*). The manipulative experiments in which single *L. coquilletti* males or loudspeakers were maintained on 12 preferred and 12 avoided *Larrea* shrubs in a 0.25 ha plot indicated that presence of a calling male or broadcasting loudspeaker was necessary to attract and retain females on a shrub (Shelly *et al.* 1987; Shelly and Greenfield 1991). That is, females failed to accumulate or arrive on preferred shrubs that were intentionally kept devoid of males or loudspeakers. It is also likely that a male's call amplitude influences females. An additional playback experiment in which loudspeakers broadcasting clicks at either 56 or 64 dB (SPL at 1 m) were maintained at preferred shrubs showed that louder clicks attracted more females (Fig. 4-2b) (see Shelly and Greenfield 1991). Click amplitude did not influence the length of time that a female remained on a shrub once attracted there; however, under natural circumstances a male's call loudness might determine whether a female accepts or rejects him during courtship by influencing her local movement (cf. Riede 1983). *Ligurotettix coquilletti* calls differ in parameters other than loudness, such as click rate, incidence of serial or double clicks, etc., but the repeatability of these parameters within individual males and influences of parameter values on females are unknown (see Butlin *et al.* 1985; Butlin and Hewitt 1986a, 1988; Kriegbaum 1989; Eiríksson 1993; von Helversen and von Helversen 1994 for various cases of stabilizing and directional selection imposed by female choice on song parameters of male gomphocerine grasshoppers; see also Butlin and Hewitt 1986b for a caveat concerning interpretations of phono-responses and phonotaxes of female grasshoppers).

The importance that females may attach to vigorous male calling can be witnessed in the morning and evening choruses of *L. coquilletti*, phenomena that arise owing to peculiar features of the diel movements of females in this species (see Greenfield 1992). Females typically depart *Larrea* shrubs at dawn, search the ground for oviposition sites, and then return to host shrubs, though not necessarily the same ones, in mid- to late morning. Click rates in male calling peak at this time, probably because some females can be persuaded to change shrubs – or to return to the same ones – during these movements. However, as many as 20% of the females do not return to *Larrea* shrubs during the morning. These individuals remain on non-host shrubs throughout the day and return to *Larrea* during a 30 min interval at dusk. Presumably, this occurs because dusk represents a temporal 'gate' (see Walker 1983a on this general phenomenon), the last opportunity for these visually orienting insects to move towards food and water before the next day. Click rates in male calling peak again at dusk and may influence the return movements of some females. Courtship and mating rarely occur during the evening or night, however. Thus, any increase in mating success that a male could obtain via his evening chorusing would be delayed until the following day or later.

In species with acoustic signaling it is common for females to choose males based on signal characters that reliably reflect energy either available for or allocated to signaling (Ryan and Keddy-Hector 1992; Andersson 1994). Most characters listed above probably fall into this category (see Prestwich and Walker 1981 on the energetic demands of orthopteran stridulation), but other characters subjected to female choice may be unrelated to energy. The precedence effect, in which preference is shown for the leading one of two or more calls that occur close in time (Wyttenbach and Hoy 1993), is an example of non-energy-based choice (see Greenfield 1994). The phonotaxes of *Ligurotettix* females toward male calls are influenced by precedence effects (Minckley and Greenfield 1995); it is believed that these effects have selected for mechanisms of interactive signaling in males that cause individuals to refrain from calling shortly after their nearest neighbors (Minckley *et al.* 1995; see Otte and Loftus-Hills 1979 on the possibility of a precedence effect in the gomphocerine grasshopper *Syrbula admirabilis*).

Female choice in acoustic species may also be based on morphology or behavioral features other than calling. This potential exists in *Ligurotettix*, because females have ample time to assess males at close range during courtship. Prospective characters evaluated include the form and persistence of visual displays, posture and tactile cues during mounting, and size.

Direct selection

If *Ligurotettix* females do move toward louder calling males, the attraction may serve as a 'short-cut' by which females quickly find high-quality resource patches when relocating (see Stamps 1987 on the use of conspecific cues in finding suitable territories). None the less, if the females mate with these males, female choice may still function. Here, females choosing loud callers could benefit from a reduction in the risk, time, and energy expended while searching for high-quality resources – and males – provided that it is significantly easier to orient toward these calls. Such direct, material benefits would be augmented if values of male call parameters were influenced by quality of the resources (food) in the patch. To answer these questions, thorough data on the costs incurred by females while searching for and orienting toward males and resources and the effect of diet on male signaling will be needed.

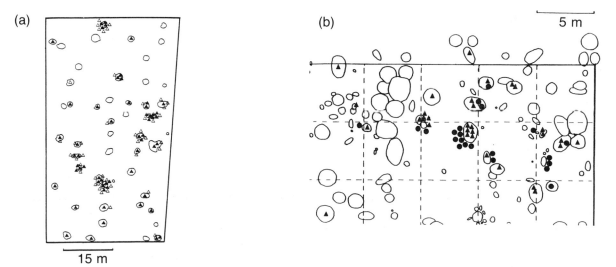

Fig. 4-3. (a) Distribution of male (▲) and female (△) *Ligurotettix coquilletti* in a 0.25 ha study plot at Deep Canyon, California, during a census day (19 July) in the middle of the 1983 season. The elliptical outlines represent *Larrea* shrubs; most remaining area is bare ground. (b) Distribution of male (▲) and female (●) *Ligurotettix planum* in a study plot at Rodeo, New Mexico, during a census day (12 August) in the middle of the 1985 season. The elliptical outlines represent *Flourensia* shrubs; most remaining area is bare ground.

Choice of congregated males?

The dispersion of *Ligurotettix* males is usually skewed such that certain host shrubs harbor several individuals whereas other host shrubs remain vacant (Greenfield and Shelly 1990). Some of the excess males at congregation sites are subordinate individuals, but often several codominant males actively signal on the same shrub. These congregations can remain stable for many days despite a high frequency of aggression between their members.

Systematic monitoring of *Ligurotettix* showed that congregated signaling males enjoyed higher female encounter rates and mating success, on a per individual basis, than solitary signalers (Fig. 4-3) (see, for example, Greenfield and Shelly 1985; Shelly and Greenfield 1989; see Doolan and MacNally 1981 and Sivinski 1989 on similar findings in Cicadidae and Tephritidae, respectively). This finding could indicate female choice for congregated males *per se*. Choice for congregants may be expected because it would afford a female the opportunity to effect a simultaneous, as opposed to sequential, comparison of males and thereby allow selection of the 'best' available mate quickly and accurately (see Alexander 1975; Alexander *et al.*, this volume; Janetos 1980). However, congregation is confounded with resource patch quality and with male dominance: congregations do not form on low-quality host shrubs, and solitary signalers are often subordinate males who call, possibly in an impotent fashion, on low-quality shrubs as an alternative to satellite behavior.

In *L. coquilletti*, the potential influences noted above were separated again via a playback experiment monitoring the orientation of females toward loudspeakers deployed at specific *Larrea* shrubs in a 0.25 ha plot (Fig. 4-2c) (see Shelly and Greenfield 1991). Here, single loudspeakers were placed at each of eight preferred shrubs, and three loudspeakers were placed at each of another eight preferred shrubs. All loudspeakers broadcast identical clicks at $30\,min^{-1}$. Clicks were timed by a central computer such that any two loudspeakers in the plot never broadcast clicks simultaneously. Therefore, clicks emanated from the shrubs supplied with three loudspeakers at a rate three times that emanating from the shrubs with single loudspeakers. The click rate from grouped loudspeakers was planned to reflect the elevated click rate produced by male congregations. Four replicates of this experiment were conducted, each continuing for 7–10 days. Females released into the plot neither moved in greater numbers toward nor remained for longer stays on shrubs with three loudspeakers, suggesting that congregation *per se* was not an attractive male feature (see Otte and Loftus-Hills 1979; Cade 1981; Doolan and MacNally 1981; Aiken 1982; and Walker 1983b for similar negative findings in other acoustic insects; only Morris *et al.* 1978 report positive results).

Observations of natural *L. coquilletti* populations, however, appear to contradict the above experimental results. On several occasions a proportion of the females on a shrub harboring a congregation of signaling males suddenly abandoned the shrub following the departure of several signalers (Shelly and Greenfield 1991). Whereas these observations may be artifacts, they may also be attributed to female assessment of features other than the congregation's overall click rate. Males call louder and produce more serial and double clicks when in a group than when solitary (see Aiken 1982 on analogous effects in Corixidae) and it is possible that females are heavily influenced by such call features. A full explanation would have to account for the apparent failure of males to exploit this preference when solitary by simply calling louder and producing serial and double clicks, however. Therefore, it is more likely that females somehow assess the number of males present on a shrub directly and remain or depart accordingly.

If females in a resource defense polygyny system do prefer males in congregations, some signaling males may be expected to congregate actively via mutual attraction and form 'resource-based leks' (Alexander 1975). Males may also orient toward conspecific males to circumvent a difficult trial-and-error search for a high-quality resource patch. These possibilities were examined in *L. coquilletti* via an experiment testing the mutual attraction (phonotaxis?) of males (Fig. 4-2d) (see Shelly and Greenfield 1991). A 0.25 ha plot was kept devoid of local males, and naive males – individuals collected from sites sufficiently distant that they would not have known the locations of high- and low-quality shrubs in the plot – of various ages were introduced singly at positions dispersed regularly throughout the plot. Six replicates of this experiment failed to reveal any indications of mutual phonotaxis or attraction, which if found might be construed as evidence that adult males prefer to settle on shrubs already harboring other signaling males. However, in playback experiments conducted recently (Muller 1995) males did show some positive phonotaxis, implying that male congregations may represent, in part, either resource-based leks or short-cuts to finding high-quality resources. In addition, congregations may simply form as 'sinks', sites where males accumulate because resource quality is sufficiently high that fortunate individuals who arrive do not readily depart when confronted by aggression from other males sharing the site.

Unlike males, *L. coquilletti* females on the same *Larrea* shrub do not exhibit mutual aggression, and they accumulate on high-quality shrubs without apparent restriction. Possibly, this results because neither mating opportunities nor food resources are limiting. This accumulation of females may yield distribution patterns among host plant shrubs even more skewed than those of males (Fig. 4-3) and it may reinforce the sink effect retaining males on high-quality shrubs. As in other purported relationships, tests of these hypotheses may be accomplished with the appropriate manipulations of insects and loudspeakers.

DISCUSSION

Mating systems of the Acridoidea

The mating systems of most acridids and other acridoids can be described as scramble competition between males (see, for example, Otte 1970, 1977; Uvarov 1977; Whitman and Orsak 1985; Wickler and Seibt 1985; Riede 1987; Bland 1991; von Helversen and von Helversen 1994). Even within Gomphocerinae, the subfamily that includes *Ligurotettix*, this characterization generally remains valid (see Otte 1970). Thus, resource defense polygyny in *Ligurotettix* represents an anomaly. It may result from unusually specialized – for Gomphocerinae (see Otte and Joern 1977; Gangwere *et al.* 1989) – feeding habits on regularly spaced desert shrubs, which males can defend economically (Otte and Joern 1975). This hypothesis is supported by the observation that the only other territorial acridids reported are species with similar host-plant associations. Males of *Bootettix argentatus*, a monophagous gomphocerine restricted to *Larrea tridentata* in the deserts of southwestern North America, occasionally defend their host shrubs (Schowalter and Whitford 1979). Likewise, males of *Hylopedetes* sp. (Rhytidochrotinae) defend fronds of their host ferns in the mountains of Costa Rica (C. H. F. Rowell, personal communication). However, in the absence of an accurate phylogeny of the Gomphocerinae and thorough field studies on mating systems, sex ratios, and dispersion of resources and of insects in a diversity of species, a robust analysis of the determinants of grasshopper mating systems is presently impossible.

Despite its atypical mating systems, the basic signaling and courtship behaviors in *Ligurotettix* do not differ fundamentally from those in other gomphocerine grasshoppers (Otte 1970; Uvarov 1977; Riede 1987; von Helversen and von Helversen 1994). Both acoustic and visual signaling are highly developed and prevalent in this subfamily (see, for example, Otte 1972). Signal interactions may occur between neighboring males searching for and displaying to females. Pair formation often includes female phonotaxis towards male advertisement calls and an exchange of short-range signals between the male and the female (see, for example, von Helversen and von Helversen 1983). Therefore, it appears as if elements of typical gomphocerine signaling and pair formation protocols have merely been modified and coopted in *Ligurotettix* for long-range mate attraction, aggression, and territorial behaviors in a resource defense polygyny system.

Sexual conflict and confluence in resource defense polygyny

At first analysis, resource defense mating systems would not seem especially prone to the sexual conflicts (see Alexander *et al.*, this volume) that characterize courtship and copulation in many insect species. Both sexes should benefit from association with the resources at encounter sites, although males may benefit more than females do because of reinforcement and positive feedback loops. In systems where the resource is limited, however, a male could benefit by controlling the amount of resource obtained by individual females; i.e. the resources are treated as if they were spermatophores (W. A. Snedden, personal communication). This controlled amount is likely to be less than optimum for the female and development of her eggs and offspring, and a conflict thereby arises. In *Ligurotettix*, however, food resources are seldom if ever limiting, and males do not, and probably cannot, control the access of females to food.

Specific consideration of the *Ligurotettix* mating system none the less does suggest various ways in which male and female interests in resource defense polygyny may not be confluent. Individuals undoubtedly assess conditions and resources favorable to functions such as oviposition, thermoregulation, and predator avoidance when deciding whether to settle on a patch. Consequently, some may settle on patches containing suboptimum food resources if other compensation exists there. For example, females may opt for shrubs whose nutritional quality is merely

adequate, provided that the adjacent soil is highly conducive to egg development. This will generate a conflict if,
owing to heightened nutrition, males compete and signal
best when on high-quality shrubs. Similarly, aggressive,
territorial males may maximize their mating success
through regular spacing among host shrubs, whereas
females may benefit most when males congregate, as opportunities thereby arise for simultaneous assessment of
males. Such conflicts might be resolved at evolutionarily
stable 'compromises', but endless 'arms races' between the
sexes are also conceivable. These general problems invite
both theoretical and experimental analyses.

ACKNOWLEDGEMENTS

The work and ideas presented here were developed in col-
laboration with many people. Most of all, I acknowledge
the long-term efforts of Bob Minckley, Todd Shelly and
Guang-yu Wang. Field and laboratory assistance of Ed
Alkaslassy, Marco Barzman, Marc Branham, David Coy,
Shawn Daley, Jack Der-Sarkissian, Ellen Engelke, Victor
Gian, Michael Gong, Jim Hogue, David Holway, Yikweon
Jang, Judy Jolly, David Ludwig, David Menken, Alan
Molumby, Kevin Toal and Ethel Villalobos and reports on
field research by Katherine Muller were invaluable. The
administrations and staff of the Deep Canyon Reserve
(University of California Natural Reserve System), the
Southwestern Research Station (American Museum of
Natural History) and Big Bend National Park (U.S.
National Park Service) provided logistical help and
granted permission for most of the field studies. Jae Choe,
Bernard Crespi, Bob Minckley, Katherine Muller, Todd
Shelly and several anonymous referees provided helpful
criticisms of previous versions of this manuscript. Finan-
cial support was provided by the National Science Founda-
tion (grants BSR 83–05824, BSR 86–00606, BSR
86–12325, and IBN 91–96177), the National Geographical
Society (2666–83), and the University of California
(U.C.L.A. grant no. 3721).

LITERATURE CITED

Aiken, R. B. 1982. Effects of group density on call rate, phono-
kinesis, and mating success in *Palmacorixa nana* (Heteroptera:
Corixidae). *Can. J. Zool.* **60**: 1665–1672.

Alcock, J. 1990. Oviposition resources, territoriality and male
reproductive tactics in the dragonfly *Paltothemis lineatipes* (Odonata: Libellulidae). *Behaviour* **113**: 251–263.

Alexander, R. D. 1975. Natural selection and specialized chorusing
behavior in acoustical insects. In *Insects, Science, and Society*.
D. Pimentel, ed., pp. 35–77. New York: Academic Press.

Andersson, M. 1994. *Sexual Selection*. Princeton: Princeton University Press.

Bailey, W. J., M. D. Greenfield and T. E. Shelly. 1993. Transmission
and perception of acoustic signaling in the desert clicker
(Orthoptera: Acrididae). *J. Insect Behav.* **6**: 141–154.

Ball, E. D., E. R. Tinkham, R. Flock and C. T. Vorhies. 1942. The
grasshoppers and other Orthoptera of Arizona. *Tech. Bull. Ariz.
Coll. Agric.* **93**: 255–373.

Bland, R. G. 1991. Mating behaviour of *Phaulacridium vittatum*
(Orthoptera: Acrididae). *J. Austr. Entomol. Soc.* **30**: 221–229.

Blondheim, S. and M. Broza. 1970. Stridulation by *Dociostaurus cur-
vicercus* (Orthoptera: Acrididae) in relation to hormonal termination of reproductive diapause. *Ann. Entomol. Soc. Am.* **63**: 896–897.

Butlin, R. K. and G. M. Hewitt. 1986a. Heritability estimates for
characters under sexual selection in the grasshopper *Chorthippus brunneus*. *Anim. Behav.* **34**: 1256–1261.

–. 1986b. The response of female grasshoppers to male song. *Anim.
Behav.* **34**: 1896–1898.

–. 1988. The structure of grasshopper song in relation to mating
success. *Behaviour* **104**: 152–161.

Butlin, R. K., G. M. Hewitt and S. F. Webb. 1985. Sexual selection
for intermediate optimum in *Chorthippus brunneus* (Orthoptera:
Acrididae). *Anim. Behav.* **33**: 1281–1292.

Cade, W. H. 1981. Field cricket spacing and the phonotaxis of crickets and parasitoid flies to clumped and isolated cricket songs.
Z. Tierpsychol. **55**: 365–375.

Chapman, R. F., E. A. Bernays and T. Wyatt. 1988. Chemical
aspects of host-plant specificity in three *Larrea*-feeding grasshoppers. *J. Chem. Ecol.* **14**: 561–579.

Choe, J. C. 1994. Sexual selection and mating system in *Zorotypus
gurneyi* Choe (Insecta: Zoraptera): II. determinants and
dynamics of dominance. *Behav. Ecol. Sociobiol.* **34**: 233–237.

Darwin, C. 1871. *The Descent of Man and Selection in Relation to Sex*.
London: John Murray.

Doolan, J. M. and R. C. MacNally. 1981. Spatial dynamics and
breeding ecology in the cicada *Cystosoma saundersii*: the interaction between distributions of resources and intraspecific behavior. *J. Anim. Ecol.* **50**: 925–940.

Eiriksson, T. 1992. Density dependent song duration in the grasshopper *Omocestus viridulus*. *Behaviour* **122**: 121–132.

–. 1993. Female preference for specific pulse duration of male
songs in the grasshopper, *Omocestus viridulus*. *Anim. Behav.* **45**:
471–477.

Fincke, O. M. 1992. Consequences of larval ecology for territoriality and reproductive success of a neotropical damselfly. *Ecology*
73: 449–462.

Gangwere, S. K., M. C. Muralirangan and M. Muralirangan. 1989.
Food selection and feeding in acridoids: a review. *Contrib. Am.
Entomol. Inst.* **25**(5): 1–56.

Grafen, A. 1987. The logic of divisively asymmetric contests: respect for ownership and the desperado effect. *Anim. Behav.* **35**: 462–467.

Greenfield, M. D. 1992. The evening chorus of the desert clicker, *Ligurotettix coquilletti* (Orthoptera: Acrididae): mating investment with delayed returns. *Ethology* **91**: 265–278.

–. 1994. Cooperation and conflict in the evolution of signal interactions. *Annu. Rev. Ecol. Syst.* **25**: 97–126.

Greenfield, M. D., E. Alkaslassy, G.-Y. Wang and T. E. Shelly. 1989a. Long-term memory in territorial grasshoppers. *Experientia* **45**: 775–777.

Greenfield, M. D. and R. L. Minckley. 1993. Acoustic dueling in tarbush grasshoppers: settlement of territorial contests via alternation of reliable signals. *Ethology* **95**: 309–326.

Greenfield, M. D. and M. P. Pener. 1992. Alternative schedules of male reproductive diapause in the grasshopper *Anacridium aegyptium* (L.): effects of the corpora allata on sexual behavior (Orthoptera: Acrididae). *J. Insect Behav.* **5**: 245–261.

Greenfield, M. D. and T. E. Shelly. 1985. Alternative mating strategies in a desert grasshopper: evidence for density dependence. *Anim. Behav.* **33**: 1192–1210.

–. 1990. Territory-based mating systems in desert grasshoppers: effects of host plant distribution and variation. In *Biology of Grasshoppers*. R. F. Chapman and A. Joern, eds., pp. 315–335. New York: Wiley.

Greenfield, M. D., T. E. Shelly and K. R. Downum. 1987. Variation in host plant quality: implications for territoriality in a desert grasshopper. *Ecology* **68**: 828–838.

Greenfield, M. D., T. E. Shelly, and A. Gonzalez-Coloma. 1989b. Territory selection in a desert grasshopper: the maximization of conversion efficiency on a chemically defended shrub. *J. Anim. Ecol.* **58**: 761–771.

Harvey, P. H. and J. W. Bradbury. 1991. Sexual selection. In *Behavioural Ecology, An Evolutionary Approach*, 3rd edn. J. R. Krebs and N. B. Davies, eds., pp. 203–233. Oxford: Blackwell.

Hedwig, B. 1990. Modulation of auditory responsiveness in stridulating grasshoppers. *J. Comp. Physiol.* A **167**: 847–856.

Helversen, D. von and O. von Helversen. 1983. Species recognition and acoustic localization in acridid grasshoppers: a behavioral approach. In *Neuroethology and Behavioral Physiology*. F. Huber and H. Markl, eds., pp. 95–107. Berlin: Springer-Verlag.

Helversen, O. von and D. von Helversen. 1994. Forces driving coevolution of song and song recognition in grasshoppers. In *Neural Basis of Behavioural Adaptations*. K. Schildberger and N. Elsner, eds., pp. 253–284. Stuttgart: Gustav Fischer Verlag.

Janetos, A. C. 1980. Strategies of female choice: a theoretical analysis. *Behav. Ecol. Sociobiol.* **7**: 107–112.

Kirkpatrick, M. and M. J. Ryan. 1991. The paradox of the lek and the evolution of mating preferences. *Nature (Lond.)* **350**: 33–38.

Kriegbaum, H. 1989. Female choice in the grasshopper *Chorthippus biguttulus*. *Naturwissenschaften* **76**: 81–82.

Kriegbaum, H. and O. von Helversen. 1992. Influence of male songs on female mating behavior in the grasshopper *Chorthippus biguttulus* (Orthoptera: Acrididae). *Ethology* **91**: 248–254.

Marden, J. H. and J. K. Waage. 1990. Escalated damselfly territorial contests are energetic wars of attrition. *Anim. Behav.* **39**: 954–959.

Minckley, R. L. and M. D. Greenfield. 1995. Psychoacoustics of female phonotaxis and the evolution of male signal interactions. *Ethol. Ecol. Evol.* **7**: 235–243.

Minckley, R. L., M. D. Greenfield and M. K. Tourtellot. 1995. Chorus structure in tarbush grasshoppers: inhibition, selective phonoresponse, and signal competition. *Anim. Behav.* **50**: 579–594.

Morris, G. K., G. E. Kerr and J. H. Fullard. 1978. Phonotactic preferences of female meadow katydids (Orthoptera: Tettigoniidae: *Conocephalus nigropleurum*). *Can. J. Zool.* **56**: 1479–1487.

Muller, K. L. 1995. Habitat settlement in territorial species: the effects of habitat quality and conspecifics. Ph.D. dissertation, University of California, Davis.

Otte, D. 1970. A comparative study of communicative behavior in grasshoppers. *Misc. Publ. Univ. Mich. Mus. Zool.* **141**: 1–168.

–. 1972. Simple versus elaborate behaviour in grasshoppers: an analysis of communication in the genus *Syrbula*. *Behaviour* **42**: 291–322.

–. 1977. Communication in Orthoptera. In *How Animals Communicate*. T. A. Sebeok, ed., pp. 334–361. Bloomington: Indiana University Press.

–. 1981. *The North American Grasshoppers. I. Acrididae, Gomphocerinae and Acridinae.* Cambridge, Massachusetts: Harvard University Press.

Otte, D. and A. Joern. 1975. Insect territoriality and its evolution: population studies of desert grasshoppers on creosote bushes. *J. Anim. Ecol.* **44**: 29–54.

–. 1977. On feeding patterns in desert grasshoppers and the evolution of specialized diets. *Proc. Acad. Nat. Sci. Philad.* **128**: 89–126.

Otte, D. and J. Loftus-Hills. 1979. Chorusing in *Syrbula* (Orthoptera: Acrididae). cooperation, interference competition, or concealment? *Entomol. News* **90**: 159–165.

Pener, M. P. 1986. Endocrine aspects of mating behavior in acridids. *Proc. 4th Triennial Meeting Pan Am. Acridol. Soc*, Saskatoon, Canada. D. Nickle, ed., pp. 9–26. Detroit, Michigan: PAAS.

Prestwich, K. N. and T. J. Walker. 1981. Energetics of singing in crickets: effect of temperature in three trilling species (Orthoptera: Gryllidae). *J. Comp. Physiol.* B **143**: 199–212.

Rehn, J. A. G. 1923. North American Acrididae (Orthoptera). 3. A study of the Ligurotettigi. *Trans. Am. Entomol. Soc.* **49**: 43–92.

Rhoades, D. F. 1977. Integrated antiherbivore, antidesiccant and ultraviolet screening properties of creosote bush resin. *Biochem. Syst. Ecol.* **5**: 281–290.

Riede, K. 1983. Influence of the courtship song of the acridid grasshopper *Gomphocerus rufus* L. on the female. *Behav. Ecol. Sociobiol.* **14**: 21–27.

–. 1987. A comparative study of mating behaviour in some Neotropical grasshoppers (Acridoidea). *Ethology* **76**: 265–296.

song, of lower intensity than the calling song and produced by rubbing the tegmina together while flattened against the dorsal surface of the abdomen (Alexander 1961; Boake 1983). Copulation consists of the female mounting the male, followed by transfer of a single spermatophore from the male to the external genitalia of the female. Sperm flows from the spermatophore into the female's body over a period ranging from several minutes to an hour, depending on the species (Khalifa 1949; Simmons 1986a; Loher and Dambach 1989). Immediately after copulation, females often remove or eat the spermatophore and males may attempt to prevent them from doing so in a series of stereo-typed behaviors often termed postcopulatory guarding (Loher and Dambach 1989).

Variation in this pattern is seen at all stages; a few species are silent, with the apparent secondary loss of either the calling song or both the calling and courtship songs (Boake and Capranica 1982; Loher and Dambach 1989). One of these, *Amphiacusta maya* (Phalangopsinae), is gregar-ious, with females apparently encountering males in the aggregation rather than orienting acoustically to potential mates (Fig. 5-1) (Boake and Capranica 1982; Boake 1984b). With few exceptions (see, for example, Alexander and Otte 1967; Sakaluk 1984) most field crickets (Gryllinae) do not produce significant amounts of nutritive material along with the spermatophore. This absence of an energetically costly spermatophylax is one of the most marked distinc-tions between the field crickets and other ensiferan orthop-tera (Gwynne, this volume).

The absence of the spermatophylax, along with the lack of any other substantive paternal investment, makes male choice less likely than female choice in the field crickets.

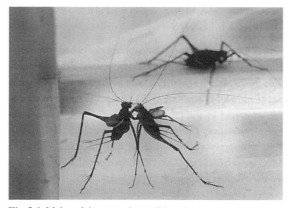

Fig. 5-1. Males of the gregarious cricket *Amphiacousta maya* engage in frequent fights over access to females.

In the tettigoniids, many of which have high paternal investment, males may only inseminate females that are most likely to have high fecundity (see, for example, Gwynne 1981, 1984). Indeed, some authors have pointed out the similarity between field cricket mating systems and those of lekking birds such as sage grouse, in which males likewise make negligible contributions to the off-spring (Alexander 1975; Bradbury and Gibson 1983; Gwynne and Morris 1983; Hoglund and Alatalo 1995), although insect and vertebrate leks may also show impor-tant distinctions in the nature of male aggregations (Hoglund and Alatalo 1995; Shelly and Whittier, this volume).

MALE MATING STRATEGIES

Crickets generally call in the easily audible (to human ears) range of 4000–6000 Hz (Bennet-Clark 1989), although at least a few species are capable of detecting ultrasonic sounds (Pollack and Hoy 1989). This sensitivity to very high frequencies may represent an adaptation to detect the signals of predatory bats (Hoy 1991). Acoustic signals mainly function in male–male interactions and in long-distance attraction of females, each of which is discussed below.

Male–male interactions and territorial spacing

The extreme aggression of male field crickets has been noted for centuries (White 1788; Fabre 1917). Laboratory populations of field crickets often exhibit dominance hier-archies (Alexander 1961; Boake and Capranica 1982); aggres-sive interactions between males, often accompanied by distinctive aggressive chirps, have been observed in both the laboratory and the field. The calling song degrades with distance, and may serve as both an agonistic signal directed towards rival males and to attract receptive females (Cade 1981; Simmons 1988a). Nevertheless, play-backs of recorded calling songs often attract males as well as females (Walker 1979). The consequence of both attrac-tive and repulsive properties of male calls is a pattern of male spacing characterized by aggregations within which males are evenly spaced (Campbell and Clark 1971; Campbell 1990). Aggregations of field crickets and other singing insects are one of the few demonstrations of con-specific attraction leading to territoriality (Thornhill and Alcock 1983). Contrary to the suggestion of Alexander *et al.* (this volume) that females promote the formation of aggregations, Cade (1981) found that solitary male crickets attracted as many females per capita as did clumped individuals.

Alexander (1961) described the levels of aggression and stereotyped behavior patterns associated with overt male fighting in a variety of field crickets. Success at winning encounters is correlated with several male attributes. As with many vertebrate species, previous experience appears to influence the outcome of aggressive interactions, with previous winners being more likely to win subsequent fights in *Teleogryllus oceanicus* (Burk 1983), *G. integer* (Dixon and Cade 1986), *G. veletis* and *G. pennsylvanicus* (Alexander 1961), and *G. bimaculatus* (Simmons 1986b). Possession of a burrow also appears to predispose a male towards winning an encounter (Alexander 1961; Burk 1983; Simmons 1986b); the addition of burrows to a population of *G. bimaculatus* increased the level of aggression (Simmons 1986b).

Not surprisingly, large body size is often correlated with winning fights (*G. bimaculatus*, Simmons 1986b; *G. integer*, Dixon and Cade 1986; *G. pennsylvanicus*, Souroukis and Cade 1993). Although sexually mature male *G. integer* were more likely to initiate and win aggressive encounters than were immature males, age did not predict the outcome of fights after maturity (Dixon and Cade 1986).

A somewhat different situation exists in the gregarious phalangopsid *A. maya*. The same type of chirp is used in both courtship and aggression in this species; experimentally silenced males are both less able to achieve high dominance rank and less successful at mating, although the reduction in mating occurs not because females are less receptive but because intruding males interrupt silent courtships (Boake and Capranica 1982; Boake 1983, 1984a). Escalated fights are probably less common in such gregarious species than in the more widely dispersed field crickets.

Both population density and habitat type influence the spatial distribution of calling males, and the likelihood that males will call at all. Despite the demonstration of dominance hierarchies and extended fights between males under laboratory conditions, the degree to which such encounters influence male spacing, calling, and mating success in the field is unclear. Alexander (1961) predicted that aggression and territoriality should diminish with increasing population density because of the increased costs of frequent fighting. Instead, males should benefit by silently searching for females. Aggression and calling is diminished at high population densities in *G. bimaculatus* (Simmons 1986b); *G. integer* and *G. campestris* males spend more time calling and less time searching when densities are low (Hissmann 1990; Cade and Cade 1992). Further, the relationship between body size and calling duration changed at different population densities in *G. integer*, with larger males calling for longer periods of time at low densities but not at high densities (Cade and Cade 1992).

That habitat structure can also influence male spacing is indicated by the observation of male *G. pennsylvanicus* uniformly distributed in tall-grass and mixed tall- and short-grass habitats, but clumped in a short-grass meadow (French *et al.* 1986). These differences may arise from the transmission properties of male calls in different habitats, although this aspect of cricket communication is largely unstudied (but see Schatral *et al.* 1984; Michelsen 1985).

Attraction of females: intraspecific variation in call structure

Males need both to establish and maintain calling sites via dominance interactions and to attract receptive females in the area. Long-range attraction is generally assumed to occur through the calling song; both neurophysiological and behavioral work has examined the differential effectiveness of various song components in attracting females.

Several studies have determined that phonotaxis in female crickets is best elicited by calls that are closest to the species-typical song (Shuvalov and Popov 1973; Pollack and Hoy 1981b; Stout *et al.* 1983), although in *T. oceanicus* females are better attracted by a song which is simpler or otherwise 'unnatural' compared with the species-typical one (Pollack and Hoy 1981b; Doolan and Pollack 1985). Indeed, the 'behavioral coupling' between the signaler and recipient is well established, and temperature-induced changes in call parameters are closely tracked by changes in female preference, so that females are more attracted to songs produced at ambient temperature (Doherty and Hoy 1985; Doherty 1985a). Whether this coupling has evolved through genetic linkage between sender and receiver neuronal pathways or through selection acting independently on the genes for signal production and reception remains an open question (Doherty and Hoy 1985, Alexander *et al.*, this volume).

Despite the behavioral coupling between sender and receiver, call parameters may vary substantially among males within a species, and females may respond differentially to both the temporal and the structural components of the song. Doherty (1985b) and Doherty and Hoy (1985) suggested that components such as syllable length and duration compensate for each other, with females simultaneously evaluating several different call properties and orienting toward a call with highly preferred characteristics in one but perhaps not all components. The use of such trade-offs by female crickets may be similar to the complex evaluation of multiple courtship displays or sexual ornamental characters in birds such as sage grouse (Gibson and Bradbury 1985; Gibson *et al.* 1991) and jungle fowl (Zuk 1991; Zuk *et al.* 1993a), and suggests that even in insects the operation of female choice may be more complicated than is assumed in many models of sexual selection.

In some species, females have been shown to prefer more intense calls in either the laboratory or the field (Shuvalov and Popov 1973; Schmitz *et al.* 1982; Forrest 1983), although in *Anurogryllus arboreus* females did not orient toward louder males (Walker 1983b). Even if females do exhibit greater phonotaxis to more intense calls, the biological significance of such a preference is unclear; because intensity obviously decreases with increasing distance from the source, a female orienting toward a loud call in the field would simply end up at the nearest male, but not necessarily the one with a louder absolute call.

Loher and Dambach (1989) point out that temporal differences in song components are more plausible cues than song intensity for females to use in discriminating among males. Demonstration of such discrimination has been surprisingly difficult, however. Hedrick (1986) found that female *G. integer* preferred male songs with longer uninterrupted bouts, a factor predicted to be of greatest importance for female discrimination by Alexander *et al.* (this volume). Other studies, although they often establish non-random mating, have failed to identify song attributes that could be used by females in making long-distance discriminations. Crankshaw (1979), for example, found that female *Acheta domesticus* preferred the recorded calls of dominant over subordinate males, but did not analyze the two types of calls for the actual cues used by the females. Female *G. bimaculatus* prefer larger males; call parameters were correlated with body size in the population studied (Simmons 1988a). Body size was unrelated to song structure in *G. integer* (Souroukis *et al.* 1992), and a comparative study of four gryllines showed no correlations between nightly calling duration and body mass (Cade and Wyatt 1984). As Alexander *et al.* (this volume) suggest, temporal patterning may be under stabilizing rather than directional selection.

Absolute amount of time spent calling would seem to increase the likelihood that a female is exposed to the song. A positive correlation between the time spent calling and mating success was found in *G. veletis* but only at low population density (French and Cade 1989). In *G. pennsylvanicus* and *G. campestris*, however, calling time was not correlated with either the number of females attracted to a particular male or his copulation rate (Rost and Honegger 1987; Zuk 1987c). This absence of a straightforward relationship between exposure to a male's signal and mating success, together with the existence of non-random mating in several species (Walker 1983b; Zuk 1988; Simmons and Zuk 1992) further supports the idea that females are choosing among the calls of potential mates. Future research on mate choice in crickets should include an exploration of the acoustical cues used by females in choosing males.

Females in several species are more frequently paired with relatively old males in the population (Table 5-1). Again, evidence on the cue used by females to determine male age is lacking, although a pilot study of *G. pennsylvanicus* showed that older males tended to have more pulses (wingstrokes) in their songs (M. Zuk, unpublished data). Age was unrelated to call duration in the gryllines studied by Cade and Wyatt (1984) and in the *G. integer* studied by Souroukis *et al.* (1992). One might expect males to increase reproductive effort as they age, being willing to take more risks by calling more and increasing the likelihood of detection by predators as their reproductive value decreases. However, a comparison of song-bout lengths before and after male *G. pennsylvanicus* were held in the laboratory for two weeks showed no increase in the tendency to sing for longer periods of time as the males aged; indeed, a paired *t*-test showed that males actually sang shorter bouts after two weeks (M. Zuk, unpublished data, mean initial bout length = 14.91 ± 2.04 min (SE), mean final bout length = 6.92 ± 1.16 min (SE); $t = 5.18$, d.f. 17, $p < 0.001$). Layers of chitin are added daily to the insect cuticle in response to photoperiod (Neville 1963; Zuk 1987a); it is possible that this morphological change causes an alteration in some song properties. Although no studies to date have looked for such an effect, our study of *G. bimaculatus* failed to show significant temporal or tuning differences related to male age (Simmons and Zuk 1992).

The degree of phonotaxis in female crickets, and hence the perceived attractiveness of male calls, may vary depending on the testing conditions. Pollack and Hoy (1981a) found that female *T. oceanicus* were less discriminating when tested using apparatus in which females approached

Table 5-1. Characteristics of paired and calling male crickets collected from natural populations (mean ± SE)

Species		Age (d)	Size (mm)[a]	Parasite load[b]	Life history
Gryllus pennsylvanicus[c]	paired (47)	11.34 ± 0.0.62*	3.00 ± 0.04*	2.94 ± 0.02*	univoltine
	calling (56)	9.61 ± 0.40	2.94 ± 0.02	3.00 ± 0.04	
G. veletis[c]	paired (37)	18.86 ± 0.70*	3.00 ± 0.03	3.54 ± 0.21**	univoltine
	calling (52)	16.74 ± 0.64	3.05 ± 0.03	4.27 ± 0.20	
G. campestris[d]	paired (23)	13.30 ± 0.56*	7.99 ± 0.09*	0.00 ± 0.00	univoltine
	calling (23)	11.52 ± 0.58	7.66 ± 0.11	0.00 ± 0.00	
G. bimaculatus[e]	paired (26)	14.00 ± 0.51*	7.50 ± 0.12	1.50 ± 0.14	bivoltine?
	calling (71)	12.14 ± 0.42	7.30 ± 0.08	1.73 ± 0.11	
Teleogryllus oceanicus[f]	paired (27)	12.67 ± 0.51	5.63 ± 0.06	0.00 ± 0.00	univoltine
	calling (34)	12.06 ± 0.58	5.59 ± 0.06	0.00 ± 0.00	

*$p < 0.05$;

**$p < 0.01$. [a] pronotum width except for G. veletis and G. pennsylvanicus, in which pronotum length was measured; [b] rank value for infection levels of gregarine parasites; [c] Zuk 1988; [d] Simmons 1995; [e] Simmons and Zuk 1992; [f] Simmons and M. Zuk, unpublished data (samples collected on Moorea, French Polynesia).

by walking than when phonotaxis was measured in flying females. An explanation for this difference may be found in the work of Hedrick and Dill (1993), who allowed female G. integer to approach speakers broadcasting calling songs under two sets of circumstances: open, in which females had no shelter in which to walk during phonotaxis, and covered, in which females were given cover in the testing arena. Females were more choosy when cover was provided, suggesting that avoidance of risk of predation influenced mate-choice decisions (Hedrick and Dill 1993). The walking females in Pollack and Hoy's (1981a) experiments may have responded similarly to the apparent dangers of moving through an exposed area.

Diel patterning of calls

Male calls may vary in daily patterning, as well as in temporal and structural components. Surprisingly little attention has been paid to calling distributions; as Walker (1983a, p. 55) points out, 'Although millions of males of thousands of species call each evening, biologists spend little time noting exactly who is calling when, especially after midnight'. In particular, because, as noted above, there is no simple relationship between the absolute amount of time a male spends calling and his mating success, it is important to determine whether the distribution of a male's calling period influences his attractiveness, and in turn what factors affect this distribution.

Alexander (1975) and Walker (1983a) suggest that males are expected to time their calling to coincide with the period of maximum activity of receptive females. In some species, this window may be rather short; female mole crickets (Gryllotalpidae) Scapteriscus acletus and S. vicinus fly during the hours just after sunset, when predators are relatively few and the temperature is still warm enough to permit activity (Forrest 1983). Not surprisingly, male mole crickets limit their calling to this period as well (Forrest 1983). Avoidance of predators, availability of energy reserves, and temperature or weather are also expected to influence diel calling patterns (Walker 1983a). Although the distribution of calling in time is expected to follow an ideal free distribution, such a pattern has been difficult to test in the field (Walker 1983a).

Most species studied appear to have greater flexibility in their distribution of calling than do the mole crickets. In G. veletis and G. pennsylvanicus, males sing mainly at night unless the temperature is too cold, in which case daytime singing is common (Alexander and Meral 1967; Loher 1989). Calling and locomotion both show a circadian rhythm in Teleogryllus, with males responding to the onset and cessation of darkness as cues (Loher 1989).

In Teleogryllus, as well as in numerous other field crickets, individuals vary in the timing of their calling (Loher 1989). Male G. veletis, G. pennsylvanicus and G. integer all exhibit variation in calling distribution, with some individuals calling in a continuous bout after sunset, others calling sporadically throughout the night, and still others calling mainly before daylight (Cade 1979; Cade and Wyatt 1984). Male *G. campestris* show all the former patterns yet may call predominantly during the day (Rost and Honegger 1987). French and Cade (1987) monitored *G. veletis*, *G. pennsylvanicus* and *G. integer* in an outdoor arena, and noted an increase in the proportion of males calling at dawn, although the calling was of lower intensity than at other times of day. Diel distributions of calling were also density-dependent, with the denser populations exhibiting the dawn calling peak more clearly (French and Cade 1987). Mating was likewise more frequent around dawn. Although daytime copulations were common in *G. veletis* and *G. pennsylvanicus*, males mainly called at night; in these species the period of peak courtship singing corresponded with a peak in spermatophore production (Zuk 1987b). In *G. campestris*, mating occurs mainly in the late afternoon and early evening, but males regularly call throughout the night and morning. This temporal separation between calling and mating suggests that females spend the night moving and perhaps evaluating males, with copulation occurring only after a male has been selected (Zuk 1987b). Burpee and Sakaluk (1993a) compared postmating calling in the spermatophylax-providing *Gryllodes sigillatus* with *Gryllus veletis*, which does not produce a spermatophylax. Calling decreased markedly following pair formation in *G. veletis*, but remained nearly constant in *Gryllodes sigillatus*, perhaps because in the latter remating is virtually absent whereas in *Gryllus veletis* (and other species providing an easily replaced spermatophore) males may benefit by continuing to attract additional females after mating (Burpee and Sakaluk 1993a). Male *G. veletis* and *G. pennsylvanicus* have been observed in the field with more than one female at a time (Zuk 1988).

The significance of the substantial variation in male calling patterns has not been explored. In the few species studied, calling was found to be energetically expensive (Prestwich and Walker 1981), but little is known about the energetics of calling distribution. For example, imagine two males, each of which calls for two hours per night. How does a male who calls in twenty-minute segments separated by an hour or more differ, either in energy use or in mating success, from one who spends the first or last two hours of the night calling steadily? Merely calling more does not guarantee attracting more females, which suggests that a closer examination of individual calling patterns and their consequences should prove fruitful.

Alternative male reproductive strategies

Sexual signaling is usually conspicuous to its recipient; in many species these signals have also been exploited by predators and parasitoids (Burk 1982; Sakaluk 1990). Among crickets, calling males have been shown to be vulnerable to predation by domestic cats (Walker 1964), geckos (Sakaluk and Belwood 1984) and birds (Bell 1979), among others (Sakaluk 1990). Perhaps the best-documented example of phonotactic orientation in insects is the use of male calling song to localize hosts by parasitoid flies in the genus *Ormia* (Tachinidae; formerly *Euphasiopteryx*) (Cade 1975; Walker 1986; Walker and Wineriter 1991). These flies land on or near a calling host and deposit larvae, which then burrow into the host's body cavity and emerge within 7–10 days, killing the host upon emergence (Cade 1975, 1984a). The hearing apparatus of the fly exhibits a finely tuned adaptation to its host, being remarkably convergent on the same apparatus in the female cricket host (Robert *et al.* 1992).

The existence of selection by these acoustically orienting parasitoids led Cade (1975, 1979, 1984a) to propose that non-calling may have arisen as a genetically determined, evolutionarily stable, alternative mating strategy. He observed silent males of *G. integer* walking near callers, and suggested that these males intercept females attracted to callers, while avoiding the risks of parasitism (Cade 1979, 1984a). At close range, both males and females may use olfactory cues to determine each other's sex and receptivity (Cade 1979). Selection experiments on *G. integer* showed significant heritabilities for nightly calling duration in lines selected for high and low amounts of calling ($h^2 = 0.50$ and 0.53); frequency-dependent selection could maintain heritable variation (Cade 1984b). In addition, more silent males were observed in this species than in three other gryllines not known to be parasitized (Cade and Wyatt 1984). Seasonal variation in calling related to parasitoid fly abundance has been documented in *G. rubens* (Burk 1982).

Although silent males may achieve the benefit of avoiding fly parasitoids, our work with parasitized and unparasitized populations of *T. oceanicus* casts some doubt on the notion that silent males necessarily represent a 'satellite' class (Zuk *et al.* 1993b). *Teleogryllus oceanicus* is parasitized by *O. ochracea* in Hawaii, where both the fly and the cricket have been introduced, but is parasite-free in other locations within its range, including northwestern Australia and the island of Moorea in French Polynesia (Zuk *et al.* 1993b). Silent males, however, were found in all three populations, although their abundance varied among the sites. In Hawaii, 12 of 44 silent males harbored fly larvae. We suggest that these non-callers may be composed of two subclasses: (1) males that are too depleted by the para-sitoid larvae to call; and (2) males that are following an alternative strategy of searching for females rather than attracting them from a fixed calling site (Zuk *et al.* 1993b, 1995).

Non-callers may be opportunistic in their search for mates; and individual males may switch between calling and non-calling modes depending on population density. As discussed earlier, at high densities, encounter rates with females are expected to be higher, and hence more males may adopt a wandering strategy instead of stationary signaling (Alexander 1961; Simmons 1986b). Hissman (1990) demonstrated a similar effect in *G. campestris*, with the number of silent searching males changing with popu-lation density. Cade (1979) also mentions that non-callers in *G. integer* may begin to call if a neighboring male is silenced so that silently searching for females may repre-sent an evolutionarily stable strategy (ESS). A final possi-ble explanation is that silent searching may represent a conditional strategy (Cade 1979; Thornhill and Alcock 1983), with males unsuccessful in competition avoiding the attraction of superior competitors by remaining silent. Simmons (1986b) showed how the frequency of calling was positively correlated with a male's competitive ability. Burk's (1983) study of *T. oceanicus* showed how subordinate males avoided attacks by dominants by not producing courtship song when interacting with females. Males gen-erally differ greatly in their tendency to call; they also may respond differentially to environmental cues. The effect of phonotactic parasitoids on cricket mating strategies may be more subtle than simply causing some males to stop calling entirely. The parasitized Hawaiian population of *T. oceani-cus* showed both a reduction in call parameters such as chirp duration and a more abrupt onset and cessation of calling at sunset and sunrise, when the flies are expected to be more active, compared with the unparasitized popula-tions (Zuk *et al.* 1993b).

Whether a trade-off exists between song components that are more attractive to females but also more costly in terms of vulnerability to the parasitoid remains to be seen. The parasitoids may be taking advantage of a window of sensory sensitivity by using the same acoustic parameters most easily detected by females (Ryan and Rand 1990; Robert *et al.* 1992). Walker (1993) found that simulated cricket calls of *G. integer* were far less effective in attracting

O. ochracea in Florida than were calls of *G. rubens*, despite *G. integer* attracting numerous flies in Texas where *G. integer* is more abundant.

FEMALE MATING STRATEGIES

Multiple mating

With few exceptions, repeated mating by female gryllids is common (Alexander and Otte 1967). Given the time, energy and risk involved in phonotaxis and copulation, workers have tried to identify the benefits obtained by females from multiple mating. Several hypotheses for the evolution of multiple mating by females have been proposed (Thornhill and Alcock 1983; Choe, this vol-ume; Eberhard, this volume). Females may remate to replenish sperm reserves. Sakaluk and Cade's (1980, 1983) studies of *G. integer* and *A. domesticus* show that multiply mated females produce more progeny than singly mated females. These results would be consistent with the sperm-replenishment hypothesis for multiple mating if females in the two groups produced the same number of eggs and multiply mated females had a higher fertility (proportion of eggs hatched). However, they could also be explained if females in the two groups had similar fertilities but laid different numbers of eggs. Simmons (1988b) moni-tored egg production and fertility of female *G. bimaculatus* after varying numbers of matings. The proportion of eggs hatching increased from 35% for singly mated females to 45% for doubly mated females and 55% for females given constant access to males. Although the rather low hatching success may have been a result of laboratory incubation techniques, the change in relative hatching proportions across the remating treatments show that females benefit from having larger sperm reserves. Why females require a large sperm store for successful fertilization is not clear.

Multiple mating also increased the number of eggs laid by female *G. bimaculatus* (Simmons 1988b). There are two explanations for this effect. First, gryllid males transfer ovi-position stimulants within the ejaculate that increase egg production. Males incorporate prostaglandin synthetase into the seminal fluids of the ejaculate, which induces the female to produce prostaglandins (Bentur *et al.* 1977; Deste-phano and Brady 1977; Stanley-Samuelson and Loher 1986; Stanley-Samuelson *et al.* 1986). These hormones increase vitellogenesis and oviposition. In *G. bimaculatus*, the receptor sites for prostaglandins are situated in the spermatheca (Bentur *et al.* 1977). Cordero (1995) speculated that chemicals in male ejaculates may represent sexually selected signals, arising either as arbitrary traits or as honest signals of male quality (Eberhard, this volume). Males with stronger chemical signals could be favored if oviposition was a graded response to the strength of the signal. However, the increased vitellogenesis of gryllids is an all-or-nothing response. Females only respond to prostaglandin after receiving doses from multiple matings (Simmons 1988b). The concordance of increased vitellogenesis with maximum fertilization rate seen in *G. bimaculatus* suggests that females respond in an adaptive manner to the levels of sperm in their stores, and thus their fertility.

The second reason for increased egg production in multiply mated females could be the acquisition of nutrients via spermatophore consumption. In Simmons' (1988b) study, some females were allowed to consume spermatophores and others were not. There was no difference in the number of eggs produced as a result of spermatophore consumption, but females allowed to consume spermatophores each day of their lives produced heavier eggs that were more likely to hatch. Similarly, Burpee and Sakaluk (1993b) showed that daily spermatophore consumption by female *G. veletis* and *Gryllodes sigillatus* increased longevity and, as a consequence, lifetime progeny production. However, Will and Sakaluk's (1994) study of *G. sigillatus* showed that this effect is not apparent for females fed up to three spermatophores in a day. Longevity was not increased by daily spermatophore consumption in Simmons' (1988b) study of *Gryllus bimaculatus*, although multiple mating has been reported to increase the longevity of starved female *Plebiogryllus guttiventris* (Bentur and Mathad 1975). Solymar and Cade (1990) showed that mating frequency has a genetic basis in female *G. integer*.

The costs associated with sampling males can theoretically prevent the evolution of female choice (Parker 1983; Pomiankowski 1987). However, because female gryllids can benefit from consuming multiple spermatophores without accepting sperm, costs of repeated sampling may be alleviated, facilitating the evolution of female preferences.

Non-random pairing

Recent field studies have used cross-sectional sampling to compare the traits of males found paired with females, with those of males that are calling to attract females. It is often assumed from these studies that paired males have been more successful in attracting females because of differences in their song. However, there is little evidence that song variability is associated with the phenotypic traits of paired and unpaired males. The patterns observed in cross-sectional sampling may equally arise from to decisions made by females after mate attraction; females remain and mate with males that have the preferred traits. Whatever the reason, the most striking result to emerge from these studies is the consistent tendency for paired males to be older than their calling conspecifics (Table 5-1). Male age is unlikely to be the direct focus of selection. However, at any given time, the relative age of potential mates may be correlated with other fitness traits, such as early and/or rapid development (Zuk 1988). Age may thus represent an indirect cue to the genetic quality of males. However, age could only be a reliable indicator of male quality in species with discrete generations; with increasing overlap of adult generations, any correlation between male age and fitness traits would break down because potential mating partners would include both old and young males from different generations that need not differ in quality.

The data in Table 5-1 are consistent with the hypothesis that females use male age as an indication of male quality for two reasons. First, non-random pairing for male age occurs in all three species that show univoltinism. The population of *G. bimaculatus* was in southern Spain, the northern limit of its distribution, where crickets are most likely bivoltine (Masaki and Walker 1987; Walker and Masaki 1989). Second, *T. oceanicus* shows no indication of non-random pairing based on male age. *T. oceanicus* is tropical in its distribution and breeds continuously in the population sampled. Male age would not be a reliable indicator of quality and is thereby unlikely to become the focus of mate selection.

Good evidence supports the idea that males successful in attracting females are of high genetic quality. With the exception of *G. veletis*, paired males tend to be larger than their calling conspecifics (Table 5-1). Body size in gryllids shows low heritability under controlled environmental conditions (Simmons 1987c; Webb and Roff 1992), but a much larger component of variance depends on environmental conditions, such as the intensity of competition during development (Simmons 1987a). Larval competition, at least in *G. bimaculatus*, is of the contest type and highly asymmetrical (Begon 1984); some good competitors achieve large adult size whereas other poor competitors eclose as small adults (Simmons 1987a). Thus, the phenotypic expression of large body size is likely to reflect overall fitness. Indeed, larger, early-emerging *G. campestris* males also exhibit a lower degree of fluctuating asymmetry (Simmons 1995), which may indicate underlying genetic quality (Palmer and Strobeck 1986; Parsons 1990; Møller 1993). Finally, the lower intensity of gregarine infections in paired *G. veletis* and *G. pennsylvanicus* compared with their calling conspecifics (Table 5-1) is consistent with the idea of variation in male genetic quality if low-quality individuals are less able to resist infection (Hamilton and Zuk 1982).

Female choice and courtship behavior

Courtship stridulation appears to be an important stimulus that male gryllids use to reduce the female's resistance to copulation. Silenced male *A. domesticus* are unable to elicit mounts from sexually receptive females, despite other aspects of courtship behavior remaining the same (Crankshaw 1979). By broadcasting courtship stridulation, Crankshaw (1979) was able to restore the ability of silenced males to elicit mounts. However, a similar experiment with the phalangopsid *Amphiacusta maya* showed that silenced males were as likely to elicit mounts as intact males (Boake 1984a). Calls produced during courtship in *A. maya* appear to function as aggressive signals directed at other males (Boake and Capranica 1982).

Information conveyed in courtship stridulation may bias the female's decision to copulate. Burk's (1983) study of *T. oceanicus* showed that males who won fights were more likely to produce courtship stridulation and were therefore more likely to copulate. However, on the few occasions when subordinate males did produce courtship song, females readily mated, suggesting that information within the courtship song does not influence mating. Boake (1984a) performed an elegant experiment with *A. maya* to establish whether females could assess male quality based on variation in courtship behavior. She paired each of five males with five females and was thus able to partition variance in the duration of courtship required to elicit a mount between males and females. Her analysis showed that variation in the duration of courtship was due to females rather than males, suggesting that females differed in their receptivity to mating rather than males varying in their quality as mates. Similarly, Simmons (1986c) observed competitive interactions among five *G. bimaculatus* males in an arena. When each of the males was paired in isolation with each of ten females, significant variation in the duration of courtship stridulation required before the female would mount was attributable to both males and females

(Fig. 5-2). As in Boake's (1984a) study, some females required extended courtship before mounting whereas others mounted almost immediately. However, the variation in courtship duration attributable to males supports the idea that males also vary in attractiveness to females. Multiple regression revealed that a significant proportion of male variation was explained by male size but not by success in competition (Simmons 1986c). Thus, females appear to obtain information about male quality during courtship.

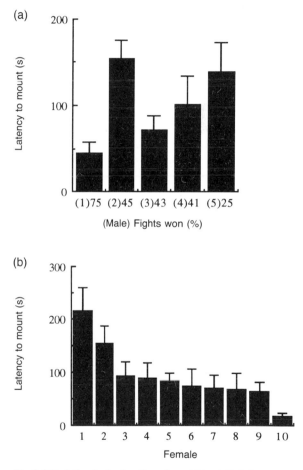

Fig. 5-2. Variation in the duration of courtship stridulation required to elicit a copulation (mean ± SE latency to mount) by female *Gryllus bimaculatus* partitioned between (a) females, $F_{(9,32)} = 3.36$, $p < 0.01$; and (b) males, $F_{(4,32)} = 4.28$, $p < 0.01$. A significant proportion of the variation due to males was explained by male size (regression, $F_{(1,44)} = 9.97$, $r^2 = 0.185$, $p < 0.01$) but not success in competition (entering the proportion of fights won into the regression increased r^2 by only 0.043, $F_{(2,43)} = 2.39$, NS) (Simmons 1986c).

Untangling the effects of intermale competition and female choice on non-random mating is a ubiquitous problem in studies of sexual selection (Halliday 1983; Partridge and Halliday 1984; Greenfield, this volume). In Burk's study of *T. oceanicus*, for example, non-random mating could be explained as a product of intermale competition without invoking active female discrimination; males avoided agonistic encounters by not producing courtship stridulation (Burk 1983). Nevertheless, studies of *G. bimaculatus* show that female choice determines non-random mating, irrespective of male competitive ability (Simmons 1986a, 1991a).

Postcopulatory guarding: sexual conflict over insemination?

After attaching a spermatophore to the female's genitalia, male crickets enter a guarding phase during which they attempt to remain in contact with the female. Postcopulatory guarding is characterized by mutual antennation and a series of aggressive responses by the male in response to female movement (Alexander and Otte 1967; Loher and Rence 1978; Simmons 1990b).

Three hypotheses have attempted to explain the significance of postcopulatory guarding in crickets. The ejaculate-protection hypothesis argues that males guard females to ensure the transfer of a full ejaculate (Gerhardt 1913). The spermatophore-replenishment hypothesis argues that males retain the female to facilitate repeated insemination (Khalifa 1950). Finally, the rival-exclusion hypothesis argues that males guard females to prevent them from copulating with additional males (Simmons 1986a; Sakaluk 1991). Although these hypotheses are not mutually exclusive, it is useful to consider them separately.

The ejaculate-protection hypothesis assumes that females will remove the spermatophore if the male is not present, and predicts a guarding duration sufficient to ensure the transfer of the ejaculate from the spermatophore. Four studies have compared spermatophore attachment times of guarded and unguarded females (Table 5-2). Three of the four studies support the ejaculate-protection hypothesis, insofar as unguarded females removed their spermatophores sooner than guarded females. Spermatophore function has been examined in *Acheta domesticus* (Khalifa 1949) and *G. bimaculatus* (Simmons 1986a). Once the spermatophore is attached, evacuating fluid passes through the inner crystalline layer and comes into contact with the pressure body, which swells, pushing the sperm mass out through the spermatophore tube

Table 5-2. *Spermatophore attachment durations (min) for guarded and unguarded female crickets (mean ± SE)pa*

Species	Guarded	Unguarded
Acheta domesticus[a,b]	36.0 ± 4.6	38.3 ± 5.5
Teleogryllus commodus[c]	71.3 ± 4.4	38.3 ± 5.5
Balamara gidya[c]	54.0 ± 9.1	36.5 ± 5.3
Gryllus bimaculatus[c]	70.4 ± 4.2	16.6 ± 6.4
Gryllodes sigillatus[c]	36.9 ± 4.4	32.5 ± 4.6

Sources: [a] Sakaluk and Cade 1983; [b] Khalifa 1950; [c] Evans 1988; [d] Simmons 1986a; [e] Sakaluk 1991.

and into the female's spermathecal duct (Fig. 5-3). For *G. bimaculatus*, the number of sperm in the female's spermatheca thereby accumulate with time. The mean duration of the guarding period appears to correspond with the time required for complete sperm transfer (Fig. 5-3).

According to the phylogeny presented by Gwynne (this volume), postcopulatory guarding in the Gryllidae had a single origin. However, a number of taxa appear to have evolved alternative means for protecting the ejaculate. In at least two South African *Teleogryllus* spp. and in *Gryllodes*

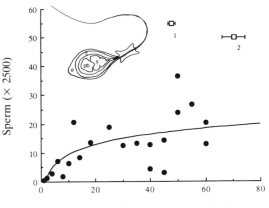

Fig. 5-3. Number of sperm transferred to the female's spermatheca in relation to the duration of spermatophore attachment for *Gryllus bimaculatus* (Sperm = 11.7 ln(attachment duration) +2.39; $F_{(1,19)} = 12.5$, $p < 0.001$; see Simmons (1986a) for more details). Inset are (1) the mean (±SE) duration of guarding, and (2) the mean spermatophore replenishment time. The internal structures of the spermatophore are shown as: c, crystalline layer; pb, pressure body; e, evacuating fluid; s, the sperm mass.

sigillatus, the male transfers a gelatinous spermatophylax attached to the ampulla of the spermatophore (Alexander and Otte 1967). Females consume the spermatophylax during insemination and at least in *G. sigillatus* the size of the spermatophylax is just sufficient to ensure complete transfer of the ejaculate (Sakaluk 1984). This type of ejaculate protection is widespread in tettigoniids (Gwynne, this volume; Wedell 1993). Other forms of nuptial feeding have also been implicated in ejaculate protection (Walker 1978; Bidochka and Snedden 1985; Loher and Dambach 1989) and in some cases appears to be associated with a secondary loss in postcopulatory guarding (Gwynne, this volume). In *Cycloptiloides canariensis* (Mogoplistinae), insemination takes only 31 s and there is no postcopulatory guarding (Dambach and Beck 1990). Alexander *et al.* (this volume) note that, within the Gryllinae, loss of postcopulatory guarding is also associated with the evolution of complex genitalia, which males use to hold onto the female after copulation. Such adaptations may prolong the mating association and perhaps spermatophore retention. Comparative data thus support the ejaculate-protection hypothesis. Interestingly, *G. sigillatus* shows nuptial feeding and has a guarding period after mating. The presence of a protective spermatophylax means that spermatophore attachment durations do not differ between guarded and unguarded females. In *G. sigillatus*, guarding appears to function to exclude rivals (Sakaluk 1991; Frankino and Sakaluk 1994).

The spermatophore-replenishment hypothesis for postcopulatory guarding was first suggested by Khalifa (1950) after he found that spermatophore attachment times for *A. domesticus* were independent of male guarding behavior (Table 5-2). In their survey of over 20 genera, Alexander and Otte (1967) concluded that no crickets have evolved a spermatophore large enough to inseminate a female fully because all species studied copulated repeatedly if left undisturbed. The spermatophore-replenishment hypothesis predicts that the time required for a male to produce a second spermatophore should be less than or equal to the duration of the guarding period. Males would thus be in a position to transfer a second spermatophore immediately after the first is evacuated. Khalifa's (1950) data provide a spermatophore-replenishment time of 33.2 ± 4.94 min, which is shorter than the duration of spermatophore attachment for guarded and unguarded females (Table 5-2) and shorter than the actual guarding duration of 52.2 ± 4.48 min. For *Gryllus bimaculatus*, spermatophore-replenishment times depend on male size (Simmons 1988c) but the mean time is less than that

required for the first spermatophore to empty (Fig. 5-3). Loher and Rence (1978) provide similar data for *T. commodus*. The data therefore support the spermatophore-replenishment hypothesis.

The spermatophore-replenishment hypothesis relies on the assumption that males benefit by copulating repeatedly with the same female. The spermatheca of gryllines is typically elastic and increases in size with successive matings (Sakaluk 1986; Simmons 1986a; Loher and Dambach 1989), suggesting that females store most of the sperm received from repeated matings. If sperm competition conforms to a 'raffle' (*sensu* Parker *et al.* 1990) then a given male's success in fertilizing eggs will increase with the proportion of his sperm in the female's spermatheca. The five studies of sperm competition in gryllids all indicate sperm mixing. When females mate twice, the expected proportion of eggs fertilized by the second male (P_2) will be 0.50 when there is no displacement and sperm mix randomly. P_2 is 0.72 for *G. integer* (Backus and Cade 1986), 0.60 for *Gryllodes sigillatus* (Sakaluk 1986), 0.62 and 0.43 for *Allenomobius fasciatus* and *A. socius*, respectively (Gregory and Howard 1994), and 0.45 for *Gryllus bimaculatus* (Simmons 1987d). In *G. bimaculatus*, sperm accumulate in the spermatheca with repeated matings (Simmons 1986a), indicating that sperm are not displaced. Formal tests of the sperm-mixing hypothesis in *G. bimaculatus* gave a significant fit to a model of random sperm mixing (Parker *et al.* 1990), whereas in *Gryllodes sigillatus* random mixing appears to be accompanied by partial displacement (Sakaluk and Eggert 1996). Simmons (1987d) also showed how male fertilization success increased with successive matings of the same female in a pattern predicted from random sperm mixing.

Most studies consider only the interests of the male in postcopulatory guarding, but why should females remove spermatophores before the ejaculate has been transferred? Thornhill and Alcock (1983) suggested that females assess male quality during postcopulatory guarding. They argued that by constantly moving and attempting to remove spermatophores, females avoid being inseminated by males of low quality, if such males are incompetent guarders. Thus, the length of the guarding period may represent the outcome of a sexual conflict over insemination, the duration of postcopulatory guarding depending on the relative abilities of males and females to exert their interests over one another (Simmons 1991b; see also Alexander *et al.*, this volume). Zuk (1987b) similarly argued that females could assess male health and vigor during postcopulatory guarding, and showed that spermatophore production rates were negatively associated with the levels of gregarine infections in *Gryllus veletis* and *G. pennsylvanicus*. By removing spermatophores, a female can ensure that uninfected males would transfer more sperm because infected males are unable to replenish spermatophores quickly. Intraspecific variation in postcopulatory guarding has been studied only in *Gryllus bimaculatus* (Simmons 1986a, 1989, 1990b, 1991a).

The sexual-conflict hypothesis for guarding as a mechanism of female choice predicts that females should treat all males equally; females assess male quality by their ability to guard successfully. Females should therefore spend more time with competent guarders, and spermatophores will remain attached for longer. In seminatural conditions, the duration of the guarding period in *G. bimaculatus* increases with male size (Fig. 5-4a). Consequently, females leave spermatophores of large males attached for longer and accept several spermatophores from large males (Simmons 1986a). However, when females were confined in small enclosures with their mates, thereby preventing them from successfully escaping, there was no relationship between male size and the frequency with which females attempted to leave (Fig. 5-4b), or the time that females spent attempting to leave (Fig. 5-4c). These data support the sexual-conflict hypothesis. Similar results were obtained from a study of the affect of gregarine infections on guarding behavior: males with heavy infections had shorter guarding periods under seminatural conditions although the behavior of females confined with males during guarding was not related to levels of infection (Simmons 1990b). The data for *G. bimaculatus* thus show that postcopulatory female choice via interference with insemination is an important factor in the evolution of mate-guarding by males. Indeed, postcopulatory guarding could be considered as a means of coercive insemination (Clutton-Brock and Parker 1995).

If females could assess quality directly, they should not attempt to interfere with insemination by high-quality males. Female *G. bimaculatus* given males matched for size but differing in their degree of relatedness to the female attempted to leave significantly less when the male was unrelated (Fig. 5-5). Accordingly, the duration of the guarding period and spermatophore attachment durations vary with the degree of relatedness between male and female (Simmons 1991a). Coupled with the greater probability of females mating with unrelated males, which they assess by using pheromones (Simmons 1989, 1990a, 1991a), the data suggest that when females can assess males directly the conflict over insemination is reduced. Females

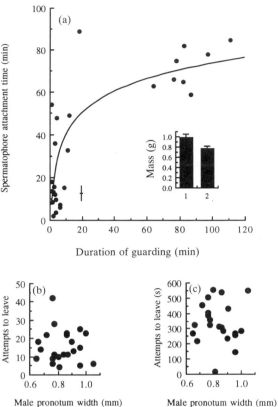

Fig. 5-4. Intraspecific variation in the guarding period of *Grylllus bimaculatus*. (a) The relationship between the time that females spend with males in their burrows and the duration of spermatophore attachment. Note the strict dichotomy between females that leave males within 15 min of mating or remain with them for the complete guarding period. Females that leave remove their spermatophores early (see Table 5-2). The guarding period does not occur when copulation is in the open (Simmons 1986a) and females remove spermatophores early (horizontal bar). Inset is the mean (±SE) mass of males that females left compared with those with whom they stayed ($t = 2.5$, df 22, $p = 0.02$). (b) The frequency with which females attempt to leave guarding males during the first 15 min after copulation in relation to male size ($r = 0.063$, NS). (c) The time spent by females attempting to leave during the first 15 min after copulation in relation to male size ($r = 0.000$, NS).

presumably benefit from mating with unrelated individuals by avoiding inbreeding.

Postcopulatory female choice

Postcopulatory female choice, originally termed 'cryptic female choice' by Thornhill (1983), can occur in two ways.

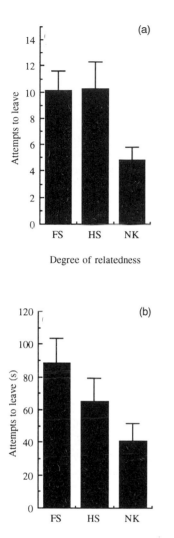

Fig. 5-5. Variation in the behavior of female *Gryllus bimaculatus* guarded by males that vary in their degree of relatedness: FS, full siblings; HS, half siblings; NK, non-kin (Simmons 1989). (a) mean (±SE) frequency with which females attempt to leave during the first 15 min after copulation; (b) mean (±SE) time spent attempting to leave during the first 15 min after copulation.

First, because mating and fertilization are temporally separated events in insects, females have the potential to use multiple mating to bias paternity in favor of preferred males by differential acceptance of ejaculates. The postcopulatory behavior of female gryllines results in longer spermatophore attachments and repeated matings for males with certain traits. Female behavior then has the effect of

biasing paternity toward preferred males because sperm stores contain a greater proportion of sperm from those males (Simmons 1987d). So far, paternity assessment in crickets, and indeed most insects, has used the sterile-male technique, which only allows the relative paternities of two males to be assessed. Nevertheless, understanding the mechanism of sperm competition in *G. bimaculatus* (Parker *et al.* 1990) allows us to calculate the relative paternities of several males based on the known rates of sperm transfer from spermatophores and the relative number of matings and spermatophore attachment durations achieved by each mate. These calculations show that behavioral aspects of female choice bias paternity towards preferred males (Simmons 1991a).

Sakuluk and Eggert (1996) recently showed that experimental manipulation of ampulla attachment in *Gryllodes sigillatus* similarly influences male paternity. The protective spermatophylax that the female must bypass before ampulla removal has, on average, evolved to a size necessary to ensure complete insemination (Sakaluk 1984) and can be considered a structure evolved for coercive insemination since it serves to promote male interests. Nevertheless, Sakaluk and Eggert (1996) argue that males who produce smaller spermatophylaxes than average will suffer premature ampulla removal so that females can still bias paternity in favor of males providing large spermatophylaxes. However, recent work by Gage and Barnard (1996) showed that spermatophylax size varies with ejaculate size, so that smaller than average spermatophylaxes will fully protect a smaller ejaculate. Variation in ejaculate size appears to be an adaptive response by males to the risks of sperm competition (Gage and Barnard 1996).

A second way in which females can exercise postcopulatory or cryptic choice is via differential reproductive investment after matings with males that vary in quality. Females faced with a trade-off between current and future reproduction (Williams 1966; Stearns 1992) should invest so as to maximize reproductive fitness. If female choice provides indirect fitness benefits, then females should invest more heavily in reproduction after mating with males that will sire fitter offspring (Burley 1988). A number of studies have reported differential patterns of reproductive investment by females following copulations with apparently preferred males (Thornhill 1983; Hughes and Hughes 1985; McLain and Marsh 1990). Female *Gryllus bimaculatus* allowed to choose their mates laid a greater total number of eggs and a greater proportion of their clutch than when allocated a mate (Fig. 5-6). Although females invested

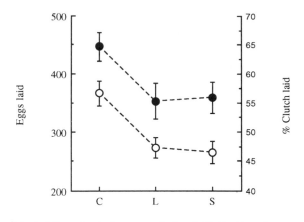

Fig. 5-6. Reproductive investment by female *Gryllus bimaculatus* allocated large (L) or small (S) males and those allowed to choose (C) their own mates. Closed symbols give the mean (±SE) total number of eggs laid ($F_{(2,55)} = 3.54, p < 0.05$) and open symbols the mean (±SE) percentage of clutch laid ($F_{(2,55)} = 7.99$, $p < 0.001$).

more heavily in oviposition when allowed to choose among males, the proportion of eggs that hatched was the same (Simmons 1987b).

These patterns of postcopulatory female choice illustrate the problems with estimating male reproductive success on the basis of mating success. At least in gryllids, and probably in many other taxa, the relationship between mating success and reproductive success is complex and more strongly influenced by female choice than generally considered (Eberhard, this volume).

Costs of female choice

Parker (1983) showed theoretically that, when the costs of mate choice are high relative to the benefits obtained, females should be indiscriminate. Female gryllids and tettigoniids suffer an increased risk of predation as a direct consequence of seeking out males (Gwynne and Dodson 1983; Sakaluk and Belwood 1984; Heller 1992). The longer a female spends searching in an open environment, the greater her risk of predation and her costs of choice. These costs can sometimes outweigh the benefits of female choice, resulting in a loss of discrimination; female *G. integer* failed to express their usual preference for males with long calling-bout lengths when searching in an open environment (Hedrick and Dill 1993). Similarly, when female *G. bimaculatus* encounter silently searching males in the open, they mate quickly and without regard for the

traits that influence mating decisions when copulations occur in the seclusion of a burrow (Simmons 1986a, 1991a). Further, they routinely leave males immediately after mating and remove spermatophores early (Fig. 5-4a). The inevitable consequence for males is that searching, as an alternative strategy, will have a relatively low pay-off in terms of fertilizations. Why females should mate at all with males in the open is a puzzle; perhaps the time costs of avoiding persistently courting males are greater than mating quickly and removing spermatophores. Moreover, females may gain long-term benefits from spermatophore consumption (Simmons 1988b).

Benefits of female choice

The two theories for the evolution of female preferences are distinguished by the mode of action of selection. Selection can be either direct, in which case females obtain some direct benefit from choosing particular males, or indirect, with the benefits of choice manifested as fitness advantages for resulting offspring (Kirkpatrick and Ryan 1991; Maynard Smith 1991).

Female choice in gryllids has been studied mainly in species where the male provides only sperm at mating and any benefits of choice must be indirect. Models that have been proposed to explain indirect selection for female preferences can be broadly classified as either Fisherian or good-genes models (Kirkpatrick and Ryan 1991). Fisherian models predict that a genetic correlation should arise between the preference in females and the chosen traits of males. Females choose males with some arbitrary trait, such as a characteristic of the call, and thereby produce sons that have attractive calls and daughters that exhibit the preference. The models rely on the extent to which traits are heritable. Some evidence is available from the gryllines for heritability of attractive male traits. Calling-bout length in *G. integer* is highly heritable in males (Hedrick 1988) and is subject to selection via female choice (Hedrick 1986). Whether genetic correlations exist between male traits and female preferences for them remains to be tested, although early work with *T. commodus* and *T. oceanicus* (Hoy and Paul 1973; Hoy 1974; Hoy *et al.* 1977) suggests that this may prove a fruitful area of research.

Good-genes models of female choice assume heritability in fitness traits. Females are proposed to choose males on the basis of traits that provide information pertaining to the genetic quality of individuals and females gain indirect benefits by increasing the fitness of their offspring. There is some evidence that good-genes female choice occurs in gryllids. The covariation between male size, timing of emergence and fluctuating asymmetry in *G. campestris* discussed above supports the idea that males vary in their underlying genetic quality. In an experiment designed to test for indirect benefits of choice in *G. bimaculatus*, Simmons (1987b) either assigned females a mate or allowed them to choose their own mating partners and measured the development rates, survival and adult sizes of resulting offspring. Females allowed to choose their own mating partners produced offspring that developed more rapidly and began their own reproduction sooner than those females allocated either large or small mates (Fig. 5-7). Females under the regime of free choice mated less frequently than those allocated mates, negating the argument that differences in offspring fitness arose from benefits obtained through multiple mating (Simmons 1987b). That females allowed to choose their mates produce offspring of superior quality supports the idea that mate choice, at least in *G. bimaculatus*, has evolved under indirect

selection, and represents one of the few demonstrations of an adaptive benefit to mate choice (see also Watt *et al.* 1986). Further, the finding that females allocated large males as mates produce offspring of inferior quality suggests that paternal size *per se* does not contribute to offspring fitness. The results from the study of *G. campestris* suggest that size is one trait that may be correlated with underlying genetic quality, but females do not choose solely on the basis of this trait (Simmons 1995).

Few studies of gryllids have examined the evolution of preferences under direct selection, although a number of species provide females with more than sperm at mating. The spermatophylax of *Gryllodes sigillatus* is consumed by the female during insemination and larger males produce larger spermatophores (Sakaluk 1985). However, the spermatophylax does not provide the female with significant nutritional benefits and there is no evidence that females prefer larger males as mates (Will and Sakaluk 1994). Some tree crickets (Oecanthinae) feed their mates from metanotal glands (Walker 1978) and some ground crickets

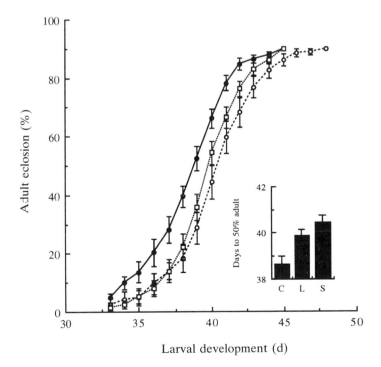

Fig. 5-7. Cumulative plot of the mean (±SE) % of total offspring reaching adult eclosion with time since hatching for female *G. bimaculatus* allocated large (L, open squares) or small (S, open circles) males compared with females allowed to choose their mates (C, closed circles). Inset is the mean time to 50% adult eclosion ($F_{(2,46)} = 5.41$, $p < 0.01$; see Simmons (1987b).

(Nemobiinae) from tibial spurs (Bidochka and Snedden 1985; Forrest *et al.* 1991). We might expect to see cases of female choice for direct benefits in these nuptial-feeding taxa. One good example of female choice for direct benefits has been documented in the mole crickets *S. acletus* and *S. vicinus*. Females orient to the loudest calls, which are an indicator of high moisture content in the soil of a male's burrow, an important determinant of egg development (Forrest 1983).

DISCUSSION

Alexander *et al.* (this volume) identify eight phases in insect mating sequences where conflict and confluences of interests between the sexes may shape mating behavior. Male gryllids provide no parental care, so that their interests should always be to achieve fertilization with any female. The interests of females, however, may not always be congruent with their male conspecifics. Evidence that female gryllids are not always willing to mate is seen in several phases of the mating sequence, although perhaps the least so at rapprochement. Despite extensive study, few elements of male song variation used by females in mate choice have been found, a result paralleled in studies of morphological correlates of mating success in lekking vertebrates (Hoglund and Alatalo 1995). Perhaps this lack exists because the costs of female choice during phonotaxis exceed the benefits of choice or perhaps, as Alexander *et al.* suggest, signaler and receiver have evolved to maximize the effective range of communication, an aspect paramount for both sexes. Alternatively, in crickets and lekking vertebrates alike, the effort of joining the lek or aggregation of callers may be sufficiently rigorous that some selection occurs before females have males from which to choose among. Once a female has been attracted, courtship songs produced by males are essential in luring the female to accept a mate.

The operational sex ratio also influences cricket mating systems; Souroukis and Murray (1995) demonstrated that female *Gryllus pennsylvanicus* were more discriminating at even or male-biased sex ratios, and moved least at a female-biased sex ratio. As with other taxa, grylline mating systems are probably more flexible than has previously been supposed, with factors such as perceived predation risk, population density, and structure of the environment providing proximate influences on male and female mating strategies (Hedrick and Dill 1993; Hissman 1990; Simmons 1988c). These findings underline the need

to study different populations of the same species under varying conditions.

Alexander *et al.* (this volume) discuss conflict and confluences of interest with regard the evolution of the copulatory phase in gryllids; in short, gryllids generally have simple genitalia that show few signs of evolution via sexual conflict. The most obvious phase of sexual conflict in gryllids occurs after copulation. Postcopulatory mate guarding by male gryllids functions to increase sperm transfer and resultant paternity. That postcopulatory guarding behavior has evolved reflects that insemination by the copulating male does not always serve female interests. Females appear to be ahead in the game: males appear unable to forcibly inseminate females by guarding. None the less, some species may have evolved more efficient means of ensuring insemination in the form of genital structures for prolonging mating associations (Alexander *et al.*, this volume) or protective spermatophylaxes attached to the spermatophore.

The final phase at which females can potentially exert their control is at fertilization. Available evidence, however, suggests that sperm are simply utilized at random from the female's spermatheca. Paternity appears to be influenced before fertilization, during the conflict over insemination. Although Alexander *et al.* (this volume) argue that sperm precedence generally favors the last male to copulate, this is not supported by the data for gryllids. They further claim that when sperm mix, females have less opportunity to choose among males. We would argue the reverse. Strong sperm precedence of the second male is more a reflection of male control, achieved either by sperm displacement or by repositioning. Once copulation is complete the female can have no further control over fertilization because she has sperm from only a single male. When sperm mix in storage, female gryllids have greater opportunity to favor particular males at fertilization by controlling the amount of sperm inseminated by different males. Further, as Alexander *et al.* (this volume) note, sperm mixing would facilitate a best-of-*n* mate choice strategy that might otherwise not be possible. If sperm competitive ability is positively associated with male quality, females may benefit considerably by staging competitions between the ejaculates of different males (Keller and Reeve 1995). Our current understanding of sperm dynamics is too limited to say whether sperm selection by females occurs in gryllids, or indeed in any other taxa. Nevertheless, there is evidence that sperm utilization is biased towards conspecific males when hybrid matings occur between *A. fasciatus* and *A. socius* (Gregory and Howard 1994). Whether homogamy is the consequence of conspecific sperm outcompeting heterospecific sperm, or whether females actively select conspecific sperm for fertilization, remains a fascinating question.

SUGGESTIONS FOR FUTURE RESEARCH

To date, field studies of non-random pairing have been cross-sectional, with the assumption that calling males are unsuccessful in attracting mates. The next step is to perform longitudinal studies in natural populations to determine whether the traits identified as important cues for non-random pairing actually correlate with the lifetime reproductive success of individuals. This will be difficult given the extent of female multiple mating and the fact that, at least in those few species that have been studied, sperm mixes in storage. Nevertheless, rapidly developing molecular techniques such as DNA fingerprinting should allow us to accurately assign paternity to a large number of putative fathers, and should thereby allow us to examine how behavioral aspects of female choice determine male reproductive success.

Individual variation is at the heart of evolutionary theory, but little is known about how male crickets differ in temporal structure and daily patterning of calling. If density affects the calling rates and timing of individuals, for example, we should be able to use current models of male clustering developed for lekking species to predict how male mating success varies with group size, amount of time spent calling, and call structure.

The largest comparative data set available to us concerned the function of postcopulatory guarding. However, this could be greatly extended. For example, the spermatophore-replenishment hypothesis assumes that there is an advantage to males for mating repeatedly with the same female. We predict that the duration of mating associations across species will be related to the degree of sperm mixing; the males of species in which sperm precedence is high should mate only once with the female, whereas the duration of mating associations should increase with increasing degree of sperm mixing. Currently we have no data on the duration of mating associations and only limited information on patterns of sperm competition. Finally, although nuptial feeding occurs in several gryllids, there are no detailed studies of female choice for immediate benefits.

ACKNOWLEDGEMENTS

We are grateful to Christine Boake for Fig. 5-1. W. H. Cade provided helpful comments on an earlier draft of this chapter.

LITERATURE CITED

Alexander, R. D. 1961. Aggressiveness, territoriality, and sexual behaviour in field crickets (Orthoptera:Gryllidae). *Behaviour* **17**: 130–223.

–. 1975. Natural selection and specialized chorusing behavior in acoustical insects. In *Insects, Science, and Society*. D. Pimentel, ed., pp. 35–47. New York: Academic Press.

Alexander, R. D. and G. H. Meral. 1967. Seasonal and daily chirping cycles in the northern spring and fall crickets, *Gryllus veletis* and *G. pennsylvanicus*. *Ohio J. Sci.* **67**: 200–209.

Alexander, R. D. and D. Otte. 1967. The evolution of genitalia and mating behavior in crickets (Gryllidae) and other Orthoptera. *Misc. Pub. Mus. Zool. Univ. Mich.* **133**: 1–59.

Backus, V. L. and W. H. Cade. 1986. Sperm competition in the field cricket *Gryllus integer* (Orthoptera: Gryllidae). *Fla. Entomol.* **69**: 722–728.

Begon, M. 1984. Density and individual fitness: asymmetric competition. In *Evolutionary Ecology*. B. Shorrocks, ed., pp. 175–194. Oxford: Oxford University Press.

Bell, P. D. 1979. Acoustic attraction of herons by crickets. *J. N. Y. Entomol. Soc.* **87**: 126–127.

Bennet-Clark, H. C. 1989. Songs and the physics of sound production. In *Cricket Behavior and Neurobiology*. F. Huber, T. E. Moore and W. Loher, eds., pp. 227–261. Ithaca, New York: Cornell University Press.

Bentur, J. S., K. Dakshayani and S. B. Mathad. 1977. Mating induced oviposition and egg production in the crickets, *Gryllus bimaculatus* De Geer and *Plebeiogryllus guttiventris* Walker. *Z. Angew. Entomol.* **84**: 129–135.

Bentur, J. S. and S. B. Mathad. 1975. Dual role of mating in egg production and survival in the cricket, *Plebeigryllus guttiventris* Walker. *Experientia* **31**: 539–540.

Bidochka, M. J. and W. A. Snedden. 1985. Effect of nuptial feeding on the mating behaviour of female ground crickets. *Can. J. Zool.* **63**: 207–208.

Boake, C. R. B. 1983. Mating systems and signals in crickets. In *Orthopteran Mating Systems: Sexual Competition in a Diverse Group of Insects*. D. T. Gwynne and G. K. Morris, eds., pp. 28–44. Boulder, Colorado: Westview Press.

–. 1984a. Male displays and female preferences in the courtship of a gregarious cricket. *Anim. Behav.* **32**: 690–697.

–. 1984b. Natural history and acoustic behaviour of a gregarious cricket. *Behaviour* **89**: 241–250.

Boake, C. R. B. and R. R. Capranica. 1982. Aggressive signal in 'courtship' chirps of a gregarious cricket. *Science (Wash., D.C.)* **218**: 580–582.

Bradbury, J. W. and R. M. Gibson. 1983. Leks and mate choice. In *Mate choice*. P. Bateson, ed., pp. 109–140. Cambridge University Press.

Burk, T. 1982. Evolutionary significance of predation on sexually signaling males. *Fla. Entomol.* **65**: 90–104.

–. 1983. Male aggression and female choice in a field cricket *(Teleogryllus oceanicus)*: the importance of courtship song. In *Orthopteran Mating Systems: Sexual Competition in a Diverse Group of Insects.* D. T. Gwynne and G. K. Morris, eds., pp. 97–119. Boulder, Colorado: Westview Press.

Burley, N. 1988. The differential allocation hypothesis: an experimental test. *Am. Nat.* **132**: 611–628.

Burpee, D. M. and S. K. Sakaluk. 1993a. The effect of pair formation on diel calling patterns in two cricket species, *Gryllus veletis* and *Gryllodes sigillatus* (Orthoptera: Gryllidae). *J. Insect Behav.* **6**: 431–440.

–. 1993b. Repeated matings offset costs of reproduction in female crickets. *Evol. Ecol.* **7**: 240–250.

Cade, W. H. 1975. Acoustically orienting parasitoids: fly phonotaxis to cricket song. *Science (Wash., D.C.)* **190**: 1312–1313.

–. 1979. The evolution of alternative male reproductive strategies in field crickets. In *Sexual Selection and Reproductive Competition in Insects.* M. S. Blum and N. A. Blum, eds., pp. 343–380. London: Academic Press.

–. 1981. Field cricket spacing, and the phonotaxis of crickets and parasitoid flies to clumped and isolated cricket songs. *Z. Tierpsychol.* **55**: 365–375.

–. 1984a. Effects of fly parasitoids on nightly duration of calling in field crickets. *Can. J. Zool.* **62**: 226–228.

–. 1984b. Genetic variation underlying sexual behavior and reproduction. *Am. Zool.* **24**: 355–366.

Cade, W. H. and E. A. Cade. 1992. Male mating success, calling and searching behaviour at high and low densities in the field cricket, *Gryllus integer. Anim. Behav.* **43**: 49–56.

Cade, W. H. and D. R. Wyatt. 1984. Factors affecting calling behaviour in field crickets, *Teleogryllus* and *Gryllus* (age, weight, density, and parasites). *Behaviour* **88**: 61–75.

Campbell, D. J. 1990. Resolution of spacial complexity in a field sample of singing crickets *Teleogryllus commodus* (Walker) (Gryllidae): a nearest-neighbour analysis. *Anim. Behav.* **39**: 1051–1057.

Campbell, D. J. and D. J. Clark. 1971. Nearest neighbour tests of significance for non-randomness in the spatial distribution of singing crickets (*Teleogryllus commodus* Walker). *Anim. Behav.* **19**: 750–756.

Campbell, D. J. and E. Shipp. 1979. Regulation of spatial pattern in populations of the field cricket *Teleogryllus commodus* (Walker). *Z. Tierpsychol.* **51**: 260–268.

Clutton-Brock, T. H. and Parker, G. A. 1995. Sexual coercion in animal societies. *Anim. Behav.* **49**: 1345–1365.

Cordero, C. 1995. Ejaculate substances that affect female insect reproductive physiology and behavior – honest or arbitrary traits. *J. Theor. Biol.* **174**: 453–461.

Crankshaw, O. S. 1979. Female choice in relation to calling and courtship songs in *Acheta domesticus. Anim. Behav.* **27**: 1274–1275.

Dambach, M. and U. Beck. 1990. Mating in the scaly cricket *Cycloptiloides canariensis* (Orthoptera: Gryllidae: Mogoplistinae). *Ethology* **85**: 289–301.

Destephano, D. B. and U. E. Brady. 1977. Prostaglandin and prostaglandin synthetase in the cricket, *Acheta domesticus. J. Insect Physiol.* **23**: 905–611.

Dixon, K. A. and W. H. Cade. 1986. Some factors influencing male-male aggression in the field cricket *Gryllus integer* (time of day, age, weight and sexual maturity). *Anim. Behav.* **34**: 340–346.

Doherty, J. A. 1985a. Temperature coupling and trade-off phenomena in the acoustic communication system of the cricket, *Gryllus bimaculatus* DeGeer (Gryllidae). *J. Exp. Biol.* **114**: 17–35.

–. 1985b. Trade-off phenomena in calling song recognition and phonotaxis in the cricket, *Gryllus bimaculatus* (Orthoptera, Gryllidae). *J. Comp. Physiol.* **156**: 787–801.

Doherty, J. A. and R. Hoy. 1985. Communication in insects. III. The auditory behaviour of crickets: some views on genetic coupling, song recognition, and predator detection. *Q. Rev. Biol.* **60**: 457–472.

Doolan, M. J. and G. S. Pollack. 1985. Phonotactic specificity of the cricket *Teleogryllus oceanicus*: intensity-dependent selectivity for temporal parameters of the stimulus. *J. Comp. Physiol.* **157**: 223–233.

Evans, A. R. 1988. Mating systems and reproductive strategies in three Australian gryllid crickets: *Bobilla victoriae* Otte, *Balamara gidya* Otte and *Teleogryllus commodus* (Walker) (Orthoptera: Gryllidae: Nemobiinae; Gryllinae). *Ethology* **78**: 21–52.

Fabre, J. H. 1917. *The Life of the Grasshopper.* London: Hodder and Stoughton.

Forrest, T. G. 1983. Calling songs and mate choice in mole crickets. In *Orthopteran Mating Systems: Sexual Competition in a Diverse Group of Insects.* D. T. Gwynne and G. K. Morris, eds., pp. 185–204. Boulder, Colorado: Westview Press.

Forrest, T. G., J. L. Sylvester, S. Testa, S. W. Smith, A. Dinep, T. L. Cupit, J. M. Huggins, K. L. Atkins and M. Eubanks. 1991. Mate choice in ground crickets (Gryllidae: Nemobiinae). *Fla. Entomol.* **74**: 74–80.

Frankino, W. A. and S. K. Sakaluk. 1994. Post-copulatory mate guarding delays promiscuous mating by female decorated crickets. *Anim. Behav.* **48**: 1479–1481.

French, B. W. and W. H. Cade. 1987. The timing of calling, movement, and mating in the field crickets *Gryllus veletis*, *G. pennsylvanicus*, and *G. integer. Behav. Ecol. Sociobiol.* **21**: 157–162.

–. 1989. Sexual selection at varying population densities in male field crickets, *Gryllus veletis* and *G. pennsylvanicus. J. Insect Behav.* **2**: 105–122.

French, B. W., E. J. McGowan and V. L. Backus. 1986. Spatial distribution of calling field crickets, *Gryllus pennsylvanicus* (Bigelow) (Orthoptera: Gryllidae). *Fla. Entomol.* **69**: 255–257.

Gage, A. R. and Barnard, C. J. 1996. Male crickets increase sperm number in relation to competition and female size. *Behav. Ecol. Sociobiol.* **38**: 349–353.

Gerhardt, U. 1913. Copulation and Spermatophoren von Grylliden und Locustiden. *Zool. Jahrb. Abt. Syst. Oekol. Geogr. Tiere* **35**: 461–531.

Gibson, R. M. and J. W. Bradbury. 1985. Sexual selection in lekking sage grouse: phenotypic correlates of male mating success. *Behav. Ecol. Sociobiol.* **18**: 117–123.

Gibson, R. M., J. W. Bradbury and S. L. Vehrencamp. 1991. Mate choice in lekking sage grouse revisited: the roles of vocal display, female site fidelity, and copying. *Behav. Ecol.* **2**: 165–180.

Gregory, P. G. and D. J. Howard. 1994. A postinsemination barrier to fertilization isolates two closely related ground crickets. *Evolution* **48**: 705–710.

Gwynne, D. T. 1981. Sexual difference theory: Mormon crickets show role reversal in mate choice. *Science (Wash., D.C.)* **213**: 779–780.

–. 1984. Sexual selection and sexual differences in Mormon crickets (Orthoptera: Tettigoniidae, *Anabrus simplex*). *Evolution* **38**: 1011–1022.

Gwynne, D. T. and G. N. Dodson. 1983. Non-random provisioning by the digger wasp *Palmodes laeviventris* (Hymenoptera: Sphecidae). *Ann. Entomol. Soc. Am.* **76**: 434–436.

Gwynne, D. T. and G. K. Morris, eds. 1983. *Orthopteran Mating Systems: Sexual Competition in a Diverse Group of Insects.* Boulder, Colorado: Westview Press.

Halliday, T. R. 1983. The study of mate choice. In *Mate Choice.* P. Bateson, ed., pp. 3–32. Cambridge University Press.

Hamilton, W. D. and M. Zuk. 1982. Heritable true fitness and bright birds: a role for parasites? *Science (Wash., D.C.)* **218**: 384–387.

Hedrick, A. V. 1986. Female preferences for male calling bout duration in a field cricket. *Behav. Ecol. Sociobiol.* **19**: 73–77.

–. 1988. Female choice and the heritability of attractive male traits: an empirical study. *Am. Nat.* **132**: 267–276.

Hedrick, A. V. and L. M. Dill. 1993. Mate choice by female crickets is influenced by predation risk. *Anim. Behav.* **46**: 193–196.

Heller, K.-G. 1992. Risk shift between males and females in the pair-forming behavior of bushcrickets. *Naturwissenschaften* **79**: 89–91.

Hissmann, K. 1990. Strategies of mate finding in the European field cricket (*Gryllus campestris*) at different population densities: a field study. *Ecol. Entomol.* **15**: 281–291.

Höglund, J. and R. V. Alatalo. 1995. *Leks.* Princeton, New Jersey: Princeton University Press.

Hoy, R. R. 1974. Genetic control of acoustic behavior in crickets. *Am. Zool.* **14**: 1067–1080.

–. 1991. Signals for survival in the lives of crickets. *Am. Zool.* **31**: 297–305.

Hoy, R. R., J. Hahn and R. C. Paul. 1977. Hybrid cricket auditory behavior: evidence for genetic coupling in animal communication. *Science (Wash., D.C.)* **195**: 82–84.

Hoy, R. R. and R. C. Paul. 1973. Genetic control of song specificity in crickets. *Science (Wash., D.C.)* **180**: 82–83.

Huber, F., T. E. Moore and W. Loher. 1989. *Cricket Behavior and Neurobiology.* Ithaca, New York: Cornell University Press.

Hughes, A. L. and M. K. Hughes. 1985. Female choice of males in a polygynous insect, the whitespotted sawyer, *Monochomus scutellatus. Behav. Ecol. Sociobiol.* **17**: 385–388.

Keller, L. and H. K. Reeve. 1995. Why do females mate with multiple males? The sexually selected sperm hypothesis. *Adv. Stud. Behav.* **24**: 291–315.

Khalifa, A. 1949. The mechanism of insemination and the mode of action of the spermatophore in *Gryllus domesticus. Q. J. Microsc. Sci.* **90**: 281–292.

–. 1950. Sexual behaviour in *Gryllus domesticus* L. *Behaviour* **2**: 264–274.

Kirkpatrick, M. and M. J. Ryan. 1991. The evolution of mating preferences and the paradox of the lek. *Nature (Lond.)* **350**: 33–38.

Kutsch, W. and F. Huber. 1989. Neural Basis of song production. In *Cricket Behavior and Neurobiology.* F. Huber, T. E. Moore and W. Loher, eds., pp. 262–309. Ithaca, New York: Cornell University Press.

Loher, W. 1989. Temporal organization of reproductive behavior. In *Cricket Behavior and Neurobiology.* F. Huber, T. E. Moore and W. Loher, eds., pp. 83–113. Ithaca, New York: Cornell University Press.

Loher, W. and M. Dambach. 1989. Reproductive behavior. In *Cricket Behavior and Neurobiology.* F. Huber, T. E. Moore and W. Loher, eds., pp. 43–82. Ithaca, New York: Cornell University Press.

Loher, W. and B. Rence. 1978. The mating behavior of *Teleogryllus commodus* (Walker) and its central and peripheral control. *Z. Tierpsychol.* **46**: 225–259.

Masaki, S. and T. J. Walker. 1987. Cricket life cycles. *Evol. Biol.* **21**: 349–423.

Maynard Smith, J. 1991. Theories of sexual selection. *Trends Ecol. Evol.* **6**: 146–151.

McLain, D. K. and N. B. Marsh. 1990. Male copulatory success: Heritability and relationship to mate fecundity in the southern green stinkbug, *Nezara viridula* (Hemiptera: Pentatomidae). *Heredity* **64**: 161–167.

Michelsen, A. 1985. Environmental aspects of sound communication in insects. In *Acoustic and Vibrational Communication in Insects.* K. Kalmring and N. Elsner, eds., pp. 1–9. Berlin: Paul Parey.

Møller, A. P. 1993. Developmental stability, sexual selection and speciation. *J. Evol. Biol.* **6**: 493–509.

Neville, A. C. 1963. Daily growth layers for determining the age of grasshopper populations. *Oikos* **14**: 1–8.

Palmer, A. C. and C. Strobeck. 1986. Fluctuating asymmetry: measurement, analysis, pattern. *Annu. Rev. Ecol. Syst.* **17**: 391–421.

Parker, G. A. 1983. Mate quality and mating decisions. In *Mate Choice.* P. Bateson, ed., pp. 141–166. Cambridge University Press.

Parker, G. A., L. W. Simmons and H. Kirk. 1990. Analysing sperm competition data: simple models for predicting mechanisms. *Behav. Ecol. Sociobiol.* **27**: 55–45.

Parsons, P. A. 1990. Fluctuating asymmetry: an epigenetic measure of stress. *Biol. Rev.* **65**: 131–145.

Partridge, L. and T. Halliday. 1984. Mating patterns and mate choice. In *Behavioural Ecology: An Evolutionary Approach*. J. R. Krebs and N. B. Davies, eds., pp. 222–250. Oxford: Blackwell Scientific Publications.

Pollack, G. S. and R. R. Hoy. 1981a. Phonotaxis in flying crickets: neural correlates. *J. Insect Physiol.* **27**: 41–45.

–. 1981b. Phonotaxis to individual rhythmic components of a complex cricket calling song. *J. Comp. Physiol.* **144**: 367–373.

–. 1989. Evasive acoustic behavior and its neurobiological basis. In *Cricket Behavior and Neurobiology*. F. Huber, T. E. Moore and W. Loher, eds., pp. 340–363. Ithaca, New York: Cornell University Press.

Pomiankowski, A. 1987. The costs of choice in sexual selection. *J. Theor. Biol.* **128**: 195–218.

Prestwich, K. N. and T. J. Walker. 1981. Energetics of singing in crickets: effect of temperature in three trilling species (Orthoptera: Gryllidae). *J. Comp. Physiol.* **143**: 199–212.

Robert, D. J. Amoroso and R. R. Hoy. 1992. The evolutionary convergence of hearing in a parasitoid fly and its cricket host. *Science (Wash., D.C.)* **258**: 1135–1137.

Rost, R. and H. W. Honegger. 1987. The timing of premating and mating behavior in a field population of the cricket *Gryllus campestris* L. *Behav. Ecol. Sociobiol.* **21**: 279–289.

Ryan, M. J. and A. S. Rand. 1990. The sensory basis of sexual selection for complex calls in the tungara frog, *Physalaemus pustulosus* (sexual selection for sensory exploitation). *Evolution* **44**: 305–314.

Sakaluk, S. K. 1984. Male crickets feed females to ensure complete sperm transfer. *Science (Wash., D.C.)* **223**: 609–610.

–. 1985. Spermatophore size and its role in the reproductive behaviour of the cricket, *Gryllodes supplicans* (Orthoptera: Gryllidae). *Can. J. Zool.* **63**: 1652–1656.

–. 1986. Sperm competition and the evolution of nuptial feeding behavior in the cricket, *Gryllodes supplicans* (Walker). *Evolution* **40**: 584–593.

–. 1990. Sexual selection and predation: Balancing reproductive and survival needs. In *Insect Defenses: Adaptive Mechanisms and Strategies of Prey and Predators*. D. L. Evans and J. O. Schmidt, eds., pp. 63–90. Albany: State University of New York Press.

–. 1991. Post-copulatory mate guarding in decorated crickets. *Anim. Behav.* **41**: 207–216.

Sakaluk, S. K. and J. Belwood. 1984. Gecko phonotaxis to cricket calling song: a case of satellite predation. *Anim. Behav.* **32**: 659–662.

Sakaluk, S. K. and W. H. Cade. 1980. Female mating frequency and progeny production in singly and doubly mated house and field crickets. *Can. J. Zool.* **58**: 404–411.

–. 1983. The adaptive significance of female multiple matings in house and field crickets. In *Orthopteran Mating Systems: Sexual Competition in a Diverse Group of Insects*. D. T. Gwynne and G. K. Morris, eds., pp. 319–336. Boulder, Colorado: Westview Press.

Sakaluk, S. K. and Eggert, A.-K. 1996. Female control of sperm transfer and intraspecific variation in sperm precedence: antecedents to the evolution of a courtship food gift. *Evolution* **50**: 694–703.

Schatral, A., W. Latimer and B. Broughton. 1984. Spatial dispersion and agonistic contacts of male bush crickets in the biotope. *Z. Tierpsychol.* **65**: 210–214.

Schmitz, B., H. Scharstein and G. Wendler. 1982. Phonotaxis in *Gryllus campestris* L. (Orthoptera, Gryllidae). I. Mechanism of acoustic orientation in intact female crickets. *J. Comp. Physiol.* **148**: 431–444.

Shuvalov, V. F. and A. V. Popov. 1973. Significance of some parameters of the calling songs of male crickets, *Gryllus bimaculatus* for phonotaxis of females. *J. Evol. Biochem. Physiol.* **9**: 177–182.

Simmons, L. W. 1986a. Female choice in the field cricket, *Gryllus bimaculatus* (De Geer). *Anim. Behav.* **34**: 1463–1470.

–. 1986b. Intermale competition and mating success in the field cricket, *Gryllus bimaculatus* (De Geer). *Anim. Behav.* **34**: 567–579.

–. 1986c. Sexual selection in the field cricket, *Gryllus bimaculatus*. Ph.D. dissertation, University of Nottingham.

–. 1987a. Competition between larvae of the field cricket, *Gryllus bimaculatus* (Orthoptera: Gryllidae) and its effects on some life-history components of fitness. *J. Anim. Ecol.* **56**: 1015–727.

–. 1987b. Female choice contributes to offspring fitness in the field cricket, *Gryllus bimaculatus* (De Geer). *Behav. Ecol. Sociobiol.* **21**: 313–321.

–. 1987c. Heritability of a male character chosen by females of the field cricket, *Gryllus bimaculatus*. *Behav. Ecol. Sociobiol.* **21**: 129–133.

–. 1987d. Sperm competition as a mechanism of female choice in the field cricket, *Gryllus bimaculatus*. *Behav. Ecol. Sociobiol.* **21**: 197–202.

–. 1988a. The calling song of the field cricket, *Gryllus bimaculatus* (De Geer): constraints on transmission and its role in intermale competition and female choice. *Anim. Behav.* **36**: 380–394.

–. 1988b. The contribution of multiple mating and spermatophore consumption to the lifetime reproductive success of female field crickets (*Gryllus bimaculatus*). *Ecol. Entomol.* **13**: 57–69.

–. 1988c. Male size, mating potential and lifetime reproductive success in the field cricket, *Gryllus bimaculatus* (De Geer). *Anim. Behav.* **36**: 372–379.

–. 1989. Kin recognition and its influence on mating preferences of the field cricket, *Gryllus bimaculatus* (De Geer). *Anim. Behav.* **38**: 68–77.

–. 1990a. Pheromonal cues for the recognition of kin by female field crickets, *Gryllus bimaculatus*. *Anim. Behav.* **40**: 192–195.

–. 1990b. Post-copulatory guarding, female choice and the levels of gregarine infections in the field cricket, *Gryllus bimaculatus*. *Behav. Ecol. Sociobiol.* **26**: 403–407.

–. 1991a. Female choice and the relatedness of mates in the field cricket, *Gryllus bimaculatus. Anim. Behav.* **41**: 493–501.

–. 1991b. On the post-copulatory guarding behaviour of male field crickets. *Anim. Behav.* **42**: 504–505.

–. 1995. Correlates of male quality in the field cricket, *Gryllus campestris* L.: age, size and symmetry determine pairing success in field populations. *Behav. Ecol.* **6**: 376–381.

Simmons, L. W. and M. Zuk. 1992. Variability in call structure and pairing success of male field crickets, *Gryllus bimaculatus*: the effects of age, size and parasite load. *Anim. Behav.* **44**: 1145–1152.

Solymar, B. D. and W. H. Cade. 1990. Heritable variation for female mating frequency in field crickets, *Gryllus integer. Behav. Ecol. Sociobiol.* **26**: 73–76.

Souroukis, K. and W. H. Cade. 1993. Reproductive competition and selection on male traits at varying sex ratios in the field cricket, *Gryllus pennsylvanicus. Behaviour* **126**: 45–42.

Souroukis, K., W. H. Cade and G. Rowell. 1992. Factors that possibly influence variation in the calling song of field crickets: temperature, time and male size, age and wing morphology. *Can. J. Zool.* **70**: 950–955.

Souroukis, K. and A.-M. Murray. 1995. Female mating behavior in the field cricket, *Gryllus pennsylvanicus* (Orthoptera: Gryllidae) at different operational sex ratios. *J. Insect Behav.* **8**: 269–279.

Stanley-Samuelson, D. W. and W. Loher. 1986. Prostaglandins in insect reproduction. *Ann. Entomol. Soc. Am.* **79**: 841–853.

Stanley-Samuelson, D. W., J. J. Peloquin and W. Loher. 1986. Egg-laying in response to prostaglandin injections in the Australian field cricket, *Teleogryllus commodus. Physiol. Entomol.* **11**: 213–219.

Stearns, S. C. 1992. *The Evolution of Life Histories.* Oxford University Press.

Stout, J. F., C. H. DeHaan and R. W. McGhee. 1983. Attractiveness of the male *Acheta domesticus* calling song to females. I. Dependence on each of the calling song features. *J. Comp. Physiol.* **153**: 509–521.

Thornhill, R. 1983. Cryptic female choice and its implications in the scorpionfly *Harpobittacus nigriceps. Am. Nat.* **122**: 765–488.

Thornhill, R. and J. Alcock. 1983. *The Evolution of Insect Mating Systems.* Cambridge, Massachusetts: Harvard University Press.

Walker, T. J. 1964. Experimental demonstration of a cat locating orthopteran prey by the prey's calling song. *Fla. Entomol.* **47**: 163–165.

–. 1978. Post-copulatory behavior of the two-spotted tree cricket, *Neoxabea bipunctata. Fla. Entomol.* **61**: 39–40.

–. 1979. Calling crickets (*Anurogryllus arboreus*) over pitfalls: females, males, and predators. *Environ. Entomol.* **8**: 441–443.

–. 1983a. Diel patterns of calling in nocturnal orthoptera. In *Orthopteran Mating Systems: Sexual Competition in a Diverse Group of Insects.* D. T. Gwynne and G. K. Morris, eds., pp. 45–42. Boulder, Colorado: Westview Press.

–. 1983b. Mating modes and female choice in short-tailed crickets (*Anurogryllus arboreus*). In *Orthopteran Mating Systems: Sexual Competition in a Diverse Group of Insects.* D. T. Gwynne and G. K. Morris, eds., pp. 240–267. Boulder, Colorado: Westview Press.

–. 1986. Monitoring the flights of field crickets (*Gryllus* sp.) and a tachinid fly (*Euphasiopteryx ochracea*) in north Florida. *Fla. Entomol.* **69**: 678–685.

–. 1993. Phonotaxis in female *Ormia ochracea* (Diptera: Tachinidae), a parasitoid of field crickets. *J. Insect Behav.* **6**: 389–410.

Walker, T. J. and S. Masaki. 1989. Natural History. In *Cricket Behavior and Neurobiology.* F. Huber, T. E. Moore and W. Loher, eds., pp. 1–42. Ithaca, New York: Cornell University Press.

Walker, T. J. and S. A. Wineriter. 1991. Hosts of a phonotactic parasitoid and levels of parasitism (Diptera: Tachinidae: *Ormia ochracea*). *Fla. Entomol.* **74**: 554–559.

Watt, W. B., P. A. Carter and K. Donohue. 1986. Females' choice of 'good genotypes' as mates is promoted by an insect mating system. *Science (Wash., D.C.)* **233**: 1187–1190.

Webb, K. L. and D. A. Roff. 1992. The quantitative genetics of sound production in *Gryllus firmus. Anim. Behav.* **44**: 823–832.

Weber, T. and J. Thorson. 1989. Phonotactic behavior of walking crickets. In *Cricket Behavior and Neurobiology.* F. Huber, T. E. Moore and W. Loher, eds., pp. 310–339. Ithaca, New York: Cornell University Press.

Wedell, N. 1993. Spermatophore size in bushcrickets: comparative evidence for nuptial gifts as sperm protection device. *Evolution* **47**: 1203–1212.

White, G. 1788. *The Natural History of Selborne* (1977 printing). London: Penguin Books.

Will, M. W. and S. K. Sakaluk. 1994. Courtship feeding in decorated crickets: is the spermatophylax a sham? *Anim. Behav.* **48**: 1309–1315.

Williams, G. C. 1966. *Adaptation and Natural Selection.* Princeton: Princeton University Press.

Zuk, M. 1987a. Age determination of adult field crickets: methodology and field application. *Can. J. Zool.* **65**: 1564–1566.

–. 1987b. The effects of gregarine parasites, body size, and time of day on spermatophore production and sexual selection in field crickets. *Behav. Ecol. Sociobiol.* **21**: 65–42.

–. 1987c. Variability in attractiviness of male field crickets (Orthoptera: Gryllidae) to females. *Anim. Behav.* **35**: 1240–1248.

–. 1988. Parasite load, body size, and age of wild-caught male field crickets (Orthoptera: Gryllidae): effects on sexual selection. *Evolution* **42**: 969–976.

–. 1991. Sexual ornaments as animal signals. *Trends Ecol. Evol.* **6**: 228–231.

Zuk, M., K. Johnson, J. D. Ligon and R. Thornhill. 1993a. Effects of experimental manipulation of male secondary sex characters on female mate preference in red jungle fowl. *Anim. Behav.* **44**: 999–1006.

Zuk, M., L. W. Simmons and L. Cupp. 1993b. Calling characteristics of parasitized and unparasitized populations of the field cricket *Teleogryllus oceanicus. Behav. Ecol. Sociobiol.* **33**: 339–343.

Zuk, M., L. W. Simmons, and J. T. Rotenberry. 1995. Acoustically-orienting parasitoids in calling and silent males of the field cricket *Teleogryllus oceanicus. Ecol. Entomol.* **20**: 380–383.

6 · The evolution of edible 'sperm sacs' and other forms of courtship feeding in crickets, katydids and their kin (Orthoptera: Ensifera)

DARRYL T. GWYNNE

ABSTRACT

Males of certain arthropods feed their mates during mating. Nowhere is the diversity of this courtship feeding greater than in the Orthopteran suborder Ensifera, the katydids, weta, and the humped-winged, wood, Jerusalem, mole, cave, camel as well as tree and other true crickets. In this chapter I discuss the origin and current utility of male glandular and body-part donations in the Ensifera. Character analysis shows that the ancestral ensiferan female removed and ate a sperm ampulla (spermatophore) that was positioned externally on her genitalia. This was followed by at least 11 origins of mate feeding, most cases of which involved female feeding on the male contribution during insemination. This sequence of events indicates that male contributions evolved through sexual selection and sexual conflict as meal size increased and prolonged the attachment of the sperm ampulla, thus increasing paternity. The most common type of male contribution in Ensifera – a gelatinous spermatophylax attachment to the sperm ampulla – evolved three times. A spermatophylax is found in virtually all katydid species (Tettigoniidae) and the other four families in the same clade. For the current utility of courtship feeding there is experimental support for different functions in different species: in a humped-winged cricket (Haglidae), two true crickets (Gryllidae) and two katydids, wing, gland or spermatophylax feeding increases paternity by preventing consumption of the sperm-ampulla. In cases of precopulatory feeding, the meal functions to obtain mates. Therefore, as in other courtship-feeding animals, mate-feeding in Ensifera appears to have evolved mainly via sexual selection and sexual conflict. One katydid is exceptional in that an evolutionary increase in spermatophylax size appears to have resulted in mutual benefits to the sexes; spermatophylax nutrients are analogous to parental care by increasing the fitness of the male's offspring.

INTRODUCTION

Reports on the mating habits of shield-backed katydids (Tettigoniidae, Decticini) were the first to describe very large spermatophores eaten by females of certain crickets, katydids and related groups (the orthopteran suborder Ensifera) (Fig. 6-1). Observations by the American explorer Captain John Feilner (1860, cited in Caudell 1908) were followed by several papers at the turn of the century (Fabre 1896, 1918; Gillette 1904; Snodgrass 1905). For example, Snodgrass observed 'a large white mass of tough albuminous matter ejected by the male (*Peranabrus*)' and eaten by the female. Rather than viewing this mass as a beneficial meal, however, he suggested that it caused the female 'much annoyance' so that she 'attempts to rid herself of it by bending her head beneath the abdomen and chewing it off'.

It was the meticulous comparative work of the pioneer French ethologist J. H. Fabre (1896, 1917) and the Russian B. T. Boldyrev (1915, 1928) that provided the details of copulation and mate-feeding in these insects. Although he studied several katydids, Fabre concentrated on *Decticus*, noting that the male transferred to the female an 'opalescent bag similar in size and colour to a mistletoe berry'. Boldyrev studied not only katydids but also camel crickets (Rhaphidophoridae) and true crickets (Gryllidae). His observations showed that elaborate spermatophores eaten by females were typical of mating in all katydids and in some species in the other groups. Boldyrev's dissections and observations revealed a dual role for the spermatophore. One role – the mating meal – was filled by a sperm-free mass that he called the spermatophylax. He noted that this mass was eaten first by the female. The other part, the sperm ampulla, clearly had an insemination role (Fig. 6-1). Because the spermatophore is transferred to the female at the end of copulation, the timing of sperm transfer in katydids and other ensiferans is quite different

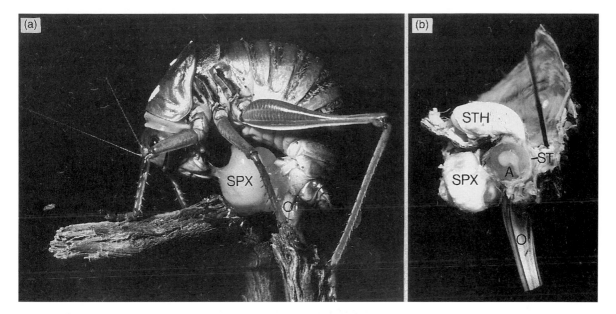

Fig. 6-1. (a) A Mormon cricket female (*Anabrus simplex*) eating the spermatophylax after mating. (b) The parts of this spermatophore shown in a dissection of the side of a female's abdomen. The spermatophylax (SPX) is attached to the rest of the spermatophore that includes a sperm ampulla (A) positioned inside the female's genital chamber. The side of the genital chamber has been removed to show the ampulla and one of two sperm cavities it contains. Sperm exits along an ejaculatory canal to the female's spermathecal tube (ST) and into her large spermatheca (STH).

from that in most other animals: it begins soon after the end of mating, thus making ejaculation a postcopulatory event. The katydid spermatophore has two sperm cavities that exit into a common ejaculatory canal. Sperm exits the canal and enters the female by passing through the spermathecal tube and into the sperm storage organ, the spermatheca.

THE FUNCTION OF THE EDIBLE SPERMATOPHORE

Boldyrev's work did more than just describe, it also discussed the adaptive significance of spermatophore anatomy, particularly how male katydids might benefit from providing their mates with a spermatophylax. Female ensiferans removed and ate the sperm ampulla some time after copulation ended, so he concluded that the spermatophylax functions to keep the female's mandibles away from the ampulla. Males transferring full ejaculates are more likely to ensure paternity through the transfer of sperm and/or of substances that turn off female receptivity (i.e. that induce a sexual refractory period) (Gwynne 1986). Boldyrev's hypothesis was a general one that he used to

explain a number of postcopulatory behaviors in Ensifera. This included dorsal gland feeding by female tree crickets (Gryllidae, Oecanthinae) and also mate-guarding in a camel cricket (Rhaphidophoridae) in which the male does not provide food but his presence and behavior prevents the female from eating the spermatophore (see also Zuk and Simmons, this volume).

Recent work on two species appeared to support Boldyrev's 'ejaculate-protection' hypothesis for the spermatophylax. When females of a katydid, *Requena verticalis* (Gwynne *et al.* 1984), and a true cricket, *Gryllodes supplicans* (Sakaluk 1984), received no spermatophylax they quickly removed and ate the sperm ampulla and little or no ejaculate was transferred. However, details of the results revealed differences between the species. The ejaculate-protection hypothesis predicts that males should provide no more of the costly spermatophylax than necessary to protect the sperm ampulla. This was the case for the cricket but not for the katydid. Male *Requena* provided a much larger spermatophylax meal than is necessary for complete transfer of ejaculate materials (Gwynne *et al.* 1984; Gwynne 1986). Furthermore, *R. verticalis* males and females might just obtain parental benefits from the extra male donation because spermatophylax nutrients were found to increase the female's reproductive output (Gwynne 1984, 1988a).

In Darwinian terms, the ejaculate-protection hypothesis argues that the spermatophylax is a sexually selected trait that could have followed an evolutionary 'arms race' resulting from sexual conflict, i.e. one of the sexes benefits while the other incurs a cost (see Alexander *et al.*, this volume; Arnqvist, this volume; Choe, this volume). Ancestral females would have gained an edge in an initial conflict as they interfered with sperm transfer, benefitting from either nutrition in the proteinaceous sperm ampulla eaten or from a direct advantage in manipulating ejaculates (such as promoting competition between the sperm of different males (Wedell 1993b)). Following its origin, the spermatophylax might have been further elaborated through sexual conflict if an increase in its size was needed to protect a larger ampulla (ejaculate) resulting from increased sperm competition (Wedell 1993b, 1994). The spermatophylax would still be no larger than necessary to protect the ejaculate (see, for example, Sakaluk 1984) because males should have been selected to minimize mating costs. An additional way males could reduce spermatophylax costs would involve a 'candy maker' strategy of decreasing the nutritional quality of the mating meal (e.g. protein content) (Wedell 1994) while maintaining its 'good taste', i.e. components that keep the female feeding, so that her mandibles do not contact the sperm ampulla. Females could benefit by receiving a spermatophylax of low nutritional quality only if the size of the meal – independent of its food quality – was somehow used to gain information on male genetic quality (a hypothesis proposed for the *Gryllodes* spermatophylax by Will and Sakaluk 1994).

If the spermatophylax evolved as an important food source for females (Thornhill 1976a; Eluwa 1978), its size could have been elaborated beyond that sufficient to protect the ejaculate as seen in the katydid *R. verticalis*. One cause of this elaboration could be additional sexual conflict if, when food-stressed, females decreased the duration of the sexual refractory period so as to obtain additional mating meals, and if the male response to this was to supply more food in a larger spermatophylax (see Simmons and Gwynne 1991 for both the hypothesis and the influence of food-stress on the refractory periods). Alternatively, the increase in the size of the mating meal could have involved mutual benefits to both sexes through natural selection on the male to provide nutrition as a 'paternal investment' benefit to his own offspring. This is an alternative hypothesis to ejaculate-protection and was proposed for the enlarged nutritious spermatophylax of *R. verticalis* (Gwynne 1984, 1988a) and for the spermatophylax of katydids in general (Gwynne 1990a).

Experiments on the function of the spermatophylax in single species (see, for example, Gwynne *et al.* 1984; Sakaluk 1984) test the *current* utility or function of the male's meal, i.e. how this food-offering trait is currently maintained by selection. The finding that the spermatophylax might have different current uses in different species, however, raises questions about the original function of this structure, both in Ensifera (Gwynne 1990a) and within Tettigoniidae (Wedell 1993b). Was it ejaculate-protection that caused the spermatophylax to evolve in the first place? Perhaps this edible elaboration of the spermatophore evolved under different selection pressures in different lineages and the original functions have not changed (Gwynne 1990a). Alternatively, the spermatophylax could have evolved only once and then changed in function in at least one of the groups. These questions about trait origins can be answered by comparing the traits in different taxa using information on the evolutionary history of the groups involved (Brooks and McLennan 1991; Harvey and Pagel 1991).

This chapter addresses the origin and adaptive significance of the spermatophylax and other forms of mate-feeding in katydids, crickets and related groups. I examine alternative hypotheses for the original and current functions of mate-feeding in Ensifera both by reviewing results of experiments and, in the spirit of Boldyrev's (1915) approach, by comparing mating in katydids and other ensiferans. First, however, I briefly review male investment in mating in ensiferans and other animals both to show that the spermatophylax is just one of a number of mechanisms by which male animals can feed their mates, and to understand the origins of the large edible sperm sac and other forms of mate-feeding in katydids, crickets and related groups.

THE DIVERSITY AND EVOLUTION OF COURTSHIP FEEDING

Males can invest in mating by providing goods and services to their offspring or to their mates (Fig. 6-2). Care of offspring by males is widespread in vertebrates but rare in insects, being confined to Isoptera and a few species of Hemiptera and Coleoptera (Wilson 1971; Smith 1980; Choe and Crespi 1997). Donation of food to mates is, however, a more common type of male nurturing behavior in

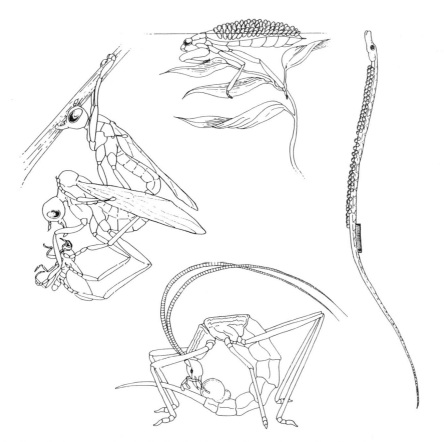

Fig. 6-2. Forms of male mating investment in animals, showing (clockwise from top) paternal care in a giant water bug that broods the eggs on its back (*Abedus herberti* (Hemiptera: Belostomatidae)), and a pipefish male (*Nerophis ophidion*, Sygnathidae) brooding eggs on its abdomen. Courtship feeding in a katydid *Kawanaphila nartee* (Tettigoniidae) and a dance fly, *Empis borealis* (Diptera: Empididae). (Reprinted with permission from *Trends in Ecology and Evolution*, vol. 6, p. 120 (1991) (drawing by Heather Proctor).)

insects and other animals. There are a variety of ways in which this occurs. 'Courtship feeding', where the male provides food to his mate before copulation, is well known in birds (Lack 1940). I use this term more broadly by including the transfer of food during as well as before copulation because courtship, when defined as sexual communication between a paired male and female, can occur at any time between first contact and the end of insemination (Thornhill and Alcock 1983; Eberhard 1985, this volume; Alexander *et al.*, this volume).

Courtship meals come in diverse forms. The donation of captured prey to the female as seen in birds is also found in certain spiders (Austad and Thornhill 1986) and insects. It is within insects and other arthropods, however, that we see the greatest diversity in the variety of substances fed to females (reviewed in Thornhill 1976a; Parker

and Simmons 1989). In addition to courtship prey donated by scorpionfly and dance fly males (Fig. 6-2), 'nuptial gifts' include other food collected by the male such as seeds used by a lygaeid bug or regurgitated nectar transferred by copulating male thynnine wasps to their wingless mates (Thornhill 1976a). Male arthropods can also feed their mates by giving up certain body parts. This varies from female humped-winged crickets (Haglidae) that feed on the fleshy hindwings of males (Fig. 6-3) (Morris *et al.* 1989) to somatic sacrifice when it is adaptive for males to allow themselves to be consumed during mating (probably rare but apparently occurring in the red-back spider, *Latrodectus hasselti* (Forster 1992; Andrade 1996)). A final category of courtship meals involves the replaceable body tissues in male glandular products. Examples are feeding on external glands in true crickets (Fig. 6-3) and the nutritious

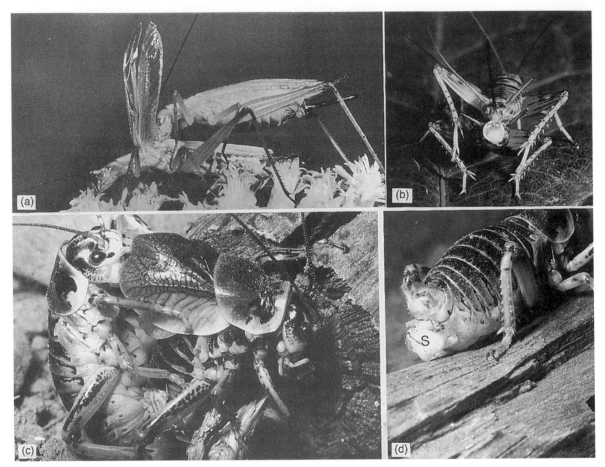

Fig. 6-3. Courtship feeding in Ensifera. True crickets (Gryllidae): (a) A female black-horned tree cricket, *Oecanthus nigricornis*, feeding on the dorsal thoracic gland of a male following copulation (photo by Maria Zorn). The arrow at the end of the female's abdomen points to the spermatophore. (b) Mating in *Gryllodes supplicans* showing the female receiving the spermatophylax from the male (photograph by Scott Sakaluk). (c, d) Humped-winged crickets (Haglidae). The two types of mating meal received by female sagebrush crickets, *Cyphoderris strepitans*. (c) A female with her head under the male's tegmina feeding on the fleshy hindwings, one of which is partly visible below the male's forewing. (d) The female *C. strepitans* also receives a spermatophylax (S) after mating, which is shown in this view of the female's abdomen. (e) A phaneropterine katydid (Tettigoniidae) and (f) a gryllacridid, showing the large spermatophylax just after mating (Fig. 6-3 f by K.-G. Heller.)

secretions from internal glands such as the enlarged salivary glands of male panorpid scorpionflies (Mecoptera) (Thornhill 1976b), the cephalic glands of certain Zoraptera (Choe, this volume) and the spermatophylax donations of ensiferan Orthoptera (Figs. 6-1 to 6-3) and dobsonflies (Megaloptera) (Hayashi 1992). Provision of substances from glands may also occur via non-oral means when ejaculated secretions are absorbed in the female's genital tract (references in Parker and Simmons 1989). This type of provisioning appears to occur in a number of arthropods and

involves not only nutrients but also specialized defensive chemicals that can contribute to female or egg survival (Eisner and Meinwold 1987) .

Orthoptera are champions of courtship-gift diversity: virtually all forms of courtship feeding are found in species of grasshoppers, katydids and various crickets (Figs. 6-3 and 6-4) (Gwynne 1983). Interestingly, the only type of courtship feeding not represented in the Orthoptera is the most widespread type in animals: the donation of prey or other food collected by the male. Examples of male

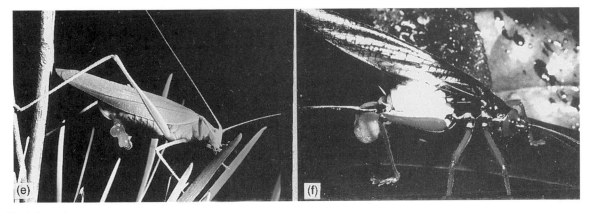

Fig. 6-3. (*cont.*)

nutrient donations in grasshoppers (suborder Caelifera) involve ejaculatory substances apparently absorbed in the genital tract (Friedel and Gillott 1977; Butlin *et al.* 1987). Ensiferans are specialists in male-derived food that is actually eaten by the female. Among crickets and katydids, these oral donations take several different forms, transferred either before or during copulation. Some species show female feeding on body parts such as legs (the shield-back katydid *Decticita* (Rentz 1963)) or wings (Fig. 6-3) during copulation. In others feeding is on glandular secretions, both before and after copulation. This is especially well-known in Gryllidae (Otte 1992). However, the most widespread courtship meal, particularly in katydids and related groups, is postcopulatory feeding by females on a spermatophylax (Figs. 6-1 to 6-3) (Gwynne 1983).

Edible additions to the spermatophore are found in six of the seven families of Ensifera in which mating has been studied (Fig. 6-4). To what degree does this represent multiple origins of the spermatophylax versus this trait being shared by common descent? I reconstructed the evolutionary history of spermatophylax and other forms of courtship feeding on a phylogeny of Ensifera (Gwynne 1995a) using parsimony methods (MacClade 3.0; Maddison and Maddison 1992). The phylogeny is independent of the characters being traced (see Brooks and McLennan 1991); the same topology – the relationships among taxa – occurs when courtship-feeding characters (i.e. spermatophore characters) are deleted from the original cladistic analysis (Gwynne 1995a). The separate phylogeny of Tettigoniidae (see Fig. 6-4, legend) and Gryllidae (DeSutter 1987) included in the more detailed tree (Fig. 6-4, bottom) are also not biassed by courtship-feeding characters (see Gwynne 1995a). Because certain subgroups within these families are represented by only one or two genera, it is likely that conclusions about the number of origins of different traits will change with the inclusion of additional taxa and courtship-feeding information.

The analysis revealed that female feeding on a spermatophylax, external glands or on body parts (legs or wings) originated 11 times within the Ensifera: three in the tettigonioid clade and eight in the grylloids (Fig. 6-4). There were four origins of feeding on metanotal glands, one on tibial glands, three on male body parts and three on a large spermatophylax (Fig. 6-4). An additional origin of an edible spermatophylax occurred in the endopterygote order Megaloptera: *Protohermes* has 'a large gelatinous part covering the sperm-containing package' that is eaten by the female (Hayashi 1992; and see also Henry, this volume, for examples with other neuropteroid insects) indicating a remarkable convergence with certain Ensifera. There are seven apparent losses of courtship feeding, comprising: a single loss of gland-feeding in Hawaiian tree crickets, *Leptogryllus* (Otte 1992; W. Brown, unpublished), a complete loss of the spermatophylax (*Deinacrida*) or its reduction to vestigial size (*Hemideina*) (giant and tree weta, respectively (Stenopelmatidae, Deinacridinae)) (Ramsey 1955; D. T. Gwynne, unpublished); a loss in the tettigoniids, *Tympanophora* (D. T. Gwynne, unpublished) and apparently *Decticita* (Rentz 1963); and two reductions to vestigial size of the spermatophylax in Phasmodinae and certain copiphorine katydids (Fig. 6-5). There may be an additional loss of spermatophylax-feeding in Jerusalem crickets (Stenopelmatidae, *Stenopelmatus*); these insects have a large spermatophylax but, in all species studied, it remains uneaten on the female's genitalia for several days (Weissman, 1995).

Because the spermatophylax is present in all families within the tettigonioid clade, character analysis reveals one origin of this trait in this group before the separation of the haglid–gryllacrid clade (five families) from the cave and camel crickets (Rhaphidophoridae). (The exact origin is equivocal because the presence of a spermatophylax is a polymorphic character in the latter family (see Fig. 6-4 bottom).) The suggestion that courtship meals evolve as benefits when males 'lure' mates (Alexander et al., this volume) is not supported in this origin of the spermatophylax meal; in the tettigonioid clade, male calling songs (of tettigoniids and haglids) followed the origin of the spermatophylax (Gwynne 1995a). It remains a possibility, however, that the ancestral spermatophylax-provider attracted females by using non-acoustic signals.

FUNCTION OF COURTSHIP FEEDING: HYPOTHESES

Copulation for males typically involves little more than the relatively trivial expense of insemination (see Alexander et al., this volume). The costly part of reproduction for most males is the effort required to battle other males and to locate receptive females (Trivers 1972). As we have seen, courtship feeding has evolved independently a number of times within the Ensifera. It can also be a costly trait. For

(a)

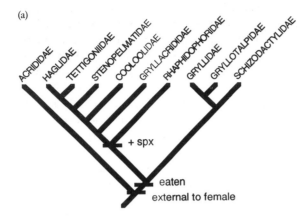

(b)

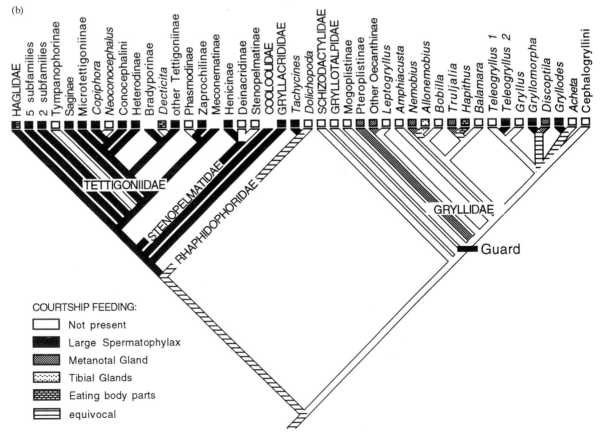

example, male katydids, *R. verticalis*, cannot produce another spermatophore for several days after mating (Gwynne 1990b). Why did costly copulations originate in the first place and what selection pressures currently maintain this male service? As mentioned previously, ejaculate-protection and paternal–investment are two explanations of elaborate edible spermatophores. There are, however, other hypotheses, some of which are more applicable to courtship feeding before and during copulation than to feeding after copulation, as is the case with the spermatophylax (Table 6-1).

Given the nature of spermatophylax and other courtship-feeding in katydids and allied groups, some of the hypotheses about the function of courtship feeding can be dismissed as unlikely for this group. Snodgrass (1905) suggested that the large katydid spermatophore 'is a plug to close the bursa copulatrix' (Table 6-1, IIC). Although the spermatophore does block the female's genital aperture, this block cannot have originated to prevent sperm loss because in all Ensifera sperm is not stored in the genital tract but in the spermatheca (Fig. 6-1). Furthermore, it is unlikely that the blocking action of the spermatophore originated via sexual selection as a device that prevents remating by the female. Such a short-term chastity device would be redundant given the days-long duration of the (probably

chemically induced; see Simmons and Gwynne 1991) sexual refractory period in females of several groups of katydids (Wedell 1993b).

The suggestion that the spermatophylax evolved by natural selection to prevent the female from eating the male is another hypothesis for the function of this trait in the few predatory katydids. This is an unlikely original function of this structure. Cannibalism, noted in some species (see, for example, Cowan 1929), appears to involve opportunistic consumption of injured or weak conspecifics. Moreover, prevention of cannibalism is an unlikely current function of the spermatophylax. Female katydids (*R. verticalis*) have been observed to eat courting males in nature (I. Dadour and L. Simmons, personal communication, 1993) but predatory attacks by females occur before copulation and thus before the putative distracting device – the spermatophylax – is produced.

An aspect of sexual selection that is likely to have been involved in the evolution of courtship feeding in Ensifera is where food offerings function in attracting females and inducing them to mate (Table 6-1, IIa,b). Courtship feeding that takes place before and during copulation probably functions in this context. Moreover, the four origins of gland and body-part feeding that occur only before or during copulation are unlikely have evolved to protect the

---

Fig. 6-4. The origins of courtship feeding and related characters in the suborder Ensifera (Orthoptera). (a) A phylogeny of families in the suborder Ensifera showing the outgroup taxon (Acrididae, suborder Caelifera) and two ensiferan clades: tettigonioids and grylloids (the most parsimonious tree from a cladistic analysis of anatomical and morphological characters (Gwynne 1995a)). The bars on the phylogeny indicate the origins of the spermatophore: being positioned externally on the female's genitalia, being eaten by the female, and acquiring a spermatophylax ('+spx') (in the tettigonioid clade). The virtual ubiquity of the spermatophylax and the lack of a single sperm cavity (Gwynne 1995a) in species in the tettigonioid clade are characters congruent with the phylogenetic separation of this group from the grylloid ensiferans. (b) The phylogeny of ensiferan families (upper-case names ending in -ae) expanded to show details of several families (lower-case names: subfamilies end in -ae, tribes end in -ini and genera are in italics). Information for Gryllidae from the phylogeny of DeSutter (1987) and Tettigoniidae from a phylogeny of Gorochov (1988) with additions from Rentz (1993; D. C. Rentz, personal communication, 1992). Because no information is available on relationships within Stenopelmatidae this family is shown as a polytomy of three subfamilies. The gryllid subfamily Gryllinae is also shown as a polytomy of four tribes (from Otte and Alexander 1983).

Rows of boxes above the phylogeny represent the occurrence of different types of courtship feeding in the taxa considered (see figure key). This information was used to trace the origins of courtship feeding (using MacClade 3.0, licensed to M. Laurin) shown in the patterned parts of the phylogeny (see the key). Also shown (bar) is the origin of male postcopulatory guarding in crickets (Gryllidae).

Information on spermatophores and other reproductive characters was obtained from Boldyrev (1915), Ander (1939) (spermatophores of Ensifera), Alexander (1962) and Evans (1988) (gryllid mating), Alexander and Otte (1967) (gryllid spermatophore, wing and gland-feeding), Carlberg (1981) (phasmatid spermatophores), Mays (1971) (nemobiine gryllid gland-feeding), Pickford and Padgham (1973) (acridid spermatophores), Dodson *et al.* (1983) (haglid wing and spermatophylax-feeding), Ramsey (1955), Richards (1973), Field and Sandlant (1983), Barrett and Ramsey (1991), Weissman (1995), G. Monteith and D. T. Gwynne (unpublished) and D. T. Gwynne (unpublished) (stenopelmatid mating and spermatophores), Rentz (1963), Gwynne (1983, 1990a; D. T. Gwynne, unpublished), Wedell (1993a) (tettigoniid spermatophores), Rentz and John (1990), D. T. Gwynne (unpublished), K.-G. Heller (personal communication, 1993) (gryllacridid mating and spermatophores), Bidochka and Snedden (1985) (nemobiine gryllid gland feeding), Evans (1988) (gryllid mating), Ono *et al.* (1989) (podoscirtine gryllid mating), and Sakaluk (1991) (guarding in Gryllidae).

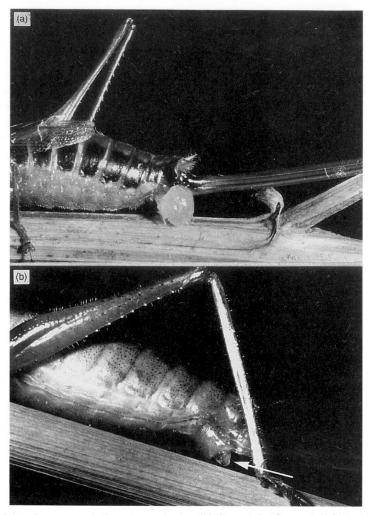

Fig. 6-5. The spermatophylax of two conocephalines, the meadow katydid, *Conocephalus* (Conocephalini) (a) and the cone-head *Belocephalus* (Copiphorini) (b). Most Conocephalinae have a large spermatophylax; it is reduced only in a few copiphorine genera.

ejaculate because they end when the spermatophore is transferred and thus before sperm transfer occurs. Conversely, obtaining copulations is an unlikely function for the spermatophylax because this male donation is not produced and eaten until after the sperm ampulla is transferred. The spermatophylax may, however, have originated (and currently functions) in obtaining inseminations, i.e. the ejaculate-protection hypothesis of Boldyrev (1915). The next section will present a preliminary comparative analysis of courtship feeding in ensiferan taxa in an attempt to understand the original function of this behavior, particularly the most widespread form, spermatophylax-feeding. Following this I will examine the

acquiring-copulations, ejaculate-protection and paternal-investment hypotheses for the current utility of this behavior.

FUNCTIONAL ORIGINS OF COURTSHIP FEEDING

Postcopulatory behavior and ejaculate protection

Boldyrev's ejaculate-protection hypothesis predicts that a vulnerable simple spermatophore evolved first, followed by female consumption of the spermatophore. Female behavior initiates sexual conflict resulting in the origin of

Table 6-1. *The adaptive significance of courtship feeding: potential benefits received by mate-feeding males*

I. Courtship feeding results from natural selection because male survival is increased; providing the female with a meal allows the male to avoid being eaten by his mate (Kessel 1955).
II. Courtship feeding results from sexual selection.
    The premating presentation of a courtship gift functions as follows.
    A. As a non-nutritive, 'symbolic' device to attract and acquire a mate (Kessel 1955).
    B. Acquiring-copulations: a meal to attract a mate and obtain copulations (Thornhill 1976a).
    C. The transfer of a courtship meal during copulation increases the probability of gaining fertilizations by acting as a plug to prevent sperm leakage from the female (Snodgrass 1905).
    D. Ejaculate-protection: the presence of a courtship meal while insemination occurs increases the probability of obtaining fertilizations by keeping the female busy with a meal (Boldyrev 1915; Fulton 1915; Thornhill 1976a).
III. Paternal investment: courtship feeding results from natural selection because nutrition provided benefits the male's own progeny (Trivers 1972; Thornhill 1976a; Gwynne 1984, 1986).

postcopulatory 'protective behavior' in which males feed or guard their mates (Gwynne 1988a, 1990a). The reconstruction of character evolution supports this evolutionary sequence (Fig. 6-4, top). (1) The single sperm cavity type of sperm-ampulla, ancestral to orthopteroid insects, was positioned externally on the genitalia of the ancestral ensiferan female (Fig. 6-4). The evolution of a vulnerable external position for the ampulla may have occurred in a more ancestral orthopteroid insect because certain grasshoppers (Acrididae) and stick insects (order Phasmatodea) have such external spermatophores (Carlberg 1981; Pickford and Padgham 1973; C. Gillott, personal communication, 1993). (2) Female consumption of the sperm ampulla is a trait ancestral to all Ensifera and originated after the Ensifera branched from other Orthoptera. Thus the vulnerability of the spermatophore was apparently exploited by ancestral females, either for nutritional purposes or possibly as a way of favoring the ejaculates of certain males (inducing sperm competition) (Wedell 1993a). These females could have obtained large nutritional benefits only by eating many ampullae; true crickets (Gryllidae) retain the ancestral type of simple spermatophore and need to eat many proteinaceous (Khalifa 1949) ampullae before there is a

significant fecundity increase (Simmons 1988). (3) Finally, following the origin of female ampulla-eating were six origins of postcopulatory feeding (three involving spermatophylax and three, gland feeding) and three origins of postcopulatory mate guarding, once each in the true crickets (Gryllidae), camel crickets (Rhaphidophoridae) and possibly certain weta (Stenopelmatidae (Barrett and Ramsey 1991; D. T. Gwynne, unpublished)) (Fig. 6-4). This sequence of evolutionary events strongly supports ejaculate protection as the original function of the spermatophylax. Although the sequence is not specifically predicted by the paternal-investment hypothesis, it does not refute this hypothesis.

If postcopulatory mate-feeding and mate-guarding function only to protect the ejaculate (Boldyrev 1915), then the two behaviors should not co-occur because the evolution of a second ampulla-protecting mechanism would be redundant. At first there appears to be some support for a lack of co-occurrence in that the origin of ampulla consumption by females was followed by the evolution of a spermatophylax in the tettigonioid clade, as noted above, and by the origin of guarding in the grylloids (Fig. 6-4, top). There were, however, subsequent gains and losses of the two characters. For a more detailed examination I counted each time that a character-state changed as an independent trial (Ridley 1983; Harvey and Pagel 1991). I deleted from the analysis two origins of gland feeding (with no guarding) because the coexistence of male guarding and female feeding on a male gland is physically impossible. Also excluded are three independent cases of absent (or vestigial) spermatophylax with no guarding in the tettigoniids *Neoconocephalus*, *Tympanophora* and *Phasmodes* (Figs. 6-4 and 6-5); in all three, the ampulla is no longer positioned outside the female but is inside the female's genital chamber in an apparently less vulnerable position. The remaining cases of postcopulatory feeding all involve the spermatophylax. The results show two origins of guarding without spermatophylax feeding, two of guarding with spermatophylax feeding, and one each of no spermatophylax with and without guarding. Although the test lacks power, the data at present provide no support for the hypothesis of no association between the two behaviors (Fisher's Exact Test, $p = 0.1$). Two origins of both spermatophylax feeding and guarding (in the henicine stenopelmatid *Hemiandrus* (D. T. Gwynne, unpublished) and the well-studied gryllid, *Gryllodes*) show that the two behaviors can co-occur. A possible explanation is that ejaculate-protection was the original function of each behavior but one or the other has changed in function. This may be the case for *G. supplicans*. The spermatophylax in this species fills an ejaculate-protection role (Sakaluk 1984) whereas guarding appears to have a different function: keeping rival males away from the female (Sakaluk 1991; see also Zuk and Simmons, this volume).

This comparative examination of courtship feeding used information on ancestral character-states to infer the direction of evolutionary change. Lacking a phylogeny of her study taxa, Wedell (1993b) used a 'non-directional' approach (Harvey and Pagel 1991) to examine the ejaculate-protection and paternal-investment hypotheses for the evolution of the katydid spermatophylax. The study examined reproductive traits at the genus, rather than species level, in an attempt to reduce the possibility of using non-independent data points. The results should be interpreted cautiously, however, because this sort of approach still lacks much control for common phylogenetic history (Harvey and Pagel 1991). Katydid genera are probably not independent; indeed, there is a suggestion of subfamily-specificity for traits such as spermatophylax size and length of the refractory period (the period in which the female remains non-receptive after mating) (see Fig. 6-6). The extent to which such specificity in these traits is constrained by common ancestry or due to similar selective histories is unknown.

Wedell (1993b) concluded that the data supported ejaculate-protection as a more general pattern for spermatophylax evolution in katydids. Under the paternal-investment but not the ejaculate-protection hypothesis, she argued that 'if all other things (are) equal' spermatophylax size and fecundity should be positively correlated because spermatophylax nutrients should be used in egg production. The absence of a positive correlation may be expected, however, for reasons other than ejaculate protection being the common function of the spermatophylax (Wedell 1993b). For example, it is doubtful whether the other variables Wedell refers to are equal across taxa. The correlation is not expected if the spermatophylax evolved 'paternally' as a replacement for a taxon-specific diet low in nutrients necessary for egg production (Wedell 1993b,c) or to provide specialized nutrients to females and/or offspring (see Gwynne 1988a). Does the spermatophylax contribution and fecundity correlate between taxa, however? The results are mixed. Wedell (1993b) showed no correlation between fecundity and spermatophylax size for genera whereas a comparison of species (Wedell 1995) does. She interprets the positive correlation as a result of species

with spermatophylax and egg masses larger than average representing those in which the spermatophylax functions as paternal investment, i.e. the nuptial meals may have increased in size beyond an ejaculate-protection role.

Another pattern was used to support the ejaculate-protection hypothesis as a more general explanation for the function of the spermatophylax in katydids. Spermatophore size should be associated with expenditures in acquiring fertilizations. Specifically, Wedell (1993b) found the size of both the ejaculate-containing ampulla, and its protective spermatophylax, correlated positively with the length of the female's refractory period (in genera; Fig. 6-6). She argued that increased effort in obtaining fertilizations due to sperm competition should result in a larger ejaculate due to the increase in sperm and in substances that induce non-receptivity in the female. An evolutionary increase in ampulla size, in turn, selected for an increase in the size of the major component of spermatophore

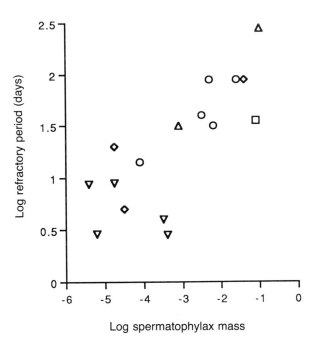

Fig. 6-6. Relationship between duration of the refractory period and mass of spermatophylax ($r = 0.78$, $p < 0.001$) for 16 genera of katydids. Redrawn from Wedell (1993b). The subfamilies of different species are represented by different symbols: diamonds for Conocephalinae, squares for Ephippigerinae, triangles for Listroscelidinae, upside-down triangles for Phaneropterinae and circles for Tettigoniinae. Refractory period also correlated with size of sperm ampulla ($r = 0.78$, $p < 0.001$)).

mass, the spermatophylax. However, the correlation between the size of the spermatophylax and level of paternity (as reflected by ampulla (ejaculate) size and length of refractory period) is also predicted by the paternal investment hypothesis. In other animals a greater male investment in offspring is commonly correlated with greater paternity assurance (Thornhill 1976a; Wedell 1993b).

Another comparative study (Wedell 1995) is used to further support the idea that the spermatophylax of some katydids functions in ejaculate protection but in others as a paternal investment. Wedell (1995) shows a roughly bimodal frequency distribution of protein content of the meals of katydid species. The most proteinaceous meals are the largest and most costly to males (the latter estimated by male remating intervals). These results are expected (assuming no strong phylogenetic biases) if the courtship gift evolved to protect ejaculates in some species and as a paternal investment in others (Wedell 1995).

Courtship feeding and diet

Tallamy's (1994) 'enhanced fecundity' hypothesis states that male investment (in care of offspring or in mating gifts) evolves if males provide resources that are limiting to female fecundity. Perhaps courtship feeding in katydids evolved to provide females with a supplement to a poor diet (a prediction consistent with both the paternal investment and acquiring copulations hypotheses for this behavior (Tallamy 1994))? Nitrogen is limiting for folivores, so leaf-eating katydids might be expected to have a greater male contribution (in terms of size and quality of the spermatophylax) than flower- and seed-eaters or predators. Although there is variation in the diet between katydid taxa (Gangwere 1961; Wedell 1994), little data is available for a character analysis and the results are therefore inconclusive: the independent origins of a greatly reduced spermatophylax in predatory Tympanophorinae, seed-eating Copiphorini (e.g. *Neoconocephalus* and *Belocephalus*) (Fig. 6-5) and flower-eating Phasmodinae appear to provide support, but the large spermatophylax in predatory Saginae and pollen-eating Zaprochilinae do not (Fig. 6-4).

A more detailed analysis of the association between diet and spermatophylax-eating in katydids is required. Wedell (1994) analyzed several reproductive variables – including relative size and protein content of the spermatophylax – in katydid species with different diets. Although the aim of the work was not to test the hypothesis that quality of diet is negatively correlated with spermatophylax quality,

Wedell did analyze the spermatophylax meals of seed- and leaf-eating species relative to the omnivorous taxa, whose diet is presumably of a higher protein (nitrogen) content (Strong *et al.* 1984; cited in Wedell 1994). Contrary to my prediction, however, it was the omnivorous and not the leaf-eating katydids that had the (significantly) largest spermatophylaces, and thus the greatest amount of protein donated. Moreover, spermatophylaces of seed-eaters had a greater percentage of protein (fresh mass) than either leaf-eaters or predacious taxa. These data remain inconclusive, however, because of uncertainty over whether differences between the three groups are due to adaptation or common ancestry; 'ecological grouping' (diet) was almost completely confounded with the subfamily grouping of the katydid species (Gwynne 1995b).

It is, however, premature to refute the fecundity-enhancement hypothesis, as acquisition (foraging) costs of acquiring seeds and prey may make protein a limiting factor for females in these diet groups. Future studies should estimate such costs.

CURRENT UTILITY OF COURTSHIP FEEDING IN ENSIFERA

Feeding before sperm transfer

Male hump-winged crickets, *Cyphoderris* (Haglidae), are unique among insects in providing a dual courtship meal to their mates; females receive a spermatophylax after copulation (Fig. 6-3c) and, during mating, they feed on the male's specialized pad-like hindwings and exuded hemolymph (Dodson *et al.* 1983) (Fig. 6-3d). Wing-eating appears to function in acquiring copulations. Morris *et al.* (1989) found male *C. strepitans* with reduced wing material due to mating were less successful in obtaining matings than were virgin males. Eggert and Sakaluk (1994) removed the hindwings and showed that without this courtship meal a male significantly reduced the chance of successful spermatophore transfer once a female had mounted. Wing-eating functioning to prevent sexual cannibalism can be ruled out, as female *Cyphoderris* confine their 'cannibalism' to the male's hindwings. A third hypothesis for wing-eating, paternal investment, has yet to be tested, however. There are no data on the effect of the male's donation of a meal of wings on female reproductive success. No study has tested the function of gland feeding before or during copulation. A species that deserves further work is a Japanese podoscirtine, *Truljalia hibicornis* (Gryllidae), in which females feed from a dorsal thoracic (metanotal) gland and receive 'unpackaged' sperm from the male; this species is unique in that it is the only ensiferan known that does not have a spermatophore (Ono *et al.* 1989).

Feeding during sperm transfer

The best-understood cases of courtship feeding in Orthoptera involve species in which females eat while being inseminated. Although thoracic-gland feeding in tree crickets (Gryllidae, Oecanthinae) may function in attracting females and inducing them to mount, most feeding occurs after the mount, during insemination, i.e. after the small spermatophore is transferred. Female black-horned tree crickets (*Oecanthus nigricornis*) feed on secretions for about two minutes before copulation but average a 12 min meal afterwards (Brown 1994). Boldyrev (1915) proposed his ejaculate-protection hypothesis to explain both postcopulatory gland feeding in tree crickets and spermatophylax-eating in katydids and certain camel crickets (Rhaphidophoridae). The alternative is that postcopulatory feeding functions as a paternal investment by increasing the fitness of the male's own offspring (Table 6-1).

Experiments with the black-horned tree cricket support an ejaculate-protection function; even though postcopulatory gland feeding had a dramatic effect on female lifetime fecundity, this effect is through increased female lifespan rather than an an increase in the rate of oviposition. Given that females remate, however, the male is unlikely to father the additional eggs laid by the female receiving additional courtship food (Brown 1994). These data, and the fact that duration of gland feeding correlates positively with the duration of ampulla attachment, favor ejaculate protection over paternal investment for the function of gland feeding in this tree cricket (Brown 1994).

Most tests of the ejaculate-protection and paternal-investment hypotheses have involved feeding on the spermatophylax. Predictions are listed in Table 6-2 and have usually been tested by manipulating the size of the spermatophylax meal to examine fecundity and the relationship between meal size and insemination. Other experiments have examined the relative paternities of two males mated to the same female. It is imortant to remember that the hypotheses are not mutually exclusive. The paternal-investment hypothesis is refuted if male-derived nutrition does not increase female or offspring fitness (Table 6-2) in a time period sufficient for the investing male to father those offspring he helps (nutritionally) to produce

(Wickler 1985). A spermatophylax larger than necessary to protect the ejaculate refutes the ejaculate-protection hypothesis (at least for the function of the additional courtship food provided). Therefore, with ejaculate protection the average time for meal consumption is expected to approximate the average time for full insemination to occur; the size of the meal should thus be directly related to the paternity of the male. Other predictions simply favor one hypothesis over the other (Table 6-2).

These predictions have been examined by D. T. Gwynne, K.-G. Heller, L. W. Simmons, N. Wedell and colleagues for the spermatophylax of four katydid species, each representing a different subfamily of Tettigoniidae (references and details in Table 6-2). For comparison there are experimental data from work by S. K. Sakaluk on a separate origin of the spermatophylax (Fig. 6-4) in the decorated cricket *G. supplicans* (Gryllidae). The results (Table 6-2, Fig. 6-7) suggest that current function varies between taxa, thus supporting the conclusion of Wedell (1993a) and refuting Gwynne's (1990a) contention that the current utility of this courtship meal of katydids in general is paternal investment (Wedell 1993a).

An ejaculate-protection function for the spermatophylax was supported for the cricket *Gryllodes*, and a katydid *Poecilimon veluchianus* (Phaneropterinae), as well as for the relatively small, low-protein offering of a tettigoniine katydid, the wart-biter *Decticus verrucivorus*. The conclusions for *Poecilimon* and *Gryllodes* are somewhat tentative as there may be an effect of spermatophylax nutrients on female fitness in both species (the relative size of hatched nymphs in *Poecilimon* (Reinhold and Heller 1993) and a non-significant, although very suggestive, effect on fecundity in *Gryllodes* (Will and Sakaluk 1994)). Wedell's studies of the ejaculate-protection function of the wart-biter spermatophylax are especially thorough (Table 6-2) and include demonstrating that the greater the male's investment, the greater his paternity (Fig. 6-8).

The function of the spermatophylax in two other katydids appears to differ from that of the wart-biter, and possibly the other species. *R. verticalis* has been studied in detail. In two studies of unmanipulated matings with either field- or laboratory-raised males, the mean size of the spermatophylax of *R. verticalis* was more than twice that necessary to protect the ejaculate (Gwynne *et al.* 1984; Gwynne 1986) (Fig. 6-7). The meal delays ampulla removal by the female for over five hours (the 3.8 h mean reported by Simmons (1995) may have resulted from some aspect of his experimental manipulation in which the spermatophylax

Table 6-2. *Predictions and tests of the ejaculate-protection (EP) and paternal-investment (PI) hypotheses for the current utility of the spermatophylax in four tettigoniids and a gryllid*

Prediction	*Requena verticalis* (Listroscelidinae)	*Decticus verrucivorus* (Tettigoniinae)	*Kawanaphila nartee* (Zaprochilinae)	*Poecilimon veluchianus* (Phaneropterinae)	*Gryllodes supplicans* (Gryllidae, Gryllinae)
(1A) Effect on fecundity or offspring-fitness: **refutes PI if not supported**. Increased number of meals: Increased size of meal:	Increases egg size and number Increases egg size but not number (1,2); other effects on offspring fitness (see text)	Experiment not done No effect on egg size, number or female lifespan (3)	Experiment not done Increases egg size and number (4)	Experiment not done Increases relative size of hatched larva but no effect on egg size or number (5)	No effect on egg number and size (6) Experiment not done
(1B) Timing of the increase in female fitness allows mating male to father the offspring he helps to produce: **refutes PI if not supported**	Paternity assured: fitness increases were measured over 30 d (2) but male nutrients incorporated into eggs (7) that are laid before females remate in nature (after about two weeks) (8)	Paternity not assured: not only do male nutrients have no effect on fecundity (3) but females also remate before nutrients are incorporated into eggs (9) and thus lose paternity in sperm competition (10)	Paternity assured: fecundity increases realized in eggs laid after 4 d, which is before the time of female remating (4)	Paternity not assured. Females remate after two days (11) and before oviposition (12) and thus lose paternity in sperm competition (12)	Paternity partially assured as paternity is proportional to spermatophylax size. Females remate quickly and before oviposition and thus paternity is mixed (13)
(2A) (Fig. 6-7). Time for female to eat spermatophylax greater than necessary for sperm and other ejaculatory substances to be transferred to female: **refutes EP**	Spermatophylax-eating time more than twice that taken for transfer of sperm and inducing a full refractory period (14)	Eating time similar to time taken to induce a full refractory period (3)	Eating time not greater than time taken to transfer sperm and induce a full refractory period (15)	Eating time not greater than time taken to transfer a full sperm complement (5)	Eating time similar to the time taken to transfer a full ejaculate (16)
(2B) Relative size of a male's food gift directly proportional to his paternity success: **refutes EP if not supported**	There is no significant correlation between the relative mass of the male's spermatophore and his paternity (spermatophylax mass not measured) (8)	Yes, there is a positive correlation between the relative mass of a male's spermatophylax and his paternity (Fig. 6-8) (9)	No data available	No data available	Relationship between spermatophylax-eating time, ampulla-attachment time and paternity (13,16) suggests that this may hold

Table 6-2. (*cont.*)

Prediction	Requena verticalis (Listroscelidinae)	Decticus verrucivorus (Tettigoniinae)	Kawanaphila nartee (Zaprochilinae)	Poecilimon veluchianus (Phaneropterinae)	Gryllodes supplicans (Gryllidae, Gryllinae)
(3) Spermatophore size as a proportion of male body size: **expected to be larger if PI**	20% (7)	9% (3)	21% (17)	25% (18)	3% (19)
(4) Cost of spermatophore (as measured by male remating interval on an *ad libitum* diet)	5 days (20)	1 day (3)	5.4 days (4)	3 days (11)	0.5 days (21)
(5) Spermatophylax quality (e.g. protein content) **expected to be higher if PI**	13.5% protein (mean fresh mass) (7)	4.2% protein (mean fresh mass) (22)	No data available	No data available	No data available

References (numbered in parentheses): (1) Gwynne (1984), (2) Gwynne (1988a), (3) Wedell and Arak (1989), (4) Simmons (1990), (5) Reinhold and Heller (1993), (6) Will and Sakaluk (1994), (7) Bowen *et al.* (1984), (8) Gwynne (1988b), Gwynne and Snedden (1995), (9) Wedell (1993b), (10) Wedell (1991), (11) Heller and von Helversen (1991), (12) Achmann *et al.* (1992), (13) Sakaluk (1986), (14) Gwynne (1986), Gwynne *et al.* (1984), (15) Simmons and Gwynne (1991), (16) Sakaluk (1984), (17) Gwynne and Bailey (1988), (18) Heller & Reinhold (1994), (19) Sakaluk (1985), (20) Gwynne (1990a), (21) Burpee and Sakaluk (1993), (22) Wedell (1993a).

was removed before consumption, weighed and returned to the female). The *Requena* spermatophylax is much higher in protein content than in the wart-biter; this is reflected in the strong effects of its nutrients on both fecundity and the fitness of the male's offspring. There is a positive influence on both egg size (which correlates with overwintering survival) (Fig. 6-9) and the growth rate, and possibly adult size, of sons (see Gwynne 1988a). The effects of a male's donations on his offspring is not a necessary prediction of the sexual-conflict hypothesis for the enlargement of the spermatophylax (Simmons and Gwynne 1991).

The conclusion that the *R. verticalis* spermatophylax is an enlarged paternal contribution in all cases (Gwynne *et al.* 1984; Gwynne 1986) was, however, premature. Just as male birds decrease parental care (feeding of chicks) when fitness returns from the mating are low (see, for example, Trivers 1985; Davies 1992), male *Requena* have been shown in laboratory experiments to reduce the size of their courtship meal when expected paternity returns are low, i.e. in matings with old females (Simmons *et al.* 1993). Males also

reduce offering size when potential mating rate is experimentally increased (Simmons *et al.* 1992; Simmons 1996). However, in contrast to other studied Ensifera (Fig. 6-7), ejaculate transfer in *R. verticalis* is not compromised even with the smallest spermatophylaces produced (Simmons 1995). It is worthwhile noting also that, when mated, field-collected males produced very large spermatophores (Gwynne 1986). Thus, the enlarged size of the meal in this species (relative to other Ensifera) appears to be currently maintained mainly through its effect on the fitness of the male's offspring (Gwynne 1988b).

The spermatophylax of the zaprochiline katydid *Kawanaphila nartee* is likely to be maintained by both natural and sexual selection pressures. The meal is not enlarged in size like that of *R. verticalis* (Fig. 6-7) but its nutrition has rapid postmating effects on the production of eggs that are probably fertilized by the mating male's sperm (Table 6-2).

Male animals are expected to realize the greatest reproductive success by increasing their fertilization of gametes (Alexander *et al.*, this volume). This conclusion is not only supported in studies for the male mating offerings of most ensiferans but also in studies of other arthropods in which males feed their mates. This includes prey-feeding by male bittacid scorpionflies (Thornhill 1976b) and dance flies (Svensson *et al.* 1990) and even a case in which the male donation represents his entire soma, i.e. male complicity in sexual cannibalism in red-back spiders (Andrade 1996). The parental contribution of the katydid *Requena verticalis* appears to be an exception; there are probably others. Likely examples are males of certain Lepidoptera and Coleoptera in which males can synthesize or ingest 'defensive' chemicals such as alkaloids. Evidence that these substances can enhance the fitness of the male's own offspring is that most of the chemical load of a male can be passed in the ejaculate in a single mating and the substances are incorporated into eggs, thus protecting them from predators (Eisner and Meinwold 1987, 1995; Eisner 1988; Dussourd *et al.* 1988).

CONCLUSIONS AND FUTURE RESEARCH

Differences between the four katydids returns us to the question of the original function of the spermatophylax, particularly in the tettigonioid clade (Fig. 6-4). As this meal was an ancestral addition to a simple spermatophore, it is reasonable to suggest that the small size of the newly evolved spermatophylax probably meant it had a sexually selected, ejaculate-protection function. This meal therefore appears to be a result of a history of sexual conflict. The paternal meal of *Requena* appears to be a derived feature (Gwynne 1986, 1990a). The outgroup to katydids, Haglidae (*Cyphoderris*), has a small spermatophylax (Fig. 6-3d) so is likely also to serve in ejaculate protection. The independent origin of the spermatophylax in *Gryllodes* as well as another origin – gland feeding in tree crickets (*Oecanthus*, Brown 1994) – appears also to function in ejaculate protection. The ejaculate-protection function of the wart-biter (*Decticus*) spermatophylax may represent the retention of the primitive state. Alternatively the paternal

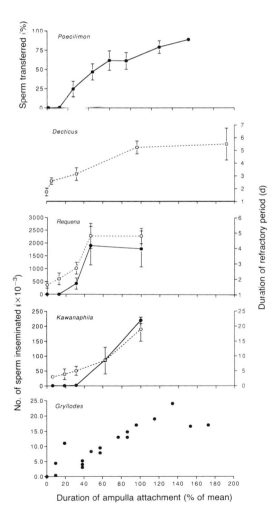

Fig. 6-7. The relationship between size of spermatophylax-meal and ejaculate transfer in a true cricket, *Gryllodes supplicans* (Gryllidae) (from Sakaluk 1984 at bottom) and four katydids (Tettigoniidae): *Poecilimon veluchianus* (Phaneropterinae) from Reinhard and Heller (1993), *Decticus verrucivorus* (Tettigoniidae) from Wedell and Arak (1989), *Requena verticalis* (Listroscelidinae) from Gwynne (1986), and *Kawanaphila nartee* (Zaprochilinae) from Simmons and Gwynne (1991) (Simmons and Gwynne presented refractory period data for several treatments. To be consistent with other studies, the data here are for females on an *ad libitum* diet that ate their spermatophylax). In order to compare taxa the duration of ampulla attachment (abscissa) is expressed as a percentage of the mean time for the female to remove the ampulla. Therefore, values in excess of 100 are mainly those in which the courtship meal was experimentally increased. Sperm inseminated (number, except for *P. veluchianus*, which was reported as a percentage of sperm in the ampulla transferred) (filled circles and lines) is on the left ordinate (note the large interspecific differences in number of sperm inseminated). Length of refractory period (open circles and dashed lines) is on the right ordinate. All data are shown as mean ± SE (except for *Gryllodes*, in which original data points are shown).

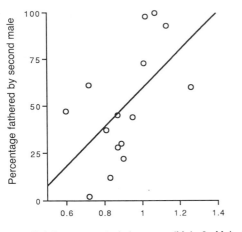

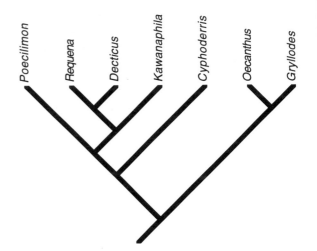

Fig. 6-8. Paternity and relative spermatophylax size in an experiment in which two male wart-biter katydids, *Decticus verrucivorus*, were mated to a single female. The plot shows the percentage of eggs fertilized by the second of the two males versus the mass of his spermatophylax relative to that of the first male ($r = 0.58$, $p < 0.025$: redrawn from Wedell 1991).

Fig. 6-10. Hypothesized relationships among the ensiferan taxa that have been subjects of experiments on the function of postcopulatory mate-feeding.

donation of *Requena* and *Kawanaphila*, taxa that branch off earlier in the clade, reflects a change in function of the spermatophylax at the base of this clade, with the ejaculate-protection role of the *Decticus* spermatophylax being secondarily derived (Fig. 6-10). These statements make the untested assumption that the function of the spermatophylax of each katydid species is representative of its subfamily. It is clear that additional detailed studies are required on the current utility of courtship feeding in both Tettigoniidae and related tettigonioid taxa (particularly the relatively unstudied Gryllacrididae, Stenopelmatidae and Rhaphidophoridae). Such studies should include experimental manipulation of the male offering to further test the predictions listed in Table 6-2 and should use phylogenetic relationships as a guide to appropriate study species. Detailed analyses of the components of the male-produced meals are also required to shed light on whether meals are designed to be highly nutritious or to serve some other function. These data will allow a better understanding of the selection pressures that initially caused ancestral camel-cricket-like males to evolve an elaborate and edible sperm sac.

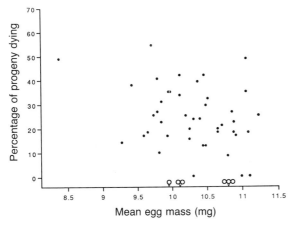

Fig. 6-9. Spermatophylax feeding and the fitness of female katydids, *Requena verticalis*. The bilobed spermatophylax was experimentally manipulated to provide three meal sizes of one, two or three lobes. Increasing size of spermatophylax eaten by females resulted in increased egg size (mass of five eggs) as shown by the three open-circle treatment symbols on the ordinate ($F = 3.9$, $p = 0.03$, covariates such as body size held constant). Egg size appears to be important to offspring fitness: larger eggs showed a significantly higher overwintering survivorship (filled circles, $r = -0.32$, $p = 0.02$). Redrawn from Gwynne (1988a).

ACKNOWLEDGEMENTS

Thanks to William Brown, Andrew Mason, Glenn Morris and especially David Rentz for discussion and information; and to Scott Sakaluk and Heather Proctor and

K.-G. Heller for their photos or artwork. Maydianne Andrade, Goran Arnqvist, George Eickwort, Mike Greenfield, Scott Sakaluk, Leigh Simmons and Nina Wedell provided important critical comments on the manuscript. Thanks to Michel Laurin and Graham Wallis for the use of MacClade. Maydianne Andrade assisted in figure preparation. This research was supported by a grant from NSERC (Canada).

LITERATURE CITED

Achmann, R., K.-G. Heller and J. T. Epplen. 1992. Last-male sperm precedence in the bushcricket *Poecilimon veluchianus* (Orthoptera: Tettigoniodea) demonstrated by DNA fingerprinting. *Molec. Ecol.* **1**: 47–54.

Alexander, R. D. 1962. The role of behavioral study in cricket classification. *Syst. Zool.* **11**: 53–72.

Alexander, R. D. and D. Otte. 1967. The evolution of genitalia and mating behavior in crickets (Gryllidae) and other Orthoptera. *Misc. Publ. Mus. Zool. Univ. Mich.* **133**: 1–62.

Ander, K. 1939. Vergleichend-anatomische und phylogenetische Studien uber die Ensifera (Saltatoria). *Opusc. Entomol. Lund (Suppl.)* **22**. (English translation by T. H. Hubbell 1971).

Andrade, M. C. B. 1996. Sexual cannibalism for male sacrifice in the Australian redback spider. *Science (Wash., D.C.)* **271**: 70–71.

Austad, S. N., and R. Thornhill. 1986. Female reproductive variation in a nuptial-feeding spider, *Pisaura mirabilis*. *Bull. Br. Arachnol. Soc.* **7**: 48–52.

Barrett, P. and G. W. Ramsey. 1991. *Keeping Wetas in Captivity*. Wellington, New Zealand: Wellington Zoological Gardens.

Bidochka, M. J. and W. A. Snedden. 1985. Effect of nuptial feeding on the mating behaviour of female ground crickets. *Can. J. Zool.* **63**: 207–208.

Boldyrev, B. T. 1915. Contributions a l' étude de la structure des spermatophores et des particularités de la copulation chez Locustodea et Gryllodea. *Horae Soc. Entomol. Rossicae* **6**: 1–245.

–. 1928. Biological studies on *Bradyporus multituberculatus* F. W. (Orth., Tettig.). *Eos* **4**: 15–57.

Bowen, B. J., C. G. Codd and D. T. Gwynne. 1984. The katydid spermatophore (Orthoptera: Tettigoniidae): male investment and its fate in the mated female. *Austr. J. Zool.* **32**: 23–31.

Brooks, D. and D. A. McLennan. 1991. *Phylogeny, Ecology and Behavior: A Research Program in Comparative Biology*. Chicago: University of Chicago Press.

Brown, W. D. 1994. Mechanisms of female mate choice in the black-horned tree cricket, *Oecanthus nigricornis* (Orthoptera: Gryllidae: Oecanthinae). Ph.D. dissertation, University of Toronto.

Burpee, D. M. and S. K. Sakaluk. 1993. Repeated matings offset costs of reproduction in female crickets. *Evol. Ecol.* **7**: 240–250.

Butlin, R. K., C. W. Woodhatch and G. M. Hewitt. 1987. Male spermatophore investment increases female fecundity in a grasshopper. *Evolution* **41**: 221–225.

Carlberg, U. 1981. Spermatophores of *Baculum extradentatum* (Brunner Von Wattenwyl), and other Phasmida. *Entomol. Mon. Mag.* **117**: 125–127.

Caudell, A. N. 1908. A old record of observations on the habits of *Anabrus*. *Entomol. News* **19**: 44–45.

Choe, J. C. and B. J. Crespi, eds. 1997. *The Evolution of Social Behavior in Insects and Arachnids*. Cambridge University Press.

Davies, N. B. 1992. *Dunnock Behaviour and Social Evolution*. Oxford University Press.

DeSutter, L. 1987. Structure et évolution du complexe phallique des Gryllidea (Orthopteres) et classification des génres néotropicaux de Grylloidea première partie. *Ann. Soc. Entomol. Fr. (N. S.)* **23**: 213–239.

Dodson, G. N., G. K. Morris and D. T. Gwynne. 1983. Mating behavior in the primitive Orthopteran genus *Cyphoderris* (Haglidae). In *Orthopteran Mating Systems: Sexual Competition in a Diverse Group of Insects*. G. K. Morris and D. T. Gwynne, eds., pp. 305–318. Boulder, Colorado: Westview Press.

Dussourd, D. E., K. Ubik, C. Harvis, J. Resch, J. Meinwald and T. Eisner. 1988. Biparental defensive endowment of eggs with acquired plant alkaloid in the moth *Utetheisia ornatrix*. *Proc. Natl. Acad. Sci. U.S.A.* **85**: 5992–5996.

Eberhard, W. G. 1985. *Sexual Selection and Animal Genitalia*. Cambridge, Mass.: Harvard University Press.

Eggert, A.-K. and S. K. Sakaluk. 1994. Sexual cannibalism and its relation to male mating success in sagebrush crickets *Cyphoderris strepitans* (Haglidae: Orthoptera). *Anim. Behav.* **47**: 1171–1177.

Eisner, T. 1988. Insekten als fürdicrsorgliche Eltern. *Verh. Dt. Zool. Ges.* **81**: 9–17.

Eisner, T. and J. Meinwold. 1987. Alakaloid-derived pheromones and sexual selection in Lepidoptera. In *Pheromone Biochemistry*, G. D. Prestwich and G. J. Blomquist, eds., pp. 251–269. Orlando, Florida: Academic Press.

–. 1995. The chemistry of sexual selection. *Proc. Natl. Acad. Sci. U.S.A.* **92**: 50–55.

Eluwa, M. C. 1978. Observations on the spermatophore and post-copulatory behaviour of three West African bush-crickets (Orth., Tettigoniidae). *Entomol. Mon. Mag.* **114**: 1–7.

Evans, A. R. 1988. Mating systems and reproductive strategies in three Australian gryllid crickets: *Bobilla victoriae* Otte, *Balamara gidya* Otte and *Teleogryllus commodus* (Walker) (Orthoptera: Gryllidae). *Ethology* **78**: 21–52.

Fabre, J. H. 1896. Étude sur les Locustiens. *Ann. Sci. Nat. Zool.* i, Ser. 8: 221–244.

–. 1918. *The Life of the Grasshopper*. (Translation by A. De Mateos.) London: Hodder and Stoughton.

Field, L. H. and G. R. Sandlant. 1983. Aggression and mating behavior in the Stenopelmatidae (Orthoptera, Ensifera), with special reference to New Zealand wetas. *In Orthopteran Mating Systems:*

Sexual Competition in a Diverse Group of Insects. D. T. Gwynne and G. K. Morris, eds., pp. 120–146. Boulder, Colorado: Westview Press.

Friedel, T. and C. Gillott. 1977. Contribution of male-produced proteins to vitellogenesis in *Melanoplus sanguinipes*. *J. Insect Physiol.* **23**: 145–151.

Forster, L. M. 1992. The stereotyped behaviour of sexual cannibalism in *Latrodectus hasselti* Thorell (Araneae: Theridiidae), the australian redback spider. *Austr. J. Zool.* **40**: 1–11.

Fulton, B. B. 1915. The Tree Crickets of New York: Life-History and Bionomics. *Tech. Bull. N. Y. Agric. Exp. Sta.*, no. 42.

Gangwere, S. K. 1961. A monograph on food selection in the Orthoptera. *Trans. Am. Entomol. Soc.* **87**: 67–203.

Gillette, C. P. 1904. Copulation and ovulation in *Anabrus simplex* Hald. *Entomol. News* **15**: 321–324.

Gorochov, A. V. 1988. The classification and phylogeny of grasshoppers (Gryllida-Orthopera, Tettigonioidea). In *The Cretaceous Biocoenotic Crisis and the Evolution of Insects*. A. Pomerenko, ed., pp. 145–190. Moscow: Hayka (in Russian).

Gwynne, D. T. 1983. Male nutritional investment and the evolution of sexual differences in the Tettigoniidae and other Orthoptera. *In Orthopteran Mating Systems: Sexual Competition in a Diverse Group of Insects*. D. Gwynne and G. Morris, eds., pp. 336–366. Boulder, Colorado: Westview Press.

Gwynne, D. T. 1984. Courtship feeding increases female reproductive success in bushcrickets. *Nature (Lond.)* **307**: 361–363.

–. 1986. Courtship feeding in katydids (Orthoptera: Tettigoniidae): investment in offspring or in obtaining fertilizations? *Am. Nat.* **128**: 342–352.

–. 1988a. Courtship feeding and the fitness of female katydids (Orthoptera: Tettigoniidae). *Evolution* **42**: 545–555.

–. 1988b. Courtship feeding in katydids benefits the mating male's offspring. *Behav. Ecol. Sociobiol.* **23**: 373–377.

–. 1990a. The katydid spermatophore: evolution of a parental investment. In *The Tettigoniidae: Biology, Systematics and Evolution*. W. J. Bailey and D. C. Rentz, eds., pp. 26–40, Bathurst, Australia: Crawford House.

–. 1990b. Testing parental investment and the control of sexual selection in katydids: the operational sex ratio. *Am. Nat.* **136**: 474–484.

–. 1995a. Phylogeny of Ensifera (Orthoptera): a hypothesis supporting multiple origins of acoustical signalling, complex spermatophores and maternal care in crickets, katydids and weta. *J. Orthopt. Res.* **4**: 203–218.

–. 1995b. Variation in bushcricket nuptial gifts may be due to common ancestry rather than ecology as taxonomy and diet are almost perfectly confounded. *Behav. Ecol.* **6**: 458.

Gwynne, D. T. and W. J. Bailey. 1988. Mating system, mate choice and ultrasonic calling in a zaprochiline katydid (Orthoptera: Tettigoniidae). *Behaviour* **105**: 202–223.

Gwynne, D. T., B. J. Bowen and C. G. Codd. 1984. The function of the katydid spermatophore and its role in fecundity and insemination. *Austr. J. Zool.* **32**: 15–22.

Gwynne, D. T. and W. A. Snedden. 1995. Paternity and female remating in *Requena verticalis* (Orthoptera: Tettigoniidae). *Ecol. Entomol.* **20**: 191–194.

Harvey, P. H. and M. D. Pagel. 1991. *The Comparative Method in Evolutionary Biology*. Oxford University Press.

Hayashi, F. 1992. Large spermatophore production and consumption in dobsonflies, *Protohermes* (Megaloptera: Corydalidae). *Jap. J. Entomol.* **60**: 59–66.

Heller, K.-G. and D. von Helversen. 1991. Operational sex ratio and individual mating frequency in two species of bushcrickets (Orthoptera: Tettigoniodea, *Poecilimon*). *Ethology* **89**: 211–228.

Heller, K.-G. and K. Reinhold. 1994. Mating effort function of the spermatophore in the bushcricket *Poecilimon veluchianus* (Orthoptera: Phaneropteridae): support from a comparison of the mating behaviour of two subspecies. *Biol. J. Linn. Soc.* **53**: 153–163.

Kessel, E. L. 1955. The mating activities of balloon flies. *Syst. Zool.* **4**: 97–104.

Khalifa, A. 1949. The mechanism of insemination and mode of action of the spermatophore in *Gryllus domesticus*. *Q. J. Microsc. Sci.* **90**: 281–294.

Lack, D. 1940. Courtship feeding in birds. *Auk* **57**: 169–178.

Maddison, W. P., and D. Maddison. 1992. *MacClade: Analysis of Phylogeny and Character Evolution*. Sunderland, Mass.: Sinauer Associates.

Mays, D. L. 1971. Mating behavior of nemobiine crickets – *Hygronemobius*, *Nemobius*, and *Pteronemobius* (Orthoptera: Gryllidae). *Fla. Entomol.* **54**: 113–126.

Morris, G. K., D. T. Gwynne, D. E. Klimas and S. K. Sakaluk. 1989. Virgin male mating advantage in a primitive acoustic insect (Orthoptera: Haglidae). *J. Insect Behav.* **2**: 173–185.

Ono, T., M. T. Siva-Jothy and A. Kato. 1989. Removal and subsequent ingestion of rivals' semen during copulation in a tree cricket. *Physiol. Entomol.* **14**: 195–202.

Otte, D. 1992. Evolution of cricket songs. *J. Orthopt. Res.* **1**: 25–48.

Otte, D. and R. D. Alexander. 1983. The Australian Crickets (Orthoptera: Gryllidae). *Academy of Natural Sciences of Philadelphia, Monograph* no. 22.

Parker, G. A. and L. W. Simmons. 1989. Nuptial feeding in insects: theoretical models of male and female interests. *Ethology* **82**: 3–26.

Pickford, R. and D. E. Padgham. 1973. Spermatophore formation and sperm transfer in the desert locust, *Schistocerca gregaria* (Orthoptera: Acrididae). *Can. Entomol.* **105**: 613–618.

Ramsey, G. W. 1955. The exoskeleton and musculature of the head, and the life-cycle of *Deinacrida rugosa* Buller, 1870. M. Sc. thesis, Victoria University of Wellington, New Zealand.

Reinhold, K. and K.-G. Heller. 1993. The ultimate function of nuptial feeding in the bushcricket *Poecilimon veluchianus* (Orthoptera: Tettigoniidae: Phaneropterinae). *Behav. Ecol. Sociobiol.* **32**: 55–60.

Rentz, D. C. 1963. Biological observations on the genus *Decticita* (Orthoptera: Tettigoniidae). *Wasmann J. Biol.* **21**: 91–94.

Rentz, D. C. F. 1993. *The Tettigoniidae of Australia, vol. 2. The Austro-saginae, Zaprochilinae and Phasmodinae.* Melbourne: CSIRO.

Rentz, D. C. F. and B. John. 1990. Studies in Australian Gryllacrididae: taxonomy, biology, ecology and cytology. *Invert. Taxon.* **3**: 1035–1210.

Richards, A. M. 1973. A comparative study of the giant wetas *Deinacrida heteracantha* and *D. fallai* (Orthoptera: Henicidae) from New Zealand. *J. Zool. (Lond.)* **169**: 195–236.

Ridley, M. 1983. *The Explanation of Organic Diversity: The Comparative Method and Adaptations for Mating.* Oxford University Press.

Sakaluk, S. K. 1984. Male crickets feed females to ensure complete sperm transfer. *Science (Wash., D.C.)* **223**: 609–610.

–. 1985. Spermatophore size and its role in the reproductive behaviour of the cricket *Gryllodes supplicans* (Orthoptera: Gryllidae). *Can. J. Zool.* **63**: 1652–1656.

–. 1986. Sperm competition and the evolution of nuptial feeding behavior in the cricket, *Gryllodes supplicans* (Walker). *Evolution* **40**: 584–593.

–. 1991. Post-copulatory mate guarding in decorated crickets. *Anim. Behav.* **41**: 207–216.

Simmons, L. W. 1988. The contribution of multiple mating and spermatophore consumption to the lifetime reproductive success of female field crickets (*Gryllus bimaculatus* De Geer). *Ecol. Entomol.* **13**: 57–69.

–. 1990. Nuptial feeding in tettigoniids: male costs and the rates of fecundity increase. *Behav. Ecol. Sociobiol.* **27**: 43–47.

–. 1995. Courtship feeding in katydids (Orthoptera: Tettigoniidae): investment in offspring and in obtaining fertilizations. *Am. Nat.* **146**: 307–315.

–. 1996. Male crickets tailor their spermatophores in relation to their remating intervals. *Funct. Ecol.* **9**: 881–886.

Simmons, L. W., M. Craig, T. Llorens, M. Schinzig and D. Hoskins. 1993. Bushcricket spermatophores vary in accord with sperm competition and parental investment theory. *Proc. R. Soc. Lond.* B **251**: 183–186.

Simmons, L. W. and D. T. Gwynne. 1991. The refractory period of female katydids (Orthoptera: Tettigoniidae): sexual conflict over the remating interval? *Behav. Ecol.* **2**: 276–282.

Simmons., L. W., R. J. Teale, M. Maier, R. J. Standish, W. J. Bailey and P. C. Withers. 1992. Some costs of reproduction for male bushcrickets, *Requena verticalis* (Orthoptera: Tettigoniidae): allocating resources to mate attraction and nuptial feeding. *Behav. Ecol. Sociobiol.* **31**: 57–62.

Smith, R. L. 1980. Evolution of exclusive postcopulatory paternal care in the insects. *Fla. Entomol.* **63**: 65–78.

Snodgrass, R. E. 1905. The coulee cricket of central Washington. (*Peranabrus scabricollis* Thomas). *J. N. Y. Entomol. Soc.* **13**: 74–82.

Strong, D. R., J. H. Lawton and R. Southwood. 1984. *Insects on Plants.* Oxford: Blackwell Scientific Publications.

Svensson, B. G., E. Petersson and M. Frisk. 1990. Nuptial gift size prolongs copulation in the dance fly *Empis borealis. Ecol. Entomol.* **15**: 225–229.

Tallamy, D. W. 1994. Nourishment and the evolution of paternal investment in subsocial arthropods. In *Nourishment and Evolution in Insect Societies.* J. H. Hunt and C. A. Nalepa, eds., pp. 21–56, Boulder, Colo.: Westview.

Thornhill, R. 1976a. Sexual selection and paternal investment in insects. *Am. Nat.* **10**: 153–163.

–. 1976b. Sexual selection and nuptial feeding behavior in Bittacus apicalis (Insecta: Mecoptera). *Am. Nat.* **110**: 529–548.

Thornhill, R. and J. Alcock. 1983. *The Evolution of Insect Mating Systems.* Cambridge, Mass.: Harvard University Press.

Trivers, R. L. 1972. Parental investment and sexual selection. In *Sexual Selection and the Descent of Man, 1871–1971.* B. Campbell, ed., pp. 1346–179, Chicago: Aldine.

–. 1985. *Social Evolution.* Menlo Park, CA: Benjamin Cummings.

Wedell, N. 1991. Sperm competition selects for nuptial feeding in a bushcricket. *Evolution* **45**: 1975–1978.

–. 1993a. Evolution of nuptial gifts in bushcrickets. Ph. D. dissertation, University of Stockholm.

–. 1993b. Spermatophore size in bushcrickets: comparative evidence for nuptial gifts as sperm protection devices. *Evolution* **47**: 1203–1212.

–. 1993c. Mating effort or paternal investment: incorporation rate and cost of male donations in the wartbiter. *Behav. Ecol. Sociobiol.* **32**: 239–246.

–. 1994. Variation in nuptial gift quality in bushcrickets. *Behav. Ecol.* **5**: 418–425.

–. 1995. Dual function of the bush cricket spematophore. *Proc. R. Soc. Lond.* B **258**: 181–188.

Wedell, N. and A. Arak. 1989. The wartbiter spermatophore and its effect on female reproductive output (Orthoptera Tettigoniidae, *Decticus verrucivorus*). *Behav. Ecol. Sociobiol.* **24**: 116–125.

Weissman, M. J. 1995. Natural history of the great sand treader camel cricket, *Daihinbaenetes giganteus* Tinkham (Orthoptera: Rhaphidophoridae) at Great Sand Dunes National Monument, Colorado. Ph. D. dissertation, Colorado State University.

Wickler, W. 1985. Stepfathers in insects and their pseudo-parental investment. *Z. Tierpsychol.* **69**: 72–78.

Will, M. W. and S. K. Sakaluk. 1994. Courtship feeding in decorated crickets: is the spermatophylax a sham? *Anim. Behav.* **48**: 1309–1315

Wilson, E. O. 1971. *The Insect Societies.* Cambridge, Mass.: Harvard University Press.

7 · The evolution of mating systems in the Zoraptera: mating variations and sexual conflicts

JAE C. CHOE

ABSTRACT

Two sympatric species of Zoraptera in central America, *Zorotypus barberi* Gurney and *Z. gurneyi* Choe, exhibit distinctively divergent mating behaviors before, during, and after copulations. The mating system of *Z. barberi* is essentially polygynandry in which both males and females mate with multiple mates. Males of *Z. barberi* perform elaborate precopulatory courtship involving courtship gifts, provide extra stimulation during copulation, and continue to court for additional copulations. Females appear to exert almost exclusive control over mating by being able to reject males at any point during the entire mating episode. In addition to deciding with whom to mate and how long each copulation lasts, *Z. barberi* females also control the frequency of copulations. They can mate with one male repeatedly, with different males, or both. Among the hypotheses examined, the material-benefit, postcopulatory female choice, sperm-supply, and fertilization-enhancement hypotheses, or some combinations of them, appear to provide good explanations for the observed mating variations in *Z. barberi*. Alternatively, repeated mating may be a result of 'parceling' of courtship gifts by males to guard females for a longer period of time. In *Z. gurneyi*, males gain considerable control over mating by establishing dominance hierarchies among themselves and thus predetermining female mating decisions to a certain degree. *Zorotypus gurneyi* males display no apparent precopulatory courtship; once genital coupling is made, they are able to prolong copulations by everting much of their internal genitalia into the female. After copulations, dominant *Z. gurneyi* males continue to protect their harems of females from other males. Consequently, body size and age (order of emergence) are important in determining dominance and reproductive success among *Z. gurneyi* males. Future studies are called to analyze the chemical composition of the cephalic secretion of *Z. barberi* males, to determine patterns of sperm transfer and the nature of parthenogenesis in *Z. gurneyi*, and most of all, to explore more species before much of the world's tropical forests disappear.

INTRODUCTION

The interests of mating females and males are often conflicting and asymmetrical (Darwin 1871; Trivers 1972; Parker 1979; West-Eberhard 1979; Eberhard 1985; Alexander *et al.*, this volume; Arnqvist, this volume; Dickinson, this volume). Although such interests may invariably overlap to a certain degree owing to a common goal, fertilization, some conflicts of interest occur at virtually every step in the mating episode. Generally, males attempt to maximize their fitness by increasing the *quantity* of matings, whereas females may benefit only by increasing the *quality* of matings. Polyandrous mating systems with paternal care are exceptions in which females may indeed increase fitness quantitatively.

Copulation is not always necessary for successful fertilization. In salamanders and collembolans, for instance, males deposit spermatophores on the substrate and females simply collect them to fertilize eggs. Copulation in many animals with internal fertilization may have evolved via the male's struggle to deliver sperm as close to eggs as possible (see also Alexander *et al.*, this volume). Although copulation may have been a male's intrusion into the female's body, having fertilization occur within their bodies may have provided females with nearly endless possibilities of 'covert' female choice since 'copulation does not always result in insemination nor does insemination always result in fertilization of eggs' (Eberhard 1991). None the less, copulation is the event of critical importance, and thus it is useful to analyze conflicts between the sexes at three distinct phases: before, during, and after copulations.

Two sympatric species of Zoraptera in central America, *Zorotypus barberi* and *Z. gurneyi*, provide excellent systems to illustrate sexual conflicts at various stages of the mating episode. These two species exhibit distinctively divergent mating strategies before, during, and after copulations. Male and female *Z. barberi* display elaborate courtship in which males present secretions from their cephalic gland as a courtship gift (Choe 1992, 1995). In contrast,

Z. gurneyi males do not possess such glands and display no apparent courtship prior to copulations. Instead, they form highly linear and stable dominance hierarchies in which the dominant male defends a harem of females (Choe 1992, 1994a,b). Following the classification of Alexander *et al.* (this volume), *Z. barberi* males show 'luring acts' whereas *Z. gurneyi* males do 'coercive' acts. Once genital coupling is made, *Z. barberi* males and females are likely to copulate repeatedly with the same mate, whereas the mating couple of *Z. gurneyi* engage in prolonged copulation. After copulations, both males and females of *Z. barberi* try to mate with additional mates, whereas dominant *Z. gurneyi* males continue to protect harems. In the following account, I describe the dynamics of conflicts between males and females of these two species of Zoraptera with respect to sexual selection, game theory, and life-history adaptation.

BIOLOGY OF THE ZORAPTERA

Zoraptera is among the least explored orders of insects. To date, only 30 species, all of which belong to a single genus, *Zorotypus* (family Zorotypidae), have been described (Choe 1990, 1992). They are pantropical, occurring in the tropical and subtropical regions of all continents except Australia (Fig. 7-1). Zorapterans are minute (2–4 mm long), soft-bodied insects that live colonially beneath the bark of rotting trees in tropical and subtropical regions. Nymphs are pale cream-colored; adults are much darker in most species. *Zorotypus hubbardi*, which occurs throughout the southeastern United States, may be an exception. In this species, there is little difference in body color between adults and nymphs. Two distinct morphs, winged and wingless, are often present in the same colony. Wingless individuals are also eyeless and much more common than winged ones. Winged individuals have a pair of darkly pigmented, fully developed compound eyes and three ocelli. Both sexes can be winged, but winged males are extremely rare. Considering that newly emerged winged females generally mate with males in the natal colony before they disperse to found new colonies (Choe 1990), selection for wingedness in males may be weak. Like reproductives in ants and termites, winged zorapterans are capable of shedding their wings. The broken-off wing-stubs persist in dealated individuals, as in termites.

Zorapteran females lay on average about two dozen eggs during their lifetime (Choe 1990). Eggs take 6–7 weeks to develop in *Z. gurneyi* and 8–9 weeks in *Z. barberi*. However, each zorapteran egg represents a large maternal investment owing to its unusually large size (nearly as large as the abdomen of the female). In other words, the general life-history traits of zorapterans are those of '*K*-selected' organisms: low reproductive output and slow development. It is counterintuitive why they have evolved *K*-selected traits, living in such ephemeral habitats as decaying logs.

Zorapterans are omnivorous (Choe 1992). They feed on fungal spores and hyphae, and dead arthropods. They are also predatory, capturing and consuming various prey

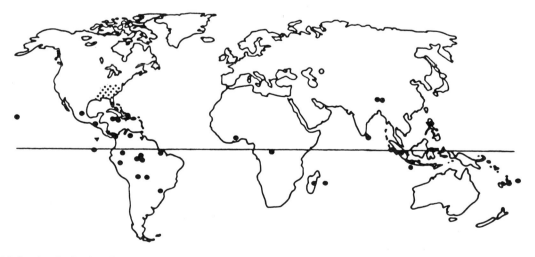

Fig. 7-1. Species distribution of the Zoraptera. The dotted area denotes the distribution range of *Z. hubbardi*; triangles denotes collection sites of *Z. barberi*.

Table 7-1. *Comparisons of mating behavior between* Zorotypus barberi *and* Z. gurneyi

Trait	*Z. barberi*	*Z. gurneyi*
Mating system	polygynandry	female defense polygyny
Sexual size dimorphism	small	arge
Microhabitat	galleries	open space
Developmental rate	faster	slower
Parthenogenesis	absent	present
Multiple mating by females	moderate	low
Copulation duration	short	long
Variance in copulation duration	large	small
Determinants of mating success	courtship feeding, courtship skills	body size, age, fighting skills
Major sexual selection force	female choice	male–male competition

such as collembolans, mites, and small nematodes. Some species are also found to ingest tiny wood fragments and harbor a variety of living protozoans and bacteria in their guts (J. C. Choe, personal observation). Although the species composition of the microorganism community appears relatively consistent between individual zorapterans, whether these microorganisms are fermentating symbionts has not been investigated.

Detailed information of mating behavior and social structure in Zoraptera comes from the studies of only two species, *Z. barberi* and *Z. gurneyi* (Choe 1990, 1994a, 1994b, 1995a). These two species are sympatric in central Panama, sometimes co-occuring under the bark of the same log (Choe 1989, 1992). Adult *Z. barberi* are about 2 mm long, measured from the head to the tip of the abdomen excluding antennae, and have an amber-colored or brownish exoskeleton. *Zorotypus gurneyi* are much darker and larger, about 2.5 mm long. In addition to differences in their external morphology (Choe 1989), the two species differ strikingly in mating behavior and life-history patterns (Table 7-1).

DIVERGENCE IN MATING BEHAVIOR

Mating behavior

Zorotypus gurneyi

The mating system of *Z. gurneyi* represents a form of female defense polygyny (Choe 1994a,b). Dominant males actively prevent other males from gaining access to female groups or harems but do not defend resources or oviposition sites (see Fig. 2 of Choe 1994a). Males establish and

maintain highly linear and stable dominance hierarchies. The most dominant male tends to walk slowly in the center of the colony, often vibrating his antennae in a rhythmic fashion, while subordinate males move swiftly along the edge of the colony arena. The dominant males constantly chase and attempt to copulate with females. Males exhibit little precopulatory courtship. After a brief antennation and/or jerking display by the male, a responsive female turns around and steps backwards toward the male. The male then quickly copulates with the female by sliding into the upside-down position. The female remains in the forward-upright position and faces the opposite direction from the male during copulation. A copulation lasts for nearly an hour on average. As found in other polygynous vertebrates, such as red deer (Clutton-Brock *et al.* 1982) and elephant seals (LeBoeuf 1974), the dominant males obtain the majority of copulations, nearly three-quarters of all copulations recorded from a colony.

Female *Z. gurneyi* can also reproduce parthenogenetically. There are two types of *Z. gurneyi* colonies in the field: all-female colonies and bisexual colonies. Sex ratios in bisexual colonies are not significantly different from 1:1 (J. C. Choe, unpublished data). Colonies on Barro Colorado Island in the Panama Canal lack males, whereas all-female and bisexual colonies coexist in the mainland along the Canal. In all-female colonies, females become more sedentary and territorial, aggressively defending small areas from other females. They are thelytokously parthenogenetic, laying unfertilized eggs, which become females. When males are made available experimentally, females appear to readily mate with males and produce offspring of both sexes (J. C. Choe, unpublished data). My preliminary data suggest that parthenogenetic females lay more eggs than bisexual females. It would be interesting to examine costs and benefits of parthenogenesis in this species, and investigate why *Z. barberi* has not evolved to be parthenogenetic.

Zorotypus barberi

The mating behavior of *Z. barberi* involves courtship feeding and repeated copulations (Choe 1995). The male displays an elaborate and rigid sequence of precopulatory courtship. When a male encounters a female, he immediately begins antennating her on the antennae and sometimes on the head. If the female antennates him back, the male then flips his antennae backward, vibrating rhythmically, and starts lowering his head while keeping the thorax in the same position. With his neck fully extended vertically, the male moves steadily towards the female (Fig. 7-2). When he moves within approximately one body-length from the female, a bulged-out skin fold protrudes from the middle of his extended neck with the tip of the bulge pointing in the direction of the female. The male can sustain the erected bulge for up to five seconds at a time. A scanning electron micrograph reveals a depressed area where the bulge appears (Fig. 7-3).

When the female takes a step toward the approaching male, he suddenly retracts the neck bulge and secretes a droplet of liquid through the opening of the cephalic gland in the middle of the head. As the female imbibes the secretion from the tufts of hairs at the gland opening

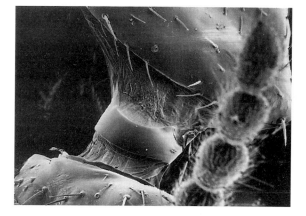

Fig. 7-3. A scanning electron micrograph of a *Z. barberi* male's extended neck. The depressed area in the center is where the bulge occurs.

(Fig. 7-4), both the male and female begin turning as if on a swivel and curl their abdomens towards each other. As soon as the male couples with the female, he slides into the upside-down position while the female maintains the forward-upright position. The male and female face in opposite directions while *in copula*. A copulation is brief, lasting less than a minute on average. Mating pairs usually copulate three or more times in succession. Before each copulation, however, the male performs the entire sequence of courtship including the presentation of cephalic secretion as a courtship gift. Females also mate with multiple mates.

Fig. 7-2. A *Z. barberi* male (left) courting a female. Notice the male's lowered head position.

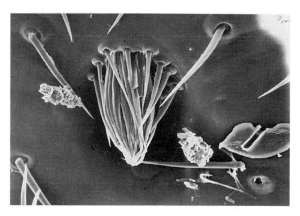

Fig. 7-4. A scanning electron micrograph of a *Z. barberi* male's head, showing the tufts of hairs above the opening of the cephalic gland.

Determinants of mating success

Relative importance of size and age in *Z. gurneyi*

Body size is probably the most commonly investigated parameter influencing the outcome of animal contests by game theorists (see, for example, Parker 1974; Hammerstein 1981) and has been found to be well correlated with the ability to win aggressive and/or ritualistic contests in a variety of insects and arachnids such as jumping spiders (see, for example, Jackson 1982), crickets (Alexander 1961), beetles (Eberhard 1979, 1980; Hughes and Hughes 1982; Johnson 1982), thrips (Crespi 1986a,b, 1988), scorpionflies (Thornhill 1984), dungflies (Borgia 1980), bees (Alcock 1979; Michener 1990) and wasps (Pardi 1948).

Body size is in general a reliable predictor of relative dominance among males in *Z. gurneyi* where male–male competition is important (Choe 1994a). It is a particularly good predictor of dominance outcome when newly emerged males from different colonies are brought together for the first time in the laboratory (Choe 1994b). In all eight of such experimentally constructed colonies, coefficients of Spearman's correlation between dominance rank and body size were highly significant.

Zorotypus gurneyi males exhibit a variety of dominance-related behavior. When two males come within one body-length of each other, both males briefly vibrate their antennae rhythmically and the subordinate often walks away, avoiding direct physical contacts. This so-called contact avoidance behavior is by far the most common interaction between males (Choe 1994a,b). Nearly as frequently as avoiding contacts, males also butt their heads against one another after a brief antennation. Head-butting is often a prelude to more escalated fights and may be a type of assessment behavior in *Z. gurneyi*. Through this relatively injury-free contact, individual zorapterans may be able to measure relative strength of the opponent. Males are sometimes engaged in a rear-to-rear hindleg-raising display and subsequently walk away from or kick at one another (Fig. 7-5). Following head-butting, hindleg-raising,

Fig. 7-5. A sketch of the rear-to-rear hindleg-raising display by two contesting *Z. gurneyi* males.

or contact avoidance behavior, a more dominant male often chases the subordinate. The most escalated form of intermale aggression is the grapple fight. During this brief but fierce fight, males grab each other's body parts using their hindlegs and roll around together. Injuries to grappling zorapterans are common; some males lose antennal segments, legs, or even their heads. Dominant males show frequent jerking movements, possibly involving stridulation, most of which are not necessarily aimed at any particular individual.

The most unusual aspect of intermale aggression is hindleg-kicking behavior. During this rear-to-rear display the hindlegs of both males, including the heavy spines on the posterior margin of the hindfemora, are pointed directly at one another. Often both males pause in this position for two or three seconds before they kick at or walk away from each other (Fig. 7-5). Although the mechanism of assessment is not completely understood, the males in the rear-to-rear position seem able to determine differences in size, dominance rank, or both, and to behave accordingly (Choe 1990). Such 'gauge-strength-by-length' types of assessment display have been observed in several insects and arachnids such as ant-mimicking Australian jumping spiders (Jackson 1982), gall aphids (Whitham 1979), thrips (Crespi 1986b), honeypot ants (Hölldobler 1976), drosophilid flies (Spieth 1981), stalk-eyed flies (McAlpine 1979; Burkhardt and de la Motte 1983; Wilkinson and Dodson, this volume) and antlered flies (Moulds 1977; Dodson 1989; Wilkinson and Dodson, this volume).

To determine why *Z. gurneyi* males sometimes kick at the opponent and other times simply walk away from one another, two separate breeding experiments were conducted. A general description of experimental procedures and methods is given elsewhere (Choe 1994b). In what I call simultaneous introduction experiments, eight experimental colonies were formed by simultaneously introducing into each colony six virgin males of varying sizes and ages from different stock colonies. In the sequential introduction experiment, six last-instar nymphs were grouped together to form an experimental colony, allowing individuals to emerge at different times. Development of dominance hierarchy was recorded by observing intermale interactions for 60 hours during five to six consecutive days. In experimental colonies where males were simultaneously introduced, males were more likely to walk away from the rear-to-rear confrontation without kicking when the difference in hindfemoral length between the opponents was equal or less than 0.002 mm (one standard deviation) than when it was greater (Table 7-2; G-test, $G = 73.93$, $p < 0.001$). Similarly, males were more likely to walk away after the display in experimental colonies where males were sequentially introduced when the difference in emergence order was equal or less than two than when it was greater (Table 7-2; $G = 8.54, 0.001 < p < 0.005$).

Game-theory models predict that realized fighting levels will depend upon the interaction of such parameters as relative fighting ability, relative resource value, ownership status, and assessment ability by both contestants (Parker and Rubenstein 1981; Maynard Smith 1982). Although most animal contests are settled by assessing relative strength of the opponent (Maynard Smith and Parker 1976), accurate information about relative fighting ability may only be acquired by trial contests rather than assessment (Parker and Rubenstein 1981). *Zorotypus gurneyi* males might have engaged in agonistic interactions more frequently owing to lack of information about one another when they were simultaneously introduced (Table 7-2). They were also more likely to escalate fights, as indicated by relatively higher frequencies of head-butting, hindleg-kicking, and grappling (Table 7-2). In the sequential experiment, however, young males were born to more or less established hierarchies and thus showed more frequent assessment and avoidance behavior (Table 7-2). Frequency distributions of intermale interactions were significantly different between the two experiments ($\chi^2 = 9.63$, $p < 0.05$).

Although higher-ranking males are generally larger in *Z. gurneyi*, the most dominant males are not always the largest (Choe 1994a). In approximately half of all field-collected colonies, dominance rank and body size were not significantly correlated (Choe 1990, 1994a). A similar trend has been found in many social insects. For instance, there is a tendency in *Polistes* wasps for dominant individuals to

Table 7-2. Relative frequencies (%) of intermale interactions in *Zorotypus gurneyi*

Type of interaction	Simultaneous introduction	Sequential introduction
Contact avoidance	8126 (33.2)	8163 (40.5)
Chasing	6617 (27.0)	6924 (34.4)
Head-butting	7760 (31.7)	4550 (22.6)
Hindleg-kicking	1749 (7.1)	115 (0.6)
Grappling	241 (1.0)	6 (0)

be larger than their subordinates (see, for example, Noonan 1981; Sullivan and Strassmann 1984), but the dominant foundresses are not always the largest females (Turillazi and Pardi 1977; Strassmann 1983). In *P. fuscatus*, successful usurpers are not necessarily larger than displaced females (Klahn 1981).

In addition to body size, a variety of other factors such as age, sex, experience, territorial familiarity, dominance status of parents, group size, and hormones may affect dynamics of dominance in group-living animals (see, for example, Schein 1975; Wilson 1975; Hand 1986). In *Z. gurneyi*, effects of age, more specifically tenure, i.e. the length of time a male has lived in a colony, sometimes override those of body size (Choe 1994b). In the sequential introduction experiment described above, body size was significantly correlated with dominance rank in less than half of the colonies. Instead, correlations between dominance rank and tenure or emergence order were significant in all but one colony. When males were simultaneously introduced to one another, however, body size was highly correlated with dominance rank in all experimental colonies whereas correlations between tenure and dominance rank were marginally significant in only two out of eight colonies. As the conditions in the sequential introduction experiment more closely simulate the natural process of colony development, males emerge throughout the year and newly emerged males appear to start at the bottom of the hierarchy.

Age has been found to influence dominance relations in other insects. In the field cricket *Gryllus integer*, mature males tend to initiate and win aggressive encounters more frequently than immature males, although age does not always predict the outcome of fights after maturity (Dixon and Cade 1986; Zuk and Simmons, this volume). Age-based dominance hierarchies have been described in *Polistes* wasps (West-Eberhard 1969; Hughes and Strassmann 1988). In *P. instabilis* emergence order determines dominance relations among workers even though dominant workers are often smaller than their younger subordinates (Hughes and Strassmann 1988). In experimentally constructed colonies of various halictine bees, the oldest female, rather than the largest female, is more likely to become the queen (Michener 1990 and references therein). Schwarz and Woods (1994) experimentally demonstrated that first eclosed females of the allodapine bee *Exoneura bicolor* have significantly larger ovaries and become reproductively dominant. In this species, dominant females are generally larger but the actual correlation between body size and dominance rank is weak (Schwarz and O'Keefe 1991). Schal and Bell (1983) also found that age is more important than size in determining the outcome of intermale contests in the cockroach *Nauphoeta cinerea*.

Prolonged and escalated fights tend to occur more often between evenly matched contestants (Parker 1974). However, disputes may also be settled without escalation if potential costs due to injuries exceed the resource value and asymmetrical pay-offs fall below a threshold value (Hammerstein 1981). The cost of fighting in *Z. gurneyi*, grapple-fighting in particular, is considerable and often causes an injured male to descend the hierarchy (J. C. Choe, unpublished). Having intact hindlegs is crucial to *Z. gurneyi* males, because they use hindlegs and the heavy spines on them as weapons during escalated fights such as hindleg-kicking and grapple fights. In fact, none of the dominant males in field-collected colonies was without a hindleg (Choe 1994a). Evenly matched *Z. gurneyi* males may assess each other for a longer period of time but avoid potentially costly fights. There was a trend, though not significant, for longer rear-to-rear assessment displays between more closely-matched opponents. In general males paused for 2.3 ± 2.8 s ($n = 52$), before they walked away, while they paused for 2.1 ± 1.6 s ($n = 32$, Mann-Whitney U-test, $Z = 1.914$, $0.1 < p < 0.5$) before they kicked at the opponents. Crespi (1986b) observed that males of a tubuliferan thrips *Elaphrothrips tuberculatus* assess size and fighting ability of the opponent through a behavior that he called 'parallel walks'. Male thrips of similar size are more likely to engage in prolonged multiple bouts. Because most zorapterans are eyeless and live in darkness, losing antennal segments appears to impair their communication ability and possibly their fighting ability as well (J. C. Choe, unpublished). Dominant males with their antennae removed do not initiate agonistic behavior and eventually lose their high-ranking status in the cockroach *Nauphoeta cinerea* (Schal and Bell 1983).

Assortative mating in *Z. barberi*

Unlike *Z. gurneyi*, in which males are much larger than females (Choe 1994a,b), there is little sexual size dimorphism in *Z. barberi*. Comparison of female and male size among copulating pairs showed that *Z. barberi* mate assortatively with respect to body size (J. C. Choe, unpublished manuscript). Crespi (1989a) identified three hypotheses for the proximate causes of size-assortative mating in arthropods: mate choice, mate availability, and mating

constraints. Male choice of large females combined with male–male competition (Johnson 1982; Ridley 1983) or mate preference for large mates by both sexes (Burley 1983; Brown 1990) may cause assortative mating. Assortative mating may also result from varying degrees of mate availability combined with male–male competition (Crespi 1989b). Finally, assortative mating may occur because of physical difficulties associated with courtship performance or genital coupling (Price and Willson 1976; Juliano 1985; Pinto and Mayor 1986; Brown 1993).

In *Z. barberi*, neither females nor males appear to prefer large mates (J. C. Choe, unpublished manuscript). Pairs courting or initiating copulations show no assortative pattern with respect to body size. The head-to-head courtship of *Z. barberi* in the dark may limit an individual's ability to assess the body size of potential mates. During the genital coupling stage, the female curls her abdomen anterolaterally while imbibing the secretion from the gland opening located at the center of the male's head (Choe 1995). In order to achieve a successful copulation, the male has to rotate his abdomen dorsoventrally and at the same time curl it anterolaterally toward the female's rear while maintaining his head at the fixed position so that the female can have an uninterrupted access to the cephalic secretion (Fig. 7-6). In a series of experiments in which males were painted over their gland openings and allowed to court females, such impaired males were found to court

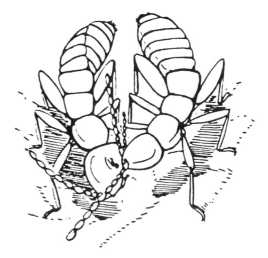

Fig. 7-6. A sketch of the feeding–coupling phase of precopulatory courtship in *Z. barberi*. The male must maintain his head at a fixed position so that the female can continue feeding on cephalic secretion, while he rotates his abdomen toward the female's rear.

with no apparent difficulties but never successfully copulated with females. Females who had begun curling their abdomens were still able to retreat when they discovered no secretion at the opening of the cephalic gland. Large males courting much smaller females often appeared not able to curl their abdomens far enough to connect with the female's genitalia. Similar difficulties were also present in copulation attempts between small males and large females. Although Ridley (1983) argued that assortative mating could not have evolved purely through mechanical constraints because selection on both sexes should be strong enough to overcome such difficulties, assortative patterns may emerge because disassortative matings are less likely (Crespi 1989a). In the absence of mate choice by size, both mate choice and mate availability hypotheses do not seem to explain assortative mating patterns in *Z. barberi*. Instead, simple physical constraints, i.e. inability of divergent size combinations to to copulate, appear to cause assortative mating in *Z. barberi*.

Sexual conflict and mating patterns

Multiple mating by females is widespread across various taxa (Birkhead *et al.* 1987; Ridley 1988; Birkhead and Møller 1992). Although males should mate with as many different females as possible (Darwin 1871; Bateman 1948; Trivers 1972), it is not obvious why females should mate multiply despite potential costs of additional matings (Daly 1978; Thornhill and Alcock 1983; Arnqvist 1989; Dickinson, this volume). Females may copulate multiply with more than one mate (multiple mating) or repeatedly with the same mate (repeated mating). Females may also prolong a copulation with any particular mate (prolonged mating). These three types of mating have different evolutionary consequences and thus different sets of hypotheses are proposed to explain them (Walker 1980; Thornhill and Alcock 1983; Hunter *et al.* 1993; Alcock 1994; Keller and Reeve 1995; Dickinson, this volume).

Female mating patterns
Females may mate with multiple mates to (1) acquire sufficient sperm supplies to achieve full fecundity (*sperm-supply hypothesis*: Birkhead *et al.* 1987; Ridley 1988), (2) ensure fertility (*fertility-insurance hypothesis*: Birkhead *et al.* 1987; Ridley 1988; Small 1990), (3) select mates based on their copulatory courtship (*postcopulatory female choice hypothesis*: Eberhard 1985; Dewsbury 1988), (4) gain genetic benefits (*genetic-benefit hypothesis*: Williams 1975; Maynard Smith

1978; Keller and Reeve 1995) by producing genetically diverse progeny, increasing the probability of being fertilized by genetically superior males, or hedging against genetic defects, (5) gain material benefits (*material-benefit hypothesis*: Hamilton 1971; Thornhill & Alcock 1983; Chen 1984; Wilcox 1984), i.e. nutritional benefits in return for mating, increased foraging success, reduced risk of predation, or protection from other males, (6) outcompete other females by reducing the likelihood of fertilization for other females or preventing other females from gaining access to high quality males (*female mate-guarding hypothesis*: Low 1979; Petrie 1992), or (7) avoid the costs of preventing additional matings (*convenience polyandry hypothesis*: Parker 1970; Thornhill and Alcock 1983; Waage 1984).

All of the above hypotheses except fertility insurance, postcopulatory female choice, and genetic benefit can explain why females mate repeatedly with the same male. In addition, repeated mating may also help females to (8) devalue the sperm from a previously accepted but inferior male by selectively and repeatedly copulating with a preferred male (*sperm-dilution hypothesis*: Simmons 1987), (9) secure paternal care by assuring paternity (*paternity-assurance hypothesis*: Davies *et al.* 1992), (10) establish and maintain a pair bond (*pair-bonding hypothesis*: Møller 1987; Birkhead & Lessells 1988), or (11) receive extra stimulation necessary to induce ovulation or oviposition (*sexual-stimulation hypothesis*: Chen 1984; Dewsbury 1984). The sperm-dilution hypothesis and sexual-stimulation hypothesis can be special cases of the postcopulatory female choice hypothesis. The material-benefit, female mate-guarding, and sexual-stimulation hypotheses may also provide some explanations for prolonged mating by females.

Male mating patterns
Reasons for repeated or prolonged mating by males are not clear. Males may mate with the same female repeatedly to (1) supply sufficient sperm to achieve full fecundity (*sperm-supply hypothesis*: Ridley 1988), (2) provide sexual stimulation to ensure fertilization (*sexual-stimulation hypothesis*: Eberhard 1985), (3) enhance fertilization rate (*fertilization-enhancement hypothesis*: Parker 1978; Otronen 1994), (4) reduce sexual receptivity of females (*receptivity-reduction hypothesis*: Chen 1984; Gillott 1988), (5) prevent other males from gaining access to females (*male mate-guarding hypothesis*: Alcock 1994), (6) numerically outcompete other males in sperm competition (*sperm-loading hypothesis*: Parker 1970), or (7) provide material benefits to increase the quantity and/or quality of offspring (*paternal-investment hypothesis*:

Gwynne 1984b). The sexual-stimulation, fertilization-enhancement, and receptivity-reduction hypotheses can be considered cases of the postcopulatory female choice hypothesis. Prolonged mating by males may also be explained by the same set of hypotheses.

Prolonged mating in Z. gurneyi and conflicts over copulation duration

Zorotypus barberi copulate for less than a minute, but *Z. gurneyi* copulate for much longer, up to nearly an hour (Table 7-3). In *Z. gurneyi*, males evert much of their internal genitalia into the female and appear to have control over copulation duration (Choe 1994a). In *Z. barberi*, however, females appear capable of terminating a copulation at nearly any time during the copulation episode (Choe 1995). The hook-shaped projection on the male's tenth tergum (Fig. 1a in Choe 1995) helps him open up the female's genital chamber but does not help him hold onto the female. Consequently, variance of copulation duration is much smaller in *Z. barberi* than in *Z. gurneyi*, although the total duration of copulation is considerably shorter in the former.

Although it is possible that female *Z. gurneyi* also benefit from prolonged mating associations, observations on copulation behavior suggest that it is the male that determines when to terminate a copulation. Why would male *Z. gurneyi* prolong a copulation? Knowledge on the timing and mechanism of sperm transfer will help determine which of the hypotheses suggested above explain the observed mating pattern. The observed pattern is especially puzzling from the viewpoints of the male mate-guarding hypothesis. Considering that the dominant male behaviorally guards a harem of females from other males, it seems equally, if not more, logical that he should mate briefly but repeatedly. While he is copulating with a female, lying on his back for nearly an hour, his harem is temporarily disbanded and subordinate males have better access to females. None the less, few subordinate males successfully copulate during the time the dominant male is *in copula*, perhaps because most females seem occupied by

the mating couple (J. C. Choe, personal observation). Mating couples are often attacked by others, the majority of whom are females (Choe 1994a). This observation provides a hint that females may also compete for the best male and guard him against other females in the harem by prolonging a copulation.

Repeated mating in Z. barberi and conflicts over copulation frequency

Female *Z. barberi* exercise both overt, precopulatory and covert, postcopulatory female choice (Choe 1995). Male *Z. barberi* present cephalic secretion as a courtship gift to entice females to copulate and continue to stimulate females even after genital intromission has begun. Mate rejection by females commonly occurs during both pre- and postcopulatory courtship (Fig. 7-7). Excluding the courtship rejections (71.1%) during the approaching phase of precopulatory courtship that is largely an indication of female sexual receptivity, receptive females reject mates during postcopulatory courtship just as frequently as during precopulatory courtship. Successful copulations are achieved only when the male's probe-like intromittent organ is able to penetrate deep inside the female, throbbing constantly, and various 'contact courtship devices' (Eberhard 1985) such as setae on the male's apical terga continually provide additional stimulation of the female. Postcopulatory female choice has also been observed from various insects, including several species of beetle (Eberhard 1990, 1991, 1992, 1993; Dickinson, this volume), the scorpionfly *Harpobittacus nigriceps* (Thornhill 1983), the mosquito *Sabethes cyaneus* (Hancock *et al.* 1990) and the fly *Dryomyza anilis* (Otronen 1990; Otronen and Siva-Jothy 1991).

Female *Z. barberi* copulate repeatedly with the same male as well as with multiple males. Among the hypotheses for female mating patterns discussed above, the paternity-assurance and pair-bonding hypotheses do not apply because *Z. barberi* neither form a pair bond nor provide parental care (Choe 1992). The sexual stimulation hypothesis may not apply either, because all uninterrupted copulations resulted in sperm transfer in the copulation-controlled breeding experiment (Choe 1995). The total duration of the mating episode between any mating partners is too short to suggest the possibility of 'mate-guarding' by females (female mate-guarding hypothesis) and females do not appear sexually harrassed (convenience polyandry hypothesis). This leaves hypotheses 1–5 and 8 for further consideration.

Table 7-3. *Variation in copulation duration of* Zorotypus gurneyi *and* Z. barberi

Species	Copulation duration ($\bar{x} \pm SD$)	n	Coefficient of variation
Zorotypus gurneyi	55.6 ± 4.9 min	76	8.8
Zorotypus barberi	50.7 ± 28.8 sec	364	56.6

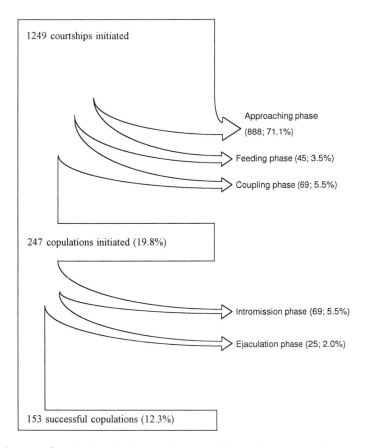

Fig. 7-7. A flow chart showing rates of copulation rejection at various stages of the mating episode in *Z. barberi*.

A series of copulation-controlled breeding experiments revealed that female *Z. barberi* that copulated three or more times laid significantly more eggs than females that copulated only once or twice (Choe 1995). One possible reason why *Z. barberi* females copulate three or more times is that they need to store sufficient sperm to achieve full fecundity. Considering that fully inseminated female *Z. barberi* lay only about two dozen eggs, however, it is still puzzling why they need such a large supply of sperm. As is often suggested by the genetic-benefit hypothesis, storing surplus sperm may also be a female strategy to weed out inferior sperm from the superior ones (Thornhill and Alcock 1983). Nearly two-thirds of *Z. barberi* females who mated with multiple mates had previously copulated with another male once or twice (Choe 1995). As predicted by the sperm-dilution hypothesis, which is essentially a subset of the genetic-benefit hypothesis, these females appear to make a choice of paternity by repeatedly copulating with a preferred male and thus devaluing the sperm from inferior

males. A purely mechanical hypothesis may also be worthy of consideration (Choe 1995). Although the mechanisms of sperm release in most insects are not fully understood, Chapman (1982) suggests that sperm are forced out from a spermatheca by contractions of compressor muscles of the spermatheca or by hemolymph pressure caused by contractions of the body musculature. A critical volume may be required for female insects to be able to squeeze out a few sperm as an egg passes down the oviduct. Although it cannot be completely ruled out, the fertility-insurance hypothesis is not likely to provide a strong explanation for multiple mating by *Z. barberi* females because all singly mated females laid fertile eggs (Choe 1995).

The material-benefit hypothesis may be supported by all three mating types: multiple, repeated, and prolonged mating. In species in which males provide courtship gifts that consist of not-so-quickly replenishable resources such as prey or a spermatophore, females should not try to mate repeatedly with the same male within a short period of time. Recently mated males of the katydid *Conocephalus nigropleurum* produce smaller spermatophores (Gwynne 1982). In the alfalfa butterfly *Colias eurytheme*, recently mated males provide a female with a smaller quantity of nutrients during copulation than males that have not recently mated, and females that mated with recently mated males showed a lower lifetime fecundity (Rutowski *et al.* 1987). By contrast, *Z. barberi* males appear capable of producing more gifts than most females could collect from an individual male. I have observed a male copulate 28 times in a row, providing a courtship gift before each copulation. After the initial copulation, males court for a much shorter time in subsequent copulations, presumably because there is no selection to expend time and energy to re-assess male quality. Such repeated mating may allow females to accumulate more gifts in a given amount of time than does multiple mating (Choe 1995).

Female *Z. barberi* appear to terminate copulation, and thus potentially control the number of sperm transferred in each copulation. Sakaluk (1986) proposed that mixed sperm utilization rather than last-male sperm precedence should be favored in species in which sperm transfer is controlled by females and males provide courtship gifts. As predicted by Walker (1980), sperm mixing is likely to occur in *Z. barberi* because its spermatheca is more or less spherical in shape. Thus, the paternity of each male may be determined numerically and females may be able to block sperm transfer altogether or reduce the number of sperm transferred by males that provide insufficient gifts, and therefore allow more generous males to fertilize more eggs. In other words, a combination of the postcopulatory female choice hypothesis and the material-benefit hypothesis may provide a more complete explanation for repeated mating in *Z. barberi*.

Why would *Z. barberi* males copulate with the same female repeatedly? All seven hypotheses suggested above for male mating variations, with the exception of the sexual-stimulation hypothesis, could provide reasonable answers. Although the sexual-stimulation hypothesis predicts that males perform one or more purely excitatory intromissions before the actual sperm transfer (Eberhard 1985), all singly mated *Z. barberi* females reproduced in copulation-controlled breeding experiments (Choe 1995). By contrast, the result of the breeding experiments may provide some support for the fertilization-enhancement hypothesis. Copulating repeatedly, at least three times in succession with the same female, increases fertilization rate significantly. The same result may also support the

sperm-supply hypothesis, as females also appear to require at least three copulations to express full fecundity. However, both the fertilization-enhancement and sperm-supply hypotheses will work only so long as males would not lose other mating opportunites by mating repeatedly with the same female (Parker 1978). Once the initial copulation is successfully completed, *Z. barberi* males do not appear to lose much time in obtaining additional copulations (Choe 1990, 1995).

Although a positive relationship between additional gifts and enhanced fertilization is generally in accord with predictions of the paternal-investment hypothesis, the result of the breeding experiments fails to support the hypothesis in *Z. barberi*. It is not clear whether cephalic secretion has any nutritional value because its chemical composition is not known. Females who copulated three or more times lay considerably more eggs than females who copulated only once or twice, and there is no difference in fecundity between females who copulated once or twice. A more linear relationship may be expected between the number of gifts and fecundity, if the gift contains critical nutrients for females (see, for example, Gwynne 1984a). Courtship feeding in *Z. barberi* as a form of prezygotic male investment may represent mating effort rather than parental effort (Alexander and Borgia 1979; Low 1979; Gwynne, this volume). None the less, some indirect evidence suggests that cephalic secretion contains some critical resources other than fecundity-enhancing material that females need. First, as discussed above, presenting a droplet of secretion is a necessary step for a successful copulation. Second, fully inseminated females show a greater tendency to cheat on males, collecting additional gifts but not sperm (Choe 1990). Finally, although *Z. barberi* eggs appear to be mechanically protected by hard shells (J. C. Choe, personal observation), they may also be chemically protected against potential egg predators such as ants and hemipterans. Such defensive chemicals might be contained in cephalic secretion. Through my long hours of observations in both the field and the laboratory, I have never seen any arthropods even touching, not to mention biting, zorapteran eggs. Although zorapterans are small insects, their eggs are fairly large (0.6 mm long and 0.4 mm wide on average) and should make decent meals for many arthropods under the bark.

In promiscuous species, repeated mating may best be viewed in terms of competition with other males. Both the receptivity-reduction and male mate-guarding hypotheses predict that males should try to reduce other males' opportunities to make genital contacts with the female. When prevention of competing males from gaining access to females is not possible, males may try to repeatedly transfer sperm so that they may numerically outcompete other males, as the sperm-loading hypothesis predicts. The sperm-loading hypothesis may explain why *Z. barberi* males copulate more than three times with the same female when there is no difference in the number of eggs laid by females that mated three, four, and five times (Choe 1995). The possibility of female receptivity reduction has not been investigated in *Z. barberi*.

I suggest *parceling* (Connor 1992) as a male strategy in species such as *Z. barberi* in which males provide a courtship gift prior to each copulation but females control copulation duration. As discussed above, female *Z. barberi* appear capable of terminating a copulation at any time and even cheating on males by taking the gift but not copulating. If males do not parcel, i.e. if they satisfy the female's need with one long bout of feeding, then females will be tempted to defect. Parceling the gift into smaller units allows males to guard the female for a longer period of time. Male *Z. barberi* appear to offer cephalic secretion for just long enough to make genital coupling. Such a discrete nature of the gift may set the stage for sexual conflict over the feeding duration of cephalic secretion.

SEXUAL SELECTION AND LIFE-HISTORY ADAPTATIONS

Zorotypus barberi and *Z. gurneyi* often coexist in central Panama and yet the two species have evolved marked differences in mating behavior and other related traits (Table 7-1). The mating system of *Z. barberi* appears to be polygynandry in which both sexes mate with multiple mates. In contrast, the mating system of *Z. gurneyi* is a rare example of female defense polygyny in arthropods (but see Kirkendall *et al.* and Saito in Choe & Crespi 1997). Little sexual dimorphism exist in *Z. barberi*, whereas males are much larger than females in *Z. gurneyi*. The association between polygyny and sexual dimorphism is well documented in birds (Selander 1972) and mammals (Alexander *et al.* 1979). Among vertebrates, polygyny is also found to be more prevalent in an open habitat such as savanna (Wittenberger 1979). Although *Z. barberi* and *Z. gurneyi* sometimes live together in a mixed colony, the former appears to prefer galleries and small spaces whereas the latter is often found in much larger spaces.

In conclusion, male–male competition is the chief selective force in *Z. gurneyi*, in which body size, age, and fighting skills are important male qualities. In *Z. barberi*, however, courtship feeding and courting skills are more important determinants of male mating success. How such strikingly different mating systems have evolved in these two closely related, sympatric species deserves further research. Whether this represents a rapid divergence in courtship behavior and related morphological features because of strong sexual selection (West-Eberhard 1984; Eberhard 1985), or character displacement of coexisting species, will be of particular importance. One aspect of life-history traits appears to be influenced by sexual selection. As discussed above, *Z. gurneyi* eggs, which are larger than *Z. barberi* eggs, develop on average two weeks faster. Because individuals that emerge sooner are more likely to become dominant over others even in the same cohort in *Z. gurneyi*, a shorter development time rather than a greater initial body size should be favored by strong sexual selection. Different modes of sexual selection can thus affect life-history strategies.

SUGGESTIONS FOR FUTURE RESEARCH

My research on the behavior and evolution of the Zoraptera has raised more questions than answers. One of the most important question has to do with the chemical composition of cephalic secretion. Without the knowledge of the exact chemical nature of the secretion, much of the arguments on the importance of courtship feeding in *Z. barberi* is admittedly speculative, although a fair amount of indirect bioassays have been conducted. Recently developed high-resolution techniques in analytical chemistry may provide proper tools for this investigation.

Patterns of sperm transfer and use in *Z. gurneyi* are critically important in understanding why males in that species exhibit prolonged mating associations rather than repeated mating patterns. If males transfer sperm only in the beginning and subsequently provide addition stimulation, one or combinations of the sexual-stimulation, fertilization-enhancement, and male mate-guarding hypotheses may be supported. By contrast, the sperm-supply, sperm-loading, and paternal-investment hypotheses may be in contention if the transfer of sperm and possibly other accessory-gland secretions occur throughout their prolonged copulation. Once such information is available, additional observation and manipulation can be made to determine the exact function of prolonged mating.

The fact that much of the information in this chapter comes from my own studies of a pair of species in central America offers a great promise for future research involving additional species. Looking at the morphology of specimens from various parts of the world, I can almost predict that entirely different mating systems await to be discovered. Professor William D. Hamilton once collected three different species in a locality near Manaus, Brazil. Genitalic structures of these three species are quite distinct (New 1978) and it would be very interesting to study their mating behavior. With the exception of *Z. hubbardi*, which often inhabits decaying sawdust piles in the United States, most zorapterans have been found only in relatively undisturbed tropical rain forests. Before much of the world's rain forests disappear, more rigorous collecting and research are desperately needed to unveil the secret lives of this obscure but remarkable group of insects.

ACKNOWLEDGEMENTS

I thank Steven Austad, William Eberhard, Bert Hölldobler, Bruce Waldman, Mary Jane West-Eberhard and Edward Wilson for sharing their expertise throughout my zorapteran project. Peter Adler, John Alcock, Göran Arnqvist, Richard Connor, Bernie Crespi, Janis Dickinson, William Eberhard, Darryl Gwynne, Leigh Simmons and Randy Thornhill helped improve this chapter tremendously. I also thank Kathy Brown-Wing for the line drawings of zorapterans. Support came from the Smithsonian Institution (Graduate and Predoctoral Fellowship), American Museum of Natural History (Theodore Roosevelt Memorial Fund), Sigma Xi (Grants-in-Aids), Organization for Tropical Studies (Noyes Fund), Harvard University (Richmond Fund and DeCuevas Fund) and Michigan Society of Fellows (Junior Fellowship).

LITERATURE CITED

Alcock, J. 1979. The evolution of intraspecific diversity in male reproductive strategies in some bees and wasps. In *Sexual Selection and Reproductive Competition in Insects*. M. S. Blum and N. A. Blum, eds., pp. 381–401. New York: Academic Press.

–. 1994. Postinsemination associations between males and females in insects: the mate-guarding hypothesis. *Annu. Rev. Entomol.* **39**: 1–21.

Alexander, R. D. 1961. Aggressiveness, territoriality and sexual behavior in field crickets (Orthoptera: Gryllidae). *Behaviour* **17**: 130–223.

Alexander, R. D. and G. Borgia. 1979. On the origin and basis of the male-female phenomenon. In *Sexual Selection and Reproductive Competition in Insects*. M. S. Blum and N. A. Blum, eds., pp. 417–440. New York: Academic Press.

Alexander, R. D., J. L. Hoogland, R. D. Howard, K. M. Noonan and P. W. Sherman. 1979. Sexual dimorphisms and breeding systems in pinnipeds, ungulates, primates and humans. In *Evolutionary Biology and Human Social Behavior: An Anthropological Perspective*. N. A. Chagnon and W. Irons, eds., pp. 402–435. North Scituate: Duxbury Press.

Arnqvist, G. 1989. Multiple mating in a water strider: mutual benefits or intersexual conflicts? *Anim. Behav.* **38**: 749–756.

Bateman, A. J. 1948. Intra-sexual selection in *Drosophila*. *Heredity* **2**: 349–368.

Birkhead, T. R. and C. M. Lessells. 1988. Copulation behaviour in ospreys *Pandion haliaetus*. *Anim. Behav.* **36**: 1672–1682.

Birkhead, T. R. and A. P. Møller. 1992. *Sperm Competition in Birds*. London: Academic Press.

Birkhead, T. R., L. Atkins, and A. P. Møller. 1987. Copulation behaviour in birds. *Behaviour* **101**: 101–138.

Borgia, G. 1980. Sexual competition in *Scatophaga stercoraria*: size- and density-related changes in male ability to capture females. *Behaviour* **75**: 185–206.

Brown, W. D. 1990. Size-assortative mating in the blister beetle *Lytta magister* (Coleoptera: Meloidae) is due to male and female preferences for larger mates. *Anim. Behav.* **40**: 901–909.

–. 1993. The cause of size-assortative mating in the leaf beetle *Trirhabda canadensis* (Coleoptera: Chrysomelidae). *Behav. Ecol. Sociobiol.* **33**: 151–157.

Burkhardt, D. and I. dela Motte. 1983. How stalk-eyed flies eye stalk-eyed flies: observations and measurements of the eyes of *Cyrtodiopsis whitei* (Diopsidae, Diptera). *J. Comp. Physiol.* A **151**: 407–421.

Burley, N. 1983. The meaning of assortative mating. *Ethol. Sociobiol.* **4**: 191–203.

Chapman, R. F. 1982. *The Insects: Structure and Function*. Cambridge, Massachusetts: Harvard University Press.

Chen, P. S. 1984. The functional morphology and biochemistry of insect male accessory glands and their secretions. *Annu. Rev. Entomol.* **29**: 233–255.

Choe, J. C. 1989. A new species *Zorotypus gurneyi* from Panama and redescription of *Z. barberi* Gurney (Zoraptera: Zorotypidae). *Ann. Entomol. Soc. Am.* **82**: 149–155.

–. 1990. The evolutionary biology of Zoraptera (Insecta). Ph.D. dissertation, Harvard University.

–. 1992. Zoraptera of Panama with a review of the morphology, systematics, and biology of the order. In *Insects of Panama and Mesoamerica: Selected Studies*. D. Quintero and A. Aiello, eds., pp. 249–256. Oxford University Press.

–. 1994a. Sexual selection and mating system in *Zorotypus gurneyi* Choe (Insecta: Zoraptera): I. Dominance hierarchy and mating success. *Behav. Ecol. Sociobiol.* **34**: 87–93.

–. 1994b. Sexual selection and mating system in *Zorotypus gurneyi* Choe (Insecta: Zoraptera): II. Determinants and dynamics of dominance. *Behav. Ecol. Sociobiol.* **34**: 233–237.

–. 1995. Courtship feeding and repeated mating in *Zorotypus barberi* (Insecta: Zoraptera). *Anim. Behav.* **49**: 1511–1520.

Choe, J. C. and B. J. Crespi, eds. 1997. *The Evolution of Social Behavior in Insects and Arachnids.* Cambridge University Press.

Clutton-Brock, T. H., F. E. Guiness and S. D. Albon. 1982. *Red Deer: Behavior and Ecology of Two Sexes.* Chicago: University of Chicago Press.

Connor, R. C. 1992. Egg-trading in simultaneous hermaphrodites: an alternative to Tit-for-Tat. *J. Evol. Biol.* **5**: 523–528.

Crespi, B. J. 1986a. Territoriality and fighting in a colonial thrips, *Hoplothrips pedicularius*, and sexual dirmorphism in Thysanoptera. *Ecol. Entomol.* **11**: 119–130.

–. 1986b. Size assessment and alternative fighting tactics in *Elaphrothrips tuberculatus* (Insecta: Thysanoptera). *Anim. Behav.* **34**: 1324–1335.

–. 1988. Risks and benefits of lethal male fighting in the colonial polygynous thrips *Hoplothrips karnyi* (Insecta: Thysanoptera). *Behav. Ecol. Sociobiol.* **22**: 293–301.

–. 1989a. Causes of assortative mating in arthropods. *Anim. Behav.* **38**: 980–1000.

–. 1989b. Sexual selection and assortative mating in subdivided populations of the thrips *Elaphrothrips tuberculatus* (Insecta: Thysanoptera). *Ethology* **83**: 265–278.

Daly, M. 1978. The cost of mating. *Am. Nat.* **112**: 771–774.

Darwin, C. 1871. *The Descent of Man and Selection in Relation to Sex.* London: John Murray.

Davies, N. B., B. J. Hatchwell, T. Robson and T. Burke. 1992. Paternity and parental effort in dunnocks *Prunella modularis*: how good are male chick-feeding rules? *Anim. Behav.* **43**: 729–745.

Dewsbury, D. A. 1984. Sperm competition in muroid rodents. In *Sperm Competition and the Evolution of Animal Mating Systems.* R. L. Smith, ed., pp. 547–571. New York: Academic Press.

–. 1988. Copulatory behavior as courtship communication. *Ethology* **79**: 218–234.

Dixon, K. A. and W. H. Cade. 1986. Some factors influencing male–male aggression in the field cricket *Gryllus integer* (time of day, age, weight and sexual maturity). *Anim. Behav.* **34**: 340–346.

Dodson, G. 1989. The horny antics of antlered flies. *Austr. Nat. Hist.* **22**: 604–611.

Eberhard, W. G. 1979. The function of horns in *Podischnus agenor* and other beetles. In *Sexual Selection and Reproductive Competition in Insects.* M. S. Blum and N. A. Blum, eds., pp. 231–258. New York: Academic Press.

–. 1980. Horned beetles. *Scient. Am.* **242**: 166–182.

–. 1985. *Sexual Selection and Animal Genitalia.* Cambridge, Mass.: Harvard University Press.

–. 1990. Animal genitalia and female choice. *Am. Sci.* **78**: 134–142.

–. 1991. Copulatory courtship and cryptic female choice in insects. *Biol. Rev.* **66**: 1–31.

–. 1992. Species isolation, genital mechanics, and the evolution of species-specific genitalia in three species of *Macrodactylus* beetles (Coleoptera, Scarabeidae, Melolonthinae). *Evolution* **46**: 1774–1783.

–. 1993. Copulatory courtship and morphology of genital coupling in seven *Phyllophaga* species (Coleoptera: Melolonthidae). *J. Nat. Hist.* **27**: 683–717.

Gillott, C. 1988. Arthropoda-Insecta. In *Reproductive Biology of Invertebrates, vol. 3. Accessory Sex Glands.* K. G. Adiyodi and R. G. Adiyodi, eds., pp. 319–471. Chichester: John Wiley & Sons.

Gwynne, D. T. 1982. Mate selection by female katydids (Orthoptera: Tettigoniidae, *Conocephalus nigropleurum*). *Anim. Behav.* **30**: 734–738.

–. 1984a. Courtship feeding increases female reproductive success in bushcrickets. *Nature (Lond.)* **307**: 361–363.

–. 1984b. Male mating effort, confidence of paternity and insect sperm competition. In *Sperm Competition and the Evolution of Insect Mating Systems.* R. L. Smith, ed., pp. 117–149. New York: Academic Press.

Hamilton, W. D. 1971. Geometry for the selfish herd. *J. Theor. Biol.* **31**: 295–311.

Hammerstein, P. 1981. The role of asymmetries in animal contests. *Anim. Behav.* **29**: 193–205.

Hancock, R. G., W. A. Foster and W. L. Yee. 1990. Courtship behavior of the mosquito *Sabethes cyaneus* (Diptera: Culicidae). *J. Insect Behav.* **3**: 401–416.

Hand, J. L. 1986. Resolution of social conflicts: dominance, egalitarianism, and spheres of dominance, and game theory. *Q. Rev. Biol.* **61**: 201–220.

Hölldobler, B. 1976. Tournaments and slavery in a desert ant. *Science (Wash., D.C.)* **192**: 912–914.

Hughes, A. L. and M. K. Hughes. 1982. Male size, mating success, and breeding habitat partitioning in the whitespotted sawyer *Monochamus scutellatus* (Say) (Coleoptera: Cerambycidae). *Oecologia (Berl.)* **55**: 258–263.

Hughes, C. R. and J. E. Strassmann. 1988. Age is more important than size in determining dominance among workers in the primitively eusocial wasp, *Polistes instabilis*. *Behaviour* **107**: 1–14.

Hunter, F. M., M. Petrie, M. Otronen, T. Birkhead and A. P. Møller. 1993. Why do females copulate repeatedly with one male. *Trends Ecol. Evol.* **8**: 21–26.

Jackson, R. R. 1982. The biology of ant-like jumping spiders: intraspecific interactions of *Myrmarachne lupata* (Araneae, Salticidae). *Zool. J. Linn. Soc.* **76**: 293–319.

Johnson, L. K. 1982. Sexual selection in a brentid weevil. *Evolution* **36**: 251–262.

Juliano, S. A. 1985. The effects of body size on mating and reproduction in *Brachinus lateralis* (Coleoptera: Carabidae). *Ecol. Entomol.* **10**: 271–280.

Keller, L. and H. K. Reeve. 1995. Why do females mate with multiple males? The sexually selected sperm hypothesis. *Adv. Stud. Behav.* **24**: 291–315.

Klahn, J. E. 1981. Alternate reproductive tactics of single foundresses of a social wasp, *Polistes fuscatus*. Ph.D. dissertation, University of Iowa.

LeBoeuf, B. J. 1974. Male-male competition and reproductive success in elephant seals. *Am. Zool.* **14**: 163–176.

Low, B. S. 1979. Sexual selection and human ornamentation. In *Evolutionary Biology and Human Social Behavior*. N. A. Chagnon and W. Irons, eds., pp. 462–487. North Scituate: Duxbury Press.

Maynard Smith, J. 1978. *The Evolution of Sex*. Cambridge University Press.

–. 1982. *Evolution and the Theory of Games*. Cambridge University Press.

Maynard Smith, J. and G. A. Parker. 1976. The logic of asymmetric contests. *Anim. Behav.* **24**: 159–175.

McAlpine, D. K. 1979. Agonistic behavior in *Achias australis* (Diptera, Platystomatidae) and the significance of eyestalks. In *Sexual Selection and Reproductive Competition in Insects*. M. S. Blum and N. A. Blum, eds., pp. 221–230. New York: Academic Press.

Michener, C. D. 1990. Reproduction and castes in social halictine bees. In *Social Insects: An Evolutionary Approach to Castes and Reproduction*. W. Engels, ed., pp. 77–121. Berlin: Springer-Verlag.

Møller, A. P. 1987. Copulation behaviour in the goshawk *Accipiter gentilis*. *Anim. Behav.* **35**: 755–763.

Moulds, M. S. 1977. Field observations on behaviour of a north Queensland species of *Phytalmia* (Diptera: Tephritidae). *J. Austr. Entomol. Soc.* **16**: 347–352.

New, T. R. 1978. Notes on neotropical Zoraptera, with descriptions of two new species. *Syst. Entomol.* **3**: 361–370.

Noonan, K. M. 1981. Individual strategies of inclusive-fitness-maximizing in *Polistes fuscatus* foundresses. In *Natural Selection and Social Behavior: Recent Research and New Theory*. R. D. Alexander and D. W. Tinkle, eds., pp. 18–44. New York: Chiron Press.

Otronen, M. 1990. Mating behaviour and sperm competition in the fly *Dryomyza anilis*. *Behav. Ecol. Sociobiol.* **26**: 349–356.

–. 1994. Repeated copulations as a strategy to maximize fertilization in the fly, *Dryomyza anilis* (Dryomyzidae). *Behav. Ecol.* **5**: 51–56.

Otronen, M. and M. T. Siva-Jothy. 1991. The effect of postcopulatory male behaviour on ejaculate distribution within the female sperm storage organs of the fly *Dryomyza anilis* (Diptera: Dryomyzidae). *Behav. Ecol. Sociobiol.* **29**: 33–37.

Pardi, L. 1948. Dominance order in *Polistes* wasps. *Physiol. Zool.* **21**: 1–13.

Parker, G. A. 1970. Sperm competition and its evolutionary consequences in the insects. *Biol. Rev.* **45**: 525–567.

–. 1974. Assessment strategy and the evolution of fighting behaviour. *J. Theor. Biol.* **47**: 223–243.

–. 1978. Searching for mates. In *Behavioral Ecology: An Evolutionary Approach*. J. R. Krebs and N. B. Davies, eds., pp. 214–244. Oxford: Blackwell Scientific Publications.

–. 1979. Sexual selection and sexual conflict. In *Sexual Selection and Reproductive Competition in Insects*. M. S. Blum and N. A. Blum, eds., pp. 123–166. New York: Academic Press.

Parker, G. A. and D. I. Rubenstein. 1981. Role assessment, reserve strategy, and the acquisition of information in asymmetric animal contests. *Anim. Behav.* **29**: 221–240.

Petrie, M. 1992. Copulation behaviour in birds: why do females copulate more than once with the same male? *Anim. Behav.* **44**: 790–792.

Pinto, J. D. and A. J. Mayor. 1986. Size, mating success, and courtship pattern in the Meloidae (Coleoptera). *Ann. Entomol. Soc. Am.* **79**: 597–604.

Price, P. W. and M. F. Willson. 1976. Some consequences for a parasitic herbivore, the milkweed longhorn beetle *Tetraopes tetraophthalmus*, of a host-shift from *Asclepias syriaca* to *A. verticillata*. *Oecologia (Berl.)* **25**: 331–340.

Ridley, M. 1983. *The Explanation of Organic Diversity: The Comparative Method and Adaptations for Mating*. Oxford: Clarendon Press.

–. 1988. Mating frequency and fecundity in insects. *Biol. Rev.* **63**: 509–549.

Rutowski, R. L., G. W. Gilchrist and B. Terkanian. 1987. Female butterflies mated with recently mated males show reduced reproductive output. *Behav. Ecol. Sociobiol.* **20**: 319–322.

Sakaluk, S. K. 1986. Is courtship feeding by male insects parental investment? *Ethology* **73**: 161–166.

Schal, C. and W. J. Bell. 1983. Determinants of dominant-subordinate interactions in males of the cockroach *Nauphoeta cinerea*. *Biol. Behav.* **8**: 117–139.

Schein, M. W. 1975. *Social Hierarchy and Dominance*. Stroudsburg: Dowden, Hutchinson & Ross.

Schwarz, M. P. and K. J. O'Keefe. 1991. Order of eclosion and reproductive differentiation in a social allodapine bee. *Ethol. Ecol. Evol.* **3**: 233–245.

Schwarz, M. P. and R. E. Woods. 1994. Oder of adult eclosion is a major determinant of reproductive dominance in the allodapine bee Exoneura bicolor. *Anim. Behav.* **47**: 373–378.

Selander, R. K. 1972. Sexual selection and dimorphism in birds. In *Sexual Selection and the Descent of Man*. B. Campbell, ed., pp. 180–230. Chicago: Aldine.

Simmons, L. W. 1987. Sperm competition as a mechanism of female choice in the field cricket, *Gryllus bimaculatus*. *Behav. Ecol. Sociobiol.* **21**: 197–202.

Small, M. F. 1990. Promiscuity in Barbary macaques (*Macaca sylvanus*). *Am. J. Primatol.* **20**: 267–282.

Spieth, H. T. 1981. *Drosophila heteroneura* and *Drosophila silvestris*: head shapes, behaviour and evolution. *Evolution* **35**: 921–930.

Strassmann, J. E. 1983. Nest fidelity and group size among foundresses of *Polistes annularis* (Hymenoptera: Vespidae). *J. Kansas Entomol. Soc.* **56**: 621–634.

Sullivan, J. D. and J. E. Strassmann. 1984. Physical variability among nest foundresses in the polygynous social wasp, *Polistes annularis*. *Behav. Ecol. Sociobiol.* **15**: 249–256.

Thornhill, R. 1983. Cryptic female choice and its implications in the scornpionfly *Harpobittacus nigriceps*. *Am. Nat.* **122**: 765–788.

–. 1984. Fighting and assessment in *Harpobittacus* scorpionflies. *Evolution* **38**: 204–214.

Thornhill, R. and J. Alcock. 1983. *The Evolution of Insect Mating Systems*. Cambridge, Massachusetts: Harvard University Press.

Trivers, R. L. 1972. Parental investment and sexual selection. In *Sexual Selection and the Descent of Man*. B. Campbell, ed., pp. 136–179. Chicago: Aldine.

Turillazi, S. and L. Pardi. 1977. Body size and hierarchy in polygynic nests of *Polistes gallicus* (L.) (Hymenoptera, Vespidae). *Monit. Zool. Ital.* (N. S.) **11**: 101–112.

Waage, J. K. 1984. Sperm competition and the evolution of odonate mating systems. In *Sperm Competition and the Evolution of Animal Mating Systems*. R. L. Smith, ed., pp. 251–290. New York: Academic Press.

Walker, W. F. 1980. Sperm utilization strategies in nonsocial insects. *Am. Nat.* **115**: 780–799.

West-Eberhard, M. J. 1969. The social biology of polistine wasps. *Misc. Publ., Mus. Zool., Univ. Mich.* **14**: 1–101.

–. 1979. Sexual selection, social competition, and evolution. *Proc. Am. Phil. Soc.* **123**: 222–234.

–. 1984. Sexual selection, competitive communication and species-specific signals in insects. In *Insect Communication*. T. Lewis, ed., pp. 283–324. London: Academic Press.

Whitham, T. G. 1979. Territorial behaviour of *Pemphigus* gall aphids. *Nature (Lond.)* **279**: 324–325.

Wilcox, R. S. 1984. Male copulatory guarding enhances female foraging in a water strider. *Behav. Ecol. Sociobiol.* **15**: 171–174.

Williams, G. C. 1975. *Sex and Evolution*. Princeton: Princeton University Press.

Wilson, E. O. 1975. *Sociobiology: The New Synthesis*. Cambridge, Mass.: Harvard University Press.

Wittenberger, J. F. 1979. The evolution of mating systems in birds and mammals. In *Handbook of Behavioral Neurobiology*, vol. 3. *Social Behavior and Communication*. P. Marler and J. Vandenbergh, eds., pp. 271–349. New York: Plenum Press.

8 · The evolution of water strider mating systems: causes and consequences of sexual conflicts

GÖRAN ARNQVIST

ABSTRACT

Water striders (Heteroptera: Gerridae) exhibit two different types of mating behavior. The most common mating system (type I) is characterized by strong apparent conflicts of interest between the sexes, and conspicuous pre- and postcopulatory struggles. Some species exhibit a mating system that involves much less apparent conflict (type II) and lack the intense copulatory struggles.

I argue that the predominant mating system in water striders is a direct consequence of sexual conflicts over mating decisions. Matings involve high costs to females (increased predation risk and energetic expenditure) but few, if any, balancing direct benefits. Sperm-displacement rates are high, and males thus gain from rematings. Mating frequencies are high; females mate multiply for reasons of convenience. In these species, males are considered to have 'won' the evolutionary conflict over the mating decision in the sense that they have made acceptance of superfluous matings 'the best of a bad job' for females, by evolving behavioral and morphological traits that make it costly for females to reject males attempting copulations. Females, however, have apparently evolved a variety of counter-adaptations to male harassment, to gain control over mating. Further, I suggest that sexual conflict may have played a crucial role in the evolution of type II matings from type I matings.

Water strider mating systems are very plastic. Females assess the rate of male harassment, and make adaptive mating decisions based on this assessment. As a consequence of variation in female mating behavior, the characteristics of the mating system vary with a number of environmental factors. Knowledge of the behavioral dynamics of water strider mating systems makes mechanistic hypotheses of sexual selection possible and provides a framework in which variations between populations in non-random mating and sexual selection can be understood and even accurately predicted. Recent insights gained from the study of water strider mating systems highlight two fruitful research avenues: (1) intraspecific variability of mating systems deserves more attention; and (2) observational and experimental studies at the population level concerned with mating patterns should be linked with experimental studies of behavioral processes at the individual level.

INTRODUCTION

In the past two decades, there has been a marked shift in the way that evolutionary ecologists view sexual behaviors. In the past, ethologists viewed courtship behaviors and mating primarily as harmonious, joint ventures in which males and females cooperated in offspring production. However, it is now widely recognized that male and female evolutionary interests in reproduction may be asymmetrical, and that sexual behavior in many cases represents the resolution of such underlying sexual conflicts (Trivers 1972, 1974; Parker 1979, 1984; Hammerstein and Parker 1987; Alexander et al., this volume). A variety of different conflicts may occur, such as conflicts over isogamous or anisogamous reproduction (Parker et al. 1972), relative parental investment (Trivers 1972, 1974), monogamous or polygamous mating systems (Alatalo et al. 1981), mating frequency and duration (Parker 1979; Hammerstein and Parker 1987; Arnqvist 1989a; Thornhill and Sauer 1991) and fertilization events (Alexander et al., this volume). This conceptual shift is obvious in the study of water strider sexual behaviors. Mating behavior in this group of bugs typically involves apparent conflicts and dramatic struggles; and several recent papers stress the role of sexual conflicts and female interests in matings (Arnqvist 1988, 1989a, 1992a; Rowe 1992, 1994; Sih and Krupa 1992; Fairbairn 1993; Krupa and Sih 1993; Rowe et al. 1994, 1996; Weigensberg and Fairbairn 1994, 1995; Jablonski and Vepsäläinen 1995; Vepsäläinen and Savolainen 1995).

Water striders (Heteroptera: Gerridae; about 500 species) form an ecologically rather homogenous group of bugs (Andersen 1982; Spence and Andersen 1994). They inhabit water surfaces of a wide range of aquatic habitats and are predators and scavengers, feeding mainly on arthropods trapped at the water surface. Mating and egg-laying continue throughout the reproductive part of the adult life, which lasts from one to three months. The daily fecundity of females ranges from 3 to 15 eggs per female per day for most species. Eggs are typically deposited on floating objects or emergent aquatic vegetation. Water striders are well suited for behavioral studies because of their open and strictly two-dimensional habitat. They can be captured, marked, manipulated and observed easily in their natural environment. They also are kept and reared easily in the laboratory, which makes them an excellent group for experimental studies. Matings are frequent and conspicuous, and there is considerable interspecific variation in mating behavior and sexual dimorphism. As a consequence, there has been a rapidly increasing interest in water strider mating systems, and a growing literature has proven water striders to be suitable for addressing many general issues about the evolution of mating systems.

The main goals of this paper are to (1) provide a comparative review of water strider mating behaviors; (2) discuss the role of sexual conflicts in the evolution of water strider mating systems; and (3) identify causal links between sexual behavior and sexual dimorphism in secondary sexual traits. Although my basic intent is to review current knowledge, I also hope that my development of the subject will prove to be provocative and hence stimulate new creative research in the field.

MATING BEHAVIOR

The literature on water strider mating behavior is rather comprehensive; I have been able to gather information on the sexual behavior of 30 species (Table 8-1). However, the information available for different species varies from single casual observational notes to thorough experimental studies. Information on temperate species is relatively satisfactory; descriptions of mating behavior of most tropical and subtropical species typically are absent or more incomplete. Even though there is considerable inter- and intraspecific variation in sexual behavior, I suggest that there is a basic dichotomy in the mating behavior of water striders and that matings can be divided into two distinct types.

Type I mating behavior

Most species of water striders studied to date exhibit type I matings, where five mating phases can be distinguished (Fig. 8-1) (Arnqvist 1988; Rowe 1992). Males may use two tactics to find potential mates; in most cases they search actively for females, but males of some species may also be territorial and use a sit-and-wait strategy (Hayashi 1985; Nummelin 1988). When females are encountered, males typically initiate matings by simply lunging at females, without persuasive courtship (cf. Alexander et al., this volume). Although males of some species use ripple signals to determine the sex of other individuals (see Wilcox 1979; Wilcox and Stefano 1991), males are generally indiscriminate in their efforts; males of many species lunge at both conspecific and heterospecific males. A male that successfully contacts a female will grasp her thorax with his forelegs and rapidly attempt to insert his genitalia. Females almost invariably respond with some form of resistance, and they may use several different behaviors in their attempts to dislodge the male. One common reluctance behavior has been termed 'somersaulting' (Fig. 8-2) (Arnqvist 1989b; Rowe 1992), in which the female rears her midlegs, which causes the pair to flip over backwards or sideways, often while she uses her forelegs to try to break the male's grasp with his forelegs. These backward somersaults are performed repeatedly, often combined with other behaviors such as jumps and/or jerks. During this precopulatory struggle females are trying to forcefully dislodge males while males are trying to endure this resistance (Andersen 1982; Arnqvist 1989b, 1992a; Krupa et al. 1990; Rowe 1992; Spence and Andersen 1994). The duration of the precopulatory struggle ranges from a few seconds to several minutes (Arnqvist 1992a; Rowe 1992; Weigensberg and Fairbairn 1994; Arnqvist and Rowe 1995). In many species, females are usually successful in dislodging males; in Gerris odontogaster and G. buenoi females have been reported to dislodge the male in approximately 80% of the struggles (Arnqvist 1989b; Rowe 1992) and more than half of the precopulatory struggles end with male dislodgment in Aquarius remigis (Weigensberg and Fairbairn 1994). However, in G. lateralis and G. lacustris males seem better able to overcome female reluctance; males of these species have been reported to 'win' more than half of the precopulatory struggles (Arnqvist 1988; Vepsäläinen and Savolainen 1995). Female A. najas appear almost incapable of dislodging males, despite vigorous struggling (Sattler 1957).

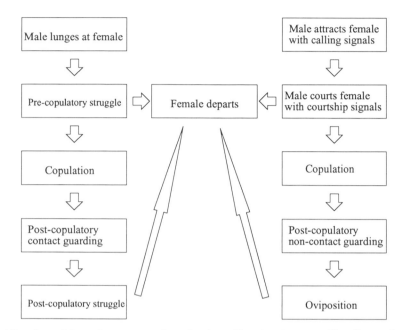

Fig. 8-1. A generalized flow chart of the mating sequence of type I and type II matings in water striders. See text for further details.

Copulation starts if female reluctance desists and the male inserts his genitalia successfully. Average copulation duration ranges from 5 to 20 minutes in most species (Table 8-1), one notable exception being *A. remigis* where the pair may remain *in copula* with joined genitalia for up to many hours (see, for example, Clark 1988; Fairbairn 1988).

Most species of water striders exhibit postcopulatory contact guarding. After retracting his genitalia, the male remains mounted on top of the female without genital contact. The duration of guarding is highly variable both within and between species, and may last from some minutes to several hours (Table 8-1). The longest postcopulatory guarding reported occurs in *A. najas*, where the pair may remain in contact for several days or even several weeks (Sattler 1957; Murray and Giller 1990)!

Type I matings are normally terminated with a postcopulatory struggle. During this phase, the female once again struggles to dislodge the male while the male attempts to remain coupled. Again, there is considerable variation between different species in the ability of the male to endure the postcopulatory struggle. Males of *G. buenoi* and *A. remigis* are almost invariably dislodged during the

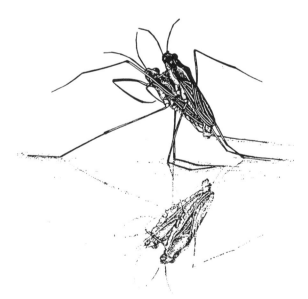

Fig. 8-2. A pair of *Gerris odontogaster* engaged in a precopulatory struggle. The male has successfully grasped the female and attempts copulation, but the female resists by rearing her midlegs, causing the pair to repeatedly 'somersault' backwards or sideways (reprinted with permission from Rowe *et al.* 1994).

first struggle (Rowe 1992; Weigensberg and Fairbairn 1994), whereas males of *G. lateralis* and *G. lacustris* often can endure several consecutive struggles (Arnqvist 1988; Jablonski and Vepsäläinen 1995). There is also considerable variation within species in how matings are terminated; males of some species may terminate matings without preceding struggle under some environmental circumstances (Erlandsson 1992; Weigensberg and Fairbairn 1994; Vepsäläinen and Savolainen 1995).

Type II mating behavior

Type II matings differ from type I matings mainly in that males 'court' females with ripple signals, there is much less apparent conflict, and fewer aggressive elements are involved (Fig. 8-1). Males are typically territorial, defending suitable oviposition sites situated in the territory. Males of several species anchor themselves to the oviposition site with the legs and produce calling signals, which apparently attract receptive females (Wilcox 1972; Hayashi 1985; Nummelin 1988). Males respond aggressively to other males, and emit repel signals and engage in fights to defend their territory. In contrast, when a female approaches, the male will switch from calling signals to different types of courtship signals. During courtship, the female often inspects the oviposition site carefully and may either (1) reject the object and skate away or (2) accept the site and allow the male to mount her without much reluctance. The male then inserts his genitalia and a comparatively brief copulation follows (copulation duration normally one or a few minutes; see Table 8-1).

In contrast to type I matings, type II matings typically involve postcopulatory non-contact guarding (Fig. 8-1). After copulation, the male dismounts and backs off from the female a few centimeters. The female will then oviposit, and the male stays in immediate proximity to the female during oviposition and guards her from disturbance of other males. Males approaching the ovipositing female will be aggressively warded off by the guarding male during the 10–20 min that oviposition may take. After ovipositing, females leave the site, and males return to producing calling signals. Hence, in type II matings there is usually a direct association between copulation and female oviposition, which is not normally the case in type I matings.

Pure type II matings as described above have been reported from species in the genus *Rhagadotarsus* (Wilcox 1972; Nummelin 1988) and in *A. elongatus* (Hayashi 1985).

In the holarctic genus *Limnoporus*, type II matings differ somewhat from the description above, and even share some characteristics of type I matings (Spence and Wilcox 1986; Wilcox and Spence 1986; Nummelin 1987; J. R. Spence, unpublished). First, territorial males do not seem to attract females actively by producing calling signals, but rather sit and wait for receptive females. Second, males generally stay mounted on top of females during oviposition, i.e. they exhibit a mate-guarding behavior much as in type I matings. Third, females sometimes show reluctance to mate even when courted by a signaling male, and females generally terminate matings.

Mating systems

Mating activity is very high in most species of water strider. Both sexes copulate repeatedly in all species studied, although the mating frequency varies with species and environmental conditions. Wilcox (1972) reports a maximum male mating frequency of five matings in one hour in *R. anomalus*. Typically, however, females as well as males have been reported to mate from 'a few' up to 'several' times per day throughout their reproductive lives, both in the field and in the laboratory (see, for example, Vepsäläinen 1974; Arnqvist 1989b, 1992b; Rowe 1992; Sih *et al.* 1990).

Female water striders are polyandrous, although there is extensive variation in female mating frequency between species. Following the classification of mating systems by Thornhill and Alcock (1983), female water strider mating systems are best described as convenience polyandry (Arnqvist 1989a; Rowe 1992; Weigensberg and Fairbairn 1994). This is true at least for mating systems characterized by type I matings. Water strider males of all species are polygynous, and there are two basic male mating systems. Mating systems where type I matings predominate are best characterized as scramble competition polygyny, since males do not monopolize either females, resources or specific sites. In cases where type II matings are involved, males provide females with oviposition sites and defense from harassment by other males, and these systems are thus best described as resource defense polygyny. *Limnoporus* species and *A. elongatus* exhibit a mixture of these two male mating systems.

In many species, the optimal behavior for acquiring matings varies with environmental conditions (e.g. sex ratio and density) and individual state (e.g. size and nutritional state), and thus males may adopt several different mating tactics

Table 8-1. *Taxonomic distribution of mating behaviors in water striders*

Subfamily	Species	Female body size (mm)	Body size dimorphism ratio (F/M)	Precopulatory struggle	Mating behavior	Copulation duration	Postcopulatory contact guarding duration	Termination of mating	References
Rhagadotarsinae	*Rhagadotarsus anomalus*	4.6	1.18	No	Type II	1 min	None[c]	Male	1
	Rhagadotarsus hutchinsoni	—	—	No	Type II	0.8 min	None[c]	Male	2
Gerrinae	*Aquarius conformis*	16.4	1.09	Yes (less frequent)	Type I	<5 min	123 min (30–790)	—	4, 35
	Aquarius elongatus	24.4	1.03	Only occasionally	Type I & II	4–5 min	None[e]	—	28
	Aquarius najas	16.5	1.28	Yes	Type I	1–4 min	hours–weeks (!)	—	9, 25, 26, 27
	Aquarius paludum	15.4	1.12	Yes	Type I	10 min[b]		Female[a,c]	9
	Aquarius remigis	13.1	1.08	Yes	Type I	20–>400 min[a]	None[f]	Female[a,c]	4, 5, 21, 22, 23, 24
	Gerris argentatus	7.4	1.13	Yes	Type I		10–20 min[a,b]	Female[a,c]	9
	Gerris buenoi	8.1	1.11	Yes	Type I	7.1–17.8 min[a]	16.4–83.0 min[a]	Female	3, 4, 5
	Gerris comatus	10.5	1.13	Yes	Type I		18.9 min[b]	Female	4, 5, 13
	Gerris gillettei	9.1	1.07	Yes	Type I	14.2 min	19.1 min	Female	37
	Gerris incognitus	9.1	1.08	Yes	Type I	10.9 min	30 min (0–110)	Female	39
	Gerris insperatus	10.4	1.10	Yes	Type I		31.6 min[b]		4, 20
	Gerris lacustris	9.7	1.11	Yes	Type I	5–20 min[a]	5–40 min[a]	Female[a,c]	9, 10, 11, 36
	Gerris lateralis	10.1	1.12	Yes	Type I	16.2 min	60–>120 min[a]	Female	8, 9
	Gerris marginatus	9.8	1.09	—	Type I		10.3 min[b]	—	4
	Gerris odontogaster	8.7	1.10	Yes	Type I	10–15 min[a]	40–90 min[a]	Female	6, 7
	Gerris pingreensis	8.9	1.02	Yes	Type I			—	13, 20
	Gerris sparkopensis	—	1.07[d]	Yes	Type I	8.2 min		Female	2
	Gerris thoracicus	11.9	1.11	Yes	Type I			—	12
	Limnoporus dissortis	14.0	1.06	Yes	Type I & II	5–25 min	5–20 min	Female	4, 17, 18, 19, 20, 36
	Limnoporus notabilis	18.1	1.04	Yes	Type I & II	5–20 min	5–20 min	Female	4, 17, 18, 19, 20, 36
	Limnoporus rufoscutellatus	15.5	1.09	Yes	Type I & II	10 min	>1 hour	Female	9, 14, 15, 16, 17, 36
	Tenagogonus albovittatus	—	1.06[d]	Yes	Type I			—	2
Cylindrostethinae	*Potamobates tridentatus*	—	—	Yes	Type I	10 min	many hours[b]	—	29
Halobatinae	*Eurymetra natalensis*	—	0.98[d]	Yes	Type I	10 min		—	2
	Halobates fijiensis	—	<1.00	—	—	0.25–2 min[b]		—	31
	Halobates robustus	5.0	1.25	—	—	many hours[b]		—	30
	Metrocoris histrio	5.8	1.02	Yes	Type I	several min[b]		Female	33, 34
	Metrocoris tenuicornis	3.9	0.81	Yes	Type I	5–10 min[b]		—	32

[a] Found to be variable, depending on environmental conditions; [b] only mating duration reported (i.e. copulation plus guarding); [c] non-contact guarding occurs; [d] size ratio based on leg lengths; [e] male reported to terminate mating occasionally; [f] species exhibits prolonged copulation.

Source: 1 Wilcox (1972); 2 Nummelin (1988); 3 Rowe (1992); 4 Fairbairn (1990); 5 Fairbairn (1988); 6 Arnqvist (1988); 7 Arnqvist (1992a); 7 Arnqvist (1992b); 8 Arnqvist (1992); 9 Erlandsson (1992); 10 Vepsäläinen (1992); 10 Vepsäläinen and Savolainen (1995); 11 Jablonski and Vepsäläinen (1995); 12 Kaitala (1991); 13 Spence (1979); 14 Vepsäläinen and Nummelin (1985b); 15 Vepsäläinen (1985); 16 Nummelin (1987); 17 Andersen and Spence (1992); 18 Wilcox and Spence (1986); 19 Spence and Wilcox (1986); 20 J. R. Spence (unpublished); 21 Wilcox (1979); 22 Sih *et al.* (1990); 23 Rubenstein (1989); 24 Clark (1988); 25 Sattler (1957); 26 Vepsäläinen and Nummelin (1985a); 27 Murray and Giller (1990); 28 Hayashi (1985); 29 Wheelwright and Wilkinson (1985); 30 Foster and Treherne (1982); 31 Foster and Treherne (1986); 32 Cheng (1966); 33 Ban *et al.* (1988); 34 Koga and Hayashi (1993); 35 J. J. Krupa (unpubl); 36 Andersen (1994); 37 G. Arnqvist (unpublished).

(Dominey 1984). This is true for water striders as well (Spence and Andersen 1994), where two main categories of alternative male mating behavior may be distinguished.

1. In some species with type I matings, males have been shown to utilize two different mate-searching tactics. Most individuals normally cruise around and search for females actively, but some males are more site-specific. These males will expel approaching males aggressively and attempt to mate with approaching females. This category of alternative mate-searching behaviors is found in *G. swakopensis*, where territorial males attack all other approaching water striders (Nummelin 1988). Attacks on intruding males and females look essentially identical, but other males are repelled whereas attacks on females result in mating attempts. Switches between similar alternative male mate-searching tactics occur in *Tenagogonus albovittatus*, *Eurymetra natalensis* and *Metrocoris histrio* (Nummelin 1988; Koga and Hayashi 1993).

2. A more complex behavioral flexibility occurs in males of some of the species that exhibit type II matings. Males adopting a type II mating behavior may not always be territorial but may actively search for females to court. If a female is encountered, the male will emit courtship signals in the same manner as a territorial male (Vepsäläinen and Nummelin 1985a,b; Spence and Wilcox 1986). These two tactics represent different means of finding potential mates, and are in that respect analogous to the behavioral dimorphism mentioned above for type I matings. However, a more profound polymorphism also occurs. Males sometimes do not signal to females with ripple signals but instead lunge at and grasp females and attempt to insert the genitalia (Hayashi 1985; Spence and Wilcox 1986; Wilcox and Spence 1986). Females will typically try to dislodge these males by intensive struggling, and males will attempt to subdue reluctant females much in the same way as in type I matings (Spence unpublished). Males adopting this tactic may either actively search for females or sit and wait (Hayashi 1985; Spence and Wilcox 1986). To summarize, males exhibiting type II matings may also attempt to grasp females without prior courtship in a way closely resembling type I matings. Irrespective of which of these two tactics a male uses to achieve a mating once a female is encountered, he may either search actively for females or use a more sedentary tactic in order to find potential mates.

Both individual state and environmental conditions have been shown to affect male choice of mating tactic. In some species, large males have been shown to more frequently be territorial (Hayashi 1985; Nummelin 1988), but age may also affect the choice of behavior (Koga and Hayashi 1993). In contrast, nutritional state does not seem to affect male choice of mating behavior (Spence and Wilcox 1986; Rowe *et al.* 1996). The presence and distribution of females in the habitat affect male mating tactics, as does the availability of suitable oviposition sites (Nummelin 1988; Hayashi 1985). However, there is a wide range of factors, both state variables (e.g. parasite load, previous mating history) and environmental factors (e.g. predator presence, population density, sex ratio, microhabitat structure) as well as interactions among these, that may influence the choice of male mating tactic but have not yet been addressed in experimental studies.

SEXUAL CONFLICTS OVER MATING

There is a basic sexual asymmetry in the relative interests in matings for animals in general. The reproductive success of males is intimately associated with the number of females they mate, whereas for females a single mating is often enough to achieve more or less maximum success. In other words, it often pays males but not females to search for additional matings, and in encounters between males and females of many species, males are under selection to mate whereas females are under selection not to mate (Parker 1979; Alexander *et al.*, this volume). This type of basic sexual conflict is thus a conflict over the mating decision. In addition to this conflict, there may also be conflicts over the mating duration (Parker 1979; Thornhill and Sauer 1991). If a female mates multiply, there may be sperm competition between males over the fertilization of the female's ova. In many arthropods, males benefit by prolonging matings to reduce the degree of sperm competition (Parker 1970; Alcock 1994). However, females may not benefit by this behavior, and in cases where the 'optimal' mating duration differs between the sexes a conflict over the mating duration will occur.

Parker (1979) stressed that it is often extremely difficult to study sexual conflicts, because of the difficulty of measuring the costs and benefits involved in matings. In water striders, there is strong apparent conflict over both the mating decision (precopulatory struggles) and the mating duration (postcopulatory struggles). Furthermore, a number of studies have addressed the relative costs and benefits involved in matings. Water striders thus provide an excellent group for evaluating costs and benefits involved in matings and for studying the dynamics of sexual conflicts.

The costs and benefits of mating for males

There may be a number of costs of mating for males, the most universal being the time and energy devoted to mating and the energetic costs of gamete production (Daly 1978; Dewsbury 1982). Although there are no direct data on the cost of ejaculate production in male water striders, two lines of evidence suggests that it is relatively low. First, water strider ejaculates consist of little more than sperm, i.e. there are no nutritious accessory substances (Andersen 1982; see below). Second, maximum male mating frequency is often high and refractory periods typically short.

In species where females mate with more than one male, one of the major factors that affects the evolution of mating systems is the sperm utilization pattern of females (Walker 1980; Smith 1984). Three different studies have addressed the degree of sperm displacement in water striders. Using reversed double mating experiments with sterilized males, Arnqvist (1988) showed that approximately 80% of a female's eggs were fertilized by the last male to mate in *G. lateralis*. Rubenstein (1989) demonstrated with a similar technique that last male advantage was in average 65% in *A. remigis*, and that the degree of sperm displacement was partly related to copulation duration. Finally, use of a genetic marker has shown sperm displacement rates of more than 90% in *Limnoporus dissortis* and *L. notabilis* (J. R. Spence, unpublished). Thus, when water strider females remate, the last male to mate will fertilize the majority of the female's eggs, even if the rate of sperm displacement may vary among species.

Because last-male sperm precedence occurs in water striders, there is strong selection in males to reduce sperm competition by other males. By guarding a female after sperm transfer, a male reduces the risk of the female mating with another male before she oviposits, but decreases his probability of finding additional mates (Parker 1974; Alcock 1994; Jablonski and Kaczanowski 1995). Although there is little field data (but see Rowe 1992), it appears as if water strider males often attempt to guard females until oviposition occurs, but that males are mostly dislodged prior to oviposition (Rowe 1992; Jablonski and Kaczanowski 1995; Jablonski and Vepsäläinen

1995). Three different mate-guarding tactics have evolved within the family.

1. In most water striders, the male will withdraw his genitalia after the relatively brief copulation but remain mounted on top of the female. The female will carry the passive non-copulating male during this postcopulatory contact guarding phase, typical for type I matings. There is considerable variation in guarding duration (both inter- and intraspecifically). The position of the male relative to the female during guarding also varies between species. In *Limnogonus* and some *Halobates* species, the male is positioned far back on the female and may even be towed rather than carried by the female (Andersen 1982). In *M. tenuicornis*, but not in the congeneric *M. histrio* (Ban *et al.* 1988; Koga and Hayashi 1993), the sex roles are reversed so that the male carries the passive female below (Cheng 1966).

2. In some species, the male dismounts after copulation but remains close to the female. The guarding male will be aggressive towards males, and repel intruders. This type of postcopulatory non-contact guarding has been described in species with typical type II matings, i.e. in the *Rhagadotarsus* species (Wilcox 1972; Nummelin 1988) and in *A. elongatus* (Hayashi 1985).

3. If males prolong copulations beyond the time necessary for insemination, males may actually be guarding females by acting as living mating plugs (Sillén-Tullberg 1981; Thornhill and Alcock 1983). This strategy has so far only been described for one species of water strider, *A. remigis*. In this species, complete sperm transfer takes approximately 20 min (Rubenstein 1989), but males typically prolong copulations and may remain *in copula* with inserted genitalia for as long as many hours (Wilcox 1984; Clark 1988; Rubenstein 1989; Sih *et al.* 1990; Fairbairn 1990).

The costs and benefits of mating for females

Females may bear a number of potential costs in matings, such as time and energy costs, increased predation risk and risk of injury or disease transmission. However, several benefits (e.g. replenishment of sperm, reception of nutrients from mating males or various genetic benefits) may balance these costs (Thornhill and Alcock 1983; Arnqvist 1989a; Dickinson, this volume).

In many insect species, females are thought to mate multiply in order to replenish their depleted sperm supplies (Thornhill and Alcock 1983). However, this is clearly not the case in water striders. Females have been found to be able to store viable sperm for extensive periods of time without reduction in fertility in several species. Sperm longevities have been reported to be over 14 d in *G. thoracicus*, over 30 d in *G. lateralis*, over 10 d in *G. odontogaster* and over 24 d in *A. remigis* (Kaitala 1987; Arnqvist 1988, 1989a; Rubenstein 1989). Neither do female water striders gain any major nutritional benefits via the ejaculate (Andersen 1982; Kaitala 1987; Arnqvist 1989a; cf. Gwynne, this volume), nor any major genetic benefits (Arnqvist 1989a), from rematings.

It is often assumed that mating activities increase the risk of predation. Water striders are particularly well studied from this point of view; three independent studies have shown that the risk of predation increases during mating. Arnqvist (1989a) showed that the risk of predation from backswimmers increased by a factor of three for female *G. odontogaster* as a consequence of mating activities. Fairbairn (1993) recorded a doubled risk of predation from frogs for mating compared with single *A. remigis* females, and Rowe (1994) demonstrated that mating *G. buenoi* females suffered two to three times greater predation risk from backswimmers than did single females. The increase in predation risk for females during mating is a combined effect of increased visibility and a decreased escape ability when attacked (Fairbairn 1993; Rowe 1994). The pre- and postcopulatory struggles are especially risky parts of mating: Rowe (1994) reported that struggles attract predators, and that predators were about five times as successful in attacks on struggling pairs compared with single individuals. Furthermore, in attacks on mating pairs, females are taken as prey more often than males (Arnqvist 1989a; Rowe 1994).

Male harassment, copulation and guarding may also be energetically costly for females (Stone 1995). Females of both *G. odontogaster* and *A. remigis* suffer from reduced mobility during mating (Arnqvist 1989a; Fairbairn 1993). Because water striders typically skate around searching for food, this will probably result in either a reduced foraging success for equal effort or a higher energetic effort for equal foraging success compared with single females (Arnqvist 1989a; Fairbairn 1993; but see below).

Form and resolution of sexual conflict

The primary interests in matings of water strider males and females are thus asymmetric, and two types of sexual conflicts occur. Males benefit from rematings in terms of paternity at little cost; females do not receive any measurable benefits from rematings but suffer energetic costs and increased predation risk. In encounters between males and females, a conflict over the control of the mating decision thus typically occurs. Similarly, males benefit from mate-guarding phases in terms of decreased sperm competition, whereas contact guardings are costly for females. Thus a conflict over the control of the mating duration also often occurs.

Thornhill and Alcock (1983) suggested that if males can evolve behavioral and/or morphological traits that make it costly for females to reject males attempting copulation, females may mate for reasons of convenience. This is exactly what has been suggested to occur in water striders with type I matings (Rubenstein 1984; Wilcox 1984; Arnqvist 1989a, 1992a; Rowe 1992; Vepsäläinen and Savolainen 1995). Arnqvist (1992a) modeled this situation and showed that water strider females making 'the best of a bad job' should resist matings, but only to a certain point representing a trade-off between the costs of rejecting harassing males and the costs of mating. Rowe (1992) arrived at similar conclusions in a verbal treatment of the situation for *G. buenoi*. This model of water strider mating dynamics has been shown to accurately predict adaptive female mating behavior in several species (Arnqvist 1992a; Rowe 1992; Sih and Krupa 1992; Krupa and Sih 1993; Vepsäläinen and Savolainen 1995; Weigensberg and Fairbairn 1994); when the rate of male harassment is high, single females have little to gain from resisting matings since they will be repeatedly harassed if the male is dislodged, and consequently females do not struggle as intensively during the precopulatory struggle before accepting a copulation. As a result, mating frequencies and mating durations increase. In contrast, when harassment rates are low, a female that dislodges a male will enjoy a relatively long period before the next harassment, and females struggle more intensively to dislodge males attempting copulations.

In water strider species with type I matings, females carrying males are not harassed as frequently or intensively as are single females; males will occasionally lunge at mating pairs but make no serious mating attempts. Wilcox (1984) showed that *A. remigis* females could even forage more effectively when carrying a male compared than when single, as a consequence of reduced harassment rates on mating pairs; he suggested that females in this way may actually 'benefit' from matings (see also Rowe *et al.* 1996). Thus, even though there is a basic sexual conflict over mating (females would be better off foraging singly given that they could effectively avoid males or reject matings at a very low cost), male water striders have in effect inflicted an apparent harmony between the sexes on females by frequent and effective harassment. Males thus seem to have made acceptance of superfluous matings 'the best of a bad job' for females in many situations (Arnqvist 1992a).

In addition to the conflict over the mating decision, there is likely to be a related conflict over the mating duration. These two conflicts are reflected in the two conspicuous fights typical for type I matings: the pre- and postcopulatory struggles. Although the various female costs of long vs. short matings have not been formally modeled, some obvious predictions can be made (Wilcox 1984; Arnqvist 1992a; Rowe 1992; Vepsäläinen and Savolainen 1995). The most intuitive prediction is that females should accept longer matings (before dislodging the male) when harassment rates are high. This prediction has been confirmed in a number of experimental studies, where mating duration has been found to increase with increased sex ratio and/or density (Clark 1988; Arnqvist 1992a; Rowe 1992; Erlandsson 1992; Jablonski and Vepsäläinen 1995; Vepsäläinen and Savolainen 1995).

Although the options for a female during the pre- and postcopulatory struggles are essentially the same, i.e. dislodging the male means avoiding the costs of carrying the male around and accepting the male means that these costs will be experienced for a certain time, the consequences of 'winning or losing' are different for males. In the precopulatory struggle, males struggle to achieve a copulation or not. In the postcopulatory struggle, males struggle to reduce sperm competition (i.e. to increase the quality of the already achieved copulation). Thus, males should always choose higher persistent time and effort than females during the precopulatory struggle (i.e. males should never withdraw, but may nevertheless be dislodged), and female interests alone should determine the outcome of precopulatory struggles (Arnqvist 1992a). This expectation seems to be consistent with the struggling behavior in type I matings (Rowe 1992). The conflict over the mating duration is theoretically more complex than the conflict over the mating decision, because both male and female interests have to be considered (Vepsäläinen and Savolainen 1995). For example, it is possible to envision situations (i.e. strongly female-biassed sex ratios) where males should not guard females from sperm competition, but rather invest time in searching for additional mates (Parker 1974; Alcock 1994). However, because harassment rates also

decrease in such situations, female interests act in the same direction; females should accept only short matings. In such situations, the conflict over the mating duration may diminish and males are predicted to terminate matings shortly after copulation without struggle. Such a behavioral pattern has been observed (Erlandsson 1992; Vepsäläinen and Savolainen 1995). Thus, when harassment rates are low both sexes should favor short matings and vice versa, and it is not obvious when and why the optimal mating duration should differ between the sexes. This issue clearly needs a formal theoretical treatment, especially as a number of factors other than sex ratio should affect the pros and cons of various mating durations for females and males (Sih et al. 1990; Arnqvist 1992a; Rowe et al. 1994).

It is tempting to view the sexual conflict over mating decisions in water striders as evolutionary 'wars of attrition' (Hammerstein and Parker 1987; Clutton-Brock and Parker 1995; Alexander et al., this volume). Even though females in one sense may 'control' matings by their reluctance (Rowe 1992), males may be said to have won the 'battle of the sexes' in the sense that they have apparently succeeded in making it costly for females to resist matings. I suggest that two major factors have made the evolution of this mating system possible. First, male behavioral and morphological traits make it costly for females to resist matings. Water striders are well adapted to determine the direction and distance of objects at the water surface (Jamieson and Scudder 1979; Andersen 1982), and males often succeed in lunging at females from distances of up to 10–15 times their own body length (Arnqvist 1989b). Males of many species have also evolved various morphological adaptations that increase their grasping ability (see below), which makes it more costly for females to dislodge them. Second, habitat characteristics make it difficult or even impossible for females to avoid males. The water surface is a strictly two-dimensional habitat with a relatively low degree of complexity. Furthermore, females have to be active in highly predictable areas in order to forage and oviposit. Thus, in contrast to many other insects (see, for example, Wiklund 1982; Wickman 1986), avoiding males is often difficult for female water striders (Krupa et al. 1990) and males are typically found searching for females in areas where females forage (Wilcox 1984; Krupa et al. 1990) or oviposit (Hayashi 1985; Foster and Treherne 1986). Under this scenario, the resolution of the sexual conflict will be dynamic and depend on behavioral, morphological and constitutional adaptations within each sex as well as on the environment. The relative ability of the sexes to control matings should then differ within and between species, which is exactly what we observe.

EVOLUTION OF WATER STRIDER MATING SYSTEMS

The least specialized and most common water strider mating system, female convenience polyandry associated with type I mating behavior, is most likely the plesiomorphic mating system within the Gerridae, as shown primarily by comparisons with outgroups (Andersen 1982). A simple mating system where the male jumps on the back of the female without prior courtship, grasps her and attempts copulation while the female attempts to dislodge the male has been described for several other families in the infraorder Gerromorpha (i.e. Veliidae (Andersen 1982, 1989) and Hebridae (Heming-van Battum and Heming 1989)).

Type II matings appear to have evolved independently at least three times within the Gerridae: in the genus *Rhagadotarsus* in the subfamily Rhagadotarsinae and in the *Limnoporus rufoscutellatus* group and the *Aquarius conformis* group within the subfamily Gerrinae (Andersen 1982, 1990, 1994; Wilcox and Spence 1986; Andersen and Spence 1992). This trend indicates that fairly general selective mechanisms are responsible for the evolution of type II from type I matings.

Finding environmental or constitutional correlates of interspecific variation in a trait is crucial in evaluating adaptive hypotheses about the evolution of the trait (Harvey and Pagel 1991; Rutowski, this volume). For the evolution of type II mating behavior from type I in water striders, it is difficult to find any such correlates. Species with type II matings occur in various kinds of habitats, e.g. in both lotic and lentic as well as in both temporary and permanent habitats, and there is no reason to believe that resources (e.g. oviposition sites) are more limited, predictable or clumped in type II species (Spence and Wilcox 1986; G. Arnqvist, personal observations). Spence and Wilcox (1986) suggested that egg parasitoids may play a role, but these are also known to infest eggs of many species with type I matings and more recent studies indicate that territoriality may actually increase parasitism (Henriquez and Spence 1993). Further, species with type I and type II matings do not differ markedly in general morphology, size or in sexual size dimorphism (Table 8-1).

I suggest that sexual conflict has played a crucial role in the evolution of type II systems from type I systems in water striders. If females can evolve attributes that in any way increase their control of mating initiation by reducing the effectiveness of male harassment, it may be more profitable for males to be sedentary, less aggressive, and harass females primarily at oviposition sites. As in other insects (Martens and Rehfeldt 1989), female water striders are particularly sensitive to male harassment during oviposition, because escaping harassment means interrupting oviposition. *Limnoporus* females have actually been found to lay more eggs if protected from harassment and interference during oviposition by a guarding male (Spence and Wilcox 1986). Thus, although females may be said to trade matings against temporary protection from further harassment during foraging and normal activity in type I matings, type II systems may have evolved by females trading matings against protection from harassment during oviposition. There are a number of potential factors that may promote such an evolutionary pathway, of which at least two have some empirical support. (1) Females may evolve morphological attributes that decrease the cost of rejecting unwanted mates, and thus shift the resolution of the mating conflict towards female interests. This seem to have occurred in *G. incognitus* (Arnqvist and Rowe 1995), and, interestingly enough, the mating system is affected: males of this species seem to be less aggressive and harass females primarily at oviposition sites. (2) A high degree of sperm displacement generally selects for an equally high degree of temporal proximity between mating and oviposition (Parker 1974; Walker 1980). If sperm precedence is very high, waiting for and/or harassing females at the oviposition sites may be a better option for males than searching and mating more randomly in space and time. Available data on sperm-displacement rates is very limited, but is at least in accordance with this hypothesis: the estimate of sperm precedence is higher in *Limnoporus* (>90%) than in species with type I matings (50–80%). The evolution of type II matings from type I matings in water striders may thus be an example of Walker's (1980) suggestion that females determine optimal male mating behavior, by the structure and/or function of their reproductive tract. However, more comparative data on sperm-displacement rates in water striders is needed to evaluate this latter hypothesis.

When discussing phylogenetic patterns of mating systems, one important aspect is to assess the degree of phylogenetic constraints (*sensu* Arnold 1992) on mating behavior. In water striders, some observations indicate a very low degree of phylogenetic constraints on behavior. One is the enormous diversity of mating behaviors found within the single genus *Aquarius*. In this genus, some species (*A. remigis*, *A. najas*, *A. conformis* and *A. paludum*) exhibit type I matings, whereas typical type II matings occur in *A. elongatus*. Further, mating durations range from very short (*A. elongatus*) to the longest ever observed within the Gerridae (*A. najas*). Males of some species show alternative mating tactics, whereas others do not; all three of the male mate-guarding tactics that have evolved in Gerridae are found within this single genus. The intensity of precopulatory conflicts also varies markedly between *Aquarius* species with type I matings (J. J. Krupa, unpublished). High variation in mating behavior has also been observed in the ocean skaters *Halobates* (Foster and Treherne 1982, 1986), in *Metrocoris* (cf. Cheng 1966; Ban *et al.* 1988; Koga and Hayashi 1993), in *Gerris* (Nummelin 1988) and in *Limnoporus* (Klingenberg and Spence 1994; J. R. Spence, unpublished). Put together, these high levels of intrageneric variability in behavior suggest that phylogenetic constraints on sexual behavior are not very strong within Gerridae, and that other factors shape mating systems adaptively (Spence and Andersen 1994; Andersen 1994).

SEXUAL DIMORPHISM AND SEXUAL SELECTION

The most consistently dimorphic trait in water striders is size: females are larger than males in most species (Andersen 1982, 1994; Fairbairn 1990; Spence and Andersen 1994; Table 8-1). The average female : male size ratio is about 1.1 within the family (range 0.81–1.32). There is considerable size-ratio variation within subfamilies and even within genera. In some species the size distribution of males is distinctly bimodal, some males being larger and some smaller than females (Andersen 1982; Chen and Nieser 1993). The function of this male dimorphism is unfortunately unknown.

Water strider males also exhibit three other types of sexual dimorphism: (1) modifications of the genital segments, (2) modifications of the legs and (3) modification of body shape. (1) The genital and pregenital abdominal segments are modified in most species of water strider, and are often provided with outgrowths of various shape and form (Andersen 1982, 1991; Arnqvist 1989b). These may be minor modifications, but in many species or genera male genital segments are large and conspicuous.

For example, in *Ptilomera* the male suranal plate is enlarged, flat and conspicuous and often provided with remarkable lateral projections (Hungerford and Matsuda 1965). Male *Halobates* typically have very conspicuous genitalia; the eighth abdominal segment is provided with various processes and the proctiger is broad and flattened (Andersen 1991; Andersen and Foster 1992). In *G. odontogaster*, males are provided with ventral processes on the seventh abdominal segment (Arnqvist 1989b). (2) Male water striders also generally have thicker and more powerful forelegs than females, despite their smaller body size (Sattler 1957; Andersen 1982; Rubenstein 1984). In addition, the male forefemora are equipped with pegs, spines, bristles and/or hair tufts in many species (for example in the genera *Amemboa*, *Metrocoris*, *Ptilomera*, *Asclepios* and *Halobates*). However, not only the forelegs may be sexually dimorphic. The most extreme male modifications of appendages occur in *Rheumatobates* (Hungerford 1954), where the antennae are modified to powerful leg-like structures and the hindlegs are curved and provided with the most bizarre outgrowths, pegs, spines and tufts of hair (see Fig. 8-3 and figures in Hungerford 1954). None of these structures occurs in females. (3) In many species there is also sexual dimorphism in body shape. Typically, the ventral surface of the abdomen is flattened, thinned or even depressed in males. This is obvious in the *Aquarius najas* group and in the *G. gillettei* group (Andersen 1990, 1993).

There are relatively few studies where sexual selection has been estimated in natural populations of water striders. Those that have been published have all been cross-sectional studies (but see Preziosi and Fairbairn 1996), where mating and single males have been compared in various ways. Several male traits have been claimed to be subjected to sexual selection in different species. Selection for large male size has been found in *G. buenoi*, *G. lateralis*, *G. lacustris*, *A. remigis* and *A. elongatus* (Hayashi 1985; Fairbairn 1988; Sih and Krupa 1992; Fairbairn and Preziosi 1994; Krupa and Sih 1993; Preziosi and Fairbairn 1996; Rowe and Arnqvist 1996), whereas selection for small male size has been reported in *G. buenoi*, *G. odontogaster* and *A. remigis* (Fairbairn 1988; Arnqvist 1992c; Kaitala and Dingle 1993; Blanckenhorn *et al.* 1995). Sexual selection for wide male forefemora and long genital segments have been demonstrated in *A. remigis* males (Rubenstein 1984; Weigensberg and Fairbairn 1995; Preziosi and Fairbairn 1996). In *G. odontogaster*, sexual selection for long male pregenital claspers has been demonstrated in several natural populations (Arnqvist 1989b, 1992b,c). Finally, pronotal

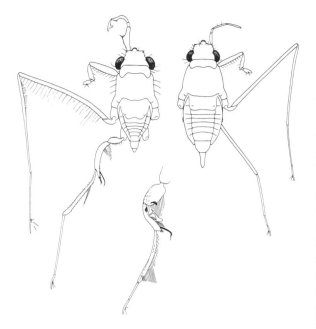

Fig. 8-3. A male (left) and a female (right) of *Rheumatobates rileyi* (reproduced with permission from Andersen 1982), and the left hindleg of a male *Rheumatobates bergrothi* (middle) (after Hungerford 1954). Note the bizarre sexual dimorphism in antennae and hindlegs, as well as the more powerful forelegs of the male.

size, gut parasite load and body mass has been found to be under sexual selection in *G. odontogaster* (Arnqvist 1992c).

Sexual conflict and female choice of male grasping ability

Most evolutionary modifications of male morphology (except body size) investigated so far are designed to grasp females efficiently during mating in one way or another (Andersen 1982, 1990). In a study of *Rheumatobates rileyi* (Fig. 8-3), Silvey (1931) noted that the male antennae were used during mating to grasp the female firmly anteriorly and the hindlegs curved around those of the female thus providing posterior attachment. The relatively powerful and modified forelegs typical of males of most species are adaptations for grasping; the forelegs are used by males during mating to hold the female thorax in a firm grip (Sattler 1957; Andersen 1982, 1989, 1990; Wilcox 1984; Arnqvist 1988, 1989b; Rowe 1992). Further, genitalic modifications in the form of outgrowths and/or processes function to grasp or pinch the female posteriorly (Arnqvist 1989b; Andersen 1991; Preziosi and Fairbairn 1996); the

flattened abdomen found in males of many species allows a tighter union to the female (Andersen 1990). It may thus be concluded that there is a variety of male adaptations that serve to increase male grasping ability during mating in water striders. This conclusion is confirmed by the direct observations of sexual selection for traits related to male grasping ability, such as powerful forelegs and long genital processes or segments (Rubenstein 1984; Arnqvist 1989b, 1992b,c; Kaitala and Dingle 1993; Weigensberg and Fairbairn 1995; Preziosi and Fairbairn 1996).

Male grasping traits, which serve to keep the sexes together during mating, are traditionally seen as adaptations to reduce sperm competition by preventing takeovers (Parker 1970, 1984). By grasping a female efficiently, the male avoids being displaced by another male, which may inseminate the same female. However, two lines of evidence strongly suggest that this is not the case in water striders (cf. Thornhill 1984). First, there are no published accounts of true takeovers of mates in water striders. Encroaching males occasionally contribute to the interruption of a mating, but the encroaching male will rarely if ever actually take over the mate; such instances are most likely due to mistaken identity rather than true takeover attempts (Wilcox 1984; Arnqvist 1989b; Rowe 1992; Krupa and Sih 1993). Second, detailed studies of one species suggest that grasping traits have another function (Arnqvist 1989a,b, 1992a,b,c, 1994). The male genital grasping apparatus of *G. odontogaster* has been shown to provide males with posterior attachment to females during the precopulatory struggle. The grasping apparatus is critical for males in enduring the struggle; males with inoperative claspers are easily dislodged by females. Furthermore, males with long claspers are better able to endure female reluctance during the precopulatory struggle, and hence achieve more matings both in the laboratory and in natural populations. In this species, there is sexual selection for long claspers, and female reluctance to mate is thus the mechanism of this selection.

All available information thus suggests that grasping morphologies in water striders are the result of intersexual selection rather than intrasexual selection. Females attempt to dislodge males for reasons of convenience, which selects for male ability to grasp and subdue the females as a side effect. This mechanism corresponds exactly with definitions of sexual selection by female choice, which is typically defined as cases where females have a behavior that tends to bias matings towards certain males (see, for example, Maynard Smith 1987). Male traits that increase grasping ability in water striders should thus generally be seen as the result of female choice for high male persistence during the pre- and postcopulatory struggles (Arnqvist 1992c; Rowe 1994). This is true irrespective of whether the traits function to grasp females during the pre- or postcopulatory struggle, or both.

Given that there is female choice of male grasping ability in water striders, the system corresponds in several ways to a current controversy as to the evolution of female choice. The current view of sexual selection by female choice may be divided into two major schools, with differing views of causes of female mate preferences (see Bradbury and Andersson 1987; Andersson 1994; Johnstone 1995). The 'good genes' school postulates that female preferences evolve under selection for females to mate with ecologically adaptive male genotypes. In contrast, the 'non-adaptive' school holds that preferences evolve for other reasons, and that selection may cause males to evolve maladaptively with respect to their ecological environment. In water striders, females struggle primarily to avoid superfluous and costly copulations rather than to choose males with good genes by testing male vigor (Arnqvist 1992a; Rowe 1992, 1994; Arnqvist and Rowe 1995; but see Weigensberg and Fairbairn 1994), and female choice of male grasping ability thus appear to be a side effect of natural selection acting on female behavior. The characteristics of female choice in water striders parallels those of some other thoroughly studied mate choice systems in frogs, fishes and lizards (Rowe *et al.* 1994).

Secondary sexual traits and coevolutionary arms races

Several authors have proposed that, given sexual conflicts over mating, male and female genitalia and secondary sexual traits may be involved in a coevolutionary arms race (Parker 1979, 1984; Alexander *et al.*, this volume). These hypotheses predict that females should evolve counteradaptations to cope with sexual harassment and gain increased control over matings. Given the conspicuous sexual conflicts in water striders and male morphological adaptations to forcefully grasp females, this group offers ideal systems for studying such mechanical conflicts of interest. For example, the highly modified abdominal spines of *G. incognitus* females have been found to function to reject harassing males: by manipulating the length of these spines, Arnqvist and Rowe (1995) were able to show that the spines increase female ability to dislodge males during

the precopulatory struggle, and thus to gain increased control over mating decisions. A number of other traits found exclusively in females are potential candidates for a similar function: females of many species have evolved various sorts of bizarre modifications such as various spines, genital processes and hooks, shortening of the abdomen, and lobes that cover the genital opening (Hungerford and Matsuda 1965; Andersen 1982, 1993; Andersen and Chen 1995). These traits probably also represent 'counteradaptations' that enable females to control matings by making it more difficult for males to establish genital contact; species exhibiting such female traits offer ideal systems to future studies of the coevolutionary dynamics of sexual conflict (Arnqvist and Rowe 1995).

Sexual size dimorphism

Sexual size dimorphism in water striders has been associated with two patterns or processes: allometry for size dimorphism and sexual selection. Comparative studies and phylogenetic analyses indicate that the evolution of sexual size dimorphism in water striders is very complex, but there are no obvious phylogenetic constraints on size dimorphism (Fairbairn 1990; Erlandsson 1992; Fairbairn and Preziosi 1994; Spence and Andersen 1994; Andersen 1994). Even though a number of adaptive mechanisms have been suggested (Andersen 1982; Vepsäläinen 1985; Fairbairn 1990, 1993; Spence and Andersen 1994), direct empirical evidence of sexual selection on body size is scarce and partly contradictory and it is still unclear, for example, why long matings are associated with a high degree of sexual size dimorphism in water striders. Most studies show sexual selection for large males (Sih and Krupa 1992; Fairbairn and Preziosi 1994; Arnqvist *et al.* 1996; Preziosi and Fairbairn 1996; Rowe and Arnqvist 1996). It might be that different mechanisms are interacting in producing apparently conflicting results: Sih and Krupa (1992) and Preziosi and Fairbairn (1996) argued that large males are better able to subdue reluctant females; Fairbairn (1993) suggested that females should allow small males to mate longer due to the costs of loading; and Blanckenhorn *et al.* (1995) suggested that small males may be favored because of lower food requirements. Further, Rowe and Arnqvist (1996) found that small males copulated longer but large males guarded longer in three *Gerris* species. Opposing mechanisms such as these could potentially interact to produce the observed inconsistency and complexity in the effect of male size on mating success between different studies and environmental conditions. At present, it is unclear how sexual conflict and sexual selection contributes to sexual size dimorphism; detailed empirical studies to evaluate this are needed.

Sexual conflict and the intensity of sexual selection

During the past decades, much effort has been spent labeling mating systems of different species. Unfortunately, this has contributed to a somewhat typological view of mating systems, which may have hindered the insights that might be gained from acknowledging that mating systems are plastic and that mating behavior varies with environmental conditions (Zeh and Zeh, this volume). Variability is the hallmark of studies of water strider mating systems; there are few other systems where there is such a thorough understanding of the mechanisms and dynamics of this variability. Recent experimental studies have shown that a number of environmental factors affect various aspects of water strider mating behavior. Factors such as sex ratio, population density, food availability and predation risk all have dynamic effects on female resistance, mating frequency, copulation duration, mate-guarding duration and general mating activity. In a review of the plasticity of water strider mating systems, Rowe et al. (1994) showed that most of this behavioral variation can be understood in terms of the sexual conflicts over matings; females seem to 'sample' the rate of male harassment (which in turn is determined by the environment) and adjust their behavior accordingly. These adaptive mating decisions, based on trade-offs between the different costs of mating, can be seen primarily as a variation in the level of female reluctance and as consequences of that variation (Rowe et al. 1994).

Mating patterns and sexual selection are population-level manifestations of the sexual behavior of individuals. Because water strider mating systems are so variable, it is not surprising that mating patterns and sexual selection have been found to vary between populations and between different sampling occasions in a given population (Fairbairn 1988; Arnqvist 1989b, 1992c; Fairbairn and Preziosi 1994; Krupa and Sih 1993; Arnqvist et al. 1996; Preziosi and Fairbairn 1996). Sexual selection for male size, for example, varies between populations or subpopulations of several species, and there is considerable variation between populations in sexual selection regimes for genital traits related to male grasping ability. There is also a remarkable intraspecific variation in the patterns of

assortative mating by size between populations in several species (Fairbairn 1988; Arnqvist et al. 1996).

The study of phenotypic selection in natural populations has been given considerable general attention in recent years, and several authors have stressed that a correlational approach is insufficient for inference of causal patterns of selection, and that a more mechanistic approach (i.e. a thorough knowledge of the ecology and behavior of the study organism in concert with studies of selection) may provide new perspectives in evolutionary ecology (see, for example, Endler 1986; Wade and Kalisz 1990; Arnqvist 1992c). In the study of water strider mating systems, the relatively thorough understanding of the mechanisms of selection provides a framework to which we can relate observations of variations in pattern of non-random mating between populations (Sih and Krupa 1992; Krupa and Sih 1993; Rowe et al. 1994). It also allows a priori predictions about the intensity of sexual selection in different natural populations to be made (Arnqvist 1992b,c, 1994).

The expected variance in reproductive success in males should increase when the operational sex ratio is biased towards males (Andersson 1994). In water striders however, the opposite pattern should be expected since male harassment increases with increased sex ratio, which causes female reluctance and thus the intensity of selection to decrease (Arnqvist 1992a; Rowe 1992, 1994). In accordance with this prediction, Arnqvist (1992b) demonstrated experimentally that the intensity of sexual selection on male claspers in G. odontogaster decreased with increased sex ratio, and Krupa and Sih (1993) showed that non-random mating by size in A. remigis males was drastically reduced in male-biased situations. The relation between population density and the intensity of selection in water striders is analogous to that between sex ratio and selection. When population density increases, female reluctance decreases as a result of increased male harassment. In G. odontogaster, patterns of sexual selection in natural populations are in accordance with this prediction; sexual selection is relaxed under high-density conditions (Arnqvist 1992c, 1994). The same is true for A. remigis, where high-density tended to decrease large male mating advantage (Krupa and Sih 1993). Thus, the predicted as well as the observed relation between sex ratio or density and the intensity of sexual selection in water striders are opposite to general theoretical predictions (Arnqvist 1992b,c; Krupa and Sih 1993; Andersson 1994), which illustrates that a thorough understanding of the mechanisms of selection is essential to predict variations in the intensity of selection and population level effects of such variations (Arnqvist 1994; Zeh and Zeh, this volume).

Because female mating behavior, and in particular female resistance to mating, is affected by a wide range of environmental factors, the intensity of sexual selection should vary accordingly. Results available so far indicate that patterns of non-random mating follow predictions based on our knowledge of female mating behavior. Experimental and observational studies of *A. remigis* have shown that non-random mating by size is also affected by hunger level and predation risk (Sih and Krupa 1992; Rowe *et al.* 1996) as well as by habitat structure (Krupa and Sih 1993).

In conclusion, there is prominent intraspecific variation in mating behavior, mating pattern and sexual selection in water striders. Much of this variation can be understood and even predicted based on our knowledge of the dynamics of sexual conflict, in particular the relation between environment, male harassment rate and female mating behavior.

ACKNOWLEDGEMENTS

This contribution is the result of many people's research, ideas and discussions. To all of these people I express my sincere gratitude. I am also most grateful to N. M. Andersen, J. C. Choe, B. J. Crespi, D. J. Fairbairn, J. J. Krupa, L. Rowe, A. Sih, J. R. Spence, K. Vepsäläinen and R. S. Wilcox for thoughtful comments on previous versions of this paper. Further, I thank J. C. Choe and B. J. Crespi for inviting me to write this chapter, N. M. Andersen, J. J. Krupa and J. R. Spence for kindly providing me with unpublished data, N. M. Andersen for allowing publication of Fig. 8-3, and G. Marklund for drawing Fig. 8-2. The Swedish Natural Science Research Council, The Craaford Foundation, The Fulbright Commission and The Swedish Institute provided generous financial support.

LITERATURE CITED

Alatalo, R. V., A. Carlson, A. Lundberg and S. Ulfstrand. 1981. The conflict between male polygamy and female monogamy: the case of the pied flycatcher, *Ficedula hypoleuca*. *Am. Nat.* 117: 738–753.

Alcock, J. 1994. Postinsemination associations between males and females in insects: the male-guarding hypothesis. *Annu. Rev. Entomol.* 39: 1–21.

Andersen, N. M. 1982. *The Semiaquatic Bugs (Hemiptera, Gerromorpha): Phylogeny, Adaptations, Biogeography and Classification.* (Entomonograph 3.) Klampenborg (Denmark): Scandinavian Science Press.

–. 1989. The coral bugs, genus *Halovelia* Bergroth (Hemiptera, Veliidae). II. Taxonomy of the *H. malaya* – group, cladistics, ecology, biology, and biogeography. *Entomol. Scand.* 20: 179–227.

–. 1990. Phylogeny and taxonomy of water striders, genus *Aquarius* Schellenberg (Insecta, Hemiptera, Gerridae), with a new species from Australia. *Steenstrupia* 16: 37–81.

–. 1991. Marine insects: genital morphology, phylogeny and evolution of sea skaters, genus *Halobates* (Hemiptera: Gerridae). *Zool. J. Linn. Soc.* 103: 21–60.

–. 1993. Classification, phylogeny, and zoogeography of the pond skater genus *Gerris* Fabricius (Hemiptera: Gerridae). *Can. J. Zool.* 71: 2473–2508.

–. 1994. The evolution of sexual size dimorphism and mating systems in water striders (Hemiptera: Gerridae): a phylogenetic approach. *Ecoscience* 1: 208–214.

Andersen, N. M. and P. P. Chen. 1995. A taxonomic revision of the Ptilomerine genus *Rhyacobates* Esaki (Hemiptera: Gerridae), with five new species from China and adjacent countries. *Tijdschr. Entomol.* 138: 51–67.

Andersen, N. M. and W. A. Foster. 1992. Sea skaters of India, Sri Lanka, and the Maldives, with a new species and a revised key to Indian Ocean species of *Halobates* and *Asclepios* (Hemiptera, Gerridae). *J. Nat. Hist.* 26: 533–553.

Andersen, N. M. and J. R. Spence. 1992. Classification and phylogeny of the Holarctic water strider genus *Limnoporus* Ståhl (Hemiptera, Gerridae). *Can. J. Zool.* 70: 753–785.

Andersson, M. 1994. *Sexual Selection.* Princeton: Princeton University Press.

Arnold, S. J. 1992. Constraints on phenotypic evolution. *Am. Nat.* 140: S85–S107.

Arnqvist, G. 1988. Mate guarding and sperm displacement in the water strider *Gerris lateralis* Schumm. (Heteroptera: Gerridae). *Freshwa. Biol.* 19: 269–274.

–. 1989a. Multiple mating in a water strider: mutual benefits or intersexual conflict? *Anim. Behav.* 38: 749–756.

–. 1989b. Sexual selection in a water strider: the function, nature of selection and heritability of a male grasping apparatus. *Oikos* 56: 344–350.

–. 1992a. Precopulatory fighting in a water strider: intersexual conflict or mate assessment? *Anim. Behav.* 43: 559–567.

–. 1992b. The effects of operational sex ratio on the relative mating success of extreme male phenotypes in the water strider *Gerris odontogaster* (Zett.) (Heteroptera: Gerridae). *Anim. Behav.* 43: 681–683.

–. 1992c. Spatial variation in selective regimes: sexual selection in the water strider, *Gerris odontogaster. Evolution* 46: 914–929.

–. 1994. The cost of male secondary sexual traits: developmental constraints during ontogeny in a sexually dimorphic water strider. *Am. Nat.* 144: 119–132.

Arnqvist, G., and L. Rowe. 1995. Sexual conflict and arms races between the sexes: a morphological adaptation for control of mating in a female insect. *Proc. R. Soc. Lond.* B 261: 123–127.

Arnqvist, G., L. Rowe, J. J. Krupa and A. Sih. 1996. Assortative mating by size: a meta-analysis of mating patterns in water striders. *Evol. Ecol.* **10**: 265–284.

Ban, Y., S. Shibata and M. Ishikawa. 1988. Life history of the water strider *Metrocoris histrio* B. White (Hemiptera: Gerridae) in Aichi Prefecture, Japan. *Verh. Int. Ver. Limnol.* **23**: 2145–2151.

Blanckenhorn, W. U., R. F. Preziosi and D. J. Fairbairn. 1995. Time and energy constraints and the evolution of sexual size dimorphism – to eat or to mate? *Evol. Ecol.* **9**: 369–375.

Bradbury, J. W. and M. B. Andersson, eds. 1987. *Sexual Selection: Testing the Alternatives.* Chichester: John Wiley.

Chen, P. P. and N. Nieser. 1993. A taxonomic revision of the Oriental water strider genus *Metrocoris* Mayr (Hemiptera, Gerridae). Part I–II. *Steenstrupia* **19**: 1–82.

Cheng, L. 1966. Studies on the biology of the Gerridae (Hem., Heteroptera) II: The life history of *Metrocoris tenuicornis* Esaki. *Entomol. Mon. Mag.* **102**: 273–282.

Clark, S. J. 1988. The effects of operational sex ratio and food deprivation on copulation duration in the water strider (*Gerris remigis* Say). *Behav. Ecol. Sociobiol.* **23**: 317–322.

Clutton-Brock, T. H. and G. A. Parker. 1995. Sexual coercion in animal societies. *Anim. Behav.* **49**: 1345–1365.

Daly, M. 1978. The cost of mating. *Am. Nat.* **112**: 771–774.

Dewsbury, D. A. 1982. Ejaculate cost and male choice. *Am. Nat.* **119**: 601–610.

Dominey, W. J. 1984. Alternative mating tactics and evolutionary stable strategies. *Am. Zool.* **24**: 385–396.

Eberhard, W. G. 1985. *Sexual Selection and Animal Genitalia.* Cambridge, Mass.: Harvard University Press.

Endler, J. A. 1986. *Natural Selection in the Wild.* Princeton: Princeton University Press.

Erlandsson, A. 1992. Life on the water surface: behaviour and evolution in semiaquatic insects. Ph.D. dissertation, Lund University, Sweden.

Fairbairn, D. J. 1988. Sexual selection for homogamy in the Gerridae: an extension of Ridley's comparative approach. *Evolution* **42**: 1212–1222.

–. 1990. Factors influencing sexual size dimorphism in temperate water striders. *Am. Nat.* **136**: 61–86.

–. 1993. The costs of loading associated with mate-carrying in the water strider *Gerris remigis*. *Behav. Ecol.* **4**: 224–231.

Fairbairn, D. J. and R. F. Preziosi. 1994. Sexual selection and the evolution of allometry for sexual size dimorphism in the water strider, *Aquarius remigis*. *Am. Nat.* **144**: 101–118.

Foster, W. A. and J. E. Treherne. 1982. Reproductive behaviour of the ocean skater *Halobates robustus* (Hemiptera: Gerridae) in the Galapagos Islands. *Oecologia (Berl.)* **55**: 202–207.

–. 1986. The ecology and behaviour of a marine insect, *Halobates fijiensis* (Hemiptera: Gerridae). *Zool. J. Linn. Soc.* **86**: 391–412.

Hammerstein, P. and G. A. Parker. 1987. Sexual selection: games between the sexes. In *Sexual Selection: Testing the Alternatives.* J. W. Bradbury and M. B. Andersson, eds., pp. 119–142. Chichester: John Wiley.

Harvey, P. H. and M. D. Pagel. 1991. *The Comparative Method in Evolutionary Biology.* Oxford University Press.

Hayashi, K. 1985. Alternative mating strategies in the water strider *Gerris elongatus* (Heteroptera, Gerridae). *Behav. Ecol. Sociobiol.* **16**: 301–306.

Heming-van Battum, K. E. and B. S. Heming. 1989. Structure, function and evolutionary significance of the reproductive system in males of *Hebrus ruficeps* and *H. pusillus*. *J. Morphol.* **202**: 281–323.

Henriquez, N. P. and J. R. Spence. 1993. Host location by the gerrid egg parasitoid *Tiphodytes gerriphagus* (Marchal) (Hymenoptera: Scelionidae). *J. Insect Behav.* **6**: 455–466.

Hungerford, H. B. 1954. The genus *Rheumatobates* Bergroth (Hemiptera-Gerridae). *Univ. Kans. Sci. Bull.* **36**: 529–588.

Hungerford, H. B. and R. Matsuda. 1965. The genus *Ptilomera* Amyott and Seville (Gerridae: Hemiptera). *Univ. Kans. Sci. Bull.* **45**: 397–515.

Jablonski, P. and S. Kaczanowski. 1995. Influence of mate-guarding duration on male reproductive success: an experiment with irradiated waterstrider (*Gerris lacustris*) males. *Ethology* **98**: 312–317.

Jablonski, P. and K. Vepsäläinen. 1995. Conflict between sexes in the water strider *Gerris lacustris*: a test of two hypotheses for male guarding behaviour. *Behav. Ecol.* **6**: 388–392.

Jamieson, G. S. and G. G. E. Scudder. 1979. Predation in *Gerris* (Hemiptera): reactive distances and locomotion rates. *Oecologia (Berl.)* **44**: 13–20.

Johnstone, R. A. 1995. Sexual selection, honest advertisement and the handicap principle: reviewing the evidence. *Biol. Rev.* **70**: 1–65.

Kaitala, A. 1987. Dynamic life-history strategy of the water strider *Gerris thoracicus* as an adaptation to food and habitat variation. *Oikos* **48**: 125–131.

–. 1991. Phenotypic plasticity in reproductive behaviour of water striders: trade-offs between reproduction and longevity during food stress. *Funct. Ecol.* **5**: 12–18.

Kaitala, A. and H. Dingle. 1993. Wing morph, territoriality and mating frequency of the water strider *Aquarius remigis*. *Ann. Zool. Fenn.* **30**: 163–168.

Klingenberg, C. P. and J. R. Spence. 1994. Heterochrony and allometry: lessons from the water strider genus *Limnoporus*. *Evolution* **47**: 1834–1853.

Koga, T. and K. Hayashi. 1993. Territorial behavior of both sexes in the water strider *Metrocoris histrio* (Hemiptera: Gerridae) during the mating season. *J. Insect Behav.* **6**: 65–77.

Krupa, J. J., W. R. Leopold and A. Sih. 1990. Avoidance of male giant water striders by females. *Behaviour* **115**: 247–253.

Krupa, J. J. and A. Sih. 1993. Experimental studies on water strider mating dynamics: spatial variation in density and sex ratio. *Behav. Ecol. Sociobiol.* **33**: 107–120.

Martens, A. and G. Rehfeldt. 1989. Female aggregation in *Platycypha caligata* (Odonata: Chlorocyphidae): a tactic to evade male interference during oviposition. *Anim. Behav.* **38**: 369–374.

Maynard Smith, J. 1987. Sexual selection: a classification of models. In *Sexual Selection: Testing the Alternatives*. J. W. Bradbury and M. B. Andersson, eds., pp. 9–20. Chichester: John Wiley.

Murray, A. M. and P. S. Giller. 1990. The life-history of *Aquarius najas* De Geer (Hemiptera: Gerridae) in Southern Ireland. *Entomologist* **109**: 53–64.

Nummelin, M. 1987. Ripple signals of the water strider *Limnoporus rufoscutellatus* (Heteroptera: Gerridae). *Ann. Entomol. Fenn.* **53**: 17–22.

–. 1988. The territorial behaviour of four Ugandan water strider species (Heteroptera, Gerridae): a comparative study. *Ann. Entomol. Fenn.* **54**: 121–134.

Parker, G. A. 1970. Sperm competition and its evolutionary consequences in the insects. *Biol. Rev.* **45**: 525–567.

–. 1974. Courtship persistence and female-guarding as male time investment strategies. *Behaviour* **48**: 157–184.

–. 1979. Sexual selection and sexual conflict. In *Sexual Selection and Reproductive Competition in Insects*. M. S. Blum and N. A. Blum, eds., pp. 123–166. New York: Academic Press.

–. 1984. Sperm competition and the evolution of animal mating strategies. In *Sperm Competition and the Evolution of Animal Mating Systems*. R. L. Smith, ed., pp. 1–60. London: Academic Press.

Parker, G. A., R. R. Baker and V. G. F. Smith. 1972. The origin and evolution of gamete dimorphism and the male-female phenomenon. *J. Theor. Biol.* **36**: 529–553.

Preziosi, R. F. and D. F. Fairbairn. 1996. Sexual size dimorphism and selection in the wild in the water strider *Aquarius remigis*: body size, components of body size and male mating success. *J. Evol. Biol.* **9**: 317–336.

Rowe, L. 1992. Conveniance polyandry in a water strider: foraging conflicts and female control of copulation frequency and guarding duration. *Anim. Behav.* **44**: 189–202.

–. 1994. The costs of mating and mate choice in water striders. *Anim. Behav.* **48**: 1049–1056.

Rowe, L. and G. Arnqvist. 1996. Analysis of the causal components of assortative mating in water striders. *Behav. Ecol. Sociobiol.* **38**: 279–286.

Rowe, L., G. Arnqvist, A. Sih and J. J. Krupa. 1994. Sexual conflict and the evolutionary ecology of mating patterns: water striders as a model system. *Trends Ecol. Evol.* **9**: 289–293.

Rowe, L., J. J. Krupa and A. Sih. 1996. An experimental test of condition dependent mating behaviour and habitat use by water striders in the wild. *Behav. Ecol.* **7**: 474–479.

Rubenstein, D. I. 1984. Resource acquisition and alternative mating strategies in water striders. *Am. Zool.* **24**: 345–353.

–. 1989. Sperm competition in the water strider *Gerris remigis*. *Anim. Behav.* **38**: 631–636.

Sattler, W. 1957. Beobachtungen zur fortpflantzung von *Gerris najas* deGeer (Heteroptera). *Z. Morphol. Oekol. Tiere* **45**: 411–428.

Sih, A. and J. J. Krupa. 1992. Predation risk, food deprivation and non-random mating by size in the stream water strider, *Aquarius remigis*. *Behav. Ecol. Sociobiol.* **31**: 51–56.

Sih, A., J. J. Krupa and S. Travers. 1990. An experimental study on the effects of predation risk and feeding regime on the mating behaviour of the water strider. *Am. Nat.* **135**: 284–290.

Sillén–Tullberg, B. 1981. Prolonged copulation: a male postcopulatory strategy in a promiscuous species, *Lygaeus equestris* (Heteroptera: Lygaeidae). *Behav. Ecol. Sociobiol.* **9**: 283–289.

Silvey, J. K. G. 1931. Observations on the life-history of *Rheumatobates rileyi* (Berg.)(Hemiptera: Gerridae). *Papers Mich. Acad. Sci.* **13**: 433–446.

Smith, R. L., ed. 1984. *Sperm Competition and the Evolution of Animal Mating Systems*. London: Academic Press.

Spence, J. R. 1979. Microhabitat selection and regional coexistence of water-striders (Heteroptera: Gerridae). Ph.D. dissertation, University of British Columbia, Canada.

Spence, J. R. and N. M. Andersen. 1994. Biology of water striders: interactions between systematics and ecology. *Annu. Rev. Entomol.* **39**: 97–124.

Spence, J. R. and R. S. Wilcox. 1986. The mating system of two hybridizing species of water striders (Gerridae). II. Alternative tactics of males and females. *Behav. Ecol. Sociobiol.* **19**: 87–95.

Stone, G. N. 1995. Female foraging responses to sexual harassment in the solitary bee *Anthophora plumipes*. *Anim. Behav.* **50**: 405–412.

Thornhill, R. 1984. Alternative hypotheses for traits believed to have evolved by sperm competition. In *Sperm Competition and the Evolution of Animal Mating Systems*. R. L. Smith, ed., pp. 151–179. London: Academic Press.

Thornhill, R. and J. Alcock. 1983. *The Evolution of Insect Mating Systems*. Cambridge, Mass.: Harvard University Press.

Thornhill, R. and K. P. Sauer. 1991. The notal organ of the scorpionfly (*Panorpa vulgaris*): an adaptation to coerce mating duration. *Behav. Ecol.* **2**: 156–164.

Trivers, R. L. 1972. Parental investment and sexual selection. In *Sexual Selection and the Descent of Man*. B. Campbell, ed., pp. 1871–1971. Chicago: Aldine.

–. 1974. Parent-offspring conflict. *Am. Zool.* **14**: 249–264.

Vepsäläinen, K. 1974. Determination of wing length and diapause in water-striders (*Gerris* Fabr., Heteroptera). *Hereditas* **77**: 163–176.

–. 1985. Exclusive female vs. male territoriality in two water strider (Gerridae) species: hypotheses of function. *Ann. Entomol. Fenn.* **51**: 45–49.

Vepsäläinen, K., and M. Nummelin. 1985a. Female territoriality in the water striders *Gerris najas* and *G. cinereus*. *Ann. Zool. Fenn.* **22**: 433–439.

–. 1985b. Male territoriality in the water strider *Limnoporus rufoscutellatus*. *Ann. Zool. Fenn.* **22**: 441–448.

Vepsäläinen, K. and R. Savolainen. 1995. Operational sex ratios and mating conflict between the sexes in the water strider *Gerris lacustris*. *Am. Nat.* **146**: 869–880.

Wade, M. J. and S. Kalisz. 1990. The causes of natural selection. *Evolution* **44**: 1947–1955.

Walker, W. F. 1980. Sperm utilization strategies in nonsocial insects. *Am. Nat.* **115**: 780–799.

Weigensberg, I. and D. J. Fairbairn. 1994. Conflict of interest between the sexes: a study of mating interactions in a semiaquatic bug. *Anim. Behav.* **48**: 893–901.

–. 1995. The sexual arms race and phenotypic correlates of mating success in the water strider, *Aquarius remigis*. *J. Insect Behav.* **9**: 307–319.

Wheelwright, N. T. and G. S. Wilkinsson. 1985. Space use by a Neotropical water strider (Hemiptera: Gerridae): sex and age-class differences. *Biotropica* **17**: 165–169.

Wickman, P.-O. 1986. Courtship solicitation by females of the small heath butterfly, *Coenonympha pamphilus* (L.) (Lepidoptera: Satyridae) and their behaviour in relation to male territories before and after copulation. *Anim. Behav.* **34**: 153–157.

Wiklund, C. 1982. Behavioural shift from courtship solicitation to mate avoidance in female ringlet butterflies (*Aphantopus hyperanthus*) after copulation. *Anim. Behav.* **30**: 790–793.

Wilcox, R. S. 1972. Communication by surface waves: mating behaviour of a water strider (Gerridae). *J. Comp. Physiol.* **80**: 255–266.

–. 1979. Sex discrimination in *Gerris remigis*: role of a surface wave signal. *Science (Wash., D.C.)* **206**: 1325–1327.

–. 1984. Male copulatory guarding enhances female foraging in a water strider. *Behav. Ecol. Sociobiol.* **15**: 171–174.

Wilcox, R. S. and J. R. Spence. 1986. The mating system of two hybridizing species of water striders (Gerridae). I. Ripple signal functions. *Behav. Ecol. Sociobiol.* **19**: 79–85.

Wilcox, R. S. and J. D. Stefano. 1991. Vibratory signals enhance mate-guarding in a water strider (Heteroptera: Gerridae). *J. Insect Behav.* **4**: 43–50.

9 · Multiple mating, sperm competition, and cryptic female choice in the leaf beetles (Coleoptera: Chrysomelidae)

JANIS L. DICKINSON

ABSTRACT

In the leaf beetles (Coleoptera: Chrysomelidae) males and females mate with multiple partners and form prolonged associations that last for hours or even days. In some species, males exhibit elaborate courtship behaviors while riding on females' backs before, during, or after insemination. Females often respond to copulation attempts by positioning their abdomens out of reach of the male's genitalia or by kicking males with their legs. Examination of internal events suggests that females control the ability of males to fully evert their genitalia and inseminate. Females of some species also emit sperm during or after copulation. Experimental work indicates that the spermathecal muscle plays a role in sperm retention and fertilization of eggs, suggesting that females control emission and retention of the male's sperm. Comprehensive analysis of the fitness consequences of prolonged and multiple mating has not been attempted for any species in the Chrysomelidae. Although mate-guarding appears to be one important function of prolonged pairing, prolonged pairing can increase proportional paternity even when it does not reduce the likelihood of remating by the female. An intriguing body of evidence suggests that cryptic female choice may occur in the leaf beetles, where females accept, retain, and use sperm of males of preferred phenotypes. The hypothesis that females exert postmating female choice by manipulating retention and use of sperm should occupy significant numbers of behavioral ecologists in the coming decade.

INTRODUCTION

Although studies of animal mating patterns have focussed primarily on the interests of males, it is widely recognized that conflict of interest between the sexes determines many aspects of reproduction, including when to mate, how long mating associations should last, patterns of sperm use when females have multiple partners, and parental investment (Walker 1980; Davies 1989; Alexander et al., this volume; Arnqvist, this volume; Choe, this volume). Males derive immediate benefits from mating with multiple females because they gain an incremental increase in reproductive success with each new mate. However, in most species of insects, females also mate with multiple partners and, in many cases, the distributions of mating frequencies are similar for the two sexes (Fincke 1982; Koenig and Albano 1986; Dickinson 1992a). Female reproductive success is usually not mate-limited, but is limited by the ability of females to survive and garner resources for offspring production (Trivers 1972). Because remating is probably costly, we expect females mating with multiple males to receive significant phenotypic or genetic gains.

The evolutionary maintenance of multiple mating by females can be understood only with experimental manipulation and careful measurement of the costs and benefits for both sexes (Tables 9-1 and 9-2). Hypothesized benefits of multiple mating for females are not mutually exclusive (Choe, this volume). By mating with multiple males, females may insure their ability to fertilize eggs or they may receive material contributions that enhance their fecundity, survival, or the viability of their offspring (Gwynne 1984, this volume; Ridley 1988). Females may benefit genetically if they exercise sequential mate choice, if multiple mating results in sequential competition between males for fertilizations, or if increasing the genetic diversity of their offspring gives females an advantage in the face of environmental variability (Walker 1980). However, females may suffer energetic costs, increased predation, or greater risk of disease and parasitism due to mating (Watson 1993). As an evolutionary problem, the question of why females mate with multiple males is difficult, because the costs and benefits for each sex depend, to some extent, on what the other sex is doing. For example, even if females derive phenotypic or genetic benefits by remating, they

Table 9-1. *Costs and benefits of multiple (M) and prolonged (P) mating for females*

Phenotypic benefits

(a) Fertility insurance (M)

(b) Acquisition of nutrients for reproduction or somatic maintenance (M, P)

(c) Acquisition of defensive chemicals for eggs or for the female herself (M, P)

(d) Protection from predators (M, P)

(e) Protection from harassment by males seeking to mate with the female (M, P)

(f) Increased foraging efficiency (M, P)

(g) Phoresy: if males carry females (M, P)

Phenotypic costs

(a) Increased exposure to parasites or disease (M)

(b) Loss of time for feeding (M, P)

(c) Increased predation due to increased conspicuousness or inability to escape as quickly (M, P)

(d) Increased energetic expenditure due to carrying the male if males ride on the female's back (M, P)

(e) Energetic costs of copulation (M, P)

Genetic Benefits

(a) Increased genetic diversity of offspring (M)

(b) Benefits due to sequential mate choice (M, P)

Genetic costs

(a) Loss of control over paternity of offspring (M, P)

Table 9-2. *Costs and benefits of prolonged and multiple mating for males*

Phenotypic benefits

(a) Incremental increase in offspring production with each new mate (M)

(b) Increased probability of survival of at least one mate (M)

(c) Reduced risk of predation if pairs can detect predators sooner, if they confuse predators, or if there is augmented chemical defense when paired (M, P)

(d) Phoresy: if females carry males (P)

(e) Increased paternity derived from investment in paternity insurance (P)

Phenotypic costs

(a) Reduced investment of time in paternity insurance (M)

(b) Increased exposure to parasites or disease (M)

(c) Loss of time spent feeding or increased predation due to searching for mates (M)

(d) Reduced mating frequency due to reduction in time to locate and fertilize multiple females (P)

(e) Reduced survival due to loss of time spent feeding, increased conspicuousness to predators, or reduced ability to escape from predators (P)

Genetic benefits

(a) Production of genetically diverse offspring (M)

(b) Benefits due to sequential mate choice (M)

Genetic costs

(a) None

---

may be 'making the best of a bad job' owing to the costs of resistance imposed on them by mate-seeking males (Thornhill and Alcock 1983; Arnqvist, this volume).

Verbal and mathematical models indicate that mating patterns reflect trade-offs among paternity assurance, mate location, mating costs, and parental effort (Davies 1989). In spite of their importance, such trade-offs have rarely been measured in the field (but see Travers and Sih 1991; Dickinson 1996). Both nutrient contribution and prolonged mating may increase a male's paternity at the cost of additional mating opportunities. However, nutrient investment and prolonged mating may also have survival costs if males are less able to feed themselves or incur a greater risk of predation while riding on females' backs. By seeking additional matings, males may suffer energetic costs of actively searching for mates, increased predation due to greater mobility, and reduced foraging time. As a consequence of these costs, males of some species do not live as long as females in the field (Thornhill 1979; Watson 1988).

In this chapter, I discuss the fitness consequences of multiple mating for females and trade-offs among mate location, paternity assurance, and maintenance for males in the leaf beetle family Chrysomelidae. There are an estimated 35 000 species of leaf beetle in nine to nineteen subfamilies (Jolivet 1988; Suzuki 1988). Only a few species have been studied in detail with respect to multiple mating and sperm competition, so a phylogenetic analysis of chrysomelid mating systems is not currently possible. However, existing studies indicate that we could learn much from comparative analysis; there is considerable variation in copulatory behavior, modes of insemination, and sperm-use mechanisms. In the leaf beetles, success in fertilizing eggs is the result of a complex of male competitive interactions, female behaviors that appear to thwart mounting and copulation attempts, paternity assurance mechanisms on the parts of males, and mechanisms that enable females to control sperm retention and use. I use information from existing studies to illustrate how further study of sexual conflict over multiple mating in the Chrysomelidae may provide important insights into the evolution and maintenance of polyandrous mating systems.

CHRYSOMELID LIFE HISTORY AND REPRODUCTIVE BEHAVIOR

Leaf beetles feed on a diversity of plant taxa; most are oligophagous, specializing on a single family or genus of plants. In some species both larvae and adults feed on plant foliage, whereas in others leaf feeding is restricted to the adult stage and larvae burrow underground to feed on the roots. Females usually cluster their eggs and lay multiple egg clutches throughout their lives. For example, females of the blue milkweed beetle (*Chrysochus cobaltinus*, Eumolpinae) lay a mean of 37 clutches in the field (Dickinson 1995). In some species of chrysomelid, females produce a fecal covering that protects eggs from natural enemies (Eisner *et al.* 1967; Damman and Cappuccino 1991), whereas in others, females simply glue their eggs to the undersides of leaves (Dickinson 1992b).

The number of eggs per clutch varies among and within species. Female *Microrhopala vittata* (Hispinae) lay a mean of three eggs per clutch, whereas mean clutch sizes of female *Labidomera clivicollis* (Chrysomelinae) range from 12.2 to 16.5 (Damman and Cappuccino 1991; Dickinson 1992b). Clustering of eggs often coincides with cannibalism and larval gregariousness (Breden and Wade 1985; Dickinson 1992b). Extended parental care is provided through prolonged retention of fertile eggs in viviparous species in the subfamily Chrysomelinae (Bontems 1988); females of some species of cassidines and chrysomelines guard eggs and immatures (Windsor 1987; Choe 1989; Windsor and Choe 1994).

In the field, leaf beetles are relatively long-lived; in some populations, adults survive to breed more than one season (Dickinson 1992a). Because of their sedentary habits, it is possible to follow a high proportion of marked adults throughout their reproductive lives. The predominant mate-locating behavior of leaf beetles is scramble competition, in which males search competitively for females (Dickinson 1992a, 1995). Studies of marked individuals have shown that both males and females mate with multiple partners in the field (Dickinson 1992a, 1995; Eberhard *et al.* 1993). In the two species where the distributions of mating frequencies have been examined for both sexes, they are similar (Dickinson 1992a, 1995). In *Chrysochus cobaltinus*, which lives on the milkweed *Asclepias eriocarpa* (Asclepiadaceae), mating frequencies are truly astonishing. Males and females mate a mean of 26 times with 13 different mates in the field and often remate with the same individuals when they encounter them on subsequent occasions (Fig. 9-1).

The prevalence of multiple mating in the Chrysomelidae provides an arena for cryptic female choice, which is defined as biasing of paternity by the female once copulation has begun (Thornhill 1983; Eberhard and Cordero 1995). Such biasing can occur if females favor some male phenotypes over others by refusing to remate. However, cryptic female choice can also occur when females appear to remate at random with respect to male phenotype. Chrysomelid females often use sperm from more than one male to fertilize eggs within a single clutch (Dickinson 1986, 1988; McCauley *et al.* 1988; Rodriguez 1993a; Stevens and McCauley 1993). In some species that diapause as adults, females store sperm from mating in the fall, before entering diapause, and use these sperm to fertilize eggs when they become reproductively active in the spring (Tauber 1988; Stevens and McCauley 1993). Cryptic female choice may occur if females bias sperm storage in favor of males with preferred phenotypes.

Prolonged mating associations, where males ride on the backs of females for periods of a few hours to days, are common in the Chrysomelidae. During these associations,

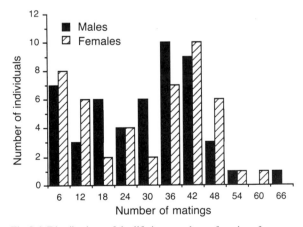

Fig. 9-1. Distributions of the lifetime numbers of matings for marked male and female *Chrysochus cobaltinus* (Eumolpinae) based on censusing nine times per day for 31 consecutive days in the field. Individuals often remated with the same partners on subsequent encounters, so although the mean numbers of mating events were 26.6 ± 2.1 ($n = 49$ males) and $25.9 \pm .4$ ($n = 47$ females) the mean numbers of different partners were 12.8 ± 1.1 and 12.4 ± 1.1, respectively.

males may inseminate females repeatedly or just once, shortly after mounting. In general, males do not court prior to mounting, but exhibit highly stereotyped courtship-like behaviors just after mounting, during copulation, after copulation, or in the interval between repeated copulations (Dickinson 1986; Eberhard 1991, 1994; Rodriguez 1993a,b, 1994a). Eberhard (1991, this volume) refers to this behavior as 'copulatory courtship' and hypothesizes that females choose the genetic sires of their offspring on the basis of courtship behaviors by exerting control over insemination and sperm retention. This is an extension of his hypothesis that cryptic female choice has resulted in elaboration of male genitalic characters that are involved in stimulating females (Eberhard 1985).

Chrysomelid mating systems are characterized by prolonged pairing, copulatory courtship, multiple mating, and mixed paternity of broods. In this chapter, I use multiple mating in the sense of Parker (1970) to mean multi-male mating by females and multi-female mating by males. I differentiate the number of matings from the number of different mates only where the influence of multiple mating on offspring genotypes is concerned; otherwise remating with a prior mate at a later date is probably equivalent to mating with a new partner. Finally, I refer to prolonged mating associations, where pairs intersperse repeated copulations with periods during which there is no genital contact, as repeated or intermittent copulation.

REPRODUCTIVE MORPHOLOGY AND THE PROCESS OF INSEMINATION

Multiple mating has been studied in the Chrysomelidae by examining both the ultimate fitness consequences and proximate mechanisms governing insemination and sperm use. A rudimentary knowledge of the reproductive morphology of leaf beetles and the processes of copulation and fertilization is necessary for understanding the mechanisms governing the outcome of multiple mating. Below, I describe the general layout of the male and female reproductive systems.

The female reproductive tract

In females, the genital pore opens into a swelling of the common oviduct, called the genital chamber or bursa copulatrix ('bursa') (Fig. 9-2). In spermatophore-producing species, the spermatophore is deposited in the bursa, just ventral and posterior to the spermathecal duct. The vagina and bursa open into a common oviduct, through which

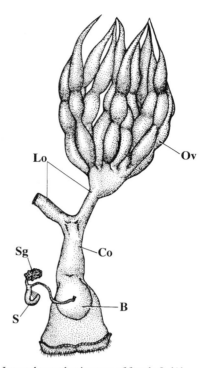

Fig. 9-2. Internal reproductive tract of female *Labidomera clivicollis* (Chrysomelinae) showing the ovaries, consisting of groups of ovarioles (Ov), the lateral oviducts (Lo), the common oviduct (Co), the bursa copulatrix (bursa) (B), the spermatheca (S) and the spermathecal gland (Sg). (Drawing by J. L. Dickinson.)

eggs pass during their journey from the ovaries. Most species have a spermatheca, which is the primary site of sperm storage. The site of fertilization is generally thought to be just below the entrance of the spermathecal duct (Rodriguez 1994b), but in some chrysomelines, sperm are also stored in the terminal portions of the ovarioles and fertilization takes place shortly after eggs are released into the ovary (Bontems 1988). Sperm storage in the pedicels of the ovarioles seems to occur in both viviparous and oviparous Chrysomelinae and is not associated with the presence or absence of a spermatheca. However, viviparous species do appear to have reduced spermathecae (Bontems 1988).

In *Chelymorpha alternans* (Cassidinae), both fertilization and the proportion of sperm in the lumen of the spermatheca relative to its duct are controlled, in part, by the spermathecal muscle (Rodriguez 1994b). Twenty-four hours after mating, a mean of 36% more sperm were found outside the lumen of the spermatheca when the spermathecal muscle was cut than when females received sham operations. This indicates that the spermathecal muscle facilitates uptake of sperm into the spermatheca. Females with cut spermathecal muscles laid just as many eggs, but did not fertilize the eggs as efficiently as sham-operated females. After three ovipositions, female *C. alternans* with cut spermathecal muscles had retained more sperm in their reproductive tracts than sham-operated controls, indicating that the spermathecal muscle may also function to control release of sperm during fertilization. The spermathecal muscle has been found in most of the 19 generally accepted subfamilies of the Chrysomelidae (Suzuki 1988) as well as in *Callosobruchus maculatus* (Bruchidae) (Eady 1994), and the boll weevil, *Anthonomus grandis* (Curculionidae) (Villavaso 1975a). In the boll weevil, the function of the spermathecal muscle is similar, but not identical to that in *Chelymorpha alternans*; it does not influence sperm uptake or retention, but controls fertilization of eggs and sperm priority.

Rodriguez's (1994b) experiments indicate that additional mechanisms must be acting in concert with the spermathecal muscle to control sperm uptake, retention, and release from the spermatheca. How sperm enter the spermatheca is not known, but possibilities include sperm motility and chemotaxis (Villavaso 1975b), positive pressure created by the male during ejaculation (Dickinson 1988), positive pressure in the spermatophore (Eberhard *et al.* 1993), negative pressure created by absorption of fluid already in the spermatheca (Linley and Simmons 1981), muscular deformation of the spermatheca to create a negative pressure (Rodriguez 1994b), and direct placement of sperm in the spermatheca by the male (Rodriguez 1994b).

The spermatheca of *L. clivicollis* has a small gland associated with it (Fig. 9-2), which may function to provide nutrients for long-term sperm storage (Chapman 1982). Extirpation of the spermathecal gland in the boll weevil demonstrated that it aids migration of spermatozoa to the spermatheca and maintains viability and motility of sperm (Villavaso 1975b). A link between the spermathecal gland and long-term sperm storage has yet to be investigated in the leaf beetles.

Male genitalia and reproductive organs

The internal reproductive tract of the male consists of paired testes; in *L. clivicollis* each has two large lobes that contain sperm in various degrees of maturation (Fig. 9-3). As is generally true for insects, the sperm in the testes are in bundles, but the seminal vesicles, which in *L. clivicollis* are enlargements of the vasa deferentia, contain single, mature spermatozoa.

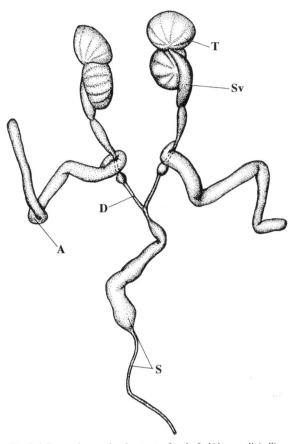

Fig. 9-3. Internal reproductive tract of male *Labidomera clivicollis* (Chrysomelinae) showing the paired testes (T), the seminal vesicles in which sperm are stored (Sv), the paired accessory glands (A), the ejaculatory duct duplex (D) and the ejaculatory duct simplex (S), which is differentiated into at least three different glandular regions (Drawing by J. L. Dickinson.)

Sperm transfer occurs by copulation. In *L. clivicollis* sperm in the seminal vesicles are vigorously motile *in vitro* and males have a mean of $229\,600 \pm 43\,200$ SE ($n = 7$ males, range $199\,000$–$302\,700$) sperm in their seminal vesicles after their first copulation with a female. Females have a mean of only $35\,000 \pm 3700$ SE ($n = 10$ females) sperm after mating repeatedly with a male during a 45 h period. Unless sperm are discharged or removed from the female's reproductive tract during repeated copulations, virgin males have enough sperm, on average, to inseminate six females. This is probably an underestimate of the number of sperm available, because the testes contain additional sperm and spermatogenesis is ongoing.

A single copulation reduces sperm counts to 55% of virgin levels in the seminal vesicles of male *C. alternans*, but 30 min later males have more than enough sperm in their seminal vesicles to inseminate a second female (Rodriguez 1993a). Males show no reduction in sperm counts relative to virgins when allowed two copulations one to three minutes apart; this indicates that sperm are released into the seminal vesicles in response to mating. This suggests that males are not limited by their ability to produce sperm. However, these sperm counts were conducted on males mated to virgin females. If more semen is needed to dilute and displace the sperm of a female's prior mates than is required to fully inseminate a virgin female,

males may deplete their sperm supplies significantly during a single mating with a non-virgin female (Parker and Simmons 1991).

The male's penis has three parts: a heavily sclerotized median lobe or aedeagus, a membraneous or partially sclerotized internal sac, and a sclerotized but flexible flagellum or virga (Fig. 9-4a). In *L. clivicollis*, the median lobe lies on its side in the male's body cavity and is rotated upright when everted. Only when the median lobe is inside the female are the internal sac and flagellum everted. In some species, sclerites of the internal sac are associated with spermatophore formation (Crowson 1981; W. Eberhard and S. Kariko, personal communication).

(a)

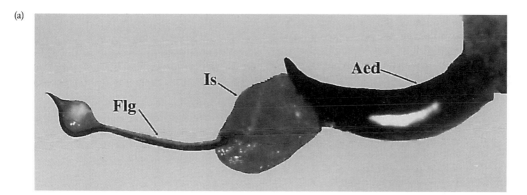

(b)

(c)

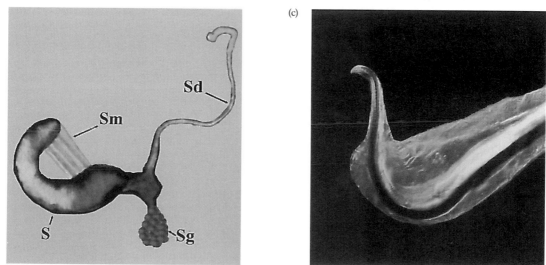

Fig. 9-4. (a) Penis of *Labidomera clivicollis* (Chrysomelinae) with the flagellum everted. The aedeagus (Aed) is heavily sclerotized and the flagellum (Flg) somewhat less rigid. The internal sac (Is) is membraneous. (b) Spermatheca of *Labidomera clivicollis* at ×150 showing the main reservoir of the spermatheca (S), the spermathecal gland (Sg), the spermathecal muscle (Sm) and the spermathecal duct (Sd) leading to the bursa. (c) Tip of the flagellum of *Labidomera clivicollis* at ×150 (Computer-enhanced photographs by J. L. Dickinson.)

The length of the flagellum varies considerably in the Chrysomelidae. The flagellum of the chrysomeline *L. clivicollis* is intermediate in length (Fig. 9-4a). In the cassidine *C. alternans* the flagellum is three times the male's body length; in some Alticinae it is reduced or absent (Rodriguez 1993a; W. Eberhard and S. Kariko, personal communication). In the alticine *Macrohaltica jamaicensis* the male's flagellum is reduced and does not enter the spermathecal duct, but males produce a spermatophore with a nipple that fits in the entrance to the duct (W. Eberhard and S. Kariko, personal communication). In *M. jamaicensis* and *L. clivicollis*, the diameter of the flagellum appears too large to fit inside the female's spermathecal duct (Fig. 9-4b,c). When female *L. clivicollis* are dissected immediately after a single, brief copulation, 94% of the inseminated sperm are located in the spermatheca. This suggests that males place the tip of the flagellum in the entrance to the duct, which is slightly swollen. By freezing mating pairs in liquid nitrogen, Rodriguez (1993a) observed that the thin, tapering flagellum of *C. alternans* is inserted far into the spermathecal duct and is often rolled upon itself within the ampulla, a dilation of the duct approximately two-thirds of the way to the spermatheca. In one case, the tip of the flagellum reached into the spermatheca itself, indicating that sperm may be inseminated directly into the spermatheca (Rodriguez 1993a).

In the Chrysomelidae, males have one or more pairs of accessory glands and an ejaculatory duct that is often differentiated into a variable number of glandular sections, the function of which is not known (Suzuki 1988). The ejaculatory duct and paired accessory glands may produce different parts of the spermatophore in species that package sperm. However, I am aware of no study of a chrysomelid that combines observation, dissection, and labeling of secretions to determine where they end up in the spermatophore. In *L. clivicollis* (Dickinson 1986) secretion from the paired accessory glands appears as an amorphous mass in the female's vagina and does not have sperm masses associated with it. Such wads of accessory-gland secretion may function to prevent leakage of sperm or reduce sperm displacement, although there is no evidence for such a function in the leaf beetles (Parker 1970; Chapman 1982).

In butterflies and moths, males normally make spermatophores, but some species in the Coleoptera, Diptera, and Hymenoptera make spermatophores whereas others inseminate directly (Chapman 1982, p. 365). Spermatophore production is considered ancestral and direct insemination into the female's spermatheca derived (Davey 1960;

Alexander *et al.*, this volume). This pattern is thought to reflect selection for placing sperm closer to the site of sperm storage and fertilization. Within the Chrysomelidae, male *M. jamaicensis* (Alticinae), *Diabrotica virgifera* (Galerucinae) and *C. alternans* (Cassidinae) make spermatophores (Lew and Ball 1980; Eberhard *et al.* 1993; Rodriguez 1993a), but males of *L. clivicollis* and *Leptinotarsa decemlineata* (Chrysomelinae) do not (Thibout 1982; Dickinson 1986).

Unlike the large nutrient-laden spermatophores produced by some Lepidoptera and Orthoptera, spermatophores of leaf beetles tend to be quite small. Male *M. jamaicensis* produce spermatophores weighing only 1% of their body mass (Eberhard *et al.* 1993) (Fig. 9-5). Female *M. jamaicensis* sometimes discharge spermatophores before even half the sperm have exited, a situation analogous to spermatophore removal in some Orthoptera (Gwynne, this volume). Spermatophore discharge also occurs in a tenebrionid beetle (De Villiers and Hanrahan 1991). Direct insemination may make discharge of sperm difficult, particularly if sperm are placed in the spermatheca. In *C. alternans* males both make spermatophores and thread their flagellum into the spermathecal duct (Rodriguez 1993a). Females discharge droplets containing large numbers of active sperm. It is not yet clear whether male *C. alternans* inseminate directly, into the spermathecal duct, in addition to producing a spermatophore. If they do, there may be advantages of both spermatophore production and direct insemination. A phylogenetic analysis of families in the Chrysomelidae could yield information on whether there has been gradual evolution toward direct insemination with species like *C. alternans* representing intermediate stages.

Fig. 9-5. *Macrohaltica jamaicensis* (Alticinae).
(Photograph by W. G. Eberhard.)

SPERM COMPETITION AND MECHANISMS OF SPERM PRIORITY

As an evolutionary consequence of sperm storage and multiple mating by females, we expect mechanisms that inhibit females from remating and increase the success of males in superseding sperm of a female's prior mates (Parker 1970). Whether females remate for phenotypic gains or to avoid the costs of resisting harassment by mate-seeking males, they will benefit by being able to influence the genotypes of their offspring (Walker 1980). For this reason, we expect selection to favor females that control the process of fertilization.

Hypothesized mechanisms governing sperm priority include: (1) positional sperm priority, whereby the last male's sperm lie nearest the site of fertilization, either because they are stratified within the spermatheca or its duct or because sperm are placed in both the bursa and the spermatheca and sperm are more likely to be used from one of these sites than the other (Schlager 1960; Walker 1980); (2) removal and replacement, whereby a male removes or flushes away sperm of earlier males before or during insemination with his own sperm (Parker 1970; Waage 1979); (3) sperm discharge by the female or relocation of sperm to a portion of the female reproductive tract where they die (Parker 1970; Rodriguez 1993a); (4) inactivation, in which an earlier male's sperm are inactivated by substances in the ejaculate of later males (Silberglied *et al.* 1984); (5) dilution, in which a male dilutes the sperm of earlier (or later) males and derives a paternity advantage based upon the relative numbers of his sperm in the female (Parker 1970; Walker 1980); and (6) differential competitiveness of sperm based on age or phenotypic differences (Sivinski 1984). These mechanisms of sperm priority are not mutually exclusive. For example, dilution may be the primary mechanism of sperm use until the female's sperm stores are full, after which the mechanism may switch to one of constant random sperm displacement (Parker and Simmons 1991). In some cases, it is possible to use observations of the distribution of sperm and information on sperm-use patterns to infer the mechanism (Parker *et al.* 1990).

Determinants of sperm use patterns in *Labidomera clivicollis*

In *L. clivicollis*, a complex series of experiments indicates that mating duration, mating order, and the duration of the gap between matings all affect sperm use by doubly mated females, whereas body size does not (Dickinson 1986, 1988). In this species, mating involves repeated copulation over a period of up to 2.5 d in the field. Males exhibit stereotyped courtship behaviors during and between copulations they antennate rapidly, stroke the tip of the female's abdomen with their hind legs, rub the female's elytra with their tarsal pads, thrust their genitalia in and out, turn around on females' backs, evert their genitalia near the female's head, and bite the female while rocking vigorously on her back (Fig. 9-6).

All experiments involved examining paternity in the first three egg clutches laid after the second mating, but because paternity did not differ among these egg clutches, offspring from all three clutches were grouped (Dickinson 1986, 1988). Paternity analysis was based on 6.3 ± 0.9 (SE) to 10.1 ± 1.9 (SE) offspring per clutch (Dickinson 1986). Because mating associations are prolonged and variable in duration, the first experiment was designed to determine

Fig. 9-6. Mated pair of the milkweed leaf beetle, *Labidomera clivicollis* (Chrysomelinae) (Photograph by J. L. Dickinson.)

whether the duration of pairing influenced paternity (Dickinson 1986). In the first treatment, females remained with their first mate for 6 h after copulation began and their second mate for 24 h. The second treatment was the reverse of the first; females remained with the first mate for 24 h and the second mate for 6 h. Allozyme analysis showed that the 24 h males sired more offspring than the 6 h males. The surprising result was that the longer-mating male sired more offspring regardless of whether he was first or second to mate. Because the number of copulations, total copulation duration, the number of sperm transferred, and paternity all increased with mating duration, dilution appeared to be the mechanism of sperm priority.

A prediction of the dilution hypothesis is that virgin females mated to two males for equal periods of time will produce offspring sired equally by the two males (Dickinson 1988). When two consecutive matings were 15 h long, the mean paternity of the second male (P_2) did not differ statistically from 0.5, whereas when the two matings were 45 h long, the proportion of offspring sired by the second male was statistically greater than 0.5. There was also significant last-male paternity when two 15 h matings were staged five days apart. These experiments indicate that sperm priority is not simply a consequence of dilution.

Sperm counts also indicate that dilution is not the sole mechanism of sperm priority. After a single 15 h mating, the female's spermatheca was not completely full, but after 45 h, the spermatheca and its duct held the maximum capacity of 35 000 sperm (SE = 700; $n = 10$ females), more than 100 sperm per egg laid by females over a 30 d period. Although second males sired 78 ± 9% (SE) of the offspring after two consecutive 45 h matings, females had no more sperm after two 45 h matings than after one (Dickinson 1988). High P_2 is not due to insemination with an undetectable number of sperm that are more vigorous or in a better location for fertilizing eggs than the first male's sperm, because short second matings (6 h) do not result in last-male sperm priority (Dickinson 1986). Because there was no increase in the number of sperm stored by females with two versus one 45 h mating, the most likely explanation for second-male sperm priority is replacement of the first male's sperm by the second male's sperm.

Although I invoked sperm flushing by the male to explain the disappearance of sperm, I was not able to determine where the sperm went (Dickinson 1988). Sperm were not present in droplets expelled by females, nor did they disappear when incubated *in vitro* in homogenates of bursa and common oviduct (J. L. Dickinson, unpublished). This suggests that sperm are unlikely to suffer rapid degradation once they leave the spermatheca. One alternative that I did not investigate is that males remove and dispose of sperm of a female's earlier mates inside their own bodies. Male *Tenebrio molitor* (Tenebrionidae) remove sperm by trapping them on recurved spines on the aedeagus (Gage 1992). However, such spines are not apparent on the aedeagus, internal sac, or flagellum of *L. clivicollis* (Fig. 9-4a,c). How sperm of *L. clivicollis* disappear and whether their disappearance is under male or female control remain a mystery.

The proportion of offspring sired by the second male was highest when there was a 5 d gap between two 15 h matings, suggesting that latency to remating also influences patterns of sperm use in *L. clivicollis* (Dickinson 1988). Although females do not become more receptive to copulation as latency to remating increases, it is possible that they are cryptically more receptive to insemination. Inefficient use of sperm could also reduce the first male's paternity through a reduction in the number of sperm stored with latency to remating. Alternatively, stored sperm may become less competitive with time.

In *L. clivicollis*, three different mechanisms appear to determine sperm use patterns: dilution, removal and replacement, and an unknown mechanism that results in increased last-male paternity with an increase in the duration of the gap between matings. However, in a field study the mean mating duration was 18 h and the mean gap between matings was six days, raising the possibility that, in spite of the complex of mechanisms that determine which male's sperm are used, most eggs are fertilized by a female's most recent mate in the field (Dickinson 1988).

Determinants of sperm-use patterns in *C. alternans*

Rodriguez (1993a) conducted sperm-competition experiments that involved mating female *C. alternans* to two and three different males. He combined paternity experiments with sperm counts, detailed observations of copulation, and freezing of pairs to observe the internal events leading to insemination. Paternity analysis was based upon offspring reared from the first eleven egg clutches laid after termination of the second mating and involved a mean of 179 offspring. Whether females copulated with two or three males, there was nearly always one male that sired significantly more offspring than the others. The predominant sires tended to be those with long flagella (Rodriguez 1993a). However, flagellum length is significantly longer for one of the color morphs than for the other two, so that the apparent relationship between flagellum length and paternity could be due to selection on color. In a Friedman two-way analysis of variance, males with long flagella sired an increased proportion of offspring, whereas order of copulation, copula duration, color morph, body length, and the length of the median lobe of the male's genitalia did not influence paternity (V. Rodriguez, W. Eberhard and D. Windsor, personal communication).

Rodriguez (1993a) allowed virgin females to copulate with multiple males and counted sperm in different parts of the female's reproductive tract 24 h after the last copulation. He found that the number of sperm in the spermatheca at 24 h increased with the first four copulations, after which it leveled off. Females sometimes expel droplets of sperm during copulation; non-virgin females are more likely to do this than virgin females. Females that discharge droplets have fewer sperm in their spermathecae 24 h after copulation than females not discharging sperm. The increase in sperm number with successive copulations, coupled with evidence of sperm discharge, suggests numerical sperm competition, whereby males increase their paternity by having more of their sperm retained by females. If both dilution and sperm discharge play a role in determining sperm-use patterns, then males that inseminate with more sperm, induce females to retain their sperm, or induce females to discharge sperm of other males, will sire more offspring.

COSTS AND BENEFITS OF PROLONGED AND MULTIPLE MATING FOR FEMALES

Fitness consequences of multiple mating may reflect conflict or confluence of interest between the sexes (Alexander *et al.*, this volume). There is no single species of chrysomelid beetle for which the costs and benefits of multiple mating have been thoroughly examined (Tables 9-1, 9-2). Analyzing the fitness consequences of multiple mating is complicated by a tendency for pairs to form prolonged mating associations that last from several hours to days. Because some phenotypic costs and benefits depend only on the amount of time individuals spend paired, there is considerable overlap between the hypothesized fitness consequences of prolonged mating and multiple mating.

Phenotypic costs and benefits

There are three ways that multiple mating by females may increase the number, size, and fertility of eggs laid: (1) nutrients transferred to females at mating may increase female survival and fecundity (Boggs and Gilbert 1979); (2) females may remate to maintain adequate sperm supplies (Walker 1980; Ridley 1988); and (3) females may remate to guard against an infertile first mating (Walker 1980; Ridley 1988). Females can acquire additional nutrients and sperm by mating with multiple males, by remating with the same partner on a later occasion, or by copulating more than once without separating from a current partner. Empirically, benefits of acquiring nutrients and sperm can be confounded with each other if females reduce their egg production when their sperm supplies are depleted.

Ideally, experiments to examine the effect of multiple mating on fecundity and fertility should include egg counts, counts of the number of eggs that have been fertilized (counting both the number hatching and the number with embryos), and counts of sperm stored by singly and multiply mated females. If females housed in small containers with males experience significant levels of harassment, the males should be removed between matings and mating should be staged at intervals similar to those in the field.

Male leaf beetles usually transfer small quantities of accessory-gland secretions, suggesting that nutrient contribution may not be as important in the Chrysomelidae as in some species of Lepidoptera and Orthoptera (Gwynne 1984, this volume). In *M. jamaicensis*, discharged spermatophores do not appear collapsed or digested as they do in species in which nutrient contribution occurs (W. Eberhard and S. Kariko, personal communication). However, where quantities are small, it is still possible that males contribute significant amounts of certain limiting nutrients.

Mating with more than one male results in increased fecundity in *Chrysomela crotchi* (Chrysomelinae) (Smereka 1965), *M. jamaicensis* (Eberhard *et al.* 1993), and *Oulema melanopus* (Criocerinae) (Wellso *et al.* 1975). Examination of development and hatching success of eggs in the latter two species showed that fertility is also apparently enhanced by multiple mating. It is not clear whether once-mated females lay fewer eggs because they lack nutrients or whether they reduce their oviposition frequency as they begin to run out of sperm. It is possible that sperm depletion precipitates a reduction in the rate of oviposition to a level typical of virgin females. Although there appear to be phenotypic benefits of multiple mating in these species, it is not clear whether these benefits arise from improved fecundity or fertility.

Although virgin female *L. clivicollis* lay infertile eggs, mated females lay more than twice as many eggs as virgins in a 30 d period (Dickinson 1988). Providing that the initial mating is successful, multiple mating does not increase female fecundity or fertility; females consorting for 24 h with a single male lay and fertilize as many eggs in the next 30 d as females mated to six different males at five-day intervals (Dickinson 1988).

In spite of the absence of an effect of multiple mating on fecundity and fertility of female *L. clivicollis*, the relatively high proportion of infertile first matings indicates that fertility insurance may favor remating at least once. Four of 33 virgin females (12%) housed with their mates for 24 h after a first copulation laid no fertile eggs in the next 5–30 d (J. L. Dickinson, unpublished). Although this high proportion of infertile first matings supports the hypothesis that females ensure their fertility by mating with more than one partner, it fails to explain why females should mate with up to ten different males in the field (Dickinson 1988).

Prolonged or repeated association with males may have positive or negative effects on a female's foraging efficiency. For example, in the water strider *Aquarius remigis* (Heteroptera: Gerridae), foraging frequency is enhanced by multiple or prolonged copulation because copulating males protect females from harassment (Wilcox 1984; Arnqvist, this volume), whereas in the spider *Linyphia litigiosa* (Linyphiidae) females improve their foraging success by participating in aspermic precopulatory courtship because they engage males that would otherwise steal their prey (Watson 1993).

Using a combination of behavioral observations and demographic data, I have addressed the question of whether mating influences foraging efficiency of female milkweed beetles. Female leaf beetles are usually able to walk and feed with males on their backs, whereas males feed rarely, if at all, while riding on females. Table 9-3 shows identical analyses for the two milkweed chrysomelids *Chrysochus cobaltinus* and *Labidomera clivicollis*. These data are not experimental, but each female's foraging frequency when mating was compared, statistically, with her foraging frequency when alone. In the field, female *C. cobaltinus* spent 35% of their time paired and fed just as frequently when paired as when unpaired. In contrast, female *L. clivicollis* spent only 12–22% of their time paired and spent less time feeding when males were riding on their backs than when they were alone. These lifetime data on marked individuals suggest that female *L. clivicollis* experience tradeoffs between feeding and mating, whereas female *C. cobaltinus* do not.

Another potential cost of prolonged and multiple mating that has not been investigated in leaf beetles is the cost of carrying a male on the female's back. Fairbairn (1993) glued lead weights and dead male water striders to the backs of female water striders (*Aquarius remigis*; Heteroptera: Gerridae), and measured their effects on foraging success, mobility, and the risk of predation. This novel approach provides some interesting measures of the costs of carrying males, but such studies would be greatly

Table 9-3. *Field data on foraging by lone and paired females*

Study population	Mean SE proportion of observations			
	Feeding (lone)		Feeding (paired)	Paired (all)
Eumolpinae:				
Chrysochus cobaltinus				
Carmel Valley, California (1990)	0.07 ± 0.01		0.08 ± 0.01	0.35 ± 0.02
Chrysomelinae:				
Labidomera clivicollis				
Austin, Texas (spring 1985)	0.28 ± 0.04	*	0.14 ± 0.05	0.13 ± 0.02
Bridgeport, New York (summer 1985)	0.57 ± 0.38	*	0.09 ± 0.03	0.22 ± 0.04
Austin, Texas (fall 1985)	0.17 ± 0.02	*	0.06 ± 0.03	0.12 ± 0.02

* Proportion of observations when lone females fed was statistically greater than the proportion of observations when paired females fed: Wilcoxon Matched-Pairs test.

enhanced if coupled with measures of energetics. There are also problems with the manipulations because live male water striders often touch the water with their long legs, so that the weight unmanipulated females carry is less than the actual weight of the male. In leaf beetles this would not be a problem, because males usually rest wholly on the females' backs. However, leaf beetles often exhibit a defensive behavior that involves dropping from the host plant and remaining quiescent for a short period. If pairs ordinarily confuse predators by splitting apart when attacked, sticking males to females could overestimate the predation risk associated with carrying males.

Although there may be trade-offs between feeding and mating in *L. clivicollis*, there is no evidence of survival costs of multiple or prolonged mating for either species of milkweed leaf beetle. In *L. clivicollis* there was no relationship between longevity and the time females invested in mating, whereas in *C. cobaltinus*, there was a significant positive association between longevity and the proportion of time males and females were paired (Table 9-4). The positive relationship in *C. cobaltinus* may reflect a survival benefit of mating if paired females are better protected from predators than single individuals. Alternatively, the attributes that cause individuals to survive well may also confer success in attracting or retaining mates.

Experimental manipulations are needed to measure the survival costs and benefits of prolonged and multiple mating for females in the Chrysomelidae. However, no single field experiment is ideal for addressing this question. Although predation costs of multiple mating are best measured in the field, where there are opportunities for shelter and escape, marking and observing animals

requires a presence that might deter vertebrate predators. Fitness of isolated, singly mated females could be compared with that of females that have continuous access to males, but this comparison would not account for the costs of resistance females would incur by avoiding remating. In the aggregating species *C. cobaltinus*, male harassment is frequent and females sometimes carry up to three males on their backs at a time. A comprehensive study of the survival costs of multiple mating must include manipulation of key features, such as operational sex ratio, density of beetles, density of predators, and risk of parasitism or disease. Researchers have successfully manipulated predator density and sex ratio in water striders (Sih *et al.* 1990; Arnqvist, this volume); such methods could also be applied to the leaf beetles.

My observations and demographic studies of milkweed chrysomelids indicate that predation on adults is rare, so predation may not be as important in the Chrysomelidae as in the water striders (Arnqvist, this volume). Abbot (1994) examined the influence of ectoparasitic mites (*Chrysomelobia labidomerae*) on mate choice in *L. clivicollis* and, although he found an influence of mite infestation on survival of adult females and nutritionally stressed males, neither sex preferred uninfested individuals as mates. Even though there is no mate preference based on parasite load, it is likely that multiple mating increases the risk of contracting mites for both sexes.

Chemical defense may play an important role in the survival of eggs, larvae and adults in the Chrysomelidae (Blum 1994; Hilker 1994). In *L. clivicollis*, males rub females' elytra with their tarsi. This may be a form of courtship behavior, but it is also possible that males transfer

Table 9-4. *Proportion of observations paired versus days seen*

Study population	Study population selection gradient ± SE (*n*)	
	Males	Females
Eumolpinae:		
Chrysochus cobaltinus		
Carmel Valley, California (1990)	1.52 ± 0.34 (49)*	1.43 ± 0.63 (47)*
Chrysomelinae:		
Labidomera clivicollis		
Austin, Texas (spring 1985)	0.02 ± 0.37 (33)	0.23 ± 0.56 (47)
Bridgeport, New York (summer 1985)	0.93 ± 0.66 (50)	0.30 ± 0.59 (43)
Austin, Texas (fall 1985)	0.41 ± 0.77 (37)	−0.35 ± 0.26 (74)

*$p < 0.05$ by a delete-one weighted jackknife procedure (Wu 1986)

chemicals that protect females from predators. Multiple mating may also enhance chemical defense of eggs if males transfer defensive substances in seminal fluids (Sierra *et al.* 1976). The egg cases of *Chrysochus cobaltinus* contain larger concentrations of cardenolides than are found in females themselves (Sady 1994). In spite of the importance of chemical defense in the Chrysomelidae, the possibility that multiple mating enhances chemical protection of females or their eggs remains unexplored.

Genetic costs and benefits

Females may derive genetic benefits from multiple mating that are distinct from those of prolonged or repeated mating with the same partner. Genetic benefits are testable with one caveat: because of genotype–environment interactions, it may be necessary to test for genetic benefits in varying environments (Stearns 1992). Multiple mating often results in multiple paternity in leaf beetles (Dickinson 1988; McCauley *et al.* 1988). However, determining whether multiple paternity is an evolutionary cause or a simple consequence of multiple mating is difficult. The critical prediction is that offspring from clutches with multiple sires have greater survival and reproductive success in the field than offspring from clutches with just one sire. Larvae of several species of chrysomelid can be followed in the field for the first two instars. The importance of genetic diversity could be examined by measuring production of offspring by females mated with *n* different males, females mated *n* times with the same male, and females mated once to a single male. Predation, the proportion of eggs developing embryos, larval survival in both the laboratory and the field, and the phenotypes of adult offspring could be compared among the three treatments. Each of the *n* different males should be reused so that sequential mates of females in the multiple and repeated mating groups are equivalent in mating experience. A male's mating history could be important if males transfer lower quantities of accessory-gland secretions or sperm as they age or as they mate with increasing numbers of females.

The experiment I have described would separate the phenotypic and genetic effects of multiple mating on egg production, hatching success, larval survival, and offspring phenotype. However, paternity analysis is needed to discriminate benefits that arise due to increased genetic diversity among offspring from benefits that arise through male competition or female choice of good genes.

Genetic benefits of multiple mating may arise through female choice or male competition. After mounting occurs, females may terminate copulation before insemination is complete, elect to remate if a male of higher quality is encountered, or bias sperm use in favor of males of preferred behavioral or morphological phenotypes (Thornhill 1983; Rodriguez 1993a; Otronen 1994; Eberhard, this volume). Cryptic female choice occurs after the start of copulation and is an alternative to the lock-and-key hypothesis advanced to explain assortative mating as well as intraspecific variation in courtship behavior and male genitalia (Eberhard 1985; Brown 1993; Rodriguez 1993a; Eberhard, this volume). Cryptic female choice need not be the ultimate cause of multiple mating; it may simply offset the costs for females remating to avoid harassment or increase the benefits of multiple mating if females already receive phenotypic gains. For example, if the presence of a riding male reduces harassment for females, shifting selectivity to the postmounting stage would allow females to control the paternity of their offspring, without incurring the cost of resisting additional mating attempts.

Eberhard (1985, 1991) has proposed that both copulatory courtship behavior and elaborate male genitalia function to stimulate females to accept and use a conspecific male's sperm in fertilization. Stereotyped courtship-like behaviors occur before, during, and between copulations in many species of leaf beetle and may involve antennation, stroking, stridulation, rocking, or biting (Lew and Ball 1979; Dickinson 1986; Rodriguez 1993b, 1994a; Eberhard 1994). Females show dramatic refractory behaviors in which they attempt to scrape males off their backs, pull their abdomens up against their elytra to prevent intromission, and kick males' legs and genitalia with their hind legs before and during copulation (Lew and Ball 1979; Thibout 1982; Dickinson 1986; Rodriguez 1993a; W. Eberhard and S. Kariko, personal communication). Copulations with virgin females do not always lead to insemination, suggesting either that female refractory behaviors thwart insemination after copulation has begun or that males exert mate choice during copulation. Further observation of a large number of species is needed before we can determine whether copulatory courtship is favored by complex genital morphology and refractory behaviors of females. Observation of copulatory courtship in *M. jamaicensis* indicates that it is associated with female refractory behaviors and with particular stages of spermatophore transfer (W. Eberhard and S. Kariko, personal communication). However, I am aware of no evidence that variation in courtship behavior among males is correlated with success in copulation, insemination, sperm storage, or paternity.

Whether they encounter them simultaneously or sequentially, females may actively choose among males by controlling insemination and sperm use. Alternatively, they may simply have morphological or behavioral barriers that favor some males over others, resulting in indirect choice (Alexander *et al.*, this volume). Even if females treat all males equally, female control of insemination or sperm retention can act as a sorting process that increases the likelihood that females will use sperm of males of superior competitive or copulatory ability. In *L. clivicollis*, insemination occurs slowly, and full insemination is achieved sometime between 15 and 45 h (Dickinson 1988). Males increase their paternity by diluting and replacing the sperm of a female's prior mates, a process that is probably made lengthy by the continuous efforts of females to thwart males' copulation attempts. By making insemination difficult, females may select for more adept inseminators, males superior at warding off rivals, and males better at relocating females after mating is disrupted. This hypothesis suggests that females gain genetic benefits from mating with multiple males through competition among males for fertilizations and is analogous to the hypothesis of passive female choice put forth by Cox and Le Boeuf (1977). It provides an alternative to the hypothesis that females preferentially accept sperm from males on the basis of internal or external courtship behavior or preferred morphological characters (Eberhard 1985, 1991; Rodriguez 1993a).

The hypothesis that female leaf beetles exert cryptic mate choice makes specific predictions: (1) there should be differential success of males in insemination and fertilization that can be attributed to differences in male phenotype; and (2) females should control insemination, sperm retention, or fertilization in ways that favor successful male phenotypes. LaMunyon and Eisner (1993, 1994) provide compelling evidence that cryptic female choice favors production of large spermatophores in the moth *Utetheisa ornatrix* (Lepidoptera: Arctiidae). Preliminary support for the hypothesis of cryptic female choice, based on male morphology, is available for just one species in the Chrysomelidae, *Chelymorpha alternans* (Rodriguez 1993a).

Cryptic female choice in *Chelymorpha alternans*

Rodriguez's (1993a) study of *C. alternans* provides evidence of both differential fertilization success of males

and differential retention of sperm by females on the basis of male flagellum length. Male *C. alternans* produce a small spermatophore in the bursa and also insert the tip of the flagellum into the spermathecal duct. It is not clear whether they use the flagellum to place sperm in the spermatheca, to open the spermathecal duct to permit sperm migration, or to stimulate the female to accept and retain sperm.

Several observations suggest that females control sperm use by selective acceptance and retention. (1) After males mounted, 12 of 48 females (25%) pressed their abdomens against their elytra, successfully preventing intromission (Rodriguez 1993a). One-sixth of pairs terminated copulation after a short time without sperm transfer (12.6 ± 14.1 (SE) min, $n = 6$ of 36 pairs). Rodriguez (1993a) observed these pairs under a dissecting microscope and noted that the flagellum and internal sac were not everted during copulation. The simple observation that males fail to inseminate females is as consistent with male control as it is with female control. However, the entrance to the spermathecal duct was closed in the three females that were frozen in liquid nitrogen and dissected. Because the duct is not normally closed in virgin females, it is likely that the females use muscular contractions of the bursa and vagina to thwart insemination. (2) Rodriguez (1993a) observed that previously mated females discharged sperm during and after mating with much greater frequency than did females mating for the first time. Discharge of sperm could be a passive outcome of remating with full sperm stores, but evidence that all females discharged sperm droplets when the spermathecal muscle was cut (compared with 13.6% in sham-operated females) suggests that sperm retention is under female control (Rodriguez 1994b). When sperm were discharged by females mating for the first time, those females had fewer sperm in their spermathecae 24 h later than females not discharging sperm. Evidence (above) of numerical sperm priority in *C. alternans* suggests that female control of sperm retention will result in female control of paternity.

Two lines of evidence suggest that females control insemination and fertilization in ways that favor males with longer flagella. (1) Males with longer flagella sired more offspring when females mated with two or three consecutive mates (see above). Evidence that flagellum length is shorter when females discharge sperm than when they do not suggests that the difference in paternity based on male flagellum length is caused by differences in retention of sperm by females (Rodriguez 1993a). (2) Males with experimentally shortened flagella were not as successful as intact males at placing the flagellum into the ampulla of the female's spermathecal duct (Rodriguez 1993a). When mated to males with cut flagella, 87.5% of females emitted sperm in comparison with 13.1% of females mated to intact males (Rodriguez 1993a). This emission resulted in a reduction in the number of sperm in the spermathecae of females 24 h later. Twenty-four hours after mating, the number of sperm in the spermathecae of females that emitted sperm after mating with males that had cut flagella was not statistically different from the number in females that mated with normal males and emitted sperm; experimentally shortening the flagellum produces an outcome similar to normal matings involving males with short flagella.

In the absence of a suitable control for experimentally cutting the male's flagellum, the experiment is not convincing on its own. However, data indicating that flagellum length is shorter when females expel sperm than when they do not makes the argument more convincing. Males with longer flagella may be superior in two respects: (1) they may be better at placing sperm in the spermatheca where they will be retained, in which case, passive or indirect female choice is supported (Alexander et al., this volume) or (2) they may gain by using the flagellum to stimulate the female to retain more of their sperm, in which case females actively choose males on the basis of genitalic courtship (Eberhard 1985).

TRADE-OFFS BETWEEN PROLONGED AND MULTIPLE MATING FOR MALES

Prolonged mating associations

Prolonged mating associations are often assumed to prevent or delay remating by females, protecting the paternity of her current mate (Parker 1970). However, there are alternative hypotheses for how males benefit from prolonged mating associations (Alcock 1994) (Table 9-2). Prolonged mating associations typically exhibit four different forms, although combinations of these are sometimes found: (1) males ride on females' backs prior to copulation (Lew and Ball 1979); (2) males copulate with females, then ride on their backs without genital contact (Kirkendall 1984; Dodge and Price 1991; Dickinson 1995); (3) males and females remain together and maintain genital contact for a prolonged period (Lew and Ball 1979; Rodriguez 1993a); and (4) males ride on females' backs, interspersing copulation with passive riding and mounted courtship behavior

(Dickinson 1986; Eberhard et al. 1993). Although the phenotypic costs of prolonged and multiple mating may be similar, males may experience trade-offs between remaining with one mate to protect their paternity and locating additional females.

Precopulatory riding is expected to occur where receptive females are dispersed and in short supply. It is usually interpreted as precopulatory mate-guarding by males waiting to mate with the female (Thornhill and Alcock 1983). However, males sometimes court females while riding. Male C. alternans exhibit precopulatory mounted courtship for a mean of 13.9 ± 21.4 (SE) minutes, during which they palpate and antennate females' elytra with rapid vibratory movements (Rodriguez 1993a). Males of D. virgifera antennate females while riding on their backs for 10–60 min prior to copulation, a behavior interpreted as physical stimulation of the female by the male (Lew and Ball 1979). Precopulatory mounted courtship may function in species recognition or female choice of mates.

Postcopulatory riding is usually interpreted as mate-guarding, where males ride on females' backs to ward off rivals and prevent females from remating prior to oviposition (Thornhill and Alcock 1983). In Chrysochus cobaltinus, males ride passively on females' backs after a brief copulation that usually lasts fewer than ten minutes (Dickinson 1995). Females remate sooner in the field when males are removed from their backs than when males are allowed to continue riding. The hypothesis of mate-guarding is supported for C. cobaltinus because postcopulatory riding increases latency to remating by females in the field, takeovers are common, males do not exhibit elaborate courtship behaviors while on females' backs, and riding males actively fend off rivals trying to mate with the female. Alternatively, it is possible that females use male riding duration as an indicator of male quality. Females may become more refractory to remating as males ride longer, a response that would be the equivalent of cryptic female choice for postcopulatory riding. As a target of female choice, postcopulatory riding duration could be an excellent indicator of a male's ability to ward off rivals and cope with the trade-offs between remaining paired and feeding.

In species where copulation itself is prolonged, it is important to determine whether insemination is prolonged or whether prolongation of pairing involves insemination followed by a postcopulatory passive or courtship phase (Eberhard 1991). The spermatophore of M. jamaicensis is not hardened but is fully formed after the second of six series of penile thrusts by the male, indicating that the last four series of thrusts occur after insemination (W. Eberhard and S. Kariko, personal communication). Prolonged copulation can function in positioning or molding the spermatophore, transferring additional sperm or accessory-gland secretions, stimulating females to accept, retain, and use the male's sperm, or mate-guarding. These hypotheses are not mutually exclusive; prolonged copulation may have multiple functions (Dickinson 1988).

Two hypotheses have been advanced for why prolonged copulation in some species of insects takes the form of a variable number of short copulations interspersed with periods during which the genitalia are not engaged. Smith (1979) suggested that frequent copulation in a belostomatid water bug functions to ensure last-male sperm priority over interlopers that copulate with the female while she lays eggs on the male's back. Alcock (1976) suggested that if males of a buprestid beetle benefit from prolonged pairing in the first place, as would be the case if there are advantages of mate-guarding, then they will do better to inseminate gradually because females might terminate pairing as soon as they are fully inseminated. However, gradual insemination could be achieved without breaking genital contact. Alcock (1976) generated a second hypothesis to explain why males intersperse short copulations with periods of riding on the female's back without the genitalia engaged. Assuming gradual insemination is beneficial for males, intermittent copulation may afford them a better chance to escape predation if they can escape more quickly when not *in copula*. The role of the female in prolonging insemination is ignored by these hypotheses. By ceasing copulation, male *L. clivicollis* can climb higher on females' backs, antennate females' heads, rock and bite, and rub the females' elytra with their tarsi (Dickinson 1986). Males may copulate intermittently because the courtship behaviors allowed by breaking genital contact are more effective in inducing females to accept their sperm.

Fitness trade-offs for males

Although some phenotypic costs of prolonged and multiple mating are similar for the two sexes, males experience trade-offs between mate location, courtship, copulation, feeding, and mate-guarding. In *C. colbaltinus*, fights are common and pairs often form by takeover. Male and female removal experiments have shown that postcopulatory riding both delays remating by females and costs males additional mates (Dickinson 1995). The simulated fitness for two hours after removal is greater for

postcopulatory riders than for non-riders as long as the proportion of offspring sired by the last male to mate with the female is at least 0.40 (Dickinson 1995). This simple model of trade-offs between postcopulatory riding and mate location is a beginning, but the results may be quite different when other potentially important variables are included. For example, in the fly *Dryomyza anilis* (Diptera: Dryomyzidae) the paternity of a female's most recent mate depends on his body size as well as the female's recent mating history (Otronen and Siva-Jothy 1991; Otronen 1994). Matings that form by takeover yield lower last-male paternity, because females do not have the opportunity to discharge sperm.

In spite of evidence that male *C. cobaltinus* and *L. clivicollis* feed less than females and do not feed while riding on females' backs, demographic data suggest that there are no significant survival costs of mating (Dickinson 1992a, 1995; Table 9-4). Based on demographic data, male *C. cobaltinus* that spend more time mating also survive longer. However, survival costs and benefits of mate-guarding are difficult to measure without experiments. If individual characteristics influence both longevity and mating success, then males successful at mating will live longer, not because mating confers a survival advantage, but because some quality of males makes them good at both surviving and mating. In less sedentary species, one would also have to consider the possiblity that males not successful at mating are likely to disperse off the study area.

A thorough understanding of the maintenance of prolonged and multiple mating requires analysis of the conflicting demands of predator avoidance, feeding, mate-guarding, and mate-searching (Sih et al. 1990; Travers and Sih 1991). However, these behaviors are not independent: a beetle that spends more time guarding may spend less time walking and feeding. Even if paired males suffer less predation than lone males, it is difficult to know whether predation is lower on paired males because there is a direct advantage of being paired or because the lower mobility of paired males reduces their chances of encountering a sit-and-wait predator (McCauley and Lawson 1986). Experimental evidence suggests that both mate-searching and mating increase predation risk in the water strider *Aquarius remigis* (Gerridae) (Sih et al. 1990). Although, it can be difficult to tease apart the influence of predation on such correlated behaviors as feeding, mate-searching, and mate-guarding, experimental manipulations of predator density, sex ratio, beetle density and food quality are feasible with leaf beetles in the field.

CONCLUSIONS AND FUTURE DIRECTIONS

Their host specificity and sedentary habits make leaf beetles suitable for basic field experiments to examine the
effects of sex ratio, density, dispersion, resource distribution, and predation risk on mating behavior. The possibility that males increase their opportunities for
fertilizations by adjusting their behaviors based upon
their sociosexual experience has yet to be investigated in
this group. We need more information on behavioral plasticity of males in the Chrysomelidae. Males of the beetle
Tenebrio molitor (Tenebrionidae) transfer more sperm
when accompanied by another male than when unaccom-
panied (Gage and Baker 1991). In contrast, males of the
stalk-eyed fly *Curtodiopsis whitei* (Diptera: Diopsidae)
reduce sperm waste by not mating with females that
have copulated recently (Lorch *et al.* 1993). Manipulation
of factors influencing propensity to mate, mating duration, and sperm allocation should yield information on
trade-offs among different behavioral options for both
sexes.

There is need for basic comparative work on the
mating systems of the Chrysomelidae. Some species of
cassidine and chrysomeline are subsocial, with females
investing heavily in guarding of eggs and larvae (Windsor
1987; Choe 1989; Windsor and Choe 1994; Rodriguez
1994a). It is likely that maternal care has significant effects
on the costs and benefits of prolonged and multiple
mating for both sexes. First, if the sex ratio is not
female-biased, there will be relatively few females nearing
oviposition at any one time, and precopulatory mate-
guarding may be favored. Second, females that guard eggs
may be more sedentary than females that do not exhibit
maternal care. Third, because females invest heavily in
each clutch they lay (the period from egg to pupation can
be as long as six weeks), females may be extremely selective about which males father their offspring. The costs
and benefits of prolonged mating associations will be significantly altered in subsocial species; comparative work
on the mating patterns of closely related species with and
without female parental care may provide insights into
which factors determine the frequency and duration of
matings. Because mating and parental behaviors vary
within genera of the Chrysomelidae, combining behavioral studies of courtship, insemination, sperm competition, and parental care with revisionary systematic studies
will provide both a comparative framework for analysis of

the functional significance of mating behaviors within the
group and the phylogenetic information needed to examine the coevolutionary patterns of male and female reproductive traits.

In leaf beetles, interactions between paired males and
females appear to involve an intricate series of paternity
assurance measures on the parts of males and measures to
control sperm use on the parts of females. The most exciting finding in the Chrysomelidae is the extent to which
cryptic processes appear to govern patterns of insemination and sperm use. Analyzing these processes is labor-
ious, often involving sperm counts and dissection of pairs
frozen at various stages of copulation. Ultimately, techniques that allow us greater resolution of these internal
processes will enable us to address questions of how sperm
of different males are distributed within the female reproductive tract and which male's sperm are discharged.
Having the answers to these questions will enable us to
further examine behavioral and physical attributes of
males that make them successful in insemination and
sperm competition.

Evidence that females emit sperm clarifies the importance of considering the female's point of view in studies
of sperm competition in the Chrysomelidae. Future challenges include discovering the fitness consequences of
multiple mating for females and the generality of cryptic
female choice as a means of controlling paternity. By combining laboratory and field studies we may one day arrive at
as good an understanding of multiple mating and sexual
conflict in the Chrysomelidae as we have of multiple
mating and sexual conflict in some other species of insects
(Arnqvist, this volume), spiders (Watson 1993) and vertebrates (Davies 1989).

ACKNOWLEDGEMENTS

I dedicate this chapter to the memory of George Campbell
Eickwort, whose enthusiasm, high standards, and great
laugh I will always remember. I thank Patrick Abbot, Jae
Choe, Bernie Crespi, Bill Eberhard, Walt Koenig, Viterbo
Rodriguez, Don Windsor and two anonymous reviewers
for comments. I am especially grateful to Patrick Abbot,
Göran Arnqvist, Jae Choe, Bill Eberhard, Sara Kariko,
Viterbo Rodriguez and Don Windsor for sharing unpub-
lished material. During preparation of this manuscript,
J.L.D. was supported on a Postdoctoral Fellowship from
the National Institute of Health and NSF Grant
IBN-9507365.

LITERATURE CITED

Abbot, P. K. 1994. The effect of the podapolipid mite *Chrysomelobia labidomerae* on the sexual and social behavior of its host, the milkweed leaf beetle (*Labidomera clivicollis*). M.Sc. Thesis, Simon Fraser University, Burnaby, Canada.

Alcock, J. 1976. Courtship and mating in *Hippomelas planicosta* (Coleoptera: Buprestidae). *Coleopt. Bull.* **30**: 343–348.

–. 1994. Postinsemination associations between males and females in insects: the mate-guarding hypothesis. *Annu. Rev. Entomol.* **39**: 1–21.

Blum, M. S. 1994. Antipredator devices in larvae of the Chrysomelidae: a unified synthesis for defensive eclecticism. In *Novel Aspects of the Biology of the Chrysomelidae*. P. Jolivet, M. L. Cox and E. Petitpierre eds., pp. 277–288. Dordrecht: Kluwer.

Boggs, C. L. and L. E. Gilbert 1979. Male contribution to egg production in butterflies: evidence for transfer of nutrients at mating. *Science (Wash., D.C.)* **206**: 83–84.

Bontems, C. 1988. Localization of spermatozoa inside viviparous and oviparous females of Chrysomelinae. In *Biology of the Chrysomelidae*. P. Jolivet, E. Petitpierre and T. H. Hsiao, eds., pp. 299–315. Dordrecht: Kluwer.

Breden, F. and M. J. Wade. 1985. The effect of group size and cannibalism rate on larval growth and survivorship in *Plagiodera versicolora*. *Entomography* **3**: 455–463.

Brown, W. D. 1993. The cause of size-assortative mating in the leaf beetle *Trirhabda canadensis* (Coleoptera: Chrysomelidae). *Behav. Ecol. Sociobiol.* **33**: 151–157.

Chapman, R. F. 1982. *The Insects: Structure and Function*. Cambridge, Mass.: Harvard University Press.

Choe, J. C. 1989. Maternal care in *Labidomera suterella* Chevrolat (Coleoptera: Chrysomelidae: Chrysomelinae) from Costa Rica. *Psyche* **96**: 63–67.

Cox, C. R. and B. J. Le Boeuf. 1977. Female incitation of male competition: a mechanism in sexual selection. *Am. Nat.* **111**: 317–335.

Crowson, R. A. 1981. *The Biology of the Coleoptera*. New York: Academic Press.

Damman, H. and N. Cappuccino. 1991. Two forms of egg defence in a chrysomelid beetle: egg clumping and excrement cover. *Ecol. Entomol.* **16**: 163–167.

Davey, K. G. 1960. The evolution of spermatophores in insects. *Proc. R. Entomol. Soc. Lond.* A **35**: 107–113.

Davies, N. B. 1989. Sexual conflict and the polygamy threshold model. *Anim. Behav.* **38**: 226–234.

De Villiers, P. S. and S. A. Hanrahan 1991. Sperm competition in the Namib Desert Beetle, *Onymacris unguicularis*. *J. Insect Physiol.* **37**: 1–8.

Dickinson, J. L. 1986. Prolonged mating in the milkweed leaf beetle *Labidomera clivicollis clivicollis* (Coleoptera: Chrysomelidae): a test of the 'sperm loading' hypothesis. *Behav. Ecol. Sociobiol.* **18**: 331–338.

–. 1988. Determinants of paternity in the milkweed leaf beetle. *Behav. Ecol. Sociobiol.* **23**: 9–19.

–. 1992a. Scramble competition polygyny in the milkweed leaf beetle: combat, mobility, and the importance of being there. *Behav. Ecol.* **3**: 32–41.

–. 1992b. Egg cannibalism by larvae and adults of the milkweed leaf beetle (*Labidomera clivicollis*, Coleoptera: Chrysomelidae). *Ecol. Entomol.* **17**: 209–218.

–. 1995. Trade-offs between postcopulatory riding and mate location in the blue milkweed beetle. *Behav. Ecol.* **6**: 280–286.

Dodge, K. L. and P. W. Price. 1991. Life history of *Disonycha pluriligata* (Coleoptera: Chrysomelidae) and host plant relationships with *Salix exigua* (Salicaceae). *Ann. Entomol. Soc. Am.* **84**: 248–254.

Eady, P. E. 1994. Sperm transfer and storage in relation to sperm competition in *Callosobruchus maculatus*. *Behav. Ecol. Sociobiol.* **35**: 123–129.

Eberhard, W. G. 1985. *Sexual Selection and Animal Genitalia*. Cambridge, Mass.: Harvard University Press.

–. 1991. Copulatory courtship and cryptic female choice in insects. *Biol. Rev.* **66**: 1–31.

–. 1994. Evidence for widespread courtship during copulation in 131 species of insects and spiders and implications for cryptic female choice. *Evolution* **48**: 711–733.

Eberhard, W., R. Achoy, M. C. Marin and J. Ugalde. 1993. Natural history and behavior of two species of *Macrohaltica* (Coleoptera: Chrysomelidae). *Psyche* **100**: 93–119.

Eberhard, W. and C. Cordero. 1995. Sexual selection by cryptic female choice on male seminal products – a new bridge between sexual selection and reproductive physiology. *Trends Ecol. Evol.* **10**: 493–496.

Eisner, T., E. van Tassell, and J. E. Carrell. 1967. Defensive use of a 'fecal shield' by a beetle larva. *Science (Wash., D.C.)* **158**: 1471–1473.

Fairbairn, D. J. 1993. Costs of loading associated with mate-carrying in the waterstrider, *Aquarius remigis*. *Behav. Ecol.* **4**: 224–231.

Fincke, O. M. 1982. Lifetime mating success in a natural population of the damselfly, *Enallagma hageni* (Walsh) (Odonata: Coenagrionidae). *Behav. Ecol. Sociobiol.* **10**: 293–302.

Gage, M. J. 1992. Removal of rival sperm during copulation in a beetle *Tenebrio molitor*. *Anim. Behav.* **44**: 587–589.

Gage, M. J. G. and R. R. Baker. 1991. Ejaculate size varies with socio-sexual situation in an insect. *Ecol. Entomol.* **16**: 331–337.

Gwynne, D. T. 1984. Male mating effort, confidence of paternity, and insect sperm competition. In *Sperm Competition and the Evolution of Animal Mating Systems*. R. L. Smith, ed., pp. 117–149. New York: Academic Press.

Hilker, M. 1994. Egg deposition and protection of eggs in Chrysomelidae. In *Novel Aspects of the Biology of the Chrysomelidae*. P. H. Jolivet, M. L. Cox and E. Petitpierre, eds., pp. 263–276. Dordrecht: Kluwer.

Jolivet, P. H. 1988. Food habits and food selection of Chrysomelidae. Bionomic and Evolutionary Perspectives. In *Biology of the Chrysomelidae*. P. Jolivet, E. Petitpierre and T. H. Hsiao, eds., pp. 1–24. Dordrecht: Kluwer.

Kirkendall, L. R. 1984. Long copulations and post-copulatory 'escort' behaviour in the locus leaf miner, *Odontota dorsalis* (Coleoptera: Chrysomelidae). *J. Nat. Hist.* **18**: 905–919.

Koenig, W. D. and S. S. Albano. 1986. Lifetime reproductive success, selection, and the opportunity for selection in the white-tailed skimmer *Plathemis lydia* (Odonata: Libellulidae). *Evolution* **41**: 22–36.

LaMunyon, C. W. and T. Eisner. 1993. Postcopulatory sexual selection in an arctiid moth (*Utetheisa ornatrix*). *Proc. Natl. Acad. Sci. U.S.A.* **90**: 4689–4692.

–. 1994. Spermatophore size as a determinant of paternity in an arctiid moth (*Utetheisa ornatrix*). *Proc. Natl. Acad. Sci. U.S.A.* **91**: 7081–7084.

Lew, A. C. and H. J. Ball. 1979. The mating behavior of the western corn rootworm *Diabrotica virgifera* (Coleoptera: Chrysomelidae). *Ann. Entomol. Soc. Am.* **72**: 391–393.

–. 1980. Effect of copulation time on spermatozoan transfer of *Diabrotica virgifera* (Coleoptera: Chrysomelidae). *Ann. Entomol. Soc. Am.* **73**: 360–361.

Linley, J. R. and K. R. Simmons. 1981. Sperm motility and spermathecal filling in lower Diptera. *Int. J. Invert. Reprod.* **4**: 137–146.

Lorch, P. D., G. S. Wilkinson and P. R. Reillo. 1993. Copulation duration and sperm precedence in the stalk-eyed fly *Cyrtodiopsis whitei* (Diptera: Diopsidae). *Behav. Ecol. Sociobiol.* **32**: 303–311.

McCauley, D. E. and E. C. Lawson. 1986. Mating reduces predation on male milkweed beetles. *Am. Nat.* **127**: 112–117.

McCauley, D. E., M. J. Wade, F. Breden and M. Wohltman. 1988. Spatial and temporal variation in group relatedness: evidence from the imported willow leaf beetle. *Evolution* **42**: 184–192.

Otronen, M. 1994. Fertilisation success in the fly *Dryomyza anilis* (Dryomyzidae): effects of male size and the mating situation. *Behav. Ecol. Sociobiol.* **35**: 33–38.

Otronen, M. and M. T. Siva-Jothy. 1991. The effects of postcopulatory male behaviour on ejaculate distribution within the female sperm storage organs of the fly *Dryomyza anilis* (Diptera: Dryomyzidae). *Behav. Ecol. Sociobiol.* **29**: 33–37.

Parker, G. A. 1970. Sperm competition and its evolutionary consequences in the insects. *Biol. Rev.* **45**: 525–568.

Parker, G. A. and L. W. Simmons. 1991. A model of constant random sperm displacement during mating: evidence from *Scatophaga*. *Proc. R. Soc. Lond.* B **246**: 107–115.

Parker, G. A., L. W. Simmons and H. Kirk. 1990. Analysing sperm competition data: simple models for predicting mechanisms. *Behav. Ecol. Sociobiol.* **27**: 55–65.

Ridley, M. 1988. Mating frequency and fecundity in insects. *Biol. Rev.* **63**: 509–549.

Rodriguez, V. 1993a. Fuentes de variación en la precedencia de espermatozoides de *Chelymorpha alternans* Boheman 1854 (Coleoptera: Chrysomelidae: Cassidinae). M.Sc. Thesis, Universidad de Costa Rica.

–. 1993b. Sexual behavior in *Charidotella* sp. nr. *sexpunctata* (Coleoptera: Chrysomelidae: Cassidinae). *Coleopt. Bull.* **47**: 37–38.

–. 1994a. Sexual behavior in *Omaspides convexicollis* and *Omaspides bistriata* (Coleoptera: Chrysomelidae: Cassidinae), with notes on maternal care of eggs and young. *Coleopt. Bull.* **48**: 140–144.

–. 1994b. Function of the spermathecal muscle in *Chelymorpha alternans* Boheman (Coleoptera: Chrysomelidae: Cassidinae). *Physiol. Entomol.* **19**: 198–202.

Sady, M. B. 1994. Survey of the blue milkweed beetle, *Chrysochus cobaltinus* LeConte (Coleoptera: Chrysomelidae), in western Nevada. *Coleopt. Bull.* **48**: 229.

Schlager, G. 1960. Sperm precedence in the fertilization of eggs in *Tribolium castaneum*. *Ann. Entomol. Soc. Am.* **53**: 557–560.

Sierra, J. R., W. D. Woggon and H. Schmid. 1976. Transfer of cantharidin during copulation from adult male to female *Lytta vesicatoria*. *Experientia* **32**: 142–144.

Sih, A., J. Krupa and S. Travers. 1990. An experimental study of the effects of predation risk and feeding regime on the mating behavior of the water strider. *Am. Nat.* **135**: 284–290.

Silberglied, R. E., J. G. Shepherd and J. L. Dickinson. 1984. Eunuchs: the role of apyrene sperm in Lepidoptera? *Am. Nat.* **123**: 255–265.

Sivinski, J. 1984. Sperm in competition. In *Sperm Competition and the Evolution of Animal Mating Systems*. R. L. Smith, ed., pp. 86–115. New York: Academic Press.

Smereka, E. P. 1965. The life history and habits of *Chrysomela crotchi* Brown (Coleoptera: Chrysomelidae) in northwestern Ontario. *Can. Entomol.* **97**: 541–549.

Smith, R. L. 1979. Repeated copulation and sperm precedence: paternity assurance for a male brooding water bug. *Science (Wash., D.C.)* **205**: 1029–1031.

Stearns, S. C. 1992. *The Evolution of Life Histories*. Oxford University Press.

Stevens, L. and D. E. McCauley. 1993. Mating prior to overwintering in the imported willow leaf beetle, *Plagiodera versicolora* (Coleoptera: Chrysomelidae). *Ecol. Entomol.* **14**: 219–223.

Suzuki, K. 1988. Comparative morphology of the internal reproductive system of the Chrysomelidae (Coleoptera). In *Biology of the Chrysomelidae*. P. Jolivet, E. Petitpierre, and T. H. Hsiao, eds., pp. 317–355. Dordrecht: Kluwer.

Tauber, M. J. 1988. Geographical variation in responses to photoperiod and temperature by *Leptinotarsa decemlineata* (Coleoptera: Chrysomelidae) during and after dormancy. *Ann. Entomol. Soc. Am.* **81**: 764–773.

Thibout, E. 1982. Le comportement sexuel du Doryphore, *Leptinotarsa decemlineata* Say, et son possible controle par l'hormone juvenile et les corps allates. *Behaviour* **80**: 199–217.

Thornhill, R. 1979. Adaptive female-mimicking behavior in a scorpionfly. *Science (Wash., D. C.)* **295**: 412–414.

–. 1983. Cryptic female choice and its implications in the scorpionfly *Harpobittacus nigriceps*. *Am. Nat.* **122**: 765–788.

Thornhill, R. and J. Alcock. 1983. *The Evolution of Animal Mating Systems*. Cambridge, Mass.: Harvard University Press.

Travers, S. E. and A. Sih. 1991. The influence of starvation and predators on the mating behavior of a semiaquatic insect. *Ecology* **72**: 2123–2136.

Trivers, R. L. 1972. Parental investment and sexual selection. In *Sexual Selection and the Descent of Man*. B. G. Campbell, ed., pp. 136–179 . Chicago: Aldine.

Villavaso, E. J. 1975a. Function of the spermathecal muscle in the boll weevil, *Anthonomus grandis*. *J. Insect Physiol.* **21**: 1275–1278.

–. 1975b. The role of the spermathecal gland of the boll weevil, *Antonomus grandis*. *J. Insect. Physiol.* **21**: 1457–1462.

Waage, J. K. 1979. Dual function of the damselfly penis: sperm removal and transfer. *Science (Wash., D.C.)* **203**: 916–918.

Walker, W. F. 1980. Sperm utilization in nonsocial insects. *Am. Nat.* **115**: 780–799.

Watson, P. J. 1988. The adaptive function of sequential polyandry in the spider *Linyphia litigiosa* (Linyphiidae). Ph.D. dissertation, Cornell University, Ithaca, New York.

Watson, P. 1993. Foraging advantage of polyandry for female sierra dome spiders (*Linyphia litigiosa*: Linyphiidae) and assessment of alternative direct benefit hypotheses. *Am. Nat.* **141**: 440–465.

Wellso, S. G., W. G. Ruesink and S. H. Gage. 1975. Cereal leaf beetle: relationships between feeding, oviposition, mating, and age. *Ann. Entomol. Soc. Am.* **68**: 663–668.

Wilcox, R. S. 1984. Male copulatory guarding enhances female foraging success in a water strider. *Behav. Ecol. Sociobiol.* **15**: 171–174.

Windsor, D. 1987. Natural history of a subsocial tortoise beetle, *Acromis sparsa* Boheman (Chrysomelidae, Cassidinae) in Panama. *Psyche* **94**: 127–150.

Windsor, D. M. and J. C. Choe. 1994. Origins of parental care in Chrysomelid beetles. In *Novel Aspects of the Biology of Chrysomelidae*. P. H. Jolivet, M. L. Cox and E. Petitpierre, eds., pp. 111 117. Dordrecht: Kluwer.

Wu, C. F. J. 1986. Jackknife, bootstrap, and other resampling methods in regression analysis. *Ann. Stat.* **14**: 1261–1295.

10 · Firefly mating ecology, selection and evolution

JAMES E. LLOYD

ABSTRACT

The mating signals and ecology of fireflies are diverse and the biology of exceptional species can be a source of historical information. Such species are functioning theoretical models and as working surrogates can be used for observational and experimental studies on the selection pressures, population divergences, and trajectories of history. Transitions between signaling modes and ecology, the impact of signal-focusing predators, and the influence of a species' unique ecology on its sexual biology are among phenomena that extant fireflies may illuminate. This paper describes idiosyncratic elements in the mating biology of several lampyrids, and then outlines some basic patterns in the mating biology of *Photuris* fireflies, themselves firefly predators and important agents of selection for many other fireflies.

INTRODUCTION

Fireflies initially caught my interest because they provided an opportunity to work taxonomically with a little-known group of attractive organisms, in the fashion of naturalists and curators of the past. Early observations by F. A. McDermott and H. S. Barber (1910–1951; review in Lloyd 1990) showed possibilities that existed. The renewed discussion of Darwin's sexual selection (Campbell 1972; Otte 1979), with the fresh perspectives for firefly systematics that were revealed (Lloyd 1979) and the Byzantine signal complexities and confusion of the genus *Photuris* and their interactions with species that occur with them, have sustained pursuit (Barber and McDermott 1951; McDermott 1967; Lloyd 1969a, 1980, 1981a,b, 1984a,b, 1986, J. E. Lloyd, taxonomic monograph in preparation).

This chapter first describes peculiarities in the biology of selected fireflies that can be used for exploring and understanding selection that may have acted upon lampyrids of the past. Such species are living theoretical models, working surrogates of lost histories. Next,

consideration is given to select species of the genus *Photuris* and their enigmatic flashing behavior, which in the 1920s stimulated Barber's thinking and criticism of how taxonomists viewed and treated species-level taxa, and subsequently were used as examples of sibling species apropos the biological species concept (Mayr 1963, p. 52). The predatory females of this genus, which deceive and attack signaling firefly males, seem to have been responsible for major changes in the mating signals and biology of their own and prey species. For *Photuris* and other species discussed, of interest are signaling behaviors that suggest how evolutionary transitions between signaling modes and systems may have occurred. A number of opportunities for the molecular-genetic analysis of diversity at many stages and levels will be apparent.

IDIOSYNCRASY AND OPPORTUNITY

There are two fields of opportunity I distinguish here. The first derives from the remarkable ability of many lampyrids to emit bright light, sometimes in pulsed and coded patterns. Fireflies, having acquired their light a long time ago, have built much of their life and mating systems around it and are tuned to its physical properties, including instantaneous line-of-sight transmission and structural simplicity, and to the ecological consequences of being conspicuous to many members of their communities. Their signals permit close observation of them and quantification of their sexual biology, which typically is fast-moving in North American species. Seasonal and local occurrence of adults and hotspots and moments of increased sexual activity at sites can be recognized and monitored; mate seeking signals and flights of individual males, calling and response signals of individual females, and courtship and competition sequences can be closely observed, recorded, measured, and simulated; many subtle and often obscure elements, including adjustments to ecological and competitive conditions, predation and operational sex ratios can be quantified; and comparisons among local populations under varying environmental circumstances are facilitated (Lewis and Wang 1993; Lloyd 1979, 1983, J. E. Lloyd, unpublished; Wing 1984, 1985, 1988, 1991).

At a second level of idiosyncracy there are the unique features of a species biology that appear as unexpected deviations from established patterns or current understandings. Among lampyrids, some are exceptional consequences of the general features and limitations of luminescence mentioned above, but a special class of them arose secondarily, only after their simplicity had made them vulnerable to mimicry predation by *Photuris* firefly females and the directionality of luminescent signals had permitted fast action and aerial attack by these same predators (Lloyd 1983, 1984a; Lloyd and Wing 1983).

EVOLUTION OF SIGNALING MODES AND PATTERNS

North American fireflies use three basic signaling modes: (1) diurnal fireflies fly during daylight and use pheromones (examples, *Ellychnia, Pyropyga, Lucidota*, about 25 U.S. and Canadian species); (2) glowworm males fly after dark, females are larviform and burrowing and they glow near burrow entrances, thus attracting their non-luminous males (examples, *Microphotus, Phausis, Pleotomodes*, about 20 U.S. and Canadian species); (3) flying males of flashing fireflies emit species-typical flash patterns, females answer them with luminescent emissions, and males approach the females while a dialogue continues (*Photinus, Photuris, Pyractomena*, about 125 U.S. and Canadian species).

Species using a combination of these modes demonstrate the feasibility of intermediate stages of evolution, and may permit discovery and analysis of what selection pressures might have been historically. Males of *Pleotomus* glowworms are unusual in that they have large and branched antennae, suggesting that they may detect female pheromones at long range before their glows are visible. Additionally, these glowworms may incorporate another dimension of male competition. Although they may serve for chemical detection and permit use of a signal mode combination, male antennae are not the delicately branched plumes of male glowworm beetles (Phengodidae, a related family) or those of several Neotropical lampyrids. They are curiously robust, suggesting that males might use them for jousting combat.

Males of the Appalachian glowworm (*Phausis reticulata*) differ from most other glowworm males in that they emit a bright glow as they search, triggering females to glow. This would be expected to reduce female exposure to danger except during moments of mate availability, and at the same time to increase developmental costs in males, because a light organ must be produced, and the costs and risks of mate search that males must pay. Because glowworm fireflies are more 'primitive' than flashers, *P. reticulata*'s intermediacy suggests one evolutionary route to the flash-response system that is used by flashing fireflies. This species with its unlit-male congeners provides an opportunity to examine the role that intersexual manipulation, in which females 'force' males to light up, might play in a signal evolution that has led into a completely new adaptive zone, that of flash then flash–response communication. Such flashed signals are used by hundreds of species worldwide.

The origin of flashes (short, sharp emissions) from glows, with the many complex lantern refinements necessitated, is lost in antiquity, although phylogenetic studies may make educated guesses possible. This transition might have happened more than once, in different lampyrid lineages, and under different ecological pressures. Species in the genus *Pyractomena* may hold possibilities for understanding the biology of transitional species. Idiosyncratically, the species-specific signal pattern of male *Pyractomena angustata*, a rare Gulf Coast species, is a long-continued glow that is indistinguishable from the glows of *Phausis reticulata*. However, *Pyractomena* has been regarded as a flashing genus because males of most species emit flashes or flickers, thus exercising rapid and precise temporal control of their emissions. The species of this genus share many characters and comprise a distinct, presumably monophyletic taxon. Is *P. angustata*'s glow primitive (plesiomorphic) or is it a reversal, a secondary return and derivative (apomorphic) within the genus? A case can be made for the first possibility by *ad hoc* arrangement of the male emissions of members of the genus to produce a logical sequence: the species-specific emission pattern of the rare *P. sinuata* appears to the eye to be a short glow (2–6 s) with the pause between glows of about the same duration, but their glows are subtly modulated at about the same rate as are the separate pulses of the next and flashing species; the flash pattern of *P. dispersa* is a group of 5–7 short flashes with a rate of about 4 Hertz (at 15 °C); and the flash patterns of two species, *P. angulata* and *P. barberi*, are flickers of 7–15 modulations at nearly twice *P. dispersa*'s rate. The patterns of other species in the genus, including short and long flashes, which sometimes are followed by dim pulses, can be attached to an evolutionary tree to produce a logical scheme. These emission patterns illustrate a simple evolutionary scheme from glows, through broken glows and flickers, to groups of flashes. But if the patterns of *P. angustata* and *P. sinuata* are apomorphic, secondarily simplified, it is reasonable to ask what sexual and ecological circumstances have driven this. These questions are probably oversimplifications; within the genus, sequences of evolutionary change may run both ways.

Further adding to the mystery, unlike all but one other *Pyractomena* species, the color of *P. angustata*'s luminescence is not yellow-orange but green, like the original firefly bioluminescence color (this conclusion having been reached on the basis of color sensitivity of superposition insect eyes in general (Seliger *et al.* 1982)). That is, *P. angustata* has in combination two seemingly independent characters, glows and color, that are presumably plesiomorphic at some basic level of lampyrid evolution.

Flashes emitted by fireflies are under neural control, with the neuronal connections to lantern cells occurring at complex structures called tracheal end-organs, which play a key role in the process (Herring 1978). However, more than a century ago, it was suggested that flashing might be due to the hiding and showing of a continuously lit lantern (for example, via telescoping segments). The notion was criticized at the time and I have not seen it discussed since, but hiding and showing or tail-waving could reasonably be a beginning, a transitional step producing modulation and pulsing that was subsequently enhanced and then supplanted by neural control of photocyte emission. Perhaps it could be argued that such a mechanical intermediate stage in the evolution from glowing to pulsing would even be necessary to make the transition.

The anatomies of a Caribbean and a Colombian firefly suggest curious and crude mechanisms that might permit modulated emissions from continually emissive lanterns. Although there is no evidence to suggest that they resemble early stages in the evolution of any extant flashing species, it is the selective contexts in which they are employed today that are invitations to examination (J. E. Lloyd and L. A. Ballantyne, unpublished). In a Caribbean species, possibly in the genus *Robopus,* apical segments of the abdomen can be rolled longitudinally, ventrally. The surface of the lantern can be rolled inward on itself somewhat, so that the tergites enclose the lantern in their black (opaque) tube. Further, the lantern segment apparently can be partly or fully withdrawn under the next anterior black sternite. Pinned specimens are found dried in varying degrees of rolling, folding, and withdrawal. In the most extreme example it is obvious that little light must be able to escape, and even foliate extensions of one sclerite apparently form side shades. I have never seen such rolling and folding, or any such structure, in any of thousands of museum specimens belonging to hundreds of other species. Might these males hide their glows from the eyes of rivals as they approach and mount females, or enfold female terminalia to conceal their attractive luminescence from the eyes passing males? Or, by folding and withdrawing and unfolding and extending the abdomen and light while in flight, might males produce modulated or pulsed advertisement signals enhancing their conspicuousness or encoding their specificity?

The peculiar Colombian species belongs to a new and yet unnamed subfamily, whose affinities are with an Eastern Hemisphere subfamily, the Luciolinae. I found several slowly flashing individuals perched in a natural arbor of vines over a narrow mountain-ledge trail in a montane rainforest west of Cali. Only later, when pinned specimens were examined with a microscope, did I discover that the lantern hangs down from the ventral surface of the abdomen like a pocket, and realize that the pulsing that I observed might have been produced by the lantern being alternately lowered and raised, swung down to send a pulse and then up and pressed against the sternum to dim or occlude its light.

Lucidota is a primarily Neotropical genus that includes an assortment of luminescent and non-luminescent species, and is represented in the North American fauna by three diurnal fireflies that use pheromones for sexual attraction (Lloyd 1972; Lall and Lloyd 1989). *Lucidota luteicollis* lives in very dry habitats in Florida, including central-highland scrub and the longleaf-pine – turkey-oak sandhill community. Although they can be induced to glow from their orange pronota and the terminal segments of their abdomens, they are not seen glowing in nature and furthermore males fly and search only during daylight, at night remaining perched, dark, and stationary with antennae folded against their sides. In the single case in which a male was observed to find a female, he flew to her location, landed, walked about, and then thrust his abdomen and pronotum into the sand where she was hidden from view, and where he found and apparently coupled with her (Wing 1988; W. Prince *et al.*, in Lall and Lloyd 1989). Is this where the male's lights function, under the sand where ambient light is reduced, and where it is safer and less desiccating for broadcasting females? This circumstance could 'preselect for' a transition from daytime to crepuscule activity, when a glow could be used at greater range, and thereby open the window to nocturnal mating. The female of this species had remained unknown for more than 100 years.

Photuris fireflies are typically drab, dusky yellows and dilute brown, gray, or olive. One from the mountains near Cali, Colombia, is wasp-like with purple-black elytra and a bright yellow-orange pronotum. A few isolated males were seen emitting their signaling flickers from perches atop bushes along a narrow dirt roadway, across from a cutaway bank that rose at a steep angle up the mountain. On the exposed wet soil of the bank hundreds of *Photuris* larvae glowed, their lights looking like a city seen from an airplane. At dusk the next evening dozens of males of this firefly flew low over this wet bank, emitting their flickers as they swept repeatedly over the berm and up the lower bank, seemingly in a communal swarming display. In a few minutes all had disappeared, except for a few remaining atop the bushes, signaling as seen the previous night. This photurine differs from known *Photuris* in its distinctive color, sedentary signaling, 'communal display' and the structure of female tarsi, a character that taxonomists have long used to distinguish among genera of this subfamily. Perhaps the wasp–like color is protective, as males maintain territorial perches in a swarm even during daylight?

CONSEQUENCES OF *PHOTURIS* PREDATION FOR FIREFLY MATING

Many firefly puzzles seem to have developed as a consequence of the unusual predation of *Photuris*, which can influence the mating biology of prey and the predators themselves. Indeed, the Photurinae may have influenced the character of most fireflies of the Western Hemisphere, and been responsible for the disappearance of sedentary aggregating species. As a corollary to this, fireflies in regions lacking *Photuris* or that are immune to their attack (Cicero 1983; Lloyd 1986; Lloyd *et al.* 1989) may appear exceptional for that reason.

Most adult fireflies apparently eat little or nothing, with trivial plant-sipping exceptions, but females of photurine genera, *Photuris*, *Bicellonycha*, and the unnamed genus mentioned immediately above, are specialized predators of other fireflies (Lloyd 1984a; V. Viviani, personal communication). Predaceous females take perches in the activity spaces of other species, render mating flashes like those of females of resident males, attract the males and eat them, a phenomenon first suspected by Barber (Barber and McDermott 1951). With this mimicry, the predators are inside the targeted fireflies' sexual selection programs. In their attacks on flashing males in the air (Lloyd and Wing 1983), the predators affect what could otherwise merely be coded and redundant advertisements, and potentially put pressure on males to alter their signals to make them evasive, deceptive, intimidating or less conspicuous. Thus the predators directly influence the signals that the males' females monitor, recognize and accept. Further, because the signal codes of prey species that occur together may in combination cause predator confusion and error, and because there probably is mutual predation among predator species, there is reason to expect that the signaling ecology of fireflies may be very complex and difficult to understand.

Because resident prey and predator species composition, and predation levels and modes, will differ among firefly sites, one would expect that signaling codes and behavior could sometimes diverge rapidly among local populations. A male immigrant might have lowered mating success first because of his increased susceptibility to predation, and then because of his reduced acceptibility to discriminating females that monitor a signal parameter that is linked with a vulnerable feature of his flashing behavior. It would seem that in such a situation, sexual selection and the formation of reproductive isolation mechanisms among demes of a single species are a single phenomenon.

A general consideration regarding the significance that *Photuris* predation could have on prey species' mating systems concerns the balance of male–female mate choice. Although it is a baseline expectation that males are eager and females coy, these tendencies can be modulated by local population and ecological conditions. Lewis and Wang (1993) found that the relative abundance of the two sexes seemed to influence their mating tendency. When the cost of a mating is considerable, as measured in loss of future reproduction, then a potential mate might be approached with greater discrimination. Because of *Photuris* predation, a bad courtship decision can be fatal. That is, male risk is as mating effort, and females of high-risk species have less opportunity for choice, and may not exercise as much control over their mating interactions as might otherwise be expected. The mating systems of such species may develop unusual elements. How would a clue to such a reversal be recognized?

The typical single-response flashes of females may tentatively be attributed to the sexual asymmetry, a notion I first heard expressed by J. Alcock. Females of the *Photinus ardens* species group (all *Photinus* groups are those recognized by Green 1956) differ from other *Photinus* in that their flash responses to males are often multiple flashes, with rhythmic or conspicuously arhythmic timings and bimodal flashes (Lloyd 1966). I would look for elements of their mating biology that would require females to make this extra mating effort, and examine local variations and associations with individual age, sex ratios, male and female competition, and in particular, *Photuris* predation on the males.

That male signaling can be influenced by predaceous behaviors of predatory females is demonstrated by the incorporation of predator-like flashes and flash interactions into the mate competition of Florida's *Photinus macdermotti* (Lloyd 1981, J. E. Lloyd, unpublished). When male *P. macdermotti* approach response flashes of their females or their predators they often land at some distance (<50 cm) and approach slowly with long pauses between their signals. Such delays permit other searching, rival males to discover the female and land nearby. When a male in such a gathering competative group emits the *macdermotti* (species-typical) flash pattern other males emit a variety of flashes, including those that are timed like female responses and others that are timed like those predators often emit. These flashes clearly appear to be deceptions to misinform rivals. There is an element of irony when rival males so compete but the respondent, instead of being one of their own, is a predaceous *Photuris* female.

Photuris predation may also have been responsible for the shift to an early twilight-mating window and concomitant changes in luminescence color and eye-filtering for species of *Photinus* of Division I; yellow tuning enhances signal contrast in the green-dusk environment (Lall *et al.* 1980). Interestingly, *Photinus cookii* of this group has lost its lantern and luminescence and is day-active. A shift in eye and luminescence tuning has also occurred in species of the *Photinus pyralis* group (Lall 1981, 1994), and males of *Photinus scintillans* search in shady woodlands with their orange-yellow flashes as early as one or more hours before sunset (Lloyd 1966). Because all species that have shifted their activity periods and signal color have simple one-flash patterns, one is led to ask whether limited variability in their signaling programs has forced twilight and daylight solutions on them. However, both *Photinus pyralis* and *Photinus umbratus* are twilight-active species in which the male flash pattern is a single flash (Lloyd 1966). In these allopatric and distantly related species females emit (code) their response flashes with a 2–4 s delay. Such delays may have more significance in tricking responding predators to disclose their true identity than in functioning as reproductive isolating mechanisms, as previously interpreted.

Species with multiple-flash patterns have other options. Early studies of members of the *Photinus consanguineus* group disclosed sympatric sibling species with distinctive differences in the timing of the flashes in the male flash pattern (Lloyd 1966, 1969a). These codes are important in species recognition, but could have initially been connected with predator avoidance (Lloyd 1984b,c, 1986). Extensive study of these species throughout the eastern half of the United States has revealed considerable geographic variation in flash timing, such that although it is possible to recognize distinct entities (biological species) locally on a region-by-region basis, it often is not evident what species comprise beyond each region of field study. That is, an understanding of the connections, disjunctions, and shared histories of geographically isolated populations seem no longer accessible through fieldwork alone but will require data from molecular genetic analysis.

Barber's *Photuris*

Although fireflies in general are excellent subjects for studies on sexual selection – because of the accessibility and simplicity of their mating signals, the ease with which their signals may be observed, recorded and simulated, their ecological and taxonomic diversity, and the variations seen in their mating signals and systems (Lloyd 1969b, 1979, 1983; Ohba 1983) – it is fireflies of the genus *Photuris* and their relatives that perhaps present the greatest opportunities for such studies, especially in the exploration of connections between mating signals and systems and predation ecology and speciation. The patterns of signaling behavior sketched below have emerged during field study directed primarily towards resolving *Photuris* alpha-level taxonomy (J. E. Lloyd, taxonomic monograph in preparation).

From careful observation of the flashing and morphology of *Photuris* fireflies in the 1920s, Barber described several new species, removed old ones from synonomy, and pointed out the extreme complexity and specific problems that existed, and in doing so suggested the way for future students (Barber and McDermott 1951). First among the problems he drew attention to was the fact that two of his new species, *Photuris lucicrescens* and *P. tremulans*, emitted two distinctively different flash patterns, a surprising and inexplicable discovery. I have found that not only are there

many more *Photuris* species than could have been ima-
gined, even exercising the most conservative taxonomic
judgement, but that males of more than half of them use
two or more distinctive flash patterns. In some species a
large proportion or an entire population changes predictably from one pattern to the other through the evening
(Lloyd 1990). In others, individual males switch back and
forth, and in some cases entire populations may use one
of their patterns predominantly or exclusively over a
period of time, irrespective of time of night, and even for
several days (nights) at a time.

In species with multiple patterns, in nearly all cases at
least one pattern is virtually identical, except for color, to
the patterns of coactive species in other genera. In match-
combinations from well-studied areas, matched patterns
belong to known prey of the *Photuris* males' own females.
Thus, among possible explanations that were initially suggested was that prey-mimicking males could deceive,
locate, and forcefully inseminate their hunting females
(Lloyd 1980). Such is not the case. When a matched pattern
is answered by a female, or a decoy light is flashed in simulation of the female response, males switch (default) to their
non matching (their species' own) flash pattern. Males
always switch away from matched patterns. Thus,
males always identify themselves as conspecifics and thus
females are not possibly inseminated contrary to their mate
choice (from their positions at a distance and in the dark).
This is not to say that the origin of male pattern-matching
was not in the context of deceiving hunting females and
forced insemination; I keep this notion as a working model.

After the prey-species pattern was incorporated into the
repertoire of a *Photuris* species as a deceptive tactic, there is
one additional step that a classical ethologist would antici-
pate. A matched pattern might be completely ritualized
and become the sole flash pattern and identifier of a *Photuris* species. There are two known examples that would
suggest that this has happened. First, *Photuris* 'B' in Florida is one of several *Photuris* species in the *Photuris versicolor* group, including Barber's *P. tremulans*, that uses a flicker
pattern similar or identical to that of *Pyractomena angulata*'s
flash pattern (the A-flicker, *ca.* 7.5 Hz at 16 °C and *ca.* 12 Hz
at 25 °C). *Photuris* 'B' appears to be a Florida isolate of *P.
tremulans*, but its males always search with the
A-flicker, never with the short flash pattern that Barber
and I have noted is by far the more commonly used
by *P. tremulans*. When *Photuris* 'B' males are answered
with a decoy light, sometimes they continue using the
A-flicker, approach, land, and go to the decoy without ever

defaulting; but sometimes they pause several seconds, then
begin emitting a series of very dim, short flashes, timed
like the default flashes of *P. tremulans*. Has 'B' been intercepted in a transitional state?

In the southern Appalachian Mountains *Photuris* 'ST'
emits a flash pattern comprising 4–8 short flashes, identical to that of the local *Photinus ardens*-group species. This
clearly appears to be a matched pattern, but when males
are answered they continue emitting it as they approach
the decoy. During three evenings of observation and
experimentation with decoys, during which many males
approached and landed near the answering decoy, males
never emitted any other flash pattern. *Photuris* 'ST' is morphologically indistinguishable from *Photuris tremulans*. If
Photuris 'ST' evolved from *P. tremulans*, then its ancestors
sometimes searched with the A-flicker but defaulted to a
short flash. Since that origin, 'ST' has acquired its *ardens*-
group pulsing pattern and apparently stopped using its
ancestral default pattern. Although I did not observe the
use of a second or default pattern during my observations
of this species, another pattern and defaulting may occur at
other times. Perhaps isolated populations of 'ST' in remote
mountain coves and marshes will show intermediate or
other signaling repertoires, and provide clues.

DISCUSSION

As comparative biology has long taught, we may learn
about the evolution of mating signals, their historical ecology and the selection that has produced them, from studies
of living species. Comparative studies of populations of
prey *Photinus* species that are in contact with different
Photuris may reveal specific signaling differences for the
study of divergence and speciation, and these differences
may then be understood as specific counter-measures
against predation. However, so far with each additional
piece of information on *Photuris* mating behavior, the
meaning of specific aspects of their signal patterns has
seemed less knowable. It is necessary to seek significance,
but one asks whether, for example, the significance of
Photuris' matched patterns could be as unsatisfyingly simple
as an ultrasensitivity in the female visual and nervous systems for a prey's flash configuration, making such patterns
effective for search, but merely luminous counterparts to
Eberhard's (1985, this volume) titillators on animal genitalia?

When males search for females, their behavioral and
signaling tactics in different sites, habitats, and seasons
may be influenced by: predictions of mean and variance in

female reproductive value; prey species abundance, combinations, and activity spaces; female predispositions in hunting, including preferred prey species and hunting perches; and competitor numbers, tactics, origins, and ages. Females, under constraints themselves from varying prey and mate shortages, aging, and ecological isolation and deteriorations, potentially view males as egg triggers, composed of nutritional materials for sustenance and yolk, possibly containing poisonous nutritional elements to be assimilated for their own and egg defenses, and as contributors of genes for producing ecologically competent and competitive progeny. But there could be females that produce no males, that merely use them, even facultatively and at intervals longer than a generation, and we would not yet know this.

Photuris mating is politics, with temporary coalitions between adversaries; females may be holding the option of a pocket veto. On three occasions I have found walking, twinkling *Photuris* males with their genitalia extruded and dragging behind them, and questioned whether mates had turned on them and they had nearly been victims of cannibalism. Should a female fertilize eggs with sperm from a mate she could catch? Unlike genitalia in many other groups of lampyrids, which gadgets are among the best taxonomic characters discovered for many of them, male genitalia of *Photuris* are virtually identical and have long, extremely delicate, fishbone-like fingers that remain outside and alongside female terminalia during copulation, as though proprioreceptors for alarm and to trigger a bail-out.

SUGGESTIONS FOR FUTURE RESEARCH

The visual signals of fireflies are well suited for the analysis of the influence of the environment on sexual selection, communication and associated behavior. Although it is obvious that certain physical properties of luminescence (e.g. line-of-sight propagation) and conditions of the environment (e.g. many opaque objects, sun- and moonlight noise) have been important factors in shaping firefly behavior, it is the ecological interactions among organisms that have been the more complex in their influences. These include the transfers and manipulations of information leading to mate acquisition, the exploitation of the mating signals of prey species by Photurinae predators, the (probable) synergistic and detrimental interactions among prey and among predator species, and the interplay of all of these.

Although experiments that involve photurine predators must be carried out in the field with free subjects, it is relatively simple to set up controlled interactions among *Photinus* fireflies and artificial computer-controlled lights. Based on experiments with *Photinus macdermotti* (Lloyd 1981a,b; J. E. Lloyd, unpublished), it should be relatively simple to employ computer programs to control artificial lights that will interact with live subjects in an 'extemporaneous, ongoing' fashion. The only negative consideration, which is a research problem in itself, is that *Photuris* females respond poorly in captivity. The background information that is needed for any study is a general knowledge of species that are present in a study area and that may interact, and their space, hours and season of occurrence.

Four examples of studies

1. Flying, flashing males can often be attracted to land near, beside or upon a decoy that is flashed in simulation of female responses. Incidental observations and expectation would suggest that there may be more to this attraction than simple 'flash–answer codes'. During close approach males may introduce subtle timing changes into their patterns to induce errors by signal-mimicking predators. During close approach, males are also especially vulnerable to aerial attack. Some *Pyractomena* fireflies seem to have countered this. Upon receipt of a flashed answer, males of *P. angulata* and *P. barberi* drop from the air to the substrate and wait in the dark three or four minutes, thus introducing a long and perhaps coded timing element, before emitting another flash pattern.

2. Male flash patterns in the *Photinus ardens* group consist of several separate pulses (Lloyd 1966). Males vary the number of pulses per pattern within a range that is characteristic of their species. For example, the pattern of one species in the *P. consimilis* complex consists of 4–9 pulses. Sometimes most males in an active population will emit patterns with few pulses (say 4–6) and at other times, most emit patterns with more (e.g. 6–9). This could be related to vegetation density, competition intensity, presence of species with intrusive flashes or high ambient (moon) light levels, or seasonal mate availability. Males that emit longer patterns are more vulnerable to aerial attack than those emitting fewer pulses per pattern, so what is the advantage of the longer patterns?

3. Males of species in the *Photuris versicolor* group emit patterns of variable pulse number, and there is a tendency for males of an active population to use similar numbers at a given time. In small populations of Florida *Photuris versicolor* males emit 3–4 pulse patterns, but in populations of 20 or more males, 5–6 pulse patterns prevail. Males of *P. v. quadrifulgens* increase pulse number during evening activity (Forrest and Eubanks 1995). This species and some others in the species group also use a flicker pattern that matches the flicker of *P. angulata*. In populations of species 'LIV' in northeastern United States males vary pulse number in their pulsing pattern through the evening; the proportion of males emitting the flicker pattern also varies through the evening. This rises rapidly from none to a maximum (60–80% of males) about an hour after onset of activity, then slowly falls across the next several hours (Lloyd 1986). Thus, during the span of an evening males of but a single species emit a slow and inexplicable kaleidoscope of patterns. What could be the significance of this? Any of the elements mentioned above are possible, and additionally, there is the matching–mimicry connection that was noted earlier for the flicker pattern.

4. *Photuris* females take perches in the sites of prey species to hunt via mimicry and aerial attack. Prey may not always be present in a *Photuris*' own breeding site, or easily found elsewhere, and once a female has located a prey population she may exploit it on successive nights during the two or three weeks she is active. Recall that prey species often have restricted times of evening activity. Do *Photuris* predators learn hunting locations and return to them (homing)? Do they learn and return to a site specifically at the appropriate time of evening (time sense)? During the attraction of a prey male, do females begin with a general 'program' (e.g. short delay, short flash) and then modify their flash delay and flash form during a male's approach (trial and error)?

ACKNOWLEDGEMENTS

I thank the editors, referees, and John Sivinski for comments during early preparation of the manuscript.

LITERATURE CITED

Barber, H. S. and F. A. McDermott. 1951. North American fireflies of the genus *Photuris*. *Smithsonian Misc. Coll.* **117**(1):1–58.

Campbell, B., ed. 1972. *Sexual Selection and the Descent Of Man 1871–1971*. London: Heinemann.

Cicero, J. M. 1983. Lek assembly and flash synchrony in the Arizona firefly *Photinus knulli* Green (Coleoptera: Lampyridae). *Coleopt. Bull.* **37**: 318–342.

Eberhard, W. G. 1985. *Sexual Selection and Animal Genitalia*. Cambridge, Mass.: Harvard University Press.

Forrest, T. G. and M. D. Eubanks. 1995. Variation in the flash-pattern of the firefly, *Photuris versicolor quadrifulgens* (Coleoptera: Lampyridae). *J. Insect Behav.* 8(1): 33–34.

Green, J. W. 1956. Revision of the nearctic species of *Photinus* (Lampyridae: Coleoptera). *Proc. Calif. Acad. Sci.* **28**: 561–613.

Herring, P. J. 1978. *Bioluminescence in Action*. New York: Academic Press.

Lall, A. B. 1981. Vision tuned to species bioluminescence emission in firefly *Photinus pyralis*. *J. Exp. Zool.* **216**: 317–319.

–. 1994. Spectral cues for the regulation of bioluminescent flashing activity in the males of twilight-active fireflies *Photinus scintillans* (Coleoptera: Lampyridae) in nature. *J. Insect Physiol.* **40**: 359–363.

Lall, A. B. and J. E. Lloyd. 1989. Spectral sensitivity of the compound eyes in two day-active fireflies (Coleoptera: Lampyridae: *Lucidota*). *J. Comp. Physiol.* A **166**: 257–260.

Lall, A. B., H. H. Seliger, W. H. Biggley, and J. E. Lloyd. 1980. Ecology of colors of firefly bioluminescence. *Science (Wash., D.C.)* **210**: 560–562.

Lewis, S. and O. T. Wang. 1993. Reproductive ecology of two species of *Photinus* fireflies (Coleoptera: Lampyridae). *Psyche* **98**: 293–307.

Lloyd, J. E. 1966. Studies on the flash communication system in *Photinus* fireflies. *Univ. Mich. Misc. Pub.* **130**: 1–95.

–. 1969a. Flashes, behavior and additional species of Nearctic *Photinus* fireflies (Coleoptera: Lampyridae). *Coleopt. Bull.* **23**: 29–40.

–. 1969b. Flashes of *Photuris* fireflies: their value and use in recognizing species. *Fla. Entomol.* **52**: 29–35.

–. 1972. Chemical communication in fireflies. *Environ. Entomol.* **1**: 265–266.

–. 1979. Sexual selection in luminescent beetles. In *Sexual Selection and Reproductive Competition in Insects*. M. S. Blum and N. A. Blum, eds., pp. 293–342. New York: Academic Press.

–. 1980. Male *Photuris* fireflies mimic sexual signals of their females' prey. *Science (Wash., D.C.)* **210**: 669–671.

–. 1981a. Firefly mate-rivals mimic their predators and vice versa. *Nature (Lond.)* **290**: 498–499.

–. 1981b. Mimicry in the sexual signals of fireflies. *Scient. Am.* **245**: 138–145.

–. 1983. Bioluminescence and communication in insects. *Annu. Rev. Entomol.* **28**: 131–160.

–. 1984a. Occurrence of aggressive mimicry in fireflies. *Fla. Entomol.* **67**: 369–376.

–. 1984b. Deception, a way of all flesh, and firefly signals and systematics. In *Oxford Surveys in Evolutionary Biology*, Vol. 1. R. Dawkins and M. Ridley, eds., pp. 48–84. Oxford: Oxford University Press.

–. 1984c. Evolution of a firefly flash code. *Fla. Entomol.* **67**: 228–239.

–. 1986. Firefly communication and deception: oh what a tangled web. In *Deception: Perspectives on Human and Nonhuman Deceit.* R. W. Mitchell and N. S. Thompson, eds., pp. 113–128. Albany: SUNY Press.

–. 1990. Firefly semiosystematics and predation: A history. *Fla. Entomol.* **73**: 51–66.

Lloyd, J. E. and S. A. Wing. 1983. Nocturnal aerial predation of fireflies by light-seeking fireflies. *Science (Wash., D.C.)* **222**: 634–635.

Lloyd, J. E., S. Wing and T. Hongtrakul. 1989. Ecology, flashes and behavior of congregating Thai fireflies. *Biotropica* **21**: 373–376.

McDermott, F. E. 1910. A note on the light emission of some American Lampyridae. *Can. Entomol.* **42**: 357–363.

–. 1914. The ecologic relations of the photogenic function among insects. *Z. Wiss. Insektenbiol.* **10**: 303–307.

–. 1917. Observations on the light-emission of American Lampyridae: The photogenic function as a mating adaptation. *Can. Entomol.* **49**: 53–61.

–. 1967. North American fireflies of the genus *Photuris* DeJean: A modification of Barber's key (Coleoptera: Lampyridae). *Coleopt. Bull.* **21**: 106–116.

Mayr, E. 1963. *Animal Species and Evolution.* Cambridge, Mass.: Harvard University Press.

Otte, D. 1979. Historical development of sexual selection theory. In *Sexual Selection and Reproductive Competition in Insects.* M. S. Blum and N. A. Blum, eds., pp. 1–18. New York: Academic Press.

Seliger, H. H., A. B. Lall, J. E. Lloyd and W. H. Biggley. 1982. The colors of firefly bioluminescence I.: Optimization model. *Photochem. Photobiol.* **36**: 673–680.

Wing, S. R. 1984. Female monogamy and male competition in *Photinus collustrans* (Coleoptera: Lampyridae). *Psyche* **91**: 153–160.

–. 1985. Prolonged copulation in *Photinus macdermotti* with comparative notes on *Photinus collustrans* (Coleoptera: Lampyridae). *Fla. Entomol.* **68**: 627–634.

–. 1988. Cost of mating for female insects: risk of predation in *Photinus collustrans* (Coleoptera: Lampyridae). *Am. Nat.* **131**: 139–142.

–. 1991. Timing of *Photinus collustrans* reproductive activity: finding a mate in time. (Coleoptera: Lampyridae). *Coleopt. Bull.* **45**: 57–74.

11 · Modern mating systems in archaic Holometabola: sexuality in neuropterid insects

CHARLES S. HENRY

ABSTRACT

The three orders of Neuropterida are together considered to be the basal, most plesiomorphic representatives of the Endopterygota (= Holometabola; insects with complete metamorphosis). Therefore, their mating systems are particularly interesting from a phylogenetic perspective, in that they could provide insight into the ancient past of insect behavioral evolution. However, sexual behavior is extremely diverse within the 21 families of Neuropterida, and not unlike that found in insects that are usually considered more 'advanced'. This chapter describes what little is known of sexual attraction, courtship and mating in the orders Megaloptera, Raphidioptera and Neuroptera. Each neuropterid mating system shows clear signs of having been molded by the same intense and conflicting pressures of mate attraction and intersexual and intrasexual competition that have been described for other animal and plant groups. In the most plesiomorphic taxa (Sialidae, Raphidioptera, Ithonoidea), female pheromones serve to attract multiple males; this attraction has produced synchronized swarming behavior and male scramble-competition polygyny in some species. Also plesiomorphic and nearly universal within the superorder is sperm transfer by means of a large spermatophore, which can represent a significant paternal investment. Probably as a consequence, prolonged copulation and mate-guarding have evolved wherever particularly large spermatophores are exchanged. Courtship is found in the majority of Neuropterida, and is usually mediated by sex pheromones deployed by males from an anatomically diverse array of androconia (scent glands). Courtship is visually based only in those taxa with diurnal habits and a strong commitment to visual prey-capture, such as Ascalaphidae, Nemopteridae and Mantispidae. An unusual courtship component is substrate-borne vibrational communication, which appears to be part of the ground-plan of neuropterids. This behavior has been lost in most lineages, but has been retained in a highly sophisticated form in certain Chrysopidae (green lacewings). In the holarctic genus *Chrysoperla*, identical 'silent songs', produced by abdominal oscillations, must be exchanged between the male and female prior to copulation. As in other insect taxa which duet silently, this system in *Chrysoperla* has apparently encouraged the proliferation of numerous cryptic, sibling species, separable only by their distinctive courtship songs.

INTRODUCTION

The three insect orders of the superorder Neuropterida (Kristensen 1981) form a small but extraordinarily diverse, ancient clade of holometabolous (endopterygote) insects. With about 5500 species, the clade constitutes an insignificant fraction of the world's insect fauna, yet it encompasses 21 highly distinct families (Figs. 11-1, 11-2), many of whose members are characterized by unique, peculiar habits and aberrant behavior patterns (Henry 1982a). Within the group, there are insects that trap their prey in sandy pits (Myrmeleontidae), subsist as larvae entirely on freshwater sponges (Sisyridae), live among ants and consume the colony's brood (Chrysopidae: larvae of *Italochrysa*), parasitize spiders and social insects (larval Mantispidae), kill their termite prey by lethal flatulence (larvae of Berothidae; see Johnson and Hagen 1981), or skillfully forage for aerial plankton like dragonflies (Ascalaphidae, Fig. 11-1; see Henry 1977). Their physical appearance can be equally remarkable. For example, mantispids faithfully duplicate the powerful raptorial front legs and elongated prothoraces of true mantises and are completely maggot-like as larvae; at least one coniopterygid taxon possesses strap-like, coriaceous front wings resembling the elytra of beetles (Riek 1975; Henry *et al.* 1992). Clearly, the neuropterids are among the most biologically and morphologically diverse higher taxa in all Insecta. Reproductive behavior within the order is correspondingly eclectic. Here, I will provide an overview and synthesis of our knowledge of courtship and mating patterns in Neuroptera and its plesiomorphic

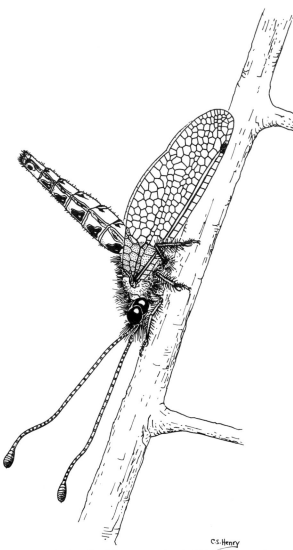

inconspicuous or 'unimportant' insects: e.g. rare, noctur-nal, or cryptic (physically indistinguishable) species; phylo-genetically archaic or basal taxa; or species considered neither harmful nor helpful to humans. It is not surpris-ing, then, that sexual behavior is best known in the green lacewings (family Chrysopidae), which have long been pro-minent as agents for biological control of populations of aphids and scales in vineyards, cotton fields, and other agri-cultural settings (New 1975). Nor is it unexpected that sexual communication has actually been found to be com-plex in this well-studied family: often when attention is finally focussed on a group, as it has been for Chrysopi-dae, hidden dimensions of the mating system are revealed. However, preliminary work on other neuropterid families indicates that diverse and intricate mating systems are also found in highly plesiomorphic taxa and probably represent the ancestral condition in the superorder; as we shall see, various families show well-developed mate-guarding, courtship feeding, reciprocal singing, modulated pheromonal release, near-lekking behavior, and territorial defense. A special courtship theme common to several taxa of both generalized and derived neuropterids is abdominal tremulation behavior (see Fig. 11-2), by which substrate-borne, vibrational songs are exchanged by court-ing individuals. Although the details of sexual behavior in most groups remain unclear, enough is now known of courtship and mating conventions in each of the ancient neuropterid families to integrate them into the mainstream of modern mating system theory, and certainly to whet the appetite for further research.

ORDER MEGALOPTERA

Megaloptera is a relict taxon characterized by retention of many primitive (generalized) features, including archaic wing venation, loosely articulated body plan, unmodified cursorial legs, and, relative to other Endopterygota, simple metamorphosis. Adults are medium-sized to very large soft-bodied insects (25–110 mm in length) and larvae are aquatic, often aggressive predators of ponds and streams. The order is geographically cosmopolitan but includes only two families: Corydalidae, the dobsonflies, and Siali-dae, the alderflies, with about 300 species altogether.

Corydalidae (dobsonflies)

Although the genus *Corydalus* is conspicuously sexually dimorphic, other genera of dobsonflies show few obvious

Fig. 11-1. Adult owlfly (Neuroptera: Ascalaphidae), in normal daytime resting position. This species, *Ululodes mexicana* (McLachlan), occurs in northern Mexico, southern Arizona and New Mexico.

relatives in the smaller orders Megaloptera and Raphidiop-tera. The three orders together form a monophyletic assem-blage, the Neuropterida (Henry 1982a). With few exceptions, this information is sketchy and anecdotal, and therefore I will have to focus on a few of the best-studied families.

Reproductive behavior in the three neuropterid orders is usually assumed to be simple, perfunctory, and stereo-typed. That view stems largely from serious ignorance of the processes of courtship and mating characterizing

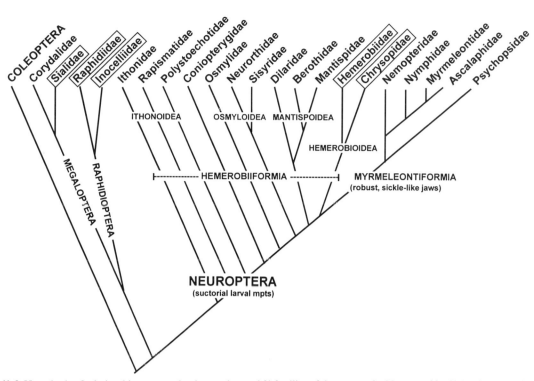

Fig. 11-2. Hypothesis of relationships among the three orders and 21 families of the superorder Neuropterida. Abdominal vibration (tremulation) has been confirmed in those taxa enclosed within boxes.

differences between the sexes. It is therefore perhaps misleading that courtship and mating is best known in the atypical genus, and particularly in *Corydalus cornutus* (Parfin 1952), which is characterized by very large males bearing greatly elongated rod-like mandibles. During courtship, the male uses these intimidating structures to 'prod' the female as he approaches from the side; if the female responds, he places them on top of her folded wings for several minutes prior to copulation. The primary function of the greatly enlarged mandibles, however, is probably in competition among males for females, in the context of strong intrasexual selection. In one case where two males were seen to court a single female, Parfin (1952) reported that the mandibles were used 'in combat', much as many large tropical beetles have been shown to use their enlarged mandibles, front legs, or antennae during intermale jousts to achieve dominance and enhanced access to females (Thornhill and Alcock 1983; Zeh *et al.* 1992). Copulation in *C. cornutus* is apparently very brief, but the male remains with the female for a much longer period of time after mating, again with mandibles resting on the female's dorsal surface. Such behavior is consistent with

mate-guarding, commonly seen in male sexual behavior throughout the animal kingdom. Presumably, it serves here as in other animals to assure paternal investment (Møller and Birkhead 1991; Sakaluk 1991; Zeigler 1991; Alcock 1992, 1994; Whittingham *et al.* 1993).

Dobsonflies transfer a spermatophore to effect fertilization, as do nearly all neuropterid insects (as well as plesiomorphic members of most other insect orders). Typically, the spermatophore is attached to the female's genitalia at copulation, and is visible externally. In *Protohermes*, the structure has a large gelatinous component and can be quite massive, approaching 20% of the male's body mass in *P. grandis* (Hayashi 1992, 1993). Dobsonfly females devote considerable time to eating the spermatophore after mating, which strongly suggests that the structure is an important source of nutrition for the newly inseminated female and that it represents a substantial paternal investment in future offspring. In species of *Protohermes* with particularly heavy spermatophores, the frequency of individual male matings is significantly reduced compared with closely related species that have smaller spermatophores (Hayashi 1993). Thus, the costs of paternity can be high in several ways (large spermatophores and reduced potential for multiple matings), providing added incentives for each male to guard its mate. This pattern parallels that seen in many other insects whose males invest heavily in the nutritional state of their mates, and where a trade-off is often seen between spermatophore mass and lifetime number of inseminations (Thornhill 1980; Sakaluk 1985; Svard 1985; Oberhauser 1988; Simmons 1990; Gwynne, this volume).

Sialidae (alderflies)

This small plesiomorphic family of only 50–60 species is best known for its dominant genus, *Sialis*, which is found throughout the northern hemisphere. Early accounts of sexual behavior in the European species *Sialis lutaria* and *S. fuliginosa* suggested that courtship and mating were simple and brief (Killington 1932; DuBois and Geigy 1935), mediated by male pheromones produced in glandular patches located basally on the wings (Geigy and DuBois 1935). It is now known, however, that copulation in *S. lutaria* and several other species is preceded by a lengthy period of rhythmic abdominal vibration in both sexes (Rupprecht 1975). This tremulation behavior produces substrate-borne signals of low frequency (120–230 Hz), remarkably similar to the vibrational songs characterizing other neuropterids such as Raphidiidae, Hemerobiidae and especially Chrysopidae (see Fig. 11-2 and later discussion). As in those other taxa, songs in alderflies allow potential mates to locate and approach one another and to choose their mates with respect to sex and species. Individuals might also be using song features to assess other important fitness-related traits of their prospective partners, including vigor, size, reproductive condition, and spermatophore mass. Such information could provide a reliable basis for female or even male choice in sialids, as has been suggested for silently tremulating katydids (Morris 1980). Appropriate experiments exploring those ideas have not been performed.

There is also evidence in alderflies of competition among males for females, probably initiated by female sex pheromones (DuBois and Geigy 1935). In *S. lutaria*, one female is usually pursued by 2–4 males, and unsuccessful suitors will even stick their spermatophores to various parts of the female's abdomen. In the scramble for success in copulation, each female is often inseminated by several males in quick succession. Even ovipositing females may be interrupted by the males and 'forced' to mate (Balduf

1939; Alexander *et al.*, this volume). Thus, questions arise concerning sexual conflict, paternity assurance, sperm precedence, sperm displacement, or even sperm (or spermatophore) removal by males (Waage 1979). However, lack of obvious sexual dimorphism in most alderfly species argues against the general pre-eminence of strong intrasexual selection in their reproductive behavior. Except for the presence of courtship, the alderfly mating system resembles that of a typical swarming insect, characterized by intense male scramble-competition. As in many mayflies, caddisflies and small Diptera, swarming male aggregations in alderflies are probably favored by the short (2–3 d) adult lifespan (Shelly and Whittier, this volume).

ORDER RAPHIDIOPTERA (SNAKEFLIES OR CAMELBACK-FLIES)

The snakeflies are another generalized, poorly understood, and probably relict neuropterid order, containing some 175 described species widely distributed in all warmer geographical regions of the northern hemisphere. Their closest phylogenetic affinities are probably with the Megaloptera (Fig. 11-2), even though their terrestrial larvae are superficially quite unlike the aquatic naiads of dobsonflies and alderflies (Achtelig and Kristensen 1973; Aspöck *et al.* 1991). Like Megaloptera, Raphidioptera includes only two extant families, Raphidiidae and Inocelliidae. Although sexual behavior is not well studied in the order, observations exist for members of both snakefly families.

A consensus has emerged that courtship in snakeflies is much like that in alderflies (Sialidae): in both groups, males and females approach each other with similar postures and orientations and vigorously vibrate their abdomens for sustained periods (Kastner 1934; Eglin 1939; Zabel 1941; Woglum and McGregor 1958; Acker 1966; Kovarik *et al.* 1991). In addition, audible wing-fluttering during courtship has been documented in both orders, and male and/or female pheromones are probably important as well (Kovarik *et al.* 1991). However, snakeflies are considerably more aggressive than sialids: females of *Agulla* have been reported to decapitate courting males in the laboratory (Acker 1966) and all published descriptions of sexual behavior in raphidiopterans emphasize mutual aggression in each courtship encounter, indicative of strong sexual conflict (Lyle 1913; Williams 1913; Kovarik *et al.* 1991). The existence of obligatory and possibly complex acoustical communication in the snakeflies reinforces the idea of a phylogenetic link between this order and both Megaloptera and Neuroptera (see later discussion). As in crickets, songs might be a source of information for females choosing mates (Boake 1983; Maynard Smith 1987; Simmons 1988; Cade and Cade 1992), but raphidiopteran species do not show the patterns of pronounced sexual dimorphism expected under acoustically-mediated intrasexual competition for mates. Additional study is needed before we will have a clear understanding of snakefly mating systems.

ORDER NEUROPTERA (— PLANIPENNIA)

Neuroptera is the largest and most diverse of the neuropterid orders, with representatives in every biogeographic realm. It is defined by a major autapomorphy: suctorial larval jaws, formed from tightly appressed mandible and maxillary lacinia. In other respects, adults and larvae of most Neuroptera show little similarity from family to family, although the order's more generalized taxa (Ithonidae, Rapismatidae and Polystoechotidae) resemble Megaloptera in overall appearance and internal anatomy. More than 4700 species in 17 families are recognized (New 1991).

Two distinct higher taxa exist within Neuroptera (Fig. 11-2), clearly defined by larval features (MacLeod 1964; Henry 1978, 1982a). One of these, the infraorder Hemerobiiformia, is the more conservative and plesiomorphic; it is characterized by larval jaws of modest size, lacking teeth or strong curvature. Hemerobiiformia is a paraphyletic assemblage, and it includes the majority of families of Neuroptera. They are generalized feeders as immatures, taking mostly small, soft-bodied arthropod prey. The other infraorder, Myrmeleontiformia, is probably monophyletic and includes five closely related families possessing an 'antlion-like' larva, characterized by a robust, heavily sclerotized head capsule and large, sharply curved, often toothed jaws. Myrmeleontiforms typically capture and subdue much larger, more aggressive prey than do hemerobiiforms.

Unfortunately (but not unexpectedly), phylogenetic patterns inferred from either larval or adult characters lend little insight into patterns of sexual behavior in the families of Neuroptera. As is true of most large insect taxa, neuropteran mating systems are evolutionarily dynamic and labile and therefore poorly correlated with higher phylogenetic clades. Each family seems to have its own independent set of rules, described below.

Infraorder Hemerobiiformia

Superfamily Ithonoidea: Ithonidae, Rapismatidae and Polystoechotidae

This paraphyletic superfamily of few and generally rare species includes the most plesiomorphic families of Neuroptera: the moth-lacewings of Australia and North America (Ithonidae; 15 species), the giant lacewings of North and South America (Polystoechotidae; 3 species), and the rapismatids of India and southeast Asia (Rapismatidae; 16 species). All are large insects, with a definite megalopteranlike ground plan but no clear synapomorphies (Henry 1982a). The entire family Rapismatidae is known from only 30 preserved adult specimens: larvae are yet to be discovered, and we are ignorant of all aspects of their general biology (Barnard 1981; Barnard and New 1986). Not much is known about the biology and life histories of the other families, either, but what little has been discovered hints at some interesting and unusual habits.

Moth-lacewings and giant lacewings are both characterized by mass emergences of adults, as some environmental cue causes the long-lived terrestrial or subterranean larvae to mature, pupate and eclose in synchrony with one another. In polystoechotids, the reproductive significance of mass emergence has not been studied (Welch 1914; Hungerford 1931), but in Ithonidae it is associated with mate attraction and copulation. The process of adult emergence and subsequent courtship and mating has been described in ithonids from opposite sides of the world, so we have some confidence that the observed patterns are typical of the family (Tillyard 1922; Riek 1974; Faulkner 1990). In both Australia (*Megalithone* and *Ithone*) and southern California (*Oliarces*), males emerge first and congregate on high ground on several consecutive days in the springtime. Each morning on those days, the males leave their shelters and gather in crowded groups on particular rock surfaces or special natural outcroppings, where they behave as though dispersing a pheromone from the tip of their abdomen (Faulkner 1990). Individual females then fly upwind into the aggregation, and are immediately surrounded by males attempting copulation. Whether true courtship exists is not known, but males compete vigorously for the opportunity to mate, and will try to dislodge any successful individual. Males *in copula* are tenacious, and have been seen to remain joined to their mates for 24 hours. The short-lived males may not remate, nor do they participate in competitions on any subsequent days. Within three days of its first appearance, the population of adult ithonids collapses, leaving a few tattered, dying males. Females have brief adult lifespans as well; they quickly deposit their eggs haphazardly and disappear (Faulkner 1990).

The mating system described above has several properties of a substrate-based lek: an integrated mix of tiny individual male 'territories', located on one or two seemingly arbitrary and unremarkable sites, which probably remain the same from year to year (Gould 1982). In addition, to be a proper lek in the classic sense (Shelly and Whittier, this volume), male courtship displays rather than resources or nest (oviposition) sites should be the major influence on male mating success, and most copulations should accrue to a mere handful of multiply mated males (Balmford 1991). The short-lived ithonids have little time to remate elsewhere, and the nature of their individual displays is not known, but highly male-biassed sex ratios at the mating sites (Riek 1974; Faulkner 1990) are at least consistent with lek-like behavior. However, such characteristics are also seen in swarming insects, where males aggregate and compete with each other for females, but do not court. Lekking is relatively uncommon but widely distributed among insects, having been identified and described in many Hawaiian and Australian *Drosophila*, tephritid flies, various other Diptera, some landmark-defending butterflies and one moth, acoustical cicadas, several fireflies (Coleoptera), a katydid (Orthoptera), and numerous solitary and social bees and wasps (Thornhill and Alcock 1983; Shelly and Whittier, this volume). Many of these examples, like the Ithonidae, seem to bend or break at least one of the cardinal rules of a true lek (Alcock 1981; Kaneshiro and Boake 1987; Shelly and Kaneshiro 1991; Gibson and Höglund 1992), largely because a continuum exists in insects between simple swarming behavior and sexually selected mating systems (Svensson and Petersson 1992). Based on existing information, it seems that the mating pattern described for Ithonidae conforms more closely to a scramble-competition model than it does to substrate-based lek polygyny. In that respect, the moth-lacewings resemble the alderflies, described earlier. Until further observations are made, ithonoid sexual behavior will remain poorly understood but provocative. At least we know now that members of these families are not so much rare as they are extremely localized, with an ephemeral adult stage.

Superfamily Coniopterygoidea: Coniopterygidae (dusty-wings)
The highly divergent family Coniopterygidae shares few synapomorphies with other Neuroptera, suggesting early origin and differentiation of the taxon and supporting its phylogenetic position in its own superfamily (New 1991). It is a globally distributed family and includes about 300 species of generally minute insects, not more than 3 mm long, that superficially resemble bark lice (Psocoptera) or even whiteflies (Hemiptera: Sternorrhyncha: Aleyrodidae).

Courtship and mating have been described in *Parasemidalis*, *Semidalis*, *Conwentzia* and *Aleuropteryx*, but only anecdotally (Withycombe 1922; Collyer 1951; Henry 1976; Johnson and Morrison 1979). Receptive females apparently release a pheromone, because multiple males have been seen clustered near single females; both sexes slowly wave the abdomen and flutter the wings in a manner suggestive of pheromonal wafting. In the most plesiomorphic of the three subfamilies, Coniopteryginae, pairing is accomplished by the male approaching its partner from behind and seizing several of her legs with his forelegs and mandibles (Johnson and Morrison 1979). Couples remain 5–15 minutes *in copula*. Unique within the neuropterid families is the absence of spermatophore: coniopterygids possess a sperm pump and transfer fluid semen (MacLeod 1962).

Superfamily Mantispoidea: Dilaridae, Berothidae and Mantispidae
Superfamily Mantispoidea has been used for different taxa by different authors; here, it is restricted to three families, the primitive members of which all possess genitalic similarities and several specializations of the larval head (Tjeder 1959, 1968; MacLeod 1964; Willmann 1990). Dilaridae (pleasing lacewings) and Berothidae (beaded lacewings) are small but widespread families, with only 35 and 60 described species, respectively (Henry *et al.* 1992). Mantispidae (mantisflies) encompasses more than 250 species, many of which have a close parasitic or predatory association as larvae with a wide range of wasps, bees, moths, scarab beetles or spiders (Lambkin 1986a,b). Essentially nothing is known of sexual behavior in Dilaridae; only a few tantalizing fragments of information exist for Mantispidae and Berothidae.

Both chemical and visual signals are probably important in courtship and mating of mantispids. One to several prominent glandular organs or sites of glandular pores, which may disperse pheromones, are variably present on the male abdomen in a number of genera of mantisflies (Eltringham 1932; Poivre 1986). In addition, some mantispids wave their legs during courtship in a manner suggestive of a stereotyped visual mating display (Eltringham 1932), which is consistent with the visual acuity of these diurnal predators. However, no good evidence exists for either pheromonal or visual mediation of sexual inter-actions (New 1989). An intricately coiled, thread-like intro-mittent organ, the penisfilum, occurs in many mantispids, as well as in some members of the other two mantispoid families. Again, the reproductive function of the penisfilum has not been documented, but its unusual morphology suggests special or even unique properties of the sperm transfer or manipulation system.

Berothid sexual behavior has attracted a little more attention. In two species of *Lomamyia* (MacLeod and Adams 1967), courtship involves rapid, alternate wing-fluttering in both sexes, followed by brief (1–5 min) copula-tion. Sperm transfer includes significant investment by the male in a large, oval spermatophore, which protrudes from the female and remains largely intact for up to four days. MacLeod and Adams (1967) speculate that the thread-like, coiled penisfilum in these species, when inserted into the female's genital duct during copulation, functions as the axis along which the inner core of the spermatophore is secreted. Eventually, the female eats this paternal contri-bution to reproduction, implying that males may have to protect their parental investment in some way after mating to assure paternity.

Superfamily Osmyloidea: Neurorthidae, Osmylidae and Sisyridae

Osmyloidea is considered by some to be a grouping of convenience, and therefore either para- or polyphyletic (Fig. 11-2). Its members share several venational features that may or may not be synapomorphies, and their larvae are semi- or subaquatic. Neurorthidae is an archaic family with a highly disjunct geographic distribution, consisting of only nine described species; little is known about their adult or larval habits (Zwick 1967). Osmylidae includes about 160 species, many of which are large, handsome lacewings, up to 80 mm in wingspan. As in Neurorthidae, biogeographic patterns within Osmylidae suggest that the family is very old (New 1989). Sisyridae, the spongillaflies, are small hemerobiid-like insects that as larvae subsist entirely upon freshwater sponges (genus *Spongilla*). Adult habits resemble those of Osmylidae (Pupedis 1980). It is a cosmopolitan family, which includes 48 species in three genera.

At least in Osmylidae, courtship behavior seems to involve pheromones: a courting male *Osmylus* everts a pair of large, white, finger-like 'scent glands', located dorsally toward the tip of the abdomen (Withycombe 1923; David

1936). Females stroke these glands with antennae and palpi prior to mating. As in the berothid *Lomamyia* (see above), a very large white spermatophore is deposited in the female's copulatory bursa, which the female later devours voraciously. Spongillaflies also transfer an obvious, externally visible spermatophore, which the female removes and consumes (Withycombe 1923; Pupedis 1985). In other respects, sexual behavior in the osmyloid families is unremarkable. Some wing-fluttering or wing signalling has been reported in osmylids and sisyrids; the courting *Osmylus* male seizes the female's front legs with his jaws just before copulation. Withycombe (1923) reported that the osmylid male remains with his partner for an hour or more after mating, continually antennating her body. Withycombe construed that behavior as the male's way of preventing the female from withdrawing and eating (or discarding) the spermatophore, which would leave the female open to remating. Mate-guarding of this type often develops from sexual conflict, when the best interests of the two sexes do not coincide (Trivers 1985; Alexander *et al.*, this volume; Phelan, this volume).

Superfamily Hemerobioidea: Hemerobiidae and Chrysopidae

The brown lacewings (Hemerobiidae) and green lacewings (Chrysopidae) form a single, natural clade. Larval similari-ties between the two families are particularly striking. Containing roughly 550 and 1200 described species, respec-tively, these familiar and common Neuroptera constitute the majority of species diversity within Hemerobii-formia, and are distributed on all continents except (perhaps) Antarctica (Brooks and Barnard 1990; Oswald 1993). Almost nothing is known of sexual behavior in Hemerobiidae, but a large body of literature on chrysopid courtship and mating has emerged in recent years (Henry 1984; New 1989).

Among all the Hemerobiidae, mating has been described only in *Sympherobius* (Smith 1923; Killington 1936). The mating process in that genus has been por-trayed as 'much the same as in chrysopids' (Smith 1923), except less complex – which could mean almost anything, as we will see later. My own observations of several North American hemerobiid taxa indicate that abdominal vibra-tion is part of the courtship repertory, but such signals have not been recorded or analyzed. These observations support the suggestion made earlier, that vibrational signal-ling is an ancient symplesiomorphy of the neuropterid families.

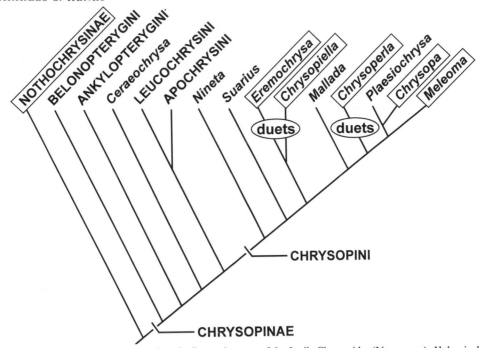

Fig. 11-3. Hypothesis of relationships among selected tribes and genera of the family Chrysopidae (Neuroptera). Abdominal vibration (tremulation) has been confirmed in those taxa enclosed within boxes.

The sexual behavior of green lacewings, family Chryso-pidae, is richly varied. In the primitive subfamily Notho-chrysinae, it resembles what one sees in Megaloptera, or in the most plesiomorphic taxa within Ithonoidea, Mantis-poidea, and Osmyloidea. However, in the other apochry-sine and chrysopine chrysopid lineages, the mating process has taken on new dimensions, including complex male-based chemical communication, stridulation, court-ship feeding, and species-specific substrate tremulation. Visual displays are less well developed in the family, per-haps because green lacewings are most active at night. The diurnal, brightly-colored members of the Hawaiian species complex *Anomalochrysa* may be conspicuous excep-tions to this generalization, but our knowledge of the behav-ior of that endemic genus is insufficient to address the issue (Zimmerman 1957; Tauber *et al*. 1991).

Nothochrysinae (= Dictyochrysinae) is a small subfam-ily of about 20 green lacewing species, characterized by relatively large size (wingspans up to 60 mm), complex wing venation, and a wide but discontinuous distribution (Adams 1967). Courtship has not been thoroughly stud-ied, but seems to involve abdominal jerking and wing-fluttering, at least in North American *Nothochrysa* (Toschi 1965). As in many primitive neuropterids, a large, whitish,

edible spermatophore is transferred and remains partly visible for several hours after copulation (Killington 1936; Toschi 1965). A whitish collar is also placed by the male on the abdomen of the female, which appears to be an addi-tional nuptial gift, because it is consumed rapidly by the recipient (Withycombe 1923). Clearly, males invest heavily in their progeny in this plesiomorphic taxon.

The balance of Chrysopidae are included in the subfam-ily Chrysopinae, whose members share a specialized tym-panal receptor in the radial vein of the forewing that is used to detect the ultrasonic sounds produced by echolocating bats (Miller and Olesen 1979). These more familiar green lacewings exhibit a broad range of mating systems. In nearly all species, chemicals probably serve an important role in attracting individuals to local regions where mating can occur. Honeydew feeders, such as *Chrysoperla*, may orient to the tryptophan component of the honeydews secreted by their aphid or scale insect hosts (Hagen *et al*. 1976). Alternatively, or perhaps as adjuncts, pheromones elaborated by any of several major exocrine glands may serve to initiate both sexual and social interactions. For example, both sexes of 'stinkfly' genera like *Chrysopa* and *Meleoma* produce a foul-smelling concoction from their large prothoracic 'repugnatorial' glands, typically when they are alarmed (Blum *et al.* 1973), but also when sexually active (Henry 1982b). Other genera such as *Chrysoperla* do not noticeably stink but possess equally large (and often sexually dimorphic) prothoracic glands, thus leaving open the possibility of important long-distance sexual attractants in olfactorially silent taxa. Chemical communication has been documented in the common European species *Chrysopa perla*, in which the male repeatedly everts a pair of prominent abdominal glands during courtship (Wattebled *et al.* 1978; Wattebled and Canard 1981). In addition, many species of chrysopids possess as a basic structural element of the male genitalia a pair of eversible, glandular scent pouches, which balloon outward during the late stages of courtship (Killington 1936; MacLeod 1962). More specialized chemical communication characterizes some members of the genus *Meleoma*, in which sexually dimorphic, glandular cavities exist on the male's head, between enlarged, prehensile antennal bases. The female feeds from this frontal cavity while the male uses his hooked antennal scapes to grasp her head (Tauber 1969). Unfortunately, chemical communication in most Chrysopidae has not received much attention.

Once aggregated by whatever means, receptive green lacewings may switch to acoustical or vibrational modes of communication. In a few species, audible sound is apparently produced by stridulation, which is the rubbing together of two or more body parts. For example, in both sexes of *Meleoma schwarzi*, a row of tubercles on the hind femora makes contact with specialized sclerites of the anterior abdominal region, which could produce a weak sound when the legs are moved (Adams 1962). Similar but independently derived peg-and-striae strigels characterize one or both sexes of other *Meleoma*, *Brinckochrysa*, and *Chrysocerca* (Brooks 1987). Additional lacewings, in many genera, possess 'presumptive stridulatory structures' on the ventral surfaces of the anal lobes of their forewings, which bear against roughened metanotal patches on the thorax (Riek 1967; Eichele and Villiger 1974). Both types of structure could easily be activated by vertical jerking of the abdomen, which has long been recognized as a component of courtship in many lacewing taxa (Smith 1922). Despite substantial circumstantial evidence, however, actual sound production by any such device has not been demonstrated. Instead, audible sounds in courting lacewings are usually percussional, and associated with relatively unspecialized structures. Males of *Mallada basalis* from Micronesia, for example, strike the reinforced front margin of their wings violently against the substrate, producing an audible buzz-

ing noise (Duelli and Johnson 1982), and various members of *Chrysopa*, *Meleoma*, *Chrysopiella*, *Chrysoperla*, and even Hawaiian *Anomalochrysa* flick, rattle or drum their wings with similar audible effect (Tauber 1969; Tauber *et al.* 1991). Even in demonstrably noisy species, though, the effects and functions of acoustical signals have not been explored experimentally or phylogenetically.

Vibrational Songs of Green Lacewings. The most interesting songs produced by Chrysopidae are the 'silent' ones that result from abdominal jerking behavior, in both sexes of many species. It is difficult to think of a less specialized, more mundane device for producing intricate signals than the lacewing abdomen, yet the substrate-borne signals that result from abdominal vibration have proven to be quite complex and crucially important in mating decisions. The process is simple, involving rapid, vigorous oscillation of the abdomen in the vertical plane, which shakes a suitably lightweight substrate of grass or leaf at the same rate as the abdominal motions (Henry 1980b). The signal is picked up by subgenual organs in the legs (tibiae) of any lacewing that chances to be standing on the same substrate. The receptors are quite unspecialized, except that they are tuned to the tremulation frequency range typical of a particular species (Devetak *et al.* 1978; Devetak 1992; Devetak and Pabst 1994). Depending upon conditions, signals can travel up to a meter through contiguous surfaces (Henry and Wells 1990). Vibrational behavior is organized into bouts (called volleys or syllables), of different duration in different taxa. Tremulation frequencies are under 200 Hz, so these are very low-pitched signals.

Tremulation is by no means universal in chrysopine Chrysopidae, nor equally well developed in those taxa which tremulate. Instead, it is scattered rather haphazardly throughout the subfamily (Fig. 11-2), having been reported from *Chrysopa*, *Meleoma*, *Mallada*, *Chrysopiella*, *Eremochrysa* and *Chrysoperla*. In some, such as *Chrysopa*, *Meleoma*, and *Mallada*, tremulation singing is generally simple, perfunctory, and more often expressed in males than in females (Smith 1922; Principi 1949; Philippe 1972; Henry 1980b, 1982b). It is considerably more important in the courtship displays of *Chrysopiella* and *Eremochrysa* (Henry and Johnson 1989), but reaches greatest development in the *carnea* species-group of the genus *Chrysoperla*. The taxonomic distribution of singing in Chrysopidae, and its presence in most basal lineages of other neuropterids, again supports the suspicion that tremulation is a plesiomorphic trait, which has been lost in most derived lineages but elaborated exquisitely in several others.

Song species in Chrysoperla. Singing in *Chrysoperla* has some unusual features (Henry 1985; Wells 1991; Wells and Henry 1992; Henry *et al.* 1993). For example, unlike the sexual calls or displays of most other animals, chrysoperlan songs are equally well developed in males and females. Not only do both sexes sing identically during courtship, but each must match the temporal and tonal pattern of the other in a precise, synchronized duet before copulation will occur. Some species have monosyllabic songs, consisting of one type of syllable (volley) which is repeated over and over for many seconds. In those species, participants in a duet exchange single syllables of vibration in a polite manner, effectively interdigitating their signals. Other species have multisyllabic songs: the shortest repeated unit (SRU) exchanged between partners in a duet includes more than one volley of abdominal vibration. Additional variables include the length of each volley, its characteristic repetition rate, its overall pitch, any frequency modulation within the volley, and the existence of more than one type of volley within the song (Fig. 11-4).

Because each species possesses a uniquely structured song (Fig. 11-4), duetting behavior has the effect of quickly disrupting all heterospecific interactions and thus curtailing potential hybridization. Both males and females are clearly exercising choice: either sex is just as likely to break off a non-synchronized exchange of vibrational

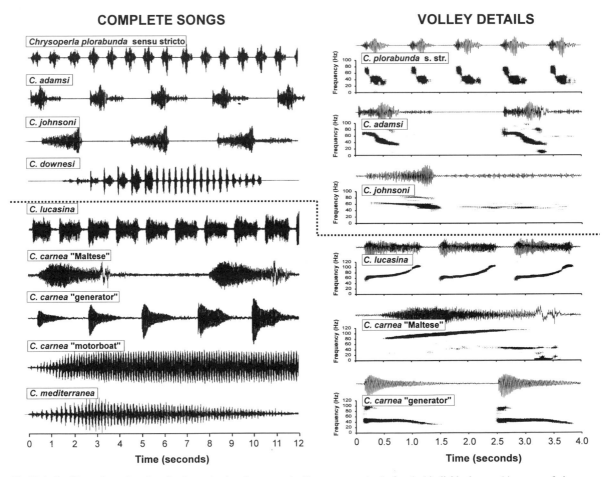

Fig. 11-4. Oscillograph tracings (amplitude vs. time) and sonographs (frequency vs. time) of typical individual courtship songs of nine biological species of the *Chrysoperla carnea* species-group. Species above the dotted line are from North America; those below the line are from Europe. Complete songs, scaled to the same 12 s time base, are shown on the left, illustrating how abdominal vibration is organized into bouts or volleys. On the right are sonographs of a 4 s section of the songs of six of the nine species, showing the frequency structure of single volleys.

signals. Consequently, the reproductively isolating effect of these courtship songs is very strong, even between closely related taxa. Controlled song playback experiments using sibling species pairs have shown conclusively that individuals choose to duet only with songs of their own species, and when given a choice of partners will mate only with a conspecific (Henry 1980a). Given no choice, under laboratory conditions, closely-related species pairs such as *C. plorabunda* × *C. downesi* or *C. plorabunda* × *C. johnsoni* will hybridize, producing progeny with intermediate song phenotypes (Henry 1985; Wells and Henry 1994). Those first-generation hybrids, like distinctive song species, discriminate strongly against the songs of their parents and mate principally among themselves (Wells 1993). Such unusually strong fidelity to one's own courtship signal should produce essentially instantaneous reproductive isolation and initiate genetic divergence between populations that differ acoustically even in small ways from one another, as predicted by runaway sexual selection models of speciation (Fisher 1958; West-Eberhard 1983). However, those models are based on choice by females only; when males choose as well, as they do in ductting species, initiation of irreversible divergence is much more likely. True duetting will be particularly common in species with 'silent' courtship songs, because female singing behavior will evolve (or persist) only when there is little risk of detection by enemies or parasites (Bennet-Clark *et al.* 1980; Henry 1994). Consequently, we should see more evidence of accelerated divergence and sibling species proliferation in taxa characterized by silent rather than audible sexual signals. This prediction is borne out by recent studies of small auchenorrhynchan hoppers belonging to several different families and superfamilies, which communicate by means of complex substrate-borne signals produced by tymbals and abdominal vibration (Ossianilsson 1949; Claridge 1990; Bailey 1991; DeWinter 1992; Hunt 1993, 1994; Wood 1993).

Hybrid studies have also shown that the genetic basis of major song differences between biological species of *Chrysoperla* green lacewings is relatively simple. In *C. plorabunda* × *C. downesi*, for example, inheritance of song phenotype is consistent with a two-locus, four-allele Mendelian system, in which each species is homozygous for a completely different set of alleles (Henry 1985). Thus we have the prerequisites for rapid speciation in *Chrysoperla*, based on chance divergence of songs: a single allelic substitution in a population is sufficient to isolate its members behaviorally from the parental type, as long as the first mutant individual can find a (reluctant) partner. There is no reason to prevent this process from occurring in sympatry, although geographical heterogeneity probably assists it. The genetic and behavioral features of this unusual mating system should result in the proliferation of cryptic, sibling species, and in fact that prediction of hidden taxonomic diversity is confirmed in North American and European *Chrysoperla* (Fig. 11-4). *Chrysoperla carnea*, which was recognized until just a few years ago as a single holarctic morphological species, is now seen to consist of swarms of reproductively isolated, cryptic song species, most of which remain to be fully described and named (Brooks, 1994). Those that are best known, such as *C. downesi*, *C. plorabunda*, *C. adamsi*, *C. johnsoni*, 'common' *C. carnea* (s. str.) and *C. mediterranea*, are nearly indistinguishable in their morphological and ecological traits; even molecular techniques have so far failed to discriminate reliably among them (Bullini and Cianchi 1983; Henry 1983; Wells 1991, 1994).

As mentioned, behavioral differences between the sexes are generally negligible in the taxa making up *Chrysoperla* species swarms. Sexual dimorphism of any other kind is also slight. However, individual males are capable of securing many more matings than are females, possibly resulting in much greater lifetime reproductive success for a certain few males than for the most prolific females (e.g. 9633 vs. 1553 offspring, in one study of *C. plorabunda* and *C. downesi* (Henry and Busher 1988)). Therefore, even in *Chrysoperla*, conditions exist favoring intrasexual competition among males for females. Songs have obvious potential for mediating male–male sexual contests or for providing data useful in female choice, but there is no evidence that they are used in those ways, and overt conflict or aggression of any sort between individuals is minor. Perhaps the opportunity for disproportionate individual male success in reproduction is seldom realized in nature, as suggested by the minimal criteria model of sexual selection (Alexander *et al.*, this volume). Chrysoperlan females will not remate until their supply of sperm is nearly exhausted, so most females encountered by a male during his lifetime will probably be unreceptive (Henry and Busher 1988).

Infraorder Myrmeleontiformia

Superfamily Myrmeleontoidea: Psychopsidae, Nemopteridae, Nymphidae, Ascalaphidae and Myrmeleontidae

Antlions and their relatives constitute a speciose group of relatively large Neuroptera, characterized by robust sit-and-wait predatory larvae and large-winged adults (New 1991). Altogether, some 2300 species have been described, accounting for almost half of the described Neuroptera (Fig. 11-2). The two dominant families, Ascalaphidae or owlflies (*ca.* 360 spp.) and Myrmeleontidae or antlions (*ca.* 1750 spp.), are badly in need of taxonomic revision. Interrelationships among the myrmeleontoid families have not been firmly resolved, either. Psychopsidae, the silky lacewings, is clearly the most plesiomorphic family of the superfamily, with 26 species conforming to an early Gondwanan distribution in South Africa, Australia, and parts of Asia. Nemopteridae contains about 130 species of bizarre thread-wing or spoon-wing lacewings; current evidence suggests that this is the sister group of all the other non-psychopsid families (Henry 1978; Mansell 1992). The Nymphidae, with 27 species confined to the Australian Region, fit somewhere between the antlions and ascalaphids phylogenetically and may be paraphyletic. Once again, little is known of the biology of any of these principally nocturnal insects.

Mate attraction, courtship, and copulation in myrmeleontoids undoubtedly involves communication by male pheromones, secreted by abdominal hairpencils in some antlions (New 1982), by Eltringham's Organ at the base of the hindwing in other antlions (Eltringham 1926; Elöfsson and Lofqvist 1974), or by eversible abdominal glands in still others (Tjeder 1954). Some acanthaclisine antlions possess both hairpencils and an Eltringham's Organ, probably indicating a particularly complex form of chemical communication in that tribe. In males of some Nemopteridae, setose swellings called bullae are present at characteristic locations on the trailing edge of the forewing and on the thread-like shaft of the long, trailing hindwing; it has been surmised that these also serve as sources of pheromones. Detailed chemical analyses of secretions from Eltringham's organ in several closely related European antlions have demonstrated the existence of species-specific mixtures of different nerol-derived compounds and monoterpenes, suggesting the possibility of reproductive isolation by unique, aerially dispersed scents (Lofqvist and Bergstrom 1980; Bergstrom *et al.* 1992). Males of New World owlflies also produce conspicuous, sweet odors when handled, and sometimes sport a prominent glandular protuberance near the base of the abdomen (Henry 1977). Restriction of putative pheromones to the male sex, as seems prevalent in myrmeleontoids, is unusual but certainly not unprecedented: many nocturnal insects, including most moths, utilize a diverse array of male chemicals, which are released

in large quantities during courtship from specialized scent organs or androconia (Phelan, this volume). Unfortunately, all evidence for chemical communication in antlions and their relatives is circumstantial, based on morphology and chemistry rather than behavior.

Although reproductive behavior in Psychopsidae is quite simple and much like that described in other plesiomorphic neuropterid taxa (Tillyard 1918), many myrmeleontoid insects court and sometimes even mate in flight. Certainly the location of Eltringham's Organ and the hirsute bullae on the wings of antlions and nemopterids is compatible with efficient dispersal of pheromones while flying. Aerial encounters and displays are common in *Ululodes*, (Fig. 11-1), a fast-flying North American owlfly with aerial foraging skills that rival the swiftest, most maneuverable dragonflies (Henry 1977). Other ascalaphids in the Americas, Europe and Africa have been seen in tandem flight, presumably mating on the wing (De Lafresnaye 1846, 1854; Fraser 1957; MacNeill 1962). Completing the analogy with dragonflies are reports in Ascalaphidae of local aggregations of presumably territorial males, patrolling sections of forest clearings or roadways and intercepting females entering their air space (Henry 1973; Covell 1989). Ascalaphids and nemopterids with day-flying habits have brightly colored, butterfly-like wings and apparently use stereotyped visual displays during courtship, much like the butterflies they resemble (Picker 1984, 1987). For such a large and conspicuous clade, it is remarkable that actual copulation has been seen so rarely in the superfamily. For example, I can find no record in the literature of any antlion observed *in copula*.

CONCLUSIONS AND SUGGESTIONS FOR FUTURE RESEARCH

As in most insects, communication by pheromones is the foundation of the courtship and mating process in nearly all neuropterids, serving to attract females to males, males to females, or large numbers of individuals to local mating sites. The ancestral condition within the superorder, retained today by plesiomorphic members of Megaloptera, Raphidioptera, Ithonoidea and Coniopterygoidea, is the attraction of males to stationary females by small concentrations of sex pheromones. This system, similar to that seen in most Trichoptera and Lepidoptera, is best illustrated today in the neuropteran families Ithonidae and Polystoechotidae, where it takes the form of synchronized adult emergence and swarming behavior. Once in place, a system based on female calling will inevitably produce male competition and intense intrasexual selection, as long as each male is capable of inseminating more than one female. Consequently, we see scramble-competition polygyny developing in these plesiomorphic taxa, with evidence of a lek-like system in a few genera, such as *Oliarces* (Faulkner 1990). Mating systems in which females call males with pheromones are probably much more widespread in the neuropterids than currently documented.

Neuropterids do more than just swarm or compete intrasexually: in most species for which information exists, a well-defined courtship follows attraction. Again, chemicals are the principal mediators of behavior in courting individuals as they approach and contact each other, but in this context pheromones are produced by males rather than females. Oftentimes, eversible glands are present, or specialized sexually dimorphic structures exist, upon which females may feed. Based on their divergent morphologies and their disparate locations on the body, male scent glands (androconia) have probably evolved many times independently in the Sialidae, Raphidioptera, Mantispidae, Osmylidae, Chrysopidae, Nemopteridae and Myrmeleontidae–Ascalaphidae. Male chemical signaling is particularly well developed in the monophyletic myrmeleontiform families of Neuroptera, perhaps because they are strong fliers with nocturnal habits. In that respect, the antlions and their relatives resemble the caddisflies and moths, which are also night-fliers and possess elaborate, labile male-pheromone systems employed during courtship (Phelan, this volume).

The large spermatophore characteristic of plesiomorphic neuropterids predisposes them to significant male investment in reproduction. The spermatophore is particularly massive in members of Corydalidae, Sialidae, Berothidae and Osmylidae, where it appears to serve a major nutritive role for the female, usually through active ingestion. Postcopulatory mate-guarding should be present in those species with heavy paternal investment, and such is universally the case in those taxa with the largest spermatophores. Hayashi's (1993) comparative study of several species of *Protohermes* (Corydalidae) elegantly demonstrates the evolutionary trade-off that must exist for males under such a mating system, between spermatophore mass and lifetime number of inseminations.

Other types of unusual courtship behavior have evolved in a few myrmeleontoid taxa, including aerial courtship, odonate-like territorial behavior, and bright visual displays. These definitely represent derived phylogenetic states, not found in the neuropterid ground plan. Because of their nocturnal habits, most neuropterids are emphatically non-visual and use other channels for communication. However, the European Ascalaphidae and several Nemopteridae are strongly diurnal, like butterflies, which helps explain the atypical occurrence of visual courtship displays in those taxa. Mantispids are also visually acute predators; they, too, have visual components in their courtship routines. Visually based territoriality in North American Ascalaphidae is just one more remarkable part of the striking evolutionary convergence that has occurred between the owlflies and the anisopteran Odonata (dragonflies). Unfortunately, the significance of territoriality in Ascalaphidae is completely unknown.

Vibrational communication through the substrate, often using abdominal oscillation or jerking, is an ancient, plesiomorphic trait within the Neuropterida (Fig. 11-2). Although the condition has been retained in the plesiomorphic Sialidae (Megaloptera), Raphidioptera, Hemerobiidae and nothochrysine Chrysopidae (Neuroptera), most clades have apparently lost the trait (although it is possible that it has been overlooked in many groups). With its distribution in the basal lineages of Neuropterida, vibrational communication during courtship links these most ancient Holometabola with the hemimetabolous order Plecoptera, which also tremulate and drum (Rupprecht 1982; Zeigler 1989). This pattern could be construed as strengthening the hypothesized link between the stoneflies, an exopterygote order, and the clade that includes all the higher endopterygote insect orders, as originally suggested by Adams (1958). However, it is difficult to establish homology between the two vibrational systems, despite their apparent similarity. Within the Chrysopidae, vibrational communication is found in both basal and terminal taxa (Fig. 11-3), again suggesting multiple loss of the trait in many genera and tribes. It has become highly embellished in the green lacewing genus *Chrysoperla*, where conspecific, male-female singing duets mediate individual mate choice and have probably facilitated the proliferation of sibling, cryptic species swarms on several continents.

Vibrational signals in chrysopids are far too labile to be of much use as characters for informing phylogeny at any taxonomic level above the species. Even within the single genus *Chrysoperla*, for example, it can be seen that closely similar songs have evolved at distant geographical locations: European 'generator' produces a song with much the same temporal and frequency characteristics as North American *C. adamsi* (Fig. 11-4), but subtle differences between the two song species argue for their independent origin. There are simply not that many ways to change a vibrational signal whose expression is controlled by just two or three gene loci (Henry 1985). However, it is possible to reconstruct relationships of closely related, cryptic species within a 'species-group' of a genus. For example, *Chrysoperla plorabunda*, *C. adamsi* and *C. johnsoni* share song features that clearly identify the three as constituting a clade, and which distinguish them from *C. downesi* and its relatives (Henry *et al.* 1993; see also Fig. 11-4). Molecular data (for example, from appropriately variable regions of mitochondrial DNA) will be especially useful for resolving systematic problems within tremulating Chrysopidae such as *Chrysoperla*. These will provide a robust, independent set of characters that can be used to generate an hypothesis of phylogenetic relationships; courtship and mating system features can then be mapped onto that phylogeny, to test different evolutionary hypotheses.

ACKNOWLEDGEMENTS

This study was supported in part by NSF Awards BSR-8508080 and DEB-9220579 to Charles S. Henry and by the Research Foundation of the University of Connecticut. I thank my colleague and frequent collaborator, Dr Marta Lucía Martínez Wells, for her contribution to much of the original research upon which this chapter is based, as well as for her constructive comments on the manuscript. Additional thanks go to Peter Duelli (Swiss Federal Institute of Forestry Research), J. B. Johnson (University of Idaho, Moscow), Phillip A. Adams (California State University, Fullerton), Stephen J. Brooks (British Museum, Natural History), Raymond J. Pupedis (Yale University), Norman D. Penny (California Academy of Sciences, San Francisco) and Julie J. Henry (University of Connecticut, Storrs), for help in collecting and maintaining living lacewings from all over the world.

LITERATURE CITED

Achtelig, M. and N. P. Kristensen. 1973. A re-examination of the relationships of the Raphidioptera (Insecta). *Z. Zool. Syst. Evol.* **11**: 268–274.

Acker, T. S. 1966. Courtship and mating behavior in *Agulla* species (Neuroptera: Raphidiidae). *Ann. Entomol. Soc. Am.* **59**: 1–6.

Adams, P. A. 1958. *Studies in the Neuroptera, with special reference to wing structure and evolution in the Osmyloidea.* Ph.D. dissertation, Harvard University.

–. 1962. A stridulatory structure in Chrysopidae (Neuroptera). *Pan-Pac. Entomol.* **38**(3): 178–180.

Adams, P. A. 1967. A review of the Mesochrysinae and Nothochrysinae (Neuroptera: Chrysopidae). *Bull. Mus. Comp. Zool. Harv.* **135**(4): 215–238.

Alcock, J. 1981. Lek territoriality in a tarantula hawk wasp *Hemipepsis ustulata* (Hymenoptera: Pompilidae). *Behav. Ecol. Sociobiol.* **8**: 309–317.

–. 1992. The duration of strong mate-guarding by males of the libellulid dragonfly *Paltothemis lineatipes*: proximate causation. *J. Insect Behav.* **5**(4): 507–515.

–. 1994. Postinsemination associations between males and females in insects: the mate-guarding hypothesis. *Annu. Rev. Entomol.* **39**: 1–21.

Aspöck, H., U. Aspöck, and H. Rausch. 1991. *Die Raphidiopteren der Erde. Eine monographische Darstellung der Systematik, Taxonomie, Biologie, Oekologie und Chorologie der rezenten Raphidiopteran der Erde, mit einer zusammenfassenden Übersicht der fossilen Raphidiopteren (Insecta: Neuropteroidea)* (Vol. 1, 370pp., vol. 2, 550pp.) Krefeld: Goecke und Evers.

Bailey, W. J. 1991. *Acoustic Behavior of Insects: An Evolutionary Perspective.* New York: Chapman & Hall.

Balduf, W. V. 1939. *The Bionomics of Entomophagous Insects*, part 2. St. Louis: John S. Swift Co., Inc.

Balmford, A. 1991. Mate choice on leks. *Trends Ecol. Evol.* **6**(3): 87.

Barnard, P. C. 1981. The Rapismatidae (Neuroptera): montane lacewings of the oriental region. *Syst. Entomol.* **6**: 121–136.

Barnard, P. C. and T. R. New. 1986. The male of *Rapisma burmanum* Navás (Neuroptera: Rapismatidae). *Neuropt. Int.* **4**(2): 125–127.

Bennet-Clark, H. C., Y. Leroy and L. Tsacas. 1980. Species and sex-specific songs and courtship behavior in the genus *Zaprionus* (Diptera: Drosophilidae). *Anim. Behav.* **28**: 230–255.

Bergstrom, G., A. B. Wassgren, H. E. Hogberg, E. Hedenstrom, A. Hefetz, D. Simon, T. Ohlsson and J. Lofqvist. 1992. Species-specific, two-component, volatile signals in two sympatric ant-lion species: *Synclysis baetica* and *Acanthaclisis occitanica* (Neuroptera, Myrmeleontidae). *J. Chem. Ecol.* **18**(7): 1177–1188.

Blum, M. S., J. B. Wallace and H. M. Fales. 1973. Skatole and tridecene: identification and possible role in a chrysopid secretion. *Insect Biochem.* **3**: 353–357.

Boake, C. R. B. 1983. *Mating systems and signals in crickets.* In *Orthopteran Mating Systems: Sexual Competition in a Diverse Group of Insects.* D. T. Gwynne and G. K. Morris, eds., pp. 28–44. Boulder, Colorado: Westview Press.

Brooks, S. J. 1987. Stridulatory structures in three green lacewings (Neuroptera: Chrysopidae). *Int. J. Insect Morphol. Embryol.* **16**(3–4): 237–244.

–. 1994. A taxonomic review of the common green lacewing genus *Chrysoperla* (Neuroptera: Chrysopidae). *Bull. Br. Mus. Nat. Hist. (Entomol.)* **63**(2): 137–210.

Brooks, S. J. and P. C. Barnard. 1990. The green lacewings of the world: a generic review (Neuroptera: Chrysopidae). *Bull. Br. Mus. Nat. Hist. (Entomol.)* **59**(2): 117–286.

Bullini, L. and R. Cianchi. 1983. Electrophoretic studies on gene-enzyme systems in chrysopid lacewings. In *Biology of Chrysopidae*. M. Canard, Y. Séméria and T. R. New, eds., pp. 48–56. The Hague: W. Junk.

Cade, W. H. and E. S. Cade. 1992. Male mating success, calling and searching behaviour at high and low densities in the field cricket, *Gryllus integer*. *Anim. Behav.* **43**(1): 49–56.

Claridge, M. F. 1990. Acoustic recognition signals: barriers to hybridization in Homoptera Auchenorrhyncha. *Can. J. Zool.* **68**(8): 1741–1746.

Collyer, E. 1951. The separation of *Conwentzia pineticola* End. from *Conwentzia psociformis* (Curt.) and notes on their biology. *Bull. Entomol. Res.* **42**: 555–564.

Covell, C. V. Jr. 1989. Aggregation behavior in a Neotropical owlfly, *Cordulecerus maclachlani* (Neuroptera: Ascalaphidae). *Entomol. News* **100**(4): 135–138.

David, K. 1936. Beiträge zur anatomie und Lebensgeschichte von *Osmylus chrysops* L. *Zeitchr. Morphol. Oekol. Tiere* **31**: 151–206.

De Lafresnaye, F. 1846. Note sur l'accouplement de l'*Ascalaphus longicornis*. *Ann. Soc. Entomol. Fr. (2)* **4**: 115–116.

–. 1854. Note sur l'accouplement de l'*Ascalaphus italicus*. *Ann. Soc. Entomol. Fr. (3)* **3**: 48.

Devetak, D. 1992. Physiology of neuropteran vibration receptors: *Chrysoperla carnea* (Stephens) as an example (Insecta: Neuroptera: Chrysopidae). In *Current Research in Neuropterology. Proceedings of the Fourth International Symposium on Neuropterology, Bagnères-de-Luchon, France, 1991*. M. Canard, H. Aspöck, and M. W. Mansell, eds., pp. 123–130. Toulouse, France: Sacco.

Devetak, D., M. Gogala and A. Cokl. 1978. Prispevek k fiziologiji vibroreceptorjev stenic iz druzine Cydnidae (Heteroptera). *Biol. Vestn. (Ljubljana)* **26**(2): 131–139.

Devetak, D. and M. A. Pabst. 1994. Structure of the subgenual organ in the green lacewing, *Chrysoperla carnea*. *Tissue Cell* **26**(2): 249–257.

DeWinter, A. J. 1992. The genetic basis and evolution of acoustic mate recognition signals in a *Ribautodelphax* planthopper (Homoptera, Delphacidae). 1. The female call. *J. Evol. Biol.* **5**(2): 249–265.

DuBois, A. and R. Geigy. 1935. Beiträge zur Okologie, Fortpflanzungsbiologie und Metamorphose von *Sialis lutaria*. Studien am Sempachersee. *Rev. Suisse Zool.* **42**: 169–248.

Duelli, P. and J. B. Johnson. 1982. Behavioral origin of tremulation, and possible stridulation, in green lacewings (Neuroptera: Chrysopidae). *Psyche* **88**(3–4): 375–381.

Eglin, W. 1939. Zur Biologie und Morphologie der Raphidien und Myrmeleoniden (Neuropteroidea) von Basel und Umgebung. *Verh. Naturforsch. Gesellsch. Basel* **50**: 163–220.

Eichele, G. and W. Villiger. 1974. Untersuchungen an den Stridulationsorganen der Florfliege, *Chrysopa carnea* (St.) (Neuroptera: Chrysopidae). *J. Insect Morphol. Embryol.* **3**(1): 41–46.

Elöfsson, R. and J. Lofqvist. 1974. The Eltringham organ and a new thoracic gland: ultrastructure and presumed pheromone function (Insecta, Myrmeleontidae). *Zool. Scripta* **3**: 31–40.

Eltringham, H. 1926. On the structure of an organ in the hindwing of *Myrmeleon nostras* Fourc. *Trans. Entomol. Soc. Lond.* **74**: 267–268.

–. 1932. On an extrusible glandular structure in the abdomen of *Mantispa styriaca* Poda. *Trans. Entomol. Soc. Lond.* **80**: 103–105.

Faulkner, D. K. 1990. Current knowledge of the biology of the moth-lacewing *Oliarces clara* Banks (Insecta: Neuroptera: Ithonidae). In *Advances in Neuropterology. Proceedings of the Third International Symposium on Neuropterology, Berg en Dal, Kruger National Park, R.S.A., 1988*. M. W. Mansell and H. Aspöck, eds., pp. 197–203. Pretoria, S. Africa: Department of Agricultural Development.

Fisher, R. A. 1958. *The Genetical Theory of Natural Selection*, 3rd edn. New York: Dover Press.

Fraser, F. C. 1957. Two new species of Ascalaphidae from Madagascar (Neuroptera). *Nat. Malgache* **9**(2): 247–250.

Geigy, R. and A. M. DuBois. 1935. Sinnesphysiologische Beobachtungen über die Begattung von *Sialis lutaria* L. *Rev. Suisse Zool.* **42**: 447–457.

Gibson, R. M. and J. Höglund. 1992. Copying and sexual selection. *Trends Ecol. Evol.* **7**(7): 229–232.

Gould, J. L. 1982. *Ethology: the Mechanisms and Evolution and Behavior*. New York: W. W. Norton & Co.

Hagen, K. S., P. Greany, E. F. Sawall Jr. and R. L. Tassan. 1976. Tryptophan in artificial honeydews as a source of an attractant for adult *Chrysopa carnea*. *Environ. Entomol.* **5**: 458–468.

Hayashi, F. 1992. Large spermatophore production and consumption in dobsonflies *Protohermes* (Megaloptera: Corydalidae). *Jap. J. Entomol.* **60**: 59–66.

–. 1993. Male mating costs in two insect species (*Protohermes*, Megaloptera) that produce large spermatophores. *Anim. Behav.* **45**(2): 343–349.

Henry, C. S. 1973. The biology of two North American ascalaphids (Neuroptera: Ascalaphidae). Ph.D. dissertation, Harvard University.

–. 1977. The behavior and life histories of two North American ascalaphids. *Ann. Entomol. Soc. Am.* **70**: 179–195.

–. 1978. An evolutionary and geographical overview of repagula (abortive eggs) in the Ascalaphidae (Neuroptera). *Proc. Entomol. Soc. Wash.* **80**: 75–86.

–. 1980a. The courtship call of *Chrysopa downesi* Banks (Neuroptera: Chrysopidae): its evolutionary significance. *Psyche* **86**: 291–297.

–. 1980b. The importance of low-frequency, substrate-borne sounds in lacewing communication (Neuroptera: Chrysopidae). *Ann. Entomol. Soc. Am.* **73**: 617–621.

–. 1982a. Neuroptera. In *McGraw Hill Synopsis and Classification of Living Organisms*. S. Parker, ed., pp. 470–482. New York: McGraw Hill.

Trivers, R. L. 1985. *Social Evolution*. Menlo Park, California: Benjamin/Cummings.

Waage, J. K. 1979. Dual function of the damselfly penis: sperm removal and transfer. *Science (Wash., D.C.)* **203**: 916–918.

Wattebled, S., J. Bitsch and A. Rousset. 1978. Ultrastructure of pheromone-producing eversible vesicles in males of *Chrysopa perla* L. (Insecta, Neuroptera). *Cell Tiss. Res.* **194**: 481–496.

Wattebled, S. and M. Canard. 1981. La parade nuptiale et l'accouplement chez *Chrysopa perla* (L.) (Insecta, Neuroptera, Chrysopidae). Rôle des vésicules exsertiles du mâle et variations de la parade en fonction de la réceptivité de la femelle. *Ann. Sci. Nat. (Paris) (Zool.)*, ser. 13, **3**: 129–140.

Welch, P. S. 1914. The early stages in the life history of *Polystoechotes punctatus* Fabr. *Bull. Brooklyn Entomol. Soc.* **9**: 1–6.

Wells, M. M. 1991. Reproductive isolation and genetic divergence among populations of green lacewings of the genus *Chrysoperla*. Ph.D. dissertation, University of Connecticut.

–. 1993. Laboratory hybridization in green lacewings (Neuroptera, Chrysopidae, *Chrysoperla*): evidence for genetic incompatibility. *Can. J. Zool.* **71**(2): 233–237.

–. 1994. Small genetic distances among populations of green lacewings of the genus *Chrysoperla* (Neuroptera: Chrysopidae). *Ann. Entomol. Soc. Am.* **87**(6): 737–744.

Wells, M. M. and C. S. Henry. 1992. The role of courtship songs in reproductive isolation among populations of green lacewings of the genus *Chrysoperla* (Neuroptera: Chrysopidae). *Evolution* **46**(1): 31–42.

–. 1994. Behavioral responses of hybrid lacewings (Neuroptera: Chrysopidae) to courtship songs. *J. Insect Behav.* **7**(5): 649–662.

West-Eberhard, M. J. 1983. Sexual selection, social competition, and speciation. *Q. Rev. Biol.* **58**(2): 155–182.

Whittingham, L. A., P. O. Dunn and R. J. Robertson. 1993. Confidence of paternity and male parental care: an experimental study in tree swallows. *Anim. Behav.* **46**(1): 139–147.

Williams, C. B. 1913. Some biological notes on *Raphidia maculicollis* Steph. *Entomologist* **46**: 6–8.

Willmann, R. 1990. The phylogenetic position of the Rhachiberothidae and the basal sister-group relationships within the Mantispidae (Neuroptera). *Syst. Entomol.* **15**: 253–265.

Withycombe, C. L. 1922. *Parasemidalis annae* Enderlein, a coniopterygid new to Britain, with notes on some other British Coniopterygidae. *Entomologist* **55**: 169–172.

–. 1923. Notes on the biology of some British Neuroptera (Planipennia). *Trans. Entomol. Soc. Lond.* 1922: 501–594.

Woglum, R. S. and E. A. McGregor. 1958. Observations on the life history and morphology of *Agulla bractea* Carpenter (Neuroptera: Raphidiodea: Raphidiidae). *Ann. Entomol. Soc. Am.* **51**: 129–141.

Wood, T. K. 1993. Diversity in the New World Membracidae. *Annu. Rev. Entomol.* **38**: 409–435.

Zabel, J. 1941. Die Kamelhals-Fliege. *Nat. Volk* **71**: 187–195.

Zeh, D. W., J. A. Zeh and G. Tavakilian. 1992. Sexual selection and sexual dimorphism in the harlequin beetle *Acrocinus longimanus*. *Biotropica* **24**(1): 86–96.

Zeigler, D. D. 1989. Drumming behavior of three Pennsylvania stonefly (Plecoptera) species. *Proc. Entomol. Soc. Wash.* **91**(4): 583–587.

–. 1991. Passive choice and possible mate guarding in the stonefly *Pteronarcella badia* (Plecoptera, Pteronarcyidae). *Fla. Entomol.* **74**(2): 335–340.

Zimmerman, E. C. 1957. *Insects of Hawaii*, vol. 6. *Ephemeroptera – Neuroptera – Trichoptera and Supplement to Volumes 1–5*. Honolulu: University of Hawaii Press.

Zwick, P. 1967. Beschreibung der aquatischen Larve von *Neurorthus fallax* (Rambur) und Errichtung der neuen Planipennierfamilie Neurorthidae fam. nov. *Gewasser und Abwasser* **44/45**: 65–86.

12 · Mating systems of parasitoid wasps

H. C. J. GODFRAY AND J. M. COOK

ABSTRACT

Parasitoid wasps are a large group of hymenopteran insects whose larvae develop by feeding on the bodies of other insects. The spatial distributions of both the hosts and the parasitoid larvae influence the mating systems found in these wasps. Specifically, there is a strong tendency towards mating at the emergence site in species whose offspring develop in gregarious clutches and in species that attack clumped hosts. The genetic and sex-determination systems of parasitoids also influence their mating systems. All sexual parasitic wasps are haplodiploid, with unfertilized eggs becoming males. Thus females can produce male offspring without mating. Some species also show complementary sex determination (CSD) with sex determined by the segregation of alleles at a highly polymorphic sex determination locus. A consequence of this is that inbreeding leads to the production of sterile diploid males. There is therefore selection for females to avoid inbreeding in species with CSD and this is achieved via a premating refractory period in *Bracon hebetor*. However, we also expect sexual conflict with respect to inbreeding as males are probably selected to take all available mating opportunities.

While laboratory studies of parasitoids have yielded considerable insights, there is a need for more field data on mating systems. Laboratory studies are also needed to help delimit the distribution of CSD, and to investigate patterns of sperm use and mate choice. We argue that a fuller consideration of the interaction between mating systems, sex allocation and sex determination may lead to a better understanding of patterns in the evolution of parasitoid wasp mating systems.

INTRODUCTION

Parasitoid wasps are a large group of insects with a relatively homogeneous natural history. Their larvae develop endo- or ectoparasitically on the bodies of other arthropods, typically larval insects. The eventual death of the host through the action of the larval parasitoid distinguishes parasitoids from true parasites. The adult parasitoid does not move the host to a prepared nest or cache, unlike certain solitary, aculeate wasps, which otherwise have a very similar natural history. Parasitoid wasps are part of the insect order Hymenoptera, and are almost certainly derived from phytophagous sawflies (Symphyta) (Gauld and Bolton 1988). There is also little doubt that the aculeate Hymenoptera – solitary and social wasps, bees and ants – are derived from parasitoids. In fact, the aculeates are probably the sister group of the parasitoid superfamily Ichneumonoidea (Rasnitsyn 1988; Downton and Austin 1994). Parasitoid Hymenoptera are distributed among approximately ten superfamilies, of which the best known are the Ichneumonoidea (Ichneumonidae + Braconidae), Chalcidoidea, Cynipoidea and Proctotrupoidea. There are approximately 50 000 described species of parasitoid Hymenoptera and another 18 000 species of parasitoid in other orders, primarily Diptera (true flies) (LaSalle and Gauld 1991; Eggleton and Belshaw 1992). However, many species are undescribed and there are probably between one half and one million species of parasitoid on earth (Godfray 1994).

Parasitoids have been subject to close scrutiny by applied entomologists because of their importance as biological control agents. A number of species have been released by man and have successfully controlled economically important pests. The heyday of descriptive parasitoid biology was in the decades leading up to the discovery of synthetic insecticides (see, for example, Clausen 1940), although there has been a renaissance in recent years as the importance of biological and integrated pest control has become more apparent. These studies furnish an invaluable database of the biology of many parasitoid species. More recently, behavioral ecologists have used parasitoids as model systems to study a variety of evolutionary questions, using an ever-growing number of different species (Godfray 1994).

The modern era of parasitoid behavioral ecology began in 1967 when Hamilton published a landmark paper explaining the correlation between female-biased sex ratios and competition among brothers for mating opportunities: the theory of local mate competition. Fisher's (1930) earlier explanation of balanced sex ratios implicitly assumed that parental fitness returns are related linearly to the production of both male and female offspring. However, in circumstances when brothers compete for matings (including those with their sisters), extra sons contribute far less than extra daughters to the production of more grandchildren (because a few sons can inseminate many females). Consequently, the parent obtains diminishing returns from the production of extra sons and the evolutionarily stable sex ratio is biased towards females. Hamilton supported his theoretical arguments with a list of species in which most matings took place between siblings, and in which parents produced highly female-biased sex ratios. No less than eighteen of the twenty-six species mentioned by Hamilton were parasitoid wasps. Since then, parasitoid wasps have played a central role in testing local mate competition theory, as well as many other aspects of modern sex-allocation theory. Because this topic has been reviewed extensively in recent years (see, for example, Charnov 1982; King 1987; Godfray 1994), we concentrate here on other aspects of parasitoid mating systems.

The success of local mate competition theory, and the rightful prominence given to this work, has led to a slight distortion of parasitoid mating systems in the minds of many behavioral ecologists. It is common to read in nonspecialist texts that most or all parasitoid wasps have female-biased sex ratios and a mating system characterized by high levels of sibling mating. In fact, only a small minority of species fall within this category. This misconception is probably exacerbated by a bias in the species chosen for study. The mating system of species with local mate competition is often easy to study in the laboratory, whereas the investigation of other mating systems requires field observations and experiments with typically very small insects. One aim of this chapter is to temper the automatic association of parasitoids and local mate competition (see also Hardy 1994a).

Our main aim, however, is to bring together two fields of study, sex determination and behavioral ecology, that have developed with very little overlap. We believe that this synthesis is valuable because recent advances in the study of hymenopteran sex determination have important

consequences for the behavioral ecology of parasitoid mating systems (see also Cook and Crozier 1995) and generate a series of new questions that parasitoid biologists should address.

We divide the chapter into three parts. In the first we describe, and place into an ecological framework, the diversity of mating systems, using the classification of Godfray (1994). However, there are many ways to categorize parasitoid mating systems and we claim no special primacy for our system. In the second part we describe the genetics of sex determination in parasitoids. Like other members of the order, hymenopteran parasitoids are haplodiploid with females developing from fertilized eggs and males developing from unfertilized eggs. In some species, sex is determined by a complementary or allelic-diversity mechanism that also involves the production of sterile diploid males. In the third section we consider how both ecology and complementary sex determination generate selection pressures (often creating sexual conflict) that influence the evolution of parasitoid mating systems.

MATING SYSTEMS

Most categorizations of mating systems stress the spatial distribution of females in the environment, and the influence of female reproductive biology (Thornhill and Alcock 1983; Davies 1991). Ours is no exception (Table 12-1). Parasitoid Hymenoptera are holometabolous insects that become sexually mature when, or soon after, they emerge from the pupal stage. If males and females pupate in the near vicinity of one another, it will often be in the males' interest to remain at the pupation site to copulate with emerging females, although females may be selected to avoid inbreeding (discussed later). Mating between males and females at their shared pupation site is our first parasitoid mating system. Where males and females emerge in different parts of the environment, they must rendezvous for mating to occur. The rendezvous may occur in part of the environment where males can expect to find females. We distinguish three separate situations: males search for female emergence sites, for female oviposition sites, or for sites where females feed. Alternatively, females may signal their presence to a male by emitting a pheromone. Finally, males may congregate in swarms or leks, which attract females.

Table 12-1. *A classification of parasitoid mating systems based on the site of mating*

Mating site	Premating dispersal	Inbreeding	Putative correlates	Example
Emergence sites	neither sex	high	male competition	*Melittobia*
Female emergence sites	males	low	first-male precedence	*Rhyssa*
Feeding sites	both sexes	low		none yet, but see Jervis *et al.* (1993)
Oviposition sites	both sexes	low	last male precedence	Some parasitoids of *Drosophila*
Arbitrary: determined by solitary females	both sexes	low	pheromones	*Cotesia rubecula*
Arbitrary: determined by groups of males	both sexes	low	female choice? High variance in male reproductive success?	*Bracon hebetor*

The second and third columns indicate which sex, if any, disperses prior to mating, and the expected consequence of dispersal for inbreeding levels. We then suggest some possible correlates of these mating systems. These are reasonable, but largely untested, speculations. Tests of their validity will be possible only when a larger comparative data set is available.

Males remain and mate at the emergence site

There are two major ecological correlates of this mating system: gregariousness and attacking gregarious hosts. A parasitoid is termed gregarious if more than one individual develops in the same host. Gregariousness is widespread among parasitoid taxa but tends to be derived from a solitary lifestyle (Godfray 1994). Typically, gregarious parasitoid larvae pupate in the vicinity of their host's cadaver, often retaining the eggshell, larval skin or pupal coat of their host as protection for the parasitoid pupae. Because all parasitoid larvae developing on a given host pupate at about the same time, the adults then hatch nearly simultaneously. This facilitates sibling mating, although selection sometimes favours inbreeding avoidance in females. A similar scenario applies to solitary parasitoids that attack gregarious hosts. Good examples are those solitary egg parasitoids that attack hosts that lay their eggs in clumps. Individual parasitoids emerge from adjacent eggs at about the same time and so males come into contact with their unmated sisters. An extreme case of mating at the pupation site occurs among some gregarious egg parasitoids in the chalcidoid family Trichogrammatidae, which mate with each other within the host egg before biting their way through the chorion (Suzuki and Hiehata 1985).

Not all the parasitoids emerging from a gregarious mass of pupae need necessarily be related. Most gregarious species are capable of superparasitism, in which a female lays a batch of eggs into a host that has previously been attacked by another wasp. Mating with unrelated individuals at the emergence site is also likely to be common among solitary egg parasitoids attacking gregarious hosts, and may be the rule among solitary parasitoids that attack hosts in patchy resources such as dung, carrion or fruit where numerous females may oviposit at the same time. Local mate competition will thus be of varying importance among species that mate at the emergence site (reviewed by Hardy 1994a).

Males of wasp species that mate at the emergence site frequently compete for access to females. It is not uncommon for males to bite open the cocoons of females in order to mate with a newly emerged insect, or to copulate with the female before she leaves the cocoon (see, for example, Mertins 1980; Suzuki and Hiehata 1985). Males of *Nasonia vitripennis*, the pteromalid parasitoid of fly pupae which has been intensively studied in the laboratory, eclose inside, and then gnaw a hole through, the host puparium. They then defend territories, marked with specific chemicals; those nearest the hole in the host puparium gain most matings with the emerging females (van den Assem *et al.* 1980). When male density is low, the hole can be successfully defended by a single male, but the competition degenerates into a scramble when many males are present.

Egg parasitoids in the family Scelionidae are almost invariably solitary but often attack hosts that lay their eggs in clumps. A male will defend the clump of host eggs against conspecifics and mate with the emerging females (surveyed by Waage 1982). Waage found some evidence that fighting behaviour was rarer in species developing in very large host egg aggregations, possibly because very large clumps of eggs are not defensible. Eberhard (1975) records the curious biology of two species of scelionid that attack the same species of host, a pentatomid bug. Single clumps of bug eggs may be attacked by both species of wasp. However, an emerging male will defend the egg mass not only against conspecifics, but also against males of the other species. In consequence, many females leave the emergence site unmated; whether they locate males elsewhere is unknown.

Male–male competition in a few species of parasitoid, and in many 'parasitic' fig wasps, takes a more sinister form. Both pollinating and 'parasitic' fig wasps are derived from chalcidoid parasitoids and some of the 'parasitic' fig wasps are fairly typical parasitoids. The fascinating mating systems and mating behaviors of pollinating and parasitoid fig wasps are discussed by Herre *et al.* (this volume) and are thus not further considered here, beyond noting that in most (but not all) species mating at the emergence site is common or the rule (Cook *et al.* 1996). The genus *Melittobia* consists of minute wasps in the family Eulophidae which are gregarious parasitoids of solitary bees and wasps (although they have, on occasion, been reared from the prey of solitary wasps). Up to one hundred *Melittobia* wasps develop on a single host. They are normally the progeny of a single mother, who lays eggs with a highly female-biased sex ratio. In the best-studied species, *M. acasta*, the few males emerge first and fight viciously among themselves for possession of the pupae. Normally all but one male are killed in the fighting, some perishing in the act of emerging from their pupae (Balfour Browne 1922; Buckell 1928; van den Assem 1986).

Sexes rendezvous at the emergence site

When males and females emerge in different places, it is often the case that males seek out female emergence sites in order to mate. Some of the best examples of this type of mating system are found in the ichneumonid genus *Rhyssa*, which attacks hosts that feed deep within wood. The female wasp has an extremely long ovipositor, which it uses to bore through wood to lay its eggs on siricoid wood wasps and perhaps other hosts in the same habitat. The wasp pupates deep within the log and then laboriously

digs itself out with its sharp mandibles. Males are attracted to the emergence site by several cues. First, chemicals emitted by the host and by the symbiotic fungus that the host uses to digest wood are attractive to males (as they are to ovipositing females, who make use of the same cues in host location) (Madden 1968; Spradbery 1970). Second, the male detects the vibrations made by the wasp as it digs its way to the surface (Heatwole *et al.* 1962). Often a number of males congregate around the site where a wasp is soon to emerge; the males may be disappointed if the wasp turns out to be another male – they have no way of predicting its sex – but if it is a female there is a frantic scramble among the males to achieve copulation.

The mating system of some related Rhyssini is subtly different (reviewed by Eggleton 1991). Species of *Megarhyssa* and *Rhyssella* also scramble to mate with emerging females but the winner is the male that succeeds in inserting his abdomen into the female's burrow, where mating occurs before she has completely emerged. Small males often lose out in competition with big males and there is some evidence that they do not engage in the scramble competition but wait in the vicinity in case a female emerges from the wood unexpectedly without being previously noticed. These species have compressed abdomens, which are morphologically adapted for insertion in the female's burrow. Finally, there is evidence that one species, *Lytarmes maculipennis*, defends a possible female emergence site from competitors and so has exclusive access to the female when she emerges (Eggleton 1990).

Sexes rendezvous at the feeding site

Both male and female parasitoids are commonly found at flowers. In a recent survey, Jervis *et al.* (1993) collected from inflorescences 250 species of parasitoid wasp from 15 different families. However, in 55% of these species only females were caught on flowers and in 26.5% of species only males. As the authors point out, the small sample sizes for many of these species inflate the number of cases where only one sex was recorded. Courtship and copulation was not observed in any of the 18.5% of species in which both sexes were recorded from flowers. These data suggest that the use of flowers as rendezvous sites for mating is not common, although it would be desirable to know the actual mating sites of these species before ruling out flowers. Parasitoids may also feed at other sites, such as deposits of honeydew, but we are unaware of any studies of the presence or absence of mating at these feeding sites.

Sexes rendezvous at the oviposition site

It is not uncommon to collect both sexes of a parasitoid at the oviposition site; if oviposition and pupation occur in different places, this observation suggests that males visit locations where hosts occur in order to find mates. As with the possibility that mating occurs at feeding sites, this mating system has yet to receive systematic study. Within parasitoids of *Drosophila*, males of some species are attracted to oviposition sites (Nadel and Luck 1992) but males of other species are not (Hardy and Godfray 1990). Myint and Walter (1990) found that male *Spalangia cameroni*, which attack fly pupae in cowsheds, were attracted to oviposition sites.

Mating occurs at arbitrary sites: females attract males by pheromones

Attraction of mates through the emission of volatile chemical pheromones is widespread among insects. In parasitoid Hymenoptera, short-range arrestant and attractant chemicals have been implicated in the courtship of many species (Gordh and DeBach 1978; van den Assem 1986). However, in comparison with other groups of insects (see, for example, Phelan, this volume), long-range pheromone attraction is relatively rare in hymenopteran parasitoids, although there is a growing list of species in which it has been found. The ability of a female to attract males chemically means that she does not have to rely on encountering a male at a rendezvous. In some cases, males are attracted by chemicals that are not produced specifically as sex pheromones but happen to be useful in mate location (Godfray 1994). This is suggested by the observation that males of several parasitoid species home in on conspecific males as well as females. In cases where males utilize chemicals that are not synthesized specifically for sex attraction, females have not necessarily been selected to produce a mate attractant *per se* (Williams 1992). There may be a further disincentive for hymenopteran females to invest in costly pheromone production: thanks to haplodiploidy, they can produce male offspring even if they fail to mate (Godfray 1990). Because males must mate to reproduce, haplodiploidy might also contribute to the general tendency for males, rather than females, to bear the costs of searching for mates. However, this pattern is also common to most other (not haplodiploid) arthropod taxa (Alexander *et al.*, this volume) and must also be linked to other selective pressures, such as greater variation in male than in female reproductive success.

The best-studied examples of long-range mate attraction using pheromones come from ichneumonid and braconid wasps (Lewis *et al.* 1971; Robacker *et al.* 1976; Robacker and Hendry 1977; Eller *et al.* 1984). Lewis *et al.* (1971) used the pheromone they discovered in the braconid *Cardiochiles nigriceps* to construct pheromone traps, which they employed to monitor parasitoid abundances. Male *Itoplectes conquisitor* (Ichneumonidae) show associative learning to devices used to dispense pheromone extracts. In the 1960s, females of the same species were the subject of an extensive series of experiments on associative learning involved in host location by Arthur (1971 and references therein). Pheromone cues can be quite specific: the chemical involved in mate attraction by the ichneumonid *Syndipnus rubiginosus* has been isolated (Eller *et al.* 1984) and appears to have no effect on the behavior of the congeneric *S. gaspesianus*, which occurs in the same habitats.

A recent study by Field and Keller (1993) dissects some of the behaviors involved in a pheromone-based mating system. The braconid wasp *Cotesia rubecula* (until recently placed in *Apanteles*) is a solitary parasitoid of larvae of a cabbage butterfly, *Pieris rapae*. Male *C. rubecula* are attracted by a pheromone released by the female. When in the vicinity of the female, the male engages in a series of stereotyped courtship behaviors, which includes vibrational signals transmitted through the leaf. Field and Keller (1993) made two interesting observations about mating behavior in this wasp. Between the arrival of the first male and its successful copulation with the female, a second male may arrive and parasitize the courtship behavior of the first, slipping in to achieve a sneaky mating. After copulation, the female remains receptive to mating for a short period of time. If the female is approached by a second male during this interval, the male that has just mated will attempt, usually successfully, to distract the behavior of the would-be cuckolder by mimicking the behaviors of the female.

Leks and swarms

There are some examples of lekking and swarming among parasitoids, although the ecological correlates and selection pressures involved are poorly known. In Europe, swarms of medium-sized Ichneumonoidea are frequently encountered and these normally (although not always) turn out to be Braconidae in the subfamily Blacinae, typically *Blacus* itself (Southwood 1957; van Achterberg 1977).

Swarms are composed nearly exclusively of males, although a few females are also often present. The mating system of another braconid, *Bracon hebetor*, which will be described in more detail below, appears to involve the formation of leks at certain sites in the wasp's habitat (Antolin and Strand 1992). Swarms of male wasps have also been reported from Dryinidae and several families of Chalcidoidea (Jervis 1979; Nadel 1987), although detailed observations of the mating systems of the species involved have yet to be made. Leks and swarms may offer the opportunity for male–male competition and/or female choice. Neither possibility has been investigated in parasitoids, where the small size of the insects does not facilitate accurate observations. If one or both sexes disperse to reach the lek or swarm, inbreeding may be avoided; in species where individuals are highly dispersed, swarms may have evolved because they are easier to locate than single individual members of the opposite sex.

SEX DETERMINATION IN PARASITOID WASPS

It is well known that hymenopterans typically have haploid males and diploid females. However, the demonstrated existence of diploid males in over 30 species of Hymenoptera (Cook 1993a) raises an interesting set of questions concerning both sex determination *per se* and its consequences for mating systems. Diploid males are a consequence of complementary sex determination (CSD), a system first described in the parasitoid wasp *Bracon hebetor* by P.W. Whiting (1939, 1943). Whiting showed that sex is determined by a single genetic locus with multiple alleles. Sex locus heterozygotes (A_iA_j) are female; hemizygotes (A_i or A_j) and homozygotes (A_iA_i or A_jA_j) develop as haploid and diploid males, respectively. Diploid males produce diploid sperm, which either fail to penetrate the egg (MacBride 1946) or produce triploid zygotes of low viability (Whiting 1961). Either way, diploid males are effectively sterile and impose a genetic load on the population. The sterility of diploid males (homozygotes) imposes frequency-dependent selection on sex alleles and causes any new allele to be favored by selection, because rare alleles occur mostly in heterozygotes (females) and haploid males. High allelic diversity is expected, with the number of alleles (k) determined by a mutation-drift balance that is strongly influenced by population size (Yokoyama and Nei 1979). Under outbreeding the sex determination load (proportion of diploids that are male (Cook 1993b)) equals $1/k$ (Laidlaw

et al. 1956), assuming that all alleles are equally common. However, this fraction increases dramatically with inbreeding and equals $1/2$ under sib-mating, *regardless* of k in the population at large. Consequently, it is unlikely that systematically inbreeding species have single-locus CSD (Whiting 1947; Crozier 1971). The dynamics of CSD are formally equivalent to those of gametophytic self-incompatibility locus (S locus) alleles in many plants (Yokoyama and Nei 1979). Ioerger *et al.* (1990) found that the DNA sequences of some S locus alleles in the family Solanacae are more closely related to some non-conspecific alleles than to other conspecific alleles, reflecting allelic differences that predate speciation. Clark and Kao (1991) further showed that the regions of the gene that displayed this high degree of shared polymorphism between species were also unusual in their manifestation of a non-synonymous base substitution rate that exceeded the synonomous. This presumably reflects the strong selection for diversification at the S locus. Comparable evidence for shared ancient polymorphisms that predate speciation has also been found for the mammalian major histocompatibility complex (MHC) class I locus (Figueroa *et al.* 1988; Lawlor *et al.* 1988); overdominance is implicated (Hughes and Nei 1988). Given the similarities between CSD and self-incompatibility systems, we predict that the hymenopteran sex locus will also prove to have ancient alleles that predate speciation, as well as an unusually high rate of non-synonymous base substitution. Tests of these ideas await location of the sex locus and give extra impetus to the need for investigations of the molecular genetics of CSD. The next step will most likely involve exploitation of the recent discovery of two different genetic markers (Hunt and Page 1994; Beye *et al.* 1994) linked to the sex locus in the honey bee.

Crozier (1971) proposed that a more general multilocus CSD mechanism applies to the Hymenoptera as a whole. This model involves multiple alleles at more than one locus with heterozygosity at one (or more) of these loci sufficient to cause female development. Crozier's multilocus CSD model subsumes both single-locus CSD and Snell's (1935) earlier suggestion of several loci each with two alleles. If there are n loci, each with k equally abundant alleles, the sex determination load equals $(1/k)^n$ under outbreeding; high diploid male production might be avoided in inbreeding taxa if occasional outcrosses restore heterozygosity (Crozier 1971).

Some predictions of CSD models are easily tested in many parasitoid wasps. For example, if single-locus CSD applies, 50% of sibling matings lead to diploid male production. This manifests itself as an increase in the off-spring sex ratio and/or a decrease in offspring survival, depending on whether diploid males have normal or reduced survival (Crozier 1977; Cook 1993b). However, there have been very few empirical studies and this is an obvious area for future work. Amongst parasitoids, single-locus CSD has only been demonstrated in two *Bracon* spp. (Whiting 1939, 1943; Speicher and Speicher 1940) and *Diadromus pulchellus* (Ichneumonidae) (Periquet *et al.* 1993). All three species belong to the Ichneumonoidea; diploid males have been detected in four other members of this superfamily (Stouthamer *et al.* 1992; Cook 1993a). In addition, single-locus CSD occurs in several aculeate social Hymenoptera. A recent phylogeny (Downton and Austin 1995) based on DNA sequence data supports Rasnitsyn's (1988) suggestion that Ichneumonoidea and Aculeata form a monophyletic group. Single-locus CSD is probably the ancestral form of sex determination in this clade.

Skinner and Werren (1980) showed that CSD does not occur in the chalcidoid wasp *Nasonia vitripennis*; a considerable amount of anecdotal information suggests that some other form of sex determination operates in the Chalcidoidea in general (Luck *et al.* 1992; Poirie *et al.* 1992; Stouthamer *et al.* 1992; Cook 1993a,b). There are virtually no data on sex determination in the several remaining parasitoid superfamilies, although it is unlikely that several typically inbreeding taxa have CSD. Bull (1981) argued that the ancestral sex-determining system in Hymenoptera is more likely to be single-locus CSD than multilocus CSD because selection opposes the transition from multilocus to single-locus control, which increases diploid male production. By a similar argument, multiple origins of single-locus CSD from any non-CSD sex-determining system that does not involve sterile diploid males are unlikely (Crozier 1977; Bull 1981). Existing empirical evidence is consistent with this view (Cook 1993a): single-locus CSD occurs in several taxonomically distant Hymenoptera, including two tenthredinoid sawflies (Smith and Wallace 1971; Naito and Suzuki 1991), which are separated from the Apocrita by one of the deepest branches in the order. Reports of diploid males in 33 disparate species of Hymenoptera (see Table 1 in Cook 1993a) reinforce the suggestion that single-locus CSD is widespread. The one bethyloid (Cook 1993b) and several chalcidoid (see Luck *et al.* 1992; Stouthamer *et al.* 1992; Cook 1993a) wasps that do not have CSD are all inbreeding species. It seems likely that selection against high diploid male production has caused a change in the sex determination system in these species.

The nature of such secondary sex-determining systems is unknown, but diploid males might be lost via a genetic imprinting mechanism (Poirie *et al.* 1992; Beukeboom 1995). The mechanism proposed involves maternal imprinting of the sex locus so that it does not bind an active product that is required for female development. Consequently, unfertilized eggs develop as males. The imprinting is erased during development and males do not imprint. Thus fertilized eggs contain bound active product from the sperm and develop as females. This new model has not been tested yet but *Nasonia vitripennis* and *Goniozus nephantidis* are obvious candidates for future experiments.

CSD has important consequences for behavioral ecology; its explicit consideration can add to our understanding of reproductive patterns in parasitoids and social Hymenoptera (see, for example, Ross and Fletcher 1986; Ratnieks 1990; Antolin and Strand 1992; Cook and Crozier 1995). Future studies should aim to determine the distribution of CSD in parasitoids. For example, we do not yet know whether single-locus CSD applies to most Ichneumonoidea and whether it applies to *any* parasitoids outside the Ichneumonoidea. In the meantime, we explore the consequences of single-locus CSD for mating systems, encouraged by the success of the few studies to date. Whereas the single-locus CSD model can be tested relatively easily in some species, demonstration or exclusion of multilocus CSD is a more formidable task (Cook 1993a). If multilocus CSD does exist, its consequences for mating systems are qualitatively similar to those of single-locus CSD, but weaker.

The effects of single-locus CSD on mating systems can be understood by considering the three possible types of mating combination. In *matched* matings ($A_1A_2 \times A_1$ or A_2), where the two sexes possess a common sex-determination allele, 50% of fertilized eggs become sterile diploid males; in *unmatched* matings ($A_1A_2 \times A_3$), involving three sex-determination alleles, all diploid offspring are female. The level of diploid male production in a population depends on the frequency of matched matings, which in turn depends on the number of sex alleles in the population and the degree of inbreeding. If mating is population wide, the probability of a matched mating equals $2/k$ (Adams *et al.* 1977). The only estimate for a wild parasitoid population is fifteen alleles for *Diadromus pulchellus* in the south of France (Periquet *et al.* 1993), implying that about 13% of matings produce diploid males and 6.5% of all diploid offspring are males. Most estimates for social insects fall in the range of 10–20 alleles (Cook and Crozier 1995), although there is one outlying estimate of 86 alleles for a fire-ant population (Ross *et al.* 1993). The third type of matings are those with diploid males. Until recently, information on the biology of parasitoid diploid males was only available for *Bracon hebetor*, where such males normally fail to reach maturity (Petters and Mettus 1980). However, recently, El Agoze and colleagues have shown that *D. pulchellus* diploid males have similar survival to maturity (El Agoze and Periquet 1993) and mating success (El Agoze *et al.* 1994) to haploid males. On a wider scale, diploid male biology also varies markedly between different non-parasitoid hymenopteran species (Cook and Crozier 1995) but low viability may be restricted to the genus *Bracon*.

THE DETERMINANTS OF MATING SYSTEMS

We have already noted that gregariousness and attacking gregarious hosts may predispose a species to have a mating system in which most matings occur at the emergence site. The most obvious advantage to the female of mating at the oviposition site is that she can immediately begin searching for hosts on which she can oviposit offspring of *either* sex (haplodiploid insects can produce sons without mating). Remaining at the emergence site to mate with their sisters has few disadvantages for males. Although they cannot search for other females, mating opportunities are relatively guaranteed at the emergence sites, unless females have evolved positive adaptations to prevent inbreeding.

However, associated with mating with siblings there may be penalties that will select for other mating systems. The genetic and sex-determination systems of parasitoids may act to either decrease or increase the disadvantages of sibling mating. On the one hand, the haplodiploid genetic system, which exposes even rare recessive alleles to natural selection in the haploid male sex, may result in a low number of recessive deleterious genes and thus a reduced penalty from the homozygosity associated with the commencement of inbreeding (Werren 1993). However, inbreeding increases diploid male production dramatically in species with single-locus CSD, and selection favors either inbreeding avoidance or a change in the sex-determining mechanism.

When the option of mating at its own emergence site is not open, it is possible for a male to locate females at their emergence sites, at sites where they feed, or at oviposition sites. Which mating system evolves will depend on the predictability and ease of location of the three types of site. Males may also be able to capitalize on the same cues used by the females to detect hosts or food, possibly over evolutionary time 'capturing' the behaviors that had evolved to assist the female in host location. Where recruitment to none of the three sites is possible or profitable, males may aggregate in swarms at arbitrary points in the environment. Alternatively, selection may operate on the female to provide cues to assist the male in host location (but see section on mating at arbitrary sites for counterarguments).

The behavior and physiology of the female may also be involved in determining a species' mating system. If females mate only once, there will be a premium on finding a female as young as possible while she is still a virgin. This behaviour might favor mating at the emergence site. Alternatively, if a female is willing to mate more than once then there may be a predisposition to mating at feeding or oviposition sites. The latter predisposition will be strengthened if the sperm of the last male to copulate with a female has precedence in fertilizing eggs. There is little information on sperm precedence in parasitoids, but first-male precedence occurs in both *Nasonia vitripennis* (Holmes 1974) and *Dahlbominus fuscipennis* (Wilkes 1966). Interestingly, in *N. vitripennis*, broods of mixed parentage were more common if the first male had mated before. If this translates to reduced ejaculate volume, it suggests that the spermatheca of female *N. vitripennis* can only hold one full ejaculate from a previously unmated male (Crozier and Bruckner 1981). More studies of sperm competition are required and can be performed in species with genetic markers such as eye-color mutants. Some wasps are not receptive to mating until well after emergence and require time to feed and mature their eggs. The delay in mating may be to avoid inbreeding or so that they have fresh sperm when they begin to oviposit. Again, these species are predisposed to have a mating system that does not involve mating at the emergence site.

Ridley (1993) has recently conducted a comparative study of mating frequency in parasitoid wasps, with data from nearly one hundred species. Using techniques that take phylogeny into account, he shows a strong association between gregariousness and multiple mating. He attributes this correlation to an advantage in having genetically heterogeneous offspring when they compete together for host resources. In a metaphor due to Williams (1975), in entering many lotteries, it does not matter if all your tickets have the same number; if entering a single lottery, one should purchase many different tickets. However, there are other possible explanations for these data (Godfray 1994). Gregarious species frequently have female-biased sex ratios so that one male inseminates a number of females. If the volume of ejaculate is low, the female may be selected to mate with several males. As discussed above, the mating system in which males remain at the emergence site tends to be associated with gregariousness. There may be a risk that a female who leaves the emergence site with insufficient sperm might never find another mate in the environment. This again will select for multiple matings among gregarious parasitoids. Hardy (1994b) suggests that quasi-gregarious parasitoids could be used to distinguish between the genetic heterogeneity and other hypotheses. Because these parasitoids lay only one egg per host but the hosts are clumped, the mating hypotheses apply but the heterogeneity argument does not.

CSD might also influence selection for multiple mating. The effect of multiple mating is to decrease the variance in diploid male production between families without changing the mean value. Consider the extremes. If a female mates with only one male, either 50% or 0% of her fertilized eggs are diploid males. On the other hand, if a female were to mate with an infinite number of males, then her fertilized eggs would contain the same fraction of diploid males as the population at large. Selection for multiple mating occurs in some social Hymenoptera, where colonies with high diploid male production have very low or even zero survival rates in some years because diploid males occur in place of presumptive workers (see, for example, Ross and Fletcher 1986; Cook and Crozier 1995). In parasitoids, it seems unlikely that the conditions will be met for a bet-hedging strategy to evolve (Gillespie 1977), as there will probably be no temporal variation in the fitness penalties of producing large broods of diploid males.

SELECTION ON THE FEMALE AND SEXUAL CONFLICT

Unmatedness in female haplodiploids

Of the six mating systems outlined above, only two require active female cooperation beyond willingness to copulate: the release of pheromones to attract males, and lekking and swarming, which require females to search out male aggregations. In Hymenoptera, unmated females are not reproductively dead but can produce offspring, albeit only males. If the assumptions necessary for the application of Fisher's sex-ratio theory apply, and if the population is at sex-ratio equilibrium, then by definition sons and daughters are of equal value to a mother. Hence, an unmated female will not be penalized if she produces only sons, provided the population stays at sex ratio equilibrium (Godfray 1990). If many females produce only sons, one of two things may occur. First, the population sex ratio may become male-biassed, and sex-ratio equilibrium lost, leading to selection on all females to possess the ability to produce daughters. Alternatively, mated females may respond to the slight bias by producing more daughters so that the overall population sex-ratio remains at equality and sex ratio equilibrium is not lost. In this second case there is no penalty attached to the continued production of all-male broods.

It might seem that this last argument would predict that haplodiploid populations should be composed of 50% of females producing daughters and the rest producing males. However, this ignores the role of males who actively seek out females. Males that fail to locate mates have zero fitness and it is likely to be their activities that determine the fraction of unmated females and thus, indirectly, the sex ratio produced by mated females. There have been few attempts to detect virgin oviposition in field populations of haplodiploid parasitoids. Hardy and Godfray (1990) and Godfray and Hardy (1992) surveyed the literature and found data from approximately 25 species. In many species, all ovipositing females appeared to be mated, but there were some cases where as many as 20% of females were virgin. Curiously, gregarious species with local mate competition were more likely to have virgin oviposition than solitary species, even though there are far greater advantages to being mated when mating opportunities for males are restricted to the natal brood or near neighbours (Godfray 1990). However, this result may be a statistical artifact of biased reporting. Alternatively, Heimpel (1994) has pointed out that male developmental mortality may lead to high levels of virginity in species that have female-biased sex ratios owing to local mate competition.

Sex determination and sexual conflict

In the above section we ignored sex determination. However, the possibility of mating with, or producing, sterile diploid males is a significant selection pressure in species with CSD. Further, sexual conflict over such mating decisions is expected to be common (see Alexander *et al.*, this volume). In both *B. hebetor* (Whiting 1961) and *Diadromus pulchellus* (El Agoze *et al.* 1994), females that mate with diploid males produce mostly haploid sons plus the occasional triploid offspring. Consequently, unless diploid males have very poor survival or mating success, species with CSD should have higher proportions of females constrained to produce only sons than species without CSD.

Diploid male matings

To illustrate the effects of diploid male matings, consider an outbreeding population with k equally abundant sex alleles. Let the fraction of diploid males among all males equal u in a given generation. Let us assume that females mate with haploid and diploid males at random and fertilize a fraction $(1 - p)$ of their eggs. Thus a fraction u of females mate with diploid males and we can now calculate the fraction (u') of diploid males in the next generation. Diploid males are produced by females that mate with haploid males $(1 - u)$ from fertilized eggs $(1 - p)$ that contain two identical sex alleles (probability $= 1/k$). Haploid males are produced from unfertilized eggs of females mated with haploid males $((1 - u)p)$ and from all eggs of females mated with diploid males (u). The diploid male frequency in the next generation is thus

$$u' = \frac{(\text{diploid males})}{(\text{diploid males} + \text{haploid males})}$$

$$= \frac{(1 - u)(1 - p)(1/k)}{(1 - u)(1 - p)(1/k) + (1 - u)p + u}. \quad \text{(Eqn 12-1)}$$

We can calculate the equilibrium fraction of diploid males by setting $u' = u$ and solving the resulting quadratic for given values of the sex ratio (strictly, the proportion of unfertilized eggs) produced by individual females (p) and the number of sex alleles (k). However, natural selection will operate on females to produce the optimum sex ratio given that some females are constrained to produce only sons. The optimum sex ratio is found as the value of p that results in a population-wide sex ratio of equality (the sex ratio is calculated *ignoring* diploid males, which have no genetic value). Thus setting

$$(1 - u)(1 - p)(1 - 1/k) = u + (1 - u)p, \quad \text{(Eqn 12-2)}$$

we find

$$p = 1 - \frac{1}{(1 - u)(2 - 1/k)}. \quad \text{(Eqn 12-3)}$$

Note that, if $k \to \infty$ so that $u \to 0$, then $p = 1/2$ as expected. We can substitute equation 12-3 into equation 12-1 to obtain the equilibrium proportion of diploid males given that females adaptively adjust their sex ratio. A little algebra leads to the equilibrium frequency (u^*):

$$u^* = 1/k. \quad \text{(Eqn 12-4)}$$

Again, note that as $k \to \infty$, $u \to 0$. Fig. 12-1 shows the relationship between the proportion of eggs left unfertilized (p) by mated females and the number of sex-determination alleles segregating in the population. Although the sex ratio is proportional to p, it does not equal p because some fertilized eggs give rise to diploid males rather than females. The equations predict optimum sex ratios only when three or more alleles are present. If there are only two alleles present, an equal population-wide sex ratio is impossible, even when mated females produce only fertilized eggs. When few sex-determination alleles are present, mated females produce strongly female-biased sex ratios, the ratio approaching 0.5 as the number of alleles increases.

In the absence of mating between females and diploid males, the loss of diploid eggs as non-functional males

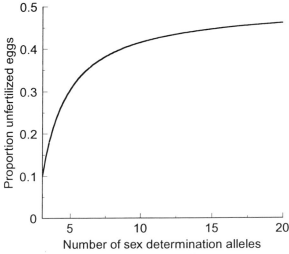

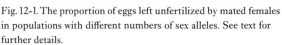

Fig. 12-1. The proportion of eggs left unfertilized by mated females in populations with different numbers of sex alleles. See text for further details.

has no influence on the sex ratio: it acts like any other differential mortality factor. In such circumstances, each female fertilizes 50% of her eggs and the equilibrium fraction of diploid males is

$$u^* = \frac{0.5(1/k)}{0.5 + 0.5(1/k)} = 1/(k+1),$$

slightly less than the sex ratio found when diploid males mate. However, as k increases, $1/(k+1) \to 1/k$ so there is little difference in populations with many sex alleles.

Selection on females to avoid mating with diploid males may be weak if diploid sperm simply fail to penetrate the egg, as is generally the case in *Bracon hebetor* (MacBride 1946). However, a recent study shows that female *D. pulchellus* mated to diploid males produce significantly fewer haploid sons than do virgin females (El Agoze *et al.* 1994) because substantial numbers of inviable triploid zygotes are formed. Females are therefore selected to avoid mating with diploid males, which, in *D. pulchellus*, closely resemble haploid males, but are bigger (El Agoze and Periquet 1993). It is not known whether females of any hymenopteran species actively avoid diploid male matings and this is clearly an area that is amenable to experimental study.

Matched matings

In parasitoids with single-locus CSD there is likely to be sexual conflict over matched matings. In general, females are under strong selection to avoid matched matings and males are selected to take all available mating opportunities (Alexander *et al.*, this volume; Cook and Crozier 1995). In principle, matched matings could be avoided by sex–allele signaling (Crozier 1987, 1988; Ratnieks 1991). However, in parasitoids male reproductive success is more likely to be limited by the number rather than the quality of matings. Nevertheless, as eye colour markers are known in both *B. hebetor* and *D. pulchellus*, simple mate-choice experiments, using individuals with known sex alleles, could be used to investigate this exciting possibility.

Inbreeding avoidance does not preclude a matched mating but provides a behavioral means to dramatically reduce the probability. *B. hebetor* provides an interesting example. As was discussed above in the section on lek mating systems, females visit male aggregations to mate, despite the fact that sib-mating opportunities are common at the site of emergence (Antolin and Strand 1992). Laboratory experiments by Ode *et al.* (1995) showed that both sexes have a refractory period (about 5 h) after

emergence, during which they do not mate. Dispersal normally occurs during this time, reducing the incidence of inbreeding. Furthermore, choice experiments showed that, for up to 5 d post-emergence, females preferred to mate with males that developed on different hosts. Egg-transplantation experiments showed that this preference was based on host cues and not on genetic relatedness *per se*, but in the wild this provides a good way of avoiding inbreeding.

MIXED MATING SYSTEMS

Some parasitoid species show a mixture of the mating systems described above. There is evidence that in several species of wasp where males typically remain at the emergence site to mate with females, a minority of males disperse into the environment to locate new emergence sites (reviewed by Hardy 1994b). This trend is taken further in other species where males show morphological dimorphism between those that remain at the emergence site and those that disperse. The most extreme examples of alternative male morphologies and mating strategies occur in some non-pollinating fig wasps, where the two morphs are so different that in the past males of one species have been placed in different subfamilies (see Herre *et al.*, this volume; Cook *et al.* 1996). Dimorphic males of scelionid egg parasitoids and bethylid larval parasitoids have been explained as mating-strategy polymorphisms (Hamilton 1979; Godfray 1994). Such polymorphisms can reflect alternative mating strategies with equal fitness (balanced polymorphisms) or may be cases where one strategy has higher fitness but some individuals are unable to adopt this strategy effectively because of constraints such as small body size (conditional polymorphisms) and must make 'the best of a bad job'. The best way to detect whether male dimorphisms are balanced or conditional is to measure the reproductive success of the two morphs, but this is often an extremely difficult task and we are not aware of any parasitoid examples. In the bethylid wasp *Cephalonomia gallicola* male dimorphism is due to two alleles at a single locus (Kearns 1934) and a balanced polymorphism would seem to be required to prevent one allele from being eliminated.

A variable mating system may also explain a curious example of wing polymorphism found in the egg parasitoid *Trichogramma semblidis*. Salt (1937, 1939) showed that males of this species reared from the alder fly, *Sialis lutraria* (Neuroptera), were wingless whereas those reared from moth eggs were winged. Alder flies lay their eggs in large clusters of 500–700 eggs. The natural lepidopteran hosts of *T. semblidis* are unknown, but are probably moths that lay their eggs singly, or at least in small clumps. It is possible that males born on a large patch of 500–700 host eggs are selected to remain on the patch to mate with females that subsequently emerge. These individuals have no need of wings and indeed may benefit from reallocating resources towards other functions. Males that emerge from solitary eggs or from small clumps of eggs, need to fly to locate other females and thus require wings (Godfray 1994).

Finally, we note that pheromone release may be involved in rendezvous mating systems at the emergence, feeding or oviposition sites. For example, we described Field and Keller's (1993) study of mating behavior in *Cotesia rubecula* under the category of mating at an arbitrary site with attraction of males by pheromone. In fact, all experiments in the field and the laboratory were conducted on cabbage plants, the food plant of the host. It is possible that both sexes rendezvous on the host plant with pheromone attraction only coming into play at short distances.

CONCLUSIONS AND SUGGESTIONS FOR FUTURE RESEARCH

Parasitoids provide numerous wonderful systems with which to test many aspects of evolutionary theory. The close connection between many aspects of parasitoid behavior and fitness, and the opportunities for facultative sex allocation afforded by a haplodiploid genetic system, are two of the main factors that make parasitoids so useful. The focus of parasitoid behavioral ecology in the past twenty years has largely, but not exclusively, been on laboratory experiments. Although many aspects of parasitoid behavioural ecology are amenable to laboratory study, field study is essential where mating systems are involved (Godfray 1994, Hardy 1994a) and there is considerable scope for valuable data collection through simple field observations of parasitoid mating behavior.

Two important aspects of mating behavior, mate choice and sperm usage, have received little study in parasitoids. Both could be studied usefully with laboratory experiments, especially where genetic markers are available. There is also an urgent need to know how widespread single-locus CSD is among parasitoids and to explore its consequences for mating behavior. Again, these questions can be addressed by laboratory exploitation of controlled matings and genetic markers.

The importance of integrating laboratory study with field observations and explicit consideration of sex determination is illustrated by the saga of *Bracon hebetor*, a species we have discussed repeatedly in this chapter. In the laboratory, mated *B. hebetor* females produce gregarious broods with female-biased sex ratios. Further, most individuals will sib-mate if kept in confined conditions. It would be easy to assume that the species is characterized by inbreeding and LMC, selecting for a female-biased sex ratio. However, *B. hebetor* has single-locus CSD (Whiting 1943), which selects against inbreeding, at least in females. Field observations by Antolin and Strand (1992) revealed a postdispersal lek mating system; judicious laboratory experiments (Ode *et al.* 1995) then showed a behavioral mechanism of inbreeding avoidance. Consequently, neither inbreeding nor LMC appears to apply in this species, and should not be assumed to explain all cases of female-biased sex ratios in gregarious parasitoids. This leaves us with the mystery of why mated female *B. hebetor* produce female-biased sex ratios. Possibly it is a response to the presence of many females in the population that are constrained to produce males, or possibly sexual asymmetry in larval competition or sex-differential dispersal is responsible (Antolin and Strand 1992; Cook *et al.* 1994; Ode *et al.* 1995). Only more study will tell.

The work on *B. hebetor*, combining genetic and behavioral ecological studies, in the field and the laboratory, is a striking example of what can be discovered about parasitoid mating systems with a multidisciplinary approach.

ACKNOWLEDGEMENTS

We thank the three reviewers of this chapter, Allen Herre, Hans van den Assem and an anonymous referee, for their constructive comments. James Cook is supported by a SERC Postdoctoral Fellowship.

LITERATURE CITED

Achterberg C. van. 1977. The function of swarming in *Blacus* species (Hymenoptera, Braconidae, Helconinae). *Entomol. Ber.* **37**: 151–152.

Adams, J., E. D. Rothman, W. E. Kerr and Z. L. Paulino. 1977. Estimation of the number of sex alleles and queen matings from diploid male frequencies in a population of *Apis mellifera*. *Genetics* **86**: 583–596.

Antolin, M. F. and M. R. Strand. 1992. Mating system of *Bracon hebetor* (Hymenoptera: Braconidae). *Ecol. Entomol.* **17**: 1–7.

Arthur, A. P. 1971. Associative learning by *Nemeritis canescens* (Hymenoptera: Ichneumonidae). *Can. Entomol.* **103**: 1137–1141.

Assem, J. van den. 1986. Mating behaviour in parasitic wasps. In *Insect Parasitoids.* J. K. Waage and D. Greathead, eds., pp. 137–167. London: Academic Press.

Assem, J. van den, M. J. Gijswijt and B. K. Nübel. 1980. Observations on courtship and mating strategies in a few species of parasitic wasps (Chalcoidea). *Netherl. J. Zool.* **30**: 208–227.

Balfour Browne, F. 1922. On the life-history of *Melittobia acasta* Walker; a chalcid parasite of bees and wasps. *Parasitology* **14**: 349–370.

Beukeboom, L. W. 1995. Sex determination in Hymenoptera: a need for genetic and molecular studies. *BioEssays* **17**: 813–817.

Buckell, E. R. 1928. Notes on the life history and habits of *Melittobia chalybii* Ashmead (Chalcoidea: Elachertidae). *Pan-Pac. Entomol.* **5**: 14–22.

Bull, J. J. 1981. Coevolution of haplodiploidy and sex determination in the Hymenoptera. *Evolution* **35**: 568–580.

Beye, M., R. F. A. Moritz and C. Epplen. 1994. Sex linkage in the honeybee *Apis mellifera* detected by multilocus DNA fingerprinting. *Naturwissenschaften* **81**: 460–462.

Charnov, E. L. 1982. *The Theory of Sex Allocation.* Princeton: Princeton University Press.

Clark, A. G. and T.-H. Kao. 1991. Excess nonsynonymous substitution at shared polymorphic sites among self-incompatibility alleles of Solanaceae. *Proc. Natl. Acad. Sci. U.S.A.* **88**: 9823–9827.

Clausen, C. P. 1940. *Entomophagous Insects.* New York: McGraw Hill.

Cook, J. M. 1993a. Sex determination in the Hymenoptera: a review of models and evidence. *Heredity* **71**: 421–435.

–. 1993b. Empirical tests of sex determination in *Goniozus nephantidis* (Hymenoptera: Bethylidae). *Heredity* **71**: 130–137.

Cook, J. M. and R. H. Crozier. 1995. Sex determination and population biology in the Hymenoptera. *Trends Ecol. Evol.* **10**: 281–286.

Cook, J. M., A. P. Rivero Lynch and H. C. J. Godfray. 1994. Sex ratio and foundress number in the parasitoid wasp *Bracon hebetor. Anim. Behav.* **47**: 687–696.

Cook, J. M., S. A. West, E. A. Herre and S. G. Compton. 1996. The evolution of extreme male dimorphism in fig wasps. *Proc. R. Soc. Lond.* B, submitted.

Crozier, R. H. 1971. Heterozygosity and sex determination in haplodiploidy. *Am. Nat.* **105**: 399–412.

–. 1977. Evolutionary genetics of the Hymenoptera. *Annu. Rev. Entomol.* **22**: 263–288.

–. 1987. Genetic aspects of kin recognition: concepts, models and synthesis. In *Kin Recognition in Animals.* D. J. C. Fletcher and C. D. Michener, eds., pp. 55–74. Chichester: Wiley.

–. 1988. Kin recognition using innate labels: a central role for piggy-backing? In *Invertebrate historecognition.* R. K. Grosberg, D. Hedgecock and K. Nelson, eds., pp. 143–156. New York: Plenum Press.

Crozier, R. H. and D. Bruckner. 1981. Sperm clumping and the population genetics of Hymenoptera. *Am. Nat.* **117**: 561–563.

Davies, N. B. 1991. Mating strategies. In *Behavioural Ecology, an Evolutionary Approach*, 3rd edn. J. R. Krebs and N. B. Davies, eds., pp. 263–294. Oxford: Blackwell Scientific Publications.

Downton, M. and A. D. Austin. 1994. A molecular phylogeny of the insect order Hymenoptera – Apocritan relationships. *Proc. Natl. Acad. Sci. U.S.A.* **91**: 9911–9915.

Eberhard, W. G. 1975. The ecology and behavior of a subsocial pentatomid bug and two scelionid wasps: strategy and counter strategy in a host and its parasite. *Smithson. Contrib. to Zool.* **205**: 1–39.

Eggleton, P. 1990. The male reproductive behaviour of *Lytarmes maculipennis* (Smith) (Hymenoptera: Ichneumonidae). *Ecol. Entomol.* **15**: 357–360.

–. 1991. Patterns in male mating strategies of the Rhyssini: a holophyletic group of parasitoid wasps (Hymenoptera: Ichneumonidae). *Anim. Behav.* **41**: 829–838.

Eggleton, P., and R. Belshaw. 1992. Insect parasitoids: an evolutionary overview. *Proc. R. Soc. Lond.* B **337**: 1–20.

El Agoze, M. and G. Periquet. 1993. Viability of diploid males in the parasitic wasp *Diadromus pulchellus* (Hym: Ichneumonidae). *Entomophaga* **38**: 199–206.

El Agoze, M., J. M. Drezen, S. Renault and G. Periquet. 1994. Analysis of the reproductive potential of diploid males in the wasp *Diadromus pulchellus* (Hymenoptera: Ichneumonidae). *Bull. Entomol. Res.* **84**: 213–218.

Eller, F. J., R. J. Bartelt, R. L. Jones and H. M. Kulman. 1984. Ethyl(Z)-9-hexadecenoate, a sex pheromone of *Syndipnus rubiginosus*, a sawfly parasitoid. *J. Chem. Ecol.* **10**: 291–300.

Field, S. A. and M. A. Keller. 1993. Alternative mating tactics and female mimicry as post-copulatory mate-guarding behaviour in the parasitic wasp *Cotesia rubecula. Anim. Behav.* **46**: 1183–1189.

Fisher, R. A. 1930. *The Genetical Theory of Natural Selection.* Oxford University Press.

Figueroa, F., G. Eberhardt and J. Klein. 1988. MHC polymorphism pre-dating speciation. *Nature (Lond.)* **335**: 265–267.

Gauld, I. and B. Bolton. 1988. *The Hymenoptera.* Oxford: Oxford University Press/ British Museum (Natural History).

Gillespie, J. H. 1977. Natural selection for variances in offspring numbers: a new evolutionary principle. *Am. Nat.* **11**: 1010–1014.

Godfray, H. C. J. 1990. The causes and consequences of constrained sex allocation in haplodiploid animals. *J. Evol. Biol.* **3**: 3–17.

–. 1994. *Parasitoids: Behavioural and Evolutionary Ecology.* Princeton: Princeton University Press.

Godfray, H. C. J. and I. C. W. Hardy. 1992. Virginity in haplodiploid animals. In *Evolution and Diversity of Sex Ratios in Insects and Mites.* D. L. Wrensch and M. A. Ebbert, eds., pp. 402–417. New York: Chapman and Hall.

Gordh, G. and P. DeBach. 1978. Courtship behavior in the *Aphytis lingnanensis* group, its potential usefulness in taxonomy, and a review of sexual behavior in the parasitic Hymenoptera (Chalc., Aphelinidae). *Hilgardia* **46**: 37–75.

Hamilton, W. D. 1967. Extraordinary sex ratios. *Science (Wash., D.C.)* **156**: 477–488.

–. 1979. Wingless and fighting males in fig wasps and other insects. In *Sexual Selection and Reproductive Competition in Insects.* M. S. Blum and N. A. Blum, eds., pp. 167–220. London: Academic Press.

Hardy, I. C. W. 1994a. Sex ratio and mating structure in the parasitoid Hymenoptera. *Oikos* **69**: 3–20.

–. 1994b. Polyandrous parasitoids: multiple mating for variety's sake. *Trends Ecol. Evol.* **9**: 202–203.

Hardy, I. C. W. and H. C. J. Godfray. 1990. Estimating the frequency of constrained sex allocation in field populations of Hymenoptera. *Behaviour* **114**: 137–147.

Heatwole, H., D. M. Davis and A. M. Wenner. 1962. The behaviour of *Megarhyssa*, a genus of parasitic hymenopterans (Ichneumonidae; Ephialtinae). *Z. Tierpsychol.* **19**: 652–664.

Heimpel, G. 1994. Virginity and the cost of insurance in highly inbred Hymenoptera. *Ecol. Entomol.* **19**: 299–302.

Holmes, H. B. 1974. Patterns of sperm competition in *Nasonia vitripennis*. *Can. J. Genet. Cytol.* **16**: 789–795.

Hughes, A. L. and M. Nei. 1988. Pattern of nucleotide substitution at major histocompatibility complex class I loci reveals overdominat selection. *Nature (Lond.).* **335**: 167–170.

Hunt, G. J. and R. E. Page. 1994. Linkage analysis of sex determination in the honeybee (*Apis mellifera*). *Molec. Gen. Genet.* **244**: 512–518.

Ioerger, T. R., A. G. Clark, and T-H. Kao. 1990. Polymorphism at the self-incompatibility locus in Solanaecae predates speciation. *Proc. Natl. Acad. Sci. U.S.A.* **87**: 9732–9735.›pa

Jervis, M. A. 1979. Courtship, mating and 'swarming' in *Aphelopus melaleucus* (Dalman) (Hymenoptera: Dryinidae). *Entomol. Gaz.* **30**: 191–193.

Jervis, M. A., N. A. C. Kidd, M. G. Fitton, T. Huddleston and H. A. Dawah. 1993. Flower-visiting by hymenopteran parasitoids. *J. Nat. Hist.* **27**: 67–105.

Kearns, C. W. 1934. Method of wing inheritance in *Cephalonomia gallicola* Ashmd. (Hymenoptera: Bethylidae). *Ann. Entomol. Soc. Am.* **27**: 533–541.

King, B. H. 1987. Offspring sex ratios in parasitoid wasps. *Q. Rev. Biol.* **62**: 367–396.

LaSalle, J. and I. D. Gauld. 1991. Parasitic Hymenoptera and the biodiversity crisis. *Redia* **74**: 315–334

Laidlaw, H. H., F. P. Gomes and W. E. Kerr. 1956. Estimation of the number of lethal alleles in a panmictic population of *Apis mellifera* L. *Genetics* **41**: 179–188.

Lawlor, D. A., F. Ward, P. Ennig, A. Jackson and P. Parham. 1988. HLA-A and B polymorphisms predate the divergence of humans and chimpanzees. *Nature (Lond.)* **335**: 268–271.

Lewis, W. J., J. W. Snow and R. L. Jones. 1971. A pheromone trap for studying populations of *Cardiochiles nigriceps*, a parasite of *Heliothis virescens. J. Econ. Entomol.* **64**: 1417–1421.

Luck, R. F., R. Stouthamer and L. P. Nunney. 1992. Sex determination and sex ratio patterns in parasitic Hymenoptera.

In *Evolution and Diversity of Sex Ratios in Insects and Mites.* D. L. Wrensch and M. A. Ebbert, eds., pp. 442–476. New York: Chapman and Hall.

MacBride, D. H. 1946. Failure of sperm of *Habrobracon* to penetrate the eggs. *Genetics* **31**: 234.

Madden, J. L. 1968. Behavioural responses of parasites to the symbiotic fungus associated with *Sirex noctilio* F. *Nature (Lond.)* **218**: 189–190.

Mertins, J. W. 1980. Life history and behaviour of *Laelius pedatus*, a gregarious bethylid ectoparasitoid of *Anthrenus verbasci*. *Ann. Entomol. Soc. Am.* **73**: 686–693.

Myint, W. W. and G. H. Walter. 1990. Behaviour of *Spalangia cameroni* males and sex ratio theory. *Oikos* **59**: 163–174.

Nadel, H. 1987. Male swarms discovered in Chalcidoidea (Hymenoptera: Encyrtidae, Pteromalidae). *Pan-Pac. Entomol.* **63**: 242–246.

Nadel, H. and R. F. Luck. 1992. Dispersal and mating structure of a parasitoid with a female-biased sex ratio: implications for theory. *Evol. Ecol.* **6**: 270–278.

Naito, T. and H. Suzuki. 1991. Sex determination in the sawfly *Athalia rosae ruficornis*: occurrence of triploid males. *J. Hered.* **82**: 101–104.

Ode, P. J., M. F. Antolin and M. R. Strand. 1995. Brood-mate avoidance in the parasitic wasp *Bracon hebetor* Say. *Anim. Behav.* **49**: 1239–1248.

Periquet, G., M. P. Hedderwick, M. El Agoze and M. Poirie. 1993. Sex determination in the Hymenoptera *Diadromus pulchellus* (Ichneumonidae): validation of the one-locus multi-allele model. *Heredity* **70**: 420–427.

Petters, R. M. and R. V. Mettus. 1980. Decreased diploid male viability in the parasitoid wasp *Bracon hebetor. J. Hered.* **71**: 353–356.

Poirie, M., G. Periquet and L. Beukeboom. 1992. The hymenopteran way of determining sex. *Semin. Devel. Biol.* **13**: 357–361.

Rasnitsyn, A. P. 1988. An outline of evolution of the hymenopterous insects (Order Vespida). *Oriental Insects* **22**: 115–145.

Ratnieks, F. L. W. 1990. The evolution of polyandry by queens in social Hymenoptera: the significance of the timing of removal of diploid males. *Behav. Ecol. Sociobiol.* **26**: 434–348.

–. 1991. The evolution of genetic odor-cue diversity in social Hymenoptera. *Am. Nat.* **137**: 202–226.

Ridley, M. 1993. Clutch size and mating frequency in parasitic Hymenoptera. *Am. Nat.* **142**: 893–910.

Robacker, D. C. and L. B. Hendry. 1977. Neral and geranial: components of the sex pheromone of the parasitic wasp *Itoplectis conquisitor. J. Chem. Ecol.* **3**: 563–577.

Robacker, D. C., K. M. Weaver and L. B. Hendry. 1976. Sexual communication and associative learning in the parasitic wasp *Itoplectis conquisitor. J. Chem. Ecol.* **2**: 39–48.

Ross, K. G. and D. J. C. Fletcher. 1986. Diploid male production – a significant colony mortality factor in the fire ant *Solenopsis invicta. Behav. Ecol. Sociobiol.* **19**: 283–291.

Ross, K. G., E. L. Vargo, L. Keller and J. C. Trager. 1993. Effect of a founder event on variation in the genetic sex-determining system of the fire ant *Solenopsis invicta*. *Genetics* **135**: 843–854.

Salt, G. 1937. The egg-parasite of *Sialis lutaria*: a study of the influence of the host upon a dimorphic parasite. *Parasitology* **29**: 539–553.

–. 1939. Further notes on *Trichogramma semblidis*. *Parasitology* **30**: 511–522.

Skinner, S. W. and J. H. Werren. 1980. The genetics of sex determination in *Nasonia vitripennis*. *Genetics* **94**: S98.

Smith, S. G. and D. R. Wallace. 1971. Allelic sex determination in a lower hymenopteran, *Neodiprion nigroscutum* Midd. *Can. J. Genet. Cytol.* **13**: 617–621.

Snell, G. D. 1935. The determination of sex in *Habrobracon*. *Proc. Natl. Acad. Sci. U.S.A.* **21**: 446–453.

Speicher, B. R. and K. G. Speicher. 1940. The occurrence of diploid males in *Habrobracon brevicornis*. *Am. Nat.* **74**: 379–382.

Stouthamer, R., R. F. Luck and J. H. Werren. 1992. Genetics of sex determination and the improvement of biological control using parasitoids. *Environ. Entomol.* **21**: 427–435.

Southwood, T. R. E. 1957. Observations on swarming in the Braconidae (Hymenoptera) and Coniopterygidae (Neuroptera). *Proc. R. Entomol. Soc. Lond.* A **32**: 80–82.

Spradbery, J. P. 1970. Host finding by *Rhyssa persuasoria*, an ichneumonid parasite of siricid woodwasps. *Anim. Behav.* **18**: 103–114.

Suzuki, Y. and K. Hiehata. 1985. Mating systems and sex ratios in the egg parasitoids, *Trichogramma dendrolimi* and *T. papilionis* (Hymenoptera:Trichogrammatidae). *Anim. Behav.* **33**: 1223–1227.

Thornhill, R. and J. Alcock. 1983. *The Evolution of Insect Mating Systems*. Cambridge, Massachusetts: Harvard University Press.

Waage, J. K. 1982. Sib-mating and sex ratio strategies in scelionid wasps. *Ecol. Entomol.* **B7**: 103–112.

Werren, J. H. 1993. The evolution of inbreeding in haplodiploid organisms. In *The Natural History of Inbreeding and Outbreeding: Theoretical and Empirical Perspectives*. N. Thornhill, ed., pp. 42–59. Chicago: University of Chicago Press.

Whiting, P. W. 1939. Sex determination and reproductive conomy in *Habrobracon*. *Genetics* **24**: 10–111.

–. 1943. Multiple alleles in complementary sex determination of *Habrobracon*. *Genetics* **28**: 365–382.

–. 1947. Some experiments with *Mellitobia* and other wasps. *J. Hered..* **38**: 11–20.

Whiting, A. R. 1961. Genetics of *Habrobracon*. *Adv. Genet.* **10**: 333–406.

Wilkes, A. 1966. Sperm utilization following multiple insemination in the wasp *Dahlbominus fuscipennis*. *Can. J. Genet. Cytol.* **8**: 451–461.

Williams, G. C. 1975. *Sex and Evolution*. Princeton: Princeton University Press.

–. 1992. *Natural Selection*. Princeton: Princeton University Press.

Yokoyama, S. and M. Nei. 1979. Population dynamics of sex-determining alleles in honey bees and self-incompatibility alleles in plants. *Genetics* **91**: 609–626.

13 · Fig-associated wasps: pollinators and parasites, sex-ratio adjustment and male polymorphism, population structure and its consequences

EDWARD ALLEN HERRE, STUART A. WEST, JAMES M. COOK,
STEVEN G. COMPTON AND FINN KJELLBERG

ABSTRACT

Fig-pollinating and fig-parasitizing wasps are integral parts of one of the most fascinating plant–insect interactions known. Moreover, studies of these wasps have been instrumental in developing and refining ideas concerning the influence of population structure and inbreeding on shaping the outcome of kin selection. We present data compiled from six studies spanning five continents that relate brood sex ratios with foundress number in 24 pollinator species. All predictions of local mate competition (LMC) and inbreeding theory are at least qualitatively supported. Additionally, the sex ratios produced by single foundresses of any given species appear to be influenced by brood size and the frequency of multiple foundress broods in that species. We then consider the assumptions underlying the testing of the specific LMC model and consider the relative merits of observational and experimental tests of the theory. Furthermore, we discuss the existing studies of the parasitic wasp species that have addressed the unusual morphological and behavioral polymorphisms for flightlessness and lethal combat that are found in the males of these species. These differences appear to be influenced by the parasites' population structure and density, although other factors are also implicated. Finally, we compare the nature of the support for LMC theory from fig-pollinating wasps with that from the parasitoid wasp *Nasonia vitripennis*, and suggest future lines of research.

INTRODUCTION

Studies of mating systems normally concern themselves with questions such as: What factors influence mate choice in females? More generally, how do females compete among themselves in order to increase their relative reproductive success? How do males compete among themselves for access to females? More fundamental questions include: What ecological and evolutionary factors influence mating systems themselves toward any of a given set of potentially recognizable forms (e.g. polygyny, serial monogamy, polyandry, etc.)? Those recognizable forms provide the context within which the first set of questions have meaning and relevance to the behaviors of the individuals that constitute them.

The ecology and the mating systems of the very diverse taxa of wasps that are associated with figs are dominated to a greater or lesser degree by the fact that all reach reproductive maturity within the cavity of a developing fig fruit. Within this highly confined environment the number and relatedness of potential mates, and potential competitors for those mates, is extremely variable both within and among species. On the one hand, the opportunities for mate choice on the part of the female wasps are severely limited in nearly all species. In the pollinators, the outcome of female–female competition is primarily based on a foundress wasp's capacity to adjust the sex ratio of her brood. On the other hand, the form that competition for mates among males may take varies dramatically from scramble competition (all pollinators) to lethal combat (some parasitic species). As a consequence of the highly female-biased sex ratios characteristic of most of these species (in some cases an average male may mate with roughly 20 females), variance in reproductive success may often be lower in males than in females. However, most of these issues hinge on population structure and the resulting sex ratios that, along with natural history, will be the primary focus of this chapter.

The wasp species that are associated with naturally occurring figs include both mutualist pollinators and true parasites (Fig. 13-1). The figs are completely dependent on the pollinators for the production of viable seeds and dispersal of pollen. This is true despite the consumption of a large

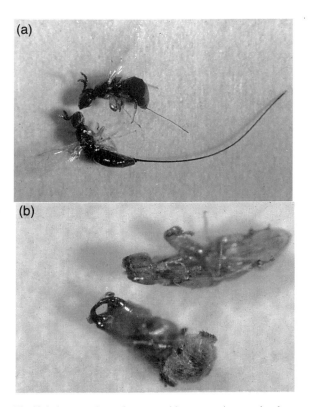

(a)

(b)

Fig. 13-1. A comparison of two agaonid wasp species associated with the New World fig, *Ficus obtusifolia*. (a) Females of the pollinator *Pegoscapus hoffmeyeri*, and the externally ovipositing parasitic wasp *Idarnes* sp. Notice the long ovipositor of the *Idarnes* female. (b) Males of the pollinator and the parasitic species. Males in the latter engage in lethal combat. Notice the very pronounced jaws in the *Idarnes* male. Magnification of the photo of males (b) is twice that of females (a).

portion of the seeds in the fig fruits during the course of the development of the pollinator offspring (Corner 1940; Galil and Eisikowitch 1968; Janzen 1979a; Herre 1989). The parasites belong to a diverse assemblege of taxa and they exhibit a wide range of ecologies (Boucek *et al.* 1981; Boucek 1993; Bronstein 1991; Compton and Hawkins 1992; West and Herre 1994; West *et al.* 1996; Machado *et al.* 1996). Some parasite species appear to compete with pollinators for available flowers, others form large galls in developing fruit, and others are true parasitoids of gall-formers (Compton 1993a; West and Herre 1994; West *et al.* 1996). However, none of the parasitic species pollinates or performs other services that hold any obvious benefit for the fig.

The pollinators are all members of the chalcidoid family Agaonidae, and all of them show similar life cycles (Wiebes 1979, 1982). Generally, some number of mated, pollen-bearing foundress wasps enter a receptive fig syconium (the enclosed inflorescence that defines the genus *Ficus* and ultimately develops into the fig fruit), pollinate the uniovulate flowers that line the interior, lay eggs in some of the flowers, and then die. Usually, the wasps die inside the one fig fruit that they pollinate, thus 'laying all of their eggs in one basket'. The offspring mature after eating the contents of one seed each. As final ripening approaches, the wingless adult males emerge from the seeds within which they matured. They crawl around the interior of the syconium, open seeds that contain females, and mate with them. There is little, if any, evidence that the winged females mate more than once. Then, the mated females emerge from their seeds, gather pollen, leave the syconium, and disperse to begin the cycle anew. After the female wasps leave, a wide range of animals eat the ripe fruit and disperse the uneaten seeds. In the vast majority of cases studied, there is a species-specific, one-to-one relationship between the pollinators and their host fig species (Wiebes 1979; Herre *et al.* 1996). Without the pollinator wasps, the figs cannot produce the fruit upon which so many species of frugivore depend. Thus, the minute wasps exert a disproportionately large influence on vertebrate populations in most tropical forests (Janzen 1979b; McKey 1989; Kalko *et al.* 1996).

In contrast to pollinator wasps, parasitic wasps species belong to many families of chalcidoid and show a strikingly wide range of life histories (Boucek *et al.* 1981; Boucek 1993; Compton 1993a; West and Herre 1994; West *et al.* 1996; Cook and Power 1996). None the less, in the vast majority of the cases studied, the parasitic wasps are specific to a given fig species (Machado *et al.* 1996). Unlike pollinators, there are usually several species of parasite associated with any particular fig species. For example, in some Old World parasite communities there are perhaps 25 species or more of parasite per host species (Boucek *et al.* 1981; Compton and Hawkins 1992). Also unlike the pollinators, the female parasitic wasps generally do not enter the syconium in order to lay their eggs. Instead, the females of many species possess long ovipositors with which they lay their eggs into the interior of the developing syconium from the outside. Often, these species are similar in size to the pollinators and, in the New World genera *Idarnes* and *Critogaster*, appear to compete with the pollinator wasps for oviposition sites (West and Herre 1994; West *et al.* 1996). Other species have relatively short ovipositors and lay their eggs in or near the fruit wall. These species (e.g. *Aepocerus*) are often much larger than the pollinators or the parasites mentioned above and develop inside true galls that they induce (Godfray 1988; Compton 1993a; West and Herre 1994; West et al. 1996). Still other species (e.g. *Physothorax*) are true parasitoids of the gall-formers (Compton 1993a; West et al. 1996). Generally the parasites lay a few eggs in each of many different developing syconia, thus 'laying a few eggs in many different baskets'.

Both across and even within species, male parasitic wasps show an extraordinary diversity in morphology and behavior (Hamilton 1979). Although many parasite species produce wingless males, as the pollinators do, other species produce winged males that can leave the syconium and mate outside it. Further, some species produce both wingless and winged males. Moreover, the wingless males in some species often show very striking polymorphisms for size, morphology and behavior. The wingless males of many of these species engage in lethal combat (Hamilton 1979; Murray 1987, 1989; Frank 1987; Vincent 1991; West et al. 1996).

The natural history of the pollinators has allowed the testing and refinement of theory that concerns population structure, local mate competition and inbreeding and their combined effects on sex-allocation and mating systems (Hamilton 1967, 1979; Kjellberg 1983; Frank 1985; Herre 1985, 1987). Further, the natural history of the parasites suggests how the form of sexual competition found among similar organisms that live side by side can hinge on such a basic consideration as 'eggs in one' vs. 'eggs in many' baskets (Hamilton 1979; Murray 1989). All of these topics are fundamentally linked with kin-selection arguments and, thereby, to the underpinnings of the theory that seems to give such powerful insight into a much wider world of animal behaviors (Hamilton 1964, 1972, 1979; Frank 1985, 1987; Herre 1993, 1995b; Herre et al. 1987). None the less, all fig wasps depend on their host fig in order to complete their life cycle; it is therefore extremely difficult, if not impossible, to understand the fig wasps' behavior outside that context.

FIG BIOGEOGRAPHY, FOSSIL RECORD DIVERSITY AND PHYLOGENY

Figs and their associated wasps are native to nearly all tropical habitats on earth (Corner 1958; Berg 1989). This global distribution suggests an ancient origin of the mutualism. That antiquity has been confirmed by the identification of fossil figs from both New and Old World strata (from currently high latitudes) that are at least 40 million years old (Collinson 1989) and by fig-pollinating wasps (along with

their parasitic nematodes (Herre 1993)) preserved in Dominican amber dating back at least 20 million years (Poinar and Herre 1991). It is clear that the mutualism between the figs and their pollinators has been successfully functioning for a long, long time.

Figs are one of the most diverse genera of flowering plants; currently, there are over 700 described species divided among four recognized subgenera (Corner 1958; Berg 1989). The division into subgenera largely reflects the fact that some figs are monoecious whereas others are functionally dioecious (Corner 1958, 1985; Wiebes 1979; Berg 1989). In the former case, individual fruit perform both female and male function, producing both viable seeds as well as pollen and the wasps that transport it to other trees (Herre 1989). In the latter case, individual trees produce fruits that are specialized to produce either viable seeds or pollen plus wasps (Corner 1940; Kjellberg et al. 1987; Patel et al. 1993).

There are over 20 recognized genera of pollinator wasp. With a few exceptions, the host species belonging to any given subsection of the figs are pollinated by species from one or a few wasp genera (Berg 1989). Although different authors have slightly different ideas of exactly how tightly linked the wasp and fig phylogenies actually are, a general congruence of fig and wasp phylogenies at these higher taxonomic levels is widely accepted (Ramirez 1974; Wiebes 1979, 1982; Corner 1985; Berg 1989; Herre et al. 1996). Moreover, with the recent publication of keys and lists of species for many parts of the world, it is now at least possible for field workers to identify figs to section and wasps to genus level (Berg and Wiebes 1992; Boucek 1993; Wiebes 1994, 1995a,b; J. Y. Rasplus, in preparation).

There have been several attempts to reconstruct the phylogenetic relationships among the higher-level fig and pollinator-wasp taxa. However, within sections of figs or genera of wasps there has not yet been any attempt to reconstruct phylogenies (but see Ulenberg 1985). The dependence of the existing morphologically based phylogenies on functional characters that are intimately related to the fig and wasp mutualism emphasizes the importance of using characters that are independent of the interactions. Presently, several studies using molecular techniques to establish phylogenetic relations are underway. For example, preliminary results from the diverse Panamanian fig-wasp fauna suggest that the phylogenies of the parasite and pollinator wasp species are congruent, implying that they have coevolved in concert with the same host fig species (Herre et al. 1996; Machado et al. 1996).

None the less, there is some evidence of non-specificity of pollinator species. Different wasp species are known to pollinate what is considered to be the same host at different parts of its range, and single trees at one site at times attract different species of pollinator (Michaloud *et al.* 1985; Compton 1993a,b). Moreover, in some instances what appears to be the same species of pollinator is known to pollinate different host species (Wiebes 1995a; J. M. Cook, unpublished results). However, non-specificity seems to be the exception that proves the general rule. In most cases, one fig species is host to one pollinator species. Similarly, despite the fact that the parasites represent very diverse taxa, with many species often reported on a single host-fig species, the parasites also appear to possess a high degree of species-specificity (Ulenberg 1985; Machado *et al.* 1996) .

POLLINATOR WASPS (LIVE FAST, DIE YOUNG, LEAVE A BEAUTIFUL CORPSE)

After emerging from the ripe fig fruit, adult pollinating wasps are very short-lived (2–4 d) and, accordingly, must encounter a receptive syconium quickly. As far as is known, adults do not feed. Not surprisingly, all species that have been studied thus far are pro-ovigenic, with all eggs ready for laying at the time of their birth. On the other hand, in some species, parasitic wasps feed as adults, are synovigenic, and may live relatively longer lives (Nefdt 1989).

Individual wasps can carry enough pollen to pollinate up to several hundred flowers. Individuals of larger species carry larger loads (Herre 1989; Kjellberg *et al.*, submitted). Some pollinator species actively gather pollen in their natal syconium and actively distribute it onto the stigmatic surfaces of the flowers in the syconium that they pollinate (Ramirez 1969; Frank 1984). In other species, this is a passive process (Galil and Neeman 1977; Okamoto and Tashiro 1981; F. Kjellberg, personal observation). Interestingly, syconia entered by wasps that do not carry pollen will sometimes produce wasp offspring. However, wasps that do not carry pollen produce fewer offspring than those that do (Compton 1993a; Nefdt 1989).

Many factors appear to affect the reproductive success of individual pollinators. Both within and among species, comparisons show a positive correlation between wasp size and fecundity (Herre 1989; Compton 1993a). The number of wasp offspring usually increases with increasing fruit size, although resource limitation in the host fig can reduce wasp fecundity (Herre 1989, 1996b). Parasitism by several different species of parasitic wasp has been shown to reduce number of pollinator offspring (Compton and Robertson 1988; Herre 1989; West and Herre 1994; West *et al.* 1996; Cook and Power 1996). Finally, parasitic nematodes usually reduce pollinator reproductive success (Herre 1993, 1995b, 1996).

Host fig density and phenology appear to affect the population densities and dynamics of the fig wasps. Many, perhaps most, species of monoecious fig are characterized by the synchronous production of receptive syconia within trees and the general asynchrony of fruiting among individual trees within a population (Milton *et al.* 1982; Kjellberg and Maurice 1989; Windsor *et al.* 1989; Bronstein 1991; Milton 1991; Herre 1996). Usually there is a more or less simultaneous arrival of would-be foundresses (Ware and Compton 1994). Pollination, fruit-ripening, and wasp development are closely synchronized so that later there is a nearly simultaneous departure of the offspring, insuring effective outcrossing of the figs. The fig fruiting cycle and the wasp generation times average roughly a month in most tropical environments. However, in more seasonal environments (e.g. France, South Africa and Australia) the fruit with developing wasps in them can spend several months on a tree during parts of the year. Moreover, crops often show pronounced asynchrony in these regions.

Most fig species naturally occur at low densities and, if they are synchronous within crowns, only a small proportion of them are producing receptive syconia at any one time. In combination with their short lifespans, this places a tremendous burden on the wasps, both pollinator and parasitic. Fortunately, the wasps have the potential to travel great distances (Nason *et al.* 1996). In some instances, distances of at least 40 km have been documented (Compton 1993b). It appears as though the phototaxic wasps fly up, get caught in the airstream, and drift downwind. As the air encounters trees, airflow is disturbed and the resulting turbulence probably brings wasps down through the leaves, fruits, and branches, rather like filter-feeding on the part of the tree. If the tree is a fig of the right species that is also receptive, the wasps drop out of the airstream and fly to the tree to seek receptive syconia. The wasp almost certainly detects the receptive trees via chemical aromatics, which appear to be released from the ostiole (Ware *et al.* 1993; Hossaert-McKey *et al.* 1994).

In Panama, direct evidence from protein electrophoresis of seedlings has shown that single trees can receive wasps from many, often dozens, of different paternal source trees

(Nason *et al.* 1996). The important implication of this observation is that, at least in these populations, not only are the different foundress wasps pollinating and laying eggs in the same syconium unlikely to have come from the same natal fruit, they are also unlikely to have come even from the same tree. However, some species growing in Florida, France, South Africa, Australia, and Panama show a pronounced asynchrony within crops, and syconia that range from receptive to producing wasps can be found on the same tree, often on the same branch. The important implication here is that in the asynchronously fruiting species wasps born in one fruit in the tree may enter another. This phenological pattern opens the possibility of related wasps cofounding broods as well as the possibility of the fig selfing.

POPULATION STRUCTURE, INBREEDING AND LOCAL MATE COMPETITION

Fisher (1930) showed that in randomly mating populations frequency-dependent selection will lead to equal parental investment in the two sexes and, generally, an even sex ratio. Later, Hamilton (1967) pointed out that in many organisms mating often takes place among the offspring of one or a few foundress mothers in isolated subpopulations (broods) from which mated females disperse to found new broods. This life cycle is characteristic of many organisms, with fig-pollinating wasps being a notable example. Hamilton went on to show that, under these conditions, local mate competition (LMC) is expected to select for female-biassed sex ratios.

Hamilton's basic model has been studied, rederived, and reinterpreted extensively (see Taylor and Bulmer 1980; Stubblefield and Seger 1990; Antolin 1993). Moreover, elaborations of that model have been tailored to the biology of fig-pollinating wasps (Kjellberg 1983; Herre 1985; Frank 1985). These models make three fundamental predictions. First, there is expected to be a female bias. Second, the degree of that female bias should depend on the number of foundresses contributing to broods. If the individual foundresses within a species are able to adjust the sex ratio of their offspring in response to the number of cofoundresses, then increasingly female-biased broods are expected with decreasing foundress number (Stubblefield and Seger 1990). Third, in haplodiploid organisms, the brood sex ratio associated with any given number of foundresses should be more female-biased in species with higher average levels of inbreeding (sib-mating) (Herre 1985; Frank 1985). This prediction follows from the higher relatedness of mothers to their offspring if the mother mates with a sibling. In haplodiploid organisms that inbreed, such as fig-pollinating wasps, the increase in mothers' relatedness is asymmetrical with respect to sons and daughters because sibmated mothers are more closely related to their daughters than they are to their sons (Herre 1985; Frank 1985).

Assuming unrelated cofoundresses, which contribute offspring equally to cofounded broods, then the predicted optimal brood sex ratio is given by:

$$p = (1 - m)(2n - 1)/(4n - 1),$$

where p is the proportion of males expected in a given brood; m is the proportion of offspring contributed to that brood by each foundress (a measure of the intensity of LMC in any given brood); and n is the proportion of sibmated females in the population (see Herre 1985; Frank 1985). Given these predictions, how do we link observation and/or experiment to theory in a meaningful way?

The number of foundresses that pollinate and lay eggs in individual syconia is variable across fruit within a tree, across different trees in the same species, and across different species (see Table 13-1). In most cases it is possible to find the bodies of foundresses present in nearly ripe fruit, count them, allow the offspring to complete their development, count male and female offspring, and then relate the associated sex ratio of the brood to foundress number. Alternatively, foundresses can be experimentally introduced.

Within species, these observations and experimental manipulations allow us to observe brood sex ratios and whether they shift with increasing foundress number, thereby testing the first two predictions. Across species, the third prediction can be tested by determining whether or not more inbred species show more female-biased sex ratios in broods with any given number of foundresses. In general we expect that species showing lower average foundress numbers will be more inbred (see Frank 1985; Herre 1988 for details of inbreeding estimates).

Fit to theory

At present, we have six studies available in which brood sex ratios of fig pollinating wasps are related to foundress number. These studies represent wasp species that span the entire range of pollinator phylogeny as it is currently understood and a geographical distribution that covers Australia, Africa, Europe, South America, and North America. Three of these studies are based on unmanipulated observations of the sex ratios of broods with different numbers of foundresses in the field (Panama (16 species), Australia (three species), Brazil (one species)) (Fig. 13-2a–c). Also there have been three studies that report brood sex ratios in which the number of foundresses has been experimentally manipulated (Africa (two species), Florida (one species), France (one species)) (Fig.13-2d–f).

Although a complete analysis of these data that takes the phylogenetic relationships of all of these species into account is desirable (Felsenstein 1985), we will treat the species as being independent. We believe that this is justified because the characters that we will be considering do not correlate with the molecularly based phylogeny of a large subset of all of the species (the 16 Panamanian species that belong to two genera, *Pegoscapus* and *Tetrapus*, the *Blastophaga* species from France, one of the *Pleistodontes* species from Australia, and an *Elisabethiella* species from South Africa). Therefore, although the preliminary phylogenetic

Table 13-1. *The distribution of foundresses found in the fig-wasp species considered in this study*

Ficus species	*n*	1	2	3	4	5	6+
Australia[a]							
macrophylla	1	0.18	0.16	0.19	0.05	0.12	0.30
obliqua	1	0.25	0.28	0.30	0.09	0.06	0.02
rubiginosa	1	0.29	0.36	0.14	0.13	0.11	0.12
Brazil[b]							
sp.	16	0.53	0.13	0.06	0.05	0.05	0.18
South Africa[c]							
burt-daveyi	1	0.77	0.15	0.05	0.03		
sur	1	0.59	0.18	0.04	0.16	0.00	0.03
Panama[d]							
perforata	12	1.0					
colubrinae	11	0.99	0.01				
paraensis	19	0.96	0.03	0.01			
pertusa	8	0.95	0.05				
obtusifolia	29	0.84	0.12	0.03	0.01		
citrifolia	31	0.82	0.16	0.02			
bullenei	12	0.74	0.17	0.05	0.02	0.02	
maxima	20	0.70	0.15	0.06	0.04	0.02	0.02
yoponensis	12	0.58	0.25	0.11	0.04	0.02	
dugandi	15	0.56	0.16	0.09	0.08	0.04	0.07
nymphaefolia	16	0.48	0.18	0.10	0.06	0.05	0.13
near *trigonata*	16	0.32	0.27	0.18	0.10	0.06	0.07
popenoei	23	0.31	0.28	0.18	0.09	0.06	0.08
insipida	33	0.27	0.21	0.18	0.14	0.08	0.12
trigonata	18	0.09	0.14	0.15	0.18	0.15	0.29
Florida[e]							
citrifolia	6	0.53	0.27	0.12	0.05	0.02	

Sources: [a] Cohen (1989); [b] Hamilton (1979); [c] Nefdt (1989); [d] Herre (1988); [e] Frank (1985).
Location and species name of the host fig is followed by the number of trees sampled to estimate the distribution, followed by the proportion of fruits found with 1, 2, 3, 4, 5, and 6+ foundresses.

analysis does not alter the broad points that we will be making here (Herre *et al.* 1996; Machado *et al.* 1996), we look forward to the additional insights that a more in-depth analysis will yield.

All species show a strong female bias. In addition, in all species the brood sex ratio shifts away from the extreme female bias associated with single foundress as the number of foundresses increases (Fig. 13-2). More-over, there is strong evidence for facultative shifts in sev-eral of the species because the number of males per foundress increases with foundress number (Herre 1987, 1988; Cohen 1989).

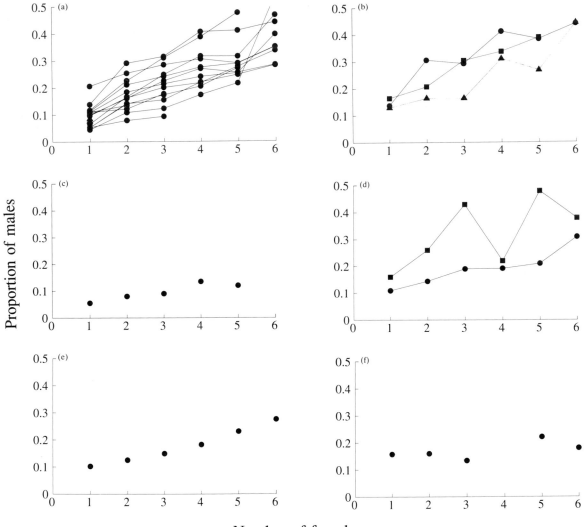

Fig. 13-2. The brood sex ratios (proportion of males) associated with different numbers of foundresses from six different studies of fig-pollinating wasps: (a) 16 species from Panama, unmanipulated field observations (from Herre 1988; E. A. Herre, unpublished data), (b) 3 species from Australia, unmanipulated field observations (from Cohen 1989), (c) 1 species from Brazil, unmanipulated field observations (from Hamilton 1979), (d) 2 species from South Africa, experimentally introduced foundresses (from Nefdt 1989), (e) 1 species from Florida, experimentally introduced foundresses (from Frank 1985), (f) 1 species from France (from Ibrahim 1985, Tayou 1991; F. Kjellberg, unpublished data). Notice that the proportion of males is very low when there is only one foundress and consistently rises with increasing foundress number (see text).

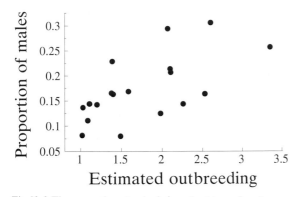

Fig. 13-3. The proportion of males in broods with two foundresses plotted against the estimated outbreeding (the reciprocal of estimated inbreeding coefficient, *n*). Notice that species with higher estimated inbreeding show more female-biased brood sex ratios.

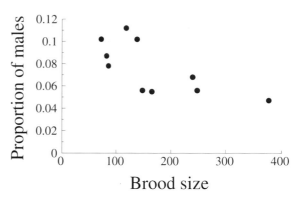

Fig. 13-4. The proportion of males in broods with one foundress plotted against the average brood size for 10 species of New World pollinator wasps in which one-foundress broods account for more than 50% of all broods. Notice that the sex ratios are more female-biased in those species with large brood sizes.

Further, those species that are characterized by the most extreme estimated opportunities for inbreeding (sib-mating) show the most female-biased sex ratios for any particular number of foundresses (Fig. 13-3). These data support the idea that the increased asymmetry of relatedness that results from inbreeding also selects for female-biassed sex ratios. It is worth noting that the influence of inbreeding is not expected to be large in comparison with the major LMC effects. Therefore, these results suggest that even the slight selection differentials that come from this second-order effect can affect sex-ratio adaptations. Overall, these results can be interpreted as supporting both a conditional facultative shift to proximate number of foundresses (or some correlate) in most species, and a population-wide response to expected inbreeding.

The predicted sex-ratio response for single-foundress broods is to produce a vanishingly small proportion of males, approaching 0 (Hamilton 1967; Hardy 1992). Taken out of context, the theory provides little guidance as to what we should expect across species. However, males are needed not only for mating but also, in most species, for cutting the exit holes needed for the mated females to leave the syconium. If all males die, then no females could exit and the result would be the complete loss of reproductive success. Therefore, we expect a reflection of risk avoidance in the sex ratio that will take the form of less female bias than would otherwise be expected if there were no risk of mortality. This effect is expected to have more influence in species with smaller brood sizes, because the sex of any individual offspring has a greater effect on brood sex

ratio (i.e. the constraint imposed by small numbers; in the extreme case, it is difficult to represent both sexes and have a female-biased sex ratio with a brood size of 2). Finally, single-foundress sex ratios are expected to be more female-biased if multiple-foundress broods are rare (Werren 1980; Herre 1987).

All species produce at least 5% males on average. Across species belonging to the New World genus *Pegoscapus*, the sex ratio for single-foundress broods is more female-biased as brood size increases (Fig. 13-4). Further, species in which the probability of cofoundresses is low show relatively more female-biased sex ratios for single-foundress broods (Fig. 13-5). Therefore, both the proportion of single-founded broods in the species and the brood size appear to influence single-foundress sex ratios. It is worth noting that without the background natural history of the individual species, and without similar information on other species with which to compare them, we would have no guide for understanding single-foundress sex ratios (Herre 1995a).

Lack of fit to theory

Although there is qualitative fit to all of the theory's basic predictions, there is considerable variability among species in the sex-ratio responses associated with different numbers of foundresses. Some species show pronounced shifts in sex ratio and others show almost none. Moreover, the sex ratios for multiple-foundress broods generally show more female-biased sex ratios than the unadjusted model

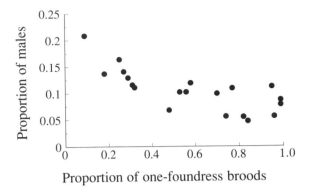

Fig. 13-5. The proportion of males in broods with one foundress for all species plotted against the proportion of all broods that have only one foundress. Notice that the proportion of males is higher in those species in which one-foundress broods are relatively rare.

predicts (Hamilton 1979; Frank 1985; Herre 1987, 1988). Conceptually, a question arises: are we using the wasps to test whether the theory is correct (essentially assuming that the wasps are adaptive and that theory should be adjusted to accomodate the observations) or are we using the theory to test how precisely adapted the wasps are (assuming the theory is correct and any discrepancy between predicted and observed results are somehow due to imperfect wasps)? One approach attempts to reconcile data and theory by adjusting parameters that are associated with various assumptions in the model (see, for example, Hamilton 1979; Frank 1985). The other approach suggests that deviations from the theory were largely due to undescribed constraints in the perfectability of adaptation (see, for example, Herre 1987). These are obviously caricatures of extreme positions. However, both approaches properly emphasize the importance of knowlege of the details of natural history of study organisms in order to make meaningful tests of theory (Herre 1995a).

The predictions of brood sex ratios for any given numbers of foundresses in the unadjusted model depend on the assumptions of equal contribution of cofoundresses and random mating within broods (Hamilton 1979; Frank 1985). Despite the fact that the predictions of optimal sex ratio are fairly robust against violations of these assumptions (Frank 1985; Herre 1988), we do not know the relative contribution of individual females to broods or the patterns of within-brood mating. That is, we do not know if m, the reciprocal of the number of foundresses, actually gives a good estimate of intensity of LMC within a given brood. Moreover, we do not

know true levels of inbreeding in these species, n. Any violations of the assumptions of the model are expected to result in selection for more female-biassed sex ratios than the unadjusted theory predicts; this is in fact what is observed. Taking this position to its logical extreme, if we are willing to assume that the wasps' responses are 'perfect', then the theory makes very specific predictions about the true levels of LMC and inbreeding. The question, then, is how far off are the estimates (of m and n) incorporated into the model that is used to make the predictions?

However, both within and across species, the deviations from theory are not random and the sex-ratio responses of the wasps generally fit theory best for the situations that the wasps encounter most frequently in nature (Herre 1987, 1988, 1995a). For example, in several instances researchers experimentally introduced a number of foundresses into figs that were never observed in nature. How should we interpret an organism's response to a situation it either never or seldom encounters in nature? It may not be surprising that in these rarely encountered foundress situations the brood sex ratios generally showed large deviations from the predictions (see Frank 1985; Nefdt 1989; see also Table 13-1, Fig. 13-2d, e). These observations suggest that the theory and even most of the assumptions incorporated into the model are fundamentally correct and that wasps appear to be limited in their ability to 'do the right thing' in response to all situations that they might encounter. This interpretation is appealing because it is consistent with a wide range of studies that demonstrate that organisms are physiologically best adapted to the environments that they encounter most frequently, and develop the most plasticity in the most variable environments (Herre 1987). Taking *this* position to its logical extreme, we would expect to be able to identify constraints that limit the wasps' capacity to produce the 'correct' sex ratios that theory predicts. For example, in the case of the wasps from *Ficus carica*, studied in southern France, the foundresses usually leave the syconia after laying eggs (Kjellberg 1983). In this species there may be little information available for the foundresses to do anything other than adjust their brood sex ratios to a global expected foundress number (see Stubblefield and Seger 1990). It is probably no coincidence that theses wasps show very little capacity to shift their brood sex ratios. The question, then, is: how precisely adapted are brood sex ratios of the wasp species over the entire range of situations that they naturally encounter?

We cannot address the tantalizing question of how precisely adapted the fig wasps' sex-ratio responses are to the range of situations that they naturally encounter until we have a clearer idea of what it is that they actually encounter. The foundress distribution data are invaluable and provide considerable explanatory power when utilized within the context of the theory. However, it is clearly desirable to have estimates of sib-mating in natural fig wasp populations that are independent of foundress distribution data and the attendant assumptions. Such estimates could be made by assaying the average levels of homozygosity by using electrophoretically variable loci (which are typically hymenopteran, and generally invariant) or perhaps regions of hypervariable DNA (Queller *et al.* 1993; S. H. Orzack and M. F. Antolin, personal communication). Given increasingly refined information we can begin addressing increasingly refined questions: Do the sex-ratio responses of different species show equivalent fit to theory? Do they appear to be equally well adapted? A number of very exciting possibilities beckon.

PARASITIC WASPS

Wing polymorphism in males

In contrast to the pollinators, the parasitic wasps are much more complex, much less studied and, as a result, much less well understood. For the vast majority of parasitic wasp species even the most basic aspects of natural history are not known (but see Compton 1993a; West and Herre 1994; West *et al.* 1996; Cook and Power 1996). For example, it is relatively difficult to determine the number of parasite females that contribute to broods in individual fig fruits; consequently, the types of sex-ratio question that can be addressed in the parasites are, at present, much less refined. However, unlike the pollinators in which all species possess only wingless males, both within and among species of parasitic wasps there is a great deal of variation in this character. In parasitic species occurring at very low population densities, winged males are produced; at high population densities, wingless males are produced; and dimorphic males consisting of both winged and wingless types are produced at intermediate population densities (Hamilton 1979) (Fig. 13-6). A tentative hypothesis is that at very low wasp population densities there is a high probability that a male may develop in a fruit with no females present, and that having wings makes it possible to search for unmated females elsewhere.

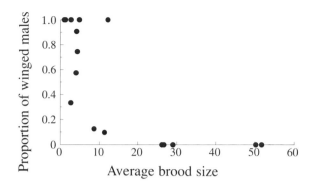

Fig. 13-6. The proportion of all parasitic wasp males that are winged plotted against average brood size. Notice that winged males are common only in those species in which the average brood size is low.

Conversely, at high population densities, as is usually found in the pollinators, a male wasp will almost always develop in a fruit that contains several females of its own species. In this case, a male should concentrate effort on the abundant mating opportunities within the fruit in which it is born. Inside a fig fruit, wings are likely to impair freedom of movement, and the resources required to grow them could more beneficially be employed otherwise (Hamilton 1979). At intermediate population densities, males have intermediate probabilities of finding conspecific females in their natal fruit. In these species, fluctuating selection pressures that result from temporal and/or spatial variability in population densities could maintain wing dimorphism (see Zeh and Zeh, this volume). At present, it is unclear to what extent these across- and within-species differences in male morphotypes are due to maternal facultative control of male morphology as opposed to fluctuating genetic polymorphisms. Certainly there is much room for future work here.

Fighting in males

There is a broad range in the level of aggression encountered between rival male fig wasps. Parasitic males often show high levels of aggression and lethal combat, whereas pollinator males usually show little aggression. Hamilton (1979) proposed that this difference was due to the different population structures of parasitic and pollinating wasps. The small number of foundresses characteristic of most species of the pollinating wasps implies that rival males are likely to be brothers. Hamilton argued that pollinator males are less likely to fight because of their generally high degree of relatedness (Hamilton 1964, 1972, 1979). In contrast, many different individual parasitic wasps can potentially oviposit in any given fig fruit, each laying only a few eggs before moving on. Consequently, rival parasite males within a fruit are unlikely to be brothers, opening a Pandora's box of adaptations leading ultimately to lethal combat.

However, Murray (1987, 1989) found no relationship between sex ratio (a predictor of relatedness of competing males (Hamilton 1979; Herre 1985, 1987; Frank 1985, 1987)) and levels of fighting in a study of 25 species of parasitic and pollinating fig wasp. Instead, he found that male density was the best predictor of fighting levels, and went on to suggest that male density constrains fighting by influencing the frequency with which rivals challenge for mating opportunities (Murray 1987, 1989). None the less, male density alone cannot explain all the variance in fighting levels between species. This is clearly emphasized by the particularly low fighting levels of the pollinators, despite what are frequently high male densities. This appears to apply as well to the parasitic species which enter the syconia to oviposit (S. G. Compton, personal observation).

In a further refinement, Vincent (1991) has shown that mating site (inside seeds, inside the fruit cavity, or outside the fruit) is related to the intensity of fighting, presumably by influencing the operational sex ratio (OSR). Parasitic species that mate inside the fruit cavity have a particularly male-biased OSR and show higher fighting levels than species that mate outside the fig or while females are still in their galls. Therefore, population structure (relatedness of interacting males), male density, and site of mating are all associated with fighting, although none of these factors accounts for it completely. Perhaps lethal combat has not evolved in any of the pollinating species because in most of them the males are further needed to cut exit holes so that the mated females can leave the syconium. However, males do not cut exit holes in all pollinator species (in some fig species the ostiole swells open, allowing the females to leave). So if there were to be cases of lethal combat in pollinators, we should expect it in these species. None the less, determining the relative importance of factors that are more variable within taxa (e.g. densities of particular parasite species through time) and those that are more fixed within taxa (e.g. mating site) across the numerous parasitic taxa provides many opportunities for future research.

CONCLUSIONS AND SUGGESTIONS FOR FUTURE RESEARCH

The sex-ratio responses of fig-pollinating wasps associated with different numbers of foundresses have provided some of the strongest and most widely accepted empirical support for specific predictions of LMC theory. Further, they are the only organisms in which it has been possible to test for the additional effects of inbreeding (via mother–offspring relatedness asymmetries). Moreover, the pattern to the deviations of fig-wasp sex ratios from predictions emphasizes the fundamental importance of the natural selective regimes for interpreting the responses of test organisms. Specifically, the pollinators appear to provide a behavioral analog to physiological studies that document greater precision of adaptation to frequently encountered situations (Herre 1987).

It is instructive to compare the opportunities that the fig wasps provide for testing theory with those offered by *Nasonia vitripennis*, perhaps the example most widely cited in support of LMC (Werren 1980, 1983; Charnov 1982; Orzack *et al.* 1991; King 1992). In contrast to the fig wasps, *Nasonia* provides excellent opportunites for experimental manipulation. In particular, the existence of eye-color mutants allows the experimenter precise knowlege of the relative contributions of different foundresses to shared broods. As in the fig-wasp studies, LMC theory's basic predictions have been broadly supported (Werren 1980, 1983; Orzack 1986). Analogously to the pattern found across fig-wasp species, different strains of *Nasonia* show varying degrees of fit to the details of theoretical expectations (Orzack & Parker 1986, 1990; Orzack *et al.* 1991). Moreover, in a particularly intriguing result, heritable variation in these responses has been demonstrated (Parker and Orzack 1985; Orzack 1990). However, the interpretation of the variation among strains in their sex-ratio response is clouded by the recognition of the fact that the natural selective regime of these strains is unknown. Is variation in response and lack of fit to the detailed predictions of theory the result of a strain's lack of experience in its natural setting to the situations to which it is exposed in the laboratory? Furthermore, the relevance of the demonstration of heritable variation to the assesment of the adaptive nature of the responses is unclear. It is well known that the estimation of heritability is dependent upon the environment in which it is measured (Falconer 1981; Lewontin 1974; see also Cook *et al.* 1994) and the heritabilities are rarely, if ever, measured in the environments in which they are naturally selected (see Antolin 1992; Molbo and Parker 1996).

Because major unresolved questions are the degree to which LMC operates in selecting for sex-ratio responses in natural settings and the degree to which the different organisms are able to respond to that selection, definitive tests of the theory must include both the characterization of natural selective regimes of test organisms and the implementation of experimental settings that reflect the natural conditions to which they are exposed (Herre 1995a). An ideal test would allow sex ratio to evolve under controlled experimental conditions. By imposing a selection regime on an organism that possesses flexibility of sex-determining mechanisms (Cook 1993), the evolutionary trajectories of sex-ratio responses could be followed and the power of the theory to predict optima unequivocally determined (Herre et al. 1987). These approaches are currently being taken with *Nasonia* and and other organisms (Molbo and Parker 1996; D. Molbo, E. D. Parker, S. H. Orzack, J. Bull and S. Skinner, personal communication). None the less, in the face of the accumulating support, it is either an ironic twist, or poetic justice, that the fig-wasp data conforming least well to Hamilton's predictions of facultative sex-ratio adjustment appear to be those collected by Hamilton himself.

ACKNOWLEGEMENTS

We thank Dave Parker, Steve Orzack, Jack Werren, Sam Skinner, Charles Godfray, Ian Hardy, Jacques Van Alphen, William Hamilton, Jim Bull, Truman Young, Jon Seger, Bill Stubblefield, Egbert Leigh, Joe Wright, Steve Frank, Roger Milkman, Koos Wiebes, Kees Berg, William Ramirez, Doyle McKey, Martine Hossaert, Marie-Charlotte Anstett, George Michaloud, Don Windsor, George Poinar, Carlos Machado, Biff Bermingham, John Nason, Katherine Cohen, Rory Nefdt, Tony Ware, Stephanie Vincent, Drude Molbo, and Elisabeth Kalko.

LITERATURE CITED

Antolin, M. F. 1992. Sex ratio variation in a parasitic wasp I. Reaction norms. *Evolution.* **46**: 1496–1510.

–. 1993. *Genetics of Biased Sex Ratios in Subdivided Populations: Models, Assumptions, and Evidence.* (*Oxford Surveys in Evolutionary Biology*). Oxford University Press.

Berg, C. C. 1989. Classification and distribution of *Ficus. Experientia* **45**: 605–611.

Berg, C. C. and J. T. Wiebes. 1992. *African Fig Trees and Fig Wasps.* Amsterdam: North-Holland.

Bronstein, J. L. 1991. The nonpollinating wasp fauna of *Ficus pertusa*: exploitation of a mutualism? *Oikos* **61**: 175–186.

Boucek, Z. 1993. The genera of chalcidoid wasps from *Ficus* fruit in the New World. *J. Nat. Hist.* **27**: 173–217.

Boucek, Z., A. Watsham and J. T. Wiebes. 1981. The fig wasp fauna of the receptacles of *Ficus thonningii. Tijdschr. Entomol.* **124**: 149–233.

Charnov, E. 1982. *The theory of sex allocation.* Princeton: Princeton University Press.

Cohen, K. 1989. Sex ratio adjustment in pollinator waps of three Australian fig species. B.Sc. thesis, University of New South Wales, Australia.

Collinson, M. E. 1989. The fossil history of the Moraceae. In *Evolution, Systematics, and Fossil History of the Hamamelidae.* P. R. Crane and S. Blackmore, eds., pp. 319–339. Oxford: Clarendon Press.

Compton, S. G. 1993a. One way to be a fig. *Afr. Entomol.* **1**(2): 151–158.

–. 1993b. A collapse of host specificity in some South African fig wasps. *S. Afr. J. Sci.* **86**: 39–40.

Compton, S. G. and B. A. Hawkins. 1992. Determinants of species richness in South African fig wasp assembleges. *Oecologia (Berl.)* **91**: 68–74.

Compton, S. G. and H. G. Robertson. 1988. Complex interactions between mutualisms: ants tending homopterans protect fig seeds and pollinators. *Ecology.* **69**: 1302–1305.

Compton, S. G., J. Y. Rasplus and A. B. Ware. 1994. African fig wasp parasitoid communities. In *Parasitoid Community Ecology.* B. A. Hawkins and W. Sheehan, eds, pp. 343–370. Oxford University Press.

Cook, J. M. 1993. Sex determination in the hymenoptera: a review of models and evidence. *Heredity.* **71**: 421–435.

Cook, J. M. and S. A. Power. 1996. Effects of within-tree flowering asynchrony on the dynamics of seed and wasp production in an Australian fig species. *J. Biogeogr.* **23**, in press.

Cook, J. M., A. P. Rivero-Lynch and H. C. J. Godfray. 1994. Sex ratio and foundress number in the parasitoid wasp, *Bracon hebetor. Anim. Behav.* **47**: 687–696.

Corner, E. J. H. 1940. *Wayside Trees of Malaya.* Singapore: Government Printing Office.

–. 1958. An introduction to the distribution of *Ficus. Reinwardtia* **4**: 325–355.

–. 1985. *Ficus* (Moraceae) and Hymenoptera (Chalcidoidea): Figs and their pollinators. *Biol. J. Linn. Soc.* **25**: 187–195.

Falconer, D. S. 1981. *Introduction to Quantitative Genetics.* 2nd edn. London: Longman.

Felsenstein, J. 1985. Phylogenies and the comparative method. *Am. Nat.* **125**: 1–15.

Fisher, R. A. 1930. *The genetic theory of natural selection.* Oxford: Clarendon Press.

Frank, S. A. 1984. The behavior and morphology of the fig wasps *Pegoscapus assetus* and *P. jiminezi*: descriptions and suggested behaviors for phylogenetic studies. *Psyche* **91**: 289–307.

–. 1985. Hierarchial selection theory and sex ratios. II. On applying the theory, and a test with fig wasps. *Evolution* **39**: 949–964.

–. 1987. Weapons and fighting in fig wasps. *Trends Ecol. Evol.* **2**: 259–260.

Galil, J. and D. Eisikowitch. 1968. Flowering cycles and fruit types of *Ficus sycomorus* in Israel. *New Phytol.* **67**: 745–758.

Galil, J. and G. Neeman. 1977. Pollen transfer and pollination in the common fig (*Ficus carica* L.). *New Phytol.* **79**: 163–171.

Godfray, H. J. C. 1988. Virginity in haplodiploid populations: a study on fig wasps. *Ecol. Entomol.* **13**: 283–291.

Hamilton, W. D. 1964. The genetical theory of social behaviour, I and II. *J. Theor. Biol.* **7**: 1–51.

–. 1967. Extraordinary sex ratios. *Science. (Wash., D.C.)* **156**: 477–488.

–. 1972. Altruism and related phenomena, mainly in social insects. *Annu. Rev. Ecol. Syst.* **3**: 193–232.

–. 1979. Wingless and fighting males in fig wasps and other insects. In *Sexual Selection and Reproductive Competition in Insects*. M. S. Blum and N. A. Blum., eds., pp. 167–220. New York: Academic Press.

Hardy, I. C. W. 1992. Nonbinomial sex allocation and brood sex ratio variances in the parasitoid hymenoptera. *Oikos.* **65**: 143–158.

Herre, E. A. 1985. Sex ratio adjustment in fig wasps. *Science. (Wash., D.C.)* **228**: 896–898.

–. 1987. Optimality, plasticity, and selective regime in fig wasp sex ratios. *Nature. (Lond.)* **329**: 627–629.

–. 1988. Sex ratio adjustment in thirteen species of Panamanian fig wasp. Ph. D. dissertation, University of Iowa.

–. 1989. Coevolution of reproductive characteristics in 12 species of New World figs and their pollinator wasps. *Experientia.* **45**: 637–647.

–. 1993. The evolution of virulence in nematode parasites of fig wasps. *Science. (Wash., D.C.)* **259**: 1442–1445.

–. 1995a. Comment on Orzack and Sober: Tests of Optimality models. *Trends Ecol. Evol.* **10**(3): 121.

–. 1995b. Factors affecting the evolution of virulence: nematode parasites of fig wasps as a case study. *Parasitology* **111** (suppl.): S179–S191.

–. 1996. Studies on a community of Panamanian figs. *J. Biogeogr.* **23**, in press.

Herre, E. A., E. G. Leigh, Jr. and E. A. Fischer. 1987. Sex allocation in animals. In *The Evolution of Sex*. S. C. Stearns., ed., pp. 219–244. Basel: Birkhauser.

Herre, E. A., C. A. Machado, E. Bermingham, S. S. McCafferty, J. D. Nason, D. Windsor, W. Van Houten and K. Bachmann. 1996. Molecular phylogenies of figs and their pollinating wasps. *J. Biogeogr.* **23**, in press.

Hossaert-McKey, M., M. Gibernau and J. E. Frey. 1994. Chemosensory attraction of fig wasps to substances producd by receptive figs. *Entomol. Exp. Appl.* **70**: 185–191.

Ibrahim, L. M. 1985. Contribution à une approche evolutive de la symbiose entre *Ficus carica* L. et *Blastophaga psenes*. M. Sc. thesis, Université des Sciences et Techniques du Languedoc, Montpellier, France.

Janzen, D. H. 1979a. How many babies do figs pay for babies? *Biotropica.* **11**: 48–50.

–. 1979b. How to be a fig. *Annu. Rev. Ecol. Syst.* **10**: 13–51.

Kalko, E. K. V., E. A. Herre and C. O. Handley, Jr. 1996. Relation of fig characteristics to fruit eating bats in the New and Old World tropics. *J. Biogeogr.* **23**, in press.

King, B. H. 1992. Sex ratio manipulation by parasitoid wasps. In *Evolution and Diversity of Sex Ratios in Insects and Mites*. D. L. Wrensch and M. Ebbert, eds., pp. 418–441. New York: Chapman and Hall.

Kjellberg, F. 1983. La strategie reproductive du figuier (*Ficus carica*) et de son pollinisateur (*Blastophaga psenes* L.), un exemple de coevolution. Ph. D. dissertation, Institute Nationale Agronomique, France.

Kjellberg, F., P.-H. Gouyon, M. Ibrahim, M. Raymond and G. Valdeyron. 1987. The stability of the symbiosis between dioecious figs and and their pollinators: a study of *Ficus carica* and *Blastophaga pnenes*. *Evolution.* **41**: 693–704.

Kjellberg, F. and S. Maurice. 1989. Seasonality in the reproductive phenology of *Ficus*: Its evolution and its consequences. *Experientia.* **45**: 653–660.

Lewontin, R. C. 1974. The analysis of variance and the analysis of causes. *Am. J. Hum. Genet.* **26**: 400–411.

Machado, C. A., E. A. Herre, S. S. McCafferty and E. Bermingham. 1996. Molecular phylogenies of fig pollinating and non-pollinating wasps and the implications for the origin and the evolution of the fig–fig wasp mutualism. *J. Biogeogr.* **23**, in press.

McKey, D. 1989. Population biology of figs: Applications for conservation. *Experientia.* **45**: 661–673.

Michaloud, G., S. Michaloud-Pelletier, J. T. Wiebes and C. C. Berg. 1985. The co-occurrence of two pollinating species of fig wasps and one species of fig. *Proc. K. Ned. Akad. Wet.*, **88**: 93–119.

Milton, K. 1991. Leaf change and fruit production in six Neotropical Moraceae species. *J. Ecol.* **79**: 1–26.

Milton, K., D. M. Windsor, D. W. Morrison, and M. A. Estribi. 1982. Fruiting phenologies of two neotropical *Ficus* species. *Ecology* **63**: 752–762.

Molbo, D. and E. D. Parker, Jr. 1996. Mating structure and sex ratio variation in a natural population of *Nasonia vitripennis*. *Proc. R. Soc. Lond.* B, in press.

Murray, M. G. 1987. The closed environment of the fig receptacle and its influence on male conflict in the Old World fig wasp, *Philotropesis pilosa*. *Anim. Behav.* **35**: 488–506.

–. 1989. Environmental constraints on fighting in flightless male fig wasps. *Anim. Behav.* **38**: 186–193.

Nason, J., E. A. Herre, and J. L. Hamrick. 1996. Paternity analysis of the breeding structure of strangler fig populations: evidence for substantial long-distance wasp dispersal. *J. Biogeogr.* **23**, in press.

Nefdt, R. J. C. 1989. Interactions between fig wasps and their host figs. M. Sc. thesis, Rhodes University, South Africa.

Okamoto, M. and M. Tashiro. 1981. Mechanism of pollen transfer and pollination in *Ficus ersta* by *Blastophaga nipponic*. *Bull. Osaka Mus. Nat. Hist.* **34**: 7–16.

Orzack, S. H. 1986. Sex ratio control in a parasitic wasp *Nasonia vitripennis*. Experimental analysis of an optimal sex ratio model. *Evolution* **40**: 341–356.

–. 1990. The comparative biology of second sex ratio evolution within a natural population of a parasitic wasp, *Nasonia vitripennis*. *Genetics* **124**: 385–396.

Orzack, S. H. and E. D. Parker. 1986. Sex ratio control in a parasitic wasp, *Nasonia vitripennis*: Genetic variation in facultative sex ratio adjustment. *Evolution* **40**: 331–340.

–. 1990. Genetic variation for sex ratio traits within a natural population of a parasitic wasp, *Nasonia vitripennis*. *Genetics* **124**: 373–384.

Orzack, S. H., E. D. Parker and J. Gladstone. 1991. The comparative biology of genetic variation for conditional sex ratio behaviour in a parasitic wasp, *Nasonia vitripennis*. *Genetics*, **127**: 583–599.

Parker, E. D. and S. H. Orzack. 1985. Genetic variation for sex ratio in *Nasonia vitripennis*. *Genetics* **110**: 93–105.

Patel, A., M. Hossaert-McKey and D. McKey. 1993. *Ficus*-pollinator research in India: Past, present, and future. *Curr. Sci.* **65**: 243–253.

Poinar, G. O. Jr. and E. A. Herre. 1991. Speciation and adaptive radiation in the fig wasp nematode, *Parasitodiplogaster* (Diplogasteridae: Rhabditida), in Panama. *Rev. Nematol.* **14**: 361–374.

Queller, D. C., J. E. Strassmann and C. R. Hughes. 1993. Microsatellites and kinship. *Trends Ecol. Evol.* **8**: 285–288.

Ramirez B. W. 1969. Fig wasps: mechanism of pollen transfer. *Science. (Wash., D.C.).* **163**: 580–581.

–. 1974. Coevolution of *Ficus* and Agaonidae. *Ann. Missouri Bot. Gard.* **61**: 770–780.

Stubblefield, J. W. and J. Seger. 1990. Local mate competition with variable fecundity: dependence of offspring sex ratios on information utilization and mode of male production. *Behav. Ecol.* **1**: 68–80.

Taylor, P. D. and M. G. Bulmer. 1980. Local mate competition and the sex ratio. *J. Theor. Biol.* **86**: 109–419.

Tayou, A. 1991. Etude du mutualisme Ficus-pollinisateur: sex ratio et transport du pollen. M. Sc. thesis, Institute National Agronomique Paris Grignon, Paris.

Ulenberg, S. A. 1985. The phylogeny of the genus *Apocrypta* Coquerel in relation to its hosts, *Ceratosolen* Mayr and *Ficus* L. In *The Systematics of the Fig Verh. K. Ned. Alcad. Wet.* C **83**: 149–176.

Vincent, S. J. 1991. Polymorphism and fighting in male fig wasps. Ph. D. dissertation, Rhodes University, South Africa.

Ware, A. B. and S. G. Compton. 1994. Dispersal of adult female fig wasps: arrivals and departures. *Entomol. Exp. Appl.* **73**: 221–229.

Ware, A. B., T. K. Perry, S. G. Compton and S. Van Noort 1993. Fig volatiles: their role in attracting pollinators and maintaining pollinator specificity. *Pl. Syst. Evol.* **186**: 147–156.

Werren, J. H. 1980. Sex ratio adaptations to local mate competition in a parasitic wasp. *Science. (Wash., D. C.).* **208**: 1157–1160.

–. 1983. Sex ratio evolution under local mate competition in a parasitic wasp. *Evolution.* **37**: 116–124.

West, S. A. and E. A. Herre. 1994. The ecology of the New World fig-parasitizing wasps *Idarnes* and the implications for the evolution of the fig-pollinator mutualism. *Proc. R. Soc. Lond.* B **258**: 67–72.

West, S. A., E. A. Herre, D. M. Windsor and P. R. S. Green. 1996. The ecology and evolution of the New World fig non-pollinating wasp communities. *J. Biogeogr.* **23**, in press.

Wiebes, J. T. 1979. Co-evolution of figs and their insect pollinators. *Annu. Rev. Ecol.Syst.* **10**: 1–12.

–. 1982. The phylogeny of the Agaonidae (Hymenoptera, Chalcidoidea). *Netherl. J. Zool.* **32**: 395–411.

–. 1994. The Indo-Australian agaonidae (pollinators of figs) *Proc. K. Ned. Akad. Wet.* **92**: 1–208.

–. 1995a. Agaonidae (Hymenoptera Chalcidoidea) and Ficus (Moraceae): fig wasps and their figs, xv (Meso-American *Pegoscapus*). *Proc. K. Ned. Akad. Wet.* **98**: 167–183.

–. 1995b. *The New World Agaonidae (Pollinators of Figs)*. Amsterdam: North-Holland.

Windsor, D. M., D. W. Morrison, M. A. Estribi and B. de Leon. 1989. Phenolgy of fruit and leaf production by 'strangler' figs on Barro Colorado Island, Panama. *Experientia* **45**: 647–653.

14 · Evolution of mate-signaling in moths: phylogenetic considerations and predictions from the asymmetric tracking hypothesis

P. LARRY PHELAN

ABSTRACT

Mate-finding in the large majority of moths is mediated by a long-distance response of males to minute quantities of pheromone emitted by females. Additionally, in many species, males may produce their own pheromone, which is employed after the sexes are brought together. More rarely, males produce acoustic signals and/or a long-distance pheromone. In the latter case, females assume the searching role. The female pheromone systems of the Heteroneura, a group that makes up 99% of extant Lepidoptera, may have had a single origin with relatively little change occurring subsequently, either in the types of chemicals used or in the glandular structures for pheromone production and release. In addition, female signaling appears to be a plesiomorphic trait for the Lepidoptera that likely was lost independently several times among the primitive groups. Recent research points to the female sternum V gland (located on the fifth abdominal sternum), found throughout primitive Lepidoptera, as the most likely evolutionary predecessor of the heteroneuran sex-pheromone system. This gland, which is shared with the sister group Trichoptera, also appears to play a defensive role in some species of the two orders. In contrast to the conservation of female pheromone systems, the incidence of male pheromones in moths is exceedingly polyphyletic and labile, suggesting intense and repeated selection for a transient function. Male acoustic signaling, although less prevalent, also appears to have had multiple origins.

I argue that the evolution of mate-signaling is determined in large measure by a sexual asymmetry in parental investment. I compare predictions from the asymmetric-tracking hypothesis with evidence from recent studies on genetic control and variation in moth pheromone systems. A wealth of studies on moth mating behaviors and pheromone chemistry, in conjunction with the extensive phylogenetic information for the Lepidoptera, provides behavioral ecologists with opportunities both to understand how
moth systems have evolved and to test hypotheses concerning the evolution of mate-signaling in general.

INTRODUCTION

Largely as a result of their significance as pests of the world's crops, we have amassed a large amount of information on the proximate mechanisms underlying moth mating in hopes of effecting pest control through population monitoring and mating disruption. During the past two decades, sex pheromones have been chemically and behaviorally characterized for hundreds of species (Arn *et al.* 1992). Building on this large database of species-specific mate-signaling systems, more recently we have seen a growth in studies characterizing the intraspecific variation in pheromone signals and response, and in the genetics controlling these systems. As a result, moth pheromones give us an exceptional opportunity to address some of the central and long-debated questions concerning the evolution of mate-signaling in animals.

Unfortunately, rather little attention has been paid to insect pheromones in evolutionary discussions of speciation and mate-signaling systems. An extreme example of this lack of cross-fertilization is demonstrated by the suggestion of Williams (1992) that female moth pheromones do not exist! His argument for the 'female pheromone fallacy' is based first on the condition (erroneous in my mind) that adaptive benefit for the sender needs to be demonstrated for a cue to be considered a signal. However, even if one accepts this condition, I doubt that there are many animal signals whose proximal mechanisms are better established than the female sex pheromones of moths. One of the reasons for Williams' questioning the existence of female pheromones is that they 'are produced in minute traces, in contrast to the abundant production of genuine pheromones, such as the occasional male sex stimulants' (Williams 1992, p. 112). However, as I shall argue below, this quantitative difference is better understood within a context of the sexual asymmetry by which individuals maximize their reproductive fitness.

The foremost objective of this chapter is to introduce a broader audience to the rich literature that exists for mate-signalling in moths. Specifically, I shall: (1) utilize our present understanding of lepidopteran phylogenetics to suggest an evolutionary history of the sex-pheromone system of moths; (2) describe some exceptional moth signaling systems that may be less familiar to the reader; and (3) use moth pheromones to illustrate the significance of sexual selection asymmetry in the evolution of animal mate-signalling.

PHYLOGENY OF FEMALE MOTH PHEROMONES

The order Lepidoptera is among the most speciose animal taxa, with more than 250 000 species named (Nielsen 1989). The monophyly of the Lepidoptera is well established and its sister-group relationship to the Trichoptera

is described as one of the best supported in all the insects (Nielsen 1989). The Lepidoptera are subdivided into three 'primitive' suborders, each of which contains a single family, with a fourth suborder (Glossata) containing 20 primitive families and all of the more advanced Ditrysia (Fig. 14-1). The Ditrysia make up 98% of the Lepidoptera (including all the butterflies), yet unlike the primitive groups it is remarkably homogeneous in morphology as well as in mating strategies. Almost universally, mate-finding in ditrysian moths is characterized by a male competitive scramble for females, who 'call' (emit pheromone) from a stationary posture. In a most extraordinarily sensitive and selective system of communication, females release picogram to nanogram levels of pheromone, yet males can locate these sources from distances of hundreds of meters, distinguishing their own females from those of related species (Baker 1985). The intensity of selection on male response is evidenced by the strong sexual dimorphism in antennae, in both the number and the tuning of chemoreceptors present. Male antennae are packed with thousands of chemoreceptors, which are narrowly tuned

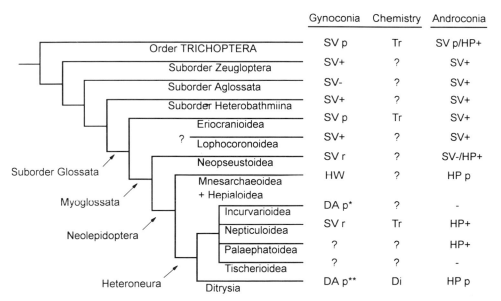

	Gynoconia	Chemistry	Androconia
Order TRICHOPTERA	SV p	Tr	SV p/HP+
Suborder Zeugloptera	SV+	?	SV+
Suborder Aglossata	SV-	?	SV+
Suborder Heterobathmiina	SV+	?	SV+
Eriocranioidea	SV p	Tr	SV+
Lophocoronoidea	SV+	?	SV+
Neopseustoidea	SV r	?	SV-/HP+
Mnesarchaeoidea + Hepialoidea	HW	?	HP p
Incurvarioidea	DA p*	?	-
Nepticuloidea	SV r	Tr	HP+
Palaephatoidea	?	?	HP+
Tischerioidea	?	?	-
Ditrysia	DA p**	Di	HP p

Fig. 14-1. Phylogeny of major lineages of the Lepidoptera and their relationship to the Trichoptera, with information on the putative scent structures of females (gynoconia) and males (androconia) and the chemistry of female pheromones. SV, Sternum V gland; HW, hindwing gland; DA, gland on distal abdominal segments; HP, hairpencils; p, pheromonal function demonstrated; +, structure present in at least some species but pheromonal function not established; r, reduced and possibly vestigial; −, not reported in any species investigated; *, suggested by observatons of mating behavior; **, except Psychidae, where pheromone is released from thorax. Pheromone chemistry is Trichopteran-like (Tr) with 6–9 carbon chains or Ditrysian-like (Di) with 12–18 carbon chains. Cladogram after Scoble (1992); information on scent structures and chemistry compiled from various sources discussed in text. See also Löfstedt and Koslov (1996) for a related phylogenetic perspective on female lepidopteran pheromones.

to components of the female pheromone blend (Priesner 1986). Once the female is located, copulation ensues quickly, sometimes preceded by a courtship sequence that may entail male-produced pheromones (see below). Of the 28 or so ditrysian superfamilies, male signaling with female search is known only in a few species of Pyralidae (e.g. *Achroia grisella*, Greenfield and Coffelt 1983; *Galleria mellonella*, Spangler 1985; and *Eldana saccharina*, Zagatti 1981; Bennett *et al.* 1991) and Arctiidae (e.g. *Estigmene acrea*, Willis and Birch 1982; *Creatonotos transiens* and *C. gangis*, Wunderer *et al.* 1986).

The evolution of female-pheromonal chemistry in the Heteroneura also has been very conservative. Most of the hundreds of species for which pheromonal components have been characterized utilize acetates, alcohols, aldehydes, or hydrocarbons with mostly even-numbered chain-lengths of 12–18 carbons (Arn *et al.* 1992). These compounds, synthesized primarily from common fatty-acyl precursors (Bjostad *et al.* 1987), are found both in the most advanced families as well as the more basal groups of the Ditrysia, such as Tineidae, Yponomeutidae and Gelechiidae (Nielsen 1989). There is no apparent phylogenetic pattern in pheromone chemistry at the higher taxonomic levels of the Ditrysia, although patterns within certain family groups have been suggested (Roelofs and Brown 1982). Despite this restricted chemical alphabet, species-specificity is realized through differences in component ratios, and in the number, position, and geometric configuration of double bonds. Similarly, uniformity is seen in female pheromone-gland morphology, usually located in the intersegmental tissue of abdominal segments VIII and IX, and only rarely outside segments VII–X (Percy-Cunningham and McDonald 1987). One notable exception is found in the Psychidae (bagworms), whose adult females remain in their pupal cases, enclosed by their protective larval 'bags'. Pheromone is produced by ventral thoracic glands, release of which is facilitated by neighboring deciduous hair-scales that are dislodged by undulating movements of the female, collect pheromone, and then accumulate in the bottom of the cocoon (Loeb *et al.* 1989).

The homogeneity of female-pheromone-based mating systems throughout the Heteroneura is strongly suggestive of a monophyletic system that has undergone relatively little change during the evolutionary history of the over 100 moth families included in this group (Nielsen 1989). In contrast, mating behaviors among the more primitive moths form a more eclectic set. An excellent phylogenetic perspective on the mating systems of primitive

Lepidoptera is provided by Wagner and Rosovsky (1991), with particular emphasis on the Hepialidae. Among the Hepialidae, mate-finding is mediated by male calling in a number of genera, but by female calling in others. Even in those species with female-produced sex pheromones, there may be significant differences relative to ditrysian mating systems. For example, mating behavior in the higher Lepidoptera typically occurs within a species-specific 2–3 h diel period, but such activity in the hepialid *Korscheltellus gracilis* is restricted to two 20–40 min periods during dawn and dusk (Wagner and Rosovsky 1991). Also unique is the site of female pheromone release, localized to the hind wings, a condition that is not known in any other family (Kuenen *et al.* 1994). In other hepialid genera, males assume the calling role by releasing pheromone from hind legs that no longer function for locomotion, but rather are highly modified for scent dissemination ('rosette organs') (Mallet 1984). In all cases, males call from aggregations or 'leks' (see Shelly and Whittier, this volume) and may do so either in stationary aggregations on vegetation, such as in '*Hepialus*' and *Sthenopis* (Wagner 1985; McCabe and Wagner 1989) or form visually dramatic aerial displays, as in *Hepialus humuli* and *Phymatopus hectus* (Mallet 1984; Wagner 1985; Turner 1988). In the latter case, swarms of males maintain station in a hovering flight while casting back and forth and releasing pheromone.

The presence of male calling in a basal lineage of Lepidoptera has led to speculation that this mating system represents the primitive condition of moths (Mallet 1984); however, Wagner and Rosovsky (1991) assert that male calling is limited to the more advanced genera. Mate finding in all known lower hepialids is mediated by female calling, and male scent structures are not found in any of these genera. Moreover, field observations by these authors of numerous groups basal to the Hepialidae suggest that mate-finding with a female-produced pheromone may be common.

Comparisons with the mating systems of the Trichoptera (caddisflies) provide further insight into the origin of moth sex pheromones, where both male- and female-based sex-pheromone systems have been demonstrated (Wood and Resh 1984; Solem 1985; Löfstedt *et al.* 1994). Identified pheromones comprise blends of 7–9 carbon ketones and alcohols produced by paired glands in abdominal sterna IV and V (Löfstedt *et al.* 1994). Such structures generally are found in both males and females throughout the Trichoptera, including the most primitive taxa (Nielsen 1980). In some species, these glands are thought to play a defensive rather than a sexual role (Duffield *et al.* 1977; Ansteeg and Dettner 1991). Significantly, homologous structures with glandular openings on sternum V also are found in many primitive moths, appearing in all the basal suborders (Fig. 14-1). In fact, the presence of the sternum V glands is considered part of the Lepidopteran ground plan and a synapomorphy of the Amphiesmenoptera (Trichoptera + Lepidoptera) (Kristensen 1984b). Sternum V glands are common in the Eriocraniidae, where the structures show sexual dimorphism, with membranous fenestrae, associated with internal globular reservoirs, appearing on sternum IV in females but only rarely in males (Davis 1978). In the most primitive moth investigated for sex pheromones to date, *Eriocrania cicatricella*, females attract males by releasing pheromone from such glands, which are reportedly absent in the males (Zhu *et al.* 1995). Similar to the Trichoptera, the pheromone of *E. cicatricella* is composed of a blend of seven-carbon alcohols and ketones (Zhu *et al.* 1995). Sternum V glands also occur in the Lophocoronidae, but in the Neopseutidae and Nepticulidae, sternum IV fenestrae occur without glandular openings on sternum V, suggesting a vestigial condition in these families (Davis 1975). A possible connection between the primitive pheromone system and that of the Ditrysia is suggested by recent studies in Nepticulidae, where females produce pheromone blends of short-chain compounds similar to that of *Eriocrania* and the Trichoptera (M. Tóth and C. Löfstedt, personal communication). Although the site of pheromone production has not been established, females of some Incurvarioidea, a possible sister group (Nielsen 1989), are reported to assume the sexual calling posture of the higher Lepidoptera, in which the ovipositor is exposed, suggesting distal abdominal glands (Wagner and Rosovsky 1991).

Based on our present knowledge of the mating behavior and phylogeny of the primitive Lepidoptera, an evolutionary path for moth sex pheromones may be hypothesized (Fig. 14-1), subject to revision as the considerable gaps in our understanding are filled. (Note: Löfstedt and Kozlov (1996) have proposed a similar phylogeny for female moth pheromones. The sternum V pheromone system of the Trichoptera and primitive Lepidoptera apparently evolved from or shared a role in chemical defense in the two orders. This glandular system was lost independently many times in one or both sexes among the primitive Lepidoptera, and likely disappeared entirely in the Myoglossata, as no complete structures have yet been found in this group. In fact, pheromone-mediated mate-finding may

have been replaced by a greater reliance on visual cues in some of these species, as suggested by diurnally active adults, sexual dimorphism in coloration, and the presence of large sexually dimorphic eyes in males (see also Rutowski, this volume). At least two changes in the site of female pheromone production occurred within the Myoglossata clade, along with shifts to male-pheromone-based mate-finding in some species. One new female-pheromone system is indicated by the hind-wing pheromone of *K. gracilis* in the Hepialoidea (putative scent scales also are found on the hindwings of female *Fraus* and other primitive hepialids (Nielsen and Kristensen 1989), and wing-fanning is commonly observed in calling females of basal hepialoids (Wagner and Rosovsky 1991)). The second innovation, the 'modern' female distal abdominal pheromone gland, probably occurred ancestral to the Ditrysia, based on its almost universal presence in this group and observations of the calling posture of some female monotrysian Heteroneura (Wagner and Rosovsky 1991). Likewise, the sex-pheromone chemistry of the Ditrysia also likely arose during the evolution of the monotrysian Heteroneura, surmised from the presence of short-chain components in the Nepticulidae, but longer-chain compounds in the pheromones of almost all Ditrysia. However, the origin of the ditrysian pheromone system remains obscure, since the phylogeny of these basal Heteroneura unfortunately is unresolved, and even with the paucity of mating studies in this group, there is evidence for a variety of mating systems, including female-pheromone-mediated mate-finding, visual mate-finding, and female attraction to male swarms in some Incurvarioidea (Luquet 1980). Thus, resolution of this key step in the evolution of mating behavior in the Lepidoptera awaits further study of the phylogeny and mate-signaling of the primitive Heteroneura.

COURTSHIP AND MALE PHEROMONES

In addition to female sex pheromones, male pheromones play a significant role in the mating behavior of many moths. These male odors and the morphological structures for their dissemination have been treated in depth previously (McColl 1969; Fitzpatrick and McNeil 1988; Birch *et al.* 1990; Phelan 1992). In contrast with female pheromones, male-pheromone systems are clearly polyphyletic in origin. Two general characteristics important to understanding their evolution are their diversity and their lability. Diversity is reflected first in the location of the scent organs (variously termed androconia, scent brushes, hair-pencils, or coremata), which can be found on virtually any part of the male body, including wings, thorax, abdomen, genitalia, legs, and even head and antennae. Similar structures have evolved independently in males of some Trichoptera as well, primarily on the head and wings (Neboiss 1991). A scheme for classifying the great diversity of androconial structures has been proposed by McColl (1969). The organs commonly consist of a hypodermal glandular cell or aggregations of such cells, each of which bears a specialized scale. The glandular scales are often hollow or possess a lattice-like internal structure that increases the effective surface area for release, and/or increases capillary action of the pheromone secretion from the gland. Alternatively, the odor-disseminating scales may be separate from the gland, but when at rest, the scales cover the glandular area and thus become loaded with pheromone. In either type of structure, the disseminating organ may be held within specialized folds or pockets, which reduce scent emission during times of sexual inactivity. Diversity is also revealed in the taxonomic distribution of these structures. Even when an apparently homologous structure occurs throughout a group, e.g. abdominal hairpencils in the Phycitinae (Pyralidae) or forewing costal hairpencils in Tortricidae, many species scattered throughout these taxa lack the structures, suggesting that they have been lost independently many times. Moreover, non-homologous androconia have appeared in other members of these taxa. For example, males of many Phycitinae possess forewing glands, in addition to or in the absence of abdominal hairpencils (Heinrich 1956).

Androconia can be found in most of the major groups of Neolepidoptera. Although hairpencils have been found on the wings of some Neopseustidae (Nielsen and Common 1991), the most primitive moths for which a pheromonal function of androconia is clearly established fall within the Hepialoidea. In addition to the rosette organs of some Hepialidae, male hairpencils are enclosed in a hindwing pocket of *Anomoses hylecoetes* (Anomosetidae) (Kristensen 1978), and hind-tibia androconia are found in some *Genustes* and *Ogygioses*. *Ogygioses caliginosa* males, like some hepialids, form mating swarms (Kuroko 1990); however, their androconia may not share a common origin with the rosette organs (Wagner and Rosovsky 1991). One of the more extreme examples of diverse origin and lability of androconia is observed in the monotrysian family Palaephatidae, where among the 28 South American species,

nine different types of androconia are concentrated in 14 species, found on the wings, legs, and/or abdomen (Davis 1986).

The function of those male scent organs that are engaged only after the sexes meet is still open to debate (Boppré 1984; Birch *et al.* 1990; Phelan 1992). However, the polyphyletic origin and loss of these structures within families across the Heteroneura is strongly suggestive of a system that has been under intense selection to satisfy a rather transient function. I have argued elsewhere that male courtship pheromones in the Lepidoptera have evolved through female preference for assortative mating to avoid hybrid matings (Phelan and Baker 1987; Phelan 1992). This hypothesis was supported by comparative studies where, in each of five ditrysian families (totalling over 800 species), androconia were more common among species that shared a host plant with congeners.

Greenfield (1981) has suggested that the low release rates of female moth pheromones represents a form of sexual selection, by which females select for males that are strong searchers. This possibility notwithstanding, there is greater potential for female choice operating during close-range courtship. For example, in *Grapholita molesta* (Tortricidae), males present terminal abdominal hairpencils in a pulsed display that provides chemical and possibly visual stimulation (Baker and Cardé 1979), which must elicit female contact with the male before the latter can attempt copulation. Similarly, male *Plodia interpunctella* (Pyralidae) approach females from behind and wait for them to turn around before attempting copulation (Grant *et al.* 1975). In other groups, females control the outcome of the mating attempt by simply denying the male access to their genitalia (for example, *Cadra figulilella* and *Ephestia elutella* (Pyralidae), Phelan and Baker 1990b; *Heliothis subflexa* (Noctuidae), Cibrian-Tovar and Mitchell 1991). To my knowledge, all cases where females play an active role in the outcome of mating involve male androconia and/or acoustic signals. *Grapholita molesta* males given access as adults to ethyl *trans*-cinnamate had higher titers of this key pheromone component in their hairpencils and attained higher mating success, owing to an increase in short-range attraction of females (Löfstedt *et al.* 1989). In a comparative study of courtship behavior in twelve phycitine (Pyralidae) moths, Phelan and Baker (1990a) found an interactive form of courtship in all species with male abdominal or wing hairpencils, whereas females played a more passive role in species without these male structures. The key behavior of the interactive courtship entails the male striking the female on the head with his abdomen, termed the head thump (Fig. 14-2). This action brings the abdominal hairpencils into contact with the female antennae, a behavior absent in the courtship of *P. interpunctella*, which has hairpencils on the wings but not on the abdomen. A quantitative analysis of phycitine courtship using information theory (Phelan and Baker 1990b) confirmed the head thump as the male behavior with the greatest communication value. Moreover, although *C. figulilella* females appeared to take a passive role in intraspecific courtship, when they were courted by *E. elutella* males they immediately rejected males after hairpencil presentation. Phycitine courtship differs from that of other insect groups where female acceptance is dependent on the male's ability to sustain courtship, as in *Drosophila* (Bastock and Manning 1955). In *E. elutella*, communication (male ability to influence female acceptance) declined rapidly with increasing courtship duration (Fig. 14-3). Thus, the phycitine courtship sequence appears to function more for

enhancing transmission of the male chemical signal than for female assessment of male courtship ability.

Female choice based on male pheromones also has been demonstrated outside the context of reproductive isolation. *E. elutella* females were more likely to mate with large males, which were found to produce larger amounts of wing-gland pheromone (Phelan and Baker 1986). This preference was adaptive in that large-male matings produced greater numbers of offspring surviving to the pupal stage, as well as greater pupal mass. In several species of arctiid and ctenuchid moths and ithomiine and danaine butterfly, males release pheromones derived from toxic pyrrolizidine alkaloids (PAs), most commonly hydroxydanaidal and/or danaidal (Boppré 1984). These compounds can occur in large quantities in male glands and it has been suggested that this signal has evolved through sexual selection as an indicator of PA quantity (Dussourd *et al.* 1991). Males of the arctiid *Utetheisa ornatrix* pass PAs with the spermatophore; females use the alkaloids for chemical protection of their eggs (Dussourd *et al.* 1988). Amounts of PA-derived pheromone (as high as 22 μg) correlated strongly with PA content in males (as high as 900 μg) (Dussourd *et al.* 1991); females showed a mating preference based on pheromone

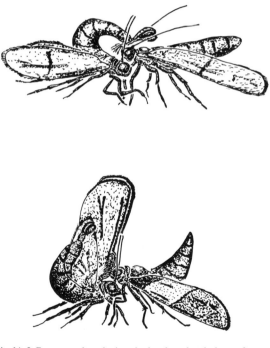

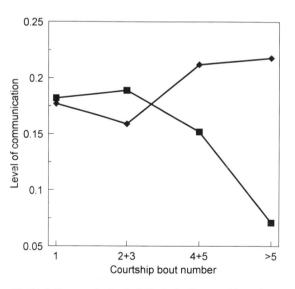

Fig. 14-2. Postures taken during the head-to-head phase of phycitine courtship just before the copulatory thrust in two species (males on left). The male *Cadra cautella* (top) displays the head thump, which positions abdominal hairpencils directly in front of the female's head. The male *Plodia interpunctella* (bottom), who possesses only vestigial abdominal hairpencils, raises the abdomen but does not use the head thump. Drawings by R. Vetter.

Fig. 14-3. Communication in *Ephestia elutella* courtships using the Informational Theoretical measure of normalized cross-covariability between male behavior and female response (squares) and female behavior and male response (diamonds). Courtship bouts are delimited by copulatory attempts. Note that as courtship is extended, female responsiveness to the male declines, whereas that of males remains high. After Phelan and Baker (1990b).

content (Conner *et al.* 1981). PA's obtained from the larval host represent a limited resource that could limit the number of matings possible for a male. Moreover, obtaining PAs by adult foraging, termed pharmacophagy (Boppré 1984), requires significant expenditures of time and energy and places the male at increased risk of predation. Thus, passing PAs to females during mating may represent significant paternal investment. It is noteworthy that *U. ornatrix* females have an exceptionally high probability of multiple mating compared with other moths. In field studies, wild-caught females of most lepidopteran species averaged 1–2 spermatophores; however, 77% of *U. ornatrix* females were multiply mated, averaging 3.8 spermatophores per female and ranging as high as 11 (Drummond 1984). Because a single mating can potentially provide enough sperm to fertilize all the female's eggs, sexual selection theory would predict that multiple matings would be uncommon if males provided only sperm.

ACOUSTIC SIGNALING AND SEXUAL ROLE REVERSALS

The important selective forces on moth sexual signaling have been reviewed previously (see, for example, Greenfield 1981; Cardé and Baker 1984; Löfstedt 1990, 1993; Phelan 1992). Particularly significant in explaining the variety of mate-signaling systems in animals has been recognition of the conflict between the optimal mating strategies of males and females. Although moth mate-finding is accomplished by a long-distance response of males to extremely low concentrations of female pheromone in the large majority of species, in exceptional cases females respond to male signals. As in other animal groups, these sexual role reversals appear to correlate with certain ecological parameters (Emlen and Oring 1977). One such case, the lesser wax moth, *Achroia grisella* (Pyralidae: Galleriinae), deviates in almost every way from the conventional moth mating system (Greenfield and Coffelt 1983). Males congregate on or near beehives and release a blend of undecanal and Z11-octadecenal from forewing glands at rates $10–10^3$ times that of female pheromones in most other species (Greenfield and Coffelt 1983). Moreover, chemical signaling is accompanied by acoustic signals produced by tymbals when males wing-fan. Initiation of chemical and/or acoustic signaling by one male stimulates signaling in other males, resulting in aggressive male–male interactions (Greenfield and Coffelt 1983). Role reversal in this pyralid moth is probably related to the patchy nature of larval

resources (honey bee brood combs). Males are more likely to find females at these oviposition sites, and mate-searching by females entails no cost beyond that for oviposition. This argument is strengthened by comparison with another galleriine, *Galleria mellonella*, whose larvae also develop in beehives. Again assuming the role of signaler, males attract females with a wing-gland pheromone and acoustic signals (Spangler 1985). As with *A. grisella*, pheromone is released at high rates, 25 μg of undecanal during the first hour of scotophase (Romel *et al.* 1992). Less amenable to an adaptive explanation is a third galleriine species, *Eldana saccharina*, where males also call with pheromone and sound, but where the ovipositional resource, sugarcane, is not particularly clumped (Zagatti 1981; Bennett *et al.* 1991). One possibility is that this species is derived from others with male signaling, and that the mating system represents phylogenetic inertia. Models by Hammerstein and Parker (1987) suggest that, although the sex with lower parental investment will most likely assume the more expensive role in mate-finding, it may be females that search if they are already the more active sex.

Although sound is presently believed to play only a minor role in sexual communication of the Lepidoptera, morphological studies indicate that sound-producing organs are widespread, suggesting that their role has been underestimated. Male-specific abdominal tymbal organs are widely but irregularly distributed across the 100 genera of Lymantriidae (Dall'Asta 1988), suggesting that this organ arose early in the group and has been lost independently many times. Particularly noteworthy is the absence of such organs in males of species with wingless or brachypterous females. The production of many moth acoustic signals in the ultrasonic range may have contributed to this oversight. Abdominal 'ears' are common in moths, playing a defensive function in avoidance of bat predators (Fenton and Fullard 1981). Chordotonal organs located on the second abdominal segment are considered part of the lepidopteran ground plan, and ventral organs found in the most primitive moth, *Micropterix*, may be the progenitors of the tympana of some higher ditrysians (Kristensen 1984a). Thus, the widespread existence of hearing, even if used in defense, would preadapt moths for acoustic communication in a sexual context. Sound production is widespread in the Arctiidae and Ctenuchidae, where ultrasonic clicking is generated by a tymbal organ located on the metepisternum (Fenton and Fullard 1981). It appears to function in this chemically protected group of moths as an aposematic signal against bats (Dunning *et al.* 1992). However, males of many species also produce sounds in the context of mating; for example, *Cycnia tenera* males induce acceptance by a combination of ultrasonic clicks and courtship phero-mones after locating the female via her pheromone (Conner 1987). In several species of the sphingid genus *Psi-logramma*, males stridulate by rubbing together specialized scales or processes on their valvae (genitalia) and eighth abdominal tergites (van Doesburg 1966; Nässig and Lütt-gen 1988).

The danger of attracting predators by acoustic commu-nication is offset in most moths by the use of sound only in courtship after primary attraction has occurred via phero-mones. One exception is the arctiid *Syntomeida epilais*, which produces sexually dimorphic ultrasonic signals by using metepisternal tymbal organs (Sanderford and Conner 1990). Although female pheromones may mediate primary mate-finding, males and females of this wasp mimic engage in extended antiphonal calling, lasting from 30 s to 12 min, while still several meters apart. Similarly, males of the pyralid *Syntonarcha iriastis* use genital stridula-tion to produce ultrasonic signals that are audible from 20 m (Gwynne and Edwards 1986). In this species, male acoustics have apparently supplanted pheromones for long-range mate-attraction.

Just as male-pheromone systems show far greater diver-sity and phylogenetic lability than those of females, so too do the sound-producing organs of male moths. In addition to those systems above, a variety of other acoustic organs have arisen on various male body segments. For example, males of several noctuids possess swollen blisters or 'fovea' on their wings, which in some species act as resonators for wing–leg mechanisms of stridulation whereas other noc-tuids strike such wing fovea together rapidly like castanets (reviewed by Scoble 1992).

Another exception to the conventional mating system of moths is the arctiid *Pyrrharctica isabella*. Females assume the calling role, attracting males with a blend of 2-methyl-heptadecane, 2-methylhexadecane and 2-methyloctade-cane (*ca.* 100 : 2 : 1) (Meyer 1984); however, they release pheromone in quantities far in excess of that of other spe-cies. At release rates averaging 240 ng min^{-1} of the major component, the pheromone is emitted as an aerosol, pro-ducing a stream visible to the eye (Krasnoff and Roelofs 1988). Aerosols have also have been seen in two other arc-tiids, *Phragmotobia fuliginosa* (Krasnoff 1987) and *Holomelina lamae* (Schal et al. 1987). In addition to large quantities of pheromone, *Py. isabella* females produce acoustic signals as males approach (Krasnoff and Yager 1988), which can

also be elicited by extracts of the male pheromone gland or by synthetic pheromone (Krasnoff and Yager 1988). How widespread female clicking is in the Arctiidae awaits a more systematic survey; however, female courtship click-ing has also been observed in the related *Ph. fuliginosa* (Krasnoff 1987).

ASYMMETRIC TRACKING AND THE EVOLUTION OF SEXUAL COMMUNICATION

A key question in the evolution of sexual communication is how signal and response are coordinated for species-speci-ficity, but shift during speciation. Paterson (1985, 1993) and others (see, for example, Lambert et al. 1987, Alexander et al., this volume) have viewed mate-signaling as a homeo-static system under strong mutually stabilizing selection. Owing to the biological significance of achieving fertiliza-tion, the system for specific-mate recognition (SMRS) is argued to be under strong selection for coordination. Thus, although some variation is allowed, 'coadapted char-acters such as those of the SMRS are not free to vary greatly' (Paterson 1993, p. 211). As a consequence, 'a popula-tion of a sexual species has greater stability over time than any group of uniparental organisms that lack the con-straints from coadaptation of signals and receiver' (Pater-son 1993). When a change in the SMRS does occur, for example as a response to changes in the environment, this change can only occur in small steps as a gradual coevolu-tionary process between signal and response. Alternatively, West-Eberhard (1983) emphasized sexual selection events in isolated populations as a mechanism for rapid shifts in a stabilized signalling system that may result in specia-tion, whereas Kaneshiro (1980) suggested that prezygotic isolation can result from the *loss* of courtship elements during a founder event. All of these hypotheses emphasize allopatric change in the SMRS. Constraints on change in a coadapted signal and response could be reduced if the two were genetically linked (Hoy et al. 1977), and particularly if signal and preference were coupled by a shared physiologi-cal mechanism (Doherty and Hoy 1985). In *Colias* butter-flies, male sexual signals and female mating preference both may be largely controlled by a 'coadapted gene com-plex' located on the X chromosome (Grula and Taylor 1979, 1980). However, in moths, pheromone production and response have consistently been found to be controlled independently, usually involving both autosomes and sex chromosomes (reviewed by Phelan 1992 and Löfstedt 1993).

Mate finding has undoubtedly evolved through the interplay of a variety of adaptive and non-adaptive forces. In addition to the commonly evoked mechanisms of natural and sexual selection, the evolution of chemical signals and response may be canalized by phylogenetic history, biosynthetic pathways, or physiochemical properties of odor emission and perception, as well as 'noise' caused by other odors in the environment (Cardé and Baker 1984; Baker 1985; Phelan 1992). The challenge is to determine the relative significance of these selective forces by deciphering the patterns in mating systems of any animal group. I will argue here that mate-signalling is best explained within a framework of sexual selection, which also predicts when other selection and non-selection processes will play a significant role.

A core component of sexual-selection theory is that variability in mating success will be driven by differential parental investment. When one sex invests significantly more than the other, the reproductive success of the former will be limited by resources that determine offspring fitness. This sex, usually the female, thus benefits by maximizing the *quality* of her matings, measured either by the genetic quality of the male or by the resources that he offers at mating. In contrast, the optimal male strategy is one that maximizes the *quantity* of matings. We can extend this argument by recognizing that such a fundamental sexual conflict should be a major determining force in the evolution of mate-signaling, as well.

Although sexual selection has previously been invoked as an important force on moth pheromones (Greenfield 1981; Cardé and Baker 1984; De Jong 1988; Phelan 1992; Löfstedt 1993), many workers argue that pheromone systems also will be under strong stabilizing selection. Thus, males will be selected that respond maximally to the pheromone blend produced by the greatest number of females, and females will be selected that produce the blend to which males are most sensitive. As I have argued elsewhere (Phelan 1992), the assumption of strong mutual stabilizing selection between signal and response is inconsistent with the predictions of sexual selection. Rather than coordination arising through a coevolutionary process of 'give and take' between signal and response, logic would predict differential selection that is inversely related to the level of sexual asymmetry in parental effort. This subtle but important distinction is formalized as the asymmetric tracking hypothesis (Phelan 1992). Although not meant to be limited to moth pheromones, the key predictions of the hypothesis are exemplified here for a pheromone system

in which the male provides only sperm with little or no parental investment.

1. Females will usually assume the least costly role in mate signalling/finding.
2. Generally, females will be under only weak stabilizing selection to produce that pheromone blend to which male response is maximized. Thus, in the absence of other selective forces, significant pheromonal variation will arise among females in a population. However, when their pheromone overlaps chemically with that of a sympatric species or different ecological race, stabilizing and/or directional selection on female signals will occur.
3. Males will show maximum sensitivity to the most common pheromone blend in the mating population; however, male response should not be so narrow as to exclude other blends that signal viable mates, even if these females belong to a different race. (Breadth of response will be limited by any trade-off that may exist with overall sensitivity.) Thus, population variance in response should be due primarily to within-male breadth of response, and variation among males should be low (De Jong 1988). Males will only experience selection against response to females of other populations when the fitness of those matings is low relative to the costs of finding a female of their own population.
4. In those species where males produce pheromone, this pheromone will be under selection to track female response.

This hypothesis thus provides specific predictions that are testable through comparative studies and genetic analyses of mate-signaling. The key distinction between this hypothesis and previous models of the evolution of signal–response resides in the mechanisms by which coordination arises, which the asymmetric tracking hypothesis views as largely the result of only one sex responding to changes in the other sex. Selection on the female is weakened by a male-biased operational sex ratio (OSR), where the number of times a male *can* mate is high relative to the average number of female matings; this increases the female's chance of mating and reduces the male's. In the absence of strong stabilizing selection, the female pheromone would be free to vary and might diverge in geographically isolated populations owing to random events. Male response would then track these pheromonal changes. Similarly, a mutation that causes a significant deviation in the female pheromone blend might nevertheless be maintained in a population by broad male response. If the change is relatively minor, male response may simply broaden further; however, if the new blend is sufficiently distinct, it could create disruptive selection on males, leading to a response polymorphism in males. By contrast, variability in female pheromone could be reduced in the presence of other populations with similar pheromones, if cross-matings produced no offspring or hybrids of reduced fitness.

Arguments of mutually stabilizing selection on signal and response are not limited to pheromone systems. For example, Alexander *et al.* (this volume) discuss the role of conflict of interest between sexes in the evolution of mating systems, but with regard to mate-signaling conclude that there is 'no reason to expect conflict of interest between males and females'. Thus, using the example of an acoustic system, '…males gain by possessing signals that are like those of all other males in the population because females gain by having hearing organs tuned to the frequency that is most prevalent of male songs…'. However, there is also no reason to believe that lack of sexual conflict of interest necessarily leads to mutually stabilizing selection between signaller and responder. More likely, this coordination results from selection on male signals to track changes in female auditory sensitivity, much as has been suggested for some frogs by Ryan and Rand (1990). Notably, Alexander *et al.* (this volume) explain the appearance of coordinated evolution between male and female genitalia by using this same logic of unilateral tracking, but seem to overlook it in the context of mate-signaling.

The sexual asymmetry of reproductive strategies has been explored with game theory by Hammerstein and Parker (1987), with two conclusions particularly instructive here. First, their models suggest that one sex can 'lose' both the parental investment game and the mate-search game, but the forces that select for male search (with low paternal investment) are stronger than selection for female search, as in prediction (1) above. Second, when individuals from two populations meet, the value of cross-matings for each sex will be determined by the relative costs of hybrid disadvantage (represented by d, e.g. reduction in number or fitness of offspring), mate encounter (s, e.g. signaling, searching, and waiting time), and reproductive investment (g, e.g. time for copulation, gamete costs, and offspring-provisioning).

Hammerstein and Parker (1987) show that a hybrid mating is advantageous for either sex when $d < s/(g + s)$.

This relationship makes clear the potential conflict between the sexes in the 'mating decision' when there is asymmetry either in the costs of searching vs. signalling or in reproductive investment (Alexander *et al.*, this volume). Because males usually invest more in mate encounter and less in reproductive investment than females, their threshold for hybrid disadvantage should be considerably higher under most conditions. This prediction is different from the conventional view that males will be under selection to avoid hybrid matings, although at an intensity lower than that on females, suggesting instead that such matings would generally be advantageous for males.

TESTING THE PREDICTIONS OF ASYMMETRIC TRACKING

The first prediction of the asymmetric tracking hypothesis is not new and is largely without controversy. Chemical signaling by female moths contrasts with male signaling in those insect orders that use acoustic calling (Greenfield 1981). Acoustic signaling is both energetically expensive and bears higher risk of predation, whereas the production of nanogram quantities of pheromones by female moths is relatively cheap and is apparently a difficult signal for predators to exploit. Also consistent is the sexual asymmetry in pheromone release, which may be orders of magnitude greater in males than in females (as in the pyralids and arctiids discussed above). Similarly, when acoustic signals are used in the context of lepidopteran mating, males almost always produce the signal. In contrast, when acoustic signals are used for defense, such as in some moth larvae or pupae, or in arctiid and ctenuchid adults, both males and females emit sound (Scoble 1992). Exceptions to the male-only acoustic mate-signaling rule were discussed above, in *Py. isabella* and *Ph. fuliginosa*, where female calls are accompanied by an inordinate release of pheromone. Sexual-selection theory predicts that females will only make such expenditures when males are providing more than just sperm. It is possible that males of these species are providing a nuptial gift, such as defensive chemicals during mating as has been demonstrated in other arctiids. However, unlike *U. ornatrix*, females of *Py. isabella* and *Ph. fuliginosa* were as likely to accept PA-deprived males with little or no pheromone as males with high pheromone titers (Krasnoff and Roelofs 1990). Thus, our present understanding of these systems runs counter to predictions, and they may represent additional examples of females losing the mate-searching game (Hammerstein and Parker 1987), owing perhaps to phylogenetic history (Krasnoff and Roelofs 1990).

The second and third predictions, which contradict the assumption of strong stabilizing selection, might appear inconsistent with those studies showing low population variation in female pheromone systems. The highly specific ratios of E- and Z-isomers in tortricid sex-pheromone blends that were the subject of some of the earliest multiple-species comparisons indicated a tightly coordinated communication system, setting expectations for moth pheromones in general (Roelofs and Brown 1982). Such examples have been used as evidence for strong reciprocal selection in mate recognition (Lambert et al. 1987). Even in studies of widely distributed species, geographical variation may be low, examples are *Pectinophora gossypiella* (Haynes and Baker 1988) and *Trichoplusia ni* (Haynes and Hunt 1990). However, as the number of studies increases, it is apparent that significant variation in moth pheromones is more common than originally thought. Pheromonal polymorphism was first recognized in *Ostrinia nubilalis* (Kochansky et al. 1975). Two bivoltine pheromone races using $97:3$ and $3:97$ Z- and $E11\text{-}14$: Ac have been found, as well as a third univoltine race that utilizes a high Z blend. Significant although apparently unidirectional gene flow between sympatric races has been measured by comparisons of frequencies of enzyme loci (Glover et al. 1991). Unidirectional gene flow, apparently due to the broader response of high-E males to blend ratios compared with high-Z males, results in hybrids with intermediate pheromone blends and responses. Reciprocal pheromone types also exist sympatrically for the larch bud moth, *Zeiraphera diniana*, with $>100:1$ and $<1:100$ blends of $E9\text{-}12$: Ac and $E11\text{-}14$: Ac (Baltensweiler and Priesner 1988 and references therein); intermediate pheromone forms suggest hybridization between these pheromone types under natural conditions. Three stable pheromone types have been discovered for the western avocado leafroller, with two of the types showing sympatric distributions (Bailey et al. 1986). At least four pheromone types appear to exist for the dingy cutworm, *Feltia jaculifera*; two sympatric populations appear to be completely isolated reproductively in the field with no cross-attraction, although the authors do not indicate whether these populations also are isolated by postmating barriers (Byers et al. 1990; Byers and Struble 1990). Other examples of high intraspecific variation in pheromone blends include the turnip moth, *Agrotis segetum* (Tóth et al. 1992; Löfstedt 1993 and references therein),

as well as some tortricids of New Zealand (Foster et al. 1989; Clearwater et al. 1991).

The forces driving the evolution of any of these variable systems can be the subject of only speculation; however, it is clear that significant variation in female pheromones and even strong polymorphism can arise either in sympatry or allopatry. Moreover, those pheromone systems with low variation do not necessarily provide evidence for mutually stabilizing selection between signal and response. In the apple leafrollers (Tortricidae), at least 12 species, many of which use pheromone blends of E- and $Z11\text{-}14$: Ac, have mating periods within 3 h after sunset (Roelofs and Brown 1982). In this case, the narrowing of the communication system in response to selection against interspecific attraction would be consistent with predictions of the asymmetric tracking hypothesis, since it would be adaptive under the model for both males and females.

Liu and Haynes (1994) have provided the first direct test of this portion of the asymmetric tracking hypothesis. In measurements of individual variation in pheromone production by the cabbage looper moth, *Trichoplusia ni*, Haynes and Hunt (1990) discovered a mutant pheromone phenotype, controlled by a single recessive gene, in which females produced a radically different pheromone blend. Whereas the normal sex pheromone is a six-component blend of $Z7\text{-}12$: Ac, 12: Ac, $Z5\text{-}12$: Ac, $11\text{-}12$: Ac, $Z7\text{-}14$: Ac, and $Z9\text{-}14$: Ac ($100:7:17:3:2:1$) (Bjostad et al. 1984), pheromone-mutant females emitted these compounds in the ratio $100:17:2:4:2:51$ (Haynes and Hunt 1990). The greatest changes in the mutant pheromone were a 3-fold decrease in the major component, $Z7\text{-}12$: Ac, a 30-fold decrease in $Z5\text{-}12$: Ac, and a 20-fold increase in $Z9\text{-}14$: Ac, resulting in a more than 40-fold reduction in the number of conspecific males attracted in the field. As in other pheromone systems, genetic control of female production was independent of male response (Haynes and Hunt 1990). Males from a colony established with mutant females had a distribution of component-specific antennal receptors similar to that of non-mutant males (Todd et al. 1992). Males from both mutant and normal colonies were initially twice as responsive to the normal pheromone as to the mutant blend; however, after 49 generations in the laboratory, males in both of two mutant cultures responded to the mutant and normal pheromone blends with equal frequency (Fig. 14-4), whereas males from the normal colony continued to show a significantly higher response to the normal blend (Liu and Haynes 1994). In contrast, the pheromone blend of mutant females did not change

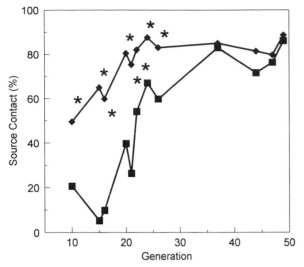

Fig. 14-4. Wind-tunnel response of *Trichoplusia ni* males from a female-pheromone-mutant colony to synthetic pheromone blends of normal (diamonds) and mutant (squares) females. Pure colonies of pheromone mutants were maintained in laboratory culture without selection on pheromone production or response. Note that although males in this culture initially showed a strong preference for the normal pheromone blend, after *ca.* 30 generations, males responded equally to mutant and normal blends. Asterisks (*) indicate significant differences in response to blends. After Liu and Haynes (1994).

over that time. Thus, even in the absence of artificial selection, there was a change in male response as predicted by the asymmetric tracking hypothesis to 'accommodate' a radical departure in the species-specific female pheromone, while no change in that signal was recorded. Although maintenance of an extreme pheromone phenotype may be easier in the laboratory, these results nevertheless indicated a sexual asymmetry in selection on the sexual-communication system.

Asymmetric selection also is suggested in the evolution of the sex-pheromone system of European *Yponomeuta*. Using a phylogeny based on allozymic and morphologic data, Löfstedt *et al.* (1991) argue that the 'primitive' *Yponomeuta* female pheromone is a blend of 14: Ac, *Z*11-14: Ac and *E*11-14: Ac, and their corresponding alcohols. Shifts in the pheromone blends of more advanced species have occurred through loss of components, accompanied either by increases in other components or by production of new compounds. Electrophysiological studies of male antennae in each of these species indicated that males have retained the ability to respond to lost compounds while evolving

sensitivity to new pheromone components. Thus, as in *T. ni*, changes in the pheromone-communication systems of *Yponomeuta* spp. appear to have been led by changes in the female pheromone, changes that males subsequently 'tracked' evolutionarily through broadening of antennal sensitivity.

The fall armyworm, *Spodoptera frugiperda*, represents a further opportunity to test the asymmetric tracking model. Two genetically differentiated strains are broadly sympatric, with one feeding on rice and various grasses and the other restricted primarily to corn (Pashley 1986). The two strains are morphologically indistinguishable; however, they can be separated on the basis of allozymic and mtDNA differences with a high degree of confidence (Pashley 1989). Genetic distinctiveness in sympatric populations of the two strains, of which cross-matings produce viable offspring, indicates the presence of prezygotic reproductive isolation. Although there was evidence of differences in seasonality and diel periodicity, the strains none the less showed considerable overlap in sexual activity (Pashley *et al.* 1992). Differences in the female pheromones appeared to play a minor role in separating the populations, as corn- and rice-strain males showed 35% and 40% cross-attraction to live females in the field, respectively (Pashley *et al.* 1992). Females showed greater discrimination when caged with males of each strain, with only 15% hybrid matings occurring. Thus, consistent with the asymmetric tracking model, *S. frugiperda* females exhibit greater intrastrain fidelity in response to mating signals than do males. Given that hybrid matings produce viable offspring, selection for male discrimination is not predicted until rejection by interstrain females is high enough to counter the costs of additional mate search or unless hybrid disadvantage becomes very large (Hammerstein and Parker 1987).

Although a full understanding of the mechanisms underlying intraspecific variation in lepidopteran female pheromones awaits further work, discovery of an increasing number of polymorphic systems forces us to reconsider previous assumptions of strong stabilizing selection between signaler and responder. Evidence now suggests that significant shifts in the female signal can arise and be maintained in a population without genetic linkage with male response, even in the face of significant gene flow between pheromone types.

We do not yet have any studies of the genetic control or heritability of male pheromones or their relationship to female response, nor is information available on geographic variation in male-pheromone blends. Thus, we cannot test the fourth prediction of the asymmetric tracking hypothesis, that male signals will follow female response, in the same manner as we have examined female pheromones. However, it is noteworthy that many male pheromones include plant volatiles. Because female moths often use odors for food-finding or ovipositional host-finding, a receptor system is already in place to respond to certain plant odors. If males were to evolve the capacity to synthesize or sequester plant compounds that elicited female response, this would be evidence of male tracking. Males of a number of noctuids emit volatile components with aromatic chemical structures. These same compounds are produced by several species of night-blooming flowers from which adult noctuids are known to feed. Phenylacetaldehyde has been identified from the flowers of *Araujia sericofera* (Cantelo and Jacobson 1979), *Abelia grandiflora* (Haynes *et al.* 1991) and night-blooming jessamine, *Cestrum nocturnum* (Heath *et al.* 1992). *A. grandiflora* also emits 2-phenylethanol, benzaldehyde and benzyl alcohol; *C. nocturnum* also releases benzaldehyde and benzyl acetate. Moreover, phenylacetaldehyde is known to attract a number of noctuid moths in the field (Creighton *et al.* 1973), and both Haynes *et al.* (1991) and Heath *et al.* (1992) found this compound, either alone or in the more complete floral blend, to elicit attraction and landing of *T. ni* females in a wind tunnel. Other examples of the connection between male odor and non-sexual female response are given elsewhere (Phelan 1992). Thus, these male pheromones may have evolved in the context of response manipulation (Krebs and Dawkins 1984), by which males produce an odor that exploits the female's food-finding response. In this manner, males could attract or at least induce arrestment in females by using volatiles to which they are already responsive, albeit in a different behavioral context. This system could evolve further by association with a female preference for the male odor through runaway sexual selection, leading to reproductive isolation (Phelan 1992). Alternatively, there may be selection for increased production of the male odor without a change in female response in a process analogous to the sensory exploitation model of Ryan and Rand (1990), proposed for auditory signaling. The possibility of unilateral evolution through sensory exploitation is particularly plausible for those groups where male pheromones may not be species-specific (Birch *et al.* 1990). Either way, the evolution of male pheromones to match a pre-existing female response would be an example of the sort of tracking envisioned in prediction (4).

SUGGESTIONS FOR FUTURE RESEARCH

The recent attention given to population variation in moth pheromone systems has been extremely useful in suggesting the role that mate-signaling plays in speciation. I have attempted to use some of these examples to suggest that the female sex-pheromone system of moths is not under strong stabilizing selection, as is normally assumed for this and other mate-signaling systems. In addition, I have tried to provide some theoretical rationale for why we should expect this not to be the case. By removing the theoretical constraint of mutual stabilizing selection, it is easier to understand how shifts in mate-signaling systems may occur that would lead to speciation. Although the emphasis here is on pheromonal systems, it should be evident to the reader that the asymmetric tracking hypothesis is applicable to all modalities.

Although recent studies have been helpful, we are still dependent on only a few examples to test various evolutionary models that have been proposed. Such a small set of studies is likely to be a biassed sample, and their conclusions may not represent the norm. Thus, we need more information on pheromone variation and stability if we are to understand the processes involved. Particularly instructive would be further work on races or sibling species that show variation in the pheromone system, such as those discussed above. Documenting patterns of geographic variation, as well as conducting empirical manipulations of these groups in the laboratory, would provide much needed information. Such information is especially lacking for male-produced pheromones. Moreover, we need more cross-fertilization between researchers in acoustic mate-signaling and those working in the chemical modality. Indeed, moth mating systems appear to offer a rich opportunity for the student of acoustic communication. Morphological studies of a number of groups suggest a greater role for this modality than presently thought; however, detailed behavioral studies are needed.

With regard to the phylogenetics of the female moth pheromone system, our present understanding of the chemistry and mating behavior of the primitive Lepidoptera is sketchy at best. Recent work in this area has provided direction for tracing this history, and I expect rapid progress in the near future. Finally, as with most evolutionary questions, the greatest advances in our understanding here have come in those groups for which phylogenetic relationships have been reasonably established. This will continue to be the case, promoting more cooperation between systematists and behavioral and chemical ecologists.

ACKNOWLEDGEMENTS

I express my appreciation for the critical comments of W. Bell, C. Löfstedt, R. Mellon and an anonymous reviewer, as well as discussions with R. Butlin and J. Wenzel. I am particularly grateful to C. Löfstedt and co-workers for sharing their unpublished results of pheromone studies with Trichoptera and primitive Lepidoptera. Finally, thanks go to Seattles' for providing office space during the writing of this work.

LITERATURE CITED

Ansteeg, O. and K. Dettner. 1991. Chemistry and possible biological significance of secretions from a gland discharging at the 5th abdominal sternite of adult caddisflies (Trichoptera). *Entomol. Gen.* **156**: 303–312.

Arn, H., H. Tóth and E. Priesner. 1992. *List of Sex Pheromones of Lepidoptera and Related Attractants*, 2nd edn. Montfavet: International Organization for Biological Control.

Bailey, J. B., L. M. McDonough and M. P. Hoffmann. 1986. Western avocado leafroller, *Amorbia cuneana* (Walsingham), (Lepidoptera: Tortricidae): Discovery of populations utilizing different ratios of sex pheromone components. *J. Chem. Ecol.* **12**: 1239–1245.

Baker, T. C. 1985. Chemical control of behavior. In *Comprehensive Insect Physiology, Biochemistry, and Pharmacology*. G. A. Kerkut and L. I. Gilbert, eds., vol. 9, pp. 621–672. Oxford: Pergamon Press.

Baker, T. C. and R. T. Cardé. 1979. Courtship behavior of the oriental fruit moth (*Grapholitha molesta*): Experimental analysis and consideration of the role of sexual selection in the evolution of courtship pheromones in the Lepidoptera. *Ann. Entomol. Soc. Am.* **72**: 173–188.

Baltensweiler, W. and E. Priesner. 1988. Studien zum Pheromon-Polymorphismus von *Zeiraphera diniana* Gn. (Lep., Tortricidae). 3. Anflugspezifität männlicher Falter zweier Wirtsrassen an synthetische Pheromonquellen. *J. Appl. Entomol.* **106**: 217–231.

Bastock, M. and A. Manning. 1955. The courtship of *Drosophila melanogaster*. *Behaviour* **8**: 85–111.

Bennett, A. L., P. R. Atchinson and N. J. S. La Croix. 1991. On communication in the African sugarcane borer, *Eldana saccharina* Walker (Lepidoptera: Pyralidae). *J. Entomol. Soc. S. Afr.* **54**: 243–259.

Birch, M. C., G. M. Poppy and T. C. Baker. 1990. Scents and eversible scent structures of male moths. *Annu. Rev. Entomol.* **35**: 25–58.

Bjostad, L. B., C. E. Linn, J.-W. Du and W. L. Roelofs. 1984. Identification of new sex pheromone components in *Trichoplusia ni*, predicted from biosynthetic precursors. *J. Chem. Ecol.* **10**: 1309–1323.

Bjostad, L. B., W. A. Wolf and W. L. Roelofs. 1987. Pheromone biosynthesis in lepidopterans: Desaturation and chain shortening. In *Pheromone Biochemistry*. G. D. Prestwich and G. J. Blomquist, eds., pp. 77–120. London: Academic Press.

Boppré, M. 1984. Chemical mediated interactions between butterflies. In *The Biology of Butterflies*. R. I. Vane-Wright and P. R. Ackery, eds., pp. 259–275. New York: Academic Press.

Byers, J. R. and D. L. Struble. 1990. Identification of sex pheromones of two sibling species in dingy cutworm complex, *Feltia jaculifera* (Gn.) (Lepidoptera: Noctuidae). *J. Chem. Ecol.* **16**: 2981–2992.

Byers, J. R., D. L. Struble, C. E. Herle, G. C. Kozub and J. D. LaFontaine. 1990. Electroannographic responses differentiate sibling species of dingy cutworm complex, *Feltia jaculifera* (Gn.) (Lepidoptera: Noctuidae). *J. Chem. Ecol.* **16**: 2969–2980.

Cantelo, W. W. and M. Jacobson. 1979. Phenylacetaldehyde attracts moths to bladder flower and to blacklight traps. *Environ. Entomol.* **8**: 444–447.

Cardé, R. T. and T. C. Baker. 1984. Sexual communication with pheromones. In *Chemical Ecology of Insects*. W. J. Bell and R. T. Cardé, eds., pp. 355–383. London: Chapman and Hall.

Cibrian-Tovar, J. and E. R. Mitchell. 1991. Courtship behavior of *Heliothis subflexa* (Gn.) (Lepidoptera: Noctuidae) and associated backcross insects obtained from hybridization with *H. virescens* (F.). *Environ. Entomol.* **20**: 419–426.

Clearwater, J. R., S. P. Foster, S. J. Muggleston, J. S. Dugdale and E. Priesner. 1991. Intraspecific variation and interspecific differences in sex pheromones of sibling species in *Ctenopseustis obliquana* complex. *J. Chem. Ecol.* **17**: 413–429.

Conner, W. E. 1987. Ultrasound: its role in the courtship of the arctiid moth, *Cycnia tenera*. *Experientia* **43**: 1029–1031.

Conner, W. E., T. Eisner, R. K. Vander Meer, A. Guerrero and J. Meinwald. 1981. *Behav. Ecol. Sociobiol.* **9**: 227–235.

Creighton, C. S., T. L. McFadden and E. R. Cuthbert. 1973. Supplemental data on phenylacetaldehyde : An attractant for Lepidoptera. *J. Econ. Entomol.* **66**: 114–115.

Dall'Asta, U. 1988. The tymbal organs of the Lymantriidae (Lepidoptera). *Nota Lepid.* **11**: 169–176.

Davis, D. R. 1975. Systematics and zoogeography of the family Neopseustidae with the proposal of a new superfamily (Lepidoptera: Neopseustoidea). *Smithson. Contrib. Zool.* **210**: 1–45.

–. 1978. A revision of the North American moths of the superfamily Eriocranioidea with the proposal of a new family Acanthopteroctetidae (Lepidoptera). *Smithson. Contrib. Zool.* **251**: 1–131.

–. 1986. A new family of monotrysian moths from austral South America (Lepidoptera: Palaephatidae) with a phylogenetic review of the Monotrysia. *Smithson. Contrib. Zool.* **434**: 1–202.

De Jong, M. C. M. 1988. Evolutionary approaches to insect communication systems. Ph. D. dissertation, University of Leiden, The Netherlands.

Doesburg, P. H., van. 1966. Ueber valväre Stridulation bei Schwärmer (Lepidoptera Sphingidae). *Zool. Med. (Leiden)* **41**: 161–170.

Doherty, J. and R. Hoy. 1985. Communication in insects. III. The auditory behavior of crickets: Some views of genetic coupling, song recognition, and predator detection. *Q. Rev. Biol.* **60**: 457–472.

Drummond, B. A. 1984. Multiple mating and sperm competition in the Lepidoptera. In *Sperm Competition and the Evolution of Animal Mating Systems*. R. L. Smith, ed., pp. 291–370. Orlando, Florida: Academic Press.

Duffield, R. M., M. S. Blum, J. B. Wallace, H. A. Lloyd and F. E. Regnier. 1977. Chemistry of the defensive secretion of the caddisfly *Pycnopsyche scabripennis* (Trichoptera: Linmnephillidae). *J. Chem. Ecol.* **3**: 649–656.

Dunning, D. C., L. Acharya, C. B. Merriman and L. Dal-Ferro. 1992. Interactions between bats and arctiid moths. *Can. J. Zool.* **70**: 2218–2223.

Dussourd, D. E., C. A. Harvis, J. Meinwald and T. Eisner. 1991. Pheromonal advertisement of a nuptial gift by a male moth (*Utethesia ornatrix*). *Proc. Natl. Acad. Sci. U.S.A.* **88**: 9224–9227.

Dussourd, D. E., K. Ubik, C. Harvis, J. Resch, J. Meinwald and T. Eisner. 1988. Biparental defensive endowment of eggs with acquired plant alkaloid in the moth *Utethesia ornatrix*. *Proc. Natl. Acad. Sci. U.S.A.* **85**: 5992–5996.

Emlen, S. T. and L. W. Oring. 1977. Ecology, sexual selection, and the evolution of mating systems. *Science (Wash., D.C.)* **197**: 215–223.

Fenton, M. B. and J. H. Fullard. 1981. Moth hearing and the feeding strategies of bats. *Am. Sci.* **69**: 266–275.

Fitzpatrick, S. M. and J. N. McNeil. 1988. Male scent in lepidopteran communication: The role of male pheromone in mating behaviour of *Pseudoletia unipuncta* (Haw.) (Lepidoptera: Noctuidae). *Mem. Entomol. Soc. Can.* **146**: 131–151.

Foster, S. P., J. R. Clearwater and S. J. Muggleston. 1989. Intraspecific variation of two components in sex pheromone gland of *Planotortrix excessana* sibling species. *J. Chem. Ecol.* **15**: 457–465.

Glover, T. J., J. J. Knodel, P. S. Robbins, C. J. Eckenrode and W. L. Roelofs. 1991. Gene flow among three races of European corn borers (Lepidoptera: Pyralidae) in New York state. *Environ. Entomol.* **20**: 1356–1362.

Grant, G. G., E. B. Smithwick and U. E. Brady. 1975. Courtship behavior of phycitid moths. II. Behavioral and pheromonal isolation of *Plodia interpunctella* and *Cadra cautella* in the laboratory. *Can. J. Zool.* **53**: 827–832.

Greenfield, M. D. 1981. Moth sex pheromones: An evolutionary perspective. *Fla. Entomol.* **64**: 4–17.

Greenfield, M. D. and J. A. Coffelt. 1983. Reproductive behaviour of the lesser waxmoth, *Achroia grisella* (Pyralidae: Galleriinae):

signalling, pair formation, male interactions, and mate guarding. *Behaviour* **84**: 287–315.

Grula, J. W. and O. R. Taylor. 1979. The inheritance of pheromone production in the sulfur butterflies, *Colias eurytheme* and *C. philodice. Heredity* **42**: 359–371.

–. 1980. The effect of X-chromosome inheritance on mate-selection behavior in the sulfur butterflies, *Colias eurytheme* and *C. philodice. Evolution* **34**: 688–695.

Gwynne, D. T. and E. D. Edwards. 1986. Ultrasound production by genital stridulation in *Syntonarcha iriastis* (Lepidoptera: Pyralidae): Long-distance signalling by male moths? *Zool. J. Linn. Soc.* **88**: 363–376.

Hammerstein, P. and G. A. Parker. 1987. Sexual selection: Games between the sexes. In *Sexual Selection: Testing the Alternatives.* J. W. Bradbury and M. B. Andersson, eds., pp. 119–142. Chichester: J. Wiley & Sons.

Haynes, K. F. and T. C. Baker. 1988. Potential for evolution of resistance to pheromones: Worldwide and local variation in chemical communication system of pink bollworm moth, *Pectinophora gossypiella. J. Chem. Ecol.* **14**: 1547–1560.

Haynes, K. F. and R. E. Hunt. 1990. A mutation in pheromonal communication system of cabbage looper moth, *Trichoplusia ni. J. Chem. Ecol.* **16**: 1249–1255.

Haynes, K. F., J. Z. Zhao and A. Latif. 1991. Identification of floral compounds from *Abelia grandiflora* that stimulate upwind flight in cabbage looper moths. *J. Chem. Ecol.* **17**: 637–646.

Heath, R. B., P. J. Landolt, B. Dueben and B. Lenczewski. 1992. Identification of floral compounds of night-blooming jessamine attractive to cabbage looper moths. *Environ. Entomol.* **21**: 854–859.

Heinrich, C. 1956. American moths of the subfamily Phycitinae. *U.S. Natl. Mus. Bull.* **207**: 1–581.

Hoy, R. R., J. Hahn and R. C. Paul. 1977. Hybrid cricket auditory behavior: Evidence for genetic coupling in animal communication. *Science (Wash., D.C.)* **195**: 82–84.

Kaneshiro, K. Y. 1980. Sexual isolation, speciation, and the direction of evolution. *Evolution* **34**: 437–444.

Kochansky, J., R. T. Cardé, J. Liebherr and W. L. Roelofs. 1975. Sex pheromone of the European corn borer. *J. Chem. Ecol.* **1**: 225–31.

Krasnoff, S. B. 1987. The chemical ecology of courtship communication in some nearctic arctiids (Lepidoptera: Arctiidae). Ph. D. dissertation, Cornell University, Ithaca, NY.

Krasnoff, S. B. and W. L. Roelofs. 1988. Sex pheromone released as an aerosol by the moth *Pyrrharctica isabella. Nature (Lond.)* **333**: 263–265.

–. 1990. Evolutionary trends in the male pheromone systems of arctiid moths: evidence from studies of courtship in *Phragmotobia fuliginosa* and *Pyrrharctia isabella* (Lepidoptera: Arctiidae). *Zool. J. Linn. Soc.* **99**: 319–338.

Krasnoff, S. B. and D. D. Yager. 1988. Acoustic response to a pheromonal cue in the arctiid moth *Pyrrharctia isabella. Physiol. Entomol.* **13**: 433–440.

Krebs, J. R. and R. Dawkins. 1984. Animal signals: Mind-reading and manipulation. In *Behavioral Ecology: An Evolutionary Approach*, 2nd edn. J. R. Krebs and N. B. Davies, eds., pp. 380–402. Sunderland, Massachusetts: Sinauer Associates.

Kristensen, N. P. 1978. Observations on *Anomoses hylecoetes* (Anomosetidae), with a key to the hepialoid families (Insecta: Lepidoptera). *Steenstrupia* 5: 1–19.

–. 1984a. The pregenital abdomen of the Zeugloptera (Lepidoptera). *Steenstrupia* 10: 113–136.

–. 1984b. Studies on the morphology and systematics of primitive Lepidoptera (Insecta). *Steenstrupia* 10: 141–191.

Kuenen, L. P. S., D. L. Wagner, W. E. Wallner and R. T. Cardé. 1994. Female sex pheromone in *Korscheltellus gracilis* (Grote) (Lepidoptera: Hepialidae). *Can. Entomol.* 126: 31–41.

Kuroko, H. 1990. Preliminary ecological notes on *Ogygioses caliginosa* Issiki and Stringer (Palaeosetidae). *Bull. Sugadaira Montane Res. Ctr.* 11: 103–104.

Lambert, D. M., B. Michaux and C. S. White. 1987. Are species self-defining? *Syst. Zool.* 36: 196–205.

Liu, Y.-B. and K. F. Haynes. 1994. Evolution of behavioral responses to sex pheromone in mutant laboratory colonies of *Trichoplusia ni. J. Chem. Ecol.* 20: 231–238.

Loeb, M. J., J. W. Neal, Jr. and J. A. Klun. 1989. Modified thoracic epithelium of the bagworm (Lepidoptera: Psychidae): site of pheromone production in adult females. *Ann. Entomol. Soc. Am.* 82: 215–219.

Löfstedt, C. 1990. Population variation and genetic control of pheromone communication systems in moths. *Entomol. Exp. Appl.* 54: 199–218.

–. 1993. Moth pheromone genetics and evolution. *Phil. Trans. R. Soc. Lond.* B 340: 167–177.

Löfstedt, C., B. S. Hansson, E. Petersson, P. Valeur and A. Richards. 1994. Pheromonal secretions from glands on the 5th abdominal sternite of hydropsychid and rhyacophilid caddisflies (Trichoptera). *J. Chem. Ecol.* 20: 153–170.

Löfstedt, C., W. M. Herrebout and S. B. J. Menken. 1991. Sex pheromones and their potential role in the evolution of reproductive isolation in small ermine moths (Yponomeutidae). *Chemoecology* 2: 20–28.

Löfstedt, C. and M. Kozlov. 1996. Phylogenetic analysis of pheromone communication in primitive moths and caddisflies. In *Pheromone Research: New Directions.* R. T. Cardé and A. K. Minks, eds. New York: Chapman and Hall, in press.

Löfstedt, C., N. J. Vickers, W. Roelofs and T. C. Baker. 1989. Diet related courtship success in the Oriental fruit moth, *Grapholita molesta* (Tortricidae). *Oikos* 55: 402–408.

Luquet, G. 1980. Observations sur l'accouplement d'*Adela reaumurella* L., espèce nouvelle pour le Vaucluse. *Alexanor* 9: 365–366.

McCabe, T. L. and D. L. Wagner. 1989. The biology of *Sthenopis auratus* (Lepidoptera: Hepialidae). *J. N. Y. Entomol. Soc.* 96: 256–273.

McColl, H. P. 1969. The sexual scent organs of male Lepidoptera. M. Sc. thesis, University College of Wales, Swansea.

Mallet, J. 1984. Sex roles in the ghost moth *Hepialus humuli* (L.) and a review of mating in the Hepialidae (Lepidoptera). *Zool. J. Linn. Soc.* 79: 67–82.

Meyer, W. L. 1984. Sex pheromone chemistry of some arctiid moths (Lepidoptera: Arctiidae): enantiomeric differences in pheromone perception. M. S. thesis, Cornell University, Ithaca, New York.

Nässig, W. A. and M. Lüttgen. 1988. Notes on genital stridulation in male hawkmoths in South East Asia (Lep., Sphingidae). *Heterocera Sumatrana* 2: 75–77.

Neboiss, A. 1991. Trichoptera. In *The Insects of Australia*, 2nd edn., vol. 2. CSIRO, ed., pp. 787–816. Ithaca, New York: Cornell University Press.

Nielsen, A. 1980. A comparative study of the genital segments and the genital chamber in female Trichoptera. *K. Dan. Vidensk. Selsk. Biol. Sk.* 23: 1–199.

Nielsen, E. S. 1989. Phylogeny of the major lepidopteran groups. In *The Hierarchy of Life.* B. Ferholm, K. Bremer, and H. Jörnvall, eds., pp. 281–294. New York: Elsevier.

Nielsen, E. S. and I. F. B. Common. 1991. Lepidoptera (moths and butterflies). In *The Insects of Australia*, 2nd edn., vol. 2, pp. 817–915. Ithaca, New York: Cornell University Press.

Nielsen, E. S. and N. P. Kristensen. 1989. Primitive ghost moths. Morphology and taxonomy of the Australian genus *Fraus* Walker (Lepidoptera: Hepialidae *s. lat.*). *Monographs on Australian Lepidoptera* 1. xvii + 286pp. Melbourne, Australia: CSIRO.

Pashley, D. P. 1986. Host-associated genetic differentiation in fall armyworm (Lepidoptera: Noctuidae): A sibling species complex? *Ann. Entomol. Soc. Am.* 79: 898–904.

–. 1989. Host-associated differentiation in armyworms: An allozymic and mtDNA perspective. In *Electrophoretic Studies on Agricultural Pests.* H. Loxdale and M. F. Claridge, eds., pp. 103–114. Oxford University Press.

Pashley, D. P., A. M. Hammond and T. N. Hardy. 1992. Reproductive isolating mechanisms in fall armyworm host strains (Lepidoptera: Noctuidae). *Ann. Entomol. Soc. Am.* 85: 400–405.

Paterson, H. E. H. 1985. The recognition concept of species. In *Species and Speciation*, E. S. Vrba, ed., pp. 21–29. (Transvaal Museum Monograph No. 4). Pretoria: Transvaal Museum.

–. 1993. Variation and the specific-mate recognition system. In *Perspectives in Ethology*. vol. 10. *Behavior and Evolution*. P. P. G. Bateson, ed., pp. 209–227. New York: Plenum Press.

Percy-Cunningham, J. E. and J. A. McDonald. 1987. Biology and ultrastructure of sex pheromone-producing glands. In *Pheromone Biochemistry.* G. D. Prestwich and G. J. Blomquist, eds., pp. 27–76. London: Academic Press.

Phelan P. L. 1992. Evolution of sex pheromones and the role of asymmetric tracking. In *Insect Chemical Ecology: An Evolutionary Approach.* B. D. Roitberg and M. B. Isman, eds., pp. 265–314. New York: Chapman and Hall.

Phelan, P. L. and T. C. Baker. 1986. Male-size-related courtship success and intersexual selection in the tobacco moth, *Ephestia elutella. Experientia* **42**: 1291–1293.

–. 1987. Evolution of male pheromones in moths: Reproductive isolation through sexual selection. *Science (Wash., D.C.)* **235**: 205–207.

–. 1990a. Comparative study of courtship in twelve phycitine moths (Lepidoptera: Pyralidae). *J. Insect Behav.* **3**: 303–326.

–. 1990b. Information transmission during intra- and interspecific courtship in *Ephestia elutella* and *Cadra figulilella. J. Insect Behav.* **3**: 589–602.

Priesner, E. 1986. Correlating sensory and behavioural responses in multichemical pheromone systems of Lepidoptera. In *Mechanisms in Insect Olfaction*. T. L. Payne, M. C. Birch and C. E. J. Kennedy, eds., pp. 225–234. Oxford: Clarendon Press.

Roelofs, W. L., and R. L. Brown. 1982. Pheromones and evolutionary relationships of Tortricidae. *Annu. Rev. Ecol. Syst.* **13**: 395–422.

Romel, K. E., C. D. Scott-Dupree and M. H. Carter. 1992. Qualitative and quantitative analyses of volatiles and pheromone gland extracts collected from *Galleria mellonella* (L.) (Lepidoptera: Pyralidae). *J. Chem. Ecol.* **18**: 1255–1268.

Ryan, M. J. and A. S. Rand. 1990. The sensory basis of sexual selection for complex calls in the túngara frog, *Physalaemus pustulosus* (Sexual selection for sensory exploitation). *Evolution* **4**: 305–314.

Sanderford, M. V. and W. E. Conner. 1990. Courtship sounds of the polka-dot wasp moth, *Syntodeida epilais. Naturwissenschaften* **77**: 345–347.

Schal, C., R. C. Charlton and R. T. Cardé. 1987. Temporal patterns of sex pheromone titers and release rates in *Holomelina lamae* (Lepidoptera: Arctiidae). *J. Chem. Ecol.* **13**: 1115–1130.

Scoble, M. J. 1992. *The Lepidoptera*. Oxford University Press.

Solem, J. O. 1985. Female sex pheromones in *Rhyacophila nubila* (Zetterstedt) (Trichoptera: Rhyacophilidae) and arrival pattern to sticky traps. *Fauna Norv. Ser.* B32: 80–82.

Spangler, H. G. 1985. Sound production and communication by the greater wax moth (Lepidoptera: Pyralidae). *Ann. Entomol. Soc. Am.* **78**: 54–61.

Todd, J. L., K. F. Haynes and T. C. Baker. 1992. Antennal neurones specific for redundant pheromone components in normal and mutant *Trichoplusia ni* males. *Physiol. Entomol.* **17**: 183–192.

Tóth, M., C. Löfstedt, B. Blair, T. Cabello, A. Farag, B. Hansson, B. Kovalev, S. Maini, E. Nesterov, I. Pajor, A. Sazonov, I. Shamshev, M. Subchev and G. Szöcs. 1992. Attraction of male turnip moths *Agrotis segetum* (Lepidoptera: Noctuidae) to sex pheromone components and their mixtures at 11 sites in Europe, Asia, and Africa. *J. Chem. Ecol.* **18**: 1337–1347.

Turner, J. R. G. 1988. Sex, leks, and fechts in swift moths *Hepialus* (Lepidoptera, Hepialidae): Evidence for the hot shot moth. *Entomologist* **107**: 90–95.

Wagner, D. L. 1985. The biosystematics of the Holarctic Hepialidae, with special emphasis on the *Hepialus californicus* species group. Ph. D. dissertation, University of California, Berkeley.

Wagner, D. L. and J. Rosovsky. 1991. Mating systems in primitive Lepidoptera, with emphasis on the reproductive behaviour of *Korscheltellus gracilis* (Hepialidae). *Zool. J. Linn. Soc.* **102**: 277–303.

West-Eberhard, M. J. 1983. Sexual selection, social competition, and speciation. *Q. Rev. Biol.* **58**: 155–183.

Williams, G. C. 1992. *Natural Selection: Domains, Levels, and Challenges*. New York: Oxford University Press.

Willis, M. A. and M. C. Birch. 1982. Male lek formation and female calling in a population of the arctiid moth, *Estigmene acrea. Science (Wash., D.C.)* **218**: 168–170.

Wood. J. R. and V. H. Resh. 1984. Demonstration of sex pheromones in caddisflies (Trichoptera). *J. Chem. Ecol.* **10**: 171–176.

Wunderer, H., K. Hansen, T. W. Bell, D. Schneider and J. Meinwald. 1986. Sex pheromones of two Asian moths (*Creatonotus transiens, C. gangis*; Lepidoptera: Arctiidae): Behavior, morphology, chemistry and electrophysiology. *Exp. Biol.* **46**: 11–27.

Zagatti, P. 1981. Comportement sexuel de la Pyrale de la Canne à sucre *Eldana saccharina* (Wlk.) lié à deux phéromones émises par le male. *Behaviour* **78**: 81–98.

Zhu, J., M. Kozlov, P. Philipp, W. Francke and C. Löfstedt. 1995. Identification of a novel moth sex pheromone in *Eriocrania cicatricella* (Lepidoptera: Eriocraniidae) and its phylogenetic implications. *J. Chem. Ecol.* **21**: 29–43.

15 · Sexual dimorphism, mating systems and ecology in butterflies

RONALD L. RUTOWSKI

ABSTRACT

Butterflies display a variety of sexual differences in size and morphology beyond those found in the gonads and genitalia. The size and nature of such differences can be explained as (1) the result of selection pressures acting on one or both sexes that are a function of these differences, or (2) as the incidental consequence of selection acting independently on each sex. This chapter reviews various attempts to relate mating-system structure and ecology to these explanations for sexual dimorphism in butterflies. Females are usually larger than males in butterflies; although there are positive size–fecundity relationships in females, the patterns of selection on male size are not understood well enough to explain this pattern. Body and wing shape reflect male mate-locating tactics, but selection pressures shaping female wing and body shape characteristics are poorly known. Sexual selection still appears to be a likely explanation for the elaborate colorations, scent-producing, and sensory structures in males, although there is clearly more work to be done. These sexual differences all appear to be the products of selection acting independently on the sexes, whereas sexual differences in eclosion patterns might be explained by selection for specific sexual differences in eclosion dates.

INTRODUCTION

A variety of sexual differences in morphology, beyond the gonads and genitalia, are common in butterflies. These include differences in wing color, wing shape, body size, body proportions, sensory structures, foreleg development, and signal-producing structures (Fig. 15-1). For all of these variables, there is interspecific variation in the magnitude of differences between the sexes. My purpose in this chapter is to determine the extent to which this variation can be explained by a knowledge of mating system structure and ecology. In other words, is the degree of sexual dimorphism in a trait evolutionarily related to the nature of the mating-system found in that species?

There is a long history of interest in ultimate explanations of sexual dimorphism in butterflies. In 1871, Charles Darwin suggested that the brilliant colors often found on the dorsal wing surfaces of males but not females evolved in the context of mate choice by females. This idea and others have often been discussed in the past 120 years (Silberglied 1984) but submitted to surprisingly few empirical tests.

A problem in empirically examining the relationship between sexual behavior and sexual dimorphism has been the difficulty of experimentally studying butterfly behavior (Rutowski 1991). Comparative approaches provide a strong alternative, but solid documentation of mating-system structure and ecology is available for a relatively small number of butterfly species (fewer than 100 out of about 12 000 species (Rutowski 1991; Wickman 1992)) and there have been relatively few efforts to describe quantitatively the patterns of variation in sexual dimorphism in butterflies. For example, there has been no thorough comparative analysis of sexual differences in color like those available for birds (see, for example, Baker and Parker 1979; Höglund 1989; Oakes 1992). None the less, the database on behavior and dimorphisms has been growing in recent years and the number of recent efforts to test hypotheses about sexual dimorphism in butterflies warrant a review at this time.

I will first briefly address some general issues surrounding the study of the evolution of sexual dimorphism, not in a full review of these issues or their history, but instead in a brief overview to indicate where arguments on these issues stand.

SOME CONCEPTUAL ISSUES

Hypotheses for the evolution of sexual dimorphism

There are two broad hypotheses for the selective circumstances that lead to sexual dimorphism that often are not clearly distinguished. The first is that there is selection

Fig. 15-1. Dorsal views of the male (above) and the female (below) of the paradise birdwing butterfly, *Ornithoptera paradisea* (adapted from D'Abrera 1971). This species displays several sexual dimorphisms found in butterflies including sexual differences in size, wing shape, color pattern, and the development of chemical signaling structures (see the presumptive scent scales on the anal margin of the male hindwing). As drawn, the stippled areas are a brilliant green and the clear areas yellow on the male. On the female, the stippled areas are brown and the clear areas cream.

due to some difference between the sexes, such as size. Such selection is expected to occur in competitive or coop‐erative interactions between males and females (Shine 1989; Mueller 1990). For example, if the members of a mating pair cooperate in some coordinated activity, the fitness con‐sequences of the activity could be a function of the direc‐tion and magnitude of the size difference between the sexes. Selection in this context could favor mate prefer‐ences that lead to assortative mating, which in turn pro‐duces the optimal size relationship, but it could also act to cause or maintain sexual differences in size or other fea‐tures.

Competitive ecological interactions might also lead to selection favoring intersexual differences in some struc‐tures. In this instance, sexual dimorphism evolves as a means of avoiding competition for resources. For exam‐ple, males and females in monogamous pairs may have

better success in provisioning the young if they do not col‐lect the same resources. This hypothesis requires that the feeding biology of one sex impinge on that of the other in a way that affects the fitness of one or both.

The second general hypothesis to explain sexual dimorphism is that selection acts independently on each sex and incidentally produces sexual differences in size or shape (Slatkin 1984; Shine 1989). The selective pressures acting on males and females are often quite different because of differences in their sexual strategies (Trivers 1972). Males are more often subject to sexual selection through female choice or intrasexual competition (see, for example, Moore 1990). Females, on the other hand, are more typically under selection favoring traits that enhance viability and fecundity. These differences in selection pres‐sures experienced by males and females may produce sexual dimorphism, depending on the nature and magni‐tude of the differences in their life styles. Unlike in the first hypothesis, differences between males and females are not in themselves the source of any selection pressure. Ver‐sions of this hypothesis about sexual differences that do not invoke sexual selection but only intrinsic differences in reproductive biology have been referred to as the ecological causes (Slatkin 1984; Shine 1989) or dimorphic niche (Hedrick and Temeles 1989) hypothesis.

When invoking the independent selection hypothesis to explain sexual dimorphisms it is important to deal with the nature of selection acting on both males and females and not just one sex (Greenwood and Adams 1987; Arak 1988; Harvey 1988). For instance, sexual selection for large size in males need not inevitably lead to males being larger than females. To make an accurate prediction about sexual size dimorphism we need to know the nature of size-related selection on both sexes. Among species within a taxon, we expect that the magnitude of sexual dimorphism will reflect the magnitude and nature of sexual differences in the action of selection.

Sexual dimorphism and comparative studies

Recently, Björklund (1991) provided the most rigorous cri‐teria to date to test by the comparative method the hypoth‐esis that large size and ornaments have evolved as a result of sexual selection. In a study focusing on hypotheses about evolutionary impact of mating systems on sexual dimorph‐ism, he suggests that in a rigorous test one must show: (1) that the trait of interest has evolved in the direction indicated; (2) that the evolution of the character follows in time the evolution of the mating system; and (3) that the evolutionary change in the trait could not be accounted for by drift. These are admirable goals. However, whereas comparative studies that do not meet these criteria do not permit strong conclusions about the origins of sexual dimorphism (as Björklund points out), they can test ideas about the role that the mating system can play in maintaining sexual dimorphism (see also Nylin and Wedell 1994). In addition, Björklund does not address the possibility that female body size has evolved in parallel with that of males but for different reasons.

None the less, as Björklund and others (e.g. Harvey and Pagel 1991) have pointed out, rigorous comparative studies of behavioral causes and effects should take into account the phylogenetic relationships among species being examined in order to control for common ancestry as an explanation for similarities between species. In this review, I will examine the extent to which controls for phylogeny have been done in comparative studies of sexual dimorphism in butterflies. In general, recent comparative studies of butterflies include careful controls for phylogeny (see, for example, Srygley and Chai 1990a,b; Marden and Chai 1991; Wickman 1992).

Sexual conflict and sexual dimorphism

Sexual conflict occurs when the optimal outcome of a sexual interaction is different for the male and the female (Parker 1979). This is likely to be the case in all sexual encounters and so sexual conflict is a pervasive feature of sexual reproduction. The usual form of sexual conflict is that it is more often advantageous for the male to mate than the female because males typically have a higher potential rate of reproduction than females. Manifestations of extreme sexual conflict include dramatic behavior patterns such as attempts by males to force copulations on females and mate-guarding by males. These sexual differences in behavior may in turn lead to sexual dimorphism in morphological structures (Clutton-Brock and Parker 1995).

The mating behavior of butterflies lays the groundwork for sexual conflict. After an initial mating, females either become unreceptive for the rest of their life or may mate again but only after after several days or even weeks (see, for example, Suzuki 1978; Rutowski and Gilchrist 1986; Oberhauser 1989). Males, on the other hand, will readily mate twice on the same day (e.g. Rutowski 1979). This sexual difference in reproductive rate has led to sexual

differences in behavior, such as mate-choice patterns and mate-acquisition behavior, that will be discussed later and are likely to have been important in the evolution of sexual dimorphism. Here I will focus briefly on the occurrence and implications in butterflies of some expected products of sexual conflict such as attempts by males to force copulation, production of mating plugs by males, and behavior of unreceptive, previously mated females. Discussion of the effects of sexual conflict on the primary sexual morphology (genitalia and gonads) of butterflies are beyond the intended scope of this review, but may be found in Drummond (1984), Eberhard (1985), and Boggs (1995).

Behavior patterns that appear to be attempts to force copulation have been observed in butterflies. In some danaids (Pliske 1975a) and papilionid butterflies (Orr and Rutowski 1991) flying males grab flying females and carry them to the ground, where they attempt to mate with them. Interestingly, not all of these interactions end in copulation, often because of effective evasive moves made by females, even when virgin (Orr and Rutowski 1991). Whether aerial takedowns are attempts to force copulation or are part of what females use normally to assess male quality is unclear. In any event, there are no special structures or modifications on males that appear to be linked to the success of these takedowns.

Mating plugs are known in butterflies (Drummond 1984) and may act as mechanical impediments to coupling (Dickinson and Rutowski 1989) or, when very large, as visual deterents to mating attempts (Orr and Rutowski 1991). Such structures are likely to be advantageous to the male because last-male precedence in sperm use appears to be the typical case (Drummond 1984). Mating plugs appear to be disadvantageous to females, but this has not been demonstrated. The structures that produce mating plugs are all part of the genitalia; no secondary sexual structures have been implicated in their production or in efforts by males to circumvent them.

Females, especially if recently mated, should evolve counterploys to such efforts by males to subvert their reproductive interests. Mate-rejection postures (Obara 1964; Wiklund and Forsberg 1985) and flight patterns (Rutowski 1978; Wiklund 1982; Wickman 1986, 1988) have been described for mated females, all of which minimize harassment by males. Once again, no special secondary sexual structures or modification have been described which are related to these behavior patterns.

To summarize, sexual conflict will favor males with characteristics that enable them to overcome female resistance to mating and mating plugs, and prevent or delay subsequent matings by females they acquire. On the other hand, sexual conflict will favor females with traits that permit them to avoid disadvantageous matings with males and circumvent efforts by males to control their post-mating behavior. Because sexual conflict means there are different sorts of selection pressures on males and females it can be expected to be an important factor shaping the evolution of mating systems and sexual dimorphism.

Sexual dimorphism and genetic correlation

Intense selection on one sex for a trait such as brilliant coloration may cause the incidental expression of this trait to some degree in the other sex. This incidental expression is explained as the result of a genetic correlation between the sexes, which in turn is caused by the many genes that conspecific males and females share in common (Lande 1987; Hedrick and Temeles 1989; Mueller 1990). The occurrence and magnitude of these correlations are difficult to predict. Their effect when they do occur, however, would be to constrain the evolution of sexual differences and result in, for example, males being less different from females than they would be if such correlations did not exist. There are no published studies of such correlations in butterflies.

MATING-SYSTEM VARIATION IN BUTTERFLIES

There are at least four axes along which mating systems vary in butterflies that might be correlated with differences between species in the degree of sexual dimorphism. All are variables that will affect the intensity of sexual selection. First, there is interspecific variation in how often females mate during their life. In some species, females mate only once, usually shortly after eclosion, whereas in others, females mate several times, usually separated by intervals of several days or weeks (Pliske 1973; Ehrlich and Ehrlich 1978). At one extreme, counts of the remains of spermatophores in the reproductive tracts of queen butterflies (*Danaus gilippus*) have indicated that females may regularly mate five or more times. In species in which females remate often, the number of potential matings a single male can acquire during his life will be higher than in species in which females typically mate only once, so the variance in male reproductive success can be greater, and, as proposed by Wiklund and Forsberg (1991), the effects of sexual selection on males could be more pronounced.

These effects might apply particularly to structures that produce courtship signals or traits important in male–male competition.

The potential for sexual conflict should also vary with whether or not or how often females remate. Attempts to force copulations by males and associated morphological adaptations and, in females, counter-adaptations might be especially advantageous if females mate only once.

Second, butterfly mating systems vary in the tactics used by males to locate females (Rutowski 1991). These tactics have several components including where and when males go to encounter females and what they do at these locations. In the majority of butterflies, males detect females visually at a distance. However, there are two primary tactics used by males to locate females: patrolling and perching. Patrolling males fly about searching widely for females, often visiting and inspecting larval foodplants. In contrast, perching males sit on the ground or on vegetation and wait for females to fly near them. In either case, females are rapidly approached and courted. These two strategies are likely to favor different sorts of morphology and associated behavior. Hence, the mate-locating tactic typical of males in a species could prove to be an important correlate of male morphology. Intraspecific variation in mate-locating tactics has been observed in butterflies (see, for example, Wickman 1985).

Third, there is variation among butterflies in the nature and complexity of behavior patterns performed by males during courtship (Rutowski 1984a). At one extreme, as in many pierids, the male flies about the female for just a few seconds before landing and attempting copulation. In other species males have complicated and stereotyped displays that are performed before attempting copulation. Examples of these displays include the hairpencilling behavior of the male queen butterfly (*Danaus gilippus*, Brower *et al.* 1965), the bowing display of the grayling (*Hipparchia semele*, Tinbergen *et al.* 1942), the proboscis display of the wood white (*Leptidea sinapis*, Wiklund 1977), and the wing-waving display of the barred sulphur (*Eurema daira*, Rutowski 1983). Receptive females perform no special displays in courtship, but the difference in male displays are likely to be due to differences in the nature of the signals used in mate choice by females. Hence, such displays are expected to be correlated with the evolution of signal-producing structures, especially those that produce chemical and visual signals.

Lastly, there is variation in the size of the spermatophores produced by males (Rutowski *et al.* 1983; Svärd and

Wiklund 1989). To the extent that these packages of sperm and nutrients constitute paternal investment in the off-spring, these can influence the nature of sexual selection and the evolution of sexual differences (see also Gwynne, this volume). In particular, if male contributions are large, then sexual selection may act more intensely on females and should produce traits in females normally associated with males. Sexual conflict might be generally more intense in species that produce small spermatophores and favor males with features that will coerce females into mat-ings and using their sperm.

The intensity and nature of selection on male morphol-ogy and behavior might vary dramatically from species to species for reasons discussed above. But what of selection acting independently on female morphology and the extent to which it produces females that are different from males? I expect that selection acting on female butterflies is likely to be more similar among species than that outlined above acting on males. As adults, female butterflies spend the greatest part of their active lives searching for suitable oviposition sites. In all species studied to date, even those in which females lay batches of eggs, females lay only part of their total expected output of eggs in a single bout of oviposition (Stamp 1980). Across species, then, the ecologi-cal demands on females are expected to be fairly similar which should be especially true within clades because of the powerful role phylogenetic history appears to play in larval foodplant preferences.

SEXUAL DIMORPHISMS IN BUTTERFLIES

Here I discuss each of several sexual dimorphisms that are known in butterflies, their potential relationship to mat-ing systems, and efforts to test for the existence of the relationship.

Body size

In the butterflies, females are typically larger than males, although the magnitude of the difference varies from spe-cies to species (Singer 1982; Opler and Krizek 1984; Wik-lund and Forsberg 1991). The size and perhaps even the direction of the difference also depends on the measure-ment of body size used. In the lycaenid *Jalmenus evagoras*, the ratio of male to female size based on mean forewing lengths is 0.976 but the same ratio based on mean body mass is 0.633 (Elgar and Pierce 1988). Wiklund and

Forsberg (1991) measured the degree of sexual size dimorphism using both forewing lengths and pupal masses of diapausing individuals. I have analyzed the data in their tables and found that, among 15 species, the ratio of male to female size based on forewing length (0.962) did not differ significantly from that based on pupal mass (0.92; paired $t = 2.08$, 14 df, $p > 0.05$), but the two measures were not significantly correlated (Spearman rank correlation, $r_s = 0.409$, $p > 0.05$). In other words, a high ratio measured by forewing length did not predict a high ratio based on pupal mass.

Explaining sexual differences in body size is difficult; there have been a number of papers and commentaries warning against studies that focus on the action of selection on only one sex. Greenwood and Adams (1987) point out that most studies of sexual dimorphism have focused solely on the role of sexual selection in the evolution of male size with-out considering the factors responsible for female size. Sexual selection on males does not necessarily produce males that are larger than females. Hence, few analyses of sexual dimorphism in body size to date make clear or unequi-vocal predictions about sexual differences in body size.

Making predictions about body-size differences is also difficult because it is not clear what a difference in body size measures. Body size is often used as an indicator of resource allocation, energy intake, or life-history strate-gies. If so, it is a poor one in that although the sexes in a species may be the same size, their bodies can be struc-tured quite differently. For example, in the black swallow-tail, female pupae contain proportionally more fat and protein than male pupae (Lederhouse *et al.* 1982). None the less, there have been several efforts to explain patterns of sexual dimorphism in body size in butterflies.

Selection on size differences

There could be selection favoring the direction or magni-tude of sexual differences in body mass. Copulation in but-terflies often lasts an hour or more; if a pair take to wing during copulation, only one of the pair flies while the other hangs suspended beneath. Within a species, only one sex flies, but the carrying sex varies from species to species (Shields and Emmel 1973). Mating pairs are at risk of predation and the carrying sex should be powerful enough to effectively evade predators while carrying its mate. This pattern leads to the prediction that the ratio of male to female size should be larger in species in which, all else being equal, males carry females. Wiklund and

Forsberg (1991) tested this prediction by comparing body-size differences with behavior during copulation in two family-level clades, the pierids (eight species) and the satyrids (five species). No formal control for phylogeny was employed. They found no significant relationship between the two variables and concluded that 'selection of large male size seems not to be associated with a tendency for males to be the carrying sex during flights in copula' (p. 377). However, their conclusion is not compelling, in that as they point out, there was little or no variation in this behavior within clades. In all but one pierid species the male carries; in all five satyrid species the female carries.

Body-size differences may also evolve in response to sexual conflict. In males, large size relative to that of females may increase a male's abilities to coerce a female to mate with him. On the other hand, a female's ability to resist male attentions may be related to how much larger she is that the typical male. There are no studies that relate the outcome of attempts by males to force copulation with females to the size difference between the sexes.

Sexual differences in selection

There are also explanations of size dimorphism that invoke sexual differences in selection. Generally, within butterfly species, fecundity in the laboratory is positively correlated with female size (Elgar and Pierce 1988 and references therein), although this relationship is stronger in some species than others (see, for example, Boggs 1986). This size-fecundity relationship may or may not result in higher reproductive success for large females in the field, but it suggests the potential for strong selection on females for large size in some species. In these species, females might tend to be larger than males unless there is similarly intense selection on males.

Wiklund and Karlsson (1988, 1990) report that, among 14 species of European satyrids, high female lifetime egg production and preference for open habitats (subjectively determined) are associated with relatively large sexual differences in forewing length. No specific control for phylogeny within this clade was employed. They propose the hypothesis that in open, and therefore warmer, environments there is more time for oviposition and so fecundity selection on females will be more intense in these environments. If this is true, then I predict that females of the three open-habitat species should be larger than females in forest-dwelling species; this is not the case. There has been no further discussion or testing of this idea since their paper.

Wiklund and Forsberg (1991) tested the idea that selection favoring large male body size will increase with the frequency with which females remate. This idea assumes that male mating success is correlated with body size through greater success in the context of either female choice or male–male competition, a reasonable assumption for some butterflies. Wiklund and Forsberg predicted that in polyandrous species, the ratio of male to female size should be larger that in monandrous species. In a study of 11 pierid and 12 satyrid species (no control for phylogeny within clades) the predicted relationship was found in each clade. However, their conclusion is weakened, especially for the pierids, by a distribution of spermatophore counts strongly skewed towards one. One or two species at the high end of the spermatophore-count spectrum make a large contribution to the positive correlation they found.

In butterfly species with short flight seasons and non-overlapping generations, males eclose on average a few days before females at the beginning of the season. This eclosion pattern is known as protandry and is viewed as a product of intrasexual competition for mating opportunities among males. Singer (1982) proposed that in protandrous species males would have shorter development times than females. He then predicted that, as an incidental effect, the ratio of male to female size should be smaller in species that are markedly protandrous. Using data from eight species that represent a mix of pierids, satyrids, danaids, and nymphalids, he found the smallest males in the three species that showed the clearest protandry. Unfortunately, these species were, as a group, more closely related to each other than to the other species, which raises questions about the independence of the data.

Wiklund and Forsberg (1991) also tested the predicted relationship between size dimorphism and protandry using their pooled data for 14 species from two families. They found the expected relationship only in diapausing generations. They concluded that this finding did not support the protandry hypothesis. Similarly, Nylin *et al.* (1993) found that sexual size dimorphism was not related to the occurrence of protandry in populations of *Pararge aegeria*. The relationship between protandry and size dimorphism appears weak at best.

Male reproductive success and body size

In addition to these efforts to test predictions about sexual dimorphism, there are a number of studies of male size (most often forewing length) and potential measures of fitness in butterflies. Some studies have found a positive correlation between body size and male ability to hold territories (Wickman 1985; Rosenberg and Enquist 1991), male mating success (Wickman 1985; Elgar and Pierce 1988), and the ability to produce both large, effective mating plugs (Matsumoto and Suzuki 1992) and large spermatophores (references in Wiklund *et al.* 1991 (4 species); Rutowski and Gilchrist 1986). In contrast, no correlation was found between body size and resident status in a territorial swallowtail (Lederhouse 1982), between body size and lifetime mating success for an Asian swallowtail (Suzuki and Matsumoto 1992), or between body size and spermatophore size in a pierid (Rutowski 1984b). Yet another study found that male mating success in a sulfur butterfly was highest for males in the middle of the size range (Rutowski 1985). With respect to sexual conflict, no studies have yet examined how male size is related to their ability to force copulations with females or to produce plugs that delay mating. However, larger males do produce larger spermatophores, which may take longer to degrade and so prolong the interval until the next mating by the female (Oberhauser 1989; Wiklund and Kaitala 1995). There is no consensus among the studies on whether male body size relates to reproductive success in butterflies generally.

Thermal consequences of sexual size differences
Two studies have addressed the body-temperature consequences of sexual differences in body size and their impact on reproductive activities. In both species, female mass was about twice that of males. Pivnick and McNeil (1986) found that the smaller body size of males permits them to warm up faster and fly in a broader range of environmental conditions than females, which may be adaptive in their searching for mates. Size-related fecundity selection is thought to outweigh the thermal benefits of small size and favor larger body sizes in females. These authors suggest that these sexual differences in selection arising from these relationships should be greater in patrolling than perching species and so produce a larger sexual size dimorphism in patrollers.

Wickman (1992) examined eight closely related butterfly clades within which mate-locating tactics varied. Within these groups he found that the forewing length of males was relatively larger than that of females in patrollers than in perchers: however, this was not the case for body mass, which is likely to be a better indicator of thermal inertia. Hence, the data do not support Pivnick and McNeil's hypothesis.

Gilchrist (1990) found similar effects of body-size differences on the activities of the sexes in a nymphalid (*Euphydryas phaeton*). However, he found that although small size permitted males to warm up faster, it also made them more susceptible to disruption of activity by short-term environmental changes such as clouds passing over the sun. He argues that sexual size dimorphism is constrained by the thermal disadvantages of large body size in females.

Body shape

The thorax and the abdomen play very different roles in the lives of butterflies. The thorax contains mostly the muscles involved in locomotion. In contrast, the abdomen contains structures for storing and processing nutrients (fat bodies and gut) and the reproductive organs. Given that resources available to larvae during development are limited, how are they allocated to the thorax and the abdomen?

Allocation to the thorax can be measured as the mass of the thorax relative to total body mass. Males consistently have a greater proportion of their body mass in the thorax (as much as 50%) than do females (Wickman 1992).

Allocation patterns should reflect several aspects of ecology and reproductive biology. Wickman (1992) tested the idea that variation among species in male mate-locating tactic will produce adaptive differences in the ratio of the thorax to body mass. The enhanced acceleration and speed that come with a large flight-muscle mass in the thorax should be a special advantage in perching males. He predicted, then, that perching males should have a higher thorax : body mass and lower abdomen : body mass ratio than patrolling males. He examined this relationship in a comparative study of 25 species that were in eight recently evolved clades. Each clade contained both patrolling and perching species. The body construction of patrollers and perchers within the sets were then compared in an analysis that also controlled for allometry and covariation between the sexes. Within the eight contrasts, perching males were significantly more likely to have larger thoraces and smaller abdomens than patrolling males.

There are at least two other important variables that might affect sexual differences in body design. The first is variation in palatability both between species and between the sexes. Srygley and Chai (1990b) and Marden and Chai (1991) found that among-species palatability is positively correlated with thoracic mass. This pattern was revealed in broad comparative studies that group taxa sharing traits by common ancestry to insure phylogenetic independence in the data set. Their hypothesis states that allocation to flight muscles used for escape from predators should be of special importance to non-mimetic, palatable species. In palatable species, predation could constrain the evolution of sexual differences in body proportions more than in those that are unpalatable. This idea predicts that sexual dimorphism could be greater in unpalatable than palatable species. In addition, sexual differences in palatability, such as those described for the monarch butterfly (Brower and Glazier 1975), could lead to sexual differences in body proportions.

Second, the number of developed eggs in the female abdomen at eclosion varies among species (Dunlap-Pianka *et al.* 1977). This variation might explain variation in the size and nature of sexual differences in body shape or proportions. In particular, the proportion of the body mass in the abdomen might be very large for females in species that eclose with all eggs present and many well along in development. In species in which females continuously develop eggs during their life, the abdominal mass at eclosion might be smaller. This variation implies that care must be taken to account for interspecific variation in female reproductive strategies and when during the life of a female mass measurements are made.

As an additional issue in the examination of body size and proportions, there is variation among butterflies in whether or not there are sexual differences in foreleg development. In most butterflies, the forelegs in both sexes are equal in size whether they are fully developed (as in pierids and papilionids) or reduced in size (as in nymphalids). However, in some groups (e.g. Libytheidae and Riodinidae) foreleg reduction is restricted to the male (Robbins 1988). Whether this sexual dimorphism in appendages is related to mating system structure or sexual conflict (as in some other insects (Crespi 1988)) is not known.

Wing shape

Sexual differences in wing shape are well known in butterflies. Perhaps the most striking difference is seen in the paradise birdwing (*Ornithoptera paradisea*) (Fig. 15-1). In this species the hindwings of the males are smaller than those of the females or of congeneric males, and each hindwing bears a thin tail. In other species in which the sexes differ in wing shape (Wickman 1993), the wings of females are larger and more rounded than those of the males.

Differences in wing shape are most likely to be a function of to sexual differences in lifestyle that make different demands on the aerodynamic performance of the wings. For example, rapid acceleration and high speeds may be especially favored in males that perch to locate females (Wickman 1992). Similarly, reliance on evasive flights to escape predators might favor wing shape that permits rapid acceleration and high maneuverability.

In his comparative study of the relationship between sexual dimorphism and mate-searching behavior discussed above, Wickman (1992) showed that males in perchers have wings that are smaller relative to body mass (higher wing loading) and narrower relative to their length (higher aspect ratio) than males in patrollers. The same result was obtained for females, so Wickman controlled for covariation between males and females. The differences between males of species with different mating systems in wing loading and aspect ratio persisted. In contrast, the differences between females in species with different mate-searching tactics disappeared when sexual covariation was taken into account. This result suggested to Wickman that 'different mating systems select for different male designs'. The high wing loading was associated with a perching tactic, as predicted, but the adaptive basis for the association of high aspect ratios with perching remains unclear.

Visual signals: wing color and pattern

When sexual dimorphism in wing color and pattern exists in butterflies, males have dorsal wing markings that, compared with those of females, are more precisely demarcated, are composed of colors of greater saturation, and sometimes have pattern elements not found in females. The colors of males are described as more brilliant, brighter, and more conspicuous than those of females. However, strong sexual dichromatism in butterflies is the exception and not the rule.

With respect to direct selection on color differences, I know of no hypothesis that proposes selection specifically favoring differences in color between the sexes. Brower *et al.* (1967) put forth the idea that in distasteful species the evolution of intraspecific differences in coloration could be constrained by selection against deviation from the color pattern that predators had learned to associate with distastefulness. This idea predicts that across species with similar mating systems, sexual dichromatism would be greater in palatable than in distasteful species. I know of no test of this idea.

Arguments about selection acting independently on the sexes have focussed on sexual selection acting on males, specifically in female choice, to explain the tendency for males to have brighter colors. There are few data supporting this view, largely because of the difficulties of setting up the necessary assays of female preference. Silberglied and Taylor (1978) showed that in *Colias eurytheme*, in which the dorsal wing surfaces of males but not females strongly reflect ultraviolet, females were less likely to mate with males whose ultraviolet reflectance had been experimentally eliminated than with control males. Rutowski (1981) showed that in *Pieris protodice*, in which the wings of males but not females absorb ultraviolet, females more often solicit courtship from ultraviolet-absorbing than ultraviolet-reflecting males presented to them on a tether. In field experiment, Wiernasz (1989) demonstrated that *P. occidentalis* females discriminate among males on the basis of certain melanic markings on the wings. In the swallowtail *Papilio glaucus*, females either mimic distasteful species or have coloration similar to that of males; females solicit courtship more readily from unaltered males than from males given a mimetic coloration (Krebs and West 1988). These studies strongly implicate female mating preferences in the evolution of male coloration. The benefits, material or genetic, of color preferences in female butterflies remain unclear.

Silberglied (1983) revitalized an alternative sexual selection explanation, first put forth by Wallace (1889), stating that bright male coloration is favored in the context of male–male competition. Rutowski (1992) examined the effect of color alteration on the ability of male *Hypolimnas bolina* to maintain residency at perching sites. The sample sizes were small but suggest no strong effect. In contrast, R. C. Lederhouse (personal communication) found that male *Papilio polyxenes* that were experimentally colored to look like females were less successful in becoming territorial residents. This idea deserves further testing but it should be kept in mind that the inter- and intrasexual hypotheses for the evolution of bright male color in butterflies are not always mutually exclusive.

These studies address intraspecific variation in male color but do not consider female coloration or permit us to understand fully why females in some species are more different from males than in others. How might selection on females, especially natural selection, differ from that on males? There may be sexual differences in the nature and level of predation. Males and females may differ in palatability. In the monarch butterfly, females are more

distasteful to birds than males are (Brower and Glazier 1975) and this may lead to sexual differences in predation by visually hunting predators and, therefore, to differences in coloration. Differences in predation between the sexes figure prominently in ideas about why mimetic color patterns are restricted to females in many species of butterfly. Codella and Lederhouse (1989) have shown that the sexual difference in coloration of *P. polyxenes*, especially on the dorsal wing surface, results in males being less effective mimics of *Battus philenor* than females. This suggests that males experience higher levels of predation than females, the costs of which are apparently countered by benefits in the context of competition for mates.

Differences in coloration might also evolve in the context of sexual differences in thermoregulatory needs, such as those described for *Euphydryas phaeton* (Gilchrist 1990) and *Thymelicus lineola* (Pivnick and McNeil 1986). Dark coloration, especially at the wing bases, increases solar heat gain (Wasserthal 1976); this might be of special advantage to females, which are generally larger than males (and therefore have greater thermal inertia) and often fly in coooler microhabitats while searching for oviposition sites. Males, whether perchers or patrollers, tend to fly about in open areas where solar radiation is not so limited. Kingsolver (1985) has suggested that in the genus *Pieris* sexual differences in the coloration of distal areas of the wing also have significant effects on solar heat gain by reflecting radiation onto the thorax and abdomen (but see Heinrich 1990).

Another apparent trend in sexual differences in coloration is that the coloration of males is less variable than that of females. This pattern is most evident in some cases of sex-limited mimicry in butterflies. For example, in the North American tiger swallowtail (*Papilio glaucus*) males are monomorphic whereas females have either a male-like coloration or a coloration that mimics that of a distasteful sympatric swallowtail, *B. philenor*. Studies of other species in which males and females are similarly colored support the the observation that females are more variable in the size of wing marking than males (Pearse and Murray 1982; Hazel 1990). The more limited variation in males is most often explained as a result of sexual selection on males.

Given the difficulties of experimental manipulations, the time is ripe for a comparative study of butterfly coloration and mating system like those that have been done recently for lekking birds (see, for example, Höglund 1989; Oakes 1992). A key prediction to test of the intrasexual competition hypothesis is that sexual dimorphism should be most striking in those species in which males occupy and

defend perching sites as a mate-locating tactic. The papilionids, lycaenids, and nymphalids provide the best opportunities for this sort of study because of the variation in male
tactics found within these families. Moreover, brilliant
male coloration may be most likely to occur in species in
which females mate more than once. Of course, these predictions depend on an understanding of the factors shaping the evolution of female color as well.

Other signal-producing structures

Chemical signals

There is variation in butterflies in the development and
location of structures that are known or thought to be
used in the production of chemical signals (Boppré 1984).
Such structures are most often found in males but not in
all species. In some, males have no obvious scent-producing structures; at the other extreme, there are species
with elaborate and complex structures such as the eversible
hairpencils of male danaids (Pliske and Salpeter 1971) and
the costal brushes of ithomiines (Pliske 1975b). Females
either lack obvious scent-producing structures or have
structures different from those found on males (Gilbert
1976; Wiklund and Forsberg 1985).

Rutowski (1980) suggested that the structure and chemistry of scent-producing devices in butterflies and other
Lepidoptera, including moths, is related to how they are
used in the courtship (see also Phelan, this volume). In
comparison with non-eversible structures, eversible structures were predicted to emit smaller, less volatile compounds and to be presented during courtship very close
or directly to the female's antennae. He found support for
this prediction in a comparison of seven diverse species of
Lepidoptera; however, there were no controls for phylogeny and the two species with non-eversible structures
were in the same family. Other than this test there has
been no effort to relate the differences between species to
differences in mating system. However, it would be of interest to see if the development of scent-producing devices is
inversely related to the brilliance of male coloration, a
trade-off like that suggested between song and coloration
in some birds.

In some species there is evidence that females produce
a scent in the pupal stage that is attractive to males
(Boppré 1984). However, the patterns of occurrence of
these signals and their precise role in the mating system
have not been examined. In species that produce such signals males detect pupae visually from a distance and but

recognize them as near eclosion by using chemical
cues (Mallet and Jackson 1980; Elgar and Pierce 1988).
It is unclear whether the production of scents by female
pupae is an evolutionary cause or consequence of the
male's behavior.

Acoustic signals

There are certain species of butterfly that produce sounds,
which may function as signals. This phenomenon is best
known in the *Hamadryas* butterflies. Males produce the
sounds and have special structures on the wings that produce the sounds (Otero 1990). *Hamadryas* males apparently
defend perches used in mate location, but no effort to relate
sound production to mating system has been undertaken.

Sensory systems

Sexual differences in the morphology of sensory systems,
especially differences in the development of antennae and
eyes, are well known in the insects. The typical case is that
males have larger and more elaborate structures than
females. Although sexual dimorphism of this sort in antennal morphology is well known for the moths (Scoble 1992),
no dramatic sexual differences have been reported for the
butterflies (Myers 1968; Grula and Taylor 1980; Odendaal
et al. 1985). As pointed out by these authors, this lack of
dimorphism in the butterflies is surprising given the documented importance of chemical signals in mate selection
by females in some of the species examined. The fact that
most chemical communication in butterflies is close-range
may explain why there is no evidence of special morphological adaptations in butterfly antennae.

Yagi and Koyama (1963) examined variation in eye structure in the Lepidoptera, including moths, to test phylogenetic hypotheses for the taxon. In doing so, they described
several sexual differences in eye morphology. First, based
on measurements made on four species (one papilionid,
two pierids and one lycaenid) using an undescribed technique, they found that males have an eye surface area that is
20–40% larger than that of females. This is surprising
given that males are generally smaller than females.
Second, eye size as measured by eye height increases with
wingspan more quickly in males than in females. Third, in
three species (*Pieris crucivora*, *Colias erate*, and *Papilio xuthus*),
they found that males have more facets or corneal lenses
than females. Finally, in the two pierids, they observed
that as eye size increases the number of facets and the area
of individual facets increases more rapidly in males than

females. These results are unfortunately presented without sample sizes for the number of specimens examined within each species, and with minimal procedural details. However, there is apparent interspecific variation in the magnitude of all these differences.

These differences in eye morphology are predicted to produce males with eyes that, compared with those of females, have a greater overall spatial resolution, a larger visual field, or both (Land and Fernald 1992; Warrant and McIntyre 1993). These enhancements of overall visual-system performance, as opposed to enhancements of the performance of specific parts of the visual-system, might be of special advantage in males that remain stationary or perch to detect mates, competitors, or predators. Similar contrasts in visual system structure are found between those odonates that perch to detect prey and those that patrol (Pritchard 1966; Sherk 1978).

In butterflies in which males patrol for mates, males and females engage in similar forward-directed search strategies: females for oviposition sites, males for mates. However, in species in which males perch, males engage in search behavior very different from that of females. Differences in selection on the visual system between the sexes and consequent differences in eye morphology are, then, expected to be greatest in butterflies in which males use perching as opposed to patrolling as a mate-locating tactic. The existing data do not permit a test of this idea.

Finally, sexual differences in the distribution of visual pigment types in the eyes have been found by Bernard and Remington (1991) in two species of lycaenid butterfly. In both sexes, the dorsal part of the eye contains visual pigments with absorbance peaks in the blue and ultraviolet. However, females have a third pigment, sensitive to red, that the authors suggest is important in recognizing the larval foodplant. From all these observations additional studies of the relationship between mating systems and sexual dimorphism in visual system structure and performance seem likely to be fruitful.

Patterns of development

Protandry, as described earlier, refers to the observation that, in many species of butterfly, males eclose from the pupa on average a few days or weeks earlier in the year than the females. This phenomenon occurs in both diapausing and non-diapausing generations that are non-overlapping. This is a context in which selection could act directly on a sexual difference. Selection might favor males

that emerge before females in the context of competition for mates (see also Choe, this volume, for a similar phenomenon in Zoraptera) and favor females that minimize the time between emergence and mating.

There have been several algebraic models put forward to describe the selective forces shaping emergence patterns in male and female butterflies. All of them include a number of mating-system features in the assumptions and in the parameters that can be varied. Most assume that females mate only once (Wiklund and Fagerstrom 1977; Fagerstrom and Wiklund 1982; Bulmer 1983; Iwasa et al. 1983; Zonneveld and Metz 1991). However, Zonneveld (1992) recently developed a model in which females can mate more than once and further matings are weighted to reflect the age-related decline in fecundity known in butterflies. This model suggests that female mating frequency will be a poor predictor of the level of protandry and that factors affecting male competition for virgins will be better. One prediction is that when mate competition at emergence is absent (i.e. there is emergence of females in reproductive diapause) there should be no protandry.

There have been four studies that have examined how female emergence patterns and behavior have influenced the evolution of protandry. Wiklund and Solbreck (1982) compared developmental patterns in diapausing and non-diapausing generations of *Leptidea sinapis* and found that in spite of differences in development the degree of protandry was the same in both generations. They argued convincingly that since the difference in emergence times is the same in both generations it is adaptive in competition for mates and not an incidental effect of males having shorter development times than females because they are smaller. Iwasa et al. (1983) used the shape of female emergence and male mortality curves in a well-studied population of *Euphydryas editha* to predict the position and shape of the male emergence curve. Contrary to the predictions, they found that in the field population males emerged earlier and did not stop emerging as abruptly. They point out several potentially incorrect assumptions that may have weakened their model. Wiklund and Forsberg (1991) studied the relationship between protandry and monogamy in females in 23 species of butterfly. Their data supported the predicted negative correlation between the degree of protandry and the level of polygamy in generations emerging from diapause. Finally, Nylin et al. (1993) have examined changes in the degree of protandry found in populations of the speckled wood butterfly, *P. aegeria*, along a north–south transect. They found that, as predicted, protandry was most likely to occur in seasonal environments in which generations did not overlap.

Two other sex differences in development have been described in butterflies and attributed in part to features of the mating system. First, Wiklund *et al.* (1991) predicted that males should have a higher growth rate than females. This prediction follows, given that protandry is beneficial and that there is selection for large size in males resulting from the benefits of producing large spermatophores and a positive correlation between body size and the size of spermatophores produced. They found a sexual difference in growth rate in *Pieris napi*. Second, Wiklund *et al.* (1992) reasoned that in species that diapause as adults males should enter diapause earlier in the season than females. Late in the flight season, they argue, males are expected to experience low gains from matings because most females are old and will display an age-related reduction in fecundity. They experimentally demonstrated that as the flight season progresses males enter dipause before than females in *P. napi*, and found strongly female-biased field populations late in the flight season in this species and two others (*Pararge aegeria* and *Polygonia c-album*).

SUGGESTIONS FOR FUTURE STUDY

Mating-system structure has shaped the evolution of sexual dimorphism in butterflies, especially through the special selection pressures brought to bear on males. Diversity in male wing and body design is related to diversity in mate-locating tactics. Elaboration of coloration, scent-producing structures, and sensory systems in males appears to be a product of female choice acting on males, while different selection pressures act on females. To date, the only evidence for selection acting directly on differences between the sexes is from studies of emergence times, in which intrasexual competition favors males that eclose some time before females.

There are many ways in which understanding of sexual dimorphism in butterflies is far from complete. Research into the aspects of the biology of butterflies suggested below might prove especially helpful in enhancing our understanding of sexual dimorphism in this taxa.

Comparative studies in their design and analysis should carefully account for phylogeny using methods appropriate to the questions being addressed (Nylin and Wedell 1994). Inadequate control for phylogeny is the most common problem with past comparative studies. This reflects the fact that careful, specific-level phylogenies are lacking for most genera and subfamilies although Martin and Pashley (1992) recently proposed a phylogeny for the subfamilies, based on molecular characters. A long-term goal would be to use Björklund's (1991) criteria to clearly establish the cause-and-effect relationships between sexual dimorphism and mating systems in butterflies.

Our understanding of mating-system diversity in butterflies is still in many respects inadequate to permit, for example, comparative tests of the hypothesis about the relationship between male mate-locating tactic and sexual dimorphism in coloration. Careful and quantitative descriptions of courtship and mate-locating behavior and especially the patterns of female behavior will help enlarge the foundation on which comparative studies can be built.

Many traits require experimental investigation of their role in butterfly mating systems and their adaptive features. This is especially true of signal-producing devices and sensory systems. Vane-Wright and Boppré (1993) should be consulted for more specific recommendations.

Studies of sexual dimorphism in a trait should examine the behavior and ecology of both sexes to get the clearest picture of how the trait in both sexes is related to reproductive success. The study by Elgar and Pierce (1988) is perhaps the best model for this sort of approach. The benefits of incorporating information on natural as well as sexual selection in explaining sexual dimorphism have been reviewed by Jennions (1993).

The genetics of sexually dimorphic traits in butterflies, except with respect to coloration (see, for example, Brakefield 1984), are poorly known. For most traits we are ignorant of the existence and magnitude of genetic correlations between the sexes and the heritability of sexually dimorphic traits. Information on these points will help refine our ability to predict the effects of selection on the evolution of sexual dimorphism.

Another issue that needs attention is how size affects butterfly behavior and physiology in butterflies. The largest species of butterflies are about 20–30 times greater in mass and about 10 greater in wingspan than the smallest. Many of the ideas presented here do not take differences in scale into account. In some birds, the degree of sexual size dimorphism is a function of body size (Webster 1992). Do such patterns exist in butterflies? The nymphalids as a group display a large diversity in body size and so might offer a good opportunity to look for and test explanations for these patterns.

Finally, understanding conflicts of interest between the sexes and how they are resolved in reproductive interactions can be helpful in understanding the evolution of sex-related morphology and, therefore, sexual dimorphism (see Arnqvist and Rowe 1995; Arnqvist, this volume). Dramatic products of sexual conflict such as forced copulations and mate-guarding are rare in butterflies. None the less, we may better understand the role of sexual conflict in the evolution of sexual dimorphism in butterflies if we attend to assessing better the costs and benefits to males and females of mating behavior in this group (see Alexander *et al.*, this volume).

ACKNOWLEDGEMENTS

Christer Wiklund and Robert Lederhouse kindly provided prepublication copies of relevant articles as well as insightful and helpful comments on an earlier draft of the manuscript. The manuscript also benefitted from the comments of John Alcock, Scott Pitnick and Per-Olof Wickman. Other helpful information and ideas were provided by Jim Collins, Diana Hews, Joel Kingsolver, Barbara Terkanian and Birgitta Tullberg. Barbara Terkanian also did the excellent artwork for Figure 1. For all this I am grateful.

LITERATURE CITED

Arak, A. 1988. Sexual dimorphism in body size: a model and a test. *Evolution* **42**: 820–825.

Arnqvist, G., and L. Rowe. 1995. Sexual conflict and arms races between the sexes – a morphological adaptation for control of mating in a female insect. *Proc. R. Soc. Lond.* B **261**: 123–127.

Baker, R. R. and G. A. Parker. 1979. The evolution of bird coloration. *Phil. Trans. R. Soc.* Lond. B **287**: 63–130.

Bernard, G. D. and C. L. Remington. 1991. Color vision in *Lycaena* butterflies: spectral tuning of receptor arrays in relation to behavioral ecology. *Proc. Natl. Acad. Sci. U.S.A.* **88**: 2783–2787.

Björklund, M. 1991. Evolution, phylogeny, sexual dimorphism, and mating system in the grackles (*Quiscalus* spp.: Icterinae). *Evolution* **45**: 608–621.

Boggs, C. L. 1986. Reproductive strategies of female butterflies: variation in and constraints on fecundity. *Ecol. Entomol.* **11**: 7–15.

–. 1995. Male nutirent donation: phenotypic consequences and evolutionary implications. In *Insect Reproduction*. S. R. Leather and J. Hardie, eds., pp. 215–242. Cleveland: CRC Press.

Boppré, M. 1984. Chemically mediated interactions between butterflies. In *The Biology of Butterflies*. R. Vanewright, P. Ackery and P. DeVries, eds., pp. 259–275. New York: Academic Press.

Brakefield, P. M. 1984. The ecological genetics of quantitative characters of *Maniola jurtina* and other butterflies. In *The Biology of Butterflies*. R. Vanewright, P. Ackery and P. DeVries, eds., pp. 167–190. New York: Academic Press.

Brower, L. P., J. V. Z. Brower and F. P. Cranston. 1965. Courtship behavior of the queen butterfly, *Danaus gilippus berenice* (Cramer). *Zoologica* **50**: 1–39.

Brower, L. P., J. V. Z. Brower and J. M. Corvino. 1967. Plant poisons in as terrestrial food chain. *Proc. Natl. Acad. Sci. U.S.A.* **57**: 893–898.

Brower, L. P. and S. C. Glazier. 1975. Localization of heart poisons in the monarch butterfly. *Science (Wash., D.C.)* **188**: 19–25.

Bulmer, M. G. 1983. Models for the evolution of protandry in insects. *Theor. Popul. Biol.* **23**: 314–322.

Clutton-Brock, T. H. and G. A. Parker. 1995. Sexual coercion in animal societies. *Anim. Behav.* **49**: 1345–1365.

Codella, S. G. Jr. and R. C. Lederhouse. 1989. Intersexual comparison of mimetic protection in the black swallowtail butterfly, *Papilio polyxenes*: experiments with captive blue jays predators. *Evolution* **43**: 410–420.

Crespi, B. J. 1988. Adaptation, compromise, and constraint: the development, morphometrics, and behavioral basis of a fighter-flyer polymorphism in male *Hoplothrips karnyi* (Insecta: Thysanoptera). *Behav. Ecol. Sociobiol.* **23**: 93–104.

D'Abrera, B. 1971. *Butterflies of the Australian Region*. Melbourne: Lansdowne Press, Ltd.

Dickinson, J. L. and R. L. Rutowski. 1989. The function of the mating plug in the chalcedon checkerspot butterfly. *Anim. Behav.* **38**: 154–162.

Drummond, B. A. III. 1984. Multiple mating and sperm competition in the Lepidoptera. In *Sperm Competion and the Evolution of Animal Mating Systems*. R. L. Smith, ed., pp. 291–370. New York: Academic Press.

Dunlap-Pianka, H., C. L. Boggs and L. E. Gilbert. 1977. Ovarian dynamics in heliconiine butterflies: programmed senescence versus eternal youth. *Science (Wash., D.C.)* **197**: 487–490.

Eberhard, W. 1985. *Sexual Selection and Animal Genitalia*. Cambridge, Mass.: Harvard University Press.

Ehrlich, A. H. and P. R. Ehrlich. 1978. Reproductive strategies in butterflies. I. Mating frequency, plugging, and egg number. *J. Kansas. Entomol. Soc.* **51**: 666–697

Elgar, M. A. and N. E. Pierce. 1988. Mating success and fecundity in an ant-tended lycaenid butterfly. In *Reproductive Success: Studies of Individual Variation in Contrasting Breeding Systems*. T. H. Clutton-Brock, ed., pp. 59–75. Chicago: University of Chicago Press.

Fagerström, T. and C. Wiklund. 1982. Why do males emerge before females? Protandry as a mating strategy in male and female butterflies. *Oecologia* **52**: 164–166.

Gilbert, L. 1976. Post-mating female odor in *Heliconius* butterflies: a male-contributed antiaphrodisiac. *Science (Wash., D.C.)* **193**: 419–420.

Gilchrist, G. W. 1990. The consequences of sexual dimorphism in body size for butterfly flight and thermoregulation. *Funct. Ecol.* **4**: 475–487.

Greenwood, P. J. and J. Adams. 1987. Sexual selection, size dimorphism, and a fallacy. *Oikos* **48**: 106–108.

Grula, J. W. and O. R. Taylor, Jr. 1980. A micromorphological and experimental study of the antennae of the sulphur butterflies, *Colias eurytheme* and *C. philodice* (Lepidoptera: Pieridae). *J. Kansas Entomol. Soc.* **53**: 476–484.

Harvey, A. W. 1988. Sexual size dimorphism and fecundity in satyrid butterflies: a comment. *Am. Nat.* **132**: 750–752.

Harvey, P. H. and M. D. Pagel. 1991. *The Comparative Method in Evolutionary Biology.* Oxford University Press.

Hazel, W. N. 1990. Sex-limited variability and mimicry in the swallowtail butterfly *Papilio polyxenes* Fabr. *Heredity* **65**: 109–114.

Hedrick, A. V. and E. J. Temeles. 1989. The evolution of sexual dimorphism in animals: hypotheses and tests. *Trends Ecol. Evol.* **4**: 136–138.

Heinrich, B. 1990. Is reflectance basking real? *J. Exp. Biol.* **154**: 31–37.

Höglund, J. 1989. Size and plumage dimorphism in lek-breeding birds: a comparative analysis. *Am. Nat.* **134**: 72–87.

Iwasa, Y., F. J. Odendaal, D. D. Murphy, P. R. Ehrlich and A. E. Launer. 1983. Emergence patterns in male butterflies: a hypothesis and a test. *Theor. Popul. Biol.* **23**: 363–379.

Jennions, M. D. 1993. Female choice in birds and the cost of long tails. *Trends Ecol. Evol.* **8**: 230–232.

Kingsolver, J. G. 1985. Thermoregulatory significance of wing melanization in *Pieris* butterflies (Lepidoptera: Pieridae): physics, posture, and pattern. *Oecologia (Berl.)* **66**: 546–553.

Krebs, R. A. and D. A. West. 1988. Female mate preference and the evolution of female-limited Batesian mimicry. *Evolution* **42**: 1101–1104.

Land, M. F. and R. D. Fernald. 1992. The evolution of eyes. *Annu. Rev. Neurosci.* **15**: 1–29.

Lande, R. 1987. Genetic correlations between the sexes in the evolution of sexual dimorphism and mating preferences. In *Sexual Selection: Testing the Alternatives.* J. W. Bradbury and M. B. Andersson, eds, pp. 83–94. New York: John Wiley and Sons, Ltd.

Lederhouse, R. C. 1982. Territorial defense and lek behavior of the black swallowtail butterfly, *Papilio polyxenes. Behav. Ecol. Sociobiol.* **10**: 109–118.

Lederhouse, R. C., M. D. Finke and J. M. Scriber. 1982. The contributions of larval growth and pupal duration to protandry in the black swallowtail butterfly, *Papilio polyxenes. Oecologia (Berl.)* **53**: 296–300.

Mallet, J. L. B. and D. A. Jackson. 1980. The ecology and social behaviour of the neotropical butterfly *Heliconius xanthocles* in Colombia. *Zool. J. Linn. Soc.* **70**: 1–13.

Marden, J. H. and P. Chai. 1991. Aerial predation and butterfly design: how palatability, mimicry, and the need for evasive flight constrain mass allocation. *Am. Nat.* **138**: 15–36.

Martin, J. A. and D. P. Pashley. 1992. Molecular systematic analysis of butterfly family and some subfamily relationships (Lepidoptera: Papilionoidea). *Ann. Entomol. Soc. Am.* **85**: 127–139.

Matsumoto, K. and N. Suzuki. 1992. Effectiveness of the mating plug in *Atrophaneura alcinous* (Lepidoptera: Papilionidae). *Behav. Ecol. Sociobiol.* **30**: 157–163.

Moore, A. J. 1990. The evolution of sexual dimorphism by sexual selection: the separate effects of intrasexual selection and intersexual selection. *Evolution* **44**: 315–331.

Mueller, H. C. 1990. The evolution of reversed sexual dimorphism in size in monogamous species of birds. *Biol. Rev.* **65**: 553–585.

Myers, J. 1968. The structure of the antennae of the Florida Queen butterfly, *Danaus gilippus berenice* (Cramer). *J. Morphol.* **125**: 315–328.

Nylin, S. and N. Wedell. 1994. Sexual size dimorphism and comparative methods. In *Phylogenetics and Ecology.* P. Eggleton and R. Vane-Wright, eds, pp. 253–280. London: Academic Press.

Nylin, S., C. Wiklund, P.-O. Wickman and E. Garcia-Barros. 1993. Absence of trade-offs in life-history evolution: the case of sexual size dimorphism and protandry in *Pararge aegeria* (Lepidoptera: Satryidae). *Ecology* **74**: 1414–1427.

Oakes, E. J. 1992. Lekking and the evolution of sexual dimorphism in birds: comparative approaches. *Am. Nat.* **140**: 665–684.

Obara, Y. 1964. Mating behaviour of the cabbage white butterfly, *Pieris rapae crucivora.* II. The 'mate-refusal' posture of the female. *Dobuts. Zasshi* **73**: 131–135.

Oberhauser, K. S. 1989. Effects of spermatophores on male and female monarch butterfly mating success. *Behav. Ecol. Sociobiol.* **25**: 237–246.

–. 1992. Rate of ejaculate breakdown and intermating intervals in monarch butterflies. *Behav. Ecol. Sociobiol.* **31**: 367–373.

Odendaal, F. J., P. R. Ehrlich and F. C. Thomas. 1985. Structure and function of the antennae of *Euphydryas editha* (Lepidoptera: Nymphalidae). *J. Morphol.* **184**: 3–22.

Opler, P. A. and G. O. Krizek. 1984. *Butterflies East of the Great Plains: An Illustrated Natural History.* Baltimore: Johns Hopkins University Press.

Orr, A. G. and R. L. Rutowski. 1991. The function of the sphragis in *Cressida cressida* (Fab.) (Lepidoptera, Papilionidae): a visual deterrent to copulation attempts. *J. Nat. Hist.* **25**: 703–710.

Otero, L. D. 1990. The stridulatory organ in *Hamadryas* (Nymphalidae): preliminary observations. *J. Lepid. Soc.* **44**: 285–288.

Parker, G. A. 1979. Sexual selection and sexual conflict. In *Sexual Selection and Reproductive Competition in Insects.* M. S. Blum and N. A. Blum, eds, pp. 123–166. New York: Academic Press.

Pearse, F. K. and N. D. Murray. 1982. Sex and variability in the common brown butterfly *Heteronympha merope merope* (Lepidoptera: Satyrinae). *Evolution* **36**: 1251–1264.

Pivnick, K. A. and J. N. McNeil. 1986. Sexual differences in the thermoregulation of *Thymelicus lineola* adults (Lepidoptera: Hesperidae). *Ecology* **67**: 1024–1035.

Pliske, T. E. 1973. Factors determining mating frequencies in some New World butterflies and skippers. *Ann. Entomol. Soc. Am.* **66**: 164–169.

–. 1975a. Courtship behavior of the monarch butterfly, *Danaus plexippus* L. *Ann. Entomol. Soc. Am.* **69**: 143–151.

–. 1975b. Courtship behavior and use of chemical communication by males of certain species of ithomiine butterflies. *Ann. Entomol. Soc. Am.* **68**: 935–942.

Pliske, T. E. and M. M. Salpeter. 1971. The structure and development of the hairpencil glands in males of the queen butterfly, *Danaus gilippus berenice*. *J. Morphol.* **134**: 215–241.

Pritchard, G. 1966. On the morphology of the compound eyes of dragonflies (Odonata: Anisoptera), with special reference to their role in prey capture. *Proc. R. Entomol. Soc.* A41 :1–8.

Robbins, R. K. 1988. Comparative morphology of the butterfly foreleg coxa and trochanter (Lepidoptera) and its systematic implications. *Proc. Entomol. Soc. Wash.* **90**: 133–154.

Rosenberg, R. H. and M. Enquist. 1991. Contest behaviour in Weidermeyer's admiral butterfly *Limenitis weidermeyerii* (Nymphalidae): the effect of size and residency. *Anim. Behav.* **42**: 805–811.

Rutowski, R. L. 1978. The form and function of ascending flights in *Colias* butterflies. *Behav. Ecol. Sociobiol.* **3**: 163–172.

–. 1979. The butterfly as a honest salesman. *Anim. Behav.* **7**: 1269–1270.

–. 1980. Male scent-producing structures in *Colias* butterflies: function, localization, and adaptive features. *J. Chem. Ecol.* **6**: 13–26.

–. 1981. Sexual discrimination using visual cues in the checkered white butterfly (*Pieris protodice*). *Z. Tierpsychol.* **55**: 325–334.

–. 1983. The wing-waving display of *Eurema daira* males (Lepidoptera: Pieridae): its structure and role in successful courtship. *Anim. Behav.* **31**: 985–989.

–. 1984a. Sexual selection and the evolution of butterfly mating behavior. *J. Res. Lepid.* **23**: 125–142.

–. 1984b. Production and use of secretions passed by males at copulation in *Pieris protodice* (Lepidoptera, Pieridae). *Psyche* **91**: 141–152.

–. 1985. Evidence for mate choice in a sulphur butterfly (*Colias eurytheme*). *Z. Tierpsychol.* **70**: 103–114.

–. 1991. The evolution of male mate-locating behavior in butterflies. *Am. Nat.* **138**: 1121–1139.

–. 1992. Male mate-locating behavior in the common eggfly, *Hypolimnas bolina* (Nymphalidae). *J. Lepid. Soc.* **46**: 24–38.

Rutowski, R. L. and G. W. Gilchrist. 1986. Copulation in *Colias eurytheme* (Lepidoptera: Pieridae): patterns and frequency. *J. Zool. (Lond.)*. **209**: 115–124.

Rutowski, R. L., M. Newton and J. Schaeffer. 1983. Interspecific variation in the size of the nutrient investment made by male butterflies during copulation. *Evolution* 37: 708–713.

Scoble, M. J. 1992. *The Lepidoptera: Form, Function, and Diversity*. New York: Oxford University Press.

Sherk, T. E. 1978. Development of the compound eyes of dragonflies (Odonata). III. Adult compound eyes. *J. Exp. Zool.* **203**: 61–80.

Shields, O. and J. F. Emmel. 1973. A review of carrying pair behavior and mating times in butterflies. *J. Res. Lepid.* **12**: 25–64.

Shine, R. 1989. Ecological causes for the evolution of sexual dimorphism: a review of the evidence. *Q. Rev. Biol.* **64**: 419–461.

Silberglied, R. E. 1984. Visual communication and sexual selection among butterflies. In *The Biology of Butterflies*. R. Vane-Wright, P. Ackery and P. DeVries, eds., pp. 207–233. New York: Academic Press.

Silberglied, R. E. and O. R. Taylor, Jr. 1978. Ultraviolet reflection and its behavioral role in the courtship of the sulfur butterflies, *Colias eurytheme* and *C. philodice*. *Behav. Ecol. Sociobiol.* **3**: 203–243.

Singer, M. C. 1982. Sexual selection for small size in male butterflies. *Am. Nat.* **119**: 440–443.

Slatkin, M. 1984. Ecological causes of sexual dimorphism. *Evolution* **38**: 622–630.

Srygley, R. B. and P. Chai. 1990a. Predation and the elevation of thoracic temperature in brightly colored neotropical butterflies. *Am. Nat.* **135**: 766–787.

–. 1990b. Flight morphology of Neotropical butterflies: palatability and distribution of mass to the thorax and abdomen. *Oecologia (Berl.)* **84**: 491–499.

Stamp, N. E. 1980. Egg deposition patterns in butterflies: why do some species cluster their eggs rather than deposit them singly? *Am. Nat.* **115**: 367–380.

Suzuki, N. and K. Matsumoto. 1992. Lifetime mating success of males in a natural population of the papilionid butterfly, *Atrophaneura alcinous* (Lepidoptera: Papilionidae). *Res. Popul. Ecol.* **34**: 397–407.

Suzuki, Y. 1978. Adult longevity and reproductive potential of the small cabbage white, *Pieris rapae crucivora* Boiduval (Lepidoptera: Pieridae). *Appl. Entomol. Zool.* **13**: 312–313.

Svärd, L. and C. Wiklund. 1989. Mass and production rate of ejaculates in relation to monandry/polyandry in butterflies. *Behav. Ecol. Sociobiol.* **24**: 395–402.

Tinbergen, N., B. J. D. Meeuse, L. K. Boerma and W. W. Varossieau. 1942. Die Balz des Samtfalters, *Eumenis* (*Satyrus*) *semele* (L.). *Z. Tierpsychol.* **5**: 182–226.

Trivers, R. L. 1972. Parental investment sexual selection. In *Sexual Selection and the Descent of Man, 1871–1971*. B. Campbell, ed., pp. 136–179. Chicago: Aldine Publishing Co.

Vane-Wright, R. I. and M. Boppré. 1993. Visual and chemical signalling in butterflies: functional and phylogenetic perspectives. *Phil. Trans. R. Soc. (Lond.* B340: 197–205.

Wallace, A. R. 1889. *Darwinism. An Exposition of the Theory of Natural Selection*. London: Macmillan.

Warrant, E. J. and P. D. McIntyre. 1993. Arthropod eye design and the physical limits to spatial resolving power. *Prog. Neurobiol.* **40**: 413–461.

Wasserthal, L. T. 1975. The role of butterfly wings in regulation of body temperature. *J. Insect Physiol.* **21**: 1921–1930.

Webster, M. S. 1992. Sexual dimorphism, mating system and body size in New World blackbirds (Icterinae). *Evolution* **46**: 1621–1641.

Wickman, P.-O. 1985. Territorial defense and mating success in males of the small heath butterfly, *Coenonympha* L. (Lepidoptera: Satyridae). *Anim. Behav.* **33**: 1162–1168.

–. 1986. Courtship solicitation by females of the small heath butterfly, *Coenonympha pamphilus* (L.)(Lepidoptera: Satyridae) and their behaviour in relation to male territories before and after copulation. *Anim. Behav.* **34**: 153–157.

–. 1988. Dynamics of mate-searching behaviour in a hilltopping butterfly, *Lasiommata megera* (L.): the effects of weather and male density. *Zool. J. Linn. Soc.* **93**: 357–377

–. 1992. Sexual selection and butterfly design – a comparative study. *Evolution* **46**: 1525–1536.

Wiernasz, D. C. 1989. Female choice and sexual selection of male wing melanin pattern in *Pieris occidentalis* (Lepidoptera). *Evolution* **43**: 1672–1682.

Wiklund, C. 1977. Courtship behaviour in relation to female monogamy in *Leptidea sinapis*. *Oikos* **29**: 275–283.

–. 1982. Behavioural shift from courtship solicitation to mate avoidance in female ringlet butterflies (*Aphantopus hyperanthus*) after copulation. *Anim. Behav.* **30**: 790–793.

Wiklund, C. and T. Fagerström. 1977. Why do males emerge before females? A hypothesis to explain the incidence of protandry in butterflies. *Oecologia (Berl.)* **31**: 153–158.

Wiklund, C. and J. Forsberg. 1985. Courtship and mate discrimination between virgin and mated females in the orange tip butterfly, *Anthocaris cardamines*. *Anim. Behav.* **34**: 328–332.

–. 1991. Sexual size dimorphism in relation to female polygamy and protandry in butterflies: a comparative study of Swedish Pieridae and Satyridae. *Oikos* **60**: 373–381.

Wiklund, C. and A. Kaitala. 1995. Sexual selection for large male size in a polyandrous butterfly: the effect of body size on male versus female reproductive success in *Pieris napi*. *Behav. Ecol.* **6**: 6–13.

Wiklund, C. and B. Karlsson. 1988. Sexual size dimorphism in relation to fecundity in some Swedish satyrid butterflies. *Am. Nat.* **131**: 132–138.

–. 1990. Sexual size dimorphism and fecundity in satyrid butterflies: a reply to Harvey's comment. *Am. Nat.* **136**: 268–269.

Wiklund, C., S. Nylin and J. Forsberg. 1991. Sex-related variation in growth rate as a result of selection for large size and protandry in a bivoltine butterfly, *Pieris napi*. *Oikos* **60**: 241–250.

Wiklund, C. and C. Solbreck. 1982. Adaptive versus incidental explanations for the occurrence of protandry in a butterfly, *Lepitidea synapsis* L. *Evolution* **36**: 56–62.

Wiklund, C., P.-O. Wickman and S. Nylin. 1992. A sex difference in the propensity to enter direct/diapause development: a result of selection for protandry. *Evolution* **46**: 519–528.

Yagi, N. and N. Koyama. 1963. *The Compound Eye of Lepidoptera: Approach from Organic Evolution*. Tokyo: Shinkyo Press.

Zonneveld, C. 1992. Polyandry and protandry in butterflies. *Bull. Math. Biol.* **54**: 957–976.

Zonneveld, C. and J. A. J. Metz. 1991. Models on butterfly protandry: virgin females are at risk to die. *Theor. Popul. Biol.* **40**: 308–321.

16 · Lek behavior of insects

TODD E. SHELLY AND TIMOTHY S. WHITTIER

The sight of a feather in a peacock's tail, whenever I gaze at it, makes me sick

Charles Darwin (Darwin, F. 1887, vol. 2, p. 296)

ABSTRACT

We provide a general overview of lek behavior in insects. Initially, we draw a distinction between substrate-based and aerial (swarming) male mating aggregations. In general, males in substrate-based groups defend territories, wait for arriving females, and perform courtship prior to mating. By contrast, males in swarms typically exhibit no intrasexual aggression before female arrivals and grasp approaching females for immediate mating without courtship. Also, compared with swarms, substrate-based aggregations tend to be small, and males are more likely to produce long-range signals to attract females. We examine the relative importance of intra- and intersexual selection in substrate-based groups with (1) male aggression but no courtship, (2) male courtship but no aggression, and (3) both male courtship and aggression. The final category, groups most closely resembling 'classical' lek species, receives most attention; data from drosophilid and tephritid fruit flies are presented to show interspecific differences in the influence of male aggression and female choice on male mating success. Few data are available that address the evolutionary origins of lek behavior in insects. Among male-initiated hypotheses, it is most likely that males cluster at transit 'hotspots' where large numbers of females are likely to pass. There is no empirical support for the idea that clustering increases signal effectiveness and hence female arrivals on a per male basis. In addition, data regarding predation and group size do not consistently reveal reduced risks with increasing group size. Among female-initiated hypotheses, the notion that females prefer clustered males because clustering facilitates mate choice is not consistently supported by field observations. Data on the adaptive benefit of female mate choice are scant; future studies should investigate both direct and indirect fitness benefits associated with female mating decisions.

INTRODUCTION

How refreshing to read, in the opening quotation to this chapter, such a sincere and simple statement from a man whose prose, though often beautifully lyrical, is characteristically so cautious and refined. How telling too of the profound difficulty that extravagant sexual ornamentation posed to Darwin and his theory of natural selection. Darwin's solution was to hypothesize an additional evolutionary force, sexual selection, to explain the bizarre and elaborate structures and behaviors used in mate procurement. Sexual selection was, according to Darwin (1871, p. 260), 'a constantly recurring struggle between the males for the possession of the females' that included both direct competition among males (intrasexual selection) and mate choice by females (intersexual selection).

Since the inception of sexual-selection theory, there has been little resistance to the notion that male weaponry evolved through reproductive competition. After all, traits promoting fighting success, such as sharp claws and agility, would likely be favored by natural selection also. By contrast, the importance of intersexual selection has been a contentious issue. As Cronin (1991) so admirably documents, Wallace and later Huxley argued successfully that forces of natural selection, such as mimicry, protection, and mate and species recognition, were sufficient to account for sexual differences in ornamentation. As a result, intersexual selection was until recently dismissed as a force of little consequence.

There were a few early advocates of sexual selection, including Edward Selous, a British ornithologist. In observing male mating aggregations of ruff and black grouse, Selous noted the clear 'power of choice' (Selous 1913, p. 96) among hens for displaying males and heralded it as 'trumpet-tongued' support (Selous 1910, p. 264) for Darwin's theory of sexual selection. It is interesting to note that

Selous (1906–07, 1909–10) was also perhaps the first to describe male mating aggregations as 'leks'.

From our current vantage, the fact that lekking species provided early evidence for sexual selection seems less an historical accident than a premonitory intersection of observation and theory. Explaining lek behavior, and female choice in particular, is now considered one of the greatest challenges to sexual-selection research (Kirkpatrick and Ryan 1991). Because lek behavior casts such bald light on sexual selection, it is not surprising that the number of articles on lek species has risen dramatically over the past several decades. However, it is also true that most of this work, both empirical and theoretical, concerns vertebrate leks (see review by Wiley 1991; Höglund and Alatalo 1995). As a result, there is a pronounced asymmetry in information flow between students of vertebrate and insect leks: data and ideas generated by the former have far greater influence on the latter than vice versa.

We provide here a general overview of lek behavior in insects. As a focal point for our discussion, we have compiled a list of insect species having lek-like mating systems, which should also serve as a useful guide to the pertinent and widely scattered literature. Three main topics are addressed in this chapter. First, we compare substrate-based and aerial aggregations with respect to the incidence of male aggression and courtship, and conclude that aerial aggregations generally do not adhere to the standard lek criteria. Second, we focus on substrate-based groups and assess the influence of male–male aggression and female choice on male mating success in species where only male aggression or courtship is evident and then in species where both operate. Finally, we review the major hypotheses on the evolution of lek behavior and evaluate them in light of available data. Before addressing these main issues, however, we examine use of the term 'lek' in studies of insect mating systems.

We acknowledge that our attempts at generalization unavoidably belie the complexity inherent in many insect mating systems. Thus, summary statements should be taken as mere starting points for examining causes of interspecific diversity. In a way, this renders our position unenviable, for (to paraphrase the mathematician Alfred North Whitehead) although we are seeking simplicity, we know and accept that it will be distrusted.

DEFINING LEKS

Despite earlier usage, the term 'lek' was not formally defined until the 1970s (Wiley 1974; Bradbury 1977, 1981).

A 'classical' lek species meets the following criteria: (1) the absence of male parental care, with males contributing nothing but gametes to the female; (2) the existence of a mating arena in which most matings occur; (3) male territories that contain no resources vital to females (or, according to a later modification (Bradbury 1985), no male-regulated access to resources that might appear in the territory); and (4) the opportunity for females to freely select a mate in the arena. This definition contrasts with earlier notions (Wilson 1975; Alexander 1975) in emphasizing the non-resource-based nature of male territoriality and the uniqueness of female mating strategies where material gains are absent.

Impressed by the great diversity of insect mating systems, Bradbury (1985), whose ideas had developed primarily through work on bats and sage grouse, questioned whether a strict typological approach is useful in studying insect lek behavior. That is, although certain insect species conform to the classical lek definition developed for vertebrates, Bradbury (1985) recognized that many insects meet some, but not all, of the criteria of classical lek species. Consequently, he proposed that all of the classical criteria, save the absence of male parental care, be considered as continuous variables, which might vary independently with ecological conditions or taxonomic affiliation. Analogous to Hutchinson's (1957) concept of an ecological niche, this approach would characterize every animal mating system according to its position in a multidimensional hypervolume. Lek (or lek-like) mating systems would then comprise a particular 'neighborhood' within this multidimensional framework.

This approach is valuable because it focusses attention away from the strict classificatory question of whether or not a certain species has a lek mating system and towards the more conceptually interesting arena of explaining variation about the archetypal lek mating system. Unfortunately, Bradbury's (1985) view may be misconstrued as advocating completely flexible criteria for lek behavior, rendering the standard definition useless. The result is that the term lek may be used to describe insect mating systems that, in our opinion, deviate dramatically from the original 'spirit' of the word (see, for example, Jarvis and Rutledge 1992; Svensson and Petersson 1992). Without doubt, the match between conceptual ideal and the empirical reality is often imperfect, i.e. not all mating systems that are perhaps best described as leks actually meet all four criteria. Still, these discrepancies do not justify abandoning the ideal; rather, they render any operational definition imprecise and idiosyncratic. Although not academically satisfying, this situation is not unique to the term lek (consider the term 'niche' for example (Vandermeer 1972)) and on the positive side this issue demands careful scrutiny of the lek criteria and explicit statement of any mismatches between the conceptual ideal and real-world observations.

SUBSTRATE-BASED VS. AERIAL AGGREGATIONS

We recognize two main types of male mating aggregation that do not involve male control of resources required by females or capture of females at emergence sites. Aggregations of the first type are substrate-based, i.e. males spend most of their time perching (e.g. cicadas, drosophilids, tephritids, satyrids) or at least perching periodically at particular stations (e.g. certain carpenter bees and wasps). Mating may be initiated on the substrate or in the air. Substrate-based aggregations that occur on the tops of hills, mountains, or ridges are referred to as hilltopping aggregations, whereas all other substrate-based aggregations are termed non-hilltopping. The second major type includes all aerial aggregations, or swarms, regardless of their location relative to hilltops. Swarming males are usually in continuous flight, and matings are typically initiated in the air (see Sivinski and Petersson, this volume). A list of species exhibiting substrate-based aggregations appears in Table 16-1, and a representative list of swarming species is given in Table 16-2. Both lists include only species with reasonably complete information on male spacing and male–male and male–female interactions.

Aside from the obvious locational difference, the distinction drawn between substrate-based and aerial aggregations emphasizes underlying differences in male territoriality and female choice. In general, males in substrate-based aggregations aggressively defend sites against conspecific males: males of 67% (35/52) of the genera listed in Table 16-1 were observed in agonistic interactions. These contests, however, vary greatly in their intensity. On one hand, males of certain species (e.g. tephritids, certain butterflies) weakly chase away intruders, and the risk of physical injury is negligible. At the other extreme, male botflies may collide violently in midair chases and incur permanent wing damage (Alcock and Schaefer 1983). Audible wing-clashing, with possible flight-threatening damage, has also been noted in wasps (Alcock 1975a) and butterflies (Alcock 1983).

The majority of males in aerial aggregations do not appear to defend specific areas or engage in any physical aggression prior to the arrival of a female. Male–male aggression was reported for only 40% (10/24) of the genera listed in Table 16-2. Descriptions of male flight within swarms usually suggest chaotic movement made independently of other males; for example, males move vertically in spiral fashion (Belkin et al. 1951), asynchronously up and down (Brodsky 1973), or in horizontal and vertical zig-zags (Chiang 1961). Thus, swarming males typically appear to search for incoming females via frequent positional changes and do not maintain a fixed location within the aggregation.

Substrate-based and aerial aggregations also tend to show a striking difference regarding the opportunity for active mate choice (sensu Parker 1983) or mate 'sampling' by females. Male courtship is present in nearly half of the genera (23/52, or 44%) that form substrate-based mating groups (Table 16-1). The Hawaiian Drosophila spp., for example, are well known for their complex male courtship behavior, which may involve semaphoring and vibrating the wings, various leg movements and vibrations, and genitalic licking (Spieth 1968). Protracted male courtship has been reported in aggregations of otitid flies (Alcock and Pyle 1979), tephritids (Dodson 1986), carpenter bees (Vinson and Frankie 1990) and cicadas (Dunning et al. 1979). Along with the occurrence of male courtship signals, females visiting substrate-based aggregations are often reported to decamp from courting males or reject copulation attempts (18/52 genera, or 34%), which suggests female sampling of potential mates. This percentage is undoubtedly an underestimate, resulting primarily from the rarity of sightings of females at many butterfly, bee and wasp leks. Excluding the Lepidoptera and Hymenoptera, mate rejection (sampling) was reported for females in over 50% (13/23) of the genera. One of the more interesting descriptions of female sampling involves the carpenter bee Xylocopa fimbriata (Vinson and Frankie 1990). In this species, females were observed to fly just downwind of several territories (males deposit a scent marking pheromone attractive to females), briefly enter each territory, and then re-enter one for mating. Data on mate-sampling have also been presented for female medflies in laboratory observations (Whittier et al. 1994). Courtship disruption by competing males at leks has rarely been reported, and females appear to exercise control over mating decisions (see Alexander et al., this volume). Shelly (1987) noted courtship disruption in a lekking Drosophila but did not observe any instances where female choice was altered by this interference.

Table 16-1. *Insects with substrate-based mating aggregations*

Column headings are as follows. Territory or perch, Area defended or, if non-territorial, perch site; Agg. size, range of male abundance in aggregations; Male signals, mode of long-range female attraction; Ct, male courtship present or absent (y/n); Sa, mate-sampling by females present or absent (y/n); Ag, Male–male aggression prior to female arrival present or absent (y/n).

Species	Location	Territory or perch	Agg. size	Male signals	Ct	Sa	Ag	Source
DIPTERA								
Bombyliidae								
Comptosia sp.	hilltop	3 m^2 ground	2–6	?	n	n	y	Yeates & Dodson 1990; Dodson & Yeates 1990
Cuterebridae								
Cuterebra austeni	hilltop	ground	3–19	n	n	n	y	Alcock & Schaefer 1983
Cuterebra grisea	non-hilltop	5–10 m^2 ground	?	n	n	n	y	Hunter & Webster 1973
Cuterebra latifrons	hilltop	8–12 m^2 ground	3–15	n	n	n	y	Catts 1967
Cuterebra lepivora	non-hilltop	5–35 m^2 ground	3–4	n	n	n	y	Meyer & Bock 1980
Cuterebra polita	non-hilltop	5 m^2 ground	2–10	n	n	n	n	Graham & Capelle 1970; Capelle 1970
Cuterebra tenebrosa	non-hilltop	30–45 m^2 ground	?	n	n	n	y	Hunter & Webster 1973
Drosophilidae								
Drosophila adunca	non-hilltop	branch	2–10	?	y	y	?	K. Y. Kaneshiro, personal observations
Drosophila comatifemora	non-hilltop	fern branch	2–6	chem	y	y	y	Spieth 1968; K. Y. Kaneshiro, personal observations
Drosophila conformis	non-hilltop	leaf	10–40	chem	y	y	y	Shelly 1987, 1989
Drosophila cnecopleura	non-hilltop	leaf	2–10	?	y	y	y	Shelly 1988
Drosophila crucigera	non-hilltop	branch	?	chem	y	y	y	Ringo 1976
Drosophila cyrtoloma	non-hilltop	branch	2–13	?	y	y	y	K. Y. Kaneshiro, personal observations
Drosophila differens	non-hilltop	branch	?	?	y	y	?	K. Y. Kaneshiro, personal observations
Drosophila engyochracea	non-hilltop	tree trunk	10–25	chem	y	y	?	Spieth 1968; K. Y. Kaneshiro, personal observations
Drosophila formella	non-hilltop	branch	?	chem	y	y	?	Ringo 1976
Drosophila grimshawi	non-hilltop	branch	?	chem	y	y	y	Droney 1992; Hodosh *et al.* 1979; Ringo 1976
Drosophila heteroneura	non-hilltop	branch	2–26	?	y	y	y	Spieth 1981; Conant 1978
Drosophila imparisetae	non-hilltop	leaf	6–12	?	y	y	y	Shelly 1990
Drosophila lasiopoda	non-hilltop	?	?	chem	y	y	?	Ringo 1976
Drosophila mimica	non-hilltop	leaf	10–15	?	y	y	y	K. Y. Kaneshiro, personal observations
Drosophila mixtura	non-hilltop	bracket fungus	?	?	y	?	?	Parsons 1977; but see Hoffman & Blows 1992
Drosophila mycetophaga	non-hilltop	bracket fungus	2–20	?	y	?	?	Parsons 1977; but see Hoffmann & Blows 1992
Drosophila nigribasis	non-hilltop	branch	2–4	?	y	y	y	K. Y. Kaneshiro, personal observations
Drosophila obscuripes	non-hilltop	branch	?	?	y	y	y	K. Y. Kaneshiro, personal observations
Drosophila percnosoma	non-hilltop	leaf	4–12	?	y	y	y	Bell & Kipp 1994
Drosophila planitibia	non-hilltop	branch	2–6	?	y	y	y	Spieth 1981; K. Y. Kaneshiro, personal observations
Drosophila polypori	non-hilltop	bracket fungus	2–16	?	y	?	?	Parsons 1976, 1977; but see Hoffmann & Blows 1992
Drosophila pullipes	non-hilltop	?	?	chem	y	y	?	Ringo 1976
Drosophila silvestris	non-hilltop	branch	2–6	?	y	y	y	Spieth 1981; Conant 1978
Gasterophilidae								
Gasterophilus intestinalis	hilltop	ground	2–10	n	n	n	y	Catts 1979; but see Cope & Catts 1991
Mydidae								
Mydas ventralis	hilltop	20–75 m^2 ground	15–75	n	n	n	y	Alcock 1989

Table 16-1. (*cont.*)

Species	Location	Territory or perch	Agg. size	Male signals	Ct	Sa	Ag	Source
Mydas xanthopterus	hilltop	bush	?	?	?	?	y	Nelson 1986
Oestridae								
Cephenemyia apicata	hilltop	tree or bush	2–15	n	n	n	n	Catts 1964
Cephenemyia jellisoni	hilltop	tree	2–15	n	n	n	n	Catts 1964
Hypoderma lineatum	non-hilltop	30–50 m² ground	2–20	n	n	n	n	Catts *et al.* 1965
Otitidae								
Physiphora demandata	non-hilltop	10 15 cm branch	'dozens'	chem	y	y	y	Alcock & Pyle 1979
Stratiomyidae								
Hermetia comstocki	non-hilltop	agave plant	2–9	?	n	n	y	Alcock 1990, 1993
Tephritidae								
Anastrepha fraterculus	non-hilltop	leaf	2–8	chem	y	?	y	Morgante *et al.* 1983; Malavasi *et al.* 1983
Anastrepha ludens	non-hilltop	leaf	2–6	chem	y	y	y	Robacker *et al.* 1991; Aluja *et al.* 1983
Anastrepha obliqua	non-hilltop	leaf	2–8	chem	y	y	y	Aluja *et al.* 1983; Burk 1991
Anastrepha pseudoparella	non-hilltop	leaf	2–5	chem	y	y	y	Burk 1991
Anastrepha serpentina	non-hilltop	leaf	2–5	chem	y	y	y	Burk 1991
Anastrepha sororcula	non-hilltop	leaf	2–5	chem	y	y	y	Burk 1991
Anastrepha suspensa	non-hilltop	leaf	8–28	chem/ acoustic	y	y	y	Burk 1983; Sivinski 1989; Nation 1972
Bactrocera cucurbitae	non-hilltop	leaf	2–20	chem	y	y	y	Kuba & Koyama 1985; Iwahashi & Majima 1986
Bactrocera dorsalis	non-hilltop	leaf	2–10	chem	n	?	y	Shelly & Kaneshiro 1991
Bactrocera tryoni	non-hilltop	leaf	?	chem	y	y	?	Tychsen 1977; Myers 1952
Ceratitis capitata	non-hilltop	leaf	2–20	chem	y	y	y	Whittier *et al.* 1992; Arita & Kaneshiro 1989
Procecidochares sp.	non-hilltop	plant stem or leaf	37–51	?	y	y	y	Dodson 1986
HYMENOPTERA								
Anthophoridae								
Centris adani	hilltop	1–3 m³ ground	2–7	chem	?	?	y	Frankie *et al.* 1980
Xylocopa caffra	non-hilltop	2–3 m³ near perch	?	?	?	?	y	Watmough 1974
Xylocopa capitata	non-hilltop	2–3 m³ near perch	?	?	?	?	y	Watmough 1974
Xylocopa fimbriata	non-hilltop	1–250 m³ near tree	?	chem	y	y	y	Vinson & Frankie 1990; Janzen 1966
Xylocopa flavorufa	non-hilltop	4 m² near bush	?	?	?	?	y	Watmough 1974; Anzenberger 1977
Xylocopa fraudulenta	non-hilltop	2–3 m² ground	?	?	?	?	y	Watmough 1974
Xylocopa gualanensis	non-hilltop	0.5–10 m³ near tree	?	chem	y	y	y	Vinson & Frankie 1990
Xylocopa nigrita	non-hilltop	bush	?	?	?	?	y	Anzenberger 1977
Apidae								
Bombus griseocollis	non-hilltop	ground	2–24	chem	n	?	y	O'Neill *et al.* 1991
Bombus nevadensis	non-hilltop	shrub	2–13	chem	n	?	y	O'Neill *et al.* 1991
Bombus rufocinctus	non-hilltop	ground	2–23	chem	n	?	y	O'Neill *et al.* 1991
Euglossa imperialis	non-hilltop	12–110 m² fallen tree	2–12	chem?	y	?	y	Kimsey 1980
Eulaema meriana	non-hilltop	12–110 m² fallen tree	2–7	chem?	y	?	y	Kimsey 1980; Stern 1991
Formicidae								
Acromyrmex versicolor	non-hilltop	shrub	?	?	n	?	n	Johnson & Rissing 1993
Formica obscuripes	non-hilltop	ground	>1000	?	n	?	n	Talbot 1972
Formica subpolita	non-hilltop	shrub	10–400	?	n	y	n	O'Neill 1994
Pogonomyrmex barbatus	non-hilltop	ground	>1000	chem	n	y	n	Hölldobler 1976; Michener 1948; Davidson 1982; Markl *et al.* 1977

Table 16-1. (*cont.*)

Species	Location	Territory or perch	Agg. size	Male signals	Ct	Sa	Ag	Source
Pogonomyrmex californicus	non-hilltop	tree	>100	chem	n	?	n	Michener 1960; Mintzer 1982
Pogonomyrmex comanche	non-hilltop	tree	>1000	?	?	?	n	Strandtmann 1942
Pogonomyrmex desertorum	non-hilltop	tree	>1000	chem	n	y	n	Hölldobler 1976; Davidson 1982
Pogonomyrmex maricopa	non-hilltop	tree	>1000	chem	n	y	n	Hölldobler 1976; Markl *et al.* 1977
Pogonomyrmex occidentalis	non-hilltop	ground	>100	?	n	?	n	Nagel & Rettenmeyer 1973
Pogonomyrmex rugosus	non-hilltop	ground	>1000	chem	n	y	n	Hölldobler 1976; Markl *et al.* 1977
Pogonomyrmex salinus	non-hilltop	shrub	25–60	?	n	?	n	Rust 1988
Masaridae								
Pseudomasaris maculifrons	hilltop	ground	3–11	?	?	?	n	Alcock 1985a
Pompilidae								
Hemipepsis ustulata	hilltop	3–5 m^2 tree	10–35	?	n	?	y	Alcock 1981
Sphecidae								
Bembix furcata	hilltop	10–15 m^2 ground	?	?	n	?	y	Dodson & Yeates 1989
Clypeadon taurulus	non-hilltop	1 m^2 shrub	8–12	chem	n	?	y	Alcock 1975a
Eucerceris arenaria	non-hilltop	1 m^2 shrub	2–15	chem	n	?	y	Alcock 1975a
Eucerceris canaliculata	non-hilltop	1 m^2 shrub	3–8	chem	n	?	y	Alcock 1975a
Eucerceris rubripes	non-hilltop	1 m^2 shrub	3–8	chem	n	?	y	Alcock 1975a
Eucerceris tricolor	non-hilltop	1 m^2 shrub	3–8	chem	n	?	y	Alcock 1975a
Philanthus barbatus	non-hilltop	75 m^2 ground	6–13	chem	n	?	y	Evans 1993
Philanthus basilaris	non-hilltop	1 m^2 ground	4–47	chem	n	?	y	O'Neill 1983
Philanthus crabroniformis	non-hilltop	1–150 m^2 ground	2–11	chem	n	?	y	Evans 1993
Philanthus multimaculatus	non-hilltop	1 m^2 grass clump	?	chem	n	?	y	Alcock 1975a, b
Philanthus triangulum	non-hilltop	1 m^2 ground	?	chem?	n	?	y	Simon-Thomas & Poorter 1972; Alcock *et al.* 1978
Vespidae								
Polistes biglumis	non-hilltop	ground	?	chem	?	?	y	Beani *et al.* 1992
Polistes canadensis	hilltop	tree	2–25	?	?	?	y	Polak 1993a,b
Polistes carnifex	hilltop	tree	2–17	chem	?	?	y	Polak 1993b
Polistes commanchus	hilltop	tree/rock	20–100	?	n	?	y	Matthes-Sears & Alcock 1986
Polistes dominulus	non-hilltop	tree branch	?	chem	?	?	y	Beani *et al.* 1992; Beani & Calloni 1991
Polistes nimpha	non-hilltop	hedge/bush	2–9	chem	?	?	y	Beani *et al.* 1992; Turrillazzi & Cervo 1982
LEPIDOPTERA								
Arctiidae								
Estigmene acrea	non-hilltop	plant stem	3–22	chem	n	?	n	Willis and Birch 1982
Hesperiidae								
Erynnis tristis	hilltop	ground	?	n	y	?	?	Shields 1967
Lycaenidae								
Atlides halesus	hilltop	tree	4–6	n	?	?	y	Alcock 1983
Callophrys dumetorum	hilltop	ground	?	n	?	?	y	Shields 1967
Celastrina argiolus	hilltop	ground	?	n	y	?	?	Shields 1967
Strymon melinus	hilltop	tree	2–7	n	?	?	y	Alcock & O'Neill 1986
Nymphalidae								
Chlosyne californica	hilltop	5–10 m^2 ground	?	n	n	?	y	Alcock 1985b
Euphydryas editha	hilltop	ground	20–50	n	?	?	?	Erlich & Wheye 1986; but see Singer & Thomas 1992

Table 16-1. (*cont.*)

Species	Location	Territory or perch	Agg. size	Male signals	Ct	Sa	Ag	Source
Vanessa atalanta	hilltop	144 m^2 ground	?	n	y	?	y	Shields 1967; Bitzer & Shaw 1979
Vanessa cardui	hilltop	ground	?	n	y	?	n	Shields 1967
Vanessa caryae	hilltop	ground	?	n	y	?	n	Shields 1967
Vanessa kershawi	hilltop	150 m^2 ground	2–11	n	?	?	y	Alcock & Gwynne 1988
Vanessa virginiensis	hilltop	plant stem	?	n	y	?	n	Shields 1967
Papillionidae								
Battus philenor	hilltop	Plant stem	2–6	n	y	y	y	Rutowski *et al.* 1989; Shields 1967
Papilio eurymedon	hilltop	plant stem	2–5	n	y	?	?	Shields 1967; Scott 1968
Papilio polyxenes	hilltop	75 m^2 of field	?	n	y	y	y	Lederhouse 1982
Papilio zelicaon	hilltop	Plant stem	2–14	n	y	y	y	Shields 1967; Scott 1968
Pieridae								
Anthocaris cethura	hilltop	ground	?	n	?	?	y	Shields 1967
Anthocaris pima	hilltop	?	?	n	y	?	n	Alcock 1984
Satyridae								
Lasiommata megera	hilltop	30 m^2 ground	4–7	n	y	?	y	Wickman 1988; Dennis 1982
Oeneis chryxus	non-hilltop	20–50 m^2 ground	2–15	n	?	?	y	Knapton 1985
HOMOPTERA								
Cicadidae								
Cystosoma saundersii	non-hilltop	1–1.5 m^3 of bush	2–37	acoustic	y	y	n	Doolan & MacNally 1981; Doolan 1981
Fidicina sericans	non-hilltop	tree branch	30–50	acoustic	?	?	n	Young 1980
Fidicina pronoe	non-hilltop	tree branch	38–60	acoustic	?	?	?	Young 1980
Magicicada cassini	non-hilltop	tree branch	2–100+	acoustic	y	y	n	Williams & Smith 1991; Karban 1983; Dunning *et al.* 1979; Alexander & Moore 1958
Magicicada decim	non-hilltop	tree branch	2–100+	acoustic	y	y	n	Williams & Smith 1991; Lloyd & Karban 1983; Dunning *et al.* 1979; Alexander & Moore 1958
Magicicada decula	non-hilltop	tree branch	2–100+	acoustic	y	y	n	Williams & Smith 1991; Alexander & Moore 1958;
Quesada gigas	non-hilltop	tree branch	17–124	acoustic	?	?	n	Young 1980
Zammara smaragdina	non-hilltop	tree branch	31–37	acoustic	?	?	n	Young 1980
HEMIPTERA								
Plastaspidae								
Megacopta punctissimum	non-hilltop	plant stem	2–26	?	y	y	n	Hibino 1985, 1986; Hibino & Ito 1983
COLEOPTERA								
Lampyridae								
Luciola cruciata	non-hilltop	bush	?	visual	y	?	y	Lloyd 1979;
Luciola obsoleta	non-hilltop	tree leaf	50–100	visual	y	y	y	Lloyd 1979, 1972
Photinus knulli	non-hilltop	patch of ground	3–200	visual	y	y	?	Cicero 1983
Pteroptyx spp.	non-hilltop	tree leaf or stem	12–1000	visual	y	y	y	Lloyd 1973, 1979; J. E. Lloyd, personal communication; Buck & Buck 1966, 1978
ORTHOPTERA								
Tettigoniidae								
Orchelimum nigripes	non-hilltop	2 m^2 of vegetation	?	acoustic	y	y	y	Feaver 1983

Table 16-2. *Insects with aerial male mating aggregations*

Column headings are as follows. Agg. size, range of male abundance in aggregations; Ct, male courtship present or absent (y/n); S, mate-sampling by females present or absent (y/n); Ag, male–male aggression prior to female arrival present or absent (y/n). For the most part, the information on swarms derived from aggregations that did not occur directly over female emergence (Thornhill 1980) or oviposition (see, for example, Savolainen 1978) sites or vertebrate hosts (see, for example, Gubler and Bhattacharya 1972).

Species	Agg. size	Ct	Sa	Ag	Source
Diptera					
Bombyliidae					
Lordotus pulchrissimus	5–67	n	n	y	Toft 1984, 1989a,b
Cecidomyiidae					
Anarete pritchardi	4–20	n	n	y	Chiang 1961; Okubo & Chiang 1974
Chironomidae					
Chironomus plumosus	100–1 000 000	n	n	n	Hilsenhoff 1966
Chironomus pseudothummi	100–10 000	n	n	n	Syrjämäki 1966
Chironomus strenzkei	20–30	n	y	n	Syrjämäki 1965
Stictochironomus crassiforceps	10–300	n	n	y	Syrjämäki 1964; Syrjämäki & Ulmanen 1970
Culicidae					
Aedes communis	15–100	n	n	?	Frohne & Frohne 1954
Aedes excrucians	200–300	n	n	n	Frohne & Frohne 1954
Anopheles franciscanus	12–5000	n	n	n	Belkin *et al.* 1951
Anopheles punctipennis	20–100	n	n	n	Knab 1907
Culex pipiens	700–4000+	n	n	n	Williams & Patterson 1969; Knab 1906
Psorophora ferox	10–60	?	y	n	Nielsen 1965
Limoniidae					
Erioptera gemina	2–15	n	n	n	Savolainen & Syrjämäki 1971
Muscidae					
Fannia canicularis	2–10	?	?	y	Zeil 1986
Ophyra leucostoma	2–7	?	?	n	Pajunen 1982
Rhagionidae					
Symphoromyia sackenia	25–100	n	n	y	Hoy & Anderson 1978
Simuliidae					
Austrosimulium pestilens	70–1000	n	n	n	Moorhouse & Colbo 1973
Syrphidae					
Syrphus ribesii	?	?	?	n	Heinrich & Pantle 1975; Gilbert 1984
Tabanidae					
Chrysops atlanticus	10–12	n	n	y	Anderson 1971
Tabanus bishoppi	2–75	n	?	y	Blickle 1959
Tipulidae					
Erioptera taenionota	10–200	n	n	?	Cuthbertson 1926
Ephemeroptera					
Ephemeridae					
Hexagenia spp.	1000+	n	n	n	Lyman 1944

Table 16-2. (cont.)

Species	Agg size	Ct	Sa	Ag	Source
Heptageniidae					
Epeorus longimanus	20–2000	n	n	y	Allan & Flecker 1989
Stenonema canadense	1000+	n	n	y	Thew 1958
Stenonema vicarium	15–25	n	n	n	Cooke 1940
Hymenoptera					
Apidae					
Apis mellifera	100+	n	n	?	Michener 1974; Strang 1970
Vespidae					
Polistes gallicus	4–15	n	n	n	Beani *et al.* 1992
Lepidoptera					
Hepialidae					
Hepialus humuli	2–1600	n	n	y	Mallet 1984
Pieridae					
Perrhybris pyrrha	<100	y	?	n	DeVries 1978

Male courtship is uniformly absent among swarming insects (see DeVries 1978 for a possible exception). Instead, males simply pounce upon females as they enter the swarm, and mating ensues (but see Tozer *et al.* 1981). Larger male mayflies (Flecker *et al.* 1988) and younger (and presumably more vigorous) male caddisflies (Petersson, 1987) may enjoy a mating advantage in swarms (see also Sivinski and Petersson, this volume). In certain midge species, however, small males are most successful at mating, owing possibly to their greater flight maneuverability (McLachlan and Allen 1987; Neems *et al.* 1990). However, in none of these instances was there an obvious association between intraswarm location and mating activity, which indicates an absence of any site-specific defense. Intense male scramble-competition is illustrated by the common observation (Knab 1906; Blickle 1959; Williams and Patterson 1969) that several males may rapidly approach a female more or less simultaneously. Although aggression among approaching males could affect mating success, no successful takeovers have been explicitly reported, and the opportunity for 'unhindered' mate choice by females seems limited. In general, males appear to control matings in insect swarms, but females of some species sequentially mate with several different males before exiting the swarm (Savolainen and Syrjämäki 1971; Michener 1974), and the possibility that these females exercise postcopulatory choice (Thornhill 1983; see also Alexander *et al.*, this volume) among different ejaculates remains open. In addition, Gullefors and Petersson (1993) have observed that small males often have difficulty in lifting females out of swarms, resulting in the pair breaking up. Females may then re-enter the swarm for another coupling attempt.

Substrate-based aggregations and swarms also differ dramatically with respect to the number of males present and the incidence of long-range attraction of females by males. Among substrate-based groups for which estimates are available (Table 16-1), median male abundance per group is less than 15 for two-thirds of the genera (26/42) and over 100 for only four genera (harvester ants, periodical cicadas and two genera of fireflies). Although quantitative estimates are difficult for most swarming species, it appears certain that swarms are generally much larger: among aerial aggregations for which data are available (Table 16-2), median group size was less than 15 in only one-fourth of the genera (6/23) and over 100 in nearly 60% of the genera (13/23). For several mosquito and mayfly species, median group size typically exceeds 1000 males. Males in substrate-based aggregations are also more likely to produce long-range signals for female attraction than those in swarms (25/52 genera vs. 3/24 genera, respectively). The value for swarming insects is perhaps an underestimate (Sullivan 1981), but the tendency of swarms to form near visual markers suggests that, like males, females are generally attracted by an environmental cue and not by conspecifics *per se* (Sivinski and Petersson, this volume).

We believe that substrate-based aggregations and aerial swarms can, in general, be distinguished by the presence of male territoriality and male courtship in the former and their absence in the latter. The incidence of male courtship further implies that the opportunity for female choice is much greater in substrate-based groups than in aerial swarms. We propose that mating systems lacking both male aggression (prior to female arrival) and courtship should not be considered leks. Based on these dual criteria, about 80% (41/52) of the genera having substrate-based aggregations are candidates for lek status (Table 16-1) compared with only 33% (8/24) of the swarming genera (Table 16-2).

RELATIVE IMPORTANCE OF INTRA- AND INTERSEXUAL SELECTION

Lekking species exhibit a remarkable diversity of mating behavior. Following the approach used above, we group species by the incidence of male aggression and courtship and consider the relative importance of intra- and intersexual selection in these groups. We consider how the opportunity for female choice might vary among species in which only male aggression or courtship is present and among species in which both activities occur. Most of our attention is directed toward this latter group, because it includes the best examples of 'classical' lek species in insects and thus represents the most justifiable and conservative listing of lekking insects.

Aggression with little or no courtship

Although male courtship is absent in swarming species, male–male aggression has been reported in some aerial aggregations ($n = 8$ genera; Table 16-2). In half of these cases, males hover and defend well-defined airspaces (0.001–$0.1\,\mathrm{m}^3$), and in the remaining instances males either patrol certain areas (Thew 1958; Zeil 1986) or chase other males while changing positions within the swarm (Toft 1984; Allan and Flecker 1989). Territorial behavior without courtship is also common among substrate-based groups, particularly among bees, wasps and brachycerous flies. Moreover, although male courtship is reported for many butterflies, this behavior sometimes appears to involve little more than a brief chase prior to copulation.

It is difficult to see how active female choice could be operating in species lacking courtship (Sullivan 1981; Alcock 1987). In the simplest scenario, females may simply mate with the first male encountered. In this case, female choice is, at best, indirect: if territorial sites are limited and certain male traits (e.g. body size, flight agility) confer an advantage in male–male contests, then females may display an indirect preference for these traits by mating only with territory-holders. Females would thus be following a 'yes/no' mating rule based strictly on a male's participation in an aggregation. Alternatively, females may base mating decisions on male position within the aggregation. In the butterfly *Chlosyne californica*, for example, male–female interactions ($n = 8$) were observed only in two territories along a ridgetop (Alcock 1985b). Here again, female choice would be indirect, and male combat, acting as a sorting or 'preselection' process (Bradbury and Gibson 1983), would be the proximate determinant of male mating success. Moreover, females might be intercepted en route to preferred territorial locations and forced to copulate (Alcock 1981). It is possible that females could gain some control in mate choice by remating with a more preferred male (particularly with last-male sperm precedence), but evidence from many solitary bees and wasps (Alcock *et al.* 1978; Eickwort and Ginsberg 1980; Evans and O'Neill 1988) and hilltopping butterflies (Drummond 1984; Alcock 1987) suggests that monandry is the rule.

Courtship with little or no aggression

Species with male aggression absent but courtship present are uncommon, being represented primarily by cicadas. Male spacing appears to be based on mutual avoidance, with site fidelity (and resident advantage) being absent (Doolan 1981; Alexander *et al.*, this volume). The lack of aggressive behavior may result from two factors: the spatial unpredictability of female arrivals (making defense of a specific site unprofitable) and the limited ability of individuals to locate members of the opposite sex over short distances within an aggregation.

Male periodical cicadas of North America, for example, adopt a 'sing–fly' searching strategy within an aggregation (Dunning *et al.* 1979). Upon detecting any non-moving, cicada-sized object, males will sing continuously from a single perch and then approach the object. If the courted object remains motionless, courtship will proceed. By contrast, males of the Australian bladder cicadas are highly sedentary, and females take an active role in eliciting court-ship. A newly arrived female produces a pheromone and displays a specialized wing-flicking behavior to advertise her presence to the nearest male (Doolan 1981). Only after receiving the female's signals does the male respond with courtship singing and eventual approach.

Male aggression with courtship

Among substrate-based aggregations, the co-occurrence of male aggression and courtship has been reported for nearly 40 species in over a dozen genera (Table 16-1). Dro-sophilids and tephritids comprise over half (23/37) of these species, and within this subset the Hawaiian *Drosophila* are numerically dominant (13 of 23 species). Because the dro-sophilids and tephritids are the best-studied taxa and are the focus of our own research, we limit the following dis-cussion to these groups.

As with many classical vertebrate leks, data on fruit flies suggest that both male–male aggression and female choice influence female mating decisions but that the relative importance of these two factors varies between species. In *D. conformis*, for example (and possibly *D. cnecopleura*; Shelly 1988), intraspecific aggression appears be the dominant force affecting female mating patterns (Shelly 1987, 1989). Males of this species defend single leaves as mating terri-tories and show a strong preference for the lowest leaves on the tree. The majority of male–male interactions occurred at low sites; following experimental removal of the resi-dent, lower territories were more quickly filled than higher ones. Moreover, because of a size advantage in aggression, larger males tended to hold the lowest sites. Continuous observations of the entire lek showed that most female encounters and matings involved large males on the lowest leaves. Because male site preference did not confer any obvious non-sexual benefits, this reflected either the tendency of females to enter the lek from the ground (where they often feed and oviposit) or active searching by females for the lowest territories (and 'best' males) in the lek. In addition to the apparent position effect, courtship disruption by intruding males was recorded for about 10% of the female visits, suggesting that female acceptance was in part dependent on a male's ability to gain exclusive control of his territory and thus to perform uninterrupted courtship.

Though perhaps less important, male sorting based on territory position appears to operate in some *Anastrepha* species as well. A position effect may act at two spatial

scales in the Mexican fruit fly, *A. ludens*, males of which defend and pheromone-call from single-leaf territories on host trees (Robacker *et al.* 1991). First, females appear to prefer territories within a certain volume of the canopy: both the total number and the proportion of successful (those ending in copulation) intersexual encounters were greater for territories in the interior of the canopy 1–2 m above ground than for other territories. Within male groups in this 'preferred volume', individuals that held the highest leaf accounted for five of the six matings observed. Males successful at mating were also dominant to unsuc-cessful males in aggressive interactions. Unlike *D. confor-mis*, neither fighting ability nor mating success was size-related in *A. ludens*, and attempts at courtship disruption were rare and unsuccessful.

The importance of male combat in these two species does not rule out the presence of active female choice. In fact, females of both *D. conformis* and *A. ludens* sometimes departed male territories without mating, both before and after the male mounted. None the less, intersexual selec-tion appears less important in these species than in another *Anastrepha* species, the Caribbean fruit fly, *A. suspensa*. While perching on leaf territories, males of *A. suspensa* (and conge-ners) evert an anal pouch and release a pheromone attrac-tive to females (so-called 'puffing' or pheromone-calling (Nation 1972)). While puffing, males fan their wings rapidly in short bursts, an action that presumably enhances pheromone dispersal but also produces an acous-tic signal (the calling song (Webb *et al.* 1976)). Upon arrival of a sexually active female, males stop calling, wave their wings in semaphore fashion, and, following a final approach by the female, mount. After mounting, males vibrate their wings rapidly for intervals ranging from a few seconds to as long as 25 minutes (the precopulatory song (Webb *et al.* 1984)).

In a series of laboratory experiments, Burk, Sivinski and Webb (Burk and Webb 1983; Webb *et al.* 1983, 1984; Sivinski *et al.* 1984; Sivinski and Webb 1986) demonstrated size-dependent female choice based partly on acoustic cues. Females presented with a choice between large and small males selected large males in over 2/3 of the mat-ings. Building on the observation that tape-recorded call-ing songs and pheromone extracts are attractive to females, these workers subsequently documented size-related varia-tion in the calling song. In addition to having a higher pro-pensity to call, larger males produce calling songs having higher intensity, lower fundamental frequency, and shorter interpulse intervals (i.e. higher pulse rates) than smaller males. Further studies showed that females showed greater movement in response to high-pulse-rate songs than to low-pulse-rate songs; in fact, pulse rates characteristic of small males failed to elicit more activity than silence. In addition to the calling song, precopulatory songs also varied with male size, with larger males producing louder precopulatory songs with narrower bandwidth than smaller males. Thus, both the calling and precopulatory song provide reliable cues to females regarding male size and therefore could be used in mating decisions.

Although female choice appears to be important in *A. suspensa*, intrasexual selection probably affects male mating success as well. Fighting ability is also size-dependent: large residents usually defend their territory successfully against small intruders, and large intruders usually usurp territories of smaller residents (Burk 1984). As a result, leks may consist primarily of large males. Sivinski (1989) also reported that all four matings he observed were achieved by large males occupying the innermost leaf (closest to the trunk), indicating once again the operation of a position effect on male mating success.

Intrasexual aggression probably has a larger (*D. conformis* and *A. ludens*) or coequal (*A. suspensa*) role than female choice in the species discussed above, but male–male combat appears relatively insignificant in leks of the Mediterranean fruit fly (or medfly), *Ceratitis capitata*. This claim, which contrasts with the earlier notion (Arita and Kaneshiro 1985) that male medflies fight for control of certain territorial positions that are preferentially visited by females, is based on recent field and laboratory observations by Whittier *et al.* (1992). During censuses of a natural population, they observed 71 matings on 70 different leaves, which suggests that the exact position of a male's territory had little impact on female choice. Of the 118 trees censused, however, ten accounted for 80% of all male sightings, and about 3/4 of the observed matings were recorded on just three trees. Although these data clearly indicate male preference for specific trees (see also Shelly and Whittier 1996), the superabundance of territories on these trees argues against interference competition for lek attendance (leks ranged from 2 to 12 males in host trees 3–4 m tall with full canopies). More importantly, males were found to only weakly defend territories only weakly, and a resident advantage was absent in male–male interactions (on the contrary, a significant intruder advantage was found (Whittier *et al.* 1992)). Although data relating male size and fighting ability are not available, Whittier *et al.* (1992) found no size difference in the field

between mating males and a random sample of non-mating males (but see Churchill-Stanland *et al.* 1986 for opposite results in laboratory-reared populations), which suggests that size-dependent fighting success is not a key determinant of male mating success. Observations of male aggregations in the laboratory (Whittier *et al.* 1994) further revealed that male–male interactions were rare and that mating success was unrelated to fighting ability.

In laboratory aggregations of *C. capitata*, mating frequency of males was non-random, with higher than expected numbers of non-mating and highly successful mating males (Whittier *et al.* 1994). Male mating success was related to their overall level of sexual activity: the total numbers of courtships performed, mountings, and females courted were all positively correlated with mating frequency (although time spent pheromone-calling was not). In addition, females appear to discriminate among males on the basis of courtship cues. Following female approach, the male curves his abdomen under the body (with the rectal epithelium still everted and releasing pheromone), vibrates his wings while fanning them toward the female, and makes oscillatory head movements. If the female remains near, the male then leaps on her back and attempts copulation (Arita and Kaneshiro 1989). Of females that were courted by two or more males, 3/4 ultimately copulated with the male that had the highest mating frequency (over the entire duration of the experiment) within the observed set of courting males. Female discrimination appeared to occur primarily after male mounting, because male mating frequency was positively related to the proportion of mountings that led to copulation but was unrelated to the proportion of courtships that led to mounting. Unfortunately, it is not known at present which male phenotypic cues are important in female choice.

In summary, studies on lekking fruit flies reveal several challenges. Distinguishing the effects of male aggression and female choice, for example, is complicated by the fact that certain male traits may function in both types of sexual competition. Moreover, although a particular trait such as body size might be expected to confer a universal advantage, this need not be the case. For example, in a laboratory study on the lekking fly *D. silvestris*, Boake (1989) found that male size was positively correlated with success in aggression but was negatively correlated with courtship success. In addition, because male courtship often involves visual, acoustic and olfactory communication, it is extremely difficult to identify exactly which cues are important to female choice. Ideally, data for all communicative modes should be gathered and analyzed simultaneously with multivariate techniques (Gibson and Bradbury 1985). To our knowledge, this approach, which would require collaboration between experts in acoustics, olfaction and behavior, has not yet been applied to any lekking fruit-fly species.

ORIGIN OF LEK BEHAVIOR

Lek mating systems appear to evolve when males are unable to monopolize and defend resources or female groups (Emlen and Oring 1977; Bradbury 1981). However, there are multiple explanations for the evolution of male clustering (as opposed, for example, to scramble competition and active mate-searching by males). These explanations fall into two categories: male-initiated, whereby males form aggregations for their own benefit, or female-initiated, whereby females favor males that cluster and thus force males to aggregate in order to achieve matings (Bradbury and Gibson 1983). Here, we briefly review several of these hypotheses in the context of insect leks, citing empirical evidence where possible. Data on insects are insufficient for comparative (phylogenetic) approaches to the question of lek evolution. However, among insect taxa, additional studies of drosophilids or tephritids (which contain both lekking and non-lekking species) may soon permit assessment of the importance of phylogenetic history in the origin of lek mating systems.

Male-initiated clustering

Four hypotheses have been proposed to explain male-initiated clustering. First, males may aggregate to increase signal range or amount of time signals are being emitted. This explanation has been used to explain clumping in a variety of insects, including ants (Hölldobler 1976), cicadas (Alexander and Moore 1958; Williams and Smith 1991) and fireflies (Lloyd 1973). Despite the popularity of this explanation, Bradbury (1981) argued convincingly that male groups at best generate a linear increase in female arrivals per male and that increased signal efficacy was therefore not a likely cause of male grouping. More importantly, data for crickets (Cade 1981), cicadas (Doolan 1981), and fruit flies (Sivinski 1989) all show that increasing group size increases the total number of females attracted but not on a per male basis.

Second, males may clump in order to reduce predation on themselves while displaying. A general function of insect congregations may be to reduce predation risks either via increased predator detection (Treherne and Foster 1980) or decreased per capita risk of predation (the 'dilution effect' (Turchin and Kareiva 1989)). However, predators may also aggregate at prey clumps, thus negating the dilution effect (Madden and Pimentel 1965).

In many insects, males, as the actively signaling sex, may suffer higher predation than the less conspicuous females (Burk 1982). Few studies have examined predation risks as a function of male spacing *per se*, and these have yielded mixed results. Based on counts of detached wings, Karban (1982) found that the intensity of avian predation was independent of cicada density at a given tree but positively related to the number of birds feeding there, suggesting that birds were consuming as many prey as possible. Although both sexes were eaten, this result none the less suggests that the birds were satiated and that the individual probability of escaping predation increased with increasing group size. In contrast to these results, Cade (1981) measured the response of acoustically orienting parasitoids to solitary vs. grouped cricket songs and found no difference between the treatments in the average number of parasitoids attracted to each male.

Third, males may cluster near hotspots through which the largest numbers of females are likely to pass. As noted above, males often aggregate near resources vital to female survival and reproduction. Tephritid leks, for example, typically occur on trees bearing both oviposition sites (fruit) and food resources (e.g. bird droppings (Whittier *et al.* 1992)). In other cases, males may aggregate at particular topographic features, which serve to funnel or channel females toward waiting males. For example, male carpenter bees often hover along gullies used by females traveling to and from ridgetops (Alcock and Smith 1987).

To the extent that females use hilltops as natural orientation guides, male hilltopping may also represent an aggregative response at a female transit route. However, unless female movement to hilltops can be demonstrated to function in some capacity other than mating, invoking the hotspot hypothesis may be a circular argument: males cluster at hilltops because females travel there, and females visit there because they can readily find mates there. This reasoning obviously cannot explain how hilltopping arose in the first place, and alternative (non-mating) mechanisms should be considered (for example, see Wickman 1988).

Fourth, males may gather near 'hotshot males' to increase their mating opportunities. Beehler and Foster (1988) proposed that male groups may result from the tendency of subordinate or less attractive males to settle in the vicinity of dominant or preferred males (the so-called hotshots). Needless to say, perhaps, little data are available to test this hypothesis for insects. Data from the Mediterranean fruit fly, however, are not consistent with one of the model's predictions. According to Beehler and Foster (1988), removal of hotshot males from a lek should result in decreased mating and male attendance. In *C. capitata*, however, repeated removal of mating pairs from a natural lek site had no noticeable effect on subsequent mating activity or lek size (Whittier *et al.* 1992).

Female-initiated clustering

Female-initiated clustering may occur through one or more of four processes. First, females may prefer clumped males to reduce predation upon them. As for males, few data are available to evaluate this hypothesis. To our knowledge, Karban's (1982) study on cicadas showing decreased predation risk with increasing lek size is the only one that specifically addresses this issue.

Second, females of species that are rare or widely dispersed may prefer males clustered at prominent landmarks in order to facilitate quick location of a mate. This explanation is frequently invoked to account for hilltopping aggregations (Thornhill and Alcock 1983) and was supported by Scott's (1968) study on butterflies, which showed that hilltopping species had lower population densities than non-hilltopping species. Landmark-based mating aggregations may also be favored in species, such as botflies, where a short adult lifespan (as little as 2–3 days (Cope and Catts 1991)) necessitates rapid mating.

Third, females may prefer males clustered at distinct landmarks that attract individuals from wide areas in order to promote outcrossing. By mating at prominent landmarks, and not at the nest site, females of social insects may prevent pairing with siblings and avoid negative effects of inbreeding, such as the production of sterile (diploid) males (Nagel and Rettenmeyer 1973; Page 1980; Matthes-Sears and Alcock 1986; Godfray and Cook, this volume).

Fourth, females may prefer clustered males because clustering facilitates mate choice. This hypothesis implies that females are more likely to mate with grouped males than with solitary males and that females gain some benefit by comparing potential mates. Despite general recognition of this hypothesis (Alexander 1975; Thornhill and Alcock 1983), few studies have directly examined either of these issues. Data on the presumed preference for grouped over solitary males are inconsistent across taxa. Hibino (1986) reported that female stink bugs were far more likely to accept courtship and matings from grouped males than solitary males. However, in the Caribbean fruit fly, *A. suspensa*, Sivinski (1989) observed that matings were equally likely to occur inside and outside male clusters. Finally, females of the cicada *Cystosoma saundersii* were less likely to mate with males in groups than with isolated males (Doolan 1981; see also Ehrlich and Wheye 1986).

Because the costs associated with locating and mating in leks generally exceed any direct benefits (Reynolds and Gross 1990), female choice has presumably evolved to increase indirect fitness benefits either through the production of 'arbitrarily sexy' progeny (by runaway or Fisherian selection) or 'adaptively vigorous' progeny (by selection for 'good genes' (Kirkpatrick and Ryan 1991)). Male traits associated with successful mating have been found to have high heritability in several insects (Cade 1981; Butlin and Hewitt 1986; Carson and Lande 1984; Hedrick 1988). However, few studies have attempted to measure the adaptive benefits of female mate choice. To our knowledge, only Whittier and Kaneshiro (1995) have addressed this issue for a lekking insect; despite non-random mating among males, they found no benefits associated with female choice (in terms of either increased fecundity or production of 'sexy' sons). However, other studies of non-lekking insects have reported enhanced offspring fitness (Partridge 1980; Taylor *et al.* 1987; McLain and Marsh 1990) through female choice (but see Boake 1985). Given these results, future work on lekking species may also provide support for the 'good genes' hypothesis, unless female choice exhibits different dynamics in leks than in other mating systems.

SUGGESTIONS FOR FUTURE RESEARCH

Insects offer several advantages over vertebrates in the study of lek behavior. Because of their small size, short lifespan, high fecundity, and adaptability to laboratory conditions, many insect species can be successfully reared in the laboratory. Thus, rigorous examination of potential benefits, direct or indirect, realized through female choice is more feasible. In addition, the sexual signals of insects can be more easily manipulated than those of vertebrates, which allows for experimental (rather than observational) tests of theoretical predictions. For instance, several studies have created male aggregations of varying size (by varying numbers of broadcast speakers (Cade 1981) or males themselves (Doolan 1981)) and monitored the resulting female settlement patterns.

Recent work by Droney (1994) provides a good example of the experimental utility of insects in studying lek mating systems. Using large, environmentally controlled rooms, Droney (1994) tested various hypotheses regarding the evolution of male mating aggregations in the Hawaiian picture-winged fly *D. grimshawi*. Among the competing hypotheses, the hotspot model offered the strongest support for lek behavior; experimental control of the distribution of food and oviposition sites revealed that males were most likely to initiate leks in close proximity to these resources. In contrast, manipulating lek size revealed decreasing rates of female arrivals and matings on a per male basis. Though an exemplary use of experimentation in lek research, this study simultaneously belies the ease in interpreting laboratory data. Quantitative field data on *D. grimshawi* leks are completely lacking, which prompts immediate doubts about the 'real world' validity of the laboratory results. For example, field collections indicate that (as Droney 1994 acknowledges) the species is very rare, and the experimental densities used were probably much higher than natural densities. Thus, the artificial regimes of lek density and size may have had unusual effects on female arrival rates and mating behavior. In addition, monitoring male and female response to positional changes of a single food/oviposition site may not have adequately mimicked nature, where a heterogeneous and temporally variable mosaic of resource patches exists. Our intent is not to unduly criticize this particular study but to remind readers of the oft-present trade-off between the precision in hypothesis testing afforded by laboratory experiments and the explanatory power of these results for natural processes.

Finally, we stress the importance of research that addresses the dynamic nature of lek mating systems. The impact of environmental conditions on the display of lek behavior undoubtedly varies among species. The medfly, for example, exhibits lek behavior under variable ecological circumstances (Hendrichs and Hendrichs 1990; Whittier *et al.* 1992), whereas other species form leks only under a specific circumstance (Wickman 1988). Information on the plasticity of mating systems will require long-term projects or concurrent studies at ecologically variable locations. Though logistically difficult, this type of research may generate valuable insight into the factors promoting lek behavior from both the male and the female perspective.

ACKNOWLEDGEMENTS

We thank J. D. Allan, E. P. Catts, R. Karban, J. Lloyd, K. M. O'Neill and R. Rutowski for sharing their expertise on insect mating systems. Comments by J. Alcock, J. Choe, C. Henry, E. Villalobos and G. Wilkinson improved the manuscript, and to all we are grateful. TES also thanks his father, G. J. Shelly, for his editorial assistance. We are especially thankful to S. Fong and C. Ihori, who helped track down many of the references cited. During the preparation of this chapter, TES and TSW were supported by United States Department of Agriculture grant 59–5320–1–189 and California Department of Food and Agriculture grant 93–0481.

LITERATURE CITED

Alcock, J. 1975a. Male mating strategies of some philanthine wasps (Hymenoptera: Sphecidae). *J. Kansas. Entomol. Soc.* **48**: 532–545.

–. 1975b. Territorial behaviour by males of *Philanthus multimaculatus* (Hymenoptera: Sphecidae) with a review of territoriality in male sphecids. *Anim. Behav.* **23**: 889–895.

–. 1981. Lek territoriality in the tarantula hawk wasp *Hemipepsis ustulata* (Hymenoptera: Pompilidae). *Behav. Ecol. Sociobiol.* **8**: 309–317.

–. 1983. Territoriality by hilltopping males of the great purple hairstreak, *Atlides halesus* (Lepidoptera, Lycaenidae): convergent evolution with a pompilid wasp. *Behav. Ecol. Sociobiol.* **13**: 57–62.

–. 1984. Convergent evolution in perching and patrolling site preferences of some hilltopping insects of the Sonoran desert. *Southwest. Nat.* **29**: 475–480.

–. 1985a. Hilltopping behavior in the wasp *Pseudomasaris maculifrons* (Fox) (Hymenoptera: Masaridae). *J. Kansas. Entomol. Soc.* **58**: 162–166.

–. 1985b. Hilltopping in the nymphalid butterfly *Chlosyne californica* (Lepidoptera). *Am. Midl. Nat.* **113**: 69–75.

–. 1987. Leks and hilltopping in insects. *J. Nat. Hist.* **21**: 319–328.

–. 1989. The mating system of *Mydas ventralis* (Diptera: Mydidae). *Psyche* **96**: 167–176.

–. 1990. A large male competitive advantage in a lekking fly, *Hermetia comstocki* Williston (Diptera: Stratiomyidae). *Psyche* **97**: 267–279.

–. 1993. The effects of male body size on territorial and mating success in the landmark-defending fly *Hermetia comstocki* (Stratiomyidae). *Ecol. Entomol.* **18**: 1–6.

Alcock, J., E. M. Barrows, G. Gordh, L. J. Hubbard, L. Kirkendall, D. W. Pyle, T. L. Ponder and F. G. Zalom. 1978. The ecology and evolution of male reproductive behavior in the bees and wasps. *J. Zool. (Lond.)* **64**: 293–326.

Alcock, J. and D. Gwynne. 1988. The mating system of *Vanessa kershawi*: Males defend landmark territories as mate encounter sites. *J. Res. Lepid.* **26**: 116–124.

Alcock, J. and K. M. O'Neill. 1986. Density-dependent mating tactics in the grey hairstreak, *Strymon melinus* (Lepidoptera: Lycaenidae). *J. Zool. (Lond.)* **209**: 105–113.

Alcock, J. and D. W. Pyle. 1979. The complex courtship behavior of *Physiphora demandata* (F.) (Diptera: Otitidae). *Z. Tierpsychol.* **49**: 352–362.

Alcock, J. and J. E. Schaefer. 1983. Hilltop territoriality in a Sonoran desert bot fly (Diptera: Cuterebridae). *Anim. Behav.* **31**: 518–525.

Alcock, J. and A. P. Smith. 1987. Hilltopping, leks and female choice in the carpenter bee *Xylocopa* (*Neoxylocopa*) *varipuncta*. *J. Zool. (Lond.)* **211**: 1–10.

Alexander, R. D. 1975. Natural selection and specialized chorusing behavior in acoustical insects. In *Insects, Science, and Society*. D. Pimentel, ed., pp. 35–77. New York: Academic Press.

Alexander, R. D. and T. E. Moore. 1958. Studies on the acoustical behavior of seventeen-year cicadas (Homoptera: Cicadidae: *Magicicada*). *Ohio J. Sci.* **58**: 107–127.

Allan, J. D. and A. S. Flecker. 1989. The mating biology of a mass-swarming mayfly. *Anim. Behav.* **37**: 361–371.

Aluja, M., J. Hendrichs and M. Cabrera. 1983. Behavior and interactions between *Anastrepha ludens* (L) and *A. obliqua* (M) on a field caged mango tree 1. Lekking behavior and male territoriality. In *Fruit Flies of Economic Importance*. R. Cavalloro, ed., pp. 122–133. Athens: Balkema.

Anderson, J. F. 1971. Autogeny and mating and their relationship to biting in the saltmarsh deer fly, *Chrysops atlanticus* (Diptera: Tabanidae). *Ann. Entomol. Soc. Am.* **64**: 1421–1424.

Anzenberger, G. 1977. Ethological study of African carpenter bees of the genus *Xylocopa* (Hymenoptera, Anthophoridae). *Z. Tierpsychol.* **44**: 337–374.

Arita, L. H. and K. Y. Kaneshiro. 1985. The dynamics of the lek system and mating success in males of the Mediterranean fruit fly, *Ceratitis capitata* (Wiedemann). *Proc. Hawaiian Entomol. Soc.* **25**: 39–48.

–. 1989. Sexual selection and lek behavior in the Mediterranean fruit fly, *Ceratitis capitata* (Diptera: Tephritidae). *Pac. Sci.* **43**: 135–143.

Beani, L. and C. Calloni. 1991. Leg tegumental glands and male rubbing behavior at leks in *Polistes dominulus* (Hymenoptera: Vespidae). *J. Insect Behav.* **4**: 449–462.

Beani, L., R. Cervo, C. M. Lorenzi, and S. Turillazzi. 1992. Landmark-based mating systems in four *Polistes* species (Hymenoptera: Vespidae). *J. Kans. Entomol. Soc.* **65**: 211–217.

Beehler, B. M. and M. S. Foster. 1988. Hotshots, hotspots, and female preferences in the organization of lek mating systems. *Am. Nat.* **131**: 203–219.

Belkin, J. N., N. Ehmann and G. Heid. 1951. Preliminary field observations on the behavior of the adults of *Anopheles franciscanus* McCracken in southern California. *Mosq. News* **11**: 23–31.

Bell, W. J. and L. R. Kipp. 1994. *Drosophila percnosoma* Hardy Lek sites: Spatial and temporal distributions of males and the dynamics of their agonistic behavior (Diptera: Drosophilidae). *J. Kansas. Entomol. Soc.* **67**: 267–276.

Bitzer, R. J. and K. C. Shaw. 1979. Territorial behavior of the Red Admiral, *Vanessa atalanta* (L.) (Lepidoptera: Nymphalidae). *J. Res. Lepid.* **18**: 36–49.

Blickle, R. L. 1959. Observations on the hovering and mating of *Tabanus bishoppi* Stone (Diptera, Tabanidae). *Ann. Entomol. Soc. Am.* **52**: 183–190.

Boake, C. R. B. 1985. Genetic consequences of mate choice: a quantitative genetic method for testing sexual selection theory. *Science (Wash., D.C.)* **227**: 1061–1063.

–. 1989. Correlations between courtship success, aggressive success, and body size in a picture-winged fly, *Drosophila silvestris*. *Ethology* **80**: 318–329.

Bradbury, J. W. 1977. Lek mating behavior in the hammer-headed bat. *Z. Tierpsychol.* **45**: 225–255.

–. 1981. The evolution of leks. In *Natural Selection and Social Behavior*. R. D. Alexander and D. W. Tinkle, eds., pp. 138–169. New York: Chiron Press.

–. 1985. Contrasts between insects and vertebrates in the evolution of male display, female choice, and lek mating. In *Experimental Ecology and Sociobiology*. B. Hölldobler and M. Lindauer, eds., pp. 273–289. New York: Gustav Fischer Verlag.

Bradbury, J. W., and R. M. Gibson. 1983. Leks and mate choice. In *Mate Choice*. P. Bateson, ed., pp. 109–138. Cambridge: Cambridge University Press.

Brodsky, A. K. 1973. The swarming behavior of mayflies (Ephemeroptera). *Entomol. Rev.* **52**: 33–39.

Buck, J. and E. Buck. 1966. Biology of synchronous flashing of fireflies. *Nature (Lond.)* **211**: 562–564.

–. 1978. Toward a functional interpretation of synchronous flashing by fireflies. *Am. Nat.* **112**: 471–492.

Burk, T. 1982. Evolutionary significance of predation on sexually signalling males. *Fla. Entomol.* **65**: 90–104.

–. 1983. Behavioral ecology of mating in the Caribbean fruit fly, *Anastrepha suspensa* (Loew) (Diptera: Tephritidae). *Fla. Entomol.* **66**: 330–344.

–. 1984. Male-male interactions in Caribbean fruit flies, *Anastrepha suspensa* (Loew) (Diptera: Tephritidae): territorial fights and signalling stimulation. *Fla. Entomol.* **67**: 542–547.

–. 1991. Sex in leks: an overview of sexual behavior in *Anastrepha* fruit flies. In *Proceedings of International Symposium on Biology and Control of Fruit Flies*. K. Kawasaki, O. Iwahashi and

K. Kaneshiro, eds., pp. 177–189. Taipei, Taiwan: The Food and Fertilizer Technology Center.

Burk, T. and J. C. Webb. 1983. Effect of male size on calling propensity, song parameters, and mating success in Caribbean fruit flies, *Anastrepha suspensa* (Loew) (Diptera: Tephritidae). *Ann. Entomol. Soc. Am.* **76**: 678–682.

Butlin, R. K. and G. M. Hewitt. 1986. Heritability estimates for characters under sexual selection in the grasshopper, *Chorthippus brunneus*. *Anim. Behav.* **34**: 1256–1261.

Cade, W. H. 1981. Field cricket spacing and the phonotaxis of crickets and parasitoid flies to clumped and isolated cricket songs. *Z. Tierpsychol.* **55**: 365–375.

Capelle, K. J. 1970. Studies on the life history and development of *Cuterebra polita* (Diptera: Cuterebridae) in four species of rodents. *J. Med. Entomol.* **7**: 320–327.

Carson, H. L. and R. Lande. 1984. Inheritance of a secondary sexual character in *Drosophila silvestris*. *Proc. Natl. Acad. Sci. U.S.A.* **81**: 6904–6907.

Catts, E. P. 1964. Field behavior of adult *Cephenemyia* (Diptera: Oestridae). *Can. Entomol.* **96**: 579–585.

–. 1967. Biology of a California rodent bot fly *Cuterebra latifrons* Coquillett (Diptera: Cuterebridae). *J. Med. Entomol.* **4**: 87–101.

–. 1979. Hilltop aggregation and mating behavior by *Gasterophilus intestinalis* (Diptera: Gasterophilidae). *J. Med. Entomol.* **16**: 461–464.

Catts, E. P., R. Garcia and J. H. Poorbaugh. 1965. Aggregation sites of males of the common cattle grub, *Hypoderma lineatum* (DeVillers) (Diptera: Oestridae). *J. Med. Entomol.* **1**: 357–358.

Chiang, H. C. 1961. Ecology of insect swarms. I. Experimental studies of the behavior of *Anarete near felti* Pritchard in artificially induced swarms (Cecidomyiidae, Diptera). *Anim. Behav.* **9**: 213–219.

Churchill-Stanland, C., R. Stanland, T. T. Y. Wong, N. Tanaka, D. O. McInnis and R. V. Dowell. 1986. Size as a factor in the mating propensity of Mediterranean fruit flies, *Ceratitis capitata*, in the laboratory. *Ann. Entomol. Soc. Am.* **79**: 614–619.

Cicero, J. M. 1983. Lek assembly and flash synchrony in the Arizona firefly *Photinus knulli* Green (Coleoptera: Lampyridae). *Coleopt. Bull.* **37**: 318–342.

Conant, P. 1978. Lek behavior and ecology of two sympatric homosequential Hawaiian *Drosophila*: *Drosophila heteroneura* and *Drosophila silvestris*. M. S. thesis, University of Hawaii at Manoa.

Cooke, H. G. 1940. Observations on mating flights of the mayfly *Stenonema vicarium* (Ephemerida). *Entomol. News* **51**: 12–14.

Cope, S. E. and E. P. Catts. 1991. Parahost behavior of adult *Gasterophilus intestinalis* (Diptera: Gasterophilidae) in Delaware. *J. Med. Entomol.* **28**: 67–73.

Cronin, H. 1991. *The Ant and the Peacock*. Cambridge University Press.

Cuthbertson, A. 1926. Studies on Clyde crane-flies: the swarming of craneflies. *Entomol. Mon. Mag.* **62**: 36–38.

Darwin, C. 1871. *The Descent of Man and Selection in Relation to Sex.* (Facsimile reproduction of the first edition. Princeton: Princeton University Press. 1981.)

Darwin, F. 1887. *The Life and Letters of Charles Darwin.* London: John Murray.

Davidson, D. W. 1982. Sexual selection in harvester ants (Hymenoptera: Formicidae: *Pogonomyrmex*). *Behav. Ecol. Sociobiol.* **10**: 245–250.

Dennis, R. L. H. 1982. Mate location strategies in the wall brown butterfly, *Lasiommata megera* (L.) (Lepidoptera: Satyridae): Wait or seek? *Entomol. Rec. J. Var.* **94**: 209–214.

DeVries, P. J. 1978. Observations on the apparent lek behavior in Costa Rican rainforest *Perrhybris pyrrha* Cramer (*Pieridae*). *J. Res. Lepid.* **17**: 142–144.

Dodson, G. 1986. Lek mating system and large male aggressive advantage in a gall-forming tephritid fly (Diptera: Tephritidae). *Ethology* **72**: 99–108.

Dodson, G. and D. K. Yeates. 1989. Male *Bembix furcata* Erichson (Hymenoptera: Sphecidae) behaviour on a hilltop in Queensland. *Pan-Pac. Entomol.* **65**: 172–175.

–. 1990. The mating system of a bee fly (Diptera: Bombyliidae). II. Factors affecting male territorial and mating success. *J. Insect Behav.* **3**: 619–636.

Doolan, J. M. 1981. Male spacing and the influence of female courtship behaviour in the bladder cicada, *Cystosoma saundersii* Westwood. *Behav. Ecol. Sociobiol.* **9**: 269–276.

Doolan, J. M. and R. C. MacNally. 1981. Spatial dynamics and breeding ecology in the cicada *Cystosoma saundersii*: the interaction between distributions of resources and intraspecific behaviour. *J. Anim. Ecol.* **50**: 925–940.

Droney, D. C. 1992. Sexual selection in a lekking Hawaiian *Drosophila*: the roles of male competition and female choice in male mating success. *Anim. Behav.* **44**: 1007–1020.

–. 1994. Tests of hypotheses for lek formation in a Hawaiian *Drosophila*. *Anim. Behav.* **47**: 351–361.

Drummond, B. A. 1984. Multiple mating and sperm competition in the Lepidoptera. In *Sperm Competition and the Evolution of Animal Mating Systems*. R. L. Smith, ed., pp. 291–370. New York: Academic Press.

Dunning, D. C., J. A. Byers and C. D. Zanger. 1979. Courtship in two species of periodical cicadas, *Magicicada septendecim* and *Magicicada cassini*. *Anim. Behav.* **27**: 1073–1090.

Ehrlich, P. R. and D. Wheye. 1986. 'Non-adaptive' hilltopping behavior in male checkerspot butterflies (*Euphydryas editha*). *Am. Nat.* **4**: 477–483.

Eickwort, G. C. and H. S. Ginsberg. 1980. Foraging and mating behavior in Apoidea. *Annu. Rev. Entomol.* **25**: 421–446.

Emlen, S. T. and L. W. Oring. 1977. Ecology, sexual selection and the evolution of mating systems. *Science (Wash., D.C.)* **197**: 215–223.

Evans, H. E. 1993. Observations on aggregations of males of two species of beewolves (Hymenoptera: Sphecidae: *Philanthus*). *Psyche* **100**: 25–33.

Evans, H. E. and K. M. O'Neill. 1988. *The Natural History and Behavior of North American Beewolves.* Ithaca: Cornell University Press.

Feaver, M. N. 1983. Pair formation in the katydid *Orchelimum nigripes* (Orthoptera: Tettigoniidae). In *Orthopteran Mating Systems.* D. T. Gwynne and G. K. Morris, eds., pp. 205–239. Boulder: Westview Press.

Flecker, A. S. , J. D. Allan and N. L. McClintock. 1988. Male body size and mating success in swarms of the mayfly *Epeorus longimanus. Holarct. Ecol.* 11: 280–285.

Frankie, G. W., S. B. Vinson and R. E. Coville. 1980. Territorial behavior of *Centris adani* and its reproductive function in the Costa Rican dry forest (Hymenoptera: Anthophoridae). *J. Kansas Entomol. Soc.* 53: 837–857.

Frohne, W. C. and R. G. Frohne. 1954. Diurnal swarms of *Culex territans* Walker, and the crepuscular swarming of *Aëdes* about a small glade in Alaska. *Mosq. News* 14: 62–64.

Gibson, R. M. and J. W. Bradbury. 1985. Sexual selection in lekking sage grouse: phenotypic correlates of male mating success. *Behav. Ecol. Sociobiol.* 18: 117–123.

Gilbert, F. S. 1984. Thermoregulation and the structure of swarms in *Syrphus ribesii* (Syrphidae). *Oikos* 42: 249–255.

Graham, C. L. and K. J. Capelle. 1970. Redescription of *Cuterebra polita* (Diptera: Cuteribridae) with notes on its taxonomy and biology. *Ann. Entomol. Soc. Am.* 63: 1569–1573.

Gubler, D. J. and N. C. Bhattacharya. 1972. Swarming and mating of *Aedes* (S.) *albopictus* in nature. *Mosq. News* 32: 219–223.

Gullefors, B. and E. Petersson. 1993. Sexual dimorphism in relation to swarming and pair formation patterns in leptocerid caddisflies (Trichoptera: Leptoceridae). *J. Insect Behav.* 6: 563–577.

Hedrick, A. V. 1988. Female choice and the heritability of attractive male traits: an empirical study. *Am. Nat.* 132: 267–276.

Heinrich, B. and C. Pantle. 1975. Thermoregulation in small flies (*Syrphus* sp.): basking and shivering. *J. Exp. Biol.* 62: 599–610.

Hendrichs, J. and M. A. Hendrichs. 1990. Mediterranean fruit fly (Diptera: Tephritidae) in nature: location and diel pattern of feeding and other activities on fruiting and non-fruiting hosts and non-hosts. *Ann. Entomol. Soc. Am.* 83: 632–641.

Hibino, Y. 1985. Formation and Maintenance of mating aggregations in a stink bug, *Megacopta punctissimum* (Montandon) (Heteroptera: Plataspidae). *J. Ethol.* 3: 123–129.

–. 1986. Female choice for male gregariousness in a stink bug, *Megacopta punctissimum* (Montandon) (Heteroptera, Plataspidae). *J. Ethol.* 4: 91–95.

Hibino, Y. and Y. Ito. 1983. Mating aggregation of a stink bug, *Megacopta punctissimum* (Montandon) (Heteroptera: Plataspidae). *Res. Pop. Ecol.* 25: 180–188.

Hilsenhoff, W. L. 1966. The biology of *Chironomus plumosus* (Diptera: Chironomidae) in Lake Winnebago, Wisconsin. *Ann. Entomol. Soc. Am.* 59: 465–473.

Hodosh, R. J., J. M. Ringo and F. T. McAndrew. 1979. Density and lek displays in *Drosophila grimshawi* (Diptera: Drosophilidae). *Z. Tierspsychol.* 49: 164–172.

Hoffmann, A. A. and M. W. Blows. 1992. Evidence that *Drosophila mycetophaga* Malloch (Diptera: Drosophilidae) is not a true 'lekking' species. *J. Austr. Entomol. Soc.* 31: 219–221.

Höglund, J. and R. V. Alatalo. 1995. *Leks.* Princeton: Princeton University Press.

Hölldobler, B. 1976. The behavioral ecology of mating in harvester ants (Hymenoptera: Formicidae: *Pogonomyrmex*). *Behav. Ecol. Sociobiol.* 1: 405–423.

Hoy, J. B. and J. R. Anderson. 1978. Behavior and reproductive physiology of blood-sucking snipe flies (Diptera: Rhagionidae: *Symphoromyia*) attacking deer in northern California. *Hilgardia* 46: 113–168.

Hunter, D. M. and J. M. Webster. 1973. Aggregation behavior of adult *Cuterebra grisea* and *C. tenebrosa* (Diptera: Cuterebridae). *Can. Entomol.* 105: 1301–1307.

Hutchinson, G. E. 1957. Concluding remarks. *Cold Spring Harbor Symp. Quant. Biol.* 22: 415–427.

Iwahashi, O. and T. Majima. 1986. Lek formation and male-male competition in the melon fly, *Dacus cucurbitae* Coquillett (Diptera: Tephritidae). *Appl. Entomol. Zool.* 21: 70–75.

Janzen, D. H. 1966. Notes on the behavior of the carpenter bee *Xylocopa fimbriata* in Mexico (Hymenoptera: Apoidea). *J. Kansas Entomol. Soc.* 39: 633–641.

Jarvis, E. K. and L. C. Rutledge. 1992. Laboratory observations on mating and leklike aggregations in *Lutzomyia longipalpis* (Diptera: Psychodidae). *J. Med. Entomol.* 29: 171–177.

Johnson, R. A. and S. W. Rissing. 1993. Breeding biology of the desert leaf- cutter and *Acromyrmex versicolor* (Pergande) (Hymenoptera: Formicidae). *J. Kansas Entomol. Soc.* 66: 127–128.

Karban, R. 1982. Increased reproductive success at high densities and predator satiation for periodical cicadas. *Ecology* 63: 321–328.

–. 1983. Sexual selection, body size and sex related mortality in the cicada *Magicicada cassini. Am. Midl. Nat.* 109: 324–330.

Kimsey, L. S. 1980. The behaviour of male orchid bees (Apidae, Hymenoptera, Insecta) and the question of leks. *Anim. Behav.* 28: 996–1004.

Kirkpatrick, M. and M. J. Ryan. 1991. The evolution of mating preferences and the paradox of the lek. *Nature (Lond.)* 350: 33–38.

Knab, F. 1906. The swarming of *Culex pipiens. Psyche* 13: 123–133.

–. 1907. The swarming of *Anopheles punctipennis* Say. *Psyche* 14: 1–4.

Knapton, R. W. 1985. Lek structure and territoriality in the Chryxus arctic butterfly, *Oeneis chryxus* (Satyridae). *Behav. Ecol. Sociobiol.* 17: 389–395.

Koenig, W. D. and S. S. Albano. 1985. Patterns of territoriality and mating success in the white-tailed skimmer *Plathemis lydia* (Odonata: Anisoptera). *Am. Midl. Nat.* 114: 1–12.

Kuba, H. and J. Koyama. 1985. Mating behavior of wild melon flies, *Dacus cucurbitae* Coquillett (Diptera: Tephritidae) in a field cage: courtship behavior. *Appl. Entomol. Zool.* 20: 365–372.

Lederhouse, R. C. 1982. Territorial defense and lek behavior of the black swallowtail butterfly, *Papilio polyxenes. Behav. Ecol. Sociobiol.* **10**: 109–118.

Lloyd, J. E. 1972. Mating behavior of a New Guinea *Luciola* firefly: a new communicative protocol (Coleoptera: Lampyridae). *Coleopt. Bull.* **26**: 155–163.

–. 1973. Fireflies of Melanesia: bioluminescence, mating behavior, and synchronous flashing (Coleoptera: Lampyridae). *Environ. Entomol.* **2**: 991–1008.

–. 1979. Sexual selection in luminescent beetles. In *Sexual Selection and Reproductive Competition in Insects.* M. S. Blum and N. A. Blum, eds., pp. 293–342. New York: Academic Press.

Lloyd, M. and R. Karban. 1983. Chorusing centers of periodical cicadas. *J. Kansas Entomol. Soc.* **56**: 299–304.

Lyman, F. E. 1944. Notes on the emergence, swarming, and mating of *Hexagenia* (Ephemeroptera). *Entomol. News* **55**: 207–210.

Madden, J. L., and D. Pimentel. 1965. Density and spatial relationships between a wasp parasite and its housefly host. *Can. Entomol.* **97**: 1031–1037.

Malavasi, A., J. S. Morgante and R. J. Prokopy. 1983. Distribution and activities of *Anastrepha fraterculus* (Diptera: Tephritidae) flies on host and nonhost trees. *Ann. Entomol. Soc. Am.* **76**: 286–292.

Mallet, J. 1984. Sex roles in the ghost moth *Hepialis humuli* (L.) and a review of mating in the Hepialidae (Lepidoptera). *Zool. J. Linn. Soc.* **79**: 67–82.

Markl, H., B. Hölldobler and T. Hölldobler. 1977. Mating behavior and sound production in harvester ants (*Pogonomyrmex*, Formicidae). *Insectes Soc.* **24**: 191–212.

Matthes-Sears, W. and J. Alcock. 1986. Hilltopping behavior of *Polistes commanchus navajoe* (Hymenoptera: Vespidae). *Ethology* **71**: 42–53.

McLachlan, A. J. and D. F. Allen. 1987. Male mating success in Diptera: advantages of small size. *Oikos* **48**: 11–14.

McLain, D. K. and N. B. Marsh. 1990. Male copulatory success: heritability and relationship to mate fecundity in the southern green stinkbug, *Nezara viridula* (Hemiptera: Pentatomidae). *Heredity* **64**: 161–167.

Meyer, R. P. and M. E. Bock. 1980. Aggregation and territoriality of *Cuterebra lepivora* (Diptera: Cuterebridae). *J. Med. Entomol.* **17**: 489–493.

Michener, C. D. 1948. Observations on the mating behavior of harvester ants. *J. N. Y. Entomol. Soc.* **56**: 239–242.

–. 1960. Treetop mating aggregations of *Pogonomyrmex* (Hymenoptera: Formicidae). *J. Kansas. Entomol. Soc.* **33**: 46.

–. 1974. *The Social Behavior of Bees.* Cambridge, Mass.: Harvard University Press.

Mintzer, A. C. 1982. Copulatory behavior and mate selection in the harvester ant, *Pogonomyrmex californica* (Hymenoptera: Formicidae). *Ann. Entomol. Soc. Am.* **75**: 323–326.

Moorhouse, D. E. and Colbo, M. H. 1973. On the swarming of *Austrosimulium pestilens* Mackerras and Mackerras (Diptera: Simuliidae). *J. Austr. Entomol. Soc.* **12**: 127–130.

Morgante, J. S., A. Malavasi and R. J. Prokopy. 1983. Mating behavior of wild *Anastrepha fraterculus* (Diptera: Tephritidae) on a caged host tree. *Fla. Entomol.* **66**: 234–241.

Myers, K. 1952. Oviposition and mating behaviour of the Queensland fruit-fly (*Dacus* (*Strumeta*) *tryoni* (Frogg.)) and the solanum fruit-fly (*Dacus* (*Strumeta*) *cacuminatus* (Hering)). *Austr. J. Biol. Sci.* **5**: 264–281.

Nagel, H. G. and C. W. Rettenmeyer. 1973. Nuptial flights, reproductive behavior and colony founding of the western harvester ant, *Pogonomyrmex occidentalis* (Hymenoptera: Formicidae). *J. Kansas Entomol. Soc.* **46**: 82–101.

Nation, J. L. 1972. Courtship behavior and evidence for a sex attractant in the male Caribbean fruit fly, *Anastrepha suspensa. Ann. Entomol. Soc. Am.* **65**: 1364–1367.

Neems, R. M., A. J. McLachlan and R. Chambers. 1990. Body size and lifetime mating success of male midges (Diptera: Chironomidae). *Anim. Behav.* **40**: 648–652.

Nelson, J. W. 1986. Ecological notes on male *Mydas xanthopterus* (Loew) (Diptera: Mydidae) and their interactions with *Hemipepsis ustulata* Dahlbohm (Hymenoptera: Pompilidae). *Pan-Pac. Entomol.* **62**: 316–322.

Nielsen, H. T. 1965. Swarming and some other habits of *Mansonia perturbans* and *Psorophora ferox* (Diptera: Culicidae). *Behaviour* **24**: 67–89.

Okubo, A. and H. C. Chiang. 1974. An analysis of the kinematics of swarming of *Anarete pritchardi* Kim (Diptera: Cecidomyiidae). *Res. Pop. Ecol.* **16**: 1–42.

O'Neill, K. M. 1983. Territoriality, body size, and spacing in males of the beewolf *Philanthus basilaris* (Hymenoptera: Sphecidae). *Behaviour* **86**: 295–332.

–. 1994. The male mating strategy of the ant *Formica subpolita* Mayr (Hymenoptera: Formicidae): Swarming, mating and predation risk. *Psyche* **101**: 93–108.

O'Neill, K. M., H. E. Evans and L. B. Bjostad. 1991. Territorial behaviour in males of three North American species of bumblebees (Hymenoptera: Apidae, *Bombus*). *Can. J. Zool.* **69**: 604–613.

Page, R. E. Jr. 1980. The evolution of multiple mating behavior by honey bee queens (*Apis mellifera* L.) *Genetics* **96**: 263–273.

Pajunen, V. I. 1982. Swarming behavior in *Ophyra leucostoma* Wied. (Diptera, Muscidae). *Ann. Zool. Fenn.* **19**: 81–85.

Parker, G. A. 1983. Mate quality and mating decisions. In *Mate Choice.* P. Bateson, ed., pp. 141–166. Cambridge University Press.

Parsons, P. A. 1976. Lek behavior in *Drosophila* (*Hirtodrosophila*) *polypori* Malloch – an Australian rainforest species. *Evolution* **31**: 223–225.

–. 1977. Lek behaviour in three species of the subgenus *Hirtodrosophila* of Australian *Drosophila. Nature (Lond.)* **265**: 48.

Partridge, L. 1980. Mate choice increases a component of offspring fitness in fruit flies. *Nature (Lond.)* **283**: 290–291.

Petersson, E. 1987. Weight-associated male mating success in the swarming caddis fly, *Mystacides azureus* L. *Ann. Zool. Fenn.* **24**: 335–339.

Polak, M. 1993a. Competition for landmark territories among male *Polistes canadensis* (L.) (Hymenoptera: Vespidae): large-size advantage and alternative mate-acquisition tactics. *Behav. Ecol.* **4**: 325–331.

–. 1993b. Landmark territoriality in the neotropical paper wasps *Polistes canadensis* (L.) and *P. carnifex* (F.) (Hymenoptera: Vespidae). *Ethology* **95**: 278–290.

Reynolds, J. D. and M. R. Gross. 1990. Costs and benefits of female mate choice: Is there a lek paradox? *Am. Nat.* **136**: 230–243.

Ringo, J. M. 1976. A communal display in Hawaiian *Drosophila* (Diptera: Drosophilidae). *Ann. Entomol. Soc. Am.* **69**: 209–214.

Robacker, D. C., R. L. Mangan, D. S. Moreno and A. M. T. Moreno. 1991. Mating behavior and male mating success in wild *Anastrepha ludens* (Diptera: Tephritidae) on a field-caged host tree. *J. Insect Behav.* **4**: 471–487.

Rust, R. 1988. Nuptial flights and mating behavior in the ant, *Pogonomyrmex salinus* Olsen (Hymenoptera: Formicidae). *J. Kansas Entomol. Soc.* **61**: 492–494.

Rutowski, R. L., J. Alcock and M. Carey. 1989. Hilltopping in the pipevine swallowtail butterfly (*Battus philenor*). *Ethology* **82**: 244–254.

Savolainen, E. 1978. Swarming in Ephermeroptera: the mechanism of swarming and the effects of illumination and weather. *Ann. Zool. Fenn.* **15**: 17–52.

Savolainen, E. and J. Syrjämäki. 1971. Swarming and mating of *Erioptera gemina* Tjeder (Dipt., Limoniidae). *Ann. Entomol. Fenn.* **37**: 79–85.

Scott, J. A. 1968. Hilltopping as a mating mechanism to aid the survival of low density species. *J. Res. Lepid.* **7**: 191–204.

Selous, E. 1906–1907. Observations tending to throw light on the question of sexual selection in birds including a day-to-day diary on the breeding habits of the ruff (*Machetes pugnax*). *Zoologist* (Ser. 4) **10**: 201–219; 285–294; 419–428; **11**: 60–65; 161–182; 367–381.

–. 1909–1910. An observational diary on the nuptial habits of the blackcock (*Tetrao tetrix*) in Scandinavia and England. *Zoologist* (Ser. 4) **13**: 401–413; **14**: 23–29; 51–56; 176–182, 248–265.

–. 1913. The nuptial habits of the blackcock. *Naturalist* **673**: 96–98

Shelly, T. E. 1987. Lek behaviour of a Hawaiian *Drosophila*: male spacing, aggression and female visitation. *Anim. Behav.* **35**: 1394–1404.

–. 1988. Lek behaviour of *Drosophila cnecopleura* in Hawaii. *Ecol. Entomol.* **13**: 51–55.

–. 1989. Waiting for mates: variation in female encounter rates within and between leks of *Drosophila conformis*. *Behaviour* **111**: 34–48.

–. 1990. Observations on the lek behavior of *Drosophila imparisetae* (Diptera: Drosophilidae) in Hawaii. *J. Kansas Entomol. Soc.* **63**: 652–655.

Shelly, T. E. and K. Y. Kaneshiro. 1991. Lek behavior of the oriental fruit fly, *Dacus dorsalis*, in Hawaii (Diptera: Tephritidae). *J. Insect Behav.* **4**: 235–241.

Shelly, T. E. and T. S. Whittier. 1995. Lek distribution in the Mediterranean fruit fly: the influence of tree size, foliage density, and neighborhood. *Proc. Hawaiian Entomol. Soc.* **32**: 113–121.

Shields, O. 1967. Hilltopping. *J. Res. Lepid.* **6**: 69–178.

Simon-Thomas, R. T. and E. P. R. Poorter. 1972. Notes on the behaviour of males of *Philanthus triangulum* (F.) (Hymenoptera, Sphecidae). *Tijdschr. Entomol.* **115**: 141–151.

Singer, M. C. and C. D. Thomas. 1992. The difficulty of deducing behavior from resource use: an example from hilltopping checkerspot butterflies. *Am. Nat.* **140**: 654–664.

Sivinski, J. 1989. Lekking and the small-scale distribution of the sexes in the Caribbean fruit fly, *Anastrepha suspensa* (Loew). *J. Insect Behav.* **2**: 3–13.

Sivinski, J., T. Burk and J. C. Webb. 1984. Acoustic courtship signals in the Caribbean fruit fly, *Anastrepha suspensa* (Loew). *Anim. Behav.* **32**: 1011–1016.

Sivinski, J. and J. C. Webb. 1986. Changes in a Caribbean fruit fly acoustic signal with social situation (Diptera: Tephritidae). *Ann. Entomol. Soc. Am.* **79**: 146–149.

Spieth, H. T. 1968. Evolutionary implications of sexual behavior in Drosophila. *Evol. Biol.* **2**: 157–193.

–. 1981. *Drosophila heteroneura* and *Drosophila silvestris*: head shapes, behavior and evolution. *Evolution* **35**: 921–930.

Stern, D. L. 1991. Male territoriality and alternative male behaviors in the Euglossine bee, *Eulaema meriana* (Hymenoptera: Apidae). *J. Kansas Entomol. Soc.* **64**: 421–437.›pa

Strandtmann, R. W. 1942. On the marriage flight of *Pogonomyrmex comanche* Wheeler (Hymenoptera: Formicidae). *Ann. Entomol. Soc. Am.* **35**: 140.

Strang, G. E. 1970. A study of honey bee drone attraction in the mating response. *J. Econ. Entomol.* **63**: 641–645.

Sullivan, R. T. 1981. Insect swarming and mating. *Fla. Entomol.* **64**: 44–65.

Svensson, B. G. and E. Petersson. 1992. Why insects swarm: testing the models for lek mating systems on swarming *Empis borealis* females. *Behav. Ecol. Sociobiol.* **31**: 253–261.

Syrjämäki, J. 1964. Swarming and mating behavior of *Allochironomus crassiforceps* Kieff. (Dipt., Chironomidae). *Ann. Zool. Fenn.* **1**: 125–145.

–. 1965. Laboratory studies on the swarming behavior of *Chironomus strenzkei* Fittkau in litt. (Dipt., Chironomidae). *Ann. Zool. Fenn.* **2**: 145–152.

–. 1966. Dusk swarming of *Chironomus pseudothummi* Strenzke (Dipt., Chironomidae). *Ann. Zool. Fenn.* **3**: 21–28.

Syrjämäki, J. and I. Ulmanen. 1970. Further experiments on male sexual behavior in *Stictochironomus crassiforceps* (Kieff.) (Diptera, Chironomidae). *Ann. Zool. Fenn.* **7**: 216–220.

Talbot, M. 1972. Flights and swarms of the ant *Formica obscuripes* Forel. *J. Kansas Entomol. Soc.* **45**: 254–258.

Taylor, C. E., A. D. Pereda and J. A. Ferrari. 1987. On the correlation between mating success and offspring quality in *Drosophila melanogaster*. *Am. Nat.* **129**: 721–729.

Thew, T. B. 1958. Studies on the mating flights of the Ephemeroptera I. The mating flights of *Ephoron album* (Say) and *Stenonema canadense* (Walker). *Fla. Entomol.* **41**: 9–12.

Thornhill, R. 1980. Sexual selection within mating swarms of the lovebug, *Plecia nearctica* (Diptera: Bibionidae). *Anim. Behav.* **28**: 405–412.

–. 1983. Cryptic female choice and its implications in the scorpionfly *Harpobittacus nigriceps. Am. Nat.* **122**: 765–788.

Thornhill, R. and J. Alcock. 1983. *The Evolution of Insect Mating Systems.* Cambridge, Mass.: Harvard University Press.

Toft, C. A. 1984. Activity budgets in two species of bee flies (*Lordotus*: Bombyliidae, Diptera): a comparison of species and sexes. *Behav. Ecol. Sociobiol.* **14**: 287–296.

–. 1989a. Population structure and mating system of a desert bee fly (*Lordotus pulchrissimus* Diptera: Bombyliidae). 1. Male demography and interactions. *Oikos* **54**: 345–358.

–. 1989b. Population structure and mating system of a desert bee fly (*Lordotus pulchrissimus*; Diptera: Bombyliidae). 2. Female demography, copulations and characteristics of swarm sites. *Oikos* **54**: 359–369.

Tozer, W. E., V. H. Resh and J. O. Solem. 1981. Bionomics and adult behavior of a lentic caddisfly, *Nectopsyche albida* (Walker). *Am. Midl. Nat.* **106**: 133–144.

Treherne, J. E. and W. A. Foster. 1980. The effects of group size on predator avoidance in a marine insect. *Anim. Behav.* **28**: 1119–1122.

Turchin, P. and P. Kareiva. 1989. Aggregation in *Aphis varians*: an effective strategy for reducing predation risk. *Ecology* **70**: 1008–1016.

Turrillazzi, S. and R. Cervo. 1982. Territorial behavior in male *Polistes nimpha* (Christ) (Hymenoptera, Vespidae). *Z. Tierpsychol.* **58**: 174–180.

Tychsen, P. H. 1977. Mating behaviour of the Queensland fruit fly, *Dacus tryoni* (Diptera: Tephritidae), in field cages. *J. Austr. Entomol. Soc.* **16**: 459–465.

Vandermeer, J. H. 1972. Niche theory. *Annu. Rev. Ecol. Syst.* **3**: 107–132.

Vinson, S. B. and G. W. Frankie. 1990. Territorial and mating behavior of *Xylocopa fimbriata* F. and *Xylocopa gualanensis* Cockerell from Costa Rica. *J. Insect Behav.* **3**: 13–32.

Watmough, R. H. 1974. Biology and behaviour of carpenter bees in southern Africa. *J. Entomol. Soc. S. Afr.* **37**: 261–281.

Webb, J. C., T. Burk and J. Sivinski. 1983. Attraction of female Caribbean fruit flies, *Anastrepha suspensa* (Diptera: Tephritidae), to the presence of males and male-produced stimuli in field cages. *Ann. Entomol. Soc. Am.* **76**: 996–998.

Webb, J. C., J. L. Sharp, D. L. Chambers, J. J. McDow and J. C. Benner. 1976. Analysis and identification of sounds produced by the male Caribbean fruit fly, *Anastrepha suspensa. Ann. Entomol. Soc. Am.* **69**: 415–420.

Webb, J. C., J. Sivinski and C. Litzkow. 1984. Acoustical behavior and sexual success in the caribbean fruit fly, *Anastrepha suspensa* (Loew) (Diptera: Tephritidae). *Environ. Entomol.* **13**: 650–656.

Whittier, T. S. and K. Y. Kaneshiro. 1995. Intersexual selection in the Mediterranean fruit fly: Does female choice enhance fitness? *Evolution*, **49**: 990–996.

Whittier, T. S., K. Y. Kaneshiro and L. D. Prescott. 1992. Mating behavior of Mediterranean fruit flies (Diptera: Tephritidae) in a natural environment. *Ann. Entomol. Soc. Am.* **85**: 214–218.

Whittier, T. S., F. Y. Nam, T. E. Shelly and K. Y. Kaneshiro. 1994. Male courtship success and female discrimination in the Mediterranean fruit fly (Diptera: Tephritidae). *J. Insect Behav.* **7**: 159–170.

Wickman, P. 1988. Dynamics of mate searching behavior in a hilltopping butterfly, *Lasiomata megera* (L.): the effects of weather and male density. *Zool. J. Linn. Soc.* **93**: 357–377.

Wiley, R. H. 1974. Evolution of Social organization and life-history patterns among grouse. *Q. Rev. Biol.* **49**: 201–227.

–. 1991. Lekking in birds and mammals: behavioral and evolutionary issues. *Adv. Stud. Behav.* **20**: 201–291.

Williams, F. M. and R. S. Patterson. 1969. Swarming and mating behavior in *Culex pipiens quinquefasciatus* Say. *Mosq. News* **29**: 662–666.

Williams, K. S. and K. G. Smith. 1991. Dynamics of periodical cicada chorus centers (Homoptera: Cicadidae: *Magicicada*). *J. Insect Behav.* **4**: 275–291.

Willis, M. A. and M. C. Birch. 1982. Male lek formation and female calling in a population of the arctiid moth *Estigmene acrea. Science (Wash., D.C.)* **218**: 168–170.

Wilson, E. O. 1975. *Sociobiology.* Cambridge, Mass.: Harvard University Press.

Yeates, D. and G. Dodson. 1990. The mating system of a bee fly (Diptera: Bombyliidae). I. Non-resource-based hilltop territoriality and a resource-based alternative. *J. Insect Behav.* **3**: 603–617.

Young, A. M. 1980. Observations on the aggregation of adult cicadas (Homoptera: Cicadidae) in tropical forests. *Can. J. Zool.* **58**: 711–722.

Zeil, J. 1986. The territorial flight of male houseflies (*Fannia canicularis* L.). *Behav. Ecol. Sociobiol.* **19**: 213–219.

17 · Mate choice and species isolation in swarming insects

JOHN M. SIVINSKI AND ERIK PETERSSON

ABSTRACT

The term 'swarm' has been applied to a variety of aerial mating systems. All are supposedly non-resource-based, although their relationship to resources is sometimes ambiguous. There are several selective contexts for mate choice. The choice of swarm markers has been implicated in species isolation. However, certain swarm markers are used by multiple species, which suggests that qualities other than specificity are important. In some instances, species appear sequentially at a marker, and there is evidence that not all times of day are equally valuable for swarming. It is not clear whether species compete for a universally best time. Swarms themselves, as well as markers, can attract participants, and a swarm's size can contribute to its attractiveness. The position of an insect within a swarm might influence its sexual opportunities. In species where male size is positively correlated with reproductive success, there is an instance where larger individuals occupy certain parts of the swarm. There are more observations of homogeneous size distributions in swarms, although in some species from several families of Diptera, the smallest males swarm very little or not at all. Presumably, they lack the energy resources to compete with larger males. Swarming insects may emit sexual signals. Pheromones seem to be rarely used, perhaps because of difficulties in determining the source of the signal. Visual signals are the easiest to trace to their sender, and furthest projected. Vision seems to be the paramount sense in swarming species. However, there is a large amount of variance in the complexity among presumed visual signals. Some of this variation may be due to environmental limitations, such as light intensity, that are imposed on swarming species. Swarming may evolve when females are uniformly dispersed and unpredictably located in space and time. A comparison of swarming insects and their parasitoids is suggested as a means of examining the roles of dispersal and phylogeny. Body size relative to resource (marker) size might influence the ability of insects to control a territory.

It is suggested that when these ratios are large, insects are less able to expel rivals. Swarm formation then becomes more likely.

INTRODUCTION

Insects often occur in groups. These groups can be accumulations of individuals at resources, and may or may not have a sexual component (for contrasting examples see Robinson and Robinson 1977 and Sivinski 1983). In other aggregations mating is the sole function of gathering. A variety of such mating systems, including swarms, are clumped into the category of 'non-resource-based' (see, for example, Borgia 1979) (Fig. 17-1).

The distinction between resource- and non-resource-based is not always simple or clear. A few examples illustrate the difficulty of categorizing aerial mating systems (see Table 17-1). Male mayflies (Ephemeroptera) often gather above the water's surface or reeds. In either case they are aggregated over aquatic oviposition sites. There are instances in mayflies, as well as in caddisflies and chironomids, where there are sequences of species swarming at different distances inland, so that some are 40–50 m from the water (Savolainen 1978). Originally, in species with short-lived adults having no or little food demands, an oviposition site might have been selected by males as an encounter site. Females would be likely to emerge in these places, and both sexes would enjoy rapid mate selection and reduced travel costs. However, in the above instances, some aggregation sites have apparently been 'pushed' away from the aquatic resource. The reasons for these progressions are unknown, though interspecific competition for signaling sites is a possibility. If so, the onshore swarms may be as close to oviposition site resources as competition allows.

Campbell and Kettle (1979) observed that males of *Culicoides brevitarsis* most frequently gathered on cattle pastures above shadows cast by clumps of fresh grass. Such markers

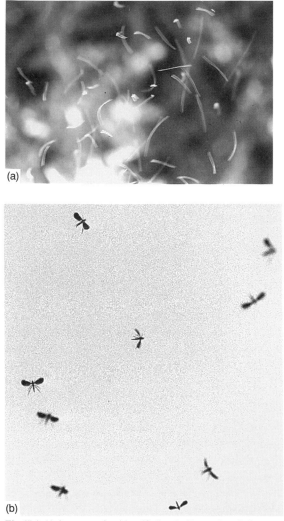

(a)

(b)

Fig. 17-1. (a) A swarm of unidentified male flies gathered above a swarm marker. The marker is not an oviposition, feeding or adult emergence site. Because of this, the aggregation is considered to be non-resource-based. We suggest that non-resource-based aggregations of flying insects are subject to unique selection pressures, and continue to make a distinction between aerial groups and those formed on a substrate (leks). The white lines in the photograph are due to the motion of the insects during the exposure period. The photograph is by J. E. Lloyd. (b) A female swarm of *Empis borealis*. Males bearing nuptial gifts choose mates from such aggregations. The photograph is by B. G. Svensson.

were common, but the swarms were not randomly distributed. The aggregates were significantly larger, closer together and more likely to contain females near cattle (female feeding site), although groups were formed in the absence of

cattle. Thus, within one species, the connection between aggregation site and a resource can vary considerably.

Finally, aggregations may also occur as side effects of mate-searching behavior. For example, male *Euphydryas* butterflies attracted to the movement of other nearby males, whom they appear to mistake for females, may attract still more males, resulting in what appears to be a classic non-resource-based aggregation (Odendaal *et al.* 1988).

Few attempts have been made to theoretically analyze the whole spectrum of non-resource-based mating aggregations. An exception is Bradbury (1985) who suggested that the best way to fit mating aggregations into a common theoretical framework is to treat characteristics such as territoriality, male investment and female choice as continuous variables instead of discrete alternatives, and to accept that they may vary independently of each other. This perspective suggests that we should not be too concerned with fitting mating systems into clearly defined categories. However, because we shall argue that location influences sexual behavior, we continue to make the commonly used distinction between 'leks' (generally aggregations touching a substratum and having a territorial component) and 'swarms' (aerial aggregations; see Shelly and Whittier, this volume). It should be kept in mind that leks and swarms may share certain characteristics in varying degrees and that the boundary between the two can be indistinct. For example, the male establishment of territories, apparent in many leks, may not distinguish them from all swarms. Maynard Smith (1974) showed theoretically that territory size should be smaller when territories are settled synchronously, as would often be the case in swarms, than when settled over a period. Pajunen (1980) suggested that swarming behavior can be derived from territoriality. Some factors, e.g. high population densities, are assumed to have decreased the aggressiveness of the participants, leading to the 'breakdown' of territories and giving rise to a 'scramble-competition' aggregation. Svensson and Petersson (1992) showed that hypotheses for the evolution of leks can successfully be adopted to insect mating swarms.

We concentrate our discussion on adaptations in swarming species to the problems of species isolation and mate choice. In particular, we address (1) the selection of location, i.e. the place at which a swarm occurs; (2) characteristics unique to swarms themselves, e.g. numerical size, that influence sexual opportunities; and (3) the advantages for and limitations on communication among members of an aerial aggregation. In our concluding section we

Table 17-1. *The distribution among "encounter sites" in insect species with aerial mating aggregations*

The table is based on 178 papers; the relevant references can be provided by E. Petersson upon request. We do not claim that this table is complete. Nevertheless, we think it gives a good view of the large variety of insect mating aggregations and how they are distributed within and among the orders. 'Classic swarms', as addressed in this chapter, do not occur over obvious resources. G, swarming above spots on ground; L, swarming close to or above landmarks; D, swarming occurs in a defined area, but no marker has been recognized; G/L, D/L, G/D, both types of behavior noted for a single species; H, swarming in vicinity of host or food plant; A, swarming in vicinity of animal host; E, swarming at (female) emergence site; O, swarming at oviposition site; S, 'summit swarmers' or hilltoppers; N, no information about orientation, but the species do aggregate.

Order/suborder		No. of species aggregating											
	C	L	D	G/L	D/L	G/D	H	A	E	O	S	N
Ephemeroptera	1	12	25	2	3	—	—	—	—	—	—	11
Odonata	—	1	—	—	—	—	—	—	—	—	—	—
Heteroptera	—	—	5	—	—	—	—	—	—	—	—	—
Neuroptera	—	1?	—	—	—	—	—	—	2	1?	—	2
Trichoptera	—	6	14	—	2	—	—	—	—	—	—	—
Lepidoptera	—	—	9	—	—	—	—	—	6	—	50	—
Coleoptera	—	—	1	—	—	—	—	—	—	—	1	—
Hymenoptera												
Symphyta	—	1	—	—	—	—	—	—	—	—	1	—
Parasitica	—	6	—	—	—	—	—	—	—	—	2	—
Other Aculeata	1	7	4	1	—	—	—	—	1	—	14	3
Diptera												
Nematocera	11	102	32	5	2	2	7	13	9	—	—	26
Brachycera	—	10	7	—	—	—	1	—	—	—	30	2

examine the evolution of swarms and consider the ecological and morphological factors that may influence their phyletic distribution (see Shelly and Whittier, this volume).

SWARM FORMATION AND ORIENTATION TO AGGREGATION SITES

Typically, but not universally, insect mating aggregations are formed by males. Swarms are usually found above contrasting spots on the ground or objects higher than their surroundings. In both cases, the object that serves as an orientation point is called a 'swarm marker'. The size and form of the marker can influence the size and form of male swarms (see, for example, Chiang 1961; Savolainen 1978; Allan and Flecker 1989) and female swarms (Svensson and Petersson 1992). Campbell and Kettle (1979) did not find such a relationship in *Culicoides brevitarsis*; the proximity to cattle was more important. There may be intraspecific variation in site preferences. In the leptocerid caddisfly *Mystacides azurea*, two types of swarm site have been reported: one over reeds, forming typical marker

swarms (Solem 1978), the other over the water surface, without any obvious landmarks (dispersed swarming, *sensu* Sullivan 1981; Petersson 1987; Gullefors and Petersson 1993). A mayfly, *Leptophlebia marginata*, swarms either over trees or horizontally over the ground (Savolainen *et al.* 1993). The latter swarm form appears to be better protected from winds. The insects that practice the different types of swarming have diverged morphologically and genetically. Speciation through specialization for particular modes of swarming may be underway. Swarming insects are often 'generalists' when choosing a marker. Many observers have reported chironomids and other nematocerans swarming above cattle, bushes, humans or any other object higher than its immediate surroundings (see Sullivan 1981). Some species are more specific in their site choices. The simuliid *Austrosimulium pestilens* is only found swarming over *Callistemon viminalis* (red bottlebrush) plants (Moorhouse and Colbo 1973). Most species use fairly simple visual cues to find an aggregation site within their habitat. Not all hilltop aggregations are the result of visual orientation; the insects may be guided uphill by wind currents (Alcock 1987).

In the following subsections we first describe the most common type of swarm, single-species swarms, and apparent adaptations for species isolation. Then we review the rarer occurrences of multiple-species swarms. Finally, we consider what the use of a site by several species reveals about the nature of swarm markers and the interspecific competition for their use.

Single-species swarms

In many insect groups, closely related species differ in their site preferences. This might contribute to species isolation. Downes (1958) reported that in Britain and Manitoba, there were seven to ten species of *Culicoides* (Diptera) and nine species of *Aedes* mosquito, respectively, that swarmed simultaneously. No mixed swarms could be observed. Hunter (1979) found three species of blackfly to have some overlap in site preferences, but they were rarely seen swarming together. Savolainen (1978) noted that mayflies in Finland swarmed at species-specific sites, and species-mixing did not seem to occur. A similar pattern is found in north European leptocerid caddisflies. Males of swarming species fly above the water surface at different distances from shore (Solem 1978, 1984; Gullefors and Petersson 1993; cf. Mori and Matubani 1953). Usually, several species are swarming simultaneously (Petersson 1989). Only during evenings with gusty wind conditions do swarms of different species occasionally become mixed.

Several species may use a single marker, but at different times. Swarm initiation is often triggered by highly species-specific illumination thresholds (see, for example, Nielsen and Nielsen 1962). Yuval and Bouskila (1993) consistently found a series of species at the same swarm markers. The focus of their study, the mosquito *Anopheles freeborni*, began to swarm 5–10 min after sunset and continued for 30–35 min until the end of twilight. Attacks by dragonflies were more numerous at the beginning of this period because the visually foraging predators became less efficient with the deepening dusk and roosted 20–25 min after sunset. Why has *A. freeborni* not moved its mating period to a later time and escaped its enemies entirely? Yuval and Bouskila argue that the visibility of swarm participants to potential mates is declining over time as well. *Anopheles freeborni* is thus forced to expose itself to both mates and foes. But the higher copulation to predation ratio in the latter part of *A. freeborni*'s swarming period also raises the question of how swarming time over a marker is partitioned among species.

If not all times are equally valuable, then how do certain species secure particularly good periods? Perhaps a robust insect species can physically displace a smaller one, as larger males displace smaller ones within species. Adults of a more abundant species might simply arrive in greater numbers at a site during the optimal time, and due to their presence *en masse*, make finding a mate a confusing process for less numerous insects. The notion of competition for a single best time may be an oversimplification. Fewer predators may be present at dusk, but there will also be less heat, light and perhaps more confusion from the swarms of other species. Poor visibility at twilight could constrain mate choice. The different physiologies of different species might divide time into a sequence of species optima. If so, the historical displacement of one species by another may never have occurred. Time-sharing of sites, however, does imply that certain sites are particularly valuable as swarm markers, a point pursued in the following subsection.

Multiple-species swarms

Some records of multiple-species swarms (Gibson 1944; Usinger 1945; Hubbard and Nagell 1976) are too scant to evaluate the degree and persistence of mixing. However, there are a few swarm systems where mixing seems to be the rule. Most have been observed in tropical or subtropical areas and contain up to nine species of mosquito or midge (Provost 1958; Cunningham-van Someren 1965, 1973). Haddow and Corbet (1961) reported on swarms that almost invariably consisted of more than one mosquito species. Mayflies and various families of Diptera, notably Tipulidae, Chironomidae and Stratiomyiidae, were also represented.

Mixed swarms might come about in at least two ways. First, for one reason or another, some sites may be better than others and in short supply. In areas with high species diversity, many species may share these limited locations, e.g. a hilltop (cf. Parker 1978). Choice habitats may be safer to swarm in. Anisopteran dragonflies prefer open spaces (cit. in Yuval and Bouskila 1993). Thus male *Anopheles freeborni* mosquitos swarming in areas sheltered by trees have higher copulation to predation ratios than those that swarm in open areas (Yuval and Bouskila 1993). Still, swarms are persistently formed in dangerous, open areas. Females that risk joining such aggregations may obtain mates that have demonstrated an ability to survive under perilous conditions.

What constitutes a favorable swarming site is some-times mysterious. Even different forms of mating system, which might be thought to have different environmental requirements, can occur in the same spot. For example, two species of 'lekking' phorid, a territorial dolichopodid and two species of swarming chironomid all shared mating sites on a Florida floodplain (Sivinski 1988b). Repla-cement of swarm marker foliage with different vegetation did not influence the extreme site specificity. In a some-what similar situation, two sympatric south European polistine wasps use the same landmarks, but have different mating systems. *Polistes gallicus* males form ordinary marker swarms, whereas *P. dominulus* males establish territories at the same site and try to chase away other males of both species (Beani and Turillazzi 1990; see also Frohne and Frohne 1952).

Second, multiple-species mating aggregations may be defensive adaptations, reducing the risk of predation through dilution (selfish-herd, cf. Hamilton 1971). The probability of predation per male in the chironomid *Chiro-nomus plumosus* decreases with increased swarm size (Neems *et al.* 1992). With more eyes for surveillance the average vigilance of birds decreases with increasing flock size (cf. Pulliam 1973; Inglis and Lazarus 1981; see, how-ever, Pulliam *et al.* 1982). In insect non-mating aggregations the distance at which a threat is discovered increases with group size (Treherne and Foster 1980, 1981, 1982; Vulinec and Miller 1989). However, it is not known whether greater awareness of predators is an advantage to individuals in multiple-species mating aggregations; i.e. whether differ-ent species can interpret each other's avoidance behavior. Predation pressure on swarms is considered to be more intense in tropical, as opposed to temperate, regions (Edmunds and Edmunds 1980). As noted, mixed swarms appear to be more common in the tropics.

ORIENTATION TO SWARMS BY POTENTIAL MEMBERS

It can be difficult to determine whether the site or the swarm attracts other insects. In some species, the visiting, non-swarming sex occasionally arrives at empty sites (Dahl 1965; Savolainen and Syrjämäki 1971; Savolainen 1978; Svensson and Petersson 1992). Savolainen and Syrjämäki (1971) reported that if a lone cranefly female flies above an artificial swarm marker, she always follows it if it is moved along the ground, i.e. she behaves as the swarming males do. Such observations suggest that both sexes are using

the same cues to find the aggregation sites and have the same site preferences ('encounter site conventions' *sensu* Parker 1978).

There are, however, aggregations for which this hypoth-esis appears to be unsatisfactory. Some *Chaoborus* (Diptera) (Downes 1958) and mayflies (Spieth 1940) form swarms above the water surface, sometimes kilometers offshore. These swarms often move as a unit. A similar pattern has been observed in lonchaeid flies (McAlpine and Munroe 1968). The swarms split and reform higher from the ground when disturbed by wind or the researchers. In such cases, females might be attracted to the swarm alone.

In the following subsections we examine the possibility that a characteristic unique to groups, i.e. the number of members, may make certain swarms more attractive than others. We also consider whether position within a swarm influences the reproductive success of its members and what influence this factor might have on the form of a swarm.

Are some swarms more attractive than others?

Particular aggregations can have characteristics that are more attractive to potential participants than others groups, regardless of whether the insects are originally attracted to the site or to the aggregation. For instance, a larger swarm represents a greater pool of potential mates. In the dance fly *Empis borealis*, females swarm and males come to swarms carrying nuptial gifts (Svensson and Petersson 1987). Once in the swarm males tend to choose the larger, more fecund, females as mates. Males stay longer in larger swarms (Svensson and Petersson 1987, 1988). The proportion of males leaving the swarms without mating decreases with swarm size (Svensson and Petersson 1992). In addition, males stay shorter times in swarms with greater variance in female size, perhaps because it is easier to discriminate among potential mates that differ remark-ably in size (Svensson *et al.* 1989). Thus, individuals of both sexes are visually attracted to the site, but the charac-teristics of the group influence their behavior within the swarm. Female *Anopheles freeborni* mosquitos are most likely to join a particular swarm at the time of its peak size (Yuval and Bouskila 1993). They may gain protection from dragonfly predation by a 'dilution effect' (Foster and Treherne 1981; Wrona and Dixon 1991).

When swarm size is an important component of its attractiveness, there might be 'runaway' growth of a group. If one aggregation is larger by chance it will recruit more members, which in turn increases its potential for further growth. Such a process would produce a highly skewed distribution of swarm sizes forming at random over various markers. Among lekking species and species aggregated at a resource, group size can also influence female arrival rates (for example, in *Drosophila conformis*, (Shelly 1989; Shelly and Whittier, this volume); and the sphaerocerid *Norrbomia frigipennis* (E. Petersson and J. M. Sivinski, unpublished)).

Position within swarms and mating success

Certain positions within an aggregation may be of different value when they are safer or because females prefer males in particular locations. For example, a central position may be attractive because it might be sheltered from predators. On the other hand, a peripheral position in a swarm could provide better access to females approaching the aggregation. If so, this would create a selective conflict in males between predator-avoidance and sexual success (Hamilton 1971). Alternatively, females might seek the relative safety of the swarm's core before coupling. Males and females would then both try to surround themselves with their neighbors. Records of predation on insect swarms contain scant information about the relative dangers of the different portions of aggregations (Banks 1919; Rao and Russel 1938; Frohne and Frohne 1952; Brickle 1959; Downes 1970, 1978; Cunningham-van Someren 1973; Chandler 1978). Most insect predators seem to enter swarms, which would even the risks to the participants.

Being in the right place at the right time could result in more contacts with opposite sex. In the dance fly *Empis borealis*, a male approaches a female from below and takes off with her in an ascending flight. Sometimes the pair couples in the air, but often the male performs the ascending flight with several or all females in the swarm before mating. He may also leave the swarm without mating (Svensson and Petersson 1987). Obviously, the males choose among females, and in such a swarm position is not critical, because males may evaluate all swarming females.

In other aggregations, with a less universal form of mate comparison, competition for position could be of enough importance to influence the form and composition of the swarm. The bibionid *Plecia nearctica* ('lovebug') swarms in pastures where adults emerge from the turf. Males search the area for females. Larger males are found closer to the ground, and hence to newly eclosed females emerging from the soil, whereas smaller males are at the top of the swarm (Thornhill 1980; see, however, Hieber and Cohen 1983). However, in other species, there are no such position effects. McLachlan and Neems (1989) made horizontal sweeps through male swarms of the chaoborid *Chaoborus flavicans*, but found no difference in male size between the tops and bottoms of swarms. Females seem to aggregate in the vegetation under the swarm, and might enter the swarm from below. McLachlan and Neems did not investigate whether males in the swarm center differed from those on the periphery. Flecker *et al.* (1988) studied the mayfly *Epeorus longimanus* and followed individual males during their nuptial display. Females probably approach the male swarms near their top (Brodsky 1973); larger males have higher mating success. Larger males were therefore predicted to be most abundant at the swarm top, but males from different sections of the swarms did not differ in body size (Flecker *et al.* 1988). It appeared that each male occupied the full vertical range encompassed by the swarm. Males of the leptocerid caddisfly *Mystacides azurea* swarm above the surface of lakes. Females approach the swarms from above, generally coming from their resting sites on the shore vegetation. The number of pair formations in different parts of the swarm do not differ (Petersson 1987). Larger male *Anopheles freeborni* mosquitoes also have greater mating success, but it is not known whether this is due to female mate choice or to the ability of large males to procure more favorable sites in the swarm (Yuval *et al.* 1993; see also Reisen *et al.* 1981). In general, position effects appear to be relatively uncommon in insect swarms.

Male mating success in the species mentioned above is mainly determined by their success in pursuing or attracting females as they approach or enter the swarms. In six species of Chironomidae, and a chaoborid, small males rarely swarm at all, but wait for females on nearby vegetation. They may exploit the attractiveness of the larger males flying above and simultaneously save energy (McLachlan and Neems 1989; cf. Arak 1988). Males of the mosquito *Anopheles freeborni* fall into three categories: (1) small, non-swarming males; (2) larger males that swarm early and for as long as their energy reserves last; and (3) the largest individuals, who fly when females are most likely to join the swarm (Yuval *et al.* 1993, 1994). In this case the energetic difficulties of swarming act as a filter. Only the larger males with the greater reserves participate in the sexual aggregation.

COMMUNICATION WITHIN SWARMS

There is a considerable range in the complexity of intraspecific communication in aggregated insects (Burk 1981; Shelly and Whittier, this volume). In particular, the often elaborate courtships of lekking species, with sometimes simultaneous acoustic, pheromonal, and visual broadcasts, have been contrasted with the seeming absence of self-advertisement in many swarming species. Yet the existence of mixed-species swarms suggests that choice of swarm site alone is not always sufficient to provide species isolation. Information is being passed between members of swarms. In this section we first address the limitations swarming might place on the type of signal produced, and conclude that vision is the most practical channel for communication. We then examine cases of apparent visual signaling within swarms and consider the role of the swarming environment on the evolution of both elaborate and relatively simple communicative organs and behaviors.

Communication channels in swarming insects

Although pheromones are important in many lekking species (e.g. tephritid fruit flies, Nation 1989), their role in swarms is less well understood. Females of the chironomid *Palpomyia brachialis* evert long glandular strings from their abdomens as they participate in sex-role-reversed female swarms (Edwards 1920). These have been interpreted as scent organs, but their bright orange color contrasting with a black body suggests a visual role as well (although the tubes are colorless in some other *Polypomyia* and species of the related *Bezyia*). Both sexes of the swarming caddisfly *Hydropsyche augustipennis* produce a chemical attractant, whereas only females of the non-swarming *Rhyacophila fasciata* emit a pheromone (Löfstedt *et al.* 1994). Thus, although pheromones do occur in swarming species, the act of swarming itself, where several insects move rapidly about in a common volume of air, would appear to limit their efficacy. It seems too difficult to track an odor plume to its producer and too simple to 'cheat' by saving the expense of producing a pheromone and exploiting the signal of a neighbor. Further exceptions tend to emphasize the dilemma. In the ghost moth *Hepialus humuli*, aggregated males hover 1–5 m apart, emitting pheromone from hind-tibial brushes (Mallet 1984). Although they bob up and down and swing 'like pendulums', they seem to be sufficiently stationary and separate to provide individual signals. Other related species are reported to loop back and forth over the same spot, though some descriptions (cit. in Mallet 1984) give the impression of overlapping flight with more possibility of pheromone plume mixture (see also the occasionally stationary flight of the mayfly *Stenonema vicarium*; (Spieth 1940)). The limoniid fly *Erioptera gemina* mates both on the substrate and in swarms. Orientation on a substrate is mediated by a female pheromone and this pheromone is suspected to function in swarms as well (Savolainen and Syrjämäki 1971). If a female joining a swarm is not immediately mated, she hovers nearly motionless over the marker, a behavior that might better allow a male to track her scent.

Acoustic displays are common in some lekking taxa (again for example, tephritid fruit flies; see Sivinski and Burk (1989 and references) and Sivinski (1988a)); 'choruses' and 'sprees' of singing Orthoptera and cicadas have been widely studied (Bailey 1991). Many nematoceran males can recognize the flight tones of conspecific females (see, for example, Duhrkopf and Hartberg 1992; Ogawa 1992). Some mosquitoes are even capable of discriminating between virgin and mated females by their sounds (cf. Sotavalta 1974). Sexually dimorphic antennae, presumably specialized in males to detect sound, are well known in swarming mosquitoes and chironomid midges (see, for example, Roth 1948). However, there is little knowledge of, or even conjecture about, sexually selected acoustic displays in swarming species. Ewing (1989) has argued that flight tones would be of little use in attraction over a long distance (but see Bailey's (1991) discussion of near-field effects), although communication among nearby insects within swarms remains a possibility. Even a particular flight pattern, with changes in wing-beat frequency, might produce a sexually selected 'song' (J. E. Lloyd, personal. communication). In tropical Mexico a large, unidentified non-swarming dipteran emits loud and dramatically modulated flight tones as it hovers (J. M. Sivinski, personal observation). The difference in intensity between pulses of sound and intervals was comparable, to my ear, to the songs of lekking tephritids (see Sivinski 1988a). However, an acoustic display, like a chemical one, might be difficult to identify with its sender in a swarm.

Visual signals seem to be the most individually recognizable, the most readily tracked, and the furthest projected of the options available to swarming insects. Not surprisingly, vision appears to be the dominant sense in swarm formation and sexual interactions within swarms. Petersson and Solem (1987) showed that males of two swarming caddisfly species are attracted to almost any object that is not too

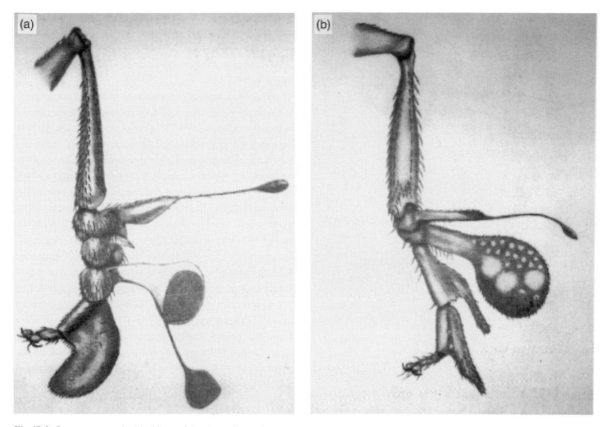

Fig. 17-2. Ornaments on the hind legs of the three allopatric species of *Calotarsa* (Platypezidae). All form swarms in widely separated regions of North America. (a) *C. pallipes*; (b) *C. calceata*; (c) (see p. 302) *C. insignis* (from Kessel and Maggioncalda 1968).

large, especially if it is moving like a conspecific female approaching the swarm. Male lonchaeid flies are attracted by any moving object about the size of a small- or medium-sized fly (McAlpine and Munroe 1968). In many tabanids, the initial detection of females by males is also visual. Males pursue small objects such as stones thrown near them (Allan *et al.* 1987). Male *Aedes triseriatus* mosquitoes show a similar behavior (Loor and DeFoliart 1970).

Many swarming males possess larger eyes than conspecific females. These are generally adaptations for mate finding (cf. Thornhill and Alcock 1983). Within Diptera, sexual dimorphism in eye size with enlarged facets in the male's upper eye is common among swarming species (cf. Downes 1969; Zeil 1983). In leptocerid caddisflies males of some species have larger eyes than the females. Gullefors and Petersson (1993) found that this sexual dimorphism is less expressed in species where the females enter the male swarms, and more expressed in species where a female is detected by a male before reaching the swarm. Larger

male eyes have also been reported for the bee *Exoneura hamulata*, where males seem to scan an area for females (Michener 1974), and in a swarming hemipteran (Kritsky 1977). In mayflies of the family Baetidae, males may possess divided eyes. The lower part functions as an ordinary eye; and the part facing upwards is adapted for scanning for females approaching the swarm. Curiously, the upper eyes are covered in the male subimago. Species of some mayfly genera, such as *Ephoron*, *Campsurs* and *Palingenia*, emerge and mate during the night. Males have undivided eyes that are relatively small (Spieth 1940). Location and recognition of mates is probably not visual.

Variance in signal production among swarming species

Once an insect has joined a group, communication with its fellows is possible. Signals could be sent to sexual rivals or attract the scrutiny of the opposite sex. There appear to be

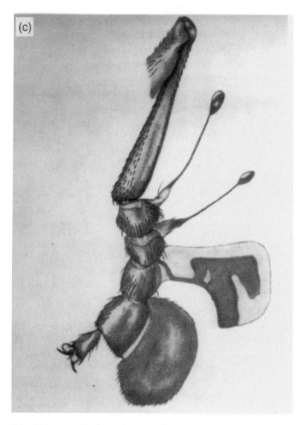

Fig. 17-2 (*cont.*). For legend see previous page.

substantial differences among species in the amount of effort and materials put into visual and other signals. Some, but not all, of this variance can be accounted for by whether or not an aggregation has occurred at a resource (see, for example, Burk 1981). Before trying to explain the distribution of investment in communication, it would be useful to have some idea of what is being communicated, particularly between the sexes.

Displays by individuals have often been interpreted as species-isolating mechanisms (see, for example, Mayr 1963; Alexander *et al.*, this volume). One problem with this idea is that often the most elaborate signals occur in taxa where selection for species isolation seems weakest (at least given the present ranges of the species) (West-Eberhard 1984). An example from the swarming Diptera are the three species of the platypezid genus *Calotarsa*. All are rare and found in widely separated North American locations: California, New Mexico, and Eastern forests. Swarming males dangle their enlarged hind legs, which

bear a variety of curious projections and glittering aluminum-colored flags (Kessel 1961, 1963) (Fig. 17-2). Snow (1884) noted how they '. . . allow their hindfeet to hang heavily downward and look as if they were carrying some heavy burden'.

There is an alternative to species isolation that seems to better explain the variety of displays and has been adapted to aggregated flies (Burk 1981; Prokopy and Roitberg 1984). Signals are hypothesized to be sexually selected advertisements of male (more rarely female) qualities directed to females (occasionally males) who are choosing a mate. There are circumstances where an individual can profitably advertise and circumstances where it cannot. One place where there is little profit in making a signal is where resources used by females are discrete, scattered, and relatively rare. Males can become closely associated with the resource, and 'guard' or 'control' it. If a female attempting to use that resource is subject to the mating attempts of a resident male or males, it might be more beneficial for her to immediately mate rather than spending time and energy choosing a particular male (note the strong selection for postcopulatory mate choice in such a situation; see Ward 1993). In the absence of female discrimination, selection for male courtship display is also absent. An unusual example of this behavior pattern occurs in the phoretic dipteran *Norrbomia frigipennis*. Females copulate frequently, often several times an hour, while riding a beetle (the resource), and there is no obvious courtship or sexual dimorphism (Sivinski 1984; see, however, Lachmann 1990).

When males do not control access to a resource, selection on signaling differs (see, for example, Headrick and Goeden 1994). For females the cost of mate choice is lower and they may be able to afford to discriminate among males; males in groups profit by appearing to be outstanding among their fellows (see, for example, Alexander 1975).

Even among non-resource-based aggregations, such as classic swarms, there is still a great deal of variation among species in displays (or markings and organs that can be interpreted as displays). At one extreme it has been noticed in both Brazil (Shannon 1931) and Africa (Haddow and Corbett 1961), that swarming mosquitoes are drab, but non-swarming species (e.g. *Toxorynchites* spp.) are often strikingly colored. It was suggested by Shannon, Haddow and Corbet that because mosquito swarms were crepuscular, or nocturnal, there was little chance for visual communication, whereas non-swarming mating systems centered over 'small containers' were diurnal and took advantage of the light to send visual signals (such as the blue leg 'paddles' of *Sabethes cyaneus* (Hancock *et al.* 1990)). There is at least one diurnal swarming mosquito, *Culex territans*, which is not unusually colored to our eye, even in the ultraviolet (although white markings on mosquitoes are often reflective in the ultraviolet (J. M. Sivinski, personal observations). However, small species of Ephemeroptera with hyaline or semihyaline wings are generally diurnal swarmers (Spieth 1940). One such species was described as shining in the sun like falling snowflakes (Morgan 1911). Large dark species usually swarm just at sunset to darkness (Spieth 1940). Spieth suggests that bulky species escape predators by swarming in the dark and small ones take advantage of their relative safety to enhance their visual apparency. Note that large termites, which may be more vulnerable to predators, tend to have their mass dispersal flight at night, and small species fly in daylight (Banks and Snyder 1920).

At the other extreme of the display spectrum among swarming flies are the previously mentioned platypezids and a number of peculiarly ornamented empidids. We argued that pheromones could be difficult to track in species that move rapidly within swarms, and so rarely evolve. The same problem may confront certain types of visual signal. Brightly colored or ornamented swarming insects often have a 'dignified' floating flight that would aid a responder in orienting to the signal (see, for example, Kessel 1955, 1963). Famous examples occur among the empidids that bear nuptial gifts. Many carry a dead insect, which they present to females. Others are more exotic, and carry flower petals or prey wrapped in silky balloons or empty balloons (cit. in Downes 1969). These 'gifts' are often bright, shining even in twilight, and can be seen bobbing against the darkened foliage in an uncanny manner.

If slow-floating swarmers were to settle to the ground or foliage, the result would be a mating system like a lek (an actual example of an insect that 'swarms' then lands to lek is *Bactrocera tryoni* (Tychsen 1977)). From a stable platform, a lekking insect can generate elaborate signals in different channels that can be easily traced back to a relatively constant point. The opportunity exists for broadcasting, and many lekking species have fulfilled the potential.

The seemingly small step between small, slow-flying swarms and leks suggests an alternative evolutionary route for elaborate displays among swarming species. Given the limitation on aerial signaling, perhaps the more ornate appendages and colorations arose in species that originally signaled from the substrate. Predation may be an important factor in the evolution of leks and the use of protected lekking sites (see, for example, Hendrichs and Hendrichs 1990). Predators may have pushed certain leks from vegetation entirely. One result could be swarms whose participants fly slowly or hover in order to better display ornaments that evolved under stationary circumstances on the 'ground'. A preswarming ancestor might have resembled certain dolichopodid flies, whose legs bear expansions similar to those illustrated in Fig. 17-2. However, these 'flags' are waved during courtships that occur on leaf surfaces (Oldroyd 1964). Dipteran taxa, such as Platypezidae, Phoridae and, to a lesser extent, Tephritidae, contain species that either swarm or lek, or display both behaviors under different circumstances, or simultaneously combine aerial and leaf-based aggregations (see, for example, Tychsen 1977; Sivinski 1988b). Examination of these groups might reveal the role of predation in the interface between leks and swarms.

The nature of sexually selected signaling information is a difficult problem that has occupied generations of biologists (see, for example, Fisher 1930; Hamilton and Zuk 1982; Alexander *et al.*, this volume). Occasionally, there are displays that offer what seems at least plausible 'translations' of the information. *Rhamphomyia longicaudata* is an empidid in which females swarm. Males provide a nuptial food-gift and may attempt to present the gift and their ejaculate to the female that can provide the most offspring. Females, when they swarm, swell up and give the impression of a flying abdomen (Steyskal 1941; Newkirk 1970). This may be an exaggerated promise of fecundity. Recently, body symmetry, as a reflection of genomic quality, has been found to play a role in female mate choice (Møller 1992; Thornhill 1992). The elaborate 'flags and feathers' that project from the legs of some swarming flies might be a difficult morphological test of the genome's ability to produce symmetry.

The plausibility of conjectures about the information content of displays depends on the mental capacity and range of experience of the choosing sex. Alexander *et al.* (this volume) propose that mate choice by insects is different from that by birds and mammals. The strongest form of their argument assumes that memory, experience with the opposite sex and individual recognition are all absent. It follows that comparisons can only be made among simultaneously encountered potential mates. There is no need for memory or experience when a choosing individual has innate minimum criteria, and the first mate to meet these standards is accepted. This, they suggest, is the typical form of mate choice in insects. Limitations on intersexual selection due to limited abilities to compare has implications about how structures used for communication are interpreted. For example, 'Fisherian runaway selection' is more likely when females 'learn about male traits so as to compare the range of variation in any group of males and choose the extreme regardless of precisely where it might fall' (Alexander et al., this volume). From this perspective, elaborate structures in insects are more likely to be due to armsraces between and within the sexes rather than 'ornamental' exaggerations that appeal to a preference for the extreme.

Are the explanations we have proposed for 'elaborate' displays (leg flags, silk balloons and air-filled abdomens) compatible with mate choice by minimum criteria? An adaptation like 'females exaggerate fecundity by swelling their abdomens' could be the result of a type of arms race. A small female that can appear larger might meet the standards of a gift-bearing male. As deceitful swelling spreads, higher minimum standards in males would become adaptive, and so on. However, a little relaxation of Alexander et al.'s assumptions in their strongest form allows Fisherian selection of 'ornaments' to occur in populations with limited mental abilities and sexual experience. Swarms are ideal places for simultaneous comparisons. This technique of mate choice requires less memory, individual recognition and experience than picking the most extreme of a sequentially encountered series of the opposite sex. A valued character, perhaps a leg plume, would be displayed by potential mates in a small area and at the same time. Individuals could be judged relative to one another, and extremes selected. One swarming insect suggests that Alexander et al.'s assumptions could be relaxed even further. As discussed earlier, males of the empidid *Empis borealis* bestow a nuptial gift and prefer large females (Svensson and Petersson 1987). They perform an 'ascending flight' with potential mates, presumably to judge their quality. A male may perform this flight with all the females in a swarm before mating; i.e. he appears to 'compare the range of variation' in female size.

CONCLUSIONS: THE EVOLUTION OF SWARMING AND SUGGESTIONS FOR FUTURE RESEARCH

We have reviewed the behaviors of insects as they approach and participate in swarms. We conclude by examining the evolution of the swarm mating system and its distribution among insect taxa. We first discuss how female unpredictability can result in the use of navigational objects as primitive swarm markers, and ask why the phyletic distribution of swarming seems to be so remarkably patchy. We then revisit Pajunen's (1980) argument that swarms result from the 'breakdown' of territoriality into 'scramble competition' and consider a simple notion based on body size of why this might occur in some taxa and not others. These discussions will suggest future research into the relationships between swarming and morphology, phylogyny and resourse distribution.

The phyletic distribution of swarming

The distribution of swarming is extremely patchy (Table 17-1). Many Diptera swarm; these species often develop in extensive, uniform habitats such as lakes and streams and their surrounding soil. Under these conditions females may be more concentrated over conspicuous objects used for navigation (swarm markers; see Parker 1978) than at emergence or oviposition sites. Males accumulating as they wait to intercept mates at navigational aids would constitute a simple type of swarm. Unlike flies, swarming beetles are seldom encountered (Table 17-1). Are there few beetles with population structures conducive to the evolution of swarming, or might the typically heavy exoskeleton of Coleoptera make swarming flight too expensive to perform? If it is robustness that constrains swarming, it is peculiar that swarming has been only sporadically described among parasitic Hymenoptera (with the exception of the braconid subfamily Blacinae (Southwood 1957; van Achterberg 1977)). There are a large number of small, lightly built species. Again, are population structures that result in female unpredictability at resources or emergence sites uncommon in parasitoids? Are there phylogenetic restraints on the evolution of swarming? These questions might be addressed through comparisons of parasitoid mating systems with those of their hosts. Adult solitary parasitoids may sometimes have distributions in time and space similar to their host's, particularly when parasitism levels approximate 50%. Their mating systems would, at least in part, be subject to similar selection pressures. An example may occur among the opiine braconid parasitoids of tephritids. *Diachasmimorpha longicaudata* and *Doryctobracon areolatus* often attack substantial proportions (*ca.* 50%) of the larvae of lekking fruit flies and also form leks in the trees infested by their hosts (J. M. Sivinski, personal observations). Similar convergence might also be possible between hosts that swarm and their parasitoids. At present, the difficulty in testing the hypothesis is finding suitable sets of insects to compare. Aquatic Diptera and the Trichoptera, which commonly swarm, are rarely parasitized (Krombein *et al.* 1979; see, however, agriotypin ichneumonids that attack pupae of caddisflies (Goulet and Huber 1993)). Swarming Blacini attack Coleoptera, but we are unaware of any information on their spacial distribution (Krombein *et al.* 1979). A swarming nearctic chalcidoid, *Bothriothorax nigripes*, attacks the larvae of Syrphidae (Nadel 1987), a dipteran family which includes species that aggregate on hilltops. Another parasitoid of syrphids, the European ichneumonid *Diplazon pectoratorius*, forms unusual male aggregations that have characteristics of swarms (Rothcray 1981). Peculiar, mostly female, swarms of a torymid, *Torymus phillyreae*, have been observed in Europe (Graham 1993). This parasitoid attacks certain Chironomidae, a family containing many swarming species (Graham 1994). However, the sexual significance of these female hymenopteran swarms has yet to be demonstrated. We suspect that opportunities for comparative studies will arise as the sexual behavior of parasitoids in the field becomes better documented (see, for example, Smith 1994). An effort to describe the little-known mating systems of dipteran parasitoids (see, for example, Toft 1989), particularly those that share hosts with or are hyperparasitized by hymenopteran parasitoids, may be especially useful in illuminating the roles of resource distribution, body size and phylogeny in the evolution of swarming.

Intrasexual conflict and swarming

For whatever reason, once females (or more rarely males) arrive over or near a marker, the air above the marker is a sexual resource (Parker 1978). Large insects, certain syrphid flies for instance, might be able to defend a territory containing a marker either by forays from a perch or by continual aerial patrol (see, for example, Fitzpatrick and Wellington 1983). Small insects, when confronted with the same marker, face a relatively larger space, which they are less able to control. Sexual rivals could accumulate without restraint. The typical absence of size distribution in swarms implies that even larger individuals in species of small size are not able to control access to locations where mating is more likely to occur. The packing together of small insects in an indefensible space might be particularly prevalent in aerial aggregations because of certain qualities

of an aerial environment. First, these spaces are relatively large compared with substrate resources. The volume over a marker is considerable larger than the area of the marker itself. A lekking tephritid would have to defend a good deal more if it fought not only for a leaf but for the air above it as well. Second, air could be a difficult medium to fight in. The energetic cost of tracking and hitting an opponent in fast-paced three-dimensional combat may be prohibitive. Smaller individuals of some species do not swarm at all, presumably because of insufficient energy reserves. Aggression might weaken an already precarious ability to remain airborne during the period of female arrival. If this argument is correct, then the ratio of insect size to swarm volume should be greater than that of insect size to the defended space of an individually held territory, be it aerial or on a substrate. Exceptions might occur in female swarms or where males bear resources (as in some Empididae). If intrasexual competition between swarmers is less intense in these aggregations, then even large insects might find it more useful to coinhabit a volume they could potentially have defended as a territory (for advantages, see Svensson and Petersson 1992).

ACKNOWLEDGEMENTS

We appreciate the many useful comments of editors and reviewers. In particular, thanks are due to James Lloyd, Bo G. Svensson and Tom Walker. We also thank Bob Sullivan for the use of his extensive library on swarming. Valerie Malcolm prepared the manuscript with her usual good-natured efficiency. The writing of this chapter was initiated during E. Petersson's postdoctoral stay at the IAB&BBRL in Gainesville, Florida, funded by the Research Council of Natural Sciences, Sweden.

LITERATURE CITED

Achterberg, I. van. 1977. The function of swarming in *Blacus* species (Hymenoptera, Braconidae, Helconinae). *Entomol. Beri.* **37**: 151–152.

Alcock, J. 1987. Leks and hilltopping in insects. *J. Nat. Hist.* **21**: 319–328.

Alexander, R. D. 1975. Natural selection and specialized chorusing behavior in acoustical insects. In *Insects, Science, and Society.* D. Pimentel, ed., pp. 35–77. New York: Academic Press.

Allan, J. D., and A. S. Flecker. 1989. The mating biology of a mass-swarming mayfly. *Anim. Behav.* **37**: 361–371.

Allan, S. A., J. F. Day and J. D. Edman. 1987. Visual ecology of biting flies. *Annu. Rev. Entomol.* **32**: 297–316.

Arak, A. 1988. Callers and satellites in the natterjack toad: evolutionary stable decision rules. *Anim. Behav.* **36**: 416–432.

Bailey, W. J. 1991. *Acoustic Behavior of Insects*. London: Chapman and Hall.

Banks, C. S. 1919. The swarming of Anopheline mosquitoes. *Phillip. J. Sci.* **15**: 283–288.

Banks, N. and T. E. Snyder. 1920. A revision of the nearctic termites. *U.S. Nat. Mus. Bull.*, no. 108.

Beani, L. and S. Turillazzi. 1990. Overlap at landmarks by lek-territorial and swarming males of two sympatric polistine wasps (Hymenoptera: Vespidae). *Ethol. Ecol. Evol.* **2**: 419–431.

Borgia, G. 1979. Sexual selection and the evolution of mating systems. In *Sexual Selection and Reproductive Competition in Insects*. M. S. Blum and N. A. Blum, eds., pp. 19–80. New York: Academic Press.

Bradbury, J. W. 1985. Contrasts between insects and vertebrates in the evolution of male display, female choice, and lek mating. In *Experimental Behavioral Ecology and Sociobiology*. B. Hölldobler and M. Lindauer, eds., pp. 286–293. New York: Springer-Verlag.

Brickle, R. L. 1959. Observations on the hovering and mating of *Tabanus bishoppi* Stone. *Ann. Entomol. Soc. Am.* **52**: 183–190.

Brodsky, A. K. 1973. The swarming behavior of mayflies (Ephemeroptera). *Entomol. Rev.* **52**: 33–39.

Burk, T. 1981. Signaling and sex in acalyptrate flies. *Fla. Entomol.* **64**: 30–43.

Campbell, M. M. and D. S. Kettle. 1979. Swarming of *Culicoides brevitarsis* Kieffer (Diptera: Ceratopogonidae) with reference to marker, swarm size, proximity of cattle, and weather. *Austr. J. Zool.* **27**: 17–30.

Chandler, P. J. 1978. Some dipterous opportunists at Windsor Forest, Berks: the attraction for flies of bonfires, wood ash and freshly cut logs. *Entomol. Gaz.* **29**: 253–257.

Chiang, H. C. 1961. Ecology of insect swarms. I. Experimental studies of the behaviour of *Anarete* sp. near *felti* Prichard in artificially induced swarms. *Anim. Behav.* **9**: 213–219.

Cunningham-van Someren, G. 1965. Male mosquito swarms: some observations in Kenya. *Proc. R. Entomol. Soc. London.* **40**: 89–91.

–. 1973. A further note on swarming of male mosquitoes and other Nematocera in Kenya. *Entomol. Mon. Mag.* **109**: 147–160.

Dahl, C. 1965. Studies on swarming activity in Trichoceridae (Diptera) in southern Sweden. *Opusc. Entomol. Suppl.*, no. 27.

Downes, J. A. 1958. Assembly and mating in the biting Nematocera. In *Proc. 10th Int. Congr. Entomol. Montreal*. E. C. Becker, ed., vol. 2, pp. 425–434.

–. 1969. The swarming and mating flight of Diptera. *Annu. Rev. Entomol.* **14**: 271–298.

–. 1970. The feeding and mating behaviour of the specialized Empidinae (Diptera); Observations on four species of *Rhamphomyia* in the arctic and a general discussion. *Can. Entomol.* **102**: 769–791.

–. 1978. Feeding and mating in the insectivorous Ceratopogoninae (Diptera). *Mem. Entomol. Soc. Can.*, no. 104.

Duhrkopf, R. E. and W. K. Hartberg. 1992. Differences in male mating response and female flight sounds in *Aedes aegypti* and *Ae. albopictus* (Diptera: Culicidae). *J. Med. Entomol.* **29**: 796–801.

Edmunds, G. F. and C. H. Edmunds. 1980. Predation, climate, and emergence and mating of mayflies. In *Advances in Ephemeroptera Biology*. J. F. Flannagan and K. E. Marshall, eds., pp. 277–285. New York: Plenum Press.

Edwards, F. W. 1920. Scent organs (?) in female midges of the *Palpomyia* group. *Annu. Mag. Nat. Hist.* ser. 9, **5**: 365–368.

Ewing, A. W. 1989. *Arthropod Bioacoustics*. Ithaca: Cornell University Press.

Fisher, R. A. 1930. *The Genetical Theory of Natural Selection*. Oxford: Clarendon Press.

Fitzpatrick, S. M. and W. G. Wellington. 1993. Contrasts in the territorial behavior of three species of Hover Fly (Diptera: Syrphidae). *Can. Entomol.* **115**: 559–566.

Flecker, A. S., J. D. Allan and N. L. McClintock. 1988. Male body size and mating success in swarms of the mayfly *Epeorus longimanus*. *Holarct. Ecol.* **11**: 280–285.

Foster, W. A. and J. E. Treherne. 1981. Evidence for the dilution effect in the selfish herd from fish predation on a marine insect. *Nature (Lond.)* **293**: 466–467.

Frohne, W. C. and R. G. Frohne. 1952. Diurnal swarms of *Culex territans* Walker, and the crepuscular swarming of *Aedes* about a small glade in Alaska. *Mosq. News* **14**: 62–64.

Gibson, N. H. E. 1944. On the mating swarms of certain Chironomidae (Diptera). *Trans. R. Entomol. Soc. Lond.* **95**: 263–294.

Goulet, H. and J. Huber. 1993. Hymenoptera of the world: An identification guide to families. *Res. Br. Ag. Can. Pub.*, no. 1894/E.

Graham, M. W. R. 1993. Swarming in chalcidoidea (Hym.) with description of a new species of *Torymus* (Hym., Torymidae) involved. *Entomol. Mon. Mag.* **129**: 15–23.

–. 1994. Further notes on a swarm-forming species, *Torymus phillyreae* (Hymenoptera: Torymidae) with new synonymy. *Entomol. Ber.* **54**: 120–122.

Gullefors, B. and E. Petersson. 1993. Sexual dimorphism in relation to swarming and pair formation behavior in Leptoceridae (Trichoptera). *J. Insect Behav.* **6**: 565–577.

Haddow, A. J. and P. S. Corbet. 1961. Entomological studies from a high tower in Mpanga Forest, Uganda. V. Swarming activity above the forest. *Trans. R. Entomol. Soc. Lond.* **113**: 284–300.

Hamilton, W. D. 1971. Geometry for the selfish herd. *J. Theor. Biol.* **31**: 295–311.

Hamilton, W. D. and M. Zuk. 1982. Heritable true fitness and bright birds: a role for parasites? *Science (Wash., D.C.)* **218**: 384–387.

Hancock, R. G., A. F. Woodbridge and W. L. Yee. 1990. Courtship behavior of the mosquito *Sabethes cyaneus* (Diptera: Culicidae). *J. Insect Behav.* **3**: 401–416.

Headrick, D. H. and R. D. Goeden. 1994. Reproductive behavior of California fruit flies and the classification and evolution of Tephritidae (Diptera) mating systems. *Stud. Dipt.* **1**: 195–252.

Hendrichs, J. and M. A. Hendrichs. 1990. Mediterranean fruit fly (Diptera: Tephritidae) in nature: Location and diel pattern of feeding and other activities on fruiting and nonfruiting hosts and nonhosts. *Ann. Entomol. Soc. Am.* **83**: 632–641.

Hieber, C. G. and J. A. Cohen. 1983. Sexual selection in the lovebug, *Plecia nearctica*: the role of male choice. *Evolution* **37**: 987–992.

Hubbard, M. D. and B. Nagell. 1976. Notes on an extraordinary high mating swarm of the ant *Myrmica laevinodes* (Hymenoptera: Formicidae, Myrmicinae). *Entomol. News* **87**: 86.

Hunter, D. M. 1979. Swarming, mating and resting behaviour of three species of black fly (Diptera: Simuliidae). *J. Austr. Entomol. Soc.* **18**: 1–6.

Inglis, I. R. and J. Lazarus. 1981. Vigilance and flock size in Brent geese: the edge effect. *Z. Tierpsychol.* **57**: 193–200.

Kessel, E. L. 1955. The mating habits of balloon flies (Diptera: Empididae). *Syst. Zool.* **4**: 97–104.

–. 1961. Observations on the mating behavior of *Platypezia Kessel* (Diptera: Platypezidae). *Wasmann J. Biol.* **19**: 295–299.

–. 1963. The genus *Calotarsa*, with special reference to *C. insignis* Aldrich (Diptera: Platypezidae). *Calif. Acad. Sci. Occas. Pap.* **39**: 1–14.

Kessel, E. L. and E. A. Maggioncalda. 1968. A revision of the genera of Platypezidae, with the descriptions of five new genera and considerations of phylogeny, circumversion and hypopygia (Diptera). *Wasmann J. Biol.* **26**: 33–106.

Kritsky, G. 1977. Observations on the morphology and behavior of the Enicocephalidae (Hemiptera). *Entomol. News* **88**: 105–110.

Krombein, K. V., P. D. Hurd, Jr, D. R. Smith and B. D. Banks. 1979. *Catalog of Hymenoptera in America North of Mexico*. Washington D.C.: Smithsonian Institution Press.

Lachmann, A. 1990. Courtship behavior and mating in two cow dung inhabiting sphaerocerid flies (Diptera: Sphaeroceridae). *Ethology* **86**: 161–169.

Loor, K. A. and G. R. DeFoliart. 1970. Field observations on the biology of *Aedes triseriatus*. *Mosq. News* **30**: 60–64.

Löfstedt, C., B. S. Hansson, E. Petersson and P. Valeur. 1994. Pheromonal secretions from glands on the 5th abdominal sterinte of hydropsychid and rhyacophilid caddisflies (Trichoptera). *J. Chem. Ecol.* **20**: 153–170.

Mallet, J. 1984. Sex roles in the ghost moth *Hepialus humali* (L.) and a review of mating in the Hepialidae (Lepidoptera). *Zool. J. Linn. Soc.* **79**: 67–82.

Maynard Smith, J. 1974. *Models in Ecology*. Cambridge University Press.

Mayr, E. 1963. *Animal species and Evolution*. Cambridge, Mass.: Harvard University Press.

McAlpine, J. F. and D. D. Munroe. 1968. Swarming of lonchaeid flies and other insects, with description of four new species of Lonchaeidae (Diptera). *Can. Entomol.* **100**: 2154–2178.

McLachlan, A. and R. Neems. 1989. An alternative mating system in small male insects. *Ecol. Entomol.* **14**: 85–91.

Michener, C. D. 1974. *The Social Behavior of the Bees: A Comparative Study*. Cambridge, Mass.: Harvard University Press.

Møller, A. P. 1992. Female swallow preference for symmetrical male sexual ornaments. *Nature (Lond.)* **357**: 238–240.

Moorhouse, D. E. and M. H. Colbo. 1973. On the swarming of *Austrosimulium pestilens* MacKerras and MacKerras (Diptera: Simuliidae). *J. Austr. Entomol. Soc.* **12**: 127–130.

Morgan, A. H. 1911. May-flies of the Fall Creek. *Ann. Entomol. Soc. Am.* **4**: 93–119.

Mori, S. and K. Matubani. 1953. Daily swarming of some caddisfly adults and their habitat segregations. *Zool. Mag. (Tokyo)* **62**: 191–198 (in Japanese with an English abstract).

Nadel, H. 1987. Male swarms discovered in Chalcidoidae (Hymenoptera: Encyrtidae, Pteromalidae). *Pan-Pac. Entomol.* **63**: 242–246.

Nation, J. L. 1989. The role of pheromones in the mating system of *Anastrepha* fruit flies. In *Fruit flies-Biology, Natural Enemies and Control*. A. S. Robinson and G. Hooper, eds., pp. 189-205. Amsterdam: Elsevier.

Neems, R., J. Lazarus and A. J. McLachlan. 1992. Swarming behavior in male chironomid midges: a cost-benefit analysis. *Behav. Ecol.* **3**: 285–290.

Nielsen, H. and E. Nielsen. 1962. Swarming of mosquitoes, laboratory experiments under controlled conditions. *Entomol. Exp. Appl.* **5**: 14–3.

Newkirk, M. R. 1970. Biology of the longtailed dance fly, *Rhamphomyia longicaudata* (Diptera: Empididae): a new look at swarming. *Ann. Entomol. Soc. Am.* **63**: 1407–1412.

Odendaal, F. J., P. Turchin and F. R. Stermitz. 1988. An incidental-effect hypothesis explaining aggregation of males in a population of *Euphydryas anicia*. *Am. Nat.* **132**: 735–749.

Ogawa, K. 1992. Field trapping of male midges *Rheotanytarsus kyotoensis* (Diptera: Chironomidae) by sound. *Jap. J. Sanit. Zool.* **43**: 77–80.

Oldroyd, H. 1964. *The Natural History of Flies*. New York: W. W. Norton & Co.

Pajunen, V. I. 1980. A note on the connection between swarming and territorial behaviour in insects. *Ann. Entomol. Fenn.* **46**: 53–55.

Parker, G. A. 1978. Evolution of competitive mate searching. *Annu. Rev. Entomol.* **23**: 173–196.

Petersson, E. 1987. Weight-associated male mating success in the swarming caddis fly, *Mystacides azureus* L. *Ann. Zool. Fenn.* **24**: 335–339.

–. 1989. Swarming activity patterns and seasonal decline in adult size in some caddis flies (Leptoceridae: Trichoptera). *Aquatic Insects* **11**: 17–28.

Petersson, E. and J. O. Solem. 1987. Male mate recognition in Leptoceridae. In *Proceedings of 5th International Symposium on Trichoptera*. M. Bournaud and H. Tachet, eds., pp. 157–160. Hingham, Massachusetts: W. Junk.

Prokopy, R. J. and B. D. Roitberg. 1984. Foraging behaviour of true fruit flies. *Am. Scient.* **72**: 41–49.

Provost, M. W. 1958. Mating and male swarming in *Psorophora* mosquitoes. In *Proc. 10th Int. Congr. Entomol., Montreal.* E. C. Becker, ed., vol. 1, pp. 553–561.

Pulliam, H. R. 1973. On the advantage of flocking. *J. Theor. Biol.* **38**: 419–422.

Pulliam, H. R., G. H. Pyke and T. Caraco. 1982. The scanning behavior of Juncos: a game-theoretical approach. *J. Theor. Biol.* **95**: 89–103.

Rao, T. R. and P. F. Russel. 1938. Some field observations on the swarming and pairing of mosquitoes, particularly *A. annularis*, in southern India. *J. Malaria Inst. India* **1**: 395–403.

Reisen, W. K., R. H. Baker, R. K. Sakai, F. Mahmood, H. R. Rathor, K. Rana and G. Toquir. 1981. *Anopheles culicifacies* Giles: Mating behavior and competitiveness in nature of chemosterilized males carrying a genetic sexing system. *Ann. Entomol. Soc. Am.* **74**: 395–401.

Robinson, M. H. and B. Robinson. 1977. Associations between flies and spiders: bibiocommensalism and dipsoparasitism. *Psyche* **84**: 150–157.

Roth, L. M. 1948. A study of mosquito behavior. An experimental laboratory study of the sexual behavior of *Aedes aegypti* (Linnaeus) *Am. Midl. Nat.* **40**: 265–352.

Rotheray, G. E. 1981. Courtship, male swarms and a sex pheromone of *Diplazon pectoratorius* (Thunberg) (Hymenoptera: Ichnuemonidae). *Entomol. Gaz.* **32**: 193–196.

Savolainen, E. 1978. Swarming in Ephemeroptera: The mechanism of swarming and the effects of illumination and weather. *Ann. Zool. Fenn.* **15**: 17–52.

Savolainen, E., A. Saura and J. Hantula. 1993. Mode of swarming in relation to reproductive isolation in mayflies. *Evolution* **47**: 1796–1804.

Savolainen, E. and J. Syrjämäki. 1971. Swarming and mating of *Erioptera gemina* Tjeder (Dipt., Limoniidae). *Ann. Entomol. Fenn.* **37**: 79–85.

Shannon, R. L. 1931. On the classification of Brazilian Culicide with special reference to those capable of harboring yellow fever virus. *Proc. Entomol. Soc. Wash.* **33**: 125–164.

Shelly, T. E. 1989. Waiting for mates: Variation in female encounter rates within and between leks of *Drosophila conformis*. *Behaviour* **111**: 34–48.

Sivinski, J. 1983. The natural history of a phoretic sphaerocerid Diptera fauna. *Ecol. Entomol.* **8**: 419–426.

–. 1984. A sexual conflict and choice in a phoretic fly, *Borborillus frigipennis* (Sphaeroceridae). *Ann. Entomol. Soc. Am.* **77**: 232–235.

–. 1988a. What do fruit fly songs mean? *Fla. Entomol.* **71**: 462–466.

–. 1988b. Unusual female-aggregated mating systems in phorid flies. *J. Insect Behav.* **1**: 123–128.

Sivinski, J. M. and T. Burk. 1989. Reproductive and mating behavior. In *Fruit Flies – Biology, Natural Enemies and Control*. A. S. Robinson and G. Hooper, eds., pp. 343–351. Amsterdam: Elsevier.

Smith, K. G. V. 1994. Swarming in *Miscogaster rufipes* Walker (Hym., Pteromalidae). *Entomol. Mon. Mag.* **130**: 130.

Snow, W. A. 1884. American Platypezidae. *Kansas Univ. Quarterly* **3**: 143–152.

Solem, J. O. 1978. Swarming and habitat segregation in the family Leptoceridae (Trichoptera). *Norw. J. Entomol.* **25**: 145–148.

–. 1984. Adult behavior of North European caddisflies. In *Proceedings of 4th International Symposium on Trichoptera*. J. C. Morse, ed., pp. 375–382. The Hague: Dr. W. Junk Publishers.

Sotavalta, O. 1974. The flight-tone (wing-stroke frequency) of insects. *Acta Entomol. Fenn.* **4**: 1–117.

Southwood, T. R. E. 1957. Observations on swarming in the Braconidae (Hymenoptera) and Coniopterygidae (Neuroptera). *Proc. R. Entomol. Soc. Lond.* A **32**: 80–82.

Spieth, H. T. 1940. Studies on the biology of Ephemeroptera. II. The nuptial flight. *J. N. Y. Entomol. Soc.* **48**: 379–390.

Steyskal, G. 1941. A curious habit of an empidid fly. *Bull. Brooklyn Entomol. Soc.* **36**: 117.

Sullivan, R. T. 1981. Insect swarming and mating. *Fla. Entomol.* **64**: 44–65.

Svensson, B. G. and E. Petersson. 1987. Sex-role reversed courtship behaviour, sexual dimorphism and nuptial gifts in the dance fly *Empis borealis* (L). *Ann. Zool. Fenn.* **24**: 323–334.

–. 1988. Non-random mating in the dance fly *Empis borealis*: The importance of male choice. *Ethology* **79**: 307–316.

–. 1992. Why insects swarm: testing the models for lek mating systems on swarming *Empis borealis* females. *Behav. Ecol. Sociobiol.* **31**: 253–261.

Svensson, B. G., E. Petersson and E. Forsgren. 1989. Why do males of the dance fly *Empis borealis* refuse to mate? The importance of female age and size. *J. Insect Behav.* **2**: 387–395.

Thornhill, R. 1980. Sexual selection within mating swarms of the lovebug, *Plecia nearctica* (Diptera: Bibionidae). *Anim. Behav.* **28**: 405–412.

–. 1992. Female preference for the pheromone of males with low fluctuating asymmetry in the Japanese scorpionfly (*Panorpa japonica*: Mecoptera). *Behav. Ecol.* **3**: 277–283.

Thornhill, R. and J. Alcock. 1983. *The Evolution of Insect Mating Systems*. Cambridge, Mass.: Harvard University Press.

Toft, C. A. 1989. Population structure and mating system of a desert bee fly (*Lordotus pulchrissimus*)(Diptera: Bombyliidae). 2. Female demography, copulations and characteristics of swarms sites. *Oikos* **54**: 359–369.

Treherne, J. E. and W. A. Foster. 1980. The effect on group size on predator avoidance in a marine insect. *Anim. Behav.* **28**: 1119–1122.

–. 1981. Group transmission of predator avoidance in a marine insect: the Trafalgar effect. *Anim. Behav.* **29**: 911–917.

–. 1982. Group size and anti-predator strategies in a marine insect. *Anim. Behav.* **30**: 536–542.

Tychsen, P. H. 1977. Mating behavior of the Queensland fruit fly, *Dacus tryoni* (Diptera: Tehpritidae), in field cages. *J. Austr. Entomol. Soc.* **16**: 459–465.

Usinger, R. L. 1945. Classifications of the Enicocephalidae (Hemiptera, Reduvioidea). *Ann. Entomol. Soc. Am.* **38**: 321–342.

Vulinec, K. and C. Miller. 1989. Aggregation and predator avoidance in whirligig beetles (Coleoptera: Gyrinidae). *J. N.Y. Entomol. Soc.* **97**: 438–447.

Ward, P. I. 1993. Females influence sperm storage and use in the yellow dung fly *Scathophaga stercoraria* (L.). *Behav. Ecol. Sociobiol.* **32**: 313–319.

West-Eberhard, M. J. 1984. Sexual selection, competitive communication and species-specific signals in insects. In *Insect Communication*. T. Lewis, ed., pp. 284–324. London: Academic Press.

Wrona, F. J. and W. J. Dixon. 1991. Group size and predation risk: a field analysis of encounter and dilution effects. *Am. Nat.* **137**: 186–201.

Yuval, B. and A. Bouskila. 1993. Temporal dynamics of mating and predation in mosquito swarms. *Oecologia (Berl.)* **95**: 65–69.

Yuval, B., J. W. Wekesa and R. K. Washino. 1993. Effect of body size on swarming behavior and mating success of male *Anopheles freeborni* (Diptera: Culicidae). *J. Insect Behav.* **6**: 333–342.

Yuval, B., M. L. Holliday-Hanson and K. K. Washino. 1994. Energy budget of swarming male mosquitos. *Ecol. Entomol.* **19**: 74–78.

Zeil, J. 1983. Sexual dimorphism in the visual system of flies: the free flight behaviour of male Bibionidae (Diptera). *J. Comp. Physiol.* A **150**: 395–412.

18 · Function and evolution of antlers and eye stalks in flies

GERALD S. WILKINSON AND GARY N. DODSON

ABSTRACT

Flies with structures protruding from the head, termed eye stalks or antlers, have long fascinated biologists. In this chapter we consider the possibility that such flies either share developmental predispositions to head modification or face similar selective pressures to augment head width. By randomizing the occurrence of eye stalks and antlers among families of acalyptrate flies over a phylogeny, we demonstrate that eye stalks and antlers have evolved especially often within one superfamily and that eye stalks and antlers exhibit correlated evolution. Because extreme sexual dimorphism of antlers and eye stalks occurs in four families, while correlated evolution of eye stalks in females occurs in others, we suggest that the degree of sexual dimorphism, but not the occurrence of head projections, has been influenced by historical genetic factors. Comparison of the ecology and mating systems of flies from the genus *Phytalmia*, all species of which possess antlers, with flies from the family Diopsidae, all species of which possess eye stalks, indicates selective similarities. All head-projection flies studied to date exhibit some form of resource-defense mating system. *Phytalmia* spp. defend rare oviposition sites, whereas dimorphic diopsids defend protected sites where females aggregate at night. In both groups, males assess the body size of competitors from the head projection, and males of similar size engage in extended physical contests for control of a resource. Females of at least two diopsid species also use eye-stalk length to select mates and can influence sperm precedence by ejecting male spermatophores. Head-structure span exhibits positive allometry with body size in *Phytalmia* and dimorphic diopsids. Because of the high correlation between head-projection span and body size, comparison of head projections provides a quick and accurate method for assessing body size. We postulate that head projections have evolved in tropical flies with resource-defense mating systems because these animals experience many mating opportunities owing to long lifespans and because competing males benefit from reduced average contest duration.

INTRODUCTION

Some of the most unusual flies in the world are those with secondary paired processes of the head capsule, often referred to as antlers or eye stalks. The term 'antler' refers to the shape of some structures, such as the cheek processes of *Phytalmia* species (Fig. 18-1), rather than to any deciduous capability; 'eye stalks' indicate that the eye bulb has been displaced laterally away from the body on peduncles. No fly yet described has both eye stalks and antlers although in a few cases, such as several species in the genus *Zygothrica* (Drosophilidae), the eye bulb is elongated into a cone, which may curl at the tip like the horns of a ram (Grimaldi 1987).

Many workers have speculated on the function of cephalic projections since the first descriptions of flies with eye stalks (Linnaeus 1775) and antlers (Gaerstecker 1860; Saunders 1860). However, until recently few observational studies on these flies had been conducted. In this chapter we summarize what is currently known regarding the mating systems of flies with antlers and eye stalks and, in conjunction with phylogenetic and allometric analyses, use this information to make inferences regarding the function and evolution of these bizarre structures.

PHYLOGENETIC RELATIONSHIP AMONG FLIES WITH HEAD PROJECTIONS

Within the order Diptera, eye stalks occur in eight families and antlers in five families (Table 18-1). The distributions of eye stalks among dipteran families (Hennig 1966; Sanderson 1991) and hypercephaly among species of drosophilids (DeSalle and Grimaldi 1993) have both been interpreted as examples of recurrent evolution. To evaluate these claims further and to determine whether eye stalks and antlers have evolved independently, we performed randomization tests using putative phylogenetic relationships among 64 acalyptrate families (Fig. 18-2) (McAlpine 1989). We restrict our analyses to acalyptrate flies because they may

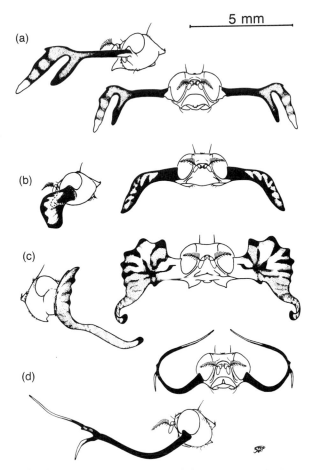

5 mm

(a)

(b)

(c)

(d)

Fig. 18-1. Head projections or 'antlers' of males from (a) *Phytalmia antilocapra*, (b) *P. cervicornis*, (c) *P. mouldsi*, and (d) *P. alcicornis*.

represent the smallest monophyletic group that includes all cases of head projections (McAlpine 1989).

The first question we ask is whether head projections have evolved independently among flies or exhibit clustered evolution on a family-level phylogeny (Fig. 18-2). Because not all members of each family have head projections (Table 18-1), one or more independent evolutionary events has occurred within each family. Consequently, comparing the number of evolutionary steps required to explain head projection evolution to some null expectation (see, for example, Sanderson 1991) is inappropriate at the family level. Instead, we estimated the probability that four families in the Tephritoidea have evolved eye stalks by assuming that eye-stalk acquisition is equiprobable among families. After shuffling the occurrence of eye stalks over 100 replicate phylogenies, only one case of four

families, and none with more families, emerged in this superfamily. The probability of such clustered evolution occurring by chance is therefore 0.01. Similarly, three families with antlers arising within the Tephritoidea is unexpected, as the probability associated with this result is also 0.01. Thus, both eye stalks and antlers are clustered non-randomly on the phylogeny.

The second question we ask is whether eye stalks and antlers have arisen independently of each other. Although antlers are known from only five families of flies, four of these families also contain species with eye stalks (Table 18-1). Using the concentrated changes test (Maddison 1990) with the occurrence of eye stalks in an acalyptrate family as the independent variable and the presence of antlers as the dependent variable, the probability of at least four gains and no more than one loss is less than 0.001.

These analyses suggest that some characteristics of these families facilitate head-projection evolution. Two non-exclusive possibilities can be envisioned. First, some developmental trait may predispose these taxa to evolve head projections. Second, these families may share an environment that favors head elaboration. To evaluate these alternatives we focus discussion on two groups, the genus *Phytalmia* (Tephritidae), all species of which possess antlers (Fig. 18-1), and the family Diopsidae, all species of which possess eye stalks.

PHYTALMIA (TEPHRITIDAE)

Distribution and morphology

Among the fly genera with cephalic projections resembling antlers, *Phytalmia* is the most speciose and their head projections are the most elaborate. When McAlpine and Schneider (1978) revised the genus they reported six species, with one additional species having been described since (Schneider 1993). All but one species are found on the mainland of New Guinea and adjacent islands (McAlpine and Schneider 1978). The single species found elsewhere is *P. mouldsi*, known only from an area of isolated rainforest in the upper Cape York Peninsula of Australia.

Across the genus, as well as throughout the other 'antlered' flies (Table 18-1), head or cheek projections are restricted to males. *Phytalmia* antlers are extensions of the cuticle arising in the cheek area, their bases forming an indentation into the compound eye (Fig. 18-1). Females have a notch at the lateral margin of the eye, but

Table 18-1. *Diptera with head projections*

Taxa	Projection	Reference
Clusiidae		
Labomyia mirabilis	cheek process	Frey 1960
Hendelia extensicornis	broad head	Frey 1960
Parahendelia latifrons	broad head	McAlpine 1975
Clusiodes gladiator	broad head	McAlpine 1975
Diopsidae (13 genera, >150 species)	eye stalks	Feijen 1989
Drosophilidae		
Chymomyza (5 species)	eye stalks	Grimaldi 1986a
Diathoneura sp.	cheek process	D. Grimaldi, personal communication
Drosophila heteroneura	eye stalks	Spieth 1981
D. (Hirtodrosophila) caputudis	eye stalks	Grimaldi 1986b
D. (Hirtodrosophila) chandleri	eye stalks	Grimaldi 1988
Mulgravea asiatica	eye stalks	Grimaldi and Fenster 1989
Zygothrica (14 species)	eye stalks or broad heads	Grimaldi 1987 Bristowe 1925
Micropezidae		
Anaeropsis guttipennis	eye stalks	McAlpine 1974
Otitidae		
Plagiocephalus (2 species)	eye stalks	Steyskal 1963
Paragorgopis (4 species)	eye stalks	G. C. Steyskal, personal communication
Ophthalmoptera spp.	eye stalks	Grimaldi and Fenster 1989
Periscelididae		
Sphyroperiscelis spp.	eye stalks	Grimaldi and Fenster 1989
Diopsosoma prima	eye stalks	G. C. Steyskal, personal communication
Platystomatidae		
Achias spp.	eye stalks	McAlpine 1994
Achiosoma (2 species)	eye stalks	McAlpine 1975
Agrochira spp.	eye stalks	McAlpine 1982
Angitula nigra	cheek process	McAlpine 1975
Apiola spp.	cheek process	McAlpine 1982
Asyntona tetyroides	broad head	McAlpine 1982
Atopognathus spp.	eye stalks	McAlpine 1982
Brea spp.	eye stalks	McAlpine 1982
Cleitamia spp.	broad head	Malloch 1939
Clitodoca spp.	cheek process	McAlpine 1982
Giraffomyia spp.	cheek process	Malloch 1940
Laglaizia spp.	eye stalks	McAlpine 1975
Mescoctenia spp.	cheek process	McAlpine 1973
Neohemigaster spp.	cheek process	McAlpine 1975
Pogonortalis doclea	broad head	McAlpine 1975
Pterogenia spp.	cheek process	McAlpine 1975
Trigonosoma spp.	eye stalks	McAlpine 1982
Zygaenula spp.	eye stalks	McAlpine 1982
Richardiidae		
Batrachophthalmum spp.	eye stalks	G. C. Steyskal, personal communication
Richardia (3 species)	eye stalks	Hennig 1937
Richardia (3 species)	cheek process	D. Grimaldi, personal communication

Table 18-1. (*cont.*)

Taxa	Projection	Reference
Tephritidae		
Diplochorda (2 species)	cheek process	McAlpine 1975
Homiothemara eurycephala	broad head	Hardy 1988
Ortaloptera callistomyia	cheek process	Hardy 1988
Pelmatops ichneumonea	eye stalks	Grimaldi and Fenster 1989
Phytalmia (7 species)	cheek process	McAlpine and Schneider 1978
Pseudopelmatops nigricostalis	eye stalks	G. C. Steyskal, personal communication
Sessilina nigrilinea	cheek process	McAlpine and Schneider 1978
Sophira limbata	cheek process	Hardy 1988
Terastiomyia lobifera	cheek process	Hardy 1986

no projections. The size of the cheek process covaries with body size in *Phytalmia* (McAlpine and Schneider 1978) (Fig. 18-3). The smallest males have nothing more than a raised area on the cheek; the largest have spans of more than 1 cm between the tips of each antler. The shapes of the cheek processes are a defining characteristic of the species and are spatulate, cylindrical, or fan-shaped, and either simple or bifurcated (Fig. 18-1).

Apart from the male head projections (and genitalia), the only conspicuous sexual difference is the relatively larger female abdomen, which terminates in a tapering oviscape. More subtle differences between the sexes include recurved spines on the male fore femora, used to grasp female wings during mating and postcopulatory guarding (Moulds 1977), and a strongly produced epistomal margin of the male face in some species (Fig. 18-1) associated with intraspecific agonistic interactions as described below and by Moulds (1977). Otherwise males and females are similar in morphology and have a wasp-like appearance.

Life history

Moulds (1977) reported on the reproductive behavior and associated biology of the Australian species, *P. mouldsi*. Subsequently, one of us (GND) has studied *P. mouldsi* and *P. alcicornis* in more detail and made preliminary observations on *P. cervicornis* and *P. biarmata*. These studies have revealed fundamental similarities in the general ecologies and mating systems of *Phytalmia* species. In particular, the larval food source appears to be limited in its distribution, enabling males to defend oviposition sites at the larval substrate, i.e. decaying wood of particular tree species or the fungus/bacteria associated with this decay. Antlers function within the context of male agonistic behavior at this resource (Moulds 1977; Dodson 1989; Dodson 1997).

Female *Phytalmia* lay eggs by inserting their ovipositors through openings in the bark of downed tree trunks or limbs. The larvae feed within the rotting sapwood. *Phytalmia* may be long-lived as adults. *P. mouldsi* adults of unknown age brought from the field into the laboratory have survived as long as 65 d following capture. Adults reared from field collected larvae have lived for up to 43 d in the laboratory (G. N. Dodson, unpublished).

Resource distribution and visitation

Evidence for the limited distribution of acceptable larval substrates comes from two sources: host suitability tests and measurements of the abundance and persistence of oviposition sites at larval substrates. In Australia and Papua New Guinea (PNG) host suitability has been assessed for fifteen tree species by systematically sampling downed or felled material for fly visitation (Moulds 1977; Dodson 1997). In all cases the trees were accessible over the same period after downing or felling in an area where adult flies were present. Three species of *Phytalmia* (*P. mouldsi*, *P. alcicornis* and *P. biarmata*) were attracted, each to a single host species. Stringent host specificity was indicated by a lack of fly attraction to either confamilial or congeneric tree species (Dodson 1997). Three species of *Phytalmia* have been reared from a single tree species (*Dysoxylum gaudichaudianum*). Inspections of innumerable trees, either naturally downed or in logged areas, have revealed only three other species as hosts (Dodson and Daniels 1988; Dodson 1997; H. Roberts, personal communication).

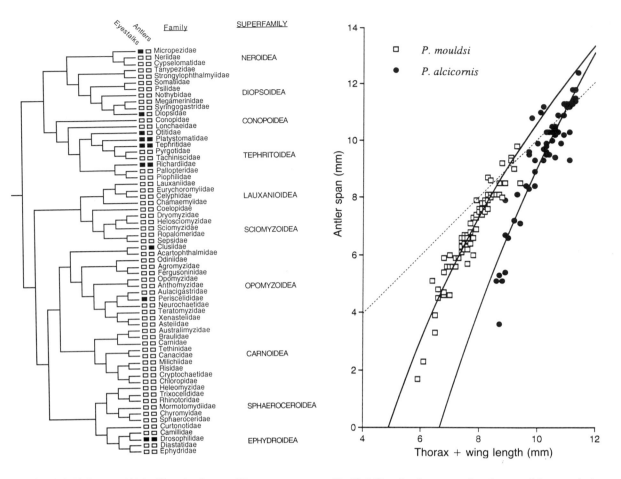

Fig. 18-2. Phylogeny of 64 families of acalyptrate Diptera slightly modified from McAlpine (1989), with the occurrence of eye stalks and antlers indicated by filled boxes adjacent to each family.

Fig. 18-3. Plot of antler span against the sum of thorax and wing length for *P. mouldsi* and *P. alcicornis*. The fitted curves represent least-squares regressions of antler span on the natural log of the sum of thorax and wing length. The dotted line represents the isometric line that passes through the origin.

The window of suitability of the oviposition substrate varied according to the size of the downed material. Tree trunks ranging in size from *ca.* 20 cm DBH to >0.5 m DBH attracted flies for maxima of 40–130 d, respectively (Dodson 1997). Regardless of the overall size of the host substrate, there were generally more males than suitable oviposition sites. For example, a decaying *D. gaudichaudianum* log monitored for nine consecutive days had at least 13.7 ± 1.1 (mean $\pm$ SE) *P. mouldsi* males visit per day, but only 4.0 ± 0.5 sites were guarded by males and used by ovipositing females each day. Individual oviposition sites

$(n = 8)$ were attractive to flies for 3.2 ± 0.5 days (Dodson 1997).

Males consistently outnumbered females at the oviposition substrates for all three *Phytalmia* species censused. Males were more numerous in 82% of 154 field censuses, with the proportion of males ranging from 0.62 to 0.85 across the three species (Dodson 1997). The male bias at the resource is apparently not attributable to a skewed sex ratio at eclosion: rearings of *P. mouldsi* (Dodson and Daniels 1988) and *P. alcicornis* (H. Roberts and G. N. Dodson, unpublished) did not differ from unity.

The mean number of days that individual *P. mouldsi* visited the oviposition site during 14 d of censusing was statistically equivalent for males and females, but the pattern of visitation differed between the sexes. Fifty-four percent of 50 marked males visited the resource on consecutive days at least once, whereas only 14% of 34 marked females had consecutive day visits. Coincidentally, 76% of the females returned to the resource after missing at least one day, whereas only 13% of the males did so. Thus, individual males tended to remain for several days and then disappear, but females were rarely present for more than one day without absences between visits (Dodson 1997).

Mating

Observations by Moulds (1977) and GND indicate that three species of *Phytalmia* exhibit similar resource-defense mating systems (Emlen and Oring 1977) with minor variation in male contest behavior. Adult flies arrive at the oviposition substrate throughout the daylight hours and depart before sundown. At the resource, males identify specific sites to which females will be attracted and then attempt to exclude other males from them. When another male moves toward a guarded oviposition site, the resident male orients directly toward the intruder with his wings held at roughly a 45° angle from the body axis and above its dorsum. From this point in an interaction, one of three events occurs: (1) the intruder male retreats without contact between them; (2) the two males push their heads against one another momentarily, remaining mainly parallel with the substrate, after which one of the males retreats; (3) the males bring their heads together and engage in a prolonged (escalated) contest until one male retreats. The proportions of non-contact, minor contact, and escalated contact interactions were similar, i.e. 32.3%, 32.6%, 36.6%, respectively, for 420 interactions between marked *P. mouldsi* males in the field (Dodson 1997).

In *P. mouldsi*, escalated interactions occurred between opponents that were the most similar in size and non-contact interactions occurred between those most disparate (Fig. 18-4). The same relationship between relative sizes of opponents and the level of interaction was obvious in *P. biarmata* and *P. alcicornis* as well (G. N. Dodson, unpublished). However, the particular form of escalated interactions differed among species. In *P. mouldsi* and *P. biarmata*, males rise up on their middle and hind legs until their bodies are perpendicular to the

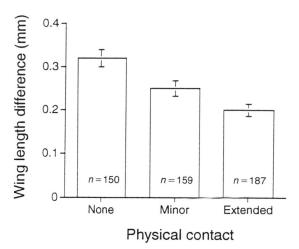

Fig. 18-4. Mean differences in the sizes of *P. mouldsi* male opponents grouped on the basis of the type of interaction (see text for descriptions). Bars indicate standard errors. ANOVA, df = 2,493, F = 16.4, $p < 0.001$.

substrate (Moulds 1977; Dodson 1989). The only point of sustained contact between opponents is the strongly produced epistomal margin on the face (McAlpine and Schneider 1978), although the antlers and forelegs make intermittent contact as well. Individual contests can last several seconds, but eventually one of the males falls backwards. This male either retreats, terminating the bout, or rarely, the same two males engage in a series of these 'stilting' contests. In contrast, the escalated contests of *P. alcicornis* (Fig. 18-5a) involve rapid bursts of pushing with less sustained contact and rarely involve the elevation of the anterior of their bodies high above the substrate. Evenly matched opponents exhibit series of bouts lasting up to 3 min, which, at their greatest intensity, include quick, jerking motions while in contact. Details of these rapid thrusts are difficult to distinguish, but it appears that little contact occurs between any facial surfaces other than the antlers.

By defending an oviposition site a male gains access to potential mates. Sexual conflict as described by Alexander *et al.* (this volume) is apparent at several points within the male–female interactions in all three species studied. The oviposition site itself is a minute point above an area of decaying sapwood, which a guarding male regularly dabs with his mouthparts and anus (Moulds 1977; Dodson 1997). Females attempt to move around males to the

Fig. 18-5. (a) Two *P. alcicornis* males engaged in a contest over an oviposition site in Papua New Guinea. (b) Two large *Cyrtodiopsis whitei* males on a rootlet approaching to compare eye spans at dusk in peninsular Malaysia.

oviposition site but are blocked by the male, who continually positions himself between the female and the site. To gain access to this site for egg-laying, a female must first allow the defending male to copulate. If a female chooses

to mate she demonstrates her receptivity by lowering her head to the substrate, at which point the male mounts her from above and secures the bases of her wings with his forelegs (Moulds 1977). Copulations are relatively brief (1–2 min) (Moulds 1977; Dodson 1997), after which the male remains astride the female as she inserts her ovipositor and begins depositing eggs. From this postcopulatory position males sometimes interact with other males who attempt to displace them from the female. Females terminate their association with a male by vigorously shaking until he dismounts (Dodson 1997).

On a spectrum of copulatory acts from luring or enticing to manipulative and coercive (Alexander *et al.*, this volume), *Phytalmia* appear to fall closer to a luring act. Males entice females to mate by controlling access to a required resource. Although females can choose whether or not to mate with a particular male, expediency to oviposit may dictate that they sometimes practice 'convenience polyandry' (Thornhill and Alcock 1983). Even so, the ability of females to exercise mate choice by storing sperm and multiply mating is expected to lead to selection on males to regain some control over fertilization (Alexander *et al.*, this volume). For *Phytalmia* this could account for the evolution of the femoral holding devices, postcopulatory guarding behavior, and the requirement that a female copulate before each oviposition bout, even when she has been only temporarily separated from a guarding male with whom she has already mated (Dodson 1997).

It should also be considered, however, that postcopulatory guarding by males might not be solely an advantage for males. Suitable, unguarded oviposition sites are rare, because there are generally excess males patrolling the substrate. Females will attempt to deposit eggs at a guarded site if a resident male is temporarily absent, but are displaced as soon as a male returns. Oviposition bouts by solitary *P. mouldsi* females were significantly shorter than bouts by guarded females (Dodson 1997). Thus, postcopulatory guarding may represent a confluence of interests to the sexes in these species.

Because copulations occur at oviposition sites monopolized by males, successful defense of a site should translate into mating success. Given that larger males win contests over oviposition sites (84% of 420 *P. mouldsi* male–male field interactions, based on wing length), mating males are expected to be larger than the overall male average. This difference was found in both *P. mouldsi* and *P. alcicornis*. Males collected while paired with a female were larger than males collected as singletons (with size estimated as the first principal component derived from four metric characters; Mann–Whitney tests: $U = 419$, $p = 0.04$, $n = 19$ paired males, $n = 33$ singletons for *P. mouldsi*; $U = 496$, $p = 0.01$, $n = 17$ paired males, $n = 41$ singletons for *P. alcicornis*).

Antler size is highly correlated with each of four metric body characters (wing, thorax, fore femur and mid femur lengths) in both *P. mouldsi* and *P. alcicornis* (Pearson $r = 0.88$ or higher for each correlation). A plot of antler size on body size reveals positive allometry for antler size in both species (Fig. 18-3). Thus, antler size is an honest indicator of overall size, which itself is a predictor of fighting success. Antler size could be used by males to assess an opponent's fighting ability, thereby avoiding unnecessary contests. At the largest body sizes, antler size appears to level off. Huxley (1932, pp. 60–61) suggested that such a diminishing relationship would be expected in a 'closed' system in which development of adult structures is dependent upon finite food reserves acquired during a larval stage. The theoretically largest heterogonic organs would fall below the sizes expected as the 'limited reserve supply would come to an end, used up by competing organs' (Huxley 1932). Whatever its cause, we suggest that this relationship would have little effect on the value of antler size as a predictor of body size for the majority of opponents because only a small proportion fall at the upper end of the distribution. Furthermore, when two of the largest males interact, an escalated fight is expected *a priori*.

Unlike the eye stalks discussed in detail below, the antlers of *Phytalmia*, and perhaps all cheek projections, are amenable to experimental investigations of function. Antlers can be excised without debilitating effects; one of us (GND) has performed preliminary experiments on the role of these structures in male–male contests in *P. mouldsi* and *P. alcicornis*. Interactions were staged between control males and treatment males whose antler size was either increased or decreased. In *P. mouldsi*, when opponents were equal in body size (as indicated by wing length), lengthening the antlers resulted in a significant increase in fighting success and shortening the antlers significantly lowered fighting success. In experiments in which body sizes differed, lengthening the antlers of small males did not increase their low success rate. Shortening the antlers of large males significantly lowered their success, but these de-antlered males still won a high percentage of fights (75–78%). In *P. alcicornis*, experiments involved differently sized males only. As in *P. mouldsi*, large males won a high percentage of their fights (80%) before the manipu-

lation and this remained unchanged after antler removal. In this species the fighting success of small males increased significantly when their antlers were lengthened. Across all experiments the proportion of interactions involving contact rose significantly for males whose antlers were shortened and did not change for males whose antlers were lengthened.

Thus, it appears that antlers are not required in order to win contests: the outcome is determined primarily by size. However, males without antlers expended more effort in escalated contests before the outcome was determined.

Sensory hairs are located on the antlers of all *Phytalmia* species and on the cheek projections of several other species, sometimes only at the apex of the projection (G. N. Dodson, personal observation). Therefore, although these manipulations may partly test for a role for antlers in visual assessment, any tactile function is probably eliminated.

DIOPSIDS

Classification and distribution

The classification of diopsids is under revision, and new species await description. In 1972, 150 species in 13 genera, including normal-headed flies in the genus *Centrioncus*, were recognized in the family (Steyskal 1972). Feijen (1983, 1989) subsequently created a new family, Centrioncidae, synonymized two genera, and created two new genera. The number of species in the Diopsidae may eventually reach 300 with 50 or more in each of three genera: *Diopsis*, *Diasemopsis* and *Teleopsis* (Feijen 1989; H. R. Feijen, personal communication).

Although most diopsids are found in the old world tropics, the genus *Sphyracephala* (Feijen 1989) is an exception, with species in eastern North America, Russia, India, Africa, and southeast Asia. Several genera – *Diopsis*, *Diasemopsis*, *Diopsina*, *Trichodiopsis*, *Chaetodiopsis*, *Cladodiopsis*, and *Cobiopsis* – are restricted to Africa with the exception of one *Diopsis* species in southeast Asia and three in the Arabian peninsula (Feijen 1989). *Cyrtodiopsis* and *Teleopsis* are limited to southeast Asia; *Eurydiopsis* is found from India to New Guinea and *Sinodiopsis* occurs in China (Feijen 1989). Fossil diopsids 38–50 million years old have been found in Baltic amber or in shale in Montana and France and placed in the genus *Prosphyracephala* (Feijen 1989) because they resemble extant species of *Sphyracephala*.

MORPHOLOGY AND
SEXUAL DIMORPHISM

Diopsids are unique in having antennae at the end of eye stalks, instead of in the facial region. Most species are 5–10 mm in body length, and often resemble ants because of their terrestrial habits (Feijen 1989; Peterson 1987). Although noted for their hypercephaly since Linnaeus' description (Linnaeus 1775), sexual dimorphism in eye span apparently was not detected until Frey (1928)

described differences in eye stalk length between the sexes of three *Teleopsis* species. Subsequently, Descamps (1957), Shillito (1971) and Burkhardt and de la Motte (1983, 1985) pointed out that males of some *Diopsis*, *Cyrtodiopsis* and *Teleopsis* species have longer eye stalks than females. Feijen (1989), however, noted that females have longer eye stalks than males in some *Sphyracephala* species.

Comparison of linear allometric relationships between eye span and body length (Fig. 18-6) reveals that the apparent reversed sexual dimorphism for eye span in

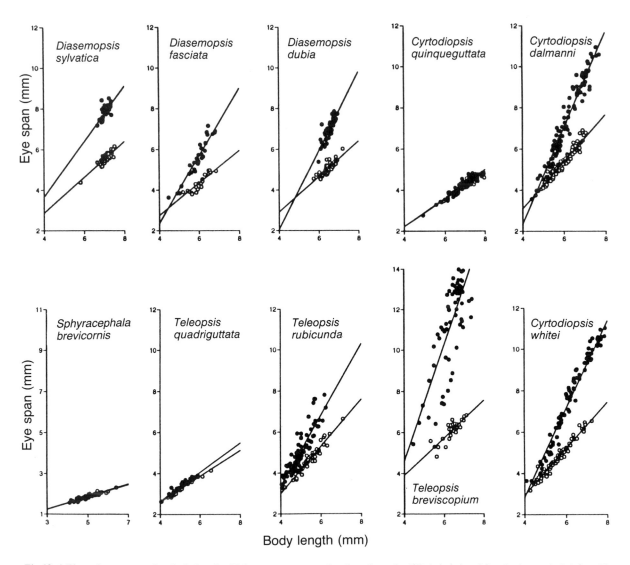

Fig. 18-6. Plots of eye span against body length with least-square regression lines for males (filled circles) and females (open circles) from 10 species of diopsids. All *Cyrtodiopsis* and *Teleopsis* species were collected in peninsular Malaysia, *Diasemopsis* in Africa, and *Sphyracephala* in Maryland.

S. brevicornis is due to females having larger bodies than males. In *S. brevicornis* both eye span ($t = 2.8$, $p = 0.007$) and body length ($t = 3.9$, $p = 0.0002$) are greater for females (Table 18-2). However, neither the slopes nor the elevations of the eye span on body length regressions differ between the sexes.

Sexual dimorphism in the eye span of diopsids is not positively correlated with body size, in contrast to enlarged mandibles among lucanid beetles (Otte and Stayman 1979). The eye span on body length regression slopes (Table 18-2) among males for each of three *Cyrtodiopsis*, *Teleopsis* and *Diasemopsis* species differ by ANCOVA (*Cyrtodiopsis*: $F = 137.9$, df $= 2,248$, $p < 0.0001$; *Teleopsis*: $F = 18.0$, df $= 2,168$, $p < 0.0001$; *Diasemopsis*: $F = 3.38$, df $= 2,111$, $p = 0.037$). In contrast, body length of males from the sexually dimorphic species does not consistently exceed that of males from sexually monomorphic species within genera, indicating that evolutionary change has occurred in male allometric relationships. This conclusion is supported by breeding studies on *C. whitei* (G. S. Wilkinson and P. R. Reillo, unpublished) and artificial selection

on the ratio of eye span to body length in *C. dalmanni* (Wilkinson 1993) demonstrating heritable variation for eye span on body length regression slope.

Sexual dimorphism in eye span to body length allometry must have evolved multiple times in the Diopsidae: at least once in *Cyrtodiopsis*, *Teleopsis* and *Diasemopsis* (Fig. 18-6) and probably in other genera. To reach this conclusion, we assume that eye-span monomorphism, as in *S. brevicornis*, *T. quadriguttata* and *C. quinqueguttata*, is plesiomorphic in the family. This assumption is supported by phylogenies derived from morphological characters (Feijen 1989; Steyskal 1972) and mitochondrial DNA sequences (R. Baker and G. S. Wilkinson, unpublished) both of which place the genus *Sphyracephala* basal to the remaining diopsids. Thus, some selective force must have favored a steeper allometric relationship between eye span and body length in males of the dimorphic species. Dimorphism could then arise if the genetic correlation between the sexes for this character was less than one, or if stabilizing selection opposed a correlated response in females, or both.

Table 18-2. *Body measurements for 10 species of diopsids*

Parentheses enclose one SE. Regression slope indicates the least-squares regression of eye span on body length. Genera as in Fig. 18-6.

Sex	Species	n	Eyespan	Body length	Thorax width	Regression slope
Male	C. dalmanni	93	749 (0.18)	623 (0.08)	167 (0.02)	231 (0.07)
	C. whitei	86	797 (0.23)	637 (0.11)	147 (0.03)	213 (0.05)
	C. quinqueguttata	102	425 (0.04)	697 (0.05)	190 (0.02)	069 (0.03)
	T. breviscopium	60	10.98 (0.28)	632 (0.08)	165 (0.03)	273 (0.30)
	T. rubicunda	89	497 (0.11)	492 (0.06)	130 (0.02)	173 (0.08)
	T. quadriguttata	29	333 (0.07)	498 (0.10)	139 (0.03)	071 (0.04)
	D. obstans	52	700 (0.07)	656 (0.03)	185 (0.01)	193 (0.16)
	D. fasciata	28	566 (0.17)	599 (0.10)	149 (0.03)	166 (0.11)
	D. sylvatica	37	790 (0.05)	709 (0.03)	199 (0.01)	116 (0.22)
	S. brevicornis	32	183 (0.03)	487 (0.08)	142 (0.03)	030 (0.02)
Female	C. dalmanni	91	521 (0.08)	585 (0.06)	160 (0.02)	114 (0.03)
	C. whitei	86	467 (0.08)	555 (0.07)	134 (0.02)	112 (0.02)
	C. quinqueguttata	66	428 (0.05)	715 (0.06)	199 (0.02)	067 (0.03)
	T. breviscopium	30	609 (0.09)	638 (0.08)	174 (0.04)	095 (0.11)
	T. rubicunda	52	465 (0.09)	544 (0.07)	148 (0.02)	115 (0.05)
	T. quadriguttata	18	337 (0.09)	521 (0.13)	143 (0.05)	063 (0.04)
	D. obstans	59	509 (0.04)	650 (0.04)	186 (0.01)	087 (0.07)
	D. fasciata	25	437 (0.08)	602 (0.09)	153 (0.03)	080 (0.09)
	D. sylvatica	62	555 (0.03)	705 (0.03)	198 (0.01)	088 (0.07)
	S. brevicornis	33	193 (0.02)	528 (0.07)	157 (0.03)	031 (0.02)

Female *C. dalmanni* exhibit a correlated response to artificial selection on male relative eye span (Wilkinson 1993). Furthermore, female *C. dalmanni* and *C. whitei* have steeper allometric relationships than female *C. quinqueguttata* (ANCOVA, $F = 37.6$, df $= 2,227$, $p < 0.0001$) and female *T. breviscopium* and *T. rubicunda* have steeper allometric relationships than female *T. quadriguttata* (ANCOVA, $F = 15.2$, df $= 2,93$, $p < 0.0001$) (Table 18-2). The differences among females in these two genera are consistent with correlated evolution of female allometric relationships in response to selection on males. By contrast, females from all three *Diasemopsis* species exhibit allometric relationships with the same slope (ANCOVA, $F = 0.38$, df $= 2,140$, $p = 0.69$, Table 18-2). Allometric data for additional *Diasemopsis* species, and a species-level phylogeny for these genera, are needed to determine whether correlated eye-span evolution is absent among females in this genus.

Life history

Despite quantitative differences in life-history traits, such as fecundity, age of sexual maturity and longevity, among species (Descamps 1957; G. S. Wilkinson, unpublished), general life-history patterns are similar among diopsids. Females lay eggs singly on either a plant stem or decaying vegetation. *C. dalmanni* females lay, on average, between 0.1 and 4 eggs per day and may continue laying eggs for many months (G. S. Wilkinson, unpublished). Eggs hatch after 1–3 d (Descamps 1957) and the larvae take from 10 d to several weeks (depending on species, temperature and food quality) to grow and pupate (Descamps 1957). Late-instar larvae typically pupate on emergent vegetation and eclose one to two weeks later. As larval food deteriorates, pupation occurs at smaller larval sizes. In dimorphic species small males exhibit eye span indistinguishable from that of small females (Fig. 18-6). Eye span is, therefore, an honest indicator of body size although at small body sizes it does not reliably signal sex.

Adult diopsids are capable of long lifespans. North American *Sphyracephala* overwinter as adults (Flint 1956; Lavigne 1962) and live 12 months or more in the laboratory (Hochberg Stasny 1985). Individual *C. dalmanni*, *C. whitei*, *C. quinqueguttata*, *T. quadriguttata*, *Diasemopsis dubia*, *D. sylvatica*, *Diopsis fumipennis*, *D. apicalis*, *E. subnotata* and *S. brevicornis* require at least two weeks after eclosion before attaining reproductive maturity (de la Motte and Burkhardt 1983; G. S. Wilkinson, unpublished) and have been kept alive for 6 months or more (G. S. Wilkinson, unpublished).

This longevity is not just due to laboratory rearing. The average adult ages of field-caught *C. dalmanni* (Wilkinson and Reillo 1994) and *C. whitei* (G. S. Wilkinson and P. R. Reillo, unpublished), estimated from pteridine eye pigments, are over 30 days with some large individuals of both sexes surviving for over 200 days.

Resource distribution and visitation

Although the larvae of some *Diopsis* species bore into monocotyledonous stems, most diopsid larvae are saprophagous and feed on decaying vegetation (de la Motte and Burkhardt 1983; Deeming 1982; Descamps 1957; Feijen 1979; 1989; Hochberg Stasny 1985; Tan 1967). Available evidence indicates that oviposition sites for most, if not all, diopsid species are widespread.

Adult diopsids ingest fungi, mold and yeast from decaying leaf litter or dead animals (de la Motte and Burkhardt 1983; Descamps 1957; Feijen 1989; Tan 1967; Wickler and Seibt 1972). During the day, individuals of most species walk on the ground or on low vegetation near streams in secondary or primary forest, although the stem-boring diopsids can be found in cultivated areas (Feijen 1989). Observations of *C. whitei*, *C. dalmanni* and *T. breviscopium* in Malaysia and of *Diasemopsis fasciata* in Kenya indicate that flies from these species typically act aggressively toward conspecifics of both sexes during the day (Lorch *et al.* 1993; G. S. Wilkinson, unpublished).

At night, adults of most dimorphic species form groups in protected sites. In Malaysia, Burkhardt and de la Motte (1985) found four out of five dimorphic species but none of four monomorphic species in nocturnal clusters. These aggregations occur on rootlets that hang underneath banks along small streams (Fig. 18-7a). In Malaysia, female *C. whitei* (Burkhardt and de la Motte 1987) and *C. dalmanni* (Wilkinson and Reillo 1994) exhibit clumped distributions on rootlets whereas males are overdispersed. In Kenya, *Diasemopsis fasciata* gather on the surface of leaves at night (Fig. 18-7b). Distributions of each sex were scored for 373 flies on 177 leaves from photographs taken at night (G. S. Wilkinson, unpublished). The number of males was not independent of the number of females ($\chi^2 = 26.5$, df $= 8$, $p = 0.001$). The ratio of variance to mean number of flies per leaf was 1.20 for females (Poisson goodness-of-fit $\chi^2 = 4.5$, df $= 4$, $p > 0.5$) and 0.79 for males ($\chi^2 = 53.7$, df $= 4$, $p < 0.001$) indicating that *D. fasciata* males were overdispersed whereas females were distributed randomly among leaves.

Monomorphic *C. quinqueguttata* do not occur along streams but are dispersed during the day on low vegetation in lowland forest (G. S. Wilkinson, unpublished). These flies do not aggregate at night in the laboratory; mating typically occurs during the day (Kotrba 1996). Nocturnal behavior has not been described for any *Diopsis* species although mating of *D. longicornis* has been observed on rice plants (Alghali 1984). More quantitative data on spatial dispersion patterns are needed for additional species to determine whether female aggregration tendency correlates with degree of sexual dimorphism.

Nocturnal aggregation sites persist for many months. Wilkinson and Reillo (1994 and unpublished) found significant positive autocorrelations in the number of *C. dalmanni* and *C. whitei* flies roosting on individually marked rootlets over a 10 month period. Only a small fraction of available rootlets were occupied, with the same ones often attracting the largest aggregations every night. Similar stable roosting preferences have been observed for *D. fasciata* in Kenya (G. S. Wilkinson, unpublished). As Fig. 18-7 indicates, flies on rootlets or leaves orient toward the end of the substrate attached to the ground. Because terrestrial predators must approach from that direction, aggregations may serve an antipredator function in addition to protecting the flies from rain.

The female bias observed in nocturnal aggregations of *C. whitei* and *C. dalmanni* is partly a consequence of a biassed adult sex ratio (de la Motte and Burkhardt 1983). This bias is caused by a sex-linked, sex-ratio-distorting gene, *sr* (G. S. Wilkinson, E. Severance and D. Presgraves, unpublished), that behaves much like sex-distorting genes in *Drosophila* (James and Jaenike 1990). Some *sr* males father only daughters. *Sr* appears to be maintained in natural populations at relatively high frequencies owing to the presence of autosomal modifiers (G. S. Wilkinson, E. Severance and D. Presgraves, unpublished).

Fig. 18-7 (a) Nocturnal aggregation of *C. whitei* on a rootlet from peninsular Malaysia. Large male has an eye span of 10 mm.
(b) (see p. 322) Nocturnal aggregation of *Diasemopsis fasciata* from the Taga National Forest, Kenya.

Mating

Obvious courtship signals do not occur in diopsids. In most species, a male approaches a female, leaps onto her back, and attempts to copulate. Species recognition is probably effected by diagnostic wing patterns, as in *Cyrtodiopsis* and *Teleopsis* species (Shillito 1971) or abdomen patterns, as in *Diasemopsis* species (Descamps 1957). Although diopsid mating resembles a coercive act as described by Alexander *et al.* (this volume), females from at least some dimorphic species can control fertilization by resisting male mating attempts (Kotrba 1996), selecting aggregation sites for mating, or influencing which sperm are stored.

In highly-dimorphic species, such as *C. whitei* and *C. dalmanni*, males fight for control of aggregation sites (Burkhardt and de la Motte 1983, 1987; Lorch *et al.* 1993). Males with similar eye span approach head-to-head (Fig. 18-4b), and if similar in size, they rise upon their mid and hind legs, spread their forelegs alongside their eye stalks, and in a series of very rapid lunging moves attempt

Fig. 18-7 (b) For legend see previous page.

to displace each other from the rootlet. In *C. whitei* 91% of all fights are won by the male with the longest eye span (Burkhardt and de la Motte 1987). In the lab, the duration of a contest between two male *C. whitei*, as measured by the amount of time spent orienting toward each other, was inversely proportional to the difference in their eye spans (G. S. Wilkinson, unpublished), as expected if these animals use sequential assessment to decide contests (Enquist and Leimar 1983).

Large male *C. whitei* can mate with over 20 females in an aggregation within 30 min after dawn; most females mate at least once a day (Burkhardt *et al.* 1994; Lorch *et al.* 1993). Small males that resemble females sometimes successfully mate in aggregations after avoiding detection by large males (Burkhardt and de la Motte 1988; Lorch *et al.* 1993). However, the average lifetime reproductive success of female

mimics is less than that of large males because small males rarely live longer than 30 d in the field (G. S. Wilkinson and P. R. Reillo, unpublished). Thus, mate selection approximates a best-of-*n* process in at least *C. whitei* and *C. dalmanni* and results in extreme polygyny (Alexander *et al.*, this volume).

Experiments with dummy male *C. whitei* and artificially selected male *C. dalmanni* have demonstrated that females can choose mates on the basis of eye-span comparison. By extending the eye span of dead males and then glueing those males on strings, Burkhardt and de la Motte (1988) found that female *C. whitei* prefer to alight in the evening on strings containing males with the longest eyespan. Moreover, Wilkinson and Reillo (1994) showed that female *C. dalmanni*, given a choice between two males separated by a perforated partition, preferred to roost with males from lines selected for increased eye span. Females from lines in which males had been selected for decreased eye span preferred males with a shorter eye span whereas females from lines selected for increased male eye span preferred males with a long eye span. Because artificial selection was exerted only on males, these results indicate that female preference is genetically correlated with male eye span: a result consistent with both arbitrary cue (Lande 1981) and good genes (Iwasa *et al.* 1991) models of sexual selection.

Female *C. whitei*, and possibly other female diopsids, can control fertilization after copulation because most diopsid males transfer sperm in a spermatophore (Kotrba 1996). Each male *C. whitei* transfers 90 sperm, on average; females hold up to 400 sperm in three spermathecae (Lorch *et al.* 1993). Sperm precedence depends on when females remate and eject spermatophores. Spermatophore ejection occurs between 5 and 60 min after mating (Kotrba 1990, 1993), sometimes before sperm is emptied from the spermatophore. First-male sperm precedence occurs when a female remates before ejecting the first male's spermatophore, whereas sperm-mixing occurs when the second male mates after the first male's spermatophore is ejected (Lorch *et al.* 1993). Paternity analysis of offspring from *C. whitei* females each paired with two males of differing eye span provides evidence consistent with postcopulatory mate choice. Males with twice the eye span of an opponent sired an average of 88% of the young even though both males mated equally often (Burkhardt *et al.* 1994). Whether this result is due to differences in sperm transfer, number, motility or utilization deserves further study.

Although females may be able to influence which sperm are stored, males control copulation duration. In *C. whitei*, sexual conflict over fertilization control is indicated by the presence of short copulations, which do not transfer sperm, shortly after a female has mated (Lorch *et al.* 1993). Males presumably abort copulations after detecting a previous male's spermatophore to reduce copulation time and avoid sperm loss. Copulation duration also varies widely across diopsid species, from 50–60 s in *C. whitei* and *C. dalmanni* (Lorch *et al.* 1993; Wilkinson and Reillo 1994) to over 40 min in *T. quadriguttata*, and correlates positively with spermatophore area (Kotrba 1996). Although some dimorphic species, such as *T. breviscopium*, have long copulation durations, within a genus dimorphic species have smaller spermatophores than monomorphic species, suggesting that selection for rapid male remating in aggregations has led to reduced spermatophore size and

copulation duration (Kotrba 1996). An exception to this pattern occurs in the genus *Sphyracephala*, where postcopulatory mate guarding occurs for up to 1 h (Kotrba 1996). These flies have short eye spans and mate when overwintering aggregations emerge in the spring (Hochberg Stasny 1985).

The consequences of these mating and sperm-precedence patterns for *C. whitei*, *C. dalmanni*, and probably for other dimorphic diopsids, are that strong sexual selection is exerted on males with long eye span. Because body length correlates highly with eye span in all diopsids (Fig. 18-6), selection on eye span is difficult to separate from selection on body length (Mitchell-Olds and Shaw 1987). Nevertheless, the selection intensity on the ratio of eyespan to body length among *C. dalmanni* males, as estimated from collections of nocturnal aggregations, exceeds half of a standard deviation (Wilkinson and Reillo 1994), and as noted above, dimorphic species do not have larger body sizes than related monomorphic species (Fig. 18-6). By contrast, no sexual selection for eye span or body length was found for two monomorphic species, *T. quadriguttata* or *C. quinqueguttata*, during mate-choice experiments conducted in the lab (G. S. Wilkinson and H. Kahler, unpublished).

DISCUSSION

Sex limitation

Comparisons of morphology and mating systems between species of *Phytalmia* and Diopsidae reveal interesting differences and similarities that shed light on the evolution of head projections. In the *Phytalmia* and all other known species with cheek processes, the structures occur only in males. By contrast, not only do both sexes of all species of diopsids exhibit eye stalks, but females of some sexually dimorphic species, e.g. *Cyrtodiopsis* and *Teleopsis*, have significantly higher eye span on body length regression slopes than monomorphic species, presumably owing to a correlated response to selection on male eye span. These differences in sex limitation of sexually selected traits may reflect a difference in the degree to which antlers and eye stalks are developmentally coupled, i.e. genetically correlated, between the sexes.

One interpretation of sex-limited head projections is that the genes that influence the growth of cells that give rise to antlers or eye stalks are located on the male-determining sex chromosome. Crosses between *Drosophila sylvestris* and *D. heteroneura* indicate, for example, that hypercephaly in *D. heteroneura* is influenced by a major sex-linked gene (Templeton 1977). The often extreme sexual dimorphism in eye stalks found in tephritid, richardiid (Grimaldi and Fenster 1989) and platystomatid (McAlpine 1979, 1994) flies might be the result, therefore, of a genetic organization shared with antler-bearing species for developmental modification of the head capsule. In contrast, crosses between selected lines of *C. dalmanni* demonstrate that genes influencing eye span exhibit no detectable sex linkage and that eye span is developmentally coupled between the sexes (Wilkinson 1993). Autosomal influences on eye-stalk expression apparently occur in other species of *Drosophila* with broad heads or eye stalks as well because females often exhibit noticeably wider heads and steeper head width to thorax length allometric regressions than monomorphic relatives from the same species group (Grimaldi and Fenster 1989). Thus, genetic constraints appear to influence the degree to which head projections are sex-limited, but cannot explain the occurrence of eye stalks or antlers because related species lack these structures.

Allometry and assessment

Antler expression in *Phytalmia* and eye-stalk development in sexually dimorphic diopsids are similar in that both traits exhibit positive allometry with body size in males (Figs. 18-3, 18-6). When the benefit to developing a trait for male–male competion or mate assessment is relatively greater for a large than for a small animal, positive allometry should be favored by sexual selection (Green 1992; Otte and Stayman 1979; Petrie 1988, 1992). We propose an additional explanation for positive allometry in flies with head projections. When projection span and body size are measured with the same degree of error and the allometric slope exceeds one, it follows that a unit change in projection span will be less than a unit change in body size, i.e. the projection-span scale will be finer than the body-length scale. Thus, the high correlation between head projection and body size, together with positive allometry, enables flies to use either antler or eye span to assess body size with greater accuracy than would be possible if body size was measured directly. Furthermore, because both antlers and eye stalks are located on the fly's head, two flies facing each other can see the extent of their opponent's process without inspecting from the side, as would be necessary to obtain a comparable perspective on body

length. By placing their eyes in close proximity during face-to-face frontal displays, stalk-eyed flies directly compare eye span (Burkhardt and de la Motte 1983; McAlpine 1979).

Frontal displays between males have been reported for all stalk-eyed flies (Bristowe 1925; Burkhardt and de la Motte 1983; Grimaldi 1987; McAlpine 1979; Spieth 1981) and antler-bearing flies (Dodson 1989; McAlpine 1975; Moulds 1977) that have been studied, strongly implicating the association of head projections with selection to assess rival body size. An alternative, but not mutually exclusive, hypothesis, that head projections function as weapons or for transmitting force in pushing contests, can be ruled out by observation in stalk-eyed flies and is inconsistent with antler-manipulation experiments in *Phytalmia*. Alexander *et al.* (this volume) suggested that extreme traits such as massive jaws and horns in insects 'are not 'ornaments' in the sense of many sexually selected traits in birds' and probably function as intraspecific weapons. Although we agree that these traits differ from bird ornaments that can vary within a lifetime, we argue that eye stalks and antlers function as ornaments to the extent that they serve as costly signals of competitive ability.

The use of head projections in male–male contests does not preclude the possibility that females may also use eye or antler span to assess males as potential mates. Female choice of head projections could be favored by direct selection (Kirkpatrick and Ryan 1991) if projection length correlated with a trait that influenced female fecundity, such as spermatophore size, or by indirect selection if head-projection length indicated either viability or sexual attractiveness of male progeny. Female choice based on eye-stalk length has been demonstrated for *C. whitei* (Burkhardt and de la Motte 1988) and for *C. dalmanni* (Wilkinson and Reillo 1994). Whether female choice in these stalk-eyed flies provides direct fitness benefits to females or genetic benefits to their progeny has not yet been determined. In order to exert choice, females must be able to visit and compare more than one male. The opportunity for female choice in any species, therefore, depends on the mating system.

Mating system

The evidence summarized above suggests that *Phytalmia* and sexually dimorphic diopsid flies have resource-defense mating systems (Emlen and Oring 1977). *Phytalmia* males defend detectable oviposition sites that are temporally and spatially clumped in the environment. Females visit fallen logs of a few tree species containing one or more potential oviposition sites and a number of males. Although females may have little choice of mates once they select an oviposition site, they clearly have an opportunity to inspect males at more than one oviposition site. In contrast, diopsids appear to have much less specific larval food preferences. Consequently, oviposition sites are dispersed and indefensible. Sexually dimorphic diopsids, however, form nocturnal aggregations in predictable locations where mating frequently occurs. Although an aggregation may be controlled by a single male, other males often roost within sight, allowing females to assess the eye span of more than one male prior to roosting. Female choice, therefore, is possible in *Phytalmia* and dimorphic diopsids.

The mating systems of other flies with head projections also appear to involve male defense of a resource or a mating site. The clumped distribution of male *D. heteroneura* display sites has been described as a lek (Spieth 1981) because males do not defend a resource and females are presumably free to choose mates (Bradbury 1985; Shelly and Whittier, this volume). Recent studies indicate that male head width, independent of body size, influences male mating success in *D. heteroneura* (C. Boake, personal communication). Male Australian stalk-eyed flies, *Achias australis* and *A. kurandanus*, form aggregations on trunks of trees (McAlpine 1979), but whether males are defending oviposition sites or courtship territories remains to be determined. Grimaldi (1987) found that up to 21 species of *Zygothrica*, many of which have broad-headed males, feed and mate on fungi (Polyporaceae). Male *Z. dispar* have been reported to butt heads with one another on fungi (Bristowe 1925; Burla 1990). Oviposition and larval development of many *Zygothrica* species occurs in flowers away from aggregations on fungi (Grimaldi 1987). Because suitable fungi are also clumped in the environment, males of these broad-headed drosophilids appear to defend feeding sites to gain access to mates.

Suggestions for future research

This review illustrates that flies with head projections exhibit resource-defense mating systems in which males attempt to mate with females at sites where feeding, ovipositing, or nocturnal resting occurs. In all cases, males compete with each other for control of these sites. Grimaldi and Fenster (1989) cited territoriality as a precondition for male hypercephaly in drosophilids. Of course, some related species compete for control of sites attractive to females,

yet have no head projections (for example, males of a temperate tephritid (*Rhagoletis completa*) defend oviposition sites on walnuts (Boyce 1934)). Another conspicuous attribute of all flies with sexually dimorphic head projections noted by McAlpine (1982) is that they live in the tropics. Perhaps all species with head projections are tropical simply because of the greater number of species in the tropics.

We suggest, however, that tropical fly species may live longer, on average, and thereby have more mating opportunities during their lifetimes than flies in temperate regions. An increase in the average number of mating opportunities in flies with resource-defense mating systems can have two consequences on male fitness. With each successive contest the degree to which each male contest improves male relative fecundity will diminish and the cost to male survival (owing to energy expenditure or injury risk) will increase. Long-lived male flies should, therefore, benefit from the evolution of traits, such as head projections, that improve assessment of competitive ability and decrease costs associated with contest settlement. In contrast, short-lived male flies should continue to fight with rivals independent of size assessment because few alternative mating possibilities exist.

Our hypothesis leads to at least four predictions that could be tested in other flies with head projections. (1) Head projection expression should exhibit positive allometry with body size. (2) Head projections should be displayed, directly compared, and used to determine the outcome of male contests over a site where mating occurs. (3) In comparison with close relatives or experimentally manipulated animals, flies without head projections should take longer to settle territorial disputes than flies with head projections that differ to the same degree in body size. (4) Flies with head projections should engage in more contests over their lifetimes than temperate relatives without head projections. Although it would be difficult to estimate the number of contests a fly has in a lifetime, it is possible to estimate age and the rate at which contests occur. Our observations on diopsids indicate that large males can participate in multiple contests each night for several hundred consecutive nights. As described above, *Phytalmia* males are known to defend territories for multiple hours over consecutive days, interacting with intruders at rates that periodically exceed $1\,min^{-1}$. Any honest signal that minimized the time spent in each contest should be favored by selection.

Rather than list alternative explanations for eye-stalk function, we recommend previous reviews (de la Motte and Burkhardt 1983; Feijen 1989; Grimaldi and Fenster 1989; Wickler and Seibt 1972) and acknowledge that our hypothesis does not explain cases of monomorphic expression of head projections. However, none of the alternative explanations does any better. Any general advantage to having eye stalks, such as improving the visual field to detect predators, fails to explain why eye stalks are not as long in monomorphic species as in males of dimorphic species. The evidence in support of sexual selection through assessment of male size as perceived by either rival males or receptive females seems sufficiently strong, in our opinion, to predict that all cases of head projections in flies will involve male size assessment. Whether our more specific predictions relating to the number of mating contests are supported awaits further study of these fascinating animals.

ACKNOWLEDGEMENTS

We thank D. Grimaldi, B. Crespi, J. Choe, M. Kotrba and an anonymous reviewer for comments on the manuscript and S. Alcorn for drawing Fig. 18-1. E. Van der Wolf, P. Reillo, F. Kelley, Y. Hoi-Sen, S. Steele and M. Kotrba assisted in collecting diopsids in Kenya, Malaysia and South Africa. R. Baker, T. Phan and P. Reillo generously shared unpublished data with us. Work on diopsids has been supported by the National Science Foundation. J. Jereb, M. Schneider, G. Daniels, T. Clarke and R. Pope were of invaluable assistance during fieldwork in Australia. None of the Papua New Guinea fieldwork would have been possible without the efforts of H. Roberts and the Papua New Guinea Institute for Forest Research in Lae. R. Chandler and B. Crespi provided statistical help. D. Yeates assisted with taxonomic information. Funding for antlered fly research has been provided by the University of Queensland, Department of Entomology, with the aid of H. Paterson; Television New Zealand with the aid of R. Morris; Ball State University; and the Indiana Academy of Science.

LITERATURE CITED

Alghali, A. M. 1984. Mating and ovipositional behavior of the stalk-eyed fly *Diopsis macrophthalma* on rice. *Entomol. Exp. Appl.* **36**: 151–157.

Boyce, A. M. 1934. Bionomics of the walnut husk fly, *Rhagoletis completa*. *Hilgardia* **8**: 363–579.

Bradbury, J. W. 1985. Contrasts between insects and vertebrates in the evolution of male display, female choice, and lek mating. In *Experimental Behavioral Ecology and Sociobiology*. B. Hölldobler and M. Lindauer, eds., pp. 273–289. Sunderland: Sinauer.

Bristowe, W. S. 1925. Notes on the habits of insects and spiders in Brazil. *Trans. R. Entomol. Soc. Lond.* **1924**: 475–504.

Burkhardt, D. and I. de la Motte. 1983. How stalk-eyed flies eye stalk-eyed flies: observations and measurements of the eyes of *Cyrtodiopsis whitei* (Diopsidae, Diptera). *J. Comp. Physiol.* **151**: 407–421.

–. 1985. Selective pressures, variability, and sexual dimorphism in stalk-eyed flies (Diopsidae). *Naturwissenschaften* **72**: 204–206.

–. 1987. Physiological, behavioural, and morphometric data elucidate the evolutive significance of stalked eyes in Diopsidae (Diptera). *Entomol. Gen.* **12**: 221–233.

–. 1988. Big 'antlers' are favoured: female choice in stalk-eyed flies (Diptera, Insecta), field collected harems and laboratory experiments. *J. Comp. Physiol.* A **162**: 649–652.

Burkhardt, D., I. de la Motte and K. Lunau. 1994. Signalling fitness: larger males sire more offspring. Studies of the stalk-eyed fly *Cyrtodiopsis whitei* (Diopsidae, Diptera). *J. Comp. Physiol.* A **174**: 61–64.

Burla, H. 1990. Lek behavior in hypercephalic *Zygothrica dispar* Wiedemann (Diptera, Drosophilidae). *Z. Zool. Syst. Evolutionsforsch.* **28**: 69–77.

de la Motte, I. and D. Burkhardt. 1983. Portrait of an Asian stalk-eyed fly. *Naturwissenschaften* **70**: 451–461.

Deeming, J. C. 1982. Hostplant records for some Nigerian *Diopsis* species (Diptera, Diopsidae). *Entomol. Mon. Mag.* **118**: 212.

DeSalle, R. and D. Grimaldi. 1993. Phylogenetic pattern and developmental process in the Drosophilidae. *Syst. Biol.* **42**: 458–475.

Descamps, M. 1957. Recherches morphologiques et biologiques sur les Diopsidae du Nord-Cameroun. *Minist. de la France d'Outre Mer, Dir. Elev. For., Sect. Tech. Agric. Trop., Bull. Sci.* **7**: 1–154.

Dodson, G. N. 1989. The horny antics of antlered flies. *Austr. Nat. Hist.* **22**: 604–611.

–. 1997. The resource defense mating system of antlered flies, *Phytalmia* spp. (Diptera: Tephritidae). *Ann. Entomol. Soc. Am.*, in press.

Dodson, G. N. and G. Daniels. 1988. Diptera reared from *Dysoxylum gaudichaudianum* (Juss.) Miq. at Iron Range, northern Queensland. *Austr. Ent. Mag.* **15**: 77–79.

Emlen, S. T. and L. W. Oring. 1977. Ecology, sexual selection, and the evolution of mating systems. *Science (Wash., D.C.)* **197**: 215–223.

Enquist, M. and O. Leimar. 1983. Evolution of fighting behaviour: decision rules and assessment of relative strength. *J. Theor. Biol.* **102**: 387–410.

Feijen, H. R. 1979. Economic importance of rice stem-borer (*Diopsis macrophthalma*) in Malawi. *Exp. Agric.* **15**: 177–186.

–. 1983. Systematics and phylogeny of Centrioncidae, a new Afromontane family of Diptera (Schizophora). *Zool. Verh.* **202**: 1–137.

–. 1989. Diopsidae. In *Flies of the Nearctic Region*. G. C. D. Griffiths, ed., pp. 1–122. Stuttgart: E. Schweizerbart'sche Verlagsbuchhandlung.

Flint, O. S. J. 1956. Hibernation of the diopsid fly, *Sphyrocephala brevicornis* Say. *Bull. Brooklyn Entoomol. Soc.* **51**: 44.

Frey, R. 1928. Philippinische Dipteren. V. Fam. Diopsidae. *Not. Entomol.* **8**: 69–77.

–. 1960. Studien uber indoaustralische Clusiiden (Dipt.) nebst Katalog der Clusiiden. *Soc. Sci. Fenn. Comment. Biol.* **22**: 1–31.

Gaerstecker, A. 1860. Beschreibung einiger ausgezeichneten neuen Dipteren aus der familie Muscariae. *Stett. Entomol. Z.* **21**: 163–202.

Green, A. J. 1992. Positive allometry is likely with mate choice, competitive display and other functions. *Anim. Behav.* **43**: 170–172.

Grimaldi, D. 1986a. The *Chymomyza aldrichii* species-group (Diptera: Drosophilidae): relationships, new neotropical species, and the evolution of some sexual traits. *J. N.Y. Entomol. Soc.* **94**: 342–371.

–. 1986b. A new *Drosophila* (*Hirtodrosophila*) from Malaysia with broad-headed males (Diptera: Drosophilidae). *J. N.Y. Entomol. Soc.* **94**: 372–376.

–. 1987. Phylogenetics and taxonomy of *Zygothrica* (Diptera: Drosophilidae). *Bull. Am. Mus. Nat. Hist.* **186**: 103–268.

–. 1988. *Drosophila* (*Hirtodrosophila*) *chandleri* (Diptera: Drosophilidae), a new species from Sri Lanka with broad-headed males. *J. N.Y. Entomol. Soc.* **96**: 323–326.

Grimaldi, D. and G. Fenster. 1989. Evolution of extreme sexual dimorphisms: structural and behavioral convergence among broad-headed Drosophilidae (Diptera). *Am. Mus. Nov.* **2939**: 1–25.

Hardy, D. E. 1973. The fruit flies (Tephritidae-Diptera) of Thailand and bordering countries. *Pac. Insects Monogr.* **31**: 1–353.

–. 1974. The fruit flies of the Philippines (Diptera: Tephritidae). *Pac. Insects Monogr.* **32**: 1–266.

–. 1986. The Adramini of Indonesia, New Guinea and adjacent islands (Diptera: Tephritidae: Trypetinae). *Proc. Hawaiian Entomol. Soc.* **27**: 53–78.

–. 1988. Fruit flies of the subtribe Gastrozonina of Indonesia, New Guinea and the Bismarck and Solomon Islands (Diptera, Tephritidae, Trypetinae, Acanthonevrini). *Zool. Script.* **17**: 77–121.

Hennig, W. 1937. Beitrage zur Systematik der Richardiiden (Dipt.). *Rev. Entomol. (Rio de Janeiro)* **7**: 21–34.

–. 1966. *Phylogenetic Systematics*. Urbana: University of Illinois Press.

Hochberg Stasny, T. A. 1985. Biology, behavior and life cycle of *Sphyrecephala brevicornis* Say (Diptera: Diopsidae). M. S. thesis, University of West Virginia.

Huxley, J. S. 1932. *Problems of Relative Growth*. London: Methuen.

Iwasa, Y., A. Pomiankowski and S. Nee. 1991. The evolution of costly mate preferences. II. The 'handicap' principle. *Evolution* **45**: 1431–1442.

James, A. C. and J. Jaenike. 1990. 'Sex ratio' meiotic drive in *Drosophila testacea*. *Genetics* **126**: 651–656.

Kirkpatrick, M. and M. J. Ryan. 1991. The evolution of mating preferences and the paradox of the lek. *Nature (Lond.)* **350**: 33–38.

Kotrba, M. 1990. Sperm transfer by spermatophore in an acalyptrate fly (Diptera: Diopsidae). *Entomol. Gen.* **15**: 181–183.

–. 1993. Das Reproduktionssystem von *Cyrtodiopsis whitei* Curran (Diopsidae, Diptera) unter besonderer Berucksichtigung der inneren weiblichen Geschlechtsorgane. *Bonn. Zool. Monogr.* **33**: 1–115.

–. 1996. Sperm transfer by spermatophore in Diptera: New results from the Diopsidae. *Zool. J. Linn. Soc.* **117**: 305–323.

Lande, R. 1981. Models of speciation by sexual selection on polygenic traits. *Proc. Natl. Acad. Sci. U.S.A.* **78**: 3721–3725.

Lavigne, R. 1962. Immature stages of the stalk-eyed fly, *Sphrycephala brevicornis* (Say)(Diptera: Diopsidae) with observations on its biology. *Bull. Brooklyn Entomol. Soc.* **57**: 5–14.

Linnaeus, C. 1775. *Dissertatio Entomologica, Bigas Insectorium Sistens, etc.* Upsaliae.

Lorch, P., G. S. Wilkinson and P. R. Reillo. 1993. Copulation duration and sperm precedence in the Malaysian stalk-eyed fly, *Cyrtodiopsis whitei* (Diptera: Diopsidae). *Behav. Ecol. Sociobiol.* **32**: 303–311.

Maddison, W. P. 1990. A method for testing the correlated evolution of two binary characters: Are gains and losses concentrated on certain branches of a phylogenetic tree? *Evolution* **44**: 539–557.

Malloch, J. R. 1939. The Diptera of the territory of New Guinea. VII. Family Otitidae (Ortalidae). *Proc. Linn. Soc. N. S. W.* **64**: 97–154.

–. 1940. The Otitidae and Phytalmidae of the Solomon Islands (Diptera). *Ann. Mag. Nat. Hist.* **11**: 66–98.

McAlpine, D. K. 1973. The Australian Platystomatidae (Diptera, Schizophora) with a revision of five genera. *Mem. Austr. Mus.* **15**: 1–256.

–. 1974. The subfamily classification of the Micropezidae and the genera of Eurybatinae (Diptera: Schizophora). *J. Entomol.* **43**: 231–245.

–. 1975. Combat between males of *Pogonortalis doclea* (Diptera, Platystomatidae) and its relation to structural modification. *Austr. Entomol. Mag.* **2**: 104–107.

–. 1979. Agonistic behavior in *Achias australis* (Diptera, Platystomatidae) and the significance of eyestalks. In *Sexual Selection and Reproductive Competition in Insects*. M. Blum and N. Blum, ed., pp. 221–230. New York: Academic Press.

–. 1982. The acalyptrate Diptera with special reference to the Platystomatidae, In *Biogeography and Ecology of New Guinea*. J. L. Gressitt, ed., pp. 659–673. The Hague: Dr W. Junk.

–. 1994. Review of the species of *Achias* (Diptera: Platystomatidae). *Invert. Taxon.* **8**: 117–281.

McAlpine, D. K. and M. A. Schneider. 1978. A systematic study of Phytalmia (Diptera, Tephritidae) with a description of a new genus. *Syst. Entomol.* **3**: 159–175.

McAlpine, J. F. 1989. Phylogeny and classification of the Musco-morpha. In *Manual of Nearctic Diptera*, vol. 3. J. F. McAlpine and D. M. Wood, eds., pp. 1397–1520. Hull: Canadian Government Publishing Center.

Mitchell-Olds, T. and R. G. Shaw. 1987. Regression analysis of natural selection: statistical inference and biological interpretation. *Evolution* **41**: 1149–1161.

Moulds, M. S. 1977. Field observations on behaviour of a North Queensland species of Phytalmia (Diptera: Tephritidae). *J. Austr. Entomol. Soc.* **16**: 347–352.

Otte, D. and K. Stayman. 1979. Beetle horns: some patterns in functional morphology. In *Sexual Selection and Reproductive Competition in Insects*. M. S. Blum and N. A. Blum, eds., pp. 259–292. New York: Academic Press.

Peterson, B. V. 1987. Diopsidae. In *Manual of Nearctic Diptera*. J. F. McAlpine, ed., pp. 675–1332. Stuttgart: Agric. Can. Res. Branch.

Petrie, M. 1988. Intraspecific variation in structures that display competitive ability: large animals invest relatively more. *Anim. Behav.* **36**: 1174–1179.

–. 1992. Are all secondary sexual display structures positively allometric and, if so, why? *Anim. Behav.* **43**: 173–175.

Sanderson, M. J. 1991. In search of homoplastic tendencies: statistical inference of topological patterns in homoplasy. *Evolution* **45**: 351–358.

Saunders, W. W. 1860. *Elaphomyia*, a genus of remarkable insects of the order Diptera. *Trans. Entomol. Soc. Lond. (N. S.)* **5**: 413–417.

Schneider, M. A. 1993. A new species of Phytalmia (Diptera: Tephritidae) from Papua New Guinea. *Austr. Entomol.* **20**: 3–8.

Shillito, J. F. 1971. The genera of Diopsidae (Insecta: Diptera). *Zool. J. Linn. Soc.* **50**: 287–295.

Spieth, H. T. 1981. *Drosophila heteroneura* and *Drosophila sylvestris*: head shapes, behavior and evolution. *Evolution* **35**: 921–930.

Steyskal, G. C. 1963. The genus *Plagiocephalus* Wiedemann (Dipt. Otitidae). *Stud. Entomol* **6**: 511–514.

–. 1972. A catalogue of species and key to the genera of the family Diopsidae. *Stuttg. Beitr. Naturkd.* A234: 1–20.

Tan, K. B. 1967. The life-histories and behaviour of some Malayan stalk-eyed flies (Diptera, Diopsidae). *Malay. Nat. J.* **20**: 31–38.

Templeton, A. R. 1977. Analysis of head shape differences between two interfertile species of Hawaiian *Drosophila*. *Evolution* **31**: 630–641.

Thornhill, R. and J. Alcock. 1983. *The Evolution of Insect Mating Systems*. Cambridge, Mass.: Harvard University Press.

Wickler, W. and U. Seibt. 1972. Zur Ethologie afrikanischer Stielaugenfliegen (Diptera, Diopsidae). *Z. Tierpsychol.* **31**: 113–130.

Wilkinson, G. S. 1993. Artificial sexual selection alters allometry in the stalk-eyed fly *Cyrtodiopsis dalmanni* (Diptera: Diopsidae). *Genet. Res.* **62**: 213–222.

Wilkinson, G. S. and P. R. Reillo. 1994. Female preference response to artificial selection on an exaggerated male trait in a stalk-eyed fly. *Proc. R. Soc. Lond.* B **255**: 1–6.

19 · Sex via the substrate: mating systems and sexual selection in pseudoscorpions

DAVID W. ZEH AND JEANNE A. ZEH

ABSTRACT

Pseudoscorpions are an ancient order of arachnids whose mating systems display an interesting mix of phylogenetic conservatism and evolutionary plasticity. A 400 million-year-old pattern of indirect sperm transfer by means of spermatophores deposited on the substrate pervades all aspects of sexual selection in pseudoscorpions. Across families, mating behavior ranges from the ancestral condition, in which males deposit structurally simple spermatophores irrespective of the presence of females (non-pairing), to a derived condition in which males engage in elaborate courtship and assist females in the uptake of structurally complex spermatophores. In non-pairing taxa, sexual selection appears to be mediated through rapid male development and prolific spermatophore production. Males are invariably the smaller sex and do not fight over access to females. Why non-pairing has persisted in six of seven superfamilies remains an enigma. Cladistic analysis suggests that pair formation has evolved independently only once. Evidence from within the most diverse family, the Chernetidae, indicates that, once pair formation evolved, sexual dimorphism became a highly variable condition. Only in productive, and hence often ephemeral and patchily distributed, micro-habitats do populations reach densities at which selection for fighting ability outweighs the costs of attaining competitive size. The harlequin-beetle-riding pseudoscorpion, *Cordylochernes scorpioides*, has provided a model system both for assessing the influence of ecological factors on the operation of sexual selection and for identifying processes that can maintain variability in male sexually selected traits. In this neotropical chernetid, extreme variability is associated with a pattern of colonization, population growth and dispersal which appears to generate regular cycles of selection alternately favoring small and then large male size.

INTRODUCTION

Pseudoscorpions are abundant, diverse and widely distributed, yet this ancient order of arachnids has been largely overlooked by behavioral ecologists. Indeed, little was known of the reproductive biology of the false scorpions prior to the 1960s when a German biologist, Peter Weygoldt, carried out extensive comparative studies of development, morphology and behavior. Weygoldt's pioneering research revealed remarkable diversity in sperm-transfer behavior and reproductive morphology within the Pseudoscorpionida (Weygoldt 1969). This diversity, coupled with novel features of their natural history (Zeh and Zeh 1990, 1992a,d, 1994a) and a developmental pattern and mating behavior amenable to experimental and observational study (Zeh 1987a,b), makes pseudoscorpions ideal subjects for investigations of mating systems and sexual selection (Thomas and Zeh 1984; Zeh and Zeh 1992a).

In this chapter, we first provide an overview of mating behavior across 13 of the 24 recently revised families of the Pseudoscorpionida (Harvey 1992). Mating behavior in the remaining 11 families remains undescribed. We then focus on Chernetidae, the largest family and the one for which the most extensive comparative data exist (Weygoldt 1969). In chernetids, ecological conditions appear to strongly affect the mating advantage to be gained from large male size and thus influence the degree to which sexual dimorphism evolves (Zeh 1987a,b).

This review of the Chernetidae provides the background for a detailed discussion of how ecological factors interact with female reproductive tactics to influence the evolution of exaggerated male traits in a highly sexually dimorphic and phenotypically variable chernetid, the harlequin-beetle-riding pseudoscorpion, *Cordylochernes scorpioides*. This pseudoscorpion has an intriguing natural history involving exploitation on a cyclical basis of two radically different environments: decaying trees and harlequin beetle abdomens (Zeh and Zeh 1992a,d).

MODES OF SPERM TRANSFER IN PSEUDOSCORPIONS: CONSTRAINT AND PLASTICITY

A fundamental goal in comparative biology is to understand how intrinsic properties of species and lineages promote or constrain adaptation. The extraordinary diversity of sperm-transfer behaviors in pseudoscorpions provides a rich resource for comparative biologists investigating the evolution of animal mating systems. All pseudoscorpions transfer sperm indirectly via a stalked spermatophore deposited on the substrate (Weygoldt 1969). However, great variability exists both in mating behavior and in the complexity of the spermatophore itself. At one extreme, the presence of females is not required for spermatophore deposition. Males scatter large numbers of small, structurally simple spermatophores over the substrate, some of which may subsequently be picked up by females. At the other extreme, males engage in elaborate courtship behaviors and directly assist females in the uptake of structurally complex spermatophores. The phylogenetic distribution of sperm transfer behavior suggests that non-pairing is the ancestral condition and that pairing has independently evolved only once in the Pseudoscorpionida (Weygoldt 1966b; Harvey 1992).

Non-pairing pseudoscorpions: a contradiction for sexual selection?

Non-pairing sperm-transfer behavior occurs in the Tridenchthoniidae, Chthoniidae, Neobisiidae, Garypidae, Pseudogarypidae, Cheiridiidae and some Olpiidae (Weygoldt 1966b, 1969, 1970; Thomas and Zeh 1984; Harvey 1992). In other olpiids (e.g. *Olpium pallipes* and *Serianus carolinensis*), an intermediate form of sperm-transfer behavior takes place in which males must physically encounter a female to trigger spermatophore deposition (Weygoldt 1969). The recent discovery of pseudoscorpion fossils from the Devonian (Schawaller *et al.* 1991), in conjunction with phylogenetic analysis (Harvey 1992), indicates that non-pairing behavior has persisted in six of seven pseudoscorpion superfamilies for at least 400 million years. The retention of this archaic and seemingly inefficient method of sperm transfer challenges the axiom in behavioral ecology that intrasexual selection should favor males that avoid sperm wastage and minimize sperm competition through close association, i.e. pairing, with females (Parker 1970, 1984; Alexander *et al.*, this volume).

Patterns in the evolution of sperm-transfer behavior in other arachnids suggest that sexual selection alone cannot bring about the evolution of pairing (copulatory) behavior from an ancestral, non-pairing condition. Some form of substantial environmental change is also involved. For example, copulation has never arisen in the terrestrial ancestors of water mites, but in water mites themselves copulation has independently evolved an estimated 91 times and is frequently associated with the evolution of swimming hairs (Proctor 1991). Using the comparative method, Proctor (1991) concluded that 'copulation evolves when behavioral dissociation from the substrate (i.e., swimming) reduces the probability of females encountering deposited spermatophores'.

In pseudoscorpions, the evolution of pairing behavior appears to be associated with more subtle environmental change. With few exceptions, non-pairing taxa are restricted to the 'cryptosphere', that is, to permanently mesic leaf-litter environments (Weygoldt 1969; Schaller 1979). Pairing taxa, by contrast, are found in more xeric and ephemeral microhabitats (Legg 1975). To account for this pattern, Thomas and Zeh (1984) proposed a cost-benefit model by which low desiccation stress on spermatophores and high risk of mate cannibalism (pseudoscorpions are voracious predators) were critical factors maintaining non-pairing behavior. The hypothesis that non-pairing can only function in mesic environments in which deposited spermatophores remain viable for prolonged periods could be tested with experiments to determine the rate of decay in spermatophore attractiveness and viability as a function of ambient humidity.

Although Thomas and Zeh's (1984) model can account for ecological associations between mating behavior and habitat utilization, it fails to adequately explain why non-pairing has been so resistant to invasion by a mutant, pairing strategy. In both mites and pseudoscorpions, non-pairing males compensate for an inherently low fertilization success rate per spermatophore by depositing hundreds or thousands of spermatophores (reviewed in Thomas and Zeh 1984). They can further increase their reproductive success by depositing spermatophores when females are most receptive which, in pseudoscorpions, is likely to occur early in the reproductive season. Females nourish embryos in an external brood sac overlying the genital atrium and are incapable of sperm uptake during the entire embryonic development period of between one and three weeks (see Weygoldt 1969). These factors are likely to generate selection for early, high reproductive effort by males, resulting in rapid development to the adult stage and smaller male size. This combination of male traits may in part explain the evolutionary stability of non-pairing. From the male standpoint, pairing behavior has clear benefits over non-pairing. It increases the probability of actually fertilizing eggs. By contrast, male–female conflict over control of fertilization (Alexander *et al.*, this volume) may render pairing disadvantageous to females. Thus, any mutation causing pairing behavior in males would be unlikely to spread in populations of small, rapidly developing males unable to seduce larger, resistant females into accepting spermatophores.

Sexual selection unleashed: the evolution of pairing behavior

With the acquisition of pairing behavior, the extreme evolutionary conservatism apparent in male sexually dimorphic traits (e.g. size relative to female, courtship behavior and spermatophore complexity) broke down completely. Pairing of males and females to achieve sperm transfer is a derived character state of the species-rich superfamily the Cheliferoidea (Harvey 1992). This superfamily consists of four families: Withiidae, Chernetidae, Atemnidae and Cheliferidae. Complex spermatophores and well-developed spermathecae are also restricted to the Cheliferoidea (Weygoldt 1969; Legg 1974). In pairing species, spermatophore production is accompanied by what Weygoldt (1969, p. 44) has described as 'spectacular mating dances that are performed with or without bodily contact'. In all cheliferoid species, once the spermatophore has been deposited on the substrate, a male grasps his mate and guides her into a position in which her gonopore region contacts the sperm packet. Sperm are then ejaculated into the female's reproductive tract. The most elaborate form of pairing behavior is exhibited by species in the family Cheliferidae. Mating is preceded by courtship in which males display ram's horn organs, a pair of erectile tubes originating from the genital atrium. In these species, males also use their modified foretarsi to assist the female in sperm uptake (see Kew 1912; Vachon 1949; Weygoldt 1969).

ECOLOGICAL DETERMINANTS OF SEXUAL DIMORPHISM IN CHERNETIDAE

Mating behavior has been most extensively studied in the family Chernetidae (Weygoldt 1970). In all the 12 genera that have been investigated, pairing occurs with little or no precopulatory courtship (Kew 1912; Levi 1953; Weygoldt 1970; our pers. obs.). Males seem unable to distinguish sex without contact; they use their pedipalpal chelae to forcefully grasp the chelae of any conspecific

encountered (Fig. 19-1). Contact between males generally results in combat (Weygoldt 1966a; Zeh 1987a); unreceptive females also respond aggressively to males. Males maintain their hold on receptive females throughout spermatophore assembly, deposition and sperm transfer. In some species, the process may be repeated, with several spermatophores deposited although not necessarily accepted by the female. In the most extreme case studied to date, mating in *Semeiochernes armiger* may last for over three hours, during which males transfer more than 100 spermatophores (D. W. Zeh and J. A. Zeh, unpublished data). Although male chernetids position the female over the spermatophore, they do not otherwise assist in sperm uptake.

In the Chernetidae, species vary greatly in the extent to which the sexes differ in the size of their conspicuous, prehensile pedipalps (Chamberlin 1931; Muchmore 1974). The pedipalps, particularly the chelae, function as fighting appendages in males (Zeh 1987a). Across chernetid species, the size (as estimated by silhouette area) of the male pedipalpal chela ranges from 60% to 150% of that of the female (Zeh 1987a). Although the evolution of enlarged male pedipalps is associated with a change from non-pairing to pairing behavior in Pseudoscorpionida, patterns within the Chernetidae indicate that sexual dimorphism is a highly variable condition, relatively unconstrained by factors associated with phylogeny (Zeh 1987a; Harvey 1992).

What accounts for the extreme variation in level of sexual dimorphism exhibited by chernetid pseudoscorpions? Male–male competition in this family appears to be a density-dependent process, tightly linked to the natural history of a species. Chernetids occupy a wide range of habitats; the characteristic population density experienced by a species is largely determined by habitat type. Many chernetids, for example, occupy leaf litter: a low-productivity, continuously distributed environment in which populations exist at relatively low density (Hoff 1959). By contrast, the ephemeral and patchily distributed habitats exploited by other species, such as rodent nests and decaying cacti, generate regular cycles of colonization, population growth to relatively high density and finally dispersal (Hoff 1959; Legg 1975; Zeh and Zeh 1992a,b,c).

Laboratory experiments carried out on several chernetid species have demonstrated that higher population density increases the intensity of sexual selection (Zeh 1987a; Zeh and Zeh 1992a). At high density, large males increase their mating success by interrupting matings and supplanting smaller rivals. This large-male advantage disappears at low density, where there is a lower

Fig. 19-1. Mating sequence in the chernetid pseudoscorpion *Dinocheirus arizonensis*. (a) Male initiates mating by grasping both chelae of the female. (b) Male leads female to mating site. (c) Male holds female over spermatophore during sperm uptake phase. (d) Female rejects male to terminate mating.

probability of encountering mating pairs. Because large male size incurs a cost in terms of prolonged development, increased exposure to predators and parasites, and delayed access to receptive females (Zeh 1987b), sexual selection should only favor the evolution of exaggerated male pedipalps in species that regularly experience high population densities. Indeed, across chernetid species, the extent of sexual dimorphism is positively correlated with estimates of population density (Zeh 1987a).

Comparative studies of sexual dimorphism in Chernetidae would greatly benefit from modern phylogenetic analysis. Although it is clear from the current Linnaean hierarchy that enlarged male size has evolved independently in a number of genera, it remains to be determined whether sexual dimorphism is concentrated in particular clades. Is enlarged male size a freely reversible characteristic or do sexually dimorphic species preferentially give rise to daughter species that are also sexually dimorphic?

OSCILLATING SEXUAL SELECTION IN *CORDYLOCHERNES SCORPIOIDES*

The harlequin-beetle-riding pseudoscorpion, *C. scorpioides*, is the largest, most sexually dimorphic member of the chernetid subfamily Lamprochernetinae. Males of this species have been described as the most phenotypically variable pseudoscorpions known (Beier 1948). For example, coefficients of variation for male pedipalpal traits are 15–20%, two to three times higher than those for females (Zeh and Zeh 1992a, 1994a). This extreme variability in exaggerated male traits, combined with a natural history that encompasses two disparate and well-defined habitats, makes *C. scorpioides* an excellent species for investigating what might best be described as the 'male variability paradox.' Selection should act to erode genetic variability (Fisher 1958) and yet, particularly among tropical terrestrial arthropods, sexual selection, far from eroding variability, appears to actually promote

Fig. 19-2 A large male *C. scorpioides* fights for control of a harlequin beetle's abdomen. Only the right pedipalp of his opponent is visible protruding from beneath the harlequin's elytrum. Meanwhile, a female (right) uses a silken thread to disembark from the beetle and colonize the new habitat.

it (see Grimaldi 1987; Zeh and Zeh 1992e). Male sexually selected traits in highly dimorphic species are often precisely those that exhibit the greatest level of variability (Cheverud *et al.* 1985; Grimaldi 1987; Alatalo *et al.* 1988; Zeh and Zeh 1992e).

The primary habitat of both *C. scorpioides* and the harlequin beetle, *Acrocinus longimanus*, is decaying trees in the families Moraceae and Apocynaceae (e.g. *Ficus* spp. and *Parahancornia fasciculata*). A combination of observations and experiments carried out in the field and laboratory has established that the pseudoscorpion is able to exploit these rich but ephemeral and patchily distributed habitats by hitch-hiking under the wings of the large cerambycid beetle. Because this association is obligate, the pseudoscorpion's opportunity to colonize new habitats is limited to the brief period when newly decaying trees draw in harlequin beetles for mating and oviposition. Pseudoscorpion populations within trees then remain marooned for three to five generations until beetle larvae complete development. Harlequin beetles emerging from their pupal chambers attract large numbers of pseudoscorpions, which climb on board and disperse *en masse* (Zeh and Zeh 1992a). This novel mode of dispersal has been exploited by males who compete to force off rivals and

remain on the beetles in order to monopolize their abdomens as strategic sites for intercepting and inseminating dispersing females. The harlequin beetle thus serves both as a dispersal agent and as a mobile mating territory (Fig. 19-2). As a consequence, sexual selection acts on *C. scorpioides* in two radically different environments: within decaying trees and on the abdomens of beetles (Zeh and Zeh 1992a, 1994a,c).

Components of fitness in the harlequin beetle environment

Dispersal on harlequin beetles can be viewed as an episode of sexual selection (Lande and Arnold 1983; Arnold and Wade 1984) with two multiplicative components contributing to variation in male reproductive success. A male must first succeed in monopolizing a beetle abdomen (W_1). He must then fertilize the eggs of females passing through his mating territory (W_2). We assessed the W_1 component of male reproductive success using multivariate morphometric comparison of beetle-riding males with males randomly sampled from predispersal tree populations (Zeh and Zeh 1992a, 1994a). Principal component analysis of

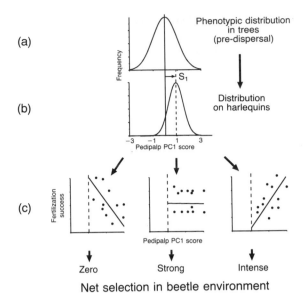

Fig. 19-3. Stages of dispersal-generated sexual selection in *C. scorpioides*. Male competition to monopolize beetle abdomens results in a shift (S1) of one standard deviation to the right in the phenotypic distribution of PC1 pedipalp score of beetle-riding males (b) relative to that of males randomly sampled from predispersal tree populations (a). The three possible relationships between male reproductive success (W_2) and male pedipalp size are shown in (c). Because overall selection in the beetle environment is the product of W_1 and W_2, only if size and reproductive success are negatively correlated can the W_1 effect be offset to result in zero selection.

seven pedipalpal traits demonstrated a shift of approximately one standard deviation unit to the right in the PC1 (a composite measure of size) distribution of males on harlequin beetles (Fig. 19-3b) relative to that of males within trees (Fig. 19-3a). This difference provides an estimate of the selection differential associated with W_1 (see Zeh and Zeh 1992a, 1994a).

Clearly, strong directional selection for large male size operates at the W_1 stage. The critical evolutionary question remains whether monopolization of mobile mating territories actually translates into enhanced reproductive success. Theoretically, the size of beetle-riding males could be negatively correlated, uncorrelated or positively correlated with reproductive success. However, as Fig. 19-3c illustrates, because overall selection is the (multiplicative) product of the two components, only a negative correlation could result in no net selection in the harlequin beetle environment.

Direct assessment of male reproductive success requires a sensitive tool for assigning paternity of offspring. Recent breakthroughs in methods for cloning minisatellite DNA (Armour *et al.* 1990) now make single-locus minisatellite DNA fingerprinting an accessible molecular tool for ecological and evolutionary studies in a wide variety of species (Bruford *et al.* 1992). To estimate the W_2 component of male reproductive success, two hypervariable, single-locus minisatellite probes were isolated

from a genomic library of *C. scorpioides* (Zeh *et al.* 1994), using the recently-developed charomid cloning technique (Bruford *et al.* 1992). Both probes display levels of heterozygosity equal to the most variable loci identified for any vertebrate species (Zeh *et al.* 1994). The two probes were then used to assign paternity in a field investigation of the relationship between male size and reproductive success in the beetle environment (a detailed report of the methods and results of this study will be presented elsewhere; see Zeh *et al.* 1997). Briefly, male and female pseudoscorpions were collected at the end of their dispersal phase, i.e. they were removed from harlequin beetles captured on newly fallen host trees. The females from these potential breeding groups were maintained in the laboratory and produced broods of offspring, which were reared to maturity. DNA fingerprinting was then carried out on these females, their offspring and the putative, beetle-riding sires.

Results of this study demonstrated a strong, positive correlation between the reproductive success of beetle-riding males (W_2) and pedipalp size. Approximately 40% of the males captured on beetles sired offspring and 70% of females produced broods that included offspring fathered by one or more of the putative sires. The selection differential (s) was positive and statistically significant for the PC1 score of the pedipalp ($s = 0.57$, $P = 0.03$). No selection on shape, as measured by PC2, was detected.

Dispersal-generated sexual selection in *C. scorpioides* thus produces a pattern in which smaller males are effectively eliminated in the first stage of competition to control beetle abdomens, thereby shifting the arena for subsequent competition to the upper end of the source (tree) population's phenotypic distribution. In the second round of male competition on harlequin beetles, this time to achieve fertilizations, single-locus DNA fingerprinting has demonstrated that the largest of these large beetle-riding males achieve the greatest reproductive success (Zeh *et al.* 1977). The overall effect of selection in the beetle environment is consequently enormous (approximately two standard deviation units). This may well represent the most extreme case of sexual selection for exaggerated male traits so far demonstrated using direct estimates of reproductive success.

The intense selection for large male size during episodes of dispersal makes the existence of extreme variability in exaggerated male traits in *C. scorpioides* even more perplexing. Purely environmental causes can be discounted. Quantitative genetic data have revealed a significant additive genetic component to this variation (J. A. Zeh and D. W. Zeh, unpublished data), as has been demonstrated for another pseudoscorpion (Zeh 1987b) and species in many other taxa (see, for example, Cade 1981, 1984; Norris 1993; Wilkinson 1993; Pomiankowski and Møller 1995). Precisely what form sexual selection takes within the decaying tree environment remains to be demonstrated in the field. Nevertheless, the major contribution of W_1 to selection in the beetle environment can be taken as indirect evidence that sexual selection must be far less intense within trees. Indeed, laboratory studies indicate that selection for a smaller optimal male phenotype may operate in the tree environment. Large size prolongs development and delays access to females, but confers no mating advantage under the relatively low-density conditions that characterize much of the population growth phase within trees (see Zeh 1987a,b; Zeh and Zeh 1992a). It seems likely, therefore, that, within trees, sexual selection favors rapid development, small male size and enhanced mate-searching ability. Testing this hypothesis will require paternity assignment to assess the relationship between the morphology and reproductive success of males from experimental tree populations. An absence of correlation or a negative correlation between reproductive success and male size would demonstrate that sexual selection in *C. scorpioides* oscillates on a cyclical basis, alternately favoring large and then small males.

If the intensity of sexual selection oscillates in a regular manner, then *C. scorpioides* might be expected to exhibit a bimodal distribution of male size, similar to the male dimorphisms found in other taxa in which variable ecological circumstances favor alternative male morphology (see, for example, Eberhard 1982; Crespi 1988a,b). However, all morphological traits in male *C. scorpioides* are distributed normally or approximately so (D. W. Zeh and J. A. Zeh, 1994a). Preliminary results from a quantitative genetic study (D. W. Zeh and J. A. Zeh, unpublished data) suggest that variation in pedipalp size has a polygenic basis and is controlled by many loci with small additive effects. Because of recombination and a lack of assortative mating, such traits may be constrained by their underlying genetic architecture to exhibit a normal distribution (Falconer 1981; Turelli and Barton 1994).

In *C. scorpioides*, the general conditions under which temporal fluctuations are predicted to augment genetic variability (Gillespie and Turelli 1989; Frank and Slatkin 1990; Gillespie 1991; Ellner and Hairston 1994) are met: coarse-grained environment; large oscillations in selection; and overlapping generations. Can oscillating sexual selection be invoked to explain the male variability paradox in general? We suggest that it can. Strong oscillations in selective pressures may be far more common than is currently appreciated (Nevo 1978; Mackay 1981; Gibbs and Grant 1987; Lynch 1987; Hairston and Dillon 1990; Gillespie 1991; Schluter *et al.* 1991). Many species, particularly among the diverse arthropod taxa that exploit ephemeral resources, experience strong, regular fluctuations in their physical and biotic environment (Hamilton 1978). As a consequence, the restricted set of ecological conditions that favors highly exaggerated male traits may occur only infrequently in the life history of these species. Such bouts of intense selection punctuate protracted periods of weak or counter-selection. Testing the generality of this mechanism will require field studies of selection and quantitative genetic investigations of trait variation. The oscillating sexual selection hypothesis predicts that within a lineage, the degree of sexual dimorphism should be positively correlated with extent of fluctuations in sexual selection and that male variability in sexually dimorphic species should be greater than in sexually monomorphic ones. These predictions could be tested in chernetid pseudoscorpions by examining patterns of male variability and sexual dimorphism in *Lustrochernes*, a diverse genus which is the likely sister group of *Cordylochernes* (Wilcox *et al.* 1997).

Polyandry and male variability in a heterogeneous environment

In *C. scorpioides*, sexual selection for large male size on beetles may not only be offset by selection for small size within trees but may also be undermined by female mating strategies and prolonged sperm storage. DNA fingerprinting has revealed extensive multiple paternity within the broods of dispersing females (Zeh *et al.* 1997). For example, in one set of ten offspring, sperm from at least four males were responsible for fertilizing the brood. Sperm-precedence experiments have demonstrated that mixed paternity results when females mate with several males and override the pattern of strong last-male sperm precedence characteristic of two-male matings (Zeh and Zeh 1994b). Although 70% of females produced broods that included some offspring of beetle-riding males, fathers could not be identified for a total of 57% of the offspring fingerprinted. Evidence from brood production studies shows that females collected while boarding beetles often carry sperm from predispersal matings within trees (Zeh *et al.* 1997). It is therefore probable that many of the offspring for whom paternity could not be assigned were sired by males within trees that had mated with the females before dispersal. If this hypothesis is correct, then female promiscuity and prolonged sperm storage enable males within trees to partly circumvent the bottleneck of dispersal-generated sexual selection and thereby perpetuate their genes in future tree populations.

The high level of mixed paternity in *C. scorpioides* broods raises the question of why females mate with several males. A growing body of molecular evidence demonstrates that the female reproductive tract represents the ultimate arena for sexual selection, with multiple paternity, and hence female promiscuity, proving to be pervasive features of natural populations (Amos *et al.* 1993; Birkhead & Møller 1993; Dunn & Lifjeld 1994). Although the importance of postcopulatory sexual selection for male reproductive success is now widely recognized, the reasons why females so commonly mate with several males remain largely unexplored. In *C. scorpioides*, female promiscuity has been shown to be a deliberate mating strategy that increases reproductive success (Zeh 1997). Several lines of evidence indicate that the lower fitness of females mated to a single male results from a small but significant risk of incompatibility between male and female genomes. Fertilizations can result in embryo failure, female-biassed sex ratios, meiotic drive and paternal determination of sex ratio (J. A. Zeh, unpublished data). In other taxa, such as mice and *Drosophila*, these phenomena are linked to reduced female reproductive success and result from an interaction between maternal and paternal haplotypes (Lyttle 1991; Hurst 1993; Silver 1993). The fitness of a female's offspring thus depends not simply on the inherent genetic quality of her mate but is a function of the interaction between maternal and paternal genomes. We propose that, by charging a female's sperm-storage organs with sperm from several males, multiple mating enables females to utilize postcopulatory mechanisms such as sperm competition, female choice of sperm, and reallocation of maternal resources from defective to viable embryos in order to minimize the risk and/or cost of fertilization by genetically incompatible sperm (Zeh and Zeh 1996, 1997).

The combination in *C. scorpioides* of female promiscuity and a temporally varying environment appears to provide precisely the conditions required for fluctuating selection to maintain genetic variability. A recent paper (Ellner and Hairston 1994) has called into question the results of discrete-generation models showing that the conditions required for fluctuating selection to maintain variability are too restrictive to be of general evolutionary importance (Hedrick 1986; Frank and Slatkin 1990). Ellner and Hairston (1994) concluded that fluctuating selection can contribute significantly to the maintenance of genetic variation when the effects of overlapping generations and age- or stage-specific selection are included in a model of temporally-varying selection. The essential result, called the 'storage effect', occurs when a genotype persists in some life-history stage that enables it to remain viable for periods long enough to experience the next favorable episode of selection (such as long-lived adults in animals and dormant seeds in annual plants). In *C. scorpioides*, sperm stored within the female may serve as the resistant life-history stage for small males which are favored in the decaying-tree environment but strongly selected against during the dispersal phase on harlequin beetles. Testing this hypothesis requires: (1) demonstrating that sexual selection within trees favors a significantly smaller optimum male phenotype than that favored in the beetle environment; and (2) establishing that a significant proportion of the offspring of dispersing females is sired by males from predispersal, within-tree matings.

SUGGESTIONS FOR FUTURE RESEARCH

Pseudoscorpions offer remarkable opportunities to address some of the most important unresolved questions in the analysis of mating systems. First, evidence for the selective pressures involved in the transition from indirect to direct pairing may be inferred from comparisons between species of Cheliferoidea, the pairing superfamily, and the non-pairing Sternophoridae, their presumed sister group. Did pairing evolve as a result of selection on males to control fertilization in an ecological context (low humidity) that selected for females to take spermatophores directly from males (Alexander *et al.*, this volume)? Does sexual selection operate in species with indirect pairing by female choice of spermatophores or novel forms of male–male competition? Second, female mating strategies in pseudoscorpions with pairing require further study. Does female choice operate in species of Cheliferoidea with complex courtship? Are multiple mating and sperm-mixing female adaptations, perhaps selected for in the context of intragenomic conflict and genetic incompatibility? Third, the evidence that selection within trees in *C. scorpioides* favors small males compels further study of the causes of selection against large size among other insects and arachnids. Finally, there is a serious need for further natural-history studies of these little-known arthropods. The cryptic habits of pseudoscorpions have hidden from view a spectacular diversity of reproductive behavior and morphology, which is only now being brought to light.

ACKNOWLEDGEMENTS

The research was supported by the Smithsonian Institution's Fellowship and Scholarly Studies programs and through an NSF/NATO Postdoctoral Fellowship to DWZ. We thank Jae Choe, Bernie Crespi and two anonymous referees for their helpful comments.

LITERATURE CITED

Alatalo, R. V., J. Hogland and A. Lunberg. 1988. Patterns of variation in tail ornament size in birds. *Biol. J. Linn. Soc.* **34**: 363–374.

Amos, W., S. Twiss, P. P. Pomeroy and S. S. Anderson. 1993. Male mating success and paternity in the grey seal, *Halichoerus grypus*: a study using DNA fingerprinting. *Proc. R. Soc. Lond.* B **252**: 199–207.

Armour, J. A. L., S. Povey, S. Jeremiah and A. J. Jeffreys. 1990. Systematic cloning of human minisatellites from ordered array charomid libraries. *Genomics* **8**: 501–512.

Arnold, S. J. and M. J. Wade. 1984. On the measurement of natural and sexual selection: theory. *Evolution* **38**: 709–719.

Beier, M. 1948. Phoresie und Phagophilie bei Pseudoscorpionen. *Oesterr. Zool. Z.* **1**: 441–497.

Birkhead, T. & A. Møller. 1993. Female control of paternity. *Trends Ecol. Evol.* **8**: 101–104.

Bruford, M. W., O. Hanotte, J. F. Y. Brookfield and T. Burke. 1992. Single-locus and multilocus DNA fingerprinting. In *Molecular Genetic Analysis of Populations. A Practical Approach.* A. R. Hoelzel, ed., pp. 225–269. Oxford: IRL Press.

Cade, W. H. 1981. Alternative male strategies: genetic differences in crickets. *Science (Wash., D.C.)* **212**: 563–564.

–. 1984. Genetic variation underlying sexual behavior and reproduction. *Am. Zool.* **24**: 355–366.

Chamberlin, J. C. 1931. The arachnid order Chelonethida. *Stanford University Publ. Biol. Sci.* **7**: 1–284.

Cheverud, J. M., M. M. Dow and W. Leutenegger. 1985. The quantitative assessment of phylogenetic constraints in comparative analyses: sexual dimorphism in body weight among primates. *Evolution* **39**: 1335–1351.

Crespi, B. J. 1988a. Risks and benefits of lethal male fighting in the colonial, polygynous thrips *Hoplothrips karnyi* (Insecta: Thysanoptera). *Behav. Ecol. Sociobiol.* **22**: 293–301.

–. 1988b. Adaptation, compromise, and constraint: the development, morphometrics, and behavioral basis of a fighter flier polymorphism in male *Hoplothrips karnyi* (Insecta: Thysanoptera). *Behav. Ecol. Sociobiol.* **23**: 93–104.

Dunn, P. O. and J. T. Lifjeld. 1994. Can extra-pair copulations be used to predict extra-pair paternity in birds? *Anim. Behav.* **47**: 983–985.

Eberhard, W. G. 1982. Beetle horn dimorphism: making the best of a bad lot. *Am. Nat.* **119**: 420–426.

Ellner, S. and N. G. Hairston Jr. 1994. Role of overlapping generations in maintaining genetic variation in a fluctuating environment. *Am. Nat.* **143**: 403–417.

Falconer, D. S. 1981. *Introduction to Quantitative Genetics*, 2nd edn. London: Longman.

Fisher, R. A. 1958. *The Genetical Theory of Natural Selection*, 2nd edn. New York: Dover.

Frank, S. A. and M. Slatkin. 1990. Evolution in a variable environment. *Am. Nat.* **136**: 244–260.

Gibbs, H. L. and P. R. Grant. 1987. Oscillating selection on Darwin's Finches. *Nature (Lond.)* **327**: 511–513.

Gillespie, J. H. 1991. *The Causes of Molecular Evolution.* Oxford University Press.

Gillespie, J. H. and M. Turelli. 1989. Genotype-environment interactions and the maintenance of polygenic variation. *Genetics* **121**: 129–138.

Grimaldi, D. A. 1987. Phylogenetics and taxonomy of *Zygothrica* (Diptera: Drosophilidae). *Bull. Am. Mus. Nat. Hist.* **186**: 103–268.

Hairston, N. G. Jr. and T. A. Dillon. 1990. Fluctuating selection and response in a population of freshwater copepods. *Evolution* **44**: 1796–1805.

Hamilton, W. D. 1978. Evolution and diversity under bark. In *Diversity of Insect Faunas*. L. A. Mound, ed., pp. 154–175. Oxford: Blackwell.

Harvey, M. S. 1992. The phylogeny and classification of Pseudoscorpionida (Chelicerata: Arachnida). *Invert. Taxon.* **6**: 1373–1435.

Hedrick, P. W. 1986. Genetic polymorphism in heterogeneous environments: a decade later. *Annu. Rev. Ecol. Syst.* **17**: 535–66.

Hoff, C. C. 1959. *The Ecology and Distribution of the Pseudoscorpions of North-Central New Mexico.* (University of New Mexico Publication in Biology, no. 8).

Hurst, L. D. 1993. The incidences, mechanisms and evolution of cytoplasmic sex ratio distorters in animals. *Biol. Rev.* **68**: 121–193.

Kew, H. W. 1912. On the pairing of pseudoscorpions. *Proc. Zool. Soc. Lond.* **25**: 376–390.

Lande, R. and S. J. Arnold.1983. The measurement of selection on correlated characters. *Evolution* **37**: 1210–1226.

Legg, G. 1974. A generalized account of the female genitalia and associated glands of pseudoscorpions (Arachnida). *Bull. Br. Arachnol. Soc.* **3**: 42–48.

–. 1975. The possible significance of spermathecae in pseudoscorpions (Arachnida). *Bull. Br. Arachnol. Soc.* **3**: 91–95.

Levi, H. W. 1953. Observations on two species of pseudoscorpions. *Can. J. Entomol.* **85**: 55– 62.

Lynch, M. 1987. The consequences of fluctuating selection for isozyme polymorphisms in *Daphnia*. *Genetics* **115**: 657–669.

Lyttle, T. W. 1991. Segregation distorters. *Annu. Rev. Genet.* **25**: 511–557.

Mackay, T. F. C. 1981. Genetic variation in varying environments. *Genet. Res. (Camb.)* **37**: 79–93.

Muchmore, W. B. 1974. Clarification of the genera *Hesperochernes* and *Dinocheirus* (Pseudoscorpionida, Chernetidae). *Fla. Entomol.* **57**: 397–407.

Nevo, E. 1978. Genetic variation in populations: patterns and theory. *Theor. Popul. Biol.* **13**: 121–177.

Norris, K. 1993. Heritable variation in a plumage indicator of variability in male great tits *Parus major*. *Nature (Lond.)* **362**: 537–539.

Parker, G. A. 1970. Sperm competition and its evolutionary consequences in the insects. *Biol. Rev.* **45**: 525–567.

–. 1984. Sperm competition and the evolution of animal mating systems. In *Sperm Competition and the Evolution of Animal Mating Systems*. R. L. Smith, ed., pp. 1–60. Orlando: Academic Press.

Pomiankowski, A. and A. P. Møller. 1995. A resolution of the lek paradox. *Proc. R. Soc. Lond.* B **260**: 21–29.

Proctor, H. C. 1991. The evolution of copulation in water mites: a comparative test for nonreversing characters.*Evolution* **45**: 558–567.

Schaller, F. 1979. Significance of sperm transfer and formation of spermatophores in arthropod phylogeny. In *Arthropod Phylogeny*. A. P. Gupta, ed., pp. 587–608. New York: Van Nostrand Reinhold.

Schawaller, W., W. A. Shear and P. M. Bonamo. 1991. The first Paleozoic pseudoscorpions (Arachnida, Pseudoscorpionida). *Am. Mus. Nov.* **3009**: 1–17.

Schluter, D., T. D. Price and L. Rowe. 1991. Conflicting selection pressures and life history tradeoffs. *Proc. R. Soc. Lond.* B **246**: 11–17.

Silver, L. M. 1993. The peculiar journey of a selfish chromosome: mouse t haplotypes and meiotic drive. *Trends Genet.* **9**: 250–254.

Thomas, R. H. and D. W. Zeh.1984. Sperm utilization strategies in arachnids: ecological and morphological constraints. In *Sperm Competition and the Evolution of Animal Mating Systems*. R. L. Smith, ed., pp. 179–221. Orlando: Academic Press.

Turelli, M. and N. H. Barton. 1990. Dynamics of polygenic characters under selection. *Theor. Popul. Biol.* **38**: 1–57.

–. 1994. Genetic and statistical analyses of strong selection on polygenic traits: what, me normal? *Genetics* **138**: 913–941.

Vachon, M. 1949. Ordre des pseudoscorpions. In *Traite de Zoologie*. P.-P. Grasse, ed., pp. 437–481. Paris: Masson.

Weygoldt, P. 1966a. Mating behavior and spermatophore morphology in the pseudoscorpion *Dinocheirus tumidus* (Banks). *Biol. Bull.* **120**: 462–467.

–. 1966b. Vergleichende Untersuchungen zur Fortpflanzungsbiologie der Pseudoscorpione: Beobachtungen über das Verhalten, die Samenubertragungsweisen und die Spermatophoren einiger einheimischer Arten. *Z. Morphol. Oekol. Tiere* **56**: 39–92.

–. 1969. *The Biology of Pseudoscorpions*. Cambridge, **MA**: Harvard University Press.

–. 1970. Vergleichende Untersuchungen zur Fortpflanzungbiologie der Pseudoscorpione II. *Z. Zool. Syst. Evol.-Forsch.* **8**: 241–259.

Wilcox, T. P., L. Hugg, J. A. Zeh and D. W. Zeh. 1997. DNA sequencing reveals extreme genetic differentiation in a cryptic species complex of neotropical pseudoscorpions. *Mol. Phylogenet. Evol.*, in press.

Wilkinson, G. S. 1993. Artificial sexual selection alters allometry in the stalk-eyed fly *Cyrtodiopsis dalmanni* (Diptera: Diopsidae). *Genet. Res. (Camb.)* **62**: 213–222.

Zeh, D. W. 1987a. Aggression, density and sexual dimorphism in chernetid pseudoscorpions (Arachnida:Pseudoscorpionida). *Evolution* **41**: 1072 –1087.

–. 1987b. Life history consequences of sexual dimorphism in a chernetid pseudoscorpion. *Ecology* **68**: 1495 1501.

Zeh, D. W. and J. A. Zeh. 1992a. Dispersal-generated sexual selection in a beetle-riding pseudoscorpion. *Behav. Ecol. Sociobiol.* **30**: 135–142.

–. 1992b. Emergence of a giant fly triggers phoretic dispersal in the neotropical pseudoscorpion, *Semeiochernes armiger* (Balzan) (Pseudoscorpionida: Chernetidae). *Bull. Br. Arachnol. Soc.* **9**: 43–46.

–. 1992c. Failed predation or transportation? Causes and consequences of phoretic behavior in the pseudoscorpion *Dinocheirus arizonensis* (Pseudoscorpionida: Chernetidae). *J. Insect Behav.* **5**: 37–49.

–. 1992d. On the function of harlequin beetle-riding in the pseudoscorpion *Cordylochernes scorpioides* (Pseudoscorpionida: Chernetidae). *J. Arachnol.* **20**: 47–51.

–. 1994a. When morphology misleads: interpopulation uniformity in sexual selection masks genetic divergence in harlequin beetle-riding pseudoscorpion populations. *Evolution* **48**: 1168–1182.

Zeh, D. W., J. A. Zeh and E. Bermingham. 1997. Polyandrous, sperm-storing females: carriers of male genotypes through episodes of adverse selection. *Proc. R. Soc. Lond.* B, in press.

Zeh, D. W., J. A. Zeh and C. A. May. 1994. Charomid cloning vectors meet the pedipalpal chelae: single-locus minisatellite DNA probes for paternity assignment in the harlequin beetle-riding pseudoscorpion. *Molec. Ecol.* **3**: 517–522.

Zeh, J. A. 1997. Polyandry and enhanced reproductive success in the harlequin beetle-riding pseudoscorpion. *Behav. Ecol. Sociobiol.*, in press.

Zeh, J. A. and D. W. Zeh. 1990. Cooperative foraging for large prey in the pseudoscorpion, *Paratemnus elongatus* (Pseudoscorpionida, Atemnidae). *J. Arachnol.* **18**: 307–311.

–. 1992e. Are sexually-selected traits reliable species characters? Implications of intra-brood variability in *Semeiochernes armiger* (Balzan) (Pseudoscorpionida: Chernetidae). *Bull. Br. Arachnol. Soc.* **9**: 61–64.

–. 1994b. Last-male sperm precedence breaks down when females mate with three males. *Proc. R. Soc. Lond.* B **257**: 287–292.

–. 1994c. Tropical liaisons on a beetle's back. *Nat. Hist.* **103**: 36–43.

–. 1996. The evolution of polyandry I: intragenomic conflict and genetic incompatilibility. *Proc. R. Soc. Lond.* B, in press.

–. 1997. The evolution of polyandry II: post-copulatory defenses against genetic incompatibility. *Proc. R. Soc. Lond.* B, in press.

20 · Jumping spider mating strategies: sex among cannibals in and out of webs

ROBERT R. JACKSON AND SIMON D. POLLARD

Who can fathom the mind of a spider

Keith McKeown (1952)

ABSTRACT

We discuss two pivotal components of salticid mating systems. The first is complexity of behavior during salticid intraspecific interactions, characterized by conditional strategies composed of distinctly different tactics and, within the context of each tactic, large repertoires of distinctive displays combined in highly variable sequences. Variability in the display repertoire of male salticids is probably the main focus of female choice. The second is the relationship of mating strategies to predation, antipredator protection and other processes in the animal's life. During conspecific interactions, there are conflicting interests because of the potential for cannibalism, and a salticid may be a potential mate or prey and rival or predator. We suggest that during these interactions, salticids may orchestrate a careful balance between stimuli that provoke (e.g. sensory exploitation) and stimuli that inhibit predatory attacks from each other. We draw parallels between salticid intraspecific interactions and the araneophagic predatory strategy of *Portia*, a genus of salticids that practice aggressive mimicry. In both systems, signaling may best be envisaged as a way of achieving dynamic fine control of another animal's behavior.

INTRODUCTION

Jumping spiders (Salticidae) are the largest family of spiders (over 4000 described species (Coddington and Levi 1991)); they are well known for excellent vision (Land 1969a,b; Blest *et al.* 1990) and elaborate visual courtship displays (Crane 1949). Our goal in this chapter is to develop a fresh perspective on salticid mating strategies. This perspective is derived from an appreciation of the predatory strategy of *Portia*, a tropical salticid with disturbingly complex and flexible behaviour.

We address two problems. The first, complexity, appears to be a special feature of salticid intraspecific interactions: individual spiders exhibit conditional strategies composed of distinctly different tactics and, within the context of each tactic, large repertoires of distinctive displays combined in highly variable sequences.

The second problem concerns the relationship of mating strategies to predation, antipredator protection and other concerns in the salticid's life. A mating strategy might be defined as how an animal solves the problem of acquiring a mate, but there is an important caveat. We obtain our data from observing behavior, not from observing a strategy directly. It is convenient to give a name such as 'courtship display' to a behavior pattern, and then to attempt to examine the animal's behavior simply in the context of mating strategies. Salticids, however, confound our attempts to do this.

PORTIA'S PREDATORY STRATEGY

Most salticids prey primarily on insects caught by actively hunting (Drees 1952; Richman and Jackson 1992), instead of using prey-ensnaring webs. However, *Portia* not only actively hunts, but builds a prey-catching web, and also invades the webs of other spiders where it feeds on the resident spider, its eggs, and insects caught in the web (Jackson 1992a).

As a web-invader, *Portia* walks directly into its prey's perceptual world because the web is an extension and critical component of the web-building spider's sensory system (Foelix 1982). Intimate contact with the prey's sensory system can be dangerous for *Portia* because the tables may be turned, and *Portia*'s intended dinner may become the diner.

After entering another spider's web, *Portia* does not simply stalk or chase down its victim but instead sends vibratory signals across the silk. The resident spider may respond to these signals from *Portia* as it would to a small insect ensnared in the web, but when the duped spider gets close, *Portia* lunges out and catches it. *Portia* joins the better-known angler fish (Wickler 1968) as an example of aggressive mimicry. However, in *Portia* aggressive mimicry is tied in with pronounced behavioral complexity.

Portia makes vibratory signals by manipulating the silk with one or any combination of its eight legs and two palps, all of which can be moved in a variety of different ways. *Portia* also makes signals by flicking its abdomen up and down, and abdominal movements can be combined with virtually any of the appendage movements. The net effect is that *Portia* seems to have at its disposal virtually an unlimited array of different signals to use on the webs of other spiders (Jackson and Blest 1982; Jackson and Hallas 1986a; Jackson and Wilcox 1993).

In this game of deceit, the stalked spiders have acute abilities to detect and discriminate between vibratory signals transmitted over the silk in their webs. How the spider interprets these web-borne vibrations varies considerably between species and also with the sex, age, previous experience and feeding state of the spider (Jackson 1986a; Masters *et al.* 1986). Yet *Portia* has been observed using aggressive mimicry to catch a highly diverse array of different web-building spiders (R. R. Jackson and Blest 1982; Jackson and Hallas 1986b; R. R. Jackson, unpublished).

How does *Portia* choose, from its large repertoire of signals, those appropriate for its diverse range of victims? This question was the impetus for a research program at the University of Canterbury, Christchurch, New Zealand, carried out in collaboration with Stimson Wilcox from the State University of New York in America, who developed a computer-based system for recording and playing back signals on webs, analogous to listening and talking to spiders in their own language. From this ongoing work, the key to *Portia*'s success at catching so many different types of spiders appears to be an interplay of two basic ploys (Jackson and Wilcox 1990, 1993): (1) using specific preprogrammed signals when cues from some of its more common prey species are detected; and (2) adjusting signals to different prey species in a flexible fashion, as a consequence of feedback from the victims.

The first ploy, using preprogrammed tactics, is consistent with the popular image of spiders as animals that are 'hard-wired' and governed simply by instinct. The spider's

brain is seen as analogous to a computer where separate instructions for different prey are programmed in. However, we might expect 'disk capacity' (i.e. the spider's brain size) to limit how many instructions can be fitted in. Yet we keep finding more and more programs (Jackson and Wilcox 1990; R. R. Jackson and R. S. Wilcox, unpublished). Surely at some point the small brain of the spider must set limits on the expansion of repertoire size for instinctive behavior, but where that point lies is far from obvious.

The second ploy is for *Portia* to solve problems and figure out, by trial and error, what to do with different victims (Jackson and Wilcox 1993), an unexpectedly flexible behavior for a spider (see Mitchell 1986). Evidently, *Portia*'s small brain does not limit it to preprogrammed tactics.

To illustrate how trial and error works, let us first look at what happens when *Portia* goes into the web of a species of web-building spider for which it does not have a preprogrammed tactic. *Portia* first presents the resident spider with a kaleidoscope of different vibratory signals. When, eventually, one of these signals elicits an appropriate response from the victim (for example, it behaves as though *Portia* were a small insect in the web (R. S. Wilcox and R. R. Jackson, unpublished data)), *Portia* ceases to vary its signals and concentrates on producing the signal that elicits the response (Jackson and Wilcox 1993). However, communication between predator and prey is often more subtle than this.

Aggressive mimicry for *Portia* is a dangerous way to make a living. When facing a large and powerful spider in a web, it would appear foolhardy for *Portia* simply to pretend to be prey and provoke a full-scale predatory attack. Instead, *Portia* appears to strive for fine control over the victim's behavior by making signals that draw the victim in slowly. Alternatively, *Portia*'s signals may keep the victim calm while moving in slowly for the kill. Calming effects appear to be achieved by monotonous repetition of a habituating signal (R. R. Jackson, unpublished), as though *Portia* were putting its victim to sleep with a vibratory lullaby derived by trial and error.

Calling *Portia*'s behavior 'aggressive mimicry' emphasizes the question of what *Portia* mimics, but this question may be less important than trying to understand how *Portia* achieves fine control of its victim's behavior. The emphasis should perhaps be on how *Portia* takes advantages of biases in the victim's nervous system, adopting a perspective akin to recent ideas (see Clark and Uetz 1992, 1993) about receiver psychology (Guilford and Dawkins 1991) and sensory exploitation (Proctor 1992; Ryan and

Keddy-Hector 1992), but with a greater emphasis on complexity, flexibility and dynamic interaction between signaler and receiver.

PORTIA'S MATING STRATEGY

When we consider *Portia*, the distinction between mating and predatory strategies is often blurred: a mate may become a meal, and what appears to be a mating tactic may also be a predatory or an antipredator tactic.

Salticid males usually approach females in rapid stop-and-go spurts of activity, punctuated with displays, while the female scrutinizes the male's displays without so actively displaying herself (Jackson 1982a). Although female animals often choose mates on the basis of the male's courtship behaviour (see Alexander *et al.*, this volume), this is not an accurate portrayal of the intersexual interactions of *Portia*.

Of three species of *Portia* studied in detail (Jackson 1982b; Jackson and Hallas 1986b), *P. labiata* and *P. schultzi* females are unconventional because they often initiate interactions with males by displaying first and because, once interaction begins, they are very active in displaying back to displaying males. Drumming (pounding on the silk with her two palps) and tugging (a sharp pull on silk with forelegs) are the dominant displays of *P. labiata* and *P. schulzi* females. These displays are never performed by females of the other species, *P. fimbriata*, or by the males of any of the three species, and they are performed only in interactions with conspecific males. During the interaction, the female tends to display, decamp a short distance, then turn back and look at the male. Intersexual interactions may take place in or outside of webs, and the female's decamping bouts normally result in her moving to higher ground (to the top of a web or up in the vegetation) with the courting male close behind.

The courtship of *Portia* males consists of various visual (e.g. posturing and waving legs) and vibratory displays (e.g. jerky walking). When the male gets close to the female, he switches to tactile displays: tapping and scraping on the female's body with his legs and palps. These behaviors are performed both prior to and simultaneous with walking over the female (mounting). Once the male is mounted, the female of *P. labiata* and *P. schultzi* drops on a dragline, with the male on board, where the pair mate.

However, the *P. labiata* or *P. schultzi* female usually tries to follow sex with a meal. She almost always makes a 'twisting lunge' in which she suddenly and violently swings around, with fangs extended while scooping toward the male with her legs, after which her unfortunate suitor may find himself impaled on her fangs. He is also at risk when approaching a female, because a passive-appearing female may make a sudden, violent lunge forward and spear the male with her fangs.

Portia labiata and *P. schultzi* females cannibalize males by using a prey-specific predatory tactic (twisting and forward lunges) designed to catch prey that deliberately approach them: courting and mating males. If the female kills the male after he begins to copulate, she chooses the male as both a sperm donor and a meal. If she kills and eats a courting male before mating, she rejects his sperm, but not his body.

The prey-specific predatory tactic *P. labiata* and *P. schultzi* females use against males appear to include more than just the lunging attacks. The female's 'courtship' displays, drumming and tugging, also appear to encourage the male to approach. If the male is killed before mating, the female's behavior might be explained as the male's failure to persuade the female to mate, but there is a viable alternative. Perhaps these prenuptially cannibalistic females use intraspecific aggressive mimicry by encouraging the male's approach when eating is their only interest.

Certainly, the implication of feminine deceit is less ambiguous when the female is immature and cannot mate. In *P. labiata* and *P. schultzi*, males apparently cannot differentiate juvenile from adult females and court them in the same way (Jackson and Hallas 1986b). The male may even mount and perform pseudocopulations, but juvenile females almost always make forward or twisting lunges and sometimes succeed in taking the male as prey.

By drumming and tugging, subadult females practice intraspecific aggressive mimicry to deceive the male into responding to them as though they were adults and thereby lure the male within range for an attack with prey-specific predatory behavior. Chemical deceit (see Stowe 1988) may also be part of the game. Pheromones from adult *Portia* females are known to draw males in and prime them to court (Pollard *et al.* 1987); subadult (i.e. penultimate-instar) females evidently simulate the chemical signal of sexually active adults (Willey and Jackson 1993).

To mate, males of *P. labiata* and *P. schultzi* are in a bind because they must approach a predator that has evolved a prey-specific predatory behavior specifically for them. However, it seems that males have made evolutionary adjustments in behavior for a defence against their rapacious mates.

Gravity assists the female should she attack with a forward lunge from a position above the male, and it is interesting that *P. labiata* and *P. schultzi*, but not *P. fimbriata*, males often make detours (see Tarsitano and Jackson 1992) and approach the female from above, although females often outmaneuver them and gain the higher ground first (Jackson and Hallas 1986b). Also, when females make even the slightest movement, *P. labiata* and *P. schultzi* males leap and run away, but *P. fimbriata* males usually stand their ground. In addition, *P. labiata* and *P. schultzi*, in contrast to *P. fimbriata*, males rarely attempt to mount before the female retracts her appendages; this makes sense because both forward and twist lunges would appear to be executed less effectively by females whose legs are retracted. Evidently, intersexual interactions of *P. labiata* and *P. schultzi* join some of *Portia*'s interspecific relations (Jackson and Wilcox 1993; R. R. Jackson and R. S. Wilcox, unpublished) as instances where the notion of predator–prey coevolution is relevant.

INTRASEXUAL CONFLICT IN *PORTIA*

Courtship is not the only intraspecific context for communication in salticids. Elaborate displays and lengthy interactions are also common among conspecifics of the same sex (Jackson 1982a). The trend in salticids is for male–male interactions to be more elaborate than female–female; male–male interactions have received greater attention, including analysis in the context of game theory (Wells 1988; Jackson and Cooper 1991; Faber and Baylis 1994). Male–male interactions may escalate in the presence of a female (Crane 1949; Wells 1988), suggesting that competition for mates and intrasexual selection are somehow involved. Yet we do not have clear evidence of territoriality or social dominance in salticids and the ultimate cause of intraspecific intolerance in salticids remains unclear (Jackson 1982a).

Again *Portia* is the exception, with female–female interactions being particularly ferocious; the resources at stake are often clearly webs and eggs (Clark and Jackson 1994). Webs spun by one female can be used by another, and the eggs in a web, the resident female's progeny, are potential food for an intruder. *Portia* males also interact, only less aggressively; possibly for *Portia* males, as for salticids in general, agonistic interactions are associated with competition for mates.

For *Portia*, a conspecific of the same sex is not just a rival for some external resource (e.g. webs, eggs or mates), but also potentially food or a predator. Envisaging intrasexual interactions not simply as competition between rivals, but also as predator–prey interactions, provides a fresh, largely unexplored perspective.

For example, a prominent display during *Portia*'s intrasexual interactions is swaying from side to side with its legs hunched (i.e. front legs highly flexed and held at the side of body), chelicerae spread apart and fangs exposed. Even to a human observer, these displays appear menacing. Why?

Surely it is no mere coincidence that these displays resemble the posture that *Portia* adopts just before attacking prey. In fact, this would be a good facsimile for the pre-attack posture of many predators. Perhaps *Portia*'s sensory landscape, to use Guilford and Dawkins' (1991) terminology, has a bias in relation to this stimulus configuration normally associated with the eminent danger of attack by a predator, a bias making *Portia* especially likely to pay attention to, and avoid, the source of stimulation. This interpretation is compatible with a hypothesis that these are displays by which rivals reveal their abilities and intentions with regard to inflicting injury or, to use prevalent terminology, resource-holding power (Parker 1974). However, even if a resource is not obvious, it may still make sense for *Portia* to exploit a bias in a conspecific's sensory landscape. For example, displaying with hunched legs may sometimes have a protective function rather like a 'pursuit invitation' (see Hasson 1991): informing another *Portia* that the element of surprise is no longer working in its favor.

'Propulsive displays' are also common in *Portia*'s intrasexual interactions (Jackson and Hallas 1986b). These consist of sudden and rapid, but usually truncated, movements toward the rival; they almost always elicit a startle response from the *Portia* on the receiving end (i.e. the rival jumps back or runs at least a short distance away). Perhaps propulsive displays take advantage of an antipredator template in the conspecific's sensory landscape.

If this interpretation of propulsive displays is correct, then *Portia* is mimicking, at least crudely, the movements of its own predators. Interestingly, these are not the kinds of movements *Portia* normally adopts against its own prey, which are approached very slowly and only attacked at close range (Jackson 1992a). These displays may be part of a Batesian mimicry system in which the mimic and the dupe are conspecifics, and the model is a heterospecific predator of both (see Hasson 1993).

Often interacting *Portia* come into physical contact, press their faces together, push back and forth, and grapple. Each grappling spider attempts to hook its fore-legs over the basal segments of its partner's legs, then pulls forcefully to remove some of the rival's legs (Jackson 1982a; Jackson and Hallas 1986b). Other salticids do not lose their legs so readily, but easy loss of legs is an advantage to *Portia* in interactions with spider prey in webs. If attacked, *Portia* can leave a leg behind and live to hunt another day, or it can turn back on the spider that took its leg, make a kill and eat the other spider plus its own leg (Jackson and Blest 1982). Loss of a leg is better than loss of life in predatory interactions, but an adult *Portia* cannot regrow the leg it loses. Grappling seems to be a fight by which a *Portia* attempts to exploit a defense mechanism of its conspecific rival and to inflict a type of injury (leg loss) to which these spiders are peculiarly susceptible. While grappling and pushing, one *Portia* often upends the other and sinks its fangs into the conspecific's soft underbelly; the *Portia* with fewer legs is the most likely to be the underdog (Jackson and Hallas 1986b).

COMPLEXITY IN THE INTRASPECIFIC INTERACTIONS OF SALTICIDS

Why salticid males court females

Intersexual selection is a consequence of mate choice (Harvey and Bradbury 1991); in salticids and animals generally, the female appears most often to do the choosing (Alexander *et al.*, this volume). When salticid males and females meet, there is usually an elaborate interaction, with the male being more persistent in displaying (Crane 1949; Richman 1982; Jackson 1992b). Females tend to alternate between watching the male and moving a short distance away. The female can accept the male's advances by allowing him to approach, mount and copulate. However, moving away until the male gives up, driving him away with threat displays and fleeing are all ways in which the female can reject her suitor (Jackson 1982b).

Female choosiness makes accounting for male behavior relatively easy: to mate, the male has to do what the female likes (Bradbury and Andersson 1987). There is no evidence that salticid males contribute significant material resources to the female, suggesting that we should look for an explanation of female choosiness in either good genes or runaway selection (Harvey and Bradbury 1991). The first step, though, is to demonstrate consistent biases for certain types of males over others.

Phidippus johnsoni females prefer males who dance

In *Phidippus johnsoni*, the salticid with the most thoroughly studied mating strategy (Jackson 1992b), there is evidence (Jackson 1981) that females prefer to mate with the males who perform a particular courtship display, called dancing. However, it is not so simple to interpret this as choice of a particular type of conspecific male because it appears that all *P. johnsoni* males can dance (Jackson 1977a). In any particular courtship sequence, the male may or may not choose to dance. Failing to dance does not mean certain rejection by the female, only a lesser chance of acceptance (Jackson 1981), but there is another variable to consider.

In virgin females, the tendency to choose males that dance is not so pronounced. Non-virgin females, who only occasionally mate again when courted, are particularly choosy, and there is no evidence that males can distinguish them from virgin females (Jackson 1977a, 1981). If using the sperm of a male who danced is advantageous, then it appears reasonable that non-virgins would be more choosy. Rejecting a non-dancing male might be costly for a virgin female because she cannot start reproducing until she has sperm, and the threat of predators and drought mean that life may be short for a *P. johnsoni* female in nature (Jackson 1976, 1978). However, a reproductively secure mated female is more free to discriminate when courted later by suitors (Jackson 1981); when a second male mates with a *P. johnsoni* female, his sperm may replace the sperm of the previous male (Jackson 1980a). This strategy may be especially common in species where females can store sperm, and therefore choose to have it displaced and replaced by that of a subsequent male (see Alexander *et al.*, this volume).

Although we do not know why *P. johnsoni* females prefer to mate with dancing males nor why males sometimes decide not to dance, a good-genes hypothesis might be appealing. For the male, dancing may be a demanding task, especially in the complex topography of stones and vegetation in nature, and the female compounds the male's problems by frequently moving away then turning to watch the courting male again. Perhaps males find it difficult not to slip off the substrate or lose sight of the female while dancing, but the kinds of motor and sensory skills that make a male a good dancer might be similar to what makes *P. johnsoni* good at catching prey and escaping from predators. If a female mates with a male that has proven himself by dancing, her offspring may also be especially good at catching prey and escaping predators. However, none of these critical suppositions have yet been investigated.

Generally, testing good-genes hypotheses is difficult. In addition, looking for a good-genes explanation for female choice may be further complicated, if intersexual selection brings about the positive feedback loop in runaway selection models (Fisher 1958; Kirkpatrick 1982, 1987). Another complication is that, in *P. johnsoni* and every salticid that has been studied in detail (Richman 1981; Jackson 1992b), courtship involves much more than any one display such as dancing. Large repertoires of distinctly different displays and exceedingly variable sequences are the rule. However, using manipulated video images of courting males (see Clark and Uetz 1990, 1992) may help unravel the morphological and behavioral characteristics of males that are important for female choice.

Copulatory courtship

Complexity and variability also describe the actual transfer of sperm from the male to the female (Jackson 1980a, 1992b). Males transfer sperm via a pair of intromittent organs on their palps and females have a pair of copulatory orifices on the ventral abdomen. Salticid males apply one palp at a time (Gerhardt and Kaestner 1938), but not in strict regular alternation. Sequences are highly variable, as are durations per palp engagement and total copulation duration (Jackson 1992b). For example, copulation duration in *P. johnsoni* varies from a few seconds to several days, with a copulation of even a few seconds being sufficient to provide for the fertilization of the female's lifetime supply of eggs (Jackson 1980a).

In *P. johnsoni*, copulation duration does not simply correlate with whether or not sperm is transferred nor with the amount of sperm transferred to the female; it functions as 'copulatory courtship' (Eberhard 1991, this volume) by affecting female fidelity. After mating longer, females are less receptive to males that court later (Jackson 1980b). The copulating male also fits the female with a mating plug, but his plug and sperm can be removed by subsequent males that the female chooses to mate with (Jackson 1980a).

How to explain complexity

With so much variability in the nature of male courtship, including copulatory courtship, looking for evidence of good genes in the details of any particular male display

may be an oversimplification. Variability in the whole display repertoire, instead of being a background complication, is probably the main focus of female choice.

A straightforward mechanism has been proposed (Jackson 1982c) for how salticid females might choose males with more variable display behavior: habituation. The evolution of enhanced preference for variability could easily be achieved by lowering the female's threshold for habituating to monotony in male courtship. However, if we want to understand the factors responsible for setting the female's threshold for habituation, hypotheses based on good-genes selection require that variability demonstrates to the female something about male fitness. How this might come about is unclear. Another problem is that in intrasexual interactions of salticids we find complexity that parallels what is seen in male–female interactions (Richman and Jackson 1992), yet mate choice has no obvious relevance here.

Lessons from *Portia's* predatory strategy

When we looked at *Portia's* intraspecific interactions, we found it useful to emphasize how individual displays might act on biases in the conspecific's nervous system, much as aggressive mimicry displays appear to act on biases in heterospecific prey's nervous system. Large repertoires of displays and exceedingly variable sequences are common to both intra- and intersexual interactions, suggesting a common agenda. Because *Portia's* agenda during interactions with heterospecific prey appears to be dynamic fine control of the other spider's behavior, it would be interesting to investigate whether similar fine control underlies variability in *Portia's* interaction with conspecifics. For example, does *Portia* attempt to draw its conspecific partner in at a certain speed, keep the conspecific quiescent in an appropriate place, or maneuver it into a particular position or orientation? Is trial and error used to achieve these goals? Perhaps these questions are relevant to salticids or animals generally, not just *Portia*.

When the other spider is heterospecific, *Portia's* goal is not only to catch prey, but also to avoid becoming prey. During conspecific interactions, the goal appears still more complex: *Portia* may be a potential mate or prey and rival or predator. However, this may be generally true for salticids, not just *Portia*.

In discussion of spider mating strategies, there is a long tradition of emphasizing sexual cannibalism, broadly defined as the eating of actual or potential mates (Elgar

1992). However, exactly what influence sexual cannibalism has had on the evolution of salticid mating strategies has never been clear.

One of the most interesting topics related to sexual cannibalism has been whether males might be programmed to feed their bodies, after mating, to their children's future mother (Buskirk *et al.* 1984). There are convincing examples of this in a few non-salticid spiders (see, for example, Forster 1992), but salticid males appear intent on surviving. In fact, a link between species identification and anticannibalism in spiders has often been emphasized as a function of male courtship (Robinson 1982). Spider females, according to conventional wisdom, are ravenous predators and males must identify themselves or else run the risk of being perceived as just another prey. However, for salticids (Jackson 1982b; Richman and Jackson 1992), and for spiders generally (Starr 1988; Jackson and Pollard 1982, 1990), this portrayal breaks down under close examination.

It breaks down in a particularly interesting way in *P. labiata* and *P. schultzi*. In these species, the female almost always tries to catch and eat the male, but because she uses a predatory tactic that is specific to the context of eating courting males, arguing mistaken identification is impossible. Instead, the female shows us by her behavior of drumming and tugging that she knows who the male is (Jackson and Hallas 1986b). It is probably generally the case in salticids that cannibalism is a danger during intraspecific interactions. Females eat males, but males also eat females, and males and females eat each other (Jackson 1982a, 1992b); but misidentification seems to have little to do with who eats whom (Jackson 1982b).

Interactions between cannibals

Encounters between salticids are encounters between cannibals, and it is unlikely that cannibalism stems simply from mistaken identity. Taking this as our starting point, we can formulate hypotheses of a type rarely considered in the past.

For example, in most salticid species studied (Richman 1981; Jackson 1992b), a prominent male display during courtship involves some variation on a posture with the first pair of legs raised and extended forward. Regardless of whether or not this posture informs the female of the male's species, sex and intentions (Drees 1952), it also puts up a barrier between a potential cannibal and the male's own body. The forward-extended legs may function partly as a physical obstruction should the female attack, but they

may also be a way of telling the female that an attack is unlikely to be successful. One way for a male to ensure that a cannibal behaves appropriately until he can persuade her to accept his advances may be to show himself and let the female recognize that he is potential food, while showing her that he is not going to be an easy catch.

Interestingly, a courting male usually does not move his legs to the side and expose his mouthparts to the female; but displays of this sort, such as the hunched-legs displays of *Portia*, are routine in the intrasexual interactions of many species. In intrasexual interactions, salticids appear to be attempting to make conspecifics decamp; showing off weapons may be an especially good way to achieve this. In addition, females tend to adopt hunched-legs and similar displays when interacting with males, regardless of whether or not they subsequently mate; it may make sense for the female to appear dangerous.

Salticids may orchestrate a careful balance between stimuli that provoke and stimuli that inhibit predatory attacks from each other. For example, in many salticids, males initiate courtship with displays involving flickering movement, such as the waving of forelegs up and down (Richman 1981; Jackson 1992b). However, flickering movements may also tend to be what attracts hungry females to potential prey such as insects (Heil 1936; Forster 1985). Clark and Uetz (1992, 1993) suggested that, by emphasizing these movements at the start of courtship, males may be exploiting reflexes of the female that evolved in the context of predation. That is, the male may be deliberately provoking the first step of a predatory sequence from the female as a way to get her attention, after which he can concentrate on trying to persuade her to mate.

In the present paper, it is not so much any particular hypotheses that we want to emphasize, but instead the different perspective that comes from recognizing that, for the salticid, eating or being eaten by a conspecific is one of the variables. Much of the complex interplay between the two spiders may start to make sense in a context similar to a lion trainer and a lion, where each salticid may be analogous to the trainer or the lion, if not both at the same time.

COURTSHIP VERSATILITY

Among animals, predatory versatility (Curio 1976), or conditional predatory strategies, are generally not as pronounced as in *Portia*, where disparate tactics are adopted depending on location (in own web, in another spider's web, or completely away from webs) and type of prey (preprogrammed prey-specific tactics). However, there is a striking parallel between *Portia*'s predatory versatility and the mating strategies of cursorial salticids (i.e. species that do not build webs).

Alternative strategies (Dominey 1984), where, for example, large males defend territories and small males wait opportunistically for undefended females, are not known in salticids. However, in almost every salticid species studied in detail (Jackson 1992b), we find pronounced conditional mating strategies (Dawkins 1980), or courtship versatility, where males adopt disparate tactics depending on the female's location and state of maturity.

Unlike *Portia*, most salticids are not web-builders, but most spin a tubular silk nest used as a refuge when quiescent, molting or ovipositing (Jackson 1979). These are also places where males court and mate with females. If the female is in a nest and she is a juvenile that will molt and reach maturity within two weeks or less, the male adjoins another chamber to the juvenile's nest, then waits for the female to mature, enters her chamber and mates (Jackson 1986b). If the female in the nest is mature, the male courts and mates without first building an adjoining chamber. Of course, if the male finds a mature female away from her nest and there is sufficient light, he performs the complex visually mediated courtship for which salticids are renowned.

Courting a female at a nest entails a problem. Salticid nests tend to be tightly woven with opaque silk and they are often situated under loose bark, inside rolled-up leaves and in similar locations where ambient light levels are low, rendering ineffective the salticid's elaborate repertoire of visually mediated displays. The salticid's solution is remarkable. It has a full alternative communication system in a different modality and with a repertoire of multiple displays distinct from those used in the open. For example, by using a battery of specialized behaviors, the salticid may tug, pull and probe with its legs and palps on the female's nest and send vibratory signals that she can perceive in total darkness (Jackson 1977b).

The salticid's vibratory communication system at nests parallels the vision-based communication system away from nests. Although the male is the more persistent at making vibratory displays, females also display to males; there are also intrasexual interactions at nests based on vibratory display. In addition, salticids combine vibratory displays in highly variable sequences (Jackson 1982a, 1992b), suggesting that dynamic fine control is relevant

not only to interactions out in the open but also to interactions at nests.

Similar vibratory communication is practiced by nest-building spiders from other families that, unlike salticids, lack acute vision (Robinson 1982; Jackson 1986a). It appears that, when vision is not appropriate, the salticid can behave like a non-salticid. Clubionids, for example, tug, pull and probe on nests with behaviors little different from those seen in various salticids (see, for example, Pollard and Jackson 1982). Repertoire size and complexity of sequences are at least comparable, if not more elaborate, in salticids.

Emphasis on vision-based communication is the tradition in literature on salticids (see, for example, Robinson 1982). However, salticids also routinely use elaborate tactile communication as a distinctive phase within both basic types of interaction: away from and at nests (Jackson 1992b). There are also examples of auditory communication in salticids (Edwards 1981; Gwynne and Dadour 1985; Maddison and Stratton 1988).

It might be tempting to suggest that pheromones and other chemical signals (Tietjen and Rovner 1982; Stowe 1988; Schultz and Toft 1993) are primarily components of the communication systems of the poor-sighted non-salticid spiders, but chemical systems are also well developed in salticids. Salticids, in fact, are probably the spiders for which we have the most information currently about chemical communication (Jackson 1987; Pollard *et al.* 1987).

Appreciation of the importance of the vibratory communication system at nests came about only when this system was investigated in detail. Likewise, appreciation of communication based on tactile, auditory and chemical cues came about only when these systems were studied in detail. If these other systems had not been investigated in detail, then on *a priori* grounds quite a different argument might have made sense. It might have been argued that development of acute vision brought about an adaptive trade-off, which moved salticids away from being proficient at communication using other sensory modalities.

The idea of adaptive trade-offs (Levins 1968; Curio 1976), coupled with an appreciation of the small size of the spider nervous system, may prime us to expect spiders with poor eyesight to be somehow better than salticids at what they can do, in compensation for what they cannot do, i.e. for not being able to communicate with vision-based signals. But where is the evidence of this? Recent studies of salticids suggest a different conclusion: that in spiders the jack of all sensory trades can be the master of all. Salticids appear not to have substituted vision for other sensory modalities; instead, they appear simply to have added on vision, setting the stage for the evolution of exceptionally complex communication systems.

CONCLUSIONS AND SUGGESTIONS FOR FUTURE RESEARCH

Mating strategies cannot evolve in isolation: they influence and are influenced by other processes of the animal's life. However, we want a detailed understanding of these inter-relationships at work. Herein may be a largely untapped advantage of using salticids as research animals. In salti-cids, it appears that mating strategies have a complex, yet tractable, relationship with other kinds of strategies, espe-cially predatory strategies.

Interrelationship of strategies was only one of two pro-blems addressed in this chapter, but the other problem, complexity, might best be viewed as a piece of the same whole because the explanation for why salticid intraspecific interactions are complex may lie largely in appreciating how conflicting interests interrelate during the exchange of salticid displays. We coin terms such as 'courtship', but what is going on for the salticid may be considerably more complex, or more subtle, than such terms prepare us to appreciate.

There was a time when, either explicitly or implicitly, dis-cussing animal communication was tightly tied up with the idea of information. This began to change especially after Dawkins and Krebs (1978) provocatively subtitled a paper on communication 'information or manipulation?' and emphasized how, when animals communicate, manipula-tion is indirect. The animal on the receiving end is not directly coerced into doing anything; instead, it perceives a stimulus and responds to it. Taking this as our starting point may not lead us entirely away from the notion of infor-mation (Markl 1985), but it does change our perspective.

In the past, we tended to look for more or less consis-tent correlations between signal and signaler, which is another way of saying 'information' (Smith 1977). The rationale was that, if we could identify what the signal is usually associated with (for example, good genes carried by the signaler), then we should be able to account for the receiving animal's response (e.g., mating with the signaler). The idea of deceit was also tied to 'information' because we tended to think of 'deceit' as instances in which the associa-tion was deliberately (adaptively) unhinged by the sender.

In salticids (Jackson 1982a, 1992b), as in many animals (Smith 1977), whatever correlations there may be between attributes of signals and the signaler tend not to be straight-forward. It may be more useful to begin by asking how the salticid's signals play directly on perceptual systems by, for example, attracting and maintaining the interest of, habitu-ating or dishabituating, and startling or calming the animal at the receiving end. This is a different perspective, with receiver psychology (Guilford and Dawkins 1991) in the foreground. The relevance of speaking of information at all may be receding into the background. This approach is not incompatible with explanations based on, for example, good-genes hypotheses for sexual selection. The starting point, however, is different.

We want 'receiver psychology' to be more than a meta-phor, and herein may lie one of the advantages of using salticids for study animals. The prospects of fathoming the mind of this particular kind of spider may be surpris-ingly good.

ACKNOWLEDGEMENTS

David Clark and Will McClintock provided valuable com-ments on the manuscript. We thank Tracey Robinson and Helen Spinks for their invaluable help in the preparation of the manuscript. Parts of the research summarized was supported by grants from the National Geographic Society (2330–81, 3226–85), the United States – New Zealand Coop-erative Program of the National Science Foundation (BNS 8617078), and the New Zealand Marsden Fund.

LITERATURE CITED

Blest, A. D., D. C. O'Carroll and M. Carter. 1990. Comparative ultrastructure of Layer I receptor mosaics in principal eyes of jumping spiders: the evolution of regular arrays of light guides. *Cell Tiss. Res.* **262**: 445–460.

Bradbury, J. W. and M. B. Andersson. 1987. Introduction. In *Sexual selection: Testing the Alternatives*. J. W. Bradbury and M. B. Anders-son, eds., pp. 1–8. New York: John Wiley & Sons Ltd.

Buskirk, R. E., C. Frohlich and K. G. Ross. 1984. The natural selec-tion of sexual cannibalism. *Am. Nat.* **123**: 612–625.

Clark, D. L. and G. W. Uetz. 1990. Video image recognition by the jumping spider, *Maevia inclemens* (Araneae: Salticidae). *Anim. Behav.* **40**: 884–890.

–. 1992. Morph-independent mate selection in a dimorphic jump-ing spider: demonstration of movement bias in female choice using video-controlled courtship behavior. *Anim. Behav.* **43**: 247–254.

–. 1993. Signal efficacy and the evolution of male dimorphism in the jumping spider, *Maevia inclemens*. *Proc. Natl. Acad. Sci. U.S.A.* **90**: 1954–1957.

Clark, R. J. and R. R. Jackson. 1994. *Portia labiata*, a cannibalistic jumping spider, discriminates between own and foreign egg-sacs. *Int. J. Comp. Psychol.* **7**: 38–43.

Coddington, J. A. and H. W. Levi. 1991. Systematics and evolution of spiders (Araneae). *Annu. Rev. Ecol. Syst.* **22**: 565–592.

Crane, J. 1949. Comparative biology of salticid spiders at Rancho Grande, Venezuela. Part IV. An analysis of display. *Zoologica (N.Y.)* **34**: 159–215.

Crompton, J. 1951. *The Life of the Spider*. Cambridge: The Riverside Press.

Curio, E. 1976. *The Ethology of Predation*. Berlin: Springer-Verlag.

Dawkins, R. 1976. *The Selfish Gene*. London: Oxford University Press.

–. 1980. Good strategy or evolutionarily stable strategy? In *Sociobiology: Beyond Nature/Nurture?* G. W. Barlow and J. Silverberg, eds., pp. 331–367. Boulder, Colorado: Westview Press.

Dawkins, R. and J. R. Krebs. 1978. Animal signals: information or manipulation? In *Behavioral Ecology: An Evolutionary Approach*, 1st edn. J. R. Krebs and N. B. Davies, eds., pp. 282–309. Sunderland, Massachusetts: Sinauer.

Dominey, W. J. 1984. Alternative mating tactics and evolutionarily stable strategies. *Am. Zool.* **24**: 385–396.

Drees, O. 1952. Untersuchungen über die angeborenen Verhaltensweisen bei Springspinnen (Salticidae). *Z. Tierpsychol.* **9**: 169–207.

Eberhard, W. G. 1991. Copulatory courtship and cryptic female choice in insects. *Biol. Rev.* **66**: 1–31.

Edwards, G. B. 1981. Sound production by courting males of *Phidippus mystaceus* (Araneae: Salticidae) *Psyche* **88**: 199–214.

Elgar, M. A. 1992. Sexual cannibalism in spiders and other invertebrates. In *Cannibalism: Ecology and Evolution among Diverse Taxa*. M. A. Elgar and B. J. Crespi, eds., pp. 128–155. Oxford University Press.

Faber, D. B. and J. R. Baylis. 1994. Effects of body size on agonistic encounters between male jumping spiders (Araneae: Salticidae). *Anim. Behav.* **45**: 289–299.

Fisher, R. A. 1958. *The Genetical Theory of Natural Selection*, 2nd edn. New York: Dover.

Foelix, R. F. 1982. *Biology of Spiders*. Cambridge, Massachusetts: Harvard University Press.

Forster, L. M. 1985. Target discrimination in jumping spiders (Araneae: Salticidae) In *Neurobiology of Arachnids*. F. G. Barth, ed., pp. 249–274. Berlin: Springer-Verlag.

–. 1992. The stereotyped behavior of sexual cannibalism in *Latrodectus hasselti* Thorell (Araneae: Theridiidae), the Australian redback spider. *Austr. J. Zool.* **40**: 1–11.

Gerhardt, U. and A. Kaestner. 1938. Araneae. In *Handbuch der Zoologie*, vol. 3. W. G. Kukenthal and T. Krumbach, eds., pp. 497–656. Berlin: De Gruyter.

Guilford, T. and M. S. Dawkins. 1991. Receiver psychology and the evolution of animal signals. *Anim. Behav.* **42**: 1–14.

Gwynne, D. T. and I. R. Dadour. 1985. A new mechanism of sound production by courting male jumping spiders (Araneae: Salticidae, *Saitis michaelseni* Simon). *J. Zool. (Lond.)* **207**: 35–42.

Harvey, P. H. and J. W. Bradbury. 1991. Sexual selection. In *Behavioral Ecology: An Evolutionary Approach*, 3rd edn. J. R. Krebs and N. B. Davies, eds., pp. 203–233. London: Blackwell Scientific Publications.

Hasson, O. 1991. Pursuit-deterrent signals: communication between prey and predator. *Trends Ecol. Evol.* **6**: 325–329.

–. 1993. Cheating signals. *J. Theor. Biol.* **167**: 223–238.

Heil, K. H. 1936. Beiträge zur Physiologie und Psychologie der Springspinnen. *Z. Vergl. Phsyiol.* **23**: 1–25.

Jackson, R. R. 1976. Predation as a selection factor in the mating strategy of the jumping spider *Phidippus johnsoni* (Salticidae, Araneae). *Psyche* **83**: 243–255.

–. 1977a. An analysis of alternative mating tactics of the jumping spider *Phidippus johnsoni* (Araneae, Salticidae). *J. Arachnol.* **5**: 185–230.

–. 1977b. Courtship versatility in the jumping spider, *Phidippus johnsoni* (Araneae: Salticidae). *Anim. Behav.* **25**: 953–957.

–. 1978. Life history of *Phidippus johnsoni* (Araneae, Salticidae). *J. Arachnol.* **6**: 1–29.

–. 1979. Nests of *Phidippus johnsoni* (Araneae, Salticidae): characteristics, pattern of occupation, and function. *J. Arachnol.* **7**: 47–58.

–. 1980a. The mating strategy of *Phidippus johnsoni* (Araneae, Salticidae): II. Sperm competition and the function of copulation. *J. Arachnol.* **8**: 217–240.

–. 1980b. The mating strategy of *Phidippus johnsoni* (Araneae, Salticidae): III. Intermale aggression and a cost-benefit analysis. *J. Arachnol.* **8**: 241–249.

–. 1981. Relationship between reproductive security and intersexual selection in a jumping spider *Phidippus johnsoni* (Araneae: Salticidae). *Evolution* **35**: 601–604.

–. 1982a. The behavior of communicating in jumping spiders (Salticidae). In *Spider Communication: Mechanisms and Ecological Significance*. P. N. Witt and J. S. Rovner, eds., pp. 213–247. Princeton: Princeton University Press.

–. 1982b. The biology of *Portia fimbriata*, a web-building jumping spider (Araneae, Salticidae) from Queensland: intraspecific interactions. *J. Zool. (Lond.)* **196**: 295–305.

–. 1982c. Habituation as a mechanism of intersexual selection. *J. Theor. Biol.* **97**: 333–335.

–. 1986a. Web building, predatory versatility, and the evolution of the Salticidae. In *Spiders: Webs, Behavior, and Evolution*. W. A. Shear, ed., pp. 232–268. Stanford: Stanford University Press.

–. 1986b. Cohabitation of males and juvenile females: a prevalent mating tactic of spiders. *J. Nat. Hist.* **20**: 1193–1210.

–. 1987. Comparative study of releaser pheromones associated with the silk of jumping spiders (Araneae, Salticidae). *N. Z. J. Zool.* **14**: 1–10.

–. 1992a. Eight-legged tricksters: spiders that specialize at catching other spiders. *BioScience* **42**: 590–598.

–. 1992b. Conditional strategies and interpopulation variation in the behavior of jumping spiders. *N. Z. J. Zool.* **19**: 99–111.

Jackson, R. R. and A. D. Blest. 1982. The biology of *Portia fimbriata*, a web-building jumping spider (Araneae, Salticidae) from Queensland: utilization of webs and predatory versatility. *J. Zool. (Lond.)* **196**: 255–293.

Jackson, R. R. and K. J. Cooper. 1991. The influence of body size and prior residency on the outcome of male-male interactions of *Marpissa marina*, a New Zealand jumping spider (Araneae: Salticidae). *Ethol. Ecol. Evol.* **3**: 79–82.

Jackson, R. R. and S. E. A. Hallas. 1986a. Capture efficiencies of web-building jumping spiders (Araneae, Salticidae): is the jack-of-all-trades the master of none? *J. Zool. (Lond.)* **209**: 1–7.

–. 1986b. Comparative biology of *Portia africana*, *P. albimana*, *P. fimbriata*, *P. labiata*, and *P. schultzi*, araneophagic web-building jumping spiders (Araneae: Salticidae): utilization of silk, predatory versatility, and intraspecific interactions. *N. Z. J. Zool.* **13**: 423–489.

Jackson, R. R. and S. D. Pollard. 1982. The biology of *Dysdera crocata* (Araneae, Dysderidae): intraspecific interactions. *J. Zool. (Lond.)* **198**: 197–214.

–. 1990. Intraspecific interactions and the function of courtship in mygalomorph spiders: a study of *Porrhothele antipodiana* (Araneae: Hexathelidae) and a literature review. *N. Z. J. Zool.* **17**: 499–526.

Jackson, R. R. and R. S. Wilcox. 1990. Aggressive mimicry, prey-specific predatory behavior and predator-recognition in the predator-prey interactions of *Portia fimbriata* and *Euryattus* sp., jumping spiders from Queensland. *Behav. Ecol. Sociobiol.* **26**: 111–119.

–. 1993. Spider flexibly chooses aggressive mimicry signals for different prey by trial and error. *Behaviour* **127**: 21–36.

Kirkpatrick, M. 1982. Sexual selection and the evolution of female choice. *Evolution* **36**: 1–12.

–. 1987. Sexual selection by female choice in polygynous animals. *Annu. Rev. Ecol. Syst.* **18**: 43–70.

Land, M. F. 1969a. Structure of the retinae of the eyes of jumping spiders (Salticidae: Dendryphantinae) in relation to visual optics. *J. Exp. Biol.* **51**: 443–470.

–. 1969b. Movements of the retinae of jumping spiders (Salticidae: Dendryphantinae) in response to visual stimuli. *J. Exp. Biol.* **51**: 471–493.

Levins, R. 1968. *Evolution in Changing Environments*. Princeton: Princeton University Press.

Maddison, W. P. and G. E. Stratton. 1988. Sound production and associated morphology of the *Habronattus agilis* species group (Araneae, Salticidae). *J. Arachnol.* **16**: 194–211.

Markl, H. 1985. Manipulation, modulation, information, cognition: some of the riddles of communication. In *Experimental Behavioral Ecology*. B. Hölldobler and M. Lindauer, eds., pp. 163–194. New York: G. Fischer Verlag.

Masters, W. M., H. S. Markl and A. M. Moffat. 1986. Transmission of vibrations in a spider's web. In *Spiders: Webs, Behavior, and Evolution*. W. A. Shear, ed., pp. 49–69. Stanford: Stanford University Press.

McKeown, K. C. 1952. *Australian Spiders: their Lives and Habits*. Sydney: Angus & Robertson.

Mitchell, R. W. 1986. A framework for discussing deception. In *Deception: Perspectives on Human and Nonhuman Deceit*. R. W. Mitchell and N. S. Thompson, eds., pp. 3–40. Albany, New York: State University of New York Press.

Parker G. A. 1974. Assessment strategy and the evolution of fighting behavior. *J. Theor. Biol.* **47**: 223–243.

Pollard, S. D. and R. R. Jackson. 1982. The biology of *Clubiona cambridgei* (Araneae, Clubionidae): intraspecific interactions. *N.Z. J. Ecol.* **5**: 44–50.

Pollard, S. D., A. M. Macnab and R. R. Jackson. 1987. Communication with chemicals: pheromones and spiders. In *Ecophysiology of Spiders*. W. Nentwig, ed., pp. 133–141. Heidelberg: Springer-Verlag.

Proctor, H. C. 1992. Sensory exploitation and the evolution of male mating behavior: a cladistic test using water mites (Acari: Parasitengona). *Anim. Behav.* **44**: 745–752.

Richman, D. B. 1981. A bibliography of courtship and agonistic display in salticid spiders. *Peckhamia* **2**: 16–23.

–. 1982. Epigamic display in jumping spiders (Araneae, Salticidae) and its use in systematics. *J. Arachnol.* **10**: 47–67.

Richman, D. B. and R. R. Jackson. 1992. A review of the ethology of jumping spiders (Araneae, Salticidae). *Bull. Br. Arachnol. Soc.* **9**: 33–37.

Robinson, M. H. 1982. Courtship and mating behavior in spiders. *Annu. Rev. Entomol.* **27**: 1–20.

Ryan, M. J. and A. Keddy-Hector. 1992. Directional patterns of female mate choice and the role of sensory biases. *Am. Nat.* **139**: S4–S35.

Schultz, S. and S. Toft. 1993. Identification of sex pheromone from a spider. *Science (Wash., D.C.)* **260**: 1635–1637.

Smith, W. J. 1977. *The Behavior of Communicating*. Cambridge, Massachusetts: Harvard University Press.

Starr, C. K. 1988. Sexual behavior in *Dictyna volucripes* (Araneae, Dictynidae). *J. Arachnol.* **16**: 321–330.

Stowe, M. K. 1988. Chemical mimicry. In *Chemical Mediation of Coevolution*. K. C. Spencer, ed., pp. 513–580. New York: Pergamon.

Tarsitano, M. S. and R. R. Jackson. 1992. Influence of prey movement on the performance of simple detours by jumping spiders. *Behaviour* **123**: 106–120.

Tietjen, W. J. and J. S. Rovner. 1982. Chemical communication in lycosids and other spiders. In *Spider Communication: Mechanisms and Ecological Significance*. P. N. Witt and J. S. Rovner, eds., pp. 249–279. Princeton: Princeton University Press.

Wells, M. S. 1988. Effects of body size and resource value on fighting behavior in a jumping spider. *Anim. Behav.* **36**: 321–326.

Wickler, W. 1968. *Mimicry in Plants and Animals*. London: Weidenfeld and Nicholson.

Willey, M. B., and R. R. Jackson. 1993. Olfactory cues from conspecifics inhibit the web-invasion behavior of *Portia*, a web-invading, araneophagic jumping spider (Araneae, Salticidae). *Can. J. Zool.* **71**: 1415–1420.

21 · Sexual conflict and the evolution of mating systems

WILLIAM D. BROWN, BERNARD J. CRESPI AND JAE C. CHOE

ABSTRACT

We review mating-systems theory and conflict theory and apply them to the analysis of intrasexual and intersexual conflicts. We distinguish three types of sexual interaction, persuasive, coercive, and forcing, that are used by members of each sex to gain control over fertilization and reproductive resources, and we discuss the nature of male–male, female–female and intersexual interactions with respect to these types of behavior. The mating sequence from pair formation to the end of parental care provides the context for analyzing sexual conflict and confluence of interest; events during any given stage in the sequence may affect the dynamics of other events. We suggest that three main sets of variables influence the evolution of mating systems: (1) resource and mate distributions; (2) the presence and extent of transferrable genetic and material benefits; and (3) the degree of control exerted by each sex over events at different stages in the mating sequence. The former two sets of variables are more or less predictable from ecology, demography and life-history adaptations not directly related to mating behavior; the latter is intimately related to the first two, but should often be unpredictable owing to the idiosyncracies of lineage-specific events.

INTRODUCTION

Insects and arachnids display a diversity of reproductive strategies unparalleled among animals; this diversity has put research on these creatures at the forefront of the study of animal mating systems (Blum and Blum 1979; Thornhill and Alcock 1983; Gwynne and Morris 1983; Smith 1984). In this chapter, we review the current status of the study of sexual selection and mating systems in insects and arachnids, focusing on the role of sexual conflict and using the chapters in this book as guides to theory and empirical knowledge. Our goals are to provide a critical overview of the state of the field, and to sketch out promising routes for the future.

We have divided this chapter into three sections. We begin with a brief review of mating-system theory (see also Otte 1979; Cronin 1991; Andersson 1994), and provide an overview of the theory related to the evolution of conflict in general and male–female (intersexual) conflict in particular. Second, we discuss the differing nature of conflicts that occur during the three types of sexual competition: male–male, female–female, and female–male. We then focus on application of sexual-conflict theory to the different stages in the sequence of mating behaviors between pair formation and the termination of parental care, drawing heavily on the hypotheses proposed by Alexander *et al.* (this volume) and Eberhard (this volume) and the examples provided by the taxa discussed in this book. Third, we discuss the formation of a cohesive theory for the evolution of mating systems that incorporates intersexual conflict among its primary dynamics.

MATING-SYSTEMS THEORY

The term 'mating system' refers to the behavioral, morphological and physiological mechanisms by which gamete union is accomplished (Emlen and Oring 1977; Davies 1991). Current mating-systems theory focusses on the role of sex differences in parental investment (Trivers 1972) and intrasexual competition by the more polygamous sex (usually males) in determining the structure of mating systems (Emlen and Oring 1977). Differences in parental investment create an asymmetry between the sexes in the fitness consequences of polygamy (Bateman 1948; Williams 1966; Trivers 1972), causing males to compete for monopolization of females and their ova (Emlen and Oring 1977; Thornhill 1986; Vehrencamp and Bradbury 1984; see also Clutton-Brock and Vincent 1991; Clutton-Brock and Parker 1992). The ability of a portion of males to monopolize access to females is thought to be the fundamental determinant of mating systems (Emlen and Oring 1977; Thornhill 1986). This ability is believed to be primarily determined by the spatial and temporal patterns of female dispersion (Crook 1965; Emlen and Oring 1977; Wittenberger 1979; Thornhill and Alcock 1983). Parker (1970, 1984) extended the theory to include opportunities for mate monopolization after mating via competition among the sperm of males that sequentially inseminate females. Differences among mating systems may therefore reflect a combination of differences in female dispersion and sperm competition.

The Emlen–Oring–Parker scheme emphasizes male–male competition, either direct or mediated through female choice, as the primary social variable causing differences within and among species in mating systems. Females have generally been treated as passive players and their effect on mating systems occurs primarily as an incidental consequence of their attraction to resources (but see Walker 1980; Knowlton and Greenwell 1984; Ahnesjö et al. 1993; Berglund et al. 1993; Gowaty 1994, 1996; Höglund and Alatalo 1995). An active role for females is mostly limited to 'default' mating systems in which the distribution of females precludes mate monopolization (see, for example, Bradbury and Gibson 1983). The Emlen and Oring (1977) classification system has proven extremely useful for many taxa (see, for example, Thornhill and Alcock 1983). For example, in many species it successfully predicts the consequences of increases in population density and skew in sex ratio on the adoption of different male mate-acquisition tactics (Alexander 1961; Fincke et al., this volume; Zuk and Simmons, this volume). The few tests of the relationship between resource dispersion and mating systems also support the predictions of Emlen and Oring (see, for example, Goldsmith 1987) as do numerous descriptive accounts (see, for example, Waage 1984a; Fincke et al., this volume). Ultimately, however, the scheme is limited by the presumption of male control over copulation and fertilization. Deviations from this pattern of male control reduce the applicability of the Emlen–Oring scheme, yet conflict between the sexes and coercion over mating and fertilization have received growing attention from behavioral ecologists (Parker 1979, 1984; Davies 1985, 1989; Eberhard 1985, 1996; West-Eberhard et al. 1987; Clutton-Brock and Parker 1995).

Reproductive conflicts between the sexes may give rise to behavioral mechanisms whereby each sex attempts to gain control over different stages in the sequence between mate attraction and the termination of parental investment (Parker 1979; West-Eberhard et al. 1987; Alexander et al., this volume). Further development of mating-systems theory therefore necessitates an understanding of under what conditions and to what extent each sex controls different stages of mating behavior, especially pair formation, copulation, fertilization, and parental investment.

CONFLICT THEORY

Conflict arises whenever the outcome of an interaction yields differing optima for different individuals or classes of individuals. Each party will be selected to manipulate the interaction in ways that bring the outcome closer to its optimum. These changes may occur via three possible types of mechanism: (1) persuasion, whereby one party increases the benefits provided to the other party by offering a fitness incentive to gain a share of control; (2) coercion, whereby one party, generally the party furthest from its optimum, attempts to gain control through actions that exert a cost to the other party; and (3) force, which involves physically, behaviorally, or physiologically taking control away (see West-Eberhard et al. 1987 for a scheme that groups force with coercion; see also Alexander et al., this volume). During the various events in the mating sequence of a given species, these processes need not be mutually exclusive, although force should seldom coincide with persuasion or coercion. We stress that both sexes may engage in each type of mechanism, to varying extents in different species.

Persuasion involves cooperative resolution of conflict, such that one party increases the benefits received by the other when it acts in its own interests. Examples of persuasive acts include nuptial gifts and attraction of mates via signaling of good genes or material resources. Persuasive interactions have been modeled in the context of cooperative breeding in animal societies (for example, the peace incentive in cooperative breeders (Keller and Reeve 1994)), and we apply the concept to sexual conflict.

Coercive and forcing interactions have been modeled by Parker (1979; Hammerstein and Parker 1987; Clutton-Brock and Parker 1995) and others (e.g. Higashi and Yamamura 1994). During coercion, opposing parties engage in a game in which costs to each party of physiological, morphological, or behavioral changes increase until for one of them the costs of further change exceed the costs of giving in to one's opponent (Higashi and Yamamura 1994). Examples of coercion include harrassment, limiting access to resources, and struggles over mating initiation or termination. Coercion may often involve a war of attrition, in which costs for each player increase with costs invested by their opponent (Parker 1979; Arnqvist and Rowe 1995). For example, a war of attrition may exist in the physical struggle between a male and female over possession of a nuptial prey item (as in scorpionflies). The effort that a female must expend to acquire the entire gift is proportional to the effort the male expends in defending the item (Thornhill 1981). Wars of attrition begin with assessment of relative competitive ability and intent, leading to either immediate settlement or struggles that eventually lead to one side giving in to the other, resolving the conflict (Parker 1979). Which party wins depends on the intensity of selection on the two players (the cost : benefit ratio of winning), on the condition of opponents, and on asymmetries in the efficacy of adaptations for competing. Over evolutionary time, wars of attrition result in improved assessment capabilities, involving quick resolution of conflict (Wilkinson and Dodson, this volume) and fewer (though possibly more extreme) escalated struggles when competitive abilities and payoffs are similar.

Forcing necessarily reduces or eliminates the benefits of control for the opponent but it may or may not incur additional costs; control is simply taken away. Examples of forcing include traumatic insemination (see, for example, Hinton 1964; Carayon 1966; Lloyd 1979a), forced copulation (see, for example, Thornhill 1980), damage to genitalia (in fishes, see Constantz 1984), sperm removal (Waage 1979a) and other forms of sperm manipulation, sensory traps (West-Eberhard et al, 1987), and infanticide (see, for example, Eggert and Müller 1997; Smith 1997). During the evolution of forcing interactions, opposing parties engage in an opponent-independent costs game (Parker 1979); one example of such a game may be the evolution of genitalic complexity, whereby each incremental change in female genitalia may occur at a cost that is incidental to the cost of genitalic complexity incurred by the male with whom she eventually mates. There is no escalation of conflict over ecological time, but given sufficient genetic variability the game can produce evolutionary 'arms races' with unpredictable outcomes and no resolutions (Parker 1979), or resolution without any struggle or assessment (i.e. when one side 'wins').

We will apply conflict theory to intrasexual and intersexual interactions, with emphasis on the object of the conflicts, when conflicts occur, and how they are resolved. We first discuss the nature of male–male and female–female conflicts, and we then focus in detail on intersexual conflict, and its interaction with intrasexual conflicts, throughout the mating sequence.

SEXUAL CONFLICT

Male–male conflict

Male–male conflict, particularly male–male competition for mates, has dominated work on sexual selection and the evolution of mating systems. Consequently, there is a tremendous literature documenting the methods males use to attempt to outcompete their rivals in obtaining matings, and the components of behavior and morphology that result in success (reviews in Blum and Blum 1979; Thornhill and Alcock 1983; Andersson 1994). However, we believe that putting male–male conflict within the wider context of conflict between and within both sexes provides novel inference on the role of male–male conflict in mating systems, and we hope to show why this is true. We begin this section with a discussion of the causes of male–male conflict and the factors determining which of these causes will predominate. Next, we discuss the factors determining when during the mating sequence male–male conflict will occur.

The primary object of male–male conflict is the limited opportunity to fertilize ova (i.e. competition for prezygotic maternal effort). Success at monopolizing females and obtaining fertilizations is the primary determinant of male lifetime reproductive success in most species and thus males commit tremendous effort towards competing for these opportunities. However, the intensity of male–male conflict over fertilizations varies considerably among species, decreasing as the ratio of male to female parental investment increases (Williams 1966; Trivers 1972), as male mating frequency is reduced owing to nuptial gifts, other forms of mating investment, or other factors (Thornhill 1976; Gwynne 1984; Clutton-Brock and Vincent 1991; Clutton-Brock and Parker 1992), and as the population sex ratio becomes increasingly female-biased (Emlen and Oring 1977), because each of these factors increases the availability of sexually receptive females relative to sexually active males (i.e. the operational sex ratio (Emlen and Oring 1977)).

Males may also compete over maternal genotypes that yield heritable benefits to offspring (i.e. viability or attractiveness genes). However, the value of good maternal genes will probably be exceeded by the fitness costs of lost mating opportunites. Male–male competition for high-quality females is probably restricted to males whose mating frequency is limited by paternal investment or mating investment and thus who cannot mate with all receptive females. Even in these cases, direct fitness benefits such as variation in fecundity of mates should usually take priority in male mate-choice decisions.

The aims of conflicting males are generally symmetrical: each competes for the benefits of fertilizing ova. The outcome of competition is thus determined by asymmetries in the benefits and costs of competition and the accuracy of assessment. Physical superiority will obviously reduce the cost of competition and will play a leading role in determining a winner. Males may also differ in intent, owing to differences in the value of winning. For example, reduced expectation of future reproduction may cause older males to invest more heavily in competition because the expense is less likely to cost them future matings (see Zuk and Simmons, this volume; see also Hansen and Price 1995).

Aggressive struggles over territories or mates fit the war-of-attrition model of conflict. The cost of competition depends on costs expended by rivals. Males assess rival competitive ability and decide whether to escalate or end the conflict. For example, male *Calopteryx maculata* that co-establish ownership of a single oviposition site engage in prolonged and energetically costly series of spiral chases (Waage 1988). Each male escalates the contest, increasing his effort until either he or his rival is spent of lipid energy stores (Marden and Waage 1990; Marden and Rollins 1994).

Opponent-independent costs games best apply to the evolution of male weaponry because costs of these traits are fixed at some point during development, and thus cannot be adjusted during competition. However, males that develop larger weapons, and possibly pay a higher price during development (as, for example, in pseudoscorpions (Zeh and Zeh, this volume; but see Crespi 1988a)), mate more successfully (see, for example, Eberhard 1979; Crespi 1988b). The model also applies to the 'weaponry' related to sperm competition and its avoidance, such as the penile sperm scoops of odonates, bush crickets, and other taxa (Waage 1979a, 1984a, 1986; Ono *et al.* 1989, 1995; von Helversen and von Helversen 1991; Fincke *et al.*, this volume).

Conflict over fertilization can potentially occur at any point prior to gamete union, but when is it most likely to occur? Male–male conflict allows only potential access to females and their ova. Direct access is determined by the outcome of male–female interactions and so the dynamics of intersexual conflict will ultimately determine male mating benefits and thus the ways in which males compete. Two key aspects of female reproductive biology will determine whether male competitive success leads to reproductive success: (1) female control over copulation;

and (2) female control over sperm use. By control, we include both the constraints of female reproductive biology, such as restricted periods of receptivity, and morphologically determined patterns of sperm use that neither the male nor female can manipulate, as well as active manipulation of copulation and sperm use by the female.

Female control over copulation and sperm influences the forms of male–male competition occurring at different stages of mating in complex ways. For example, first-male sperm precedence devalues postcopulatory defense of females and fosters precopulatory male–male competition for virgins, whereas preferential use of last-male sperm devalues precopulatory defense in favor of postcopulatory defense (Fig. 21-1). Mixed sperm use (as, for example, in crickets (Zuk and Simmons, this volume)) favors competition at all stages of the mating sequence but the value of competition (and hence the likelihood that it occurs) shifts from predominantly precopulatory to postcopulatory as sperm priority shifts from predominantly first-male to last-male. The duration of female receptivity further restricts the period over which fruitful male–male competition occurs: full loss of receptivity ends male–male competition and temporarily lost or reduced receptivity decreases the risk of female remating and hence lowers the benefits of guarding. Although males often compete for females that are not yet receptive (especially when females are monogamous), long periods of non-receptivity increase the cost of precopulatory guarding and thus reduce guarding duration (see Grafen and Ridley 1983).

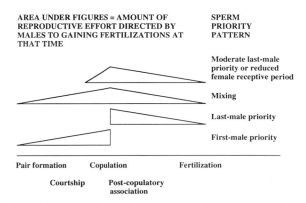

AREA UNDER FIGURES = AMOUNT OF REPRODUCTIVE EFFORT DIRECTED BY MALES TO GAINING FERTILIZATIONS AT THAT TIME

SPERM PRIORITY PATTERN

Moderate last-male priority or reduced female receptive period

Mixing

Last-male priority

First-male priority

Pair formation   Copulation   Fertilization

Courtship   Post-copulatory association

Fig. 21-1. The relationship between sperm priority patterns and the amount of reproductive effort that males direct towards gaining fertilizations at various times during the mating sequence. Areas under curves are approximate. See text for details.

Emlen and Oring (1977) predict that when resources (or females) are clustered and moderately synchronized, males will attempt to acquire matings by defending these clusters. The prediction has gained considerable support. For example, males of several odonate species defend oviposition sites along streams and ponds as a mate-acquisition tactic. These territorial species generally occur at moderate to relatively low population densities and the oviposition sites are relatively localized and easily defended. Territoriality generally breaks down, both within populations and among species, when population density becomes too high and male competition too intense for economic defense of the resource (see, for example, Uéda 1979; Conrad and Pritchard 1992; Fincke *et al.*, this volume). Similarly, in territorial butterflies males defend various sites, including bushes and topographical landmarks, as mate-encounter sites. Territoriality ranges from patrols of several sites at especially low densities, to defense of single sites at moderate density, and finally to reduced aggression and loss of territoriality at high population densities (Brown and Alcock 1990; Rutowski 1991). Many additional examples of resource-defense polygyny in arthopods exist (see, for example, Greenfield, this volume) and all provide support for Emlen and Oring's scheme. Moreover, numerous examples of female defense polygyny occur in diverse insect taxa (Orthoptera, Coleoptera, Thysanoptera, Diptera, Zoraptera), all of which support Emlen and Oring's hypothesis (Gwynne and Morris 1983; Kirkendall 1983; Crespi 1988b; Wilkinson and Dodson, this volume; Choe, this volume).

Male–male competition during the courtship and copulation stages of the mating sequence will occur when conflict is not resolved during pair formation, such as when females are not grouped into defensible clusters and females are asynchronously receptive. Under these conditions, multiple males are more likely to engage in courtship and copulation attempts with the same receptive female. Competition during courtship and copulation will also occur more frequently when these events are protracted, owing to long periods of precopulatory mate-guarding or female mate choice.

Male–male competition during courtship and copulation can occur in two ways, by takeovers or by 'competitive persuasion' (i.e. simultaneous courtship). Takeovers during precopulatory guarding and courtship are common in aquatic isopods and amphipods, such as *Asellus* and *Gammarus* (Ridley and Thompson 1979; Birkhead and Clarkson 1980; Ward 1983). We predict that this form of male competition will occur in species in which females are non-choosy or incite male competition as a mechanism of mate choice (Cox and LeBoeuf 1977; Borgia 1979; Alexander and Borgia 1979). Competitive persuasion is expected when female choice is not determined by male competition and males cannot force copulation (as, for example, in many lekking species (Shelly and Whittier, this volume)). Intrasexual conflict over mating frequency selects for signals that reach more prospective mates and provide greater sensory stimulation to the prospective mates they reach (West-Eberhard 1984).

After copulation, males compete for fertilizations by employing methods to replace the sperm of previous males and protect their own ejaculates from being replaced by rivals. Sperm competition is most likely in species in which females mate often, copulation is decoupled from fertilization, and some degree of sperm-mixing occurs. Copulatory mechanisms related to sperm competition and its avoidance by males include the spurs and plungers on the odonate aedeagus used to remove and displace sperm of a female's previous mates, flooding the female reproductive tract with sperm to displace or outnumber the sperm of other males (Eady 1995; Zuk and Simmons, this volume; Dickinson, this volume), mating plugs that block the female genital opening from futher intromission (see, for example, Drummond 1984; Devine 1984; Dickinson and Rutowski 1989; see also Orr and Rutowski 1991), and possibly specialized sperm morphs that block or kill the sperm of rivals (in humans, Baker and Bellis 1988; see also Sivinski 1984; Gage and Cook 1994). Postcopulatory mechanisms of sperm competition include guarding females until fertilizations are secured (Alcock 1994) and repeated mating with the same female to flood the genital tract with sperm (Eggert and Müller 1997; Smith 1997).

The presence and efficacy of sperm competition depend upon female morphology, physiology, and behavior (Walker 1980; Birkhead and Møller 1993; Gomendio and Roldan 1993), because female multiple mating and organs for sperm storage and mainentance create the situations that allow sperm to be in competition. Thus, females' control over mating frequency and utilization of sperm in their storage organs determine the avenues of sperm competition and its relation to male–male competition (see, for example, Austad 1984; Knowlton and Greenwell 1984; Gowaty 1994).

Males are also in conflict over the allocation of postzygotic maternal effort. In many species with mixed paternity, female postzygotic effort greatly exceeds prezygotic investment and ultimately limits male and female repro-ductive success (see Waage 1996). The selective advantage for males that monopolize postzygotic maternal effort should be substantial. However, competition for postzygo-tic care may be less prevalent than competition to fertilize ova for two reasons. First, successful competition for prezy-gotic effort will minimize competition for postzygotic investment. If one male fertilizes all of a female's eggs, there is no male–male conflict over post-zygotic maternal effort; asymmetries in the number of offspring sired by dif-ferent males create asymmetries in the value of monopoliz-ing postzygotic maternal effort, causing the primary male to invest more heavily in winning: resolving the conflict. Second, adaptations to monopolize postzygotic maternal effort are probably limited. If two or more males fertilize a female's eggs in relatively equal numbers, each male seeks a greater share of maternal effort but from the female's per-spective the optimum (*ceteris paribus*) is to invest equally. Thus, winning the conflict over postzygotic effort requires both outcompeting the rival males and manipulating the female to invest at the male optimum to her reproductive detriment. The only clear form of male competition over female parental care in insects appears to be infanticide; for example, male burying beetles kill the offspring of a rival male as a mechanism of gaining fertilizations and monopolizing the maternal investment of their mates (Eggert and Müller 1997).

Female–female conflict

Conflict and cooperation among females in mating systems has received relatively little attention because female repro-ductive success is typically limited by the availability of resources for the production of offspring, and not by access to male gametes. Thus females usually will not com-pete for sperm, unless the sperm of high-quality males is limiting. However, females will be in conflict over acquisi-tion of resources (not provided by males) and this conflict may have incidental consequences for mating systems by determining female distribution. Females will also com-pete for male-provided resources, including nuptial gifts and male parental effort (see, for example, Smith 1997), and in these cases females may compete for mates (see, for example, Gwynne 1993). Here, we discuss the causes of female–female conflict and cooperation at different stages of the mating sequence and consider its possible outcomes.

If resources are limiting, females are expected to actively compete to monopolize them. The nature of

female competition for resources will determine the envir-onmental distribution of females. Scramble competition for resources may produce an ideal-free distribution of females with respect to the resources, whereas aggressive competition will create either despotic or even distribu-tions of females (Milinski and Parker 1991). Examples of female competition for resources are abundant, ranging from defense of foraging and nesting sites to extreme reproductive suppression in some social Hymenoptera (see, for example, Choe and Crespi 1997). We expect that such female competition will reduce the environmental potential for polygyny by increasing the spacing of females, and thus it will decrease the intensity of sexual selection on males. By contrast, when resources are abun-dant and there are social or safety benefits to gregarious-ness, females may actively aggregate. Gregariousness will bring females into defendable clusters, increasing the environmental potential for polygyny and thus increasing the intensity of sexual selection on males (see also Wade 1995). Female social systems may thus have major conse-quences for mating-system structure.

Because mating is often a precondition for obtaining resources, females may also compete for direct access to males. Similarly, whenever males care for the offspring of more than one female, females may compete to monopolize male parental effort. In both cases, the intensity of female–female conflict over mating and sexual selection on females increases with the slope of the relationship between mating and fecundity (Arnold and Duvall 1994). Female–female conflict also increases with the ratio of male to female par-ental investment (Williams 1966; Trivers 1972) and the population sex ratio (Emlen and Oring 1977). These expec-tations will be modified by variance in the quality of mating benefits (Burley 1977; Parker 1983; Owens and Thompson 1994) but in general female–female conflict over mating is expected to be resolved more quickly, esca-late less frequently, and incur lower costs than male–male conflict.

Finally, females may compete for opportunities to mate with males bearing good genes. Whether conflict over good genes occurs depends on value of the heritable benefits relative to the cost of competition. Because males with good genes may generally be able to mate with many females, such competition may be rare. However, it may be more likely to occur in species where costs of mating or paternal investment limit male mating frequency.

When during the mating sequence will female–female competition occur? Because mate-allocated resources typically are conditional upon copulation, female–female conflict for these resources will intensify between pair formation and copulation. Postcopulatory conflict is expected only when females receive added benefits by mating repeatedly with a male or when the beneficial consequences of mating occur after copulation (such as access to oviposition sites). Female–female conflict over male parental effort may occur at any time before the end of parental care.

Predicting the form of female–female conflict requires an understanding of: (1) the density and distribution of males; (2) the density of females; (3) the degree of male control over mating benefits such as nuptial gifts; and (4) degree of male control over allocation of male parental effort. Examples of female competition are considerably fewer than examples of male competition. Conflict during pair formation occurs in sex-role-reversed katydids, in which females engage in scramble competition for the fecundity benefits of mating more frequently (see, for example, Gwynne and Simmons 1990). In some such katydid species, females actively displace rivals and have claspers that appear to be specialized for retaining their grasp on males, possibly to avoid takeovers (Simmons and Bailey 1990; Rentz 1993; Brown and Gwynne 1997), although they may have evolved in the context of intersexual conflict or cooperation. Additional expressions of female competition during courtship and copulation include alternation of receptivity to mating (for example, in lobsters (Cowan and Atema 1990)) which reduces competition, and suppression of reproduction via pheromones, physical aggression, or both (for example, in many social insects (Choe and Crespi 1997) and some mammals (Creel and Creel 1991)). In *Zorotypus gurneyi*, mating couples are often attacked by others, the majority of whom are females, which suggests that females compete for the best male and guard him against other females in the harem by prolonging copulation (Choe 1994a,b, this volume) and that such males provide some form of limiting resource. Similar phenomena have been observed in birds (Foster 1983; Petrie 1992) and humans (Low 1979). Finally, in a belostomatid beetle, females will attempt to kill eggs produced by other females (Ichikawa 1995), to monopolize a limiting resource, paternal care.

Male–female conflict

Individuals of both sexes must mate at least once, successfully fertilize gametes, and produce offspring that survive to reproduce, but confluence of reproductive interests of the sexes generally ends here (Alexander and Borgia 1979; West-Eberhard et al. 1987; Alexander et al., this volume). Intersexual conflict differs from conflict within the sexes because the optima for males and females are asymmetrical, such that the pay-offs of winning differ between the sexes (Parker 1979). The sexes will usually differ in optimal patterns of reproduction, with males striving to control mating in order to fertilize more (or better) offspring and females striving to control mating in order to beneficially determine offspring paternity, optimize the timing of reproduction, and reduce the costs of superfluous matings (Alexander et al., this volume). Conflict of interest between the sexes is expected at each stage between pair formation and the termination of parental care, and in all mating systems including monogamy with equal parental investment by each sex. Because disparity between optimal mating frequencies of males and females will be at its most extreme when male parental investment is much less than female parental investment, and when male potential reproductive rate is higher than that of females for other reasons such as high encounter rates (Clutton-Brock and Vincent 1991), sexual conflict over fertilization and partitioning of female parental effort will both be at their greatest in such situations (Parker 1979). Conversely, sexual conflict over the partitioning of male-provided mating benefits and male parental effort will increase with the ratio of male to female parental investment. As male parental investment increases, it becomes a limiting resource for females, strengthening sexual selection on females and weakening it for males. Thus there will be increased selection for females to compete for these benefits, both among themselves and with males, through forcing, coercing, or persuading the benefits away. By contrast, reduced sexual selection on males causes a reduction in levels of persuasion (i.e. amount of benefits or signals) required for males to achieve matings.

Sexual conflict may occur over: (1) which sex calls or moves to the other (Alexander et al, this volume); (2) whether copulation takes place; (3) the duration and frequency of copulation (see, for example, Gilbert 1976; Oberhauser 1992; Otronen 1994); (4) whether sperm are transferred; (5) how much sperm is released (see, for example, Pitnick and Markow 1994; Warner et al. 1995); (6) whether transferred sperm are used in fertilization (see, for example, Eberhard, this volume; Dickinson, this volume); (7) whether or not resources are transferred between the sexes, and the extent of any such transfer (see, for example, Oberhauser 1989; Parker and Simmons 1989;

Simmons and Parker 1989); (8) the attraction of additional mates (see, for example, Eggert and Sakaluk 1995); (9) the duration, form, and objects of parental investment (Trivers 1972); and (10) the honesty and information content of signals at any stage of the interaction (see, for example, Arak and Enquist 1995; Schluter and Price 1993).

Resolution of conflict will be a product of the selective advantage of winning, asymmetries in the effectiveness of adaptations, and the nature of these adaptations with respect to different models of conflict (Parker 1979, 1984). The ways in which these factors jointly determine the outcome of sexual conflict will often be difficult to predict because of the intricacies of measuring costs and benefits to each sex of differing reproductive interactions (Parker 1979) and the vagaries of evolutionary history. Selection may often be much stronger on males to achieve matings than on females to resist because the benefit to males of additional fertilizations will be considerably greater than the cost of mating to females. Despite this asymmetry, females may win the conflict if their adaptations for mate-rejection are more effective than male adaptations to achieve matings (Parker 1979).

The intersexual weaponry and charms available to each sex make up the strategy set upon which selection with respect to intersexual conflict can operate. Both sexes may engage in movement, calling, or courtship behavior; they may exhibit more or less refined sensory discrimination abilities; they may develop morphology to physically manipulate the other sex (such as claspers (Arnqvist and Rowe 1995; Sakaluk *et al.* 1995)); and they may produce and respond to chemicals produced either externally or within the female's genital tract to control reproductive physiology and behavior (see, for example, Cordero 1995; Eberhard and Cordero 1995; Eberhard, this volume). Exclusively female weaponry includes various primary sexual characteristics, especially spermatheca and associated morphology and physiology in many arthropods and some mammals (e.g. bats (Fenton 1984)) that allow females to manipulate sperm survival and use (Birkhead *et al.* 1993), and choice of oviposition site and rate. Exclusively male traits used in conflict include genitalia and sperm morphology that increase male control over sperm use. Conflict may involve various modalities, including physical conflict over the presence and duration of copulation, physical–chemical conflict over female detection of a full spermatheca, or sensory-response conflict over the presence and honesty of information in visual, auditory or vibratory behavior during pair formation and courtship. Whenever these

conflicts take the form of opponent-independent costs games, we expect the evolution of high complexity and diversity within and between species, as well as matches between the traits of males and females (Alexander *et al.*, this volume).

The evolution of physiological, morphological, and behavioral adaptations used in sexual conflict should depend on the strength of selection in sexual contexts versus other life-history situations (see, for example, Rowe *et al.* 1994; Sih and Krupa 1995). For example, in most species male phenotypes employed in sexual conflict may be subject to fewer trade-offs than female phenotypes, because males are selected to survive and mate whereas females also provide the bulk, or all, of investment in offspring. Thus, for females, life-history trade-offs associated with the development of sexual conflict adaptations should be more prevalent and reduce the ability of females to evolve adaptations that counter male manipulation. By contrast, males will suffer stronger trade-offs when selection in the context of male–male competition differs from selection in the context of intersexual interactions (for example, when a male's sperm must adapt to both sperm competition and female sperm manipulation). The strength of these various life-history trade-offs will depend upon to what extent different behaviors contribute to variation in lifetime reproductive success, which will vary tremendously among species with aspects of ecology, demography, genetics, and phenotypes.

Sexual interactions begin with pair formation and end sometime between fertilization and termination of parental investment. Each stage in this sequence engenders different sexual conflicts and confluences of interest; resolutions at one stage may affect the dynamics of others. We will review the expected presence and resolution of sexual conflict at each stage, to assess our ability to predict the form and outcome of sexual conflict in different lineages.

Sexual conflict and the mating sequence

Pair formation. Sexual conflict over who signals and who searches is a case in which the aims of the sexes are symmetrical (both gain from mating at least once), and the relative effectiveness of adaptations may often be similar (both can send and receive signals). Thus, resolution of conflict should be predictable from sex differences in the intensity of selection and evolutionary constraints imposed by the historical presence or absence of signaling and searching adaptations. Generally, females should assume the less risky task, forcing the expensive task on the male (Alexander and Borgia 1979; Alexander *et al.*, this volume). This prediction is upheld in the few insects where the costs of calling versus moving have been measured (Wing 1988; Gwynne 1987; Moore 1987; Heller 1992) and appears to agree well with our intuition regarding which sex has the most to gain by searching. In the vast majority of species with no obvious long-range signals, males are the searchers. The few exceptions include species such as some dance flies (Empididae) where females use male swarms as foraging patches (acquiring courtship prey items) and thus females probably achieve substantial fecundity advantages by mating (see Sivinski and Petersson, this volume).

Where long-range signaling is expected to be especially costly, largely owing to susceptibility of exploitation by predators (Sakuluk 1990), males are by far the most common callers (Greenfield, this volume; Zuk and Simmons, this volume; Lloyd, this volume). In at least one taxon where calling (with pheromones) is thought to be less costly than moving, females are the predominant callers (Alexander and Borgia 1979; Phelan, this volume). In other cases, ancestral states may limit phenotypic variation required for shifts away from male-call–female-locate systems, even when ecological conditions change the selective value of mate-searching between the sexes (West-Eberhard *et al.* 1987). For example, Sakaluk and Belwood (1984) found that predation on phonotactic female crickets of the introduced species *Gryllodes sigillatus* was higher than on calling males, despite males being subject to stronger sexual selection. Similarly, Heller (1992) found that females of one species of phonotactic katydid suffered higher predation rates than did females of a related species that answered males' calls with short, audible 'clicks' to which males moved. Male sex-role-reversed katydids (e.g. *Requena verticalis* (Gwynne 1990)) continue to signal under conditions in which the intensity of sexual selection is greater on females than males, although their calling effort is reduced.

Predictions concerning which sex calls appear to be upheld, but the information or deception content of calls is subject to a variety of selective pressures whose presence and strength may vary considerably among species. The form of calls may include aspects of: (1) species recognition; (2) mate attraction via persuasion; (3) aggressive intrasexual signaling (see, for example, Mason 1996); and (4) selection in other life-history contexts, especially predation.

Alexander *et al.* (this volume) argue that sexual conflict is expected to be relatively low over the aspects of songs that are under stabilizing selection, which may include the species-recognition components of broadcast and reception of long-range signals. Thus, sexual confluence of interest over the species-recognition function of pair-formation signals should engender stabilizing selection for tight matches between signals and receiver behavior. Evidence for such matches includes Henry and colleagues' (Henry, this volume) extensive studies of acoustic signaling in *Chrysoperla* (where especially strong selection for species recognition may have selected for duetting), plus considerable work on acoustic Orthoptera (Walker 1957; Greenfield 1996; Zuk and Simmons, this volume).

Examples of apparent mismatches between signalers and receivers are, however, common in both arthropods and vertebrates (see, for example, Ryan 1983; Mason 1991, 1996; Bailey and Simmons 1991; Phelan, this volume). For example, the ears of some female frogs and toads, including *Physalaemus pustulosus*, are tuned to the lower than average sound frequencies produced by larger than average males and these males attract more mates (Ryan 1983). Similarly, female *Oecanthus nigricornis* tree crickets prefer calls with lower than average frequencies and large males consequently mate more often (Brown *et al.* 1996). Females of some other acoustic insects, including *Cyphoderris monstrosa* (Mason 1991, 1996) and *Kwanaphila nartee* (Bailey and Simmons 1991), are apparently poorly tuned to male calls and probably respond only to nearby, loudly calling males. Similarly, the most attractive calls of *Teleogryllus oceanicus* are not the most species-typical (see Zuk and Simmons, this volume). Such apparent mismatches may be due to: (1) weakness of selection for species recognition, perhaps owing to an absence of sympatric species with similar calls; (2) interactions between species recognition and sexual selection functions (Phelan, this volume); (3) persuasion in the context of mate attraction; or (4) selection in other contexts, such as predation. None of these cases need involve females missing conspecific males that they would benefit from approaching.

Phelan (this volume) argues that, when females call, stabilizing selection is weak on call form because of strong selection on males to respond to any female signals; over evolutionary time, males should track any changes in female calls. Females presumably reject attracted heterospecific males after rapprochment; if they do not, they lose more than males to the extent that hybrids are inviable and their parental investment exceeds that of males. When males call, Phelan's hypothesis does not apply because female responses should be under strong selection for species recognition (since females generally lose more from heterospecific attraction and mating than do males), such that males are selected to provide accurate information concerning this component of the call.

Apparent mismatches between signals and receivers when males call may be due to persuasion, and may involve components of calls (such as loudness or frequency) that differ from components used in species recognition (such as temporal patterning) (see, for example, Walker 1957; Alexander *et al.*, this volume). We suggest that sexual conflict generates directional selection on signal structure, manifested in signal divergences and receiver biases. If females discriminate among (distinctively conspecific) calls based on intensity, duration, or structural variability (Zuk and Simmons, this volume; Lloyd, this volume; Phelan, this volume; Brown and Gwynne 1997), then males may achieve their interests by (1) persuading females by honestly advertising reproductive or heritable benefits, or (2) incorporating elements into their signals that deceive females into approaching (West-Eberhard 1984; Alexander *et al.*, this volume).

The most likely deceptive elements in signals are those that are easily perceived (see Guilford and Dawkins 1991), generally correlate with traits or conditions that benefit the female, and are cheap to produce. For acoustic signals, these elements may include use of baffles and resonators to bolster signal intensity, choice of perches and microhabitats to increase broadcast range, and incorporation of high-frequency harmonics to simulate calls of close proximity. Because high frequency attenuates more rapidly with distance (Ewing 1989), high-frequency harmonics or a high carrier frequency may be used by females to assess the proximity of callers (Bailey 1991) or localize them more easily (M. Greenfield, personal communication) and females may prefer songs with disproportionate power in the higher ranges to reduce the costs of moving to males. For example, male katydids, *Kwanaphila nartee*, that produce higher-frequency calls are more readily approached by females (Gwynne and Bailey 1988) and thus males that biassed their calls in favor of a high carrier frequency would probably attract more females. Call intensity (loudness) also decreases with distance; loud calls are typically more attractive to females (Forrest 1980, 1983; Greenfield, this volume; Zuk and Simmons, this volume). Preference for loud calls is often interpreted as evidence of adaptive mate choice because larger, superior males often generate more powerful calls, but whether female preference for loud calls represents choice of large males, passive attraction, or a sensory trap has apparently never been tested.

Other deceptive signals may involve exploitation of female sensory responses that evolve in other selective contexts (reviewed in Andersson 1994; Shaw 1995; Christy 1995; Christy and Blackwell 1995). For example, male *Photuris* fireflies signal with patterns that match the prey of conspecific females; after a female responds they switch back to their species-distinctive pattern. Males appear to exploit female predatory behavior to attract female attention and then engage in distinctive courtship (Lloyd 1980, this volume). Mimicry of prey to attract mates also appears to occur in the visual signals of salticid spiders (Jackson and Pollard, this volume) and the vibratory signals of aquatic water mites (Proctor 1991). The success of deception will depend in large part on the strength of the life-history or information-processing (Bernays and Wcislo 1994) tradeoffs that are exploited by the deceiver.

Alexander *et al.* (this volume) argue that persuasive mate-attraction can be maintained by the luring sex, usually males, providing females with material resources. This hypothesis appears to be supported given the association between signaling and male-derived resources, especially nuptial gifts. However, much of the work on both signaling and nuptial gifts focusses on only a few clades (e.g. Ensiferan Orthoptera (Gwynne, this volume)), so this the association may be due to deceptively few evolutionary events. Moreover, Gwynne (this volume) offers phylogenetic evidence that male calling songs evolved after the provision of courtship meals (see also Gwynne 1995). Accounts from other groups (e.g. Neuroptera (see Henry, this volume)), along with phylogenetically controlled comparative tests, are required to test this idea rigorously. Persuasive mate-attraction may also involve advertisement of genetic benefits, or some combination of genetic and material gain.

Finally, there may be sexual conflict over continued calling when a mate has already been attracted. For example, in some burying beetles females interfere with male pheromone emission because they suffer a cost if additional females arrive (Eggert and Sakuluk 1995; Eggert and Müller 1997), and in a spider males remove female-produced sex pheromone from her web (Watson 1986). Such conflicts are expected when males attract females to shareable, limiting resources, and when females benefit from male–male competition as a mechanism of mate choice.

Courtship. Courtship is defined by Alexander *et al.* (this volume) as intersexual behavior beginning when both sexes are within range of one or more sense of the other (and at least one sex is sexually motivated) and ending when the courter can no longer influence paternity. One or both sexes seek to facilitate insemination and fertilization through affecting the other's behavior, such that courtship excludes acts of forced copulation and requires some degree of control by each sex over mating.

If males reliably win sexual conflict over mating, such as via coercion, then mating frequency approaches the male optimum and there should be no courtship beyond that required for species recognition. However, when females exercise some control over mating, males may use courtship to accrue control by: (1) manipulating female receptivity through behavioral, psychological, or physiological means (a coercive act); or (2) providing females with information regarding mate quality (a persuasive act). The former presumes that females have a reaction to stimuli that is non-adaptive in the context of mating, such as sensory traps (West-Eberhard 1984), sensory biases, or physiological communication or manipulation (Eberhard 1996, this volume; Eberhard and Cordero 1995). The latter presumes adaptive female preferences but it may incorporate both honest and deceptive elements (Cordero 1995), although deception during courtship may not be evolutionarily stable (Kodric-Brown and Brown 1984). Both possibilities fall under the rubric of female choice.

Whether the outcome of sexual conflict is male manipulation or advertisement of quality depends on the intensity of selection on females for preference and against manipulation (i.e. the costs and benefits of female choice), and preconditions for achieving each alternative. For example, adaptive female choice may be more prevalent in species in which females already possess appropriate neural mechanisms for learning, remembering, and identifying individuals (especially conspecifics), such as social insects, solitary nesters, and trap-line foragers (see also Bernays and Wcislo 1994). Male manipulation may occur more frequently in species where females use the same sensory channels for finding prey or avoiding predators as for locating mates, such as visually orienting *Photuris* fireflies (Lloyd, this volume) and *Portia* spiders (Jackson and Pollard, this volume).

When females choose their mates, the mechanism of choice (Janetos 1980; Forrest and Raspet 1994; Weigmann *et al.* 1996) has considerable implications for its evolutionary dynamics. Alexander *et al.* (this volume) argue that the predominant mechanism of female choice in insects is threshold or 'minimal criteria' preferences, and that such preferences will not facilitate the evolution of sexually

selected ornaments (traits produced by female choice alone) in insects. This argument implies that Fisherian runaway selection and good-genes selection are rare in insects, at least compared with vertebrates, and that sex-limited traits are more likely the result of intrasexual or intersexual conflicts of interest (but see Eberhard, this volume). This hypothesis can be divided into two components: (1) whether preferences are 'minimal criteria'; and (2) whether sex-limited traits in insects commonly function as ornaments.

The hypothesis that insect mate preferences are minimal-criteria thresholds is based on the argument that many insects and arachnids lack either the neural capacity or the social opportunity to compare mates. Although in general this may (or may not) be true, it is worthwhile to speculate on the types of species in which comparisons among potential mates might occur. The requisite cognitive traits are individual recognition and some capacity for memory. The requisite social conditions are a frequency of interaction with potential mates that exceeds optimal female mating frequency, with sampling opportunities (and benefits) increasing with density, mobility, and range of perception. These social conditions may be common but probably reach extremes in species where males engage in long-range signalling (Alexander *et al.*, this volume), where receptive males and females congregate into large mating aggregations (see, for example, Snead and Alcock 1985; Alcock 1987), or otherwise socially come together (as in some social insects). Appropriate neural mechanisms for comparisons may be more limiting but they may be especially likely in species where learning functions in other selective contexts; for example, strong evidence for (learned) nest-mate recognition in social insects (Gadagkar 1985) plus some evidence of kin recognition in crickets (Simmons 1989, 1990) suggest that these might be fruitful taxa for the study of individual recognition functioning in mate choice (Simmons 1991). Of course, there is no reason why learned recognition of individuals or classes of individuals could not evolve solely to function in mate choice.

Minimal-criteria preferences occur in *Centris pallida*, where the female ceases to remate only if the male taps her on the head after copulating (Alcock and Buchman 1985), in *Drosophila melanogaster*, where females remate with sufficiently large males regardless of the size of previous mates (Pitnick 1991), and in cockroaches (Moore 1995). Examples of choice by comparison (i.e. relative preference) include blister beetles (Brown 1990) and tree crickets

(Brown *et al.* 1996). An example of best-of-*n* choice may be found among stalk-eyed flies (Wilkinson and Dodson, this volume, but see Alexander *et al.*, this volume) which congregate overnight, and show genetic evidence for Fisherian processes (Wilkinson and Reillo 1994). See Weigmann *et al.* (1996) for additional examples of choice mechanisms, mainly in vertebrates, and methods for testing and evaluating choice models.

Minimal-criteria preferences should restrict the evolution of ornaments, but other forms of fixed preference facilitate runaway selection (and thus ornamentation) if they are narrow within females and genetically variable among females. By contrast, 'relative' preferences (i.e. comparisons like 'best-of-*n*') are better than fixed ones for the operation of good-genes and sensory-bias models of sexual selection. All three mate-choice processes may generate ornaments, and they may often be difficult to differentiate. For example, female mole crickets, *Scapteriscus acletus* and *S. vicinus*, orient to the loudest calls, which are correlated with moisture content in the soil of a male's burrow, an important determinant of egg development (Forrest 1983). Thus female preference may be either a selective outcome of choice of oviposition substrate (see Zuk and Simmons, this volume) or proximity (see Greenfield, this volume). However, sound intensity also varies with distance, power output, orientation, and microhabitat and thus it probably holds limited information (Loher and Dambach 1989). Perceptual biases for songs of higher intensity may be a more parsimonious explanation because they require no evolutionary change in female responses.

Alternative models of ornament evolution are distinguished by analyzing the fitness consequences and the genetics of choice (Boake 1985, 1986; Heisler *et al.* 1987; Petrie 1994; Wilkinson and Reillo 1994; reviewed in Andersson 1994). For example, certain elements of courtship may falsify design criteria for selected female preferences because they yield ambiguous information or are susceptible to cheating. Temporal differences in song components could be better cues than song intensity for females to use in discriminating among males (Loher and Dambach 1989) because they tend to be more stable over distance and habitat changes, yet maintaining extended calling bouts and rhythmic patterns may be quite costly and not susceptible to cheating (Alexander *et al.*, this volume; Zuk and Simmons, this volume). For example, female *Gryllus integer* prefer male songs with longer uninterrupted bouts (Hedrick 1986). Although the information content of signal components alone cannot support or falsify

alternative models of preference evolution (uninterrupted calling bouts may be more readily perceived and thus preferences may also be due to sensory biases), together with evidence of active comparisons (relative preferences) they can support or falsify design criteria for different models of sexual selection. For example, moths that communicate information about spermatophore chemistry during courtship (Dussord *et al.* 1991) provide an exceptionally good example of honest signalling and preferences for direct benefits. By contrast, Proctor (1991) provided strong support of sensory biases by identifying the original selective context of female response (foraging). Further distinction among the alternatives requires detailed analysis of choice mechanisms, and comparative and further genetic tests of coevolution between signals and responses.

Copulation. If males normally win the conflict over mating, either by forcing copulation or via coercion at a large cost to females, then females will evolve to accept superfluous matings and convenience polyandry may result (Arnqvist, this volume). Struggles over copulation occur only when there is uncertainty over who will win and males find struggles more profitable than courtship, such as when males lack persuasive ability or have high coercive or forcing ability (as, for example, in many Mecoptera). Struggles may also represent a form of courtship (i.e. be subject to female choice) under the restrictive conditions that: (1) they are (or have been in the past) occasionally successful outside of the context of female choice; and (2) females are unable to assess males by alternative, less costly, means (see, for example, Weigensberg and Fairbairn 1994). Alexander *et al.* (this volume) suggest that these conditions are quite common.

Intersexual conflict over copulation is exemplified in waterstriders (Rowe *et al.* 1994; Arnqvist, this volume). In most species males and females struggle over copulation and the behavioral and morphological evidence suggests components of forcing (i.e. opponent-independent costs), coercion (i.e. war of attrition), and persuasion (luring to an oviposition site) under different social conditions and in different species. Both males and females have evolved morphological characteristics that appear to aid in forcing and resisting copulation, respectively. Males possess modified forelegs for grasping females, flattened abdomens for a tighter union to the female, and various genital processes that function to grasp the female posteriorly. Females of some species have abdominal spines that function to reject males, and engage in 'somersaulting' to reject males

(Arnqvist and Rowe 1995; Arnqvist, this volume). Preziosi and Fairbairn (1996) and Weigensberg and Fairbairn (1996) have shown that several of the male structures involved in such coercion are subject to strong directional selection, and have thus apparently evolved at least partly in the context of intersexual conflict.

As expected in a war of attrition, the amount of female resistance in some water striders decreases when harassment is frequent (Weigensberg and Fairbairn 1994; Arnqvist, this volume). Moreover, the rate of harassment changes the benefit of mating for females because females incur less harassment and consequently forage better when mating (Arnqvist, this volume). In odonates, males cannot force copulation but males of some species appear to coerce females with persistent harassment that reduces oviposition efficiency. Females respond by 'dislodging males, holding onto perches, shaking, curling their abdomen, and spreading their wings' (Fincke *et al.*, this volume) but nevertheless females probably often mate to avoid harassment. Moreover, whereas harassment is an effective copulatory strategy when females oviposit at the water surface (in *Calopteryx maculata*), it is absent in a related species in which females avoid harassment by ovipositing underwater (*C. dimidiata*) (Waage 1984b). However, in *C. maculata* females can and do counter this form of male control by parasitizing male guarding behavior (Waage 1979b).

Arnqvist (this volume) and Alexander *et al.* (this volume) suggest that ability to force copulation is a sign of male quality and thus female resistance may be functionally equivalent to mate choice. How does one distinguish resistance due to the cost of mating (coercion) from resistance as a test of male vigor (persuasion)? The coercion hypothesis predicts reduced resistance when rate of harassment is high, whereas the persuasion hypothesis makes two predictions: (1) constant or perhaps even greater resistance because females can presumably be more choosy; and (2) heritable benefits of resistance exceeding the cost of resistance (Arnqvist 1992). The data on harassment in water striders collected thus far, especially by Weigensberg and Fairbairn (1994), support the hypothesis of coercion. However, higher female resistance when a female has a high egg load, in both these water striders and flies (Ward *et al.* 1992), suggests that females become more discriminating nearer to fertilization.

Postcopulatory events. Female mate choice occurring after the onset of mating has accumulated a number of titles, including 'covert choice', 'cryptic choice', 'paternity control', and 'postcopulatory choice' (Thornhill 1984; Eberhard 1985; West-Eberhard *et al.* 1987; Choe, this volume). Choice prior to copulation has been distinguished by terms such as 'overt choice'. We suggest that the terms precopulatory, copulatory, and postcopulatory mate choice are most unambiguous and useful, because postcopulatory or copulatory choice may often be neither cryptic nor covert to either researchers or the males being chosen. For example, removal and consumption of spermatophores by female crickets occur after copulation, influence male insemination success (see, for example, Sakaluk 1984; Wedell 1991), and are readily observed and quantified (Fulton 1915; Gwynne, this volume; Zuk and Simmons, this volume). Moreover, some acts of precopulatory choice may be exceptionally difficult to observe (see, for example, Borgia 1981). We also note that, although acts of pre- and postcopulatory choice are generally considered separate events, certain acts of mate choice may incorporate components of both. For example, if females remate only with males that are superior to a prior mate, in one act the female is making a postcopulatory choice against her prior mate and a precopulatory choice in favor of her current mate.

Postcopulatory conflict includes conflict over mate-guarding, insemination, sperm utilization, repeated mating, resource transfer, and rate of progeny production. The male optimum is to inseminate the female to a sufficient degree to fertilize all available ova and outcompete the sperm of any prior or future males, to jockey the sperm into position to be preferentially utilized during fertilization, to stimulate oviposition during the period at which his sperm are most likely to be used and produce viable offspring, and to inhibit remating by the female during the period that his sperm are viable and fertilization may occur. Females should resist any degree of male control that causes a net reduction in fecundity or limits female ability to choose males with high heritable fitness.

Parker (1984) recognized three aspects of copulatory and post-copulatory events that may involve sexual confluence of interest: (1) postcopulatory guarding that (incidentally) reduces harassment from other males as it protects a male's paternity (see, for example, Arnqvist, this volume; Dickinson, this volume; Fincke *et al.*, this volume); (2) mechanisms to reduce the likelihood of takeover, given some cost of takeover to females and an absence of mate choice via male–male conflict at this stage (for example, male or female morphology that allows claspers of the other sex to gain a better grip; see also Eberhard, this volume); and (3) the evolution of male-produced stimuli that induce female unreceptivity for some period after mating. Although male and female interests may coincide over the length of periods of female unreceptivity, it is unclear why male-produced stimuli, especially multiple types of stimulus (see, for example, Gromko et al. 1984), need be involved in this female behavior unless there is often some degree of conflict.

Evidence for some degree of female control over post-copulatory processes includes (see also Eberhard, this volume, 1996): (1) displacement or preferential use of sperm within the genital tract (see, for example, Otronen and Siva-Jothy 1991; Ward 1993; Siva-Jothy and Hooper 1995); (2) ejection of sperm (see, for example, Davies 1985; Otronen 1994; Barnett et al. 1995; Dickinson, this volume; Wilkinson and Dodson, this volume); and (3) increased oviposition after matings with chosen males (Thornhill 1984; Zuk and Simmons, this volume), although Brown (1994) suggests that this pattern may also ensue from male control and manipulation. Females may also channel more resources to offspring of preferred males (Burley 1986) or selectively abort offspring (for example, in plants, Willson and Burley 1983; West-Eberhard et al. 1987; Marshall and Ellstrand 1988; Haig 1992; Queller 1994).

Evidence for some degree of male control over such processes includes: (1) removal of the sperm of other males; (2) the presence of invasive male-produced substances that affect female reproductive physiology and behavior in the male's interest (Gromko et al. 1984; Chapman et al. 1995; Eberhard and Cordero 1995; Eberhard, this volume); (3) increases in ejaculate size with increased risk of sperm competition (Gage and Baker 1991); (4) prolonged copulation and postcopulatory associations (see, for example, Alcock 1994; Choe, this volume; Dickinson, this volume); (5) copulatory plugs; and (6) active male suicide as adaptive mating investment (Andrade 1996), which also represents a form of sexual confluence of interest.

One of the main lines of evidence concerning the dynamics of sperm transfer and use is genitalic morphology. Among animal taxa, male genitalia exhibit highly variable degrees of complexity and species-specificity (Eberhard 1985); their morphological complexity and diversity suggest interactions with female genitalia far more involved than simple sperm transfer. Alexander et al. (this volume) challenge Eberhard's (1985) hypothesis that genitalic complexity has evolved mainly via

persuasion involving female preference for internal 'titillation' and runaway sexual selection. Instead, Alexander et al. (this volume) propose that genitalic complexity is the consequence of evolutionary arms races involving coercive and forcing attempts to control reproduction.

Arguments concerning the functional design of genitalia, and the importance of copulatory and postcopulatory mate choice, are intimately connected. Alexander et al. (this volume) argue that postcopulatory choice is unlikely to evolve as a result of selection on females, because they are much better served by precopulatory mate choice, and are more capable of choice, earlier in the sexual sequence (see also Birkhead et al. 1993). Thus, to the extent that postcopulatory choice has evolved, it probably involves female efforts to regain control over fertilization in circumstances where males control earlier events (for example, in Mecoptera with forced copulation (Thornhill 1980, 1984)). Moreover, they point out that the general absence of female adaptations to physically prevent sperm entry after copulations suggests that choice has already been affected by that stage. Eberhard (1985, 1991, 1993, 1996, this volume) argues that postcopulatory choice is common, and that although male manipulation, sensory traps, or good-genes choice may initiate changes in male genitalia, runaway sexual selection soon takes over and serves as its main evolutionary impetus. Two lines of evidence, (1) interspecific correlations between male and female genitalic complexity, especially those suggestive of evolutionary 'stand-offs' and (2) non-graded, threshold mate-choice responses rather than open-ended preferences, would support the hypotheses of Alexander et al. and weaken Eberhard's arguments. By contrast, two findings that would support Eberhard's hypothesis but weaken the hypothesis of Alexander et al. include: (1) high levels of female sensory innervation corresponding to male genitalic complexity, in the absence of morphology for coercion or forcing; and (2) coincidence of pre- and postcopulatory choice (unless precopulatory choice diminishes the value of postcopulatory choice) or concordance of the two types of choice (unless the types of benefit differed, such as material benefits for precopulatory choice and genetic benefits for postcopulatory choice). Eberhard's hypothesis also makes the basic predictions of the runaway model, including the maintenance of genetic variability (see Bakker and Pomiankowski 1995; Pomiankowski and Møller 1995) and genetic correlation (Pomiankowski and Sheridan 1994).

If the advantages of controlling paternity are equal across mating systems, then the prevalence of postcopulatory female choice should be predictable from the costs and benefits of different forms and stages of mate choice. Postcopulatory choice should be common in two situations: (1) where costs of postcopulatory choice are small compared with the costs of precopulatory choice; and (2) where males provide females with nutrients, and females benefit from both the nutrients and ability to adjust paternity. In the former situation, females of some species may be physically unable to resist forced copulations (Thornhill 1980); in others the costs of resisting male mating attempts may be prohibitively high, making it adaptive for females to mate indiscriminately (Parker 1970), i.e. 'convenience polyandry' (Thornhill and Alcock 1983; Rowe 1992; Arnqvist, this volume). In other species, females may be uncertain of future mating prospects and so they initially mate indiscriminately or with a low threshold for choice, and then choose later males more carefully, employing postcopulatory choice to preferentially use their sperm (see, for example, Watson 1990; Petrie *et al.* 1992; Brown 1996). In the latter situation, when males provide females with nuptial gifts or other mating incentives, and the quality of the benefits is not highly correlated with male quality as a sire, postcopulatory choice allows females to augment or change choice for the traits that they preferred during precopulatory mate choice in ways that confer heritable benefits on offspring. If postcopulatory choice occurs after the pair have separated, then females may be able to induce males to provide a full complement of material benefits after which they preferentially use the sperm of males that bear relatively good genes (LaMunyon and Eisner 1993; see also Bissoondath and Wiklund 1995). Females thus simultaneously enjoy the positive fitness effects of both good materials and good genes. In *Zorotypus barberi*, females reject males as frequently during the postcopulatory phase as they do during the precopulatory phase even after obtaining nuptial gifts. If a male's performance is acceptable, females allow him to copulate repeatedly, increasing the probability of fertilization by such males (Choe 1995, this volume).

If choice is expressed during mating and food is gradually received from males, females may use choice to entice larger gifts from males, possibly by allowing more sperm to be inseminated as gift size increases (see, for example, Thornhill 1983). Alternatively, males may control gift size or number, especially when females control copulation. By parceling or optimally allocating resources in each gift (Connor 1995) males may be able to keep females engaged longer and thus transfer more sperm (Choe, this volume). Therefore, feeding duration and frequency may be determined in the context of intersexual conflict.

The argument that postcopulatory female choice rewards males for providing high-quality gifts commits the sequence fallacy (see Simmons and Parker 1989); the presumptive benefits of the act occur before the act is performed and not as a consequence of the act. Postcopulatory female choice for direct benefits will only evolve when it increases future male investment, such as paternal effort. Otherwise, postcopulatory choice is restricted to (1) choice for good genes, (2) male manipulation, or (3) runaway sexual selection.

Conflicts over control of fertilization involve both competition between the sperm of different males and conflicts between males and females over sperm use (reviewed in Smith 1984; see also Birkhead and Møller 1992). Male strategies can include displacement of previous sperm and resistance to displacement (see Clark *et al.* 1995 for an analysis of genetic variation in these traits), facilitation of sperm-mixing, removal of sperm of other males, or variation in sperm number and morphology. The symmetry of such male–male conflict should lead to evolutionary arms races (Parker 1970, 1984), with unpredictable outcomes. The male optimum with respect to interaction with females is to maximize his number of total fertilizations, rather than fertilizations per female (Parker 1984). Female optima may include using the sperm of any one male, or some optimal mixture from two or more males; females may also use only particular sperm from a given male (i.e. intra-ejaculate sperm choice), which may conflict with male interests given some degree of intermale sperm competition (G. Arnqvist, personal communication; see also Haig and Bergstrom 1995).

The pattern of sperm priority found in any given species should depend upon: (1) the strength of selection on male–male vs. female–male interactions, with the former usually greater (Parker 1984); (2) the ability of males to coerce differential sperm use, which may be low given that sperm are under the control of female physiology and morphology once inseminated (Walker 1980); and (3) the magnitude of (a) male mating effort that reduces mating frequency, and (b) parental effort, with parental effort expected to be positively related to female use of sperm by investing males (Gwynne 1984). The degree of female control over sperm use will depend on the magnitude of material and genetic benefits to females of control, in relation to the strength of selection on males and the effectiveness of male and female adaptations. The high degree of both interspecific and intraspecific variation in sperm precedence (see, for example, Drummond 1984; Lewis and Austad 1990; Simmons and Parker 1992; Conner 1995) suggests that such conflicts are highly dynamic. However, the general trend toward last-male precedence may be due to a frequent confluence of interest between the sexes, whereby males gain from displacement and females gain from ability to influence fertilization via multiple mating (see also Ridley 1989).

SEXUAL CONFLICT AND MATING SYSTEMS

What are the consequences of conflict within and between the sexes for the evolution of mating systems? Mating systems represent an array of adaptations selected to maximize the reproductive rate of individuals of each sex, in the context of how conflict between the sexes is resolved. The suite of traits that form a mating system include the acts of pair formation, courtship, copulation, and postcopulatory events that determine patterns of gamete union. Certain combinations of events will be more likely than others and changes in a component of the mating sequence at one stage may facilitate or hinder changes at other stages, such that mating systems can be viewed as following a trajectory of plausible outcomes. Our main thesis is that, because mating systems are essentially the expression of a struggle between the sexes for control of reproductive resources, sexual conflict is among their primary dynamics. Male–male conflict sorts out, to varying degrees, which males engage in this conflict. Female–female conflict partly determines the effectiveness of alternative forms of male–male conflict, and sorts out which females may engage in sexual conflict over male reproductive resources. Intersexual conflict influences the utility of alternative forms of intrasexual competition: for example, female control over copulation selects for increased male displays of quality and lessened male–male aggression, and male control of sperm priority, such as via sperm scoops, selects for postcopulatory guarding.

The Emlen–Oring scheme for understanding the ecology and evolution of mating systems pivots upon the role of the spatial and temporal distribution of resources in determining the distribution of females, and the role of female distribution in determining the mating behavior of males. We are, however, interested in the spatial and temporal distribution of fertilizations rather than matings; when fertilization opportunities do not cluster with female distribution then the Emlen–Oring scheme breaks down. For example, if females suppress reproduction by rival females, then female-defense polygyny will not occur, despite groupings of females. If females do not copulate at resources, then resource-defense polygyny will not occur, despite clusters of resources (see also Birkhead and Møller 1992). And if sperm acquired away from clustered resources is used preferentially, then males will not guard these resources. Among insects and arachnids, the usual spatial and temporal uncoupling of insemination and fertilization (Eberhard 1985; Alexander *et al.*, this volume), and the recognition that females exert some control over fertilization in most taxa, indicate that considerations of sexual conflict must be incorporated into any explanatory scheme that seeks to predict mating systems from ecology and aspects of evolutionary history. In particular, female control of fertilization, evolved in the context of sexual conflict, may greatly alter the potential for polygamy (Emlen and Oring 1977) and the degree to which such potential is realized.

We propose that the evolution of a mating system depends predominantly on three sets of variables: (1) the distributions of resources and mates, with respect to economic defensibility (Emlen and Oring 1977); (2) the presence and degree of material and genetic benefits that are transferred between the sexes; and (3) who controls events during the mating sequence, and to what extent persuasion, coercion and force are involved in these events (Fig. 21-2).

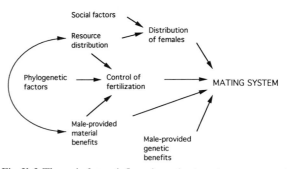

Fig. 21-2. The main factors influencing animal mating systems. 'Phylogenetic factors' refers to the historical presence or absence of adaptations related to the control of fertilization in various lineages.

The distributions of mates and resources determine the potential for different forms of competition and cooperation at different stages in the mating sequence. This axis can be characterized along a scale of economic defensibility, from spatially or temporally clumped and perfectly defensible from others of the same sex, to a uniform or random distribution that precludes any form of defense (Emlen and Oring 1977).

The presence and degree of transferrable genetic and material benefits influence the form of sexual selection (via parental-investment asymmetries, discussed above) and the extent to which sexual conflict involves persuasion, coercion or force. The importance of genetic benefits depends on the variance in mate genetic quality found in any given species, as well as the costs of transmitting, receiving, processing and acting upon the signals used to communicate genetic quality. By contrast, the importance of material benefits should depend mainly upon feeding ecology and demographics: when resources limit female reproduction, and males can transfer resources to females directly, via bodily secretions, or via the ejaculate, males may gain more from persuasion with materials, or even from paternal care (Tallamy 1994) than from other types of reproductive behavior. Male-provided resources should be especially prevalent when gathering resources is costly for females, such as when resources are dispersed or otherwise difficult to obtain (see Leimar *et al.* 1994). The environmental potential for polygamy (Emlen and Oring 1977) may thus increase with clumping of resources for three reasons: (1) decreased costs to females of gathering resources; (2) decreased provision of resources by males; and (3) increased likelihood of resource-defense polygamy by males.

Transfer of genetic or material benefits requires some degree of control by the receiver over fertilization, lest the provider forego investment in advertising or accruing resources and just force. The main consequence of a higher degree of transferrable material or genetic benefits is an increased expectation that persuasion, rather than coercion or force, will drive intersexual interactions. This view contrasts in part with that of Alexander *et al.* (this volume), who argue that in some cases coercive ability may come to be associated with quality as a mate.

In general, determining the role of control over sexual conflict in the evolution of the mating system of a given species requires specification of the optimal mating system for each sex, where and how interests differ, and how persuasion, coercion and force are involved in resolving or maintaining the conflicts. Male optima usually involve polygamy, no courtship, no transfer of resources, complete sperm precedence, and, perhaps, choice of more-fecund females of high genetic quality. By contrast, female optima involve ability to choose mates at low cost using criteria available throughout the sexual sequence, provisioning of resources by males, and control over sperm use and oviposition. For any given species, there will be some range of variation present, usually between these optima for each sex, upon which selection on each sex operates.

Male control of the mating system is most prevalent when: (1) males can control resources that increase female reproductive success (see, for example, Borgia 1979; Greenfield, this volume; Wilkinson and Dodson, this volume); (2) males can defend females directly from other males; and (3) males can harrass females and thereby coerce them into mating (see, for example, Arnqvist, this volume; Fincke *et al.*, this volume). Forced copulation as an alternative mating tactic should be relatively common where males provide or defend resources, if the resources are sufficiently expensive to accrue that some males pursue this non-resource option (Thornhill 1980).

Females tend to control the mating sequence when males cannot control resources or coerce females. Such mating sequences are typified by male persuasion and female choice; resources may or may not provide the sites for pair formation, and males may be either concentrated or dispersed (i.e. the equivalent of 'exploded leks'). Clumping of resources may thus result in either resource-defense polygyny, or resource-based female-choice (lek) mating systems, depending upon the degree to which males or females can control copulation and fertilization. Other comparative examples of the role of male versus female control in the structure of mating systems are provided by Arnqvist (this volume) and Choe (this volume). In water striders, male control produces a mating system characterized by convenience polyandry in females and scramble competition for mates in males, and female control over copulation results in presumably lower levels of polygyny and resource-defense territoriality by males. In some zorapterans, however, male control produces a low level of polyandry in females and female-defense polygyny in males, and female control results in a high level of polyandry in females, nuptial gift presentation and scramble competition in males (Choe, this volume).

One of the main difficulties in the prediction of mating systems from the three sets of variables above is intrinsic unpredictability in processes that involve arms races with various starting points, coupled with the complex interrelationships expected between resource distributions, transferrable genetic and material resources, and aspects of control.

What are the implications of the scheme described above for our understanding of the traditional characterization of mating systems, into scramble competitions, resource-defense polygyny, female defense, monogamy, swarms, leks, and others? Scramble-competition mating systems may involve male, female or mixed control over mating and fertilization, depending mainly upon population structure (for example, high density leading to high potential for polygamy) (Clutton-Brock and Vincent 1991), selection in other life-history contexts (e.g. feeding or predation), genetic or material benefits of mate choice to females, and the dynamics of genitalic interactions and sperm transfer and use.

Under resource-defense polygyny, clumped resources result in a high degree of male control over mating, via coercion or force, when the resource is limiting (as, for example, in flies with antlers (Wilkinson and Dodson, this volume)). Moreover, in such species female choice may be devalued because strong male–male competition filters male genetic quality. To the extent that female choice occurs in this mating system, it should be copulatory or postcopulatory, but in general copulation should equal fertilization and last-male sperm precedence should be common.

Female-defense mating systems should resemble resource-defense systems, except that females may control the grouping patterns that make them defensible. When males defend individual females before, during or after copulation, coercion should be common (see, for example, Arnqvist, this volume; Zuk and Simmons, this volume), by mechanisms that involve either behavior or materials (e.g. the chewy spermatophylax (Gwynne, this volume; Zuk and Simmons, this volume)).

Mating systems that involve swarming, typically by males, appear to evolve when females are uniformly dispersed and unpredictably located in space, and there are no resources for males to control (Sivinski and Petersson, this volume). Swarming may usefully be considered as a form of scramble competition where the aggregate of males and the swarm marker serve for pair formation; as noted above, males normally adopt the risky role in rapprochement (Alexander et al., this volume) and swarming appears to be a risky proposition. Sivinski and Petersson (this volume) suggest that swarming is restricted to situations where males cannot defend an aerial space economically, owing

to some combination of male size, male numbers, and the size of the encounter site. The degree to which females choose their mates in swarms is unclear; swarms certainly do not offer females easy opportunities to choose among males (as do leks), although many males are present for possible comparison (Sivinski and Petersson, this volume). Females gain from ease of finding a mate, but they may sacrifice ability to choose their mate before copulation; thus, to the extent that mate choice would benefit females, they should exhibit multiple mating and postcopulatory choice.

Leks can be categorized along two axes: exploded or not, and with or without resources present (see also Höglund and Alatalo 1995). Non-resource, non-exploded leks (as found, for example, in some lampyrid beetles (Lloyd 1979b) and euglossine bees (Kimsey 1980)) should involve dispersed resources, a high degree of female control over the mating sequence, active or passive 'hotspot' accumulation of females, and high genetic benefits to females accrued via male persuasion. Resource, non-exploded leks (for example, in some crickets and katydids (Alexander 1975; Simmons and Bailey 1990)) should involve clumped resources, female control, and genetic or material benefits (or both), depending on whether the resources serve as a gathering place or are tranferred from male to female. Exploded non-resource leks (for example, in hilltopping tarantula hawk wasps (Alcock 1981) and butterflies (Alcock 1983)) should entail dispersed females, female control, and high genetic benefits; exploded resource leks (for example, in a haglid (Mason 1996) and a butterfly (Lederhouse 1982)) should be similar but involve environmental resources and genetic benefits, transferred material resources, or both. In some vertebrates, females may mate on leks to avoid male harassment (reviewed in Höglund and Alatalo 1995) and such a mechanism could drive lek evolution; this hypothesis has apparently not been considered for insects.

Monogamous mating systems should evolve when there is selection for biparental care to the point that for either sex additional mate-searching is less beneficial than either: (1) channeling those resources into current offspring (or additional offspring with the same mate); or (2) using the resources for future reproduction (see Trivers 1972; Clutton-Brock 1991). Changes in the cost functions can instantly change the optimal tactic from monogamy to polygyny, such as when a second female arrives at a large carcass in burying beetles (Eggert and Müller 1997) or when an extra-pair female solicits copulation out of sight of the paired female in birds (see Birkhead and

Møller 1992). Both sexes should exert some control over parental investment, such that males tend to evolve elaborate paternity-assurance mechanisms.

One type of mating system with unusual forms of sexual conflict is found in wasps with subdivided population structures and high levels of local mate competition, inbreeding, and highly female-biassed sex ratios (Godfray and Cook, this volume; Herre *et al.*, this volume). In these taxa, conflict among females involves the sex ratio that they produce, and male competition involves the development of fighter or dispersal morphs, which may also involve mother–son conflict over control of morph determination (E. A. Herre, personal communication). Moreover, intersexual conflict may occur over the presence of sib-mating, because sterile, diploid males can be produced from such matings (Parker 1979; Godfray and Cook, this volume). The strength of such intersexual conflicts, how they are resolved, and whether they can result in transitions to outbred mating systems require further study. However, in *Melittobia* wasps with extreme female biases, males appear to largely control courtship and copulation (van den Assem *et al.* 1982; see also van den Assem *et al.* 1980), probably owing to the physiological rigors of inseminating so many sisters; does courtship in such taxa serve only for species recognition? Does kinship increase sexual confluence of interest?

The mating-system categorization described above provides a useful means of seeking convergences, but the boundaries between systems are not always obvious and profound differences in sexual behavior may be found within systems of the same name (see Wcislo 1997). We suggest that mating systems can also be viewed in terms of the mating sequence, as trajectories of male and female behavior involving persuasion, coercion and force. In this context, traditional categorizations represent common, albeit broad, types of trajectory; the search for homology and analogy should often involve analysis of the components of mating sequences rather than the full trajectories themselves.

SUGGESTIONS FOR FUTURE RESEARCH

One of the main goals of this volume has been to stimulate the next round of questions and studies in mating-system ecology and evolution. In this chapter, we have attempted a revision and update of the Emlen–Oring–Parker scheme, and in doing so tantalizing gaps in mating system theory and support have emerged. We have little evidence to cite

that links male courtship traits with sperm use (but see Lewis and Austad 1994), we understand the costs and benefits of multiple mating by females in only a handful of cases (see Ridley 1988; Choe, this volume; Dickinson, this volume) and our sketches of the expected interrelationships between resource distributions, transferrable genetic and material benefits, and male and female control via persuasion, coercion and force remain incomplete. Despite such difficulties, we believe that the important issues and questions are now more obvious, which is the first necessary step toward our goal.

We suggest that the most important future studies of mating systems will involve: (1) analyses of the functional design (Thornhill 1980) of traits related to intersexual interactions – such as courtship behavior and morphology, genitalia, and invasive chemicals – especially via experimental manipulations that affect male and female control or perception (see, for example, Weigensberg and Fairbairn 1994; Jablonski and Vepsäläinen 1995); and (2) comparative analyses of related species with different mating systems, to elucidate the patterns of differences in causal variables between them (see, for example, Arnqvist *et al.*, this volume; Choe, this volume; Herre *et al.*, this volume). Functional design studies should especially address mechanisms of mate choice, whether good-genes choice can compete with material benefits, and how life-history trade-offs influence mating behavior and mating sytems. Phylogenetic comparative studies may involve tests of specific hypotheses, such as male coercion evolving to occur earlier in the mating sequence (Alexander *et al.*, this volume), luring acts evolving to either forcing or provision of gifts (Alexander *et al.*, this volume), or swarms evolving to leks and vice versa (Sivinski and Petersen, this volume; Shelly and Whittier, this volume). Such studies might also include comparisons of insects and arachnids with vertebrates, to explore the effects on mating-system evolution of pair bonds, long-term parental care, and spermathecae. West-Eberhard *et al.* (1987) and Christy (1995) describe a range of other promising approaches and tests, most of which await only researchers with the dedication and insight to untangle the intricacies of mating-system politics.

ACKNOWLEDGEMENTS

We thank R. D. Alexander, G. Arnqvist, J. Cooley, D. Fairbairn, M. Greenfield, D. Marshall, R. Rutowski, L. Simmons, J. Waage and M. Zuk for helpful comments and discussion.

LITERATURE CITED

Alcock, J. 1981. Lek territoriality in the tarantula hawk wasp *Hemipepsis ustulata* (Hymenoptera: Pompilidae). *Behav. Ecol. Sociobiol.* 8: 309–317.

–. 1983. Territoriality by hilltopping males of the great purple hairstreak, *Atlides haleus* (Lepidoptera, Lycaenidae): convergent evolution with a pompilid wasp. *Behav. Ecol. Sociobiol.* 13: 57–62.

–. 1987. Leks and hilltopping in insects. *J. Nat. Hist.* 21: 319–328.

–. 1994. Postinsemination associations between males and females in insects: the mate-guarding hypothesis. *Annu. Rev. Entomol.* 39: 1–21.

Alcock, J. and Buchman, S. 1985. The significance of post-insemination display by male *Centris pallida* (Hymenoptera: Anthoporidae). *Z. Tierpsychol.* 68: 231–243.

Alexander, R. D. 1961. Aggressiveness, territoriality and sexual behavior in field crickets (Orthoptera: Gryllidae). *Behaviour* 17: 130–223.

–. 1975. Natural selection and specialized chorusing behavior in acoustical insects. In *Insects, Science and Society*. D. Pimentel, ed., pp. 35–77. New York: Academic Press.

Alexander, R. D. and G. Borgia. 1979. On the origin and basis of the male-female phenomenon. In *Sexual Selection and Reproductive Competition in Insects*. M. S. Blum and N. A. Blum, eds., pp. 417–440. New York: Academic Press.

Ahnesjö, I., A. Vincent, R. Alatalo, T. Halliday and W. J. Sutherland. 1993. The role of females in influencing mating patterns. *Behav. Ecol.* 4: 187–189.

Andersson, M. 1994. *Sexual Selection*. Princeton: Princeton University Press.

Andrade, M. C. B. 1996. Sexual selection for male sacrifice in the Australian redback spider. *Science (Wash., D.C.)* 271: 70–72.

Arak, A. and M. Enquist. 1995. Conflict, receiver bias and the evolution of signal form. *Phil. Trans. R. Soc. Lond.* B 349: 337–344.

Arnold, S. J. and D. Duvall. 1994. Animal mating systems: a synthesis based on selection theory. *Am. Nat.* 143: 317–348.

Arnqvist, G. 1992. Precopulatory fighting in a water strider: inter-sexual conflict or mate assessment? *Anim. Behav.* 43: 559–567.

Arnqvist, G. and L. Rowe. 1995. Sexual conflict and arms races between the sexes: a morphological adaptation for control of mating in a female insect. *Proc. R. Soc. Lond.* B 261: 123–127.

Assem, J. van dem, H. A. J. In Den Bosch and E. Prooy. 1982. *Melittobia* courtship behaviour: a comparative study of the evolution of a display. *Netherl. J. Zool.* 32: 427–471.

Assem, J. van dem, M. J. Gijswijt and B. K. Nübel. 1980. Observations on courtship and mating strategies in a few species of parasitic wasps (Chalcidoidea). *Netherl. J. Zool.* 30: 208–227.

Austad, S. N. 1984. Evolution of sperm priority patterns in spiders. In *Sperm Competition and the Evolution of Animal Mating Systems*. R. L. Smith, ed., pp. 223–249. New York: Academic Press.

Bailey, W. J. 1991. *Acoustic Behaviour of Insects: an Evolutionary Perspective*. New York: Chapman and Hall.

Bailey, W. J. and L. W. Simmons. 1991. Male-male behavior and sexual dimorphism of the ear of a zaprochiline tettigoniid (Orthoptera: Tettigoniidae). *J. Insect Behav.* 4: 51–65.

Baker, R. R. and M. A. Bellis. 1988. 'Kamikaze' sperm in mammals? *Anim. Behav.* 36: 936–939.

Bakker, T. C. M. and A. Pomiankowski. 1995. The genetic basis of female mate preferences. *J. Evol. Biol.* 8: 129–171.

Barnett, M., S. R. Telford and B. J. Tibbles. 1995. Female mediation of sperm competition in the millipede *Alloporus uncinatus* (Diplopoda: Spirostreptidae). *Behav. Ecol. Sociobiol.* 36: 413–419.

Bateman, A. J. 1948. Intra-sexual selection in *Drosophila*. *Heredity* 2: 349–368.

Berglund, A., C. Magnhagen, A. Basazza, B. König, and F. Huntingford. 1993. Female-female competition over reproduction. *Behav. Ecol.* 4: 184–187.

Bernays, E. A. and W. T. Wcislo. 1994. Sensory capabilities, information processing, and resource specialization. *Q. Rev. Biol.* 69: 187–204.

Birkhead, T. R. and Clarkson, K. 1980. Mate selection and precopulatory guarding in *Gammarus pulex*. *Z. Tierpsychol.* 52: 365–380.

Birkhead, T. R. and A. P. Møller. 1992. *Sperm Competition in Birds: Evolutionary Causes and Consequences*. London: Academic Press.

–. 1993. Female control of paternity. *Trends Ecol. Evol.* 8: 100–104.

Birkhead, T. R., A. P. Møller and W. J. Sutherland. 1993. Why do females make it so difficult for males to fertilize their eggs? *J. Theor. Biol.* 161: 51–60.

Bissoondath, C. J. and C. Wiklund. 1995. Protein content of spermatophores in relation to monandry/polyandry in butterflies. *Behav. Ecol. Sociobiol.* 37: 365–371.

Blum, M. S. and N. A. Blum, eds. 1979. *Sexual Selection and Reproductive Competition in Insects*. New York: Academic Press.

Boake, C. R. B. 1985. Genetic consequences of mate choice: a quantitative genetic method for testing sexual selection theory. *Science (Wash., D.C.)* 227: 1061–1063.

–. 1986. A method for testing adaptive hypotheses of mate choice. *Am. Nat.* 127: 654–666.

Borgia, G. 1979. Sexual selection and the evolution of mating systems. In *Sexual Selection and Reproductive Competition in Insects*. M. S. Blum and N. A. Blum, eds., pp. 19–80. New York: Academic Press

–. 1981. Mate selection in the fly *Scatophaga stercoraria*: female choice in a male controlled system. *Anim. Behav.* 29: 71–80.

Bradbury, J. and Gibson, R. 1983. Leks and mate choice. In *Mate Choice*. P. Bateson, ed., pp. 109–138. Cambridge University Press.

Brown, W. D. 1990. Size-assortative mating in the blister beetle *Lytta magister* (Coleoptera: Meloidae) is due to male and female preference for larger mates. *Anim. Behav.* 40: 901–909.

–. 1994. Mechanisms of female mate choice in the black-horned tree cricket *Oecanthus nigricornis* (Orthoptera: Gryllidae: Oecanthinae). Ph.D. dissertation, University of Toronto.

–. 1996. Remating and the intensity of female choice in the black-horned tree cricket, *Oecanthus nigricornis*. *Behav. Ecol.*, in press.

Brown, W. D. and J. Alcock. 1990. Hilltopping by the red admiral butterfly: mate searching alongside congeners. *J. Res. Lepid.* **29**: 1–10.

Brown, W. D. and D. T. Gwynne. 1997. Evolution of mating in crickets, katydids and wetas (Ensifera). In *Bionomics of Crickets, Katydids and their Kin.* S. K. Gangwere and M. C. Muralirangan, eds. London: CAB Press, in press.

Brown, W. D., J. Wideman, M. C. B. Andrade, A. C. Mason and D. T. Gwynne. 1996. Female mate choice for an indicator of male size in the song of the black-horned tree cricket, *Oecanthus nigricornis* (Orthoptera: Gryllidae: Oecanthinae). *Evolution* **50**: 2400–2411.

Burley, N. 1977. Parental investment, mate choice, and mate quality. *Proc. Natl. Acad. Sci. U.S.A.* **74**: 3476–3479.

–. 1986. Sexual selection for aesthetic traits in species with biparental care. *Am. Nat.* **127**: 415–445.

Carayon, J. 1966. Traumatic insemination and the paragenital system. In *Monograph of Cimicidae.* R. Usinger, ed., pp. 81–166. Baltimore: Thomas Say Foundation 7, Entomological Society of America.

Chapman, T., L. F. Liddle, J. M. Kalb, M. F. Wolfner and L. Partridge. 1995. Cost of mating in *Drosophila melanogaster* females is mediated by male accessory gland products. *Nature (Lond.)* **373**: 241–244.

Choe, J. C. 1994a. Sexual selection and mating system in *Zorotypus gurneyi* Choe (Insecta: Zoraptera): I. Dominance hierarchy and mating success. *Behav. Ecol. Sociobiol.* **34**: 87–93.

–. 1994b. Sexual selection and mating system in *Zorotypus gurneyi* Choe (Insecta: Zoraptera): II. Determinants and dynamics of dominance. *Behav. Ecol. Sociobiol.* **34**: 233–237.

–. 1995. Courtship feeding and repeated mating in *Zorotypus barberi* (Insecta: Zoraptera). *Anim. Behav.* **49**: 1511–1520.

Choe, J. C. and B. J. Crespi, eds. 1997. *The Evolution of Social Behavior in Insects and Arachnids.* Cambridge University Press.

Christy, J. H. 1995. Mimicry, mate choice, and the sensory trap hypothesis. *Am. Nat.* **146**: 171–181.

Christy, J. H. and P. R. Y. Blackwell. 1995. The sensory exploitation hypothesis. *Trends Ecol. Evol.* **10**: 417.

Clark, A. G., M. Aguadé, T. Prout, L. G. Harshman and C. H. Langley. 1995. Variation in sperm displacement and its association with accessory gland protein loci in *Drosophila melanogaster. Genetics* **139**: 189–201.

Clutton-Brock, T. H. 1991. *The Evolution of Parental Care.* Princeton: Princeton University Press.

Clutton-Brock, T. H. and G. A. Parker. 1992. Potential reproductive rates and the operation of sexual selection. *Q. Rev. Biol.* **67**: 437–456.

–. 1995. Sexual coercion in animal societies. *Anim. Behav.* **49**: 1345–1365.

Clutton-Brock, T. H. and A. C. J. Vincent. 1991. Sexual selection and the potential reproductive rates of males and females. *Nature (Lond.)* **351**: 58–60.

Conner, J. K. 1995. Extreme variability in sperm precedence in the fungus beetle, *Bolitotherus cornutus* (Coleoptera: Tenebrionidae). *Ethol. Ecol. Evol.* **7**: 277–280.

Connor, R. 1995. Altruism among non-relatives: alternatives to the 'Prisoner's Dilemma'. *Trends Ecol. Evol.* **10**: 84–86.

Conrad, K. F. and G. Pritchard. 1992. An ecological classification of odonate mating systems: the relative influence of natural, inter- and intra-sexual selection on males. *Biol. J. Linn. Soc.* **45**: 255–269.

Constantz, G. D. 1984. Sperm competition in poeciliid fishes. In *Sperm Competition and the Evolution of Animal Mating Systems.* R. L. Smith, ed., pp. 456–485. New York: Academic Press.

Cordero, C. 1995. Ejaculate substances that affect female insect reproductive physiology and behavior: honest or arbitrary traits? *J. Theor. Biol.* **174**: 453–461.

Cowan, D. F. and J. Atema. 1990. Moult staggering and serial monogamy in American lobsters, *Homarus americanus. Anim. Behav.* **39**: 1199–1206.

Cox, C. R. and B. J. LeBoeuf. 1977. Female incitation of male competition: a mechanism of mate selection. *Am. Nat.* **111**: 317–335.

Creel, S. R. and N. M. Creel. 1991. Energetics, reproductive suppression and obligate communal breeding in carnivores. *Behav. Ecol. Sociobiol.* **28**: 263–270.

Crespi, B. J. 1988a. Adaptation, compromise and constraint: the development, morphometrics and behavioral basis of a fighter-flier polymorphism in male *Hoplothrips karnyi. Behav. Ecol. Sociobiol.* **23**: 93–104.

–. 1988b. Risks and benefits of lethal male fighting in the colonial, polygynous thrips *Hoplothrips karnyi* (Insecta: Thysanoptera). *Behav. Ecol. Sociobiol.* **22**: 293–301.

Cronin, H. 1991. *The Ant and the Peacock.* Cambridge University Press.

Crook, J. H. 1965. The adaptive significance of avian social organization. *Symp. Zool. Soc. Lond.* **14**: 181–218.

Davies, N. B. 1985. Cooperation and conflict among dunnocks, *Prunella modularis* in a variable mating system. *Anim. Behav.* **33**: 628–648.

–. 1989. Sexual conflict and the polygamy threshold. *Anim. Behav.* **38**: 226–234.

–. 1991. Mating systems. In *Behavioural Ecology: an Evolutionary Approach.* J. R. Krebs and N. B. Davies, eds., pp. 263–299. Oxford University Press.

Devine, M. C. 1984. Potential for sperm competition in reptiles: behavioral and physiological consequences. In *Sperm Competition and the Evolution of Animal Mating Systems.* R. L. Smith, ed., pp. 509–521. New York: Academic Press.

Dickinson, J. L. and R. L. Rutowski. 1989. The function of the mating plug in the chalcedon checkerspot butterfly. *Anim. Behav.* **38**: 154–162.

Drummond, B. A. III. 1984. Multiple mating and sperm competition in Lepidoptera. In *Sperm Competition and the Evolution of Animal Mating Systems.* R. L. Smith, ed., pp. 291–370. New York: Academic Press.

Dussord, D. E., C. A. Harvis, J. Meinwald and T. Eisner. 1991. Pheromonal advertisement of a nuptial gift by a male moth (*Utetheisa ornatrix*). *Proc. Natl. Acad. Sci. U.S.A.* **88**: 9224–9227.

Eady, P. E. 1995. Why do male *Callosobruchus maculatus* beetles inseminate so many sperm? *Behav. Ecol. Sociobiol.* **36**: 25–32.

Eberhard, W. G. 1979. The function of horns in *Podischnus agenor* and other beetles. In *Sexual Selection and Reproductive Competition in Insects*. M. S. Blum and N. A. Blum, eds., pp. 231–258. New York: Academic Press.

–. 1985. *Sexual Selection and Animal Genitalia*. Cambridge, Mass.: Harvard University Press.

–. 1991. Copulatory courtship and cryptic female choice in insects. *Biol. Rev.* **66**: 1–31.

–. 1993. Evaluating models of sexual selection: genitalia as a test case. *Am. Nat.* **142**: 564–571.

–. 1996. *Female in Control: Sexual Selection by Cryptic Female Choice*. Princeton: Princeton University Press.

Eberhard, W. G. and C. Cordero. 1995. Sexual selection by cryptic female choice on male seminal products – a new bridge between sexual selection and reproductive physiology. *Trends Ecol. Evol.* **10**: 493–495.

Eggert, A. and A.-K. Müller. 1997. Biparental care and social evolution in burying beetles: lessons from the larder. In *The Evolution of Social Behavior in Insects and Arachnids*. J. C. Choe and B. J. Crespi, eds. Cambridge University Press.

Eggert, A.-K. and S. K. Sakaluk. 1994. Sexual cannibalism and its relation to male mating success in sagebrush crickets, *Cyphoderris strepitans* (Haglidae: Orthoptera). *Anim. Behav.* **47**: 1171–1177.

–. 1995. Female-coerced monogamy in burying beetles. *Behav. Ecol. Sociobiol.* **37**: 147–153.

Emlen, S. T. and Oring, L. W. 1977. Ecology, sexual selection, and the evolution of mating systems. *Science (Wash., D.C.)* **197**: 215–223.

Ewing, A. W. 1989. *Arthropod Bioacoustics: Neurobiology and Behaviour*. Edinburgh: Edinburgh University Press.

Fenton, M. B. 1984. Sperm competition? The case of vespertilionid and rhinolophid bats. In *Sperm Competition and the Evolution of Animal Mating Systems*. R. L. Smith, ed., pp. 573–588. New York: Academic Press.

Forrest, T. G. 1980. Phonotaxis in mole crickets: its reproductive significance. *Fla. Entomol.* **63**: 45–53.

–. 1983. Calling songs and mate choice in crickets. In *Orthopteran Mating Systems: Sexual Competition in a Diverse Group of Insects*. D. T. Gwynne and G. K. Morris, eds., pp. 185–204. Boulder: Westview Press.

Forrest, T. G. and R. Raspet. 1994. Models of female choice in acoustic communication. *Behav. Ecol.* **5**: 293–303.

Foster, M. S. 1983. Disruption, dispersion, and dominance in lek-breeding birds. *Am. Nat.* **122**: 53–72.

Fulton, B. B. 1915. The tree crickets of New York: life history and bionomics. *N.Y. Agric. Expt. Sta. Tech. Bull.* **42**: 1–47.

Gadagkar, R. 1985. Kin recognition in social insects and other animals. A review of recent findings and a considerations of their relevance for the theory of kin selection. *Proc. Indian Acad. Sci.* **94**: 587–621.

Gage, M. J. G. and R. R. Baker. 1991. Ejaculate size varies with socio-sexual situation in an insect. *Ecol. Entomol.* **16**: 331–337.

Gage, M. J. G. and P. A. Cook. 1994. Sperm size or numbers? Effects of nutritional stress upon eupyrene and apyrene sperm production strategies in the moth *Plodia interpunctella* (Lepidoptera: Pyralidae). *Funct. Ecol.* **8**: 594–599.

Gilbert, L. E. 1976. Postmating female odor in *Heliconius* butterflies: a male-contributed antiaphrodisiac? *Science (Wash., D.C.)* **193**: 419–420.

Goldsmith, S. K. 1987. Resource distribution and its effect on the mating system of a longhorn beetle, *Perarthrus linsleyi* (Coleoptera: Cerambycidae). *Oecologia (Berl.)* **73**: 317–320.

Gomendio, M., and E. R. S. Roldan. 1993. Mechanisms of sperm competition: linking physiology and behavioral ecology. *Trends Ecol. Evol.* **8**: 95–100.

Gowaty, P. A. 1994. Architects of sperm competition. *Trends Ecol. Evol.* **9**: 160–162.

–. 1996. *Feminism and Evolutionary Biology. Boundaries, Intersections, and Frontiers*. New York: Chapman and Hall.

Grafen, A. and M. Ridley. 1983. A model of mate guarding. *J. Theor. Biol.* **102**: 549–567.

Greenfield, M. 1996. Acoustic communication in Orthoptera. In *Bionomics of Crickets, Katydids and their Kin*. S. K. Gangwere and M. C. Muralirangan, eds., London: CAB Press, in press.

Gromko, M. H., D. G. Gilbert and R. C. Richmond. 1984. Sperm transfer and use in the multiple mating system of *Drosophila*. *Sperm Competition and the Evolution of Animal Mating Systems*. R. L. Smith, ed., pp. 372–427. New York: Academic Press.

Guilford, T. and M. S. Dawkins. 1991. Receiver psychology and the evolution of animal signals. *Anim. Behav.* **42**: 1–14.

Gwynne, D. T. 1984. Male mating effort, confidence of paternity, and insect sperm competition. In *Sperm Competition and the Evolution of Animal Mating Systems*. R. L. Smith, ed., pp. 117–149. New York: Academic Press.

–. 1987. Sex-biased predation and the risky mate-locating behaviour of male tick-tock cicadas (Homoptera: Cicadidae) *Anim. Behav.* **35**: 571–576.

–. 1990. Testing parental investment and the control of sexual selection in katydids: the operational sex ratio. *Am. Nat.* **136**: 474–484.

–. 1993. Food quality controls sexual selection in Mormon crickets by altering male mating investment. *Ecology* **74**: 1406–1413.

–. 1995. Phylogeny of the Ensifera (Orthoptera): a hypothesis supporting multiple origins of acoustic signalling, complex spermatophores and maternal care in crickets, katydids, and wetas. *J. Orthopt. Res.* **4**: 203–218.

Gwynne, D. T. and W. J. Bailey. 1988. Mating system, mate choice and ultasonic calling in a Zaprochiline katydid (Orthoptera: Tettigonidae). *Behaviour* **105**: 202–223.

Gwynne, D. T. and G. K. Morris, eds. 1983. *Orthopteran Mating Systems: Sexual Competition in a Diverse Group of Insects.* Boulder: Westview Press.

Gwynne, D. T. and L. W. Simmons. 1990. Experimental reversal of courtship roles in an insect. *Nature (Lond.)* **346**: 172–174.

Hammerstein, P. and G. A. Parker. 1987. Sexual selection: games between the sexes. In *Sexual Selection: Testing the Alternatives.* J. W. Bradbury and M. B. Andersson, eds., pp. 180–195. New York: John Wiley and Sons.

Haig, D. 1992. Brood reduction in gymnosperms. **In:** *Cannibalism: Ecology and Evolution among Diverse Taxa.* M. Elgar and B. J. Crespi, eds., pp. 63–84. Oxford University Press.

Haig, D. and C. T. Bergstrom. 1995. Multiple mating, sperm competition and meiotic drive. *J. Evol. Biol.* **8**: 265–282.

Hansen, T. F. and D. K. Price. 1995. Good genes and old age: do old mates provide superior genes? *J. Evol. Biol.* **8**: 759–778.

Hedrick, A. V. 1986. Female preferences for male calling bout duration in a field cricket. *Behav. Ecol. Sociobiol.* **19**: 73–77.

Heisler, L., M. B. Andersson, S. J. Arnold, C. R. Boake, G. Borgia, G. Hausfater, M. Kirkpatrick, R. Lande, J. Maynard Smith, P. O'Donald, A. R. Thornhill and F. J. Weissing. 1987. The evolution of mating preferences and sexually selected traits. In *Sexual Selection: Testing the Alternatives.* J. W. Bradbury and M. B. Andersson, eds., pp. 96–118. New York: Wiley and Sons.

Heller, K.-G. 1992. Risk shift between males and females in the pair-forming behavior of bushcrickets. *Naturwissenschaften* **79**: 89–91.

Helversen, D. von and O. von Helversen. 1991. Pre-mating sperm removal in the bushcricket *Metaplastes ornatus* Ramma 1931 (Orthoptera, Tettigonoidea, Phaneropteridae). *Behav. Ecol. Sociobiol.* **28**: 391–396.

Higashi, M. and N. Yamamura. 1994. Resolution of evolutionary conflict: a general theory and its applications. *Res. Pop. Ecol.* **36**: 15–22.

Hinton, H. E. 1964. Sperm transfer in insects and the evolution of haemocoelic insemination. In *Insect Reproduction.* K. C. Highnam, ed. (Symposium of the Royal Entomological Society of London, no. 2), pp. 95–107.

Höglund, J. and R. V. Alatalo. 1995. *Leks.* Princeton: Princeton University Press.

Ichikawa, N. 1995. Male counterstrategy against infanticide of the female giant water bug *Lethocerus deyrollei* (Hemiptera: Belostomatidae). *J. Insect Behav.* **8**: 181–188.

Jablonski, P. and K. Vepsäläinen. 1995. Conflict between the sexes in the water strider, *Gerris lacustris*: a test of two hypotheses for male guarding behavior. *Behav. Ecol.* **6**: 388–392.

Janetos, A. C. 1980. Strategies of female mate choice: a theoretical analysis. *Behav. Ecol. Sociobiol.* **7**: 107–112.

Keller, L. and H. K. Reeve. 1994. Partitioning of reproduction in animal societies. *Trends Ecol. Evol.* **9**: 98–102.

Kirkendall, L. R. 1983. The evolution of mating systems in bark and ambrosia beetles (Coleoptera: Scolytidae and Platypoddidae). *Zool. J. Linn. Soc.* **77**: 293–352.

Kimsey, L. S. 1980. The behavior of male orchid bees (Apidae, Hymenoptera, Insecta) and the question of leks. *Anim. Behav.* **28**: 996–1004.

Knowlton, N. and S. R. Greenwell. 1984. Male sperm competition avoidance mechanisms: the influence of female interests. In *Sperm Competition and the Evolution of Animal Mating Systems.* R. L. Smith, ed., pp. 62–84. New York: Academic Press.

Kodric-Brown, A. and J. H. Brown. 1984. Truth in advertising: the kinds of traits favored by sexual selection. *Am. Nat.* **124**: 309–323.

LaMunyon, C. W. and T. Eisner. 1993. Postcopulatory sexual selection in an arctiid moth, *Utetheisa ornatrix. Proc. Natl. Acad. Sci. U.S.A.* **90**: 4689–4692.

Lederhouse, R. C. 1982. Territorial defense and lek behavior of the black swallowtail butterfly, *Papilio polyxenes. Behav. Ecol. Sociobiol.* **10**: 109–118.

Leimar, O., B. Karlsson and C. Wiklund. 1994. Unpredictable food and sexual size dimorphism in insects. *Proc. R. Soc. Lond.* B **258**: 121–125.

Lewis, S. M. and S. N. Austad. 1990. Sources of intraspecific variation in sperm precedence in red flour beetles. *Am. Nat.* **135**: 351–359.

—. 1994. Sexual selection in flour beetles: the relationship between sperm precedence and male olfactory attractiveness. *Behav. Ecol.* **5**: 219–224.

Lloyd, J. E. 1979a. Mating behavior and natural selection. *Fla. Entomol.* **62**: 17–23.

—. 1979b. Sexual selection in bioluminescent beetles. In *Sexual Selection and Reproductive Competition in Insects.* M. S. Blum and N. A. Blum, eds., pp. 293–342. New York: Academic Press

—. 1980. Male *Photuris* fireflies mimic sexual signals of their female's prey. *Science (Wash., D.C.)* **210**: 669–671.

Loher, W. and M. Dambach. 1989. Reproductive behavior. In *Cricket Behavior and Neurobiology.* F. Huber, T. E. Moore and W. Loher, eds., pp. 43–82. Ithaca: Cornell University Press.

Low, B. S. 1979. Sexual selection and human ornamentation. In *Evolutionary Biology and Human Social Behavior.* N. A. Chagnon and W. Irons, eds., pp. 462–487. North Scituate, Mass.: Duxbury Press.

Marden, J. H. and J. K. Waage. 1990. Escalated damselfly territorial contests are energetic wars of attrition. *Anim. Behav.* **39**: 954–959.

Marden, J. H. and R. A. Rollins. 1994. Assessment of energy reserves by damselflies engaged in aerial contests for mating territories. *Anim. Behav.* **48**: 1023–1030.

Marshall, D. L. and N. C. Ellstrand. 1988. Effective mate choice in wild radish: evidence for selective seed abortion and its mechanism. *Am. Nat.* **131**: 736–759.

Mason, A. C. 1991. Hearing in a primative ensiferan: the auditory system of *Cyphoderris monstrosa* (Orthoptera: Haglidae). *J. Comp. Physiol.* A **168**: 351–363.

—. 1996. Territoriality and the function of song in the primitive acoustic insect *Cyphoderris monstrosa* (Orthoptera: Haglidae). *Anim. Behav.* **51**: 211–224.

Milinski, M. and G. A. Parker. 1991. Competition for resources. In *Behavioural Ecology: An Evolutionary Approach*. J. R. Krebs and N. B. Davies, eds., pp. 137–168. Oxford: Blackwell Scientific Publications.

Moore, A. J. 1995. Genetic evidence for the 'good genes' process of sexual selection. *Behav. Ecol. Sociobiol.* **35**: 235–241.

Moore, S. D. 1987. Male-biased mortality in the butterfly *Euphydryas editha*: a novel cost of mate acquisition. *Am. Nat.* **130**: 306–309.

Oberhauser, K. S. 1989. Effects of spermatophores on male and female monarch butterfly reproductive success. *Behav. Ecol. Sociobiol.* **25**: 237–246.

–. 1992. Rate of ejaculate breakdown and intermating intervals in monarch butterflies. *Behav. Ecol. Sociobiol.* **31**: 367–373.

Ono, T., M. T. Siva-Jothy and A. Kato. 1989. Removal and subsequent ingestion of rivals' semen during copulation in a tree cricket. *Physiol. Entomol.* **14**: 195–202.

Ono, T., F. Hayakawa, Y. Matsuura, M. Shiraishi, H. Yasui, T. Nakamura and M. Arakawa. 1995. Reproductive biology and function of multiple mating in the mating system of a tree cricket, *Truljalia hibinonis* (Orthoptera: Podoscritinae). *J. Insect Behav.* **8**: 813–824.

Orr, A. G. and R. L. Rutowski. 1991. The function of the sphragis in *Cressida cressida* (Fab.) (Lepidoptera, Papilionidae): a visual deterrent to copulation attempts. *J. Nat. Hist.* **25**: 703–710.

Otronen, M. 1994. Repeated copulation as a strategy to maximize fertilizations in the fly, *Dryomyza anilis* (Dryomyzidae). *Behav. Ecol.* **5**: 51–56.

Otronen, M. and M. T. Siva-Jothy. 1991. The effect of postcopulatory male behaviour on ejaculate distribution within the female sperm storage organs of the fly *Dryomyza anilis* (Diptera: Dryomyzidae). *Behav. Ecol. Sociobiol.* **29**: 33–37.

Otte, D. 1979. Historical development of sexual selection theory. In: *Sexual Selection and Reproductive Competition in Insects*. M. S. Blum and N. A. Blum, eds., pp. 1–19. New York: Academic Press.

Owens, I. P. F. and D. B. A. Thompson. 1994. Sex differences, sex ratios, and sex roles. *Proc. R. Soc. Lond.* B **258**: 93–99.

Parker, G. A. 1970. Sperm competition and its evolutionary consequences in the insects. *Biol. Rev.* **45**: 525–567.

–. 1979. Sexual selection and sexual conflict. In *Sexual Selection and Reproductive Competition in Insects*. M. S. Blum and N. A. Blum, eds., pp. 123–166. New York: Academic Press.

–. 1983. Mate quality and mating decisions. In *Mate Choice*. P. Bateson, ed., pp. 141–166. Cambridge University Press.

–. 1984. Sperm competition and the evolution of animal mating strategies. In *Sperm Competition and the Evolution of Animal Mating Systems*. R. L. Smith, ed., pp. 2–60. New York: Academic Press.

Parker, G. A. and L. W. Simmons. 1989. Nuptial feeding in insects: theoretical models of male and female interests. *Ethology* **82**: 3–26.

Petrie, M. 1992. Copulation behaviour in birds: why do females copulate more than once with the same male? *Anim. Behav.* **44**: 790–792.

–. 1994. Improved growth and survival of offspring of peacocks with more elaborate trains. *Nature (Lond.)* **371**: 598–599.

Petrie, M., M. Hall, T. Halliday, H. Budgey and C. Pierpoint. 1992. Multiple mating in a lekking bird: why do peahens mate with more than one male and with the same male more than once? *Behav. Ecol. Sociobiol.* **31**: 349–358.

Pitnick, S. 1991. Male size influences mate fecundity and remating interval in *Drosophila melanogaster*. *Anim. Behav.* **41**: 735–745.

Pitnick, S. and T. A. Markow. 1994. Male gametic strategies: sperm size, testis size, and the allocation of ejaculate among successive mates by the sperm-limited fly *Drosophila pachea* and its relatives. *Am. Nat.* **143**: 785–819.

Pomiankowski, A. and A. P. Møller. 1995. A resolution of the lek paradox. *Proc. R. Soc. Lond.* B **260**: 21–29.

Pomiankowski, A. and L. Sheridan. 1994. Linked sexiness and choosiness. *Trends Ecol. Evol.* **9**: 242–244.

Preziosi, R. F. and D. J. Fairbairn. 1996. Sexual size dimorphism and selection in the wild in the waterstrider *Aquarius remigis*: body size, components of body size and male mating success. *J. Evol. Biol.* **9**: 317–336.

Proctor, H. C. 1991. Courtship in the water mite *Neumania papillator*: males capitalize on female adaptations for predation. *Anim. Behav.* **42**: 589–598.

Queller, D. C. 1994. Male-female conflict and parent-offspring conflict. *Am. Nat.* **144**: S84–S99.

Rentz, D. C. F. 1993. *A Monograph of the Tettigoniidae of Australia*, vol. 2. *The Austrosaginae, Phasmodinae, and Zaprochilinae*. Melbourne: CSIRO.

Ridley, M. 1988. Mating frequency and fecundity in insects. *Biol. Rev.* **63**: 509–549.

–. 1989. The incidence of sperm displacement in insects: four conjectures, one corroboration. *Biol. J. Linn. Soc.* **38**: 349–367.

Ridley, M. and D. J. Thompson. 1979. Size and mating in *Asellus aquaticus* (Crustacea: Isopoda). *Z. Tierpsychol.* **51**: 380–397.

Rowe, L. 1992. Convenience polyandry in a water strider: foraging conflicts and female control of copulation frequency and guarding duration. *Anim. Behav.* **44**: 189–202.

Rowe, L., G. Arnqvist, A. Sih and J. J. Krupa. 1994. Sexual conflict and the evolutionary ecology of mating patterns: water striders as a model system. *Trends Ecol. Evol.* **9**: 289–293.

Rutowski, R. L. 1991. The evolution of male mate-locating behavior in butterflies. *Am. Nat.* **138**: 1121–1139.

Ryan, M. J. 1983. Sexual selection and communication in a Neotropical frog, *Physalaemus pustulosus*. *Evolution* **37**: 261–272.

Sakaluk, S. K. 1984. Male crickets feed females to ensure complete sperm transfer. *Science (Wash., D.C.)* **223**: 609–610.

–. 1990. Sexual selection and predation: balancing reproductive and survival needs. In *Insect Defences*. D. L. Evans and J. O. Schmidt, eds., pp. 63–90. Albany: SUNY Press.

Sakaluk, S. K., P. J. Bangert, A.-K. Eggert, C. Gack and L. W. Swanson. 1995. The gin trap as a device facilitating coercive mating in sagebrush crickets. *Proc. R. Soc. Lond.* B **261**: 65–71.

Sakaluk, S. K. and J. J. Belwood. 1984. Gecko phonotaxis to cricket calling song: a case of satellite predation. *Anim. Behav.* **32**: 659–662

Shaw, K. 1995. Phylogenetic tests of the sensory exploitation model of sexual selection. *Trends Ecol. Evol.* **10**: 117–120.

Schluter, D. and T. Price. 1993. Honesty, perception and population divergence in sexually selected traits. *Proc. R. Soc. Lond.* B **253**: 117–122.

Sih, A. and J. J. Krupa. 1995. Interacting effects of predation risk and male and female density on male/female conflicts and mating dynamics of stream water striders. *Behav. Ecol.* **6**: 316–325.

Simmons, L. W. 1989. Kin recognition and its influence on mating preferences of the field cricket, *Gryllus bimaculatus* (De Geer). *Anim. Behav.* **38**: 68–77.

–. 1990. Pheromonal cues for the recognition of kin by female field crickets, *Gryllus bimaculatus*. *Anim. Behav.* **39**: 192–195.

–. 1991. Female choice and the relatedness of mates in the field cricket, *Gryllus bimaculatus*. *Anim. Behav.* **41**: 493–501.

Simmons, L. W. and W. J. Bailey. 1990. Resource influenced sex roles of Zaprochiline tettigoniids (Orthoptera: Tettigoniidae). *Evolution* **44**: 1853–1868.

Simmons, L. W. and G. A. Parker. 1989. Nuptial feeding in insects: mating effort versus paternal investment. *Ethology* **81**: 332–343.

–. 1992. Individual variation in sperm competition success of yellow dung flies, *Scatophaga stercoraria*. *Evolution* **46**: 366–375.

Siva-Jothy, M. T. and R. E. Hooper. 1995. The disposition and genetic diversity of stored sperm in females of the damselfly *Calopteryx splendens xanthostoma* (Charpentier). *Proc. R. Soc. Lond.* B **259**: 313–318.

Sivinski, J. 1984. Sperm in competition. In *Sperm Competition and the Evolution of Animal Mating Systems*. R. L. Smith, ed., pp. 86–115. New York: Academic Press.

Smith, R. L., ed. 1984. *Sperm Competition and the Evolution of Animal Mating Systems*. New York: Academic Press.

–. 1997. Evolution of paternal care in the giant water bugs (Heteroptera: Belostomatidae). In *The Evolution of Social Behavior in Insects and Arachnids*. J. C. Choe and B. J. Crespi, eds. Cambridge University Press.

Snead, J. S. and J. Alcock. 1985. Aggregation formation and assortative mating in two meloid beetles. *Evolution* **39**: 1123–1131.

Tallamy, D. W. 1994. Nourishment and the evolution of paternal investment in subsocial arthropods. In *Nourishment and Evolution in Insect Societies*. J. H. Hunt and C. A. Nalepa, eds., pp. 21–55. Boulder: Westview Press.

Thornhill, R. 1976. Sexual selection and paternal investment in insects. *Am. Nat.* **110**: 153–163.

–. 1980. Rape in *Panorpa* scorpionflies and a general rape hypothesis. *Anim. Behav.* **28**: 52–59.

–. 1981. *Panorpa* (Mecoptera: Panorpidae) scorpionflies: systems for understanding resource-defense polygyny and alternative male reproductive efforts. *Annu. Rev. Ecol. Syst.* **12**: 355–386.

–. 1983. Cryptic female choice and its implications in the scorpionfly *Hylobittacus nigriceps*. *Am. Nat.* **122**: 765–788.

–. 1984. Alternative female choice tactics in the scorpionfly *Hylobittacus apicalis* (Mecoptera) and their implications. *Am. Zool.* **24**: 367–383.

–. 1986. Relative parental contribution of the sexes to their offspring and the operation of sexual selection. In *Evolution of Animal Behavior: Paleontological and Field Approaches*. M. H. Nitecki and J. A. Kitchell, eds., pp. 113–135. Oxford University Press.

–. 1990. The study of adaptation. In *Interpretation and Explanation in the Study of Behavior*, vol. 2. M. Bekoff and D. Jamieson, eds., pp. 1–31. Boulder: Westview Press.

Thornhill, R. and J. Alcock. 1983. *The Evolution of Insect Mating Systems*. Cambridge, Mass: Harvard University Press.

Trivers, R. L. 1972. Parental investment and sexual selection. In *Sexual Selection and the Descent of Man*. B. Campbell, ed., pp. 136–179. Chicago: Adline.

Vehrencamp, S. L. and J. W. Bradbury. 1984. Mating systems and ecology. In: *Behavioural Ecology: an Evolutionary Approach*, 2nd edn. J. R. Krebs and N. B. Davies, eds., pp. 251–278. Oxford: Blackwell Scientific Publications.

Uéda, T. 1979. Plasticity of the reproductive behaviour in a dragonfly, *Sympetrum parvulum* Barteneff, with reference to the social relationships of males and the density of males and the density of territories. *Res. Pop. Ecol.* **21**: 135–152.

Waage, J. K. 1979a. Dual function of the damselfly penis: sperm removal and transfer. *Science (Wash., D.C.)* **203**: 916–918.

–. 1979b. Adaptive significance of postcopulatory guarding of mates and nonmates by male *Calopteryx maculata* (Odonata). *Behav. Ecol. Sociobiol.* **6**: 147–154.

–. 1984a. Sperm competition and the evolution of odonate mating systems. In: *Sperm Competition and the Evolution of Animal Mating Systems*. R. L. Smith, ed., pp. 251–290. New York: Academic Press.

–. 1984b. Influence of oviposition behavior on femlae and male responses during courtship in *Calopteryx maculata* and *C. dimidiata* (Odonata: Calopterygidae). *Anim. Behav.* **32**: 400–404.

–. 1986. Evidence for widespread sperm displacement ability among Zygoptera (Odonata) and the means for predicting its presence. *Biol. J. Linn. Soc.* **28**: 285–300.

–. 1988. Confusion over residency and the escalation of damselfly territorial disputes. *Anim. Behav.* **36**: 586–595.

–. 1996. Parental investment – minding the kids or keeping control? In *Feminism and Evolutionary Biology. Boundaries, Intersections, and Frontiers*. P. Gowaty, ed. New York: Chapman and Hall.

Walker, T. J. 1957. Specificity in the response of female tree crickets (Orthoptera, Gryllidae, Oecanthinae) to calling songs of the males. *Ann. Entomol. Soc. Am.* **50**: 626–636.

Walker, W. F. 1980. Sperm utilization strategies in nonsocial insects. *Am. Nat.* **115**: 780–799.

Wade, M. 1995. The ecology of sexual selection: mean crowding of females and resource-defense polygyny. *Evol. Ecol.* **9**: 118–124.

Ward, P. I. 1983. Advantages and a disadvantage of large size for male *Gammarus pulex*. *Behav. Ecol. Sociobiol.* **14**: 69–76.

–. 1993. Females influence sperm storage and use in the yellow dung fly *Scatophaga stercoraria* (L.). *Behav. Ecol. Sociobiol.* **32**: 313–319.

Ward, P. I., J. Hemmi and T. Röösli. 1992. Sexual conflict in the dung fly *Sepsis cynipsea*. *Funct. Ecol.* **6**: 649–653.

Warner, R. R., D. Y. Shapiro, A. Marcanato and C. W. Petersen. 1995. Sexual conflict: males with highest mating success convey the lowest fertilization benefits to females. *Proc. R. Soc. Lond.* B **262**: 135–139.

Watson, P. J. 1986. Transmission of a female sex pheromone thwarted by males in the spider *Linyphia litigiosa* (Linyphiidae). *Science (Wash., D.C.)* **233**: 219–221.

–. 1990. Female-enhanced male competition determines the first mate and principal sire in the spider *Linyphia litigosa*. *Behav. Ecol. Sociobiol.* **26**: 77–90.

Wcislo, W. C. 1997. Are behavioral classifications blinders to natural variation? In *The Evolution of Social Behavior in Insects and Arachnids*. J. C. Choe and B. J. Crespi, eds. Cambridge University Press.

Wedell, N. 1991. Sperm competition selects for nuptial feeding in a bushcricket. *Evolution* **45**: 1975–1978.

Weigensberg, I. and D. J. Fairbairn 1994. Conflict of interest between the sexes: a study of mating interactions in a semiaquatic bug. *Anim. Behav.* **48**: 893–901.

–. 1996. The sexual arms race and phenotypic correlates of mating success in the waterstrider, *Aquarius remigis*. *J. Insect Behav.* **9**: 307–319.

Weigmann, D. D., L. A. Real, T. A. Capone and S. Ellner. 1996. Some distinguishing features of search behavior and mate choice. *Am. Nat.* **147**: 188–204.

West-Eberhard, M. J. 1984. Sexual selection, competitive communication and species-specific signals in insects. In *Insect Communication*. T. Lewis, ed., pp. 283–324. New York: Academic Press.

West-Eberhard, M. J., J. W. Bradbury, N. B. Davies, P.-H. Gouyon, P. Hammerstein, B. König, G. A. Parker, D. C. Queller, D. Sachser, T. Slagsvold, F. Trillmich and C. Vogel. 1987. Conflicts between and within the sexes in sexual selection. In *Sexual Selection: Testing the Alternatives*. J. W. Bradbury and M. B. Andersson, eds., pp. 180–95. New York: John Wiley and Sons.

Wilkinson, G. S. and P. F. Reillo. 1994. Female choice responds to artificial selection on an exaggerated male trait in a stalk-eyed fly. *Proc. R. Soc. Lond.* B **255**: 1–6

Williams, G. C. 1966 *Adaptation and Natural Selection: a Critique of some Current Evolutionary Thought*. Princeton: Princeton University Press.

Willson, M. F. and N. Burley. 1983. *Mate Choice in Plants: Tactics, Mechanisms, and Consequences*. Princeton: Princeton University Press.

Wing, S. R. 1988. Cost of mating for female insects: risk of predation in *Photinus collustrans* (Coleoptera: Lampyridae). *Am. Nat.* **131**: 139–142.

Wittenberger, J. F. 1979. The evolution of mating systems in birds and mammals. In *Social Behavior and Communication*. P. Marler and J. G. Vandenbergh, eds., pp. 271–349. New York: Plenum Press.

Organism index

Subject index